NUCLEAR MEDICINE PHYSICS: A HANDBOOK FOR TEACHERS AND STUDENTS

The following States are Members of the International Atomic Energy Agency:

AFGHANISTAN
ALBANIA
ALGERIA
ANGOLA
ARGENTINA
ARMENIA
AUSTRALIA
AUSTRIA
AZERBAIJAN
BAHAMAS
BAHRAIN
BANGLADESH
BELARUS
BELGIUM
BELIZE
BENIN
BOLIVIA
BOSNIA AND HERZEGOVINA
BOTSWANA
BRAZIL
BRUNEI DARUSSALAM
BULGARIA
BURKINA FASO
BURUNDI
CAMBODIA
CAMEROON
CANADA
CENTRAL AFRICAN REPUBLIC
CHAD
CHILE
CHINA
COLOMBIA
CONGO
COSTA RICA
CÔTE D'IVOIRE
CROATIA
CUBA
CYPRUS
CZECH REPUBLIC
DEMOCRATIC REPUBLIC OF THE CONGO
DENMARK
DOMINICA
DOMINICAN REPUBLIC
ECUADOR
EGYPT
EL SALVADOR
ERITREA
ESTONIA
ETHIOPIA
FIJI
FINLAND
FRANCE
GABON
GEORGIA
GERMANY
GHANA
GREECE
GUATEMALA
HAITI
HOLY SEE
HONDURAS
HUNGARY
ICELAND
INDIA
INDONESIA
IRAN, ISLAMIC REPUBLIC OF
IRAQ
IRELAND
ISRAEL
ITALY
JAMAICA
JAPAN
JORDAN
KAZAKHSTAN
KENYA
KOREA, REPUBLIC OF
KUWAIT
KYRGYZSTAN
LAO PEOPLE'S DEMOCRATIC REPUBLIC
LATVIA
LEBANON
LESOTHO
LIBERIA
LIBYA
LIECHTENSTEIN
LITHUANIA
LUXEMBOURG
MADAGASCAR
MALAWI
MALAYSIA
MALI
MALTA
MARSHALL ISLANDS
MAURITANIA, ISLAMIC REPUBLIC OF
MAURITIUS
MEXICO
MONACO
MONGOLIA
MONTENEGRO
MOROCCO
MOZAMBIQUE
MYANMAR
NAMIBIA
NEPAL
NETHERLANDS
NEW ZEALAND
NICARAGUA
NIGER
NIGERIA
NORWAY
OMAN
PAKISTAN
PALAU
PANAMA
PAPUA NEW GUINEA
PARAGUAY
PERU
PHILIPPINES
POLAND
PORTUGAL
QATAR
REPUBLIC OF MOLDOVA
ROMANIA
RUSSIAN FEDERATION
RWANDA
SAN MARINO
SAUDI ARABIA
SENEGAL
SERBIA
SEYCHELLES
SIERRA LEONE
SINGAPORE
SLOVAKIA
SLOVENIA
SOUTH AFRICA
SPAIN
SRI LANKA
SUDAN
SWAZILAND
SWEDEN
SWITZERLAND
SYRIAN ARAB REPUBLIC
TAJIKISTAN
THAILAND
THE FORMER YUGOSLAV REPUBLIC OF MACEDONIA
TOGO
TRINIDAD AND TOBAGO
TUNISIA
TURKEY
UGANDA
UKRAINE
UNITED ARAB EMIRATES
UNITED KINGDOM OF GREAT BRITAIN AND NORTHERN IRELAND
UNITED REPUBLIC OF TANZANIA
UNITED STATES OF AMERICA
URUGUAY
UZBEKISTAN
VENEZUELA, BOLIVARIAN REPUBLIC OF
VIET NAM
YEMEN
ZAMBIA
ZIMBABWE

The Agency's Statute was approved on 23 October 1956 by the Conference on the Statute of the IAEA held at United Nations Headquarters, New York; it entered into force on 29 July 1957. The Headquarters of the Agency are situated in Vienna. Its principal objective is "to accelerate and enlarge the contribution of atomic energy to peace, health and prosperity throughout the world".

NUCLEAR MEDICINE PHYSICS: A HANDBOOK FOR TEACHERS AND STUDENTS

ENDORSED BY:

AMERICAN ASSOCIATION OF PHYSICISTS IN MEDICINE,
ASIA–OCEANIA FEDERATION OF ORGANIZATIONS FOR MEDICAL PHYSICS,
AUSTRALASIAN COLLEGE OF PHYSICAL SCIENTISTS AND ENGINEERS IN MEDICINE,
EUROPEAN FEDERATION OF ORGANISATIONS FOR MEDICAL PHYSICS,
FEDERATION OF AFRICAN MEDICAL PHYSICS ORGANISATIONS,
WORLD FEDERATION OF NUCLEAR MEDICINE AND BIOLOGY

INTERNATIONAL ATOMIC ENERGY AGENCY
VIENNA, 2014

Marketing and Sales Unit, Publishing Section
International Atomic Energy Agency
Vienna International Centre
PO Box 100
1400 Vienna, Austria
fax: +43 1 2600 29302
tel.: +43 1 2600 22417
email: sales.publications@iaea.org
http://www.iaea.org/books

Printed by the IAEA in Austria
December 2014
STI/PUB/1617

IAEA Library Cataloguing in Publication Data

Nuclear medicine physics : a handbook for students and teachers. — Vienna : International Atomic Energy Agency, 2014.
p. ; 24 cm.
STI/PUB/1617
ISBN 978–92–0–143810–2

Includes bibliographical references. 1. Nuclear medicine — Handbooks, manuals, etc. 2. Medical physics handbooks. 3. Medical physics. I. International Atomic Energy Agency.

IAEAL 14–00880

FOREWORD

Nuclear medicine is the use of radionuclides in medicine for diagnosis, staging of disease, therapy and monitoring the response of a disease process. It is also a powerful translational tool in the basic sciences, such as biology, in drug discovery and in pre-clinical medicine. Developments in nuclear medicine are driven by advances in this multidisciplinary science that includes physics, chemistry, computing, mathematics, pharmacology and biology.

This handbook comprehensively covers the physics of nuclear medicine. It is intended for undergraduate and postgraduate students of medical physics. It will also serve as a resource for interested readers from other disciplines, for example, clinicians, radiochemists and medical technologists who would like to familiarize themselves with the basic concepts and practice of nuclear medicine physics.

The scope of the book is intentionally broad. Physics is a vital aspect of nearly every area of nuclear medicine, including imaging instrumentation, image processing and reconstruction, data analysis, radionuclide production, radionuclide therapy, radiopharmacy, radiation protection and biology. The authors were drawn from a variety of regions and were selected because of their knowledge, teaching experience and scientific acumen.

This book was written to address an urgent need for a comprehensive, contemporary text on the physics of nuclear medicine. It complements similar texts in radiation oncology physics and diagnostic radiology physics that have been published by the IAEA.

Endorsement of this handbook has been granted by the following international professional bodies: the American Association of Physicists in Medicine (AAPM), the Asia–Oceania Federation of Organizations for Medical Physics (AFOMP), the Australasian College of Physical Scientists and Engineers in Medicine (ACPSEM), the European Federation of Organisations for Medical Physics (EFOMP), the Federation of African Medical Physics Organisations (FAMPO), and the World Federation of Nuclear Medicine and Biology (WFNMB).

The following international experts are gratefully acknowledged for making major contributions to this handbook as technical editors: D.L. Bailey (Australia), J.L. Humm (United States of America), A. Todd-Pokropek (United Kingdom) and A. van Aswegen (South Africa). The IAEA officers responsible for this publication were S. Palm and G.L. Poli of the Division of Human Health.

EDITORIAL NOTE

Although great care has been taken to maintain the accuracy of information contained in this publication, neither the IAEA nor its Member States assume any responsibility for consequences which may arise from its use.

The use of particular designations of countries or territories does not imply any judgement by the publisher, the IAEA, as to the legal status of such countries or territories, of their authorities and institutions or of the delimitation of their boundaries.

The mention of names of specific companies or products (whether or not indicated as registered) does not imply any intention to infringe proprietary rights, nor should it be construed as an endorsement or recommendation on the part of the IAEA.

The IAEA has no responsibility for the persistence or accuracy of URLs for external or third party Internet web sites referred to in this book and does not guarantee that any content on such web sites is, or will remain, accurate or appropriate.

PREFACE

Nuclear medicine is the study and utilization of radioactive compounds in medicine to image and treat human disease. It relies on the 'tracer principle' first espoused by Georg Karl von Hevesy in the early 1920s. The tracer principle is the study of the fate of compounds in vivo using minute amounts of radioactive tracers which do not elicit any pharmacological response by the body to the tracer. Today, the same principle is used to study many aspects of physiology, such as cellular metabolism, DNA (deoxyribonucleic acid) proliferation, blood flow in organs, organ function, receptor expression and abnormal physiology, externally using sensitive imaging devices. Larger amounts of radionuclides are also applied to treat patients with radionuclide therapy, especially in disseminated diseases such as advanced metastatic cancer, as this form of therapy has the ability to target abnormal cells to treat the disease anywhere in the body.

Nuclear medicine relies on function. For this reason, it is referred to as 'functional imaging'. Rather than just imaging a portion of the body believed to have some abnormality, as is done with X ray imaging in radiology, nuclear medicine scans often depict the whole body distribution of the radioactive compound often acquired as a sequence of images over time showing the temporal course of the radiotracer in the body.

There are two main types of radiation of interest for imaging in nuclear medicine: γ ray emission from excited nuclei, and annihilation (or coincidence) radiation ($\gamma\pm$) arising after positron emission from proton-rich nuclei. Gamma photons are detected with a gamma camera as either planar (2-D) images or tomographically in 3-D using single photon emission computed tomography. The annihilation photons from positron emission are detected using a positron emission tomography (PET) camera. The most recent major development in this field is the combination of gamma cameras or PET cameras with high resolution structural imaging devices, either X ray computed tomography (CT) scanners or, increasingly, magnetic resonance imaging (MRI) scanners, in a single image device. The combined PET/CT (or PET/MRI) scanner represents one of the most sophisticated and powerful ways to visualize normal and altered physiology in the body.

It is in this complex environment that the medical physicist, along with nuclear medicine physicians and technologists/radiographers, plays a significant role in the multidisciplinary team needed for medical diagnosis. The physicist is responsible for such areas as instrumentation performance, radiation dosimetry for treatment of patients, radiation protection of staff and accuracy of the data analysis. The physicist draws on training in radiation and nuclear science, in addition to scientific rigour and attention to detail in experiments and measurements, to join forces with the other members of the multidisciplinary team in delivering optimal health care. Patients are frequently treated on the basis of the result of the scans they receive and these, therefore, have to be of the highest quality.

This handbook was conceived and written by physicists, and is intended primarily for physicists, although interested readers from medical, paramedical and other science and engineering backgrounds could find it useful. The level of understanding of the material covered will be different depending on the background of the reader. Readers are encouraged to visit the IAEA Human Health web site (http://www-naweb.iaea.org/NAHU/index.html) to discover the wealth of resources available.

The technical editors and authors, selected for their experience and in recognition of their contributions to the field, were drawn from around the world and, thus, this book represents a truly international collaboration. The technical editors travelled to the IAEA headquarters in Vienna on four occasions over three years to bring this project to fruition. We would like to thank all of the authors for their important contribution.

D.L. Bailey, J.L. Humm
A. Todd-Pokropek, A. van Aswegen

CONTENTS

CHAPTER 1. BASIC PHYSICS FOR NUCLEAR MEDICINE 1

1.1. INTRODUCTION 1
1.1.1. Fundamental physical constants 1
1.1.2. Physical quantities and units 2
1.1.3. Classification of radiation 4
1.1.4. Classification of ionizing radiation 4
1.1.5. Classification of indirectly ionizing photon radiation . . 5
1.1.6. Characteristic X rays 5
1.1.7. Bremsstrahlung 5
1.1.8. Gamma rays 6
1.1.9. Annihilation quanta 6
1.1.10. Radiation quantities and units 7
1.2. BASIC DEFINITIONS FOR ATOMIC STRUCTURE 8
1.2.1. Rutherford model of the atom 10
1.2.2. Bohr model of the hydrogen atom 10
1.3. BASIC DEFINITIONS FOR NUCLEAR STRUCTURE 10
1.3.1. Nuclear radius 12
1.3.2. Nuclear binding energy 12
1.3.3. Nuclear fusion and fission 13
1.3.4. Two-particle collisions and nuclear reactions 14
1.4. RADIOACTIVITY 16
1.4.1. Decay of radioactive parent into a stable or unstable daughter 17
1.4.2. Radioactive series decay 19
1.4.3. Equilibrium in parent–daughter activities 21
1.4.4. Production of radionuclides (nuclear activation) 22
1.4.5. Modes of radioactive decay 23
1.4.6. Alpha decay 25
1.4.7. Beta minus decay 26
1.4.8. Beta plus decay 26
1.4.9. Electron capture 27
1.4.10. Gamma decay and internal conversion 27
1.4.11. Characteristic (fluorescence) X rays and Auger electrons 28
1.5. ELECTRON INTERACTIONS WITH MATTER 29
1.5.1. Electron–orbital interactions 29
1.5.2. Electron–nucleus interactions 29

1.6. PHOTON INTERACTIONS WITH MATTER 30
1.6.1. Exponential absorption of photon beam in absorber . . . 30
1.6.2. Characteristic absorber thicknesses 31
1.6.3. Attenuation coefficients 34
1.6.4. Photon interactions on the microscopic scale 35
1.6.5. Photoelectric effect 38
1.6.6. Rayleigh (coherent) scattering. 39
1.6.7. Compton effect (incoherent scattering) 39
1.6.8. Pair production 44
1.6.9. Relative predominance of individual effects 46
1.6.10. Macroscopic attenuation coefficients 47
1.6.11. Effects following photon interactions with absorber and summary of photon interactions 48

CHAPTER 2. BASIC RADIOBIOLOGY 49

2.1. INTRODUCTION 49
2.2. RADIATION EFFECTS AND TIMESCALES 49
2.3. BIOLOGICAL PROPERTIES OF IONIZING RADIATION . . 51
2.3.1. Types of ionizing radiation 51
2.4. MOLECULAR EFFECTS OF RADIATION AND THEIR MODIFIERS 53
2.4.1. Role of oxygen 54
2.4.2. Bystander effects 54
2.5. DNA DAMAGE AND REPAIR. 55
2.5.1. DNA damage. 55
2.5.2. DNA repair 55
2.6. CELLULAR EFFECTS OF RADIATION 56
2.6.1. Concept of cell death. 56
2.6.2. Cell survival curves 56
2.6.3. Dose deposition characteristics: linear energy transfer. 57
2.6.4. Determination of relative biological effectiveness 58
2.6.5. The dose rate effect and the concept of repeat treatments 62
2.6.6. The basic linear–quadratic model 63
2.6.7. Modification to the linear–quadratic model for radionuclide therapies 64
2.6.8. Quantitative intercomparison of different treatment types. 64
2.6.9. Cellular recovery processes 65
2.6.10. Consequence of radionuclide heterogeneity 66

2.7. GROSS RADIATION EFFECTS ON TUMOURS AND TISSUES/ORGANS. 66
2.7.1. Classification of radiation damage (early versus late) . 66
2.7.2. Determinants of tumour response 67
2.7.3. The concept of therapeutic index in radiation therapy and radionuclide therapy 68
2.7.4 Long term concerns: stochastic and deterministic effects 68
2.8. SPECIAL RADIOBIOLOGICAL CONSIDERATIONS IN TARGETED RADIONUCLIDE THERAPY. 69
2.8.1. Radionuclide targeting 69
2.8.2. Whole body irradiation 69
2.8.3. Critical normal tissues for radiation and radionuclide therapies 70
2.8.4. Imaging the radiobiology of tumours 71
2.8.5. Choice of radionuclide to maximize therapeutic index. 71

CHAPTER 3. RADIATION PROTECTION. 73

3.1. INTRODUCTION 73
3.2. BASIC PRINCIPLES OF RADIATION PROTECTION 74
3.2.1. The International Commission on Radiological Protection system of radiological protection. 74
3.2.2. Safety standards. 76
3.2.3. Radiation protection quantities and units 77
3.3. IMPLEMENTATION OF RADIATION PROTECTION IN A NUCLEAR MEDICINE FACILITY 81
3.3.1. General aspects 81
3.3.2. Responsibilities 82
3.3.3. Radiation protection programme. 84
3.3.4. Radiation protection committee 84
3.3.5. Education and training. 84
3.4. FACILITY DESIGN. 85
3.4.1. Location and general layout 85
3.4.2. General building requirements 85
3.4.3. Source security and storage 86
3.4.4. Structural shielding 87
3.4.5. Classification of workplaces 87
3.4.6. Workplace monitoring 88
3.4.7. Radioactive waste 88

3.5. OCCUPATIONAL EXPOSURE 89
3.5.1. Sources of exposure. 90
3.5.2. Justification, optimization and dose limitation 91
3.5.3. Conditions for pregnant workers and young persons . . 91
3.5.4. Protective clothing. 92
3.5.5. Safe working procedures. 92
3.5.6. Personal monitoring. 94
3.5.7. Monitoring of the workplace. 95
3.5.8. Health surveillance 95
3.5.9. Local rules and supervision. 96
3.6. PUBLIC EXPOSURE 97
3.6.1. Justification, optimization and dose limitation 97
3.6.2. Design considerations 97
3.6.3. Exposure from patients 98
3.6.4. Transport of sources 98
3.7. MEDICAL EXPOSURE 99
3.7.1. Justification of medical exposure 99
3.7.2. Optimization of protection 100
3.7.3. Helping in the care, support or comfort of patients. . . . 107
3.7.4. Biomedical research 107
3.7.5. Local rules. 108
3.8. POTENTIAL EXPOSURE. 108
3.8.1. Safety assessment and accident prevention. 108
3.8.2. Emergency plans 110
3.8.3. Reporting and lessons learned. 111
3.9. QUALITY ASSURANCE 112
3.9.1. General considerations 112
3.9.2. Audit 114

CHAPTER 4. RADIONUCLIDE PRODUCTION 117

4.1. THE ORIGINS OF DIFFERENT NUCLEI. 117
4.1.1. Induced radioactivity. 118
4.1.2. Nuclide chart and line of nuclear stability. 120
4.1.3. Binding energy, Q-value, reaction threshold and nuclear reaction formalism 123
4.1.4. Types of nuclear reaction, reaction channels and cross-section 124
4.2. REACTOR PRODUCTION. 127
4.2.1. Principle of operation and neutron spectrum. 128

4.2.2. Thermal and fast neutron reactions 128
4.2.3. Nuclear fission, fission products 131
4.3. ACCELERATOR PRODUCTION........................ 132
4.3.1. Cyclotron, principle of operation, negative and positive ions 134
4.3.2. Commercial production (low and high energy)....... 136
4.3.3. In-house low energy production (PET)............. 137
4.3.4. Targetry, optimizing the production regarding yield and impurities, yield calculations 140
4.4. RADIONUCLIDE GENERATORS....................... 141
4.4.1. Principles of generators........................... 142
4.5. RADIOCHEMISTRY OF IRRADIATED TARGETS........ 143
4.5.1. Carrier-free, carrier-added systems 144
4.5.2. Separation methods, solvent extraction, ion exchange, thermal diffusion 145
4.5.3. Radiation protection considerations and hot-box facilities...................................... 147

CHAPTER 5. STATISTICS FOR RADIATION MEASUREMENT 149

5.1. SOURCES OF ERROR IN NUCLEAR MEDICINE MEASUREMENT 149
5.2. CHARACTERIZATION OF DATA...................... 153
5.2.1. Measures of central tendency and variability 153
5.3. STATISTICAL MODELS 157
5.3.1. Conditions when binomial, Poisson and normal distributions are applicable 158
5.3.2. Binomial distribution.............................. 160
5.3.3. Poisson distribution............................... 163
5.3.4. Normal distribution 165
5.4. ESTIMATION OF THE PRECISION OF A SINGLE MEASUREMENT IN SAMPLE COUNTING AND IMAGING 168
5.4.1. Assumption 168
5.4.2. The importance of the fractional σ_F as an indicator of the precision of a single measurement in sample counting and imaging 170
5.4.3. Caution on the use of the estimate of the precision of a single measurement in sample counting and imaging 171

5.5. PROPAGATION OF ERROR ... 172
5.5.1. Sums and differences ... 173
5.5.2. Multiplication and division by a constant ... 174
5.5.3. Products and ratios ... 176
5.6. APPLICATIONS OF STATISTICAL ANALYSIS ... 177
5.6.1. Multiple independent counts ... 177
5.6.2. Standard deviation and relative standard deviation for counting rates ... 178
5.6.3. Effects of background counts ... 179
5.6.4. Significance of differences between counting measurements ... 183
5.6.5. Minimum detectable counts, count rate and activity ... 184
5.6.6. Comparing counting systems ... 187
5.6.7. Estimating required counting times ... 188
5.6.8. Calculating uncertainties in the measurement of plasma volume in patients ... 189
5.7. APPLICATION OF STATISTICAL ANALYSIS: DETECTOR PERFORMANCE ... 191
5.7.1. Energy resolution of scintillation detectors ... 191
5.7.2. Intervals between successive events ... 193
5.7.3. Paralysable dead time ... 194

CHAPTER 6. BASIC RADIATION DETECTORS ... 196

6.1. INTRODUCTION ... 196
6.1.1. Radiation detectors — complexity and relevance ... 196
6.1.2. Interaction mechanisms, signal formation and detector type ... 196
6.1.3. Counting, current, integrating mode ... 197
6.1.4. Detector requirements ... 197
6.2. GAS FILLED DETECTORS ... 200
6.2.1. Basic principles ... 200
6.3. SEMICONDUCTOR DETECTORS ... 202
6.3.1. Basic principles ... 202
6.3.2. Semiconductor detectors ... 204
6.4. SCINTILLATION DETECTORS AND STORAGE PHOSPHORS ... 205
6.4.1. Basic principles ... 205
6.4.2. Light sensors ... 206
6.4.3. Scintillator materials ... 209

CHAPTER 7. ELECTRONICS RELATED TO NUCLEAR MEDICINE IMAGING DEVICES 214

7.1. INTRODUCTION 214
7.2. PRIMARY RADIATION DETECTION PROCESSES 215
7.2.1 Scintillation counters 215
7.2.2. Gas filled detection systems 216
7.2.3. Semiconductor detectors 216
7.3. IMAGING DETECTORS 217
7.3.1. The gamma camera 217
7.3.2. The positron camera 218
7.3.3. Multiwire proportional chamber based X ray and γ ray imagers 219
7.3.4. Semiconductor imagers 220
7.3.5. The autoradiography imager 221
7.4. SIGNAL AMPLIFICATION 222
7.4.1. Typical amplifier 222
7.4.2. Properties of amplifiers 224
7.5. SIGNAL PROCESSING 226
7.5.1. Analogue signal utilization 226
7.5.2. Signal digitization 226
7.5.3. Production and use of timing information........... 228
7.6. OTHER ELECTRONICS REQUIRED BY IMAGING SYSTEMS 230
7.6.1. Power supplies 230
7.6.2. Uninterruptible power supplies 231
7.6.3. Oscilloscopes 231
7.7. SUMMARY 232

CHAPTER 8. GENERIC PERFORMANCE MEASURES 234

8.1. INTRINSIC AND EXTRINSIC MEASURES 234
8.1.1. Generic nuclear medicine imagers 234
8.1.2. Intrinsic performance 236
8.1.3. Extrinsic performance 236
8.2. ENERGY RESOLUTION 237
8.2.1. Energy spectrum 237
8.2.2. Intrinsic measurement — energy resolution 238
8.2.3. Impact of energy resolution on extrinsic imager performance 239

8.3. SPATIAL RESOLUTION 240
8.3.1. Spatial resolution blurring 240
8.3.2. General measures of spatial resolution 241
8.3.3. Intrinsic measurement — spatial resolution 242
8.3.4. Extrinsic measurement — spatial resolution 242
8.4. TEMPORAL RESOLUTION 244
8.4.1. Intrinsic measurement — temporal resolution 244
8.4.2. Dead time 244
8.4.3. Count rate performance measures 246
8.5. SENSITIVITY 247
8.5.1. Image noise and sensitivity 247
8.5.2. Extrinsic measure — sensitivity 248
8.6. IMAGE QUALITY 249
8.6.1. Image uniformity 249
8.6.2. Resolution/noise trade-off 249
8.7. OTHER PERFORMANCE MEASURES 250

CHAPTER 9. PHYSICS IN THE RADIOPHARMACY 251

9.1. THE MODERN RADIONUCLIDE CALIBRATOR 251
9.1.1. Construction of dose calibrators 251
9.1.2. Calibration of dose calibrators 253
9.1.3. Uncertainty of activity measurements 254
9.1.4. Measuring pure β emitters 258
9.1.5. Problems arising from radionuclide contaminants 259
9.2. DOSE CALIBRATOR ACCEPTANCE TESTING AND QUALITY CONTROL 260
9.2.1. Acceptance tests 260
9.2.2. Quality control 262
9.3. STANDARDS APPLYING TO DOSE CALIBRATORS 262
9.4. NATIONAL ACTIVITY INTERCOMPARISONS 263
9.5. DISPENSING RADIOPHARMACEUTICALS FOR INDIVIDUAL PATIENTS 264
9.5.1. Adjusting the activity for differences in patient size and weight 264
9.5.2. Paediatric dosage charts 264
9.5.3. Diagnostic reference levels in nuclear medicine 266
9.6. RADIATION SAFETY IN THE RADIOPHARMACY 269
9.6.1. Surface contamination limits 269
9.6.2. Wipe tests and daily surveys 270
9.6.3. Monitoring of staff finger doses during dispensing . . . 270

9.7. PRODUCT CONTAINMENT ENCLOSURES 271
9.7.1. Fume cupboards 271
9.7.2. Laminar flow cabinets 272
9.7.3. Isolator cabinets 273
9.8. SHIELDING FOR RADIONUCLIDES 274
9.8.1. Shielding for γ, β and positron emitters 274
9.8.2. Transmission factors for lead and concrete 278
9.9. DESIGNING A RADIOPHARMACY 280
9.10. SECURITY OF THE RADIOPHARMACY 282
9.11. RECORD KEEPING 283
9.11.1. Quality control records 283
9.11.2. Records of receipt of radioactive materials 283
9.11.3. Records of radiopharmaceutical preparation and dispensing 284
9.11.4. Radioactive waste records 284

CHAPTER 10. NON-IMAGING DETECTORS AND COUNTERS 287

10.1. INTRODUCTION 287
10.2. OPERATING PRINCIPLES OF RADIATION DETECTORS . 287
10.2.1. Ionization detectors 288
10.2.2. Scintillation detectors 292
10.3. RADIATION DETECTOR PERFORMANCE 294
10.3.1. Sensitivity 294
10.3.2. Energy resolution 295
10.3.3. Count rate performance (‘speed’) 296
10.4. DETECTION AND COUNTING DEVICES 298
10.4.1. Survey meters 298
10.4.2. Dose calibrator 299
10.4.3. Well counter 299
10.4.4. Intra-operative probes 300
10.4.5. Organ uptake probe 302
10.5. QUALITY CONTROL OF DETECTION AND COUNTING DEVICES 305
10.5.1. Reference sources 305
10.5.2. Survey meter 306
10.5.3. Dose calibrator 307
10.5.4. Well counter 310
10.5.5. Intra-operative probe 310
10.5.6. Organ uptake probe 311

CHAPTER 11. NUCLEAR MEDICINE IMAGING DEVICES.......... 312

11.1. INTRODUCTION 312
11.2. GAMMA CAMERA SYSTEMS 312
11.2.1. Basic principles 312
11.2.2. The Anger camera 314
11.2.3. SPECT.. 341
11.3. PET SYSTEMS 353
11.3.1. Principle of annihilation coincidence detection 353
11.3.2. Design considerations for PET systems 356
11.3.3. Detector systems 362
11.3.4. Data acquisition............................. 369
11.3.5. Data corrections............................. 380
11.4. SPECT/CT AND PET/CT SYSTEMS.................. 392
11.4.1. CT uses in emission tomography 392
11.4.2. SPECT/CT..................................... 393
11.4.3. PET/CT 394

CHAPTER 12. COMPUTERS IN NUCLEAR MEDICINE............. 398

12.1. PHENOMENAL INCREASE IN COMPUTING CAPABILITIES 398
12.1.1. Moore's law.................................. 398
12.1.2. Hardware versus 'peopleware' 398
12.1.3. Future trends 399
12.2. STORING IMAGES ON A COMPUTER.................. 400
12.2.1. Number systems 400
12.2.2. Data representation 401
12.2.3. Images and volumes 403
12.3. IMAGE PROCESSING 405
12.3.1. Spatial frequencies........................... 406
12.3.2. Sampling requirements 412
12.3.3. Convolution.................................. 412
12.3.4. Filtering..................................... 414
12.3.5. Band-pass filters 416
12.3.6. Deconvolution................................ 421
12.3.7. Image restoration filters....................... 422
12.3.8. Other processing 424
12.4. DATA ACQUISITION 425
12.4.1. Acquisition matrix size and spatial resolution........ 426
12.4.2. Static and dynamic planar acquisition.............. 426

12.4.3. SPECT 427
12.4.4. PET acquisition 428
12.4.5. Gated acquisition 430
12.4.6. List-mode 431
12.5. FILE FORMAT 431
12.5.1. File format design 432
12.5.2. Common image file formats 435
12.5.3. Movie formats 437
12.5.4. Nuclear medicine data requirements 437
12.5.5. Common nuclear medicine data storage formats 442
12.6. INFORMATION SYSTEM 443
12.6.1. Database 443
12.6.2. Hospital information system 445
12.6.3. Radiology information system 445
12.6.4. Picture archiving and communication system 446
12.6.5. Scheduling 447
12.6.6. Broker 447
12.6.7. Security 447

CHAPTER 13. IMAGE RECONSTRUCTION 449

13.1. INTRODUCTION 449
13.2. ANALYTICAL RECONSTRUCTION 450
13.2.1. Two dimensional tomography 451
13.2.2. Frequency–distance relation 456
13.2.3. Fully 3-D tomography 457
13.2.4. Time of flight PET 466
13.3. ITERATIVE RECONSTRUCTION 468
13.3.1. Introduction 468
13.3.2. Optimization algorithms 473
13.3.3. Maximum-likelihood expectation-maximization 479
13.3.4. Acceleration 485
13.3.5. Regularization 488
13.3.6. Corrections 495
13.4. NOISE ESTIMATION 507
13.4.1. Noise propagation in filtered back projection 507
13.4.2. Noise propagation in maximum-likelihood expectation-maximization 508

CHAPTER 14. NUCLEAR MEDICINE IMAGE DISPLAY 512

14.1. INTRODUCTION 512
14.2. DIGITAL IMAGE DISPLAY AND VISUAL PERCEPTION. . 513
14.2.1. Display resolution 514
14.2.2. Contrast resolution. 515
14.3. DISPLAY DEVICE HARDWARE 516
14.3.1. Display controller 516
14.3.2. Cathode ray tube 517
14.3.3. Liquid crystal display panel. 519
14.3.4. Hard copy devices 521
14.4. GREY SCALE DISPLAY 521
14.4.1. Grey scale standard display function. 522
14.5. COLOUR DISPLAY 525
14.5.1. Colour and colour gamut. 528
14.6. IMAGE DISPLAY MANIPULATION 530
14.6.1. Histograms. 530
14.6.2. Windowing and thresholding. 530
14.6.3. Histogram equalization 532
14.7. VISUALIZATION OF VOLUME DATA 533
14.7.1. Slice mode 533
14.7.2. Volume mode. 534
14.7.3. Polar plots of myocardial perfusion imaging 538
14.8. DUAL MODALITY DISPLAY 540
14.9. DISPLAY MONITOR QUALITY ASSURANCE. 541
14.9.1. Acceptance testing. 542
14.9.2. Routine quality control 542

CHAPTER 15. DEVICES FOR EVALUATING IMAGING SYSTEMS . . . 547

15.1. DEVELOPING A QUALITY MANAGEMENT SYSTEM APPROACH TO INSTRUMENT QUALITY ASSURANCE. . 547
15.1.1. Methods for routine quality assurance procedures 547
15.2. HARDWARE (PHYSICAL) PHANTOMS 550
15.2.1. Gamma camera phantoms 550
15.2.2. SPECT phantoms. 558
15.2.3. PET phantoms 568
15.3. COMPUTATIONAL MODELS 575
15.3.1. Emission tomography simulation toolkits. 577

15.4. ACCEPTANCE TESTING . . . 578
15.4.1. Introduction . . . 578
15.4.2. Procurement and pre-purchase evaluations . . . 580
15.4.3. Acceptance testing as a baseline for regular quality assurance . . . 583
15.4.4. What to do if the instrument fails acceptance testing . . . 584
15.4.5. Meeting the manufacturer's specifications . . . 584

CHAPTER 16. FUNCTIONAL MEASUREMENTS IN NUCLEAR MEDICINE . . . 587

16.1. INTRODUCTION . . . 587
16.2. NON-IMAGING MEASUREMENTS . . . 588
16.2.1. Renal function measurements . . . 588
16.2.2. ^{14}C breath tests . . . 591
16.3. IMAGING MEASUREMENTS . . . 591
16.3.1. Thyroid . . . 592
16.3.2. Renal function . . . 594
16.3.3. Lung function . . . 596
16.3.4 Gastric function . . . 596
16.3.5 Cardiac function . . . 599

CHAPTER 17. QUANTITATIVE NUCLEAR MEDICINE . . . 608

17.1. PLANAR WHOLE BODY BIODISTRIBUTION MEASUREMENTS . . . 608
17.2. QUANTITATION IN EMISSION TOMOGRAPHY . . . 609
17.2.1. Region of interest . . . 609
17.2.2. Use of standard . . . 610
17.2.3. Partial volume effect and the recovery coefficient . . . 610
17.2.4. Quantitative assessment . . . 612
17.2.5. Estimation of activity . . . 616
17.2.6. Evaluation of image quality . . . 618

CHAPTER 18. INTERNAL DOSIMETRY . . . 621

18.1. THE MEDICAL INTERNAL RADIATION DOSE FORMALISM . . . 621
18.1.1. Basic concepts . . . 621
18.1.2. The time-integrated activity in the source region . . . 626

18.1.3. Absorbed dose rate per unit activity (S value). 628
18.1.4. Strengths and limitations inherent in the formalism . . . 631
18.2. INTERNAL DOSIMETRY IN CLINICAL PRACTICE 635
18.2.1. Introduction 635
18.2.2. Dosimetry on an organ level 636
18.2.3. Dosimetry on a voxel level 637

CHAPTER 19. RADIONUCLIDE THERAPY 641

19.1. INTRODUCTION 641
19.2. THYROID THERAPIES 642
19.2.1. Benign thyroid disease 642
19.2.2. Thyroid cancer. 643
19.3. PALLIATION OF BONE PAIN. 645
19.3.1. Treatment specific issues. 646
19.4. HEPATIC CANCER. 646
19.4.1. Treatment specific issues. 647
19.5. NEUROENDOCRINE TUMOURS. 647
19.5.1. Treatment specific issues. 648
19.6. NON-HODGKIN'S LYMPHOMA 649
19.6.1. Treatment specific issues. 649
19.7. PAEDIATRIC MALIGNANCIES 650
19.7.1. Thyroid cancer. 651
19.7.2. Neuroblastoma. 651
19.8. ROLE OF THE PHYSICIST 652
19.9. EMERGING TECHNOLOGY. 654
19.10. CONCLUSIONS 656

CHAPTER 20. MANAGEMENT OF THERAPY PATIENTS 658

20.1. INTRODUCTION 658
20.2. OCCUPATIONAL EXPOSURE 658
20.2.1. Protective equipment and tools 658
20.2.2. Individual monitoring 659
20.3. RELEASE OF THE PATIENT. 659
20.3.1. The decision to release the patient. 660
20.3.2. Specific instructions for releasing the radioactive patient 662
20.4. PUBLIC EXPOSURE 665
20.4.1. Visitors to patients. 665
20.4.2. Radioactive waste 665

20.5. RADIONUCLIDE THERAPY TREATMENT ROOMS AND WARDS 666
20.5.1. Shielding for control of external dose 666
20.5.2. Designing for control of contamination 668
20.6. OPERATING PROCEDURES 668
20.6.1. Transport of therapy doses 669
20.6.2. Administration of therapeutic radiopharmaceuticals 669
20.6.3. Error prevention 670
20.6.4. Exposure rates and postings 670
20.6.5. Patient care in the treating facility 672
20.6.6. Contamination control procedures 673
20.7. CHANGES IN MEDICAL STATUS 674
20.7.1. Emergency medical procedures 675
20.7.2. The radioactive patient in the operating theatre 675
20.7.3. Radioactive patients on dialysis 676
20.7.4. Re-admission of patients to the treating institution 676
20.7.5. Transfer to another health care facility 677
20.8. DEATH OF THE PATIENT 677
20.8.1. Death of the patient following radionuclide therapy 678
20.8.2. Organ donation 679
20.8.3. Precautions during autopsy 679
20.8.4. Preparation for burial and visitation 680
20.8.5. Cremation 681

APPENDIX I: ARTEFACTS AND TROUBLESHOOTING 684

APPENDIX II: RADIONUCLIDES OF INTEREST IN DIAGNOSTIC AND THERAPEUTIC NUCLEAR MEDICINE 719

ABBREVIATIONS 723
SYMBOLS 729
CONTRIBUTORS TO DRAFTING AND REVIEW 735

CHAPTER 1

BASIC PHYSICS FOR NUCLEAR MEDICINE

E.B. PODGORSAK
Department of Medical Physics,
McGill University,
Montreal, Canada

A.L. KESNER
Division of Human Health,
International Atomic Energy Agency,
Vienna

P.S. SONI
Medical Cyclotron Facility,
Board of Radiation and Isotope Technology,
Bhabha Atomic Research Centre,
Mumbai, India

1.1. INTRODUCTION

The technologies used in nuclear medicine for diagnostic imaging have evolved over the last century, starting with Röntgen's discovery of X rays and Becquerel's discovery of natural radioactivity. Each decade has brought innovation in the form of new equipment, techniques, radiopharmaceuticals, advances in radionuclide production and, ultimately, better patient care. All such technologies have been developed and can only be practised safely with a clear understanding of the behaviour and principles of radiation sources and radiation detection. These central concepts of basic radiation physics and nuclear physics are described in this chapter and should provide the requisite knowledge for a more in depth understanding of the modern nuclear medicine technology discussed in subsequent chapters.

1.1.1. Fundamental physical constants

The chapter begins with a short list of physical constants of importance to general physics as well as to nuclear and radiation physics. The data listed below were taken from the CODATA set of values issued in 2006 and are available from a web site supported by the National Institute of Science and Technology in Washington, DC, United States of America: http://physics.nist.gov/cuu/Constants

- Avogadro's number: $N_A = 6.022 \times 10^{23}$ mol^{-1} or 6.022×10^{23} atoms/mol.
- Speed of light in vacuum: $c = 2.998 \times 10^8$ m/s $\approx 3 \times 10^8$ m/s.
- Electron charge: $e = 1.602 \times 10^{-19}$ C.
- Electron and positron rest mass: $m_e = 0.511$ MeV/c^2.
- Proton rest mass: $m_p = 938.3$ MeV/c^2.
- Neutron rest mass: $m_n = 939.6$ MeV/c^2.
- Atomic mass unit: u = 931.5 MeV/c^2.
- Planck's constant: $h = 6.626 \times 10^{-34}$ J · s.
- Electric constant (permittivity of vacuum): $\varepsilon_0 = 8.854 \times 10^{-12}$ C · V^{-1} · m^{-1}.
- Magnetic constant (permeability of vacuum): $\mu_0 = 4\pi \times 10^{-7}$ V · s · A^{-1} · m^{-1}.
- Newtonian gravitation constant: $G = 6.672 \times 10^{-11}$ m^3 · kg^{-1} · s^{-2}.
- Proton mass/electron mass: $m_p/m_e = 1836.0$.
- Specific charge of electron: $e/m_e = 1.758 \times 10^{11}$ C/kg.

1.1.2. Physical quantities and units

A physical quantity is defined as a quantity that can be used in mathematical equations of science and technology. It is characterized by its numerical value (magnitude) and associated unit. The following rules apply to physical quantities and their units in general:

- Symbols for physical quantities are set in italics (sloping type), while symbols for units are set in roman (upright) type (e.g. $m = 21$ kg; $E = 15$ MeV; $K = 220$ Gy).
- Superscripts and subscripts used with physical quantities are set in italics if they represent variables, quantities or running numbers; they are in roman type if they are descriptive (e.g. N_x, λ_m but λ_{max}, E_{ab}, μ_{tr}).
- Symbols for vector quantities are set in bold italics.

The currently used metric system of units is known as the International System of Units (SI). The system is founded on base units for seven basic physical quantities. All other quantities and units are derived from the seven base quantities and units. The seven base SI quantities and their units are:

(a) Length l: metre (m).
(b) Mass m: kilogram (kg).
(c) Time t: second (s).
(d) Electric current I: ampere (A).

(e) Temperature T: kelvin (K).
(f) Amount of substance: mole (mol).
(g) Luminous intensity: candela (cd).

Examples of basic and derived physical quantities and their units are given in Table 1.1.

TABLE 1.1. BASIC QUANTITIES AND SEVERAL DERIVED PHYSICAL QUANTITIES AND THEIR UNITS IN THE INTERNATIONAL SYSTEM OF UNITS AND IN RADIATION PHYSICS

Physical quantity	Symbol	SI unit	Units commonly used in radiation physics	Conversion
Length	l	m	nm, Å, fm	1 m = 10^9 nm = 10^{10} Å = 10^{15} fm
Mass	m	kg	MeV/c^2	1 MeV/c^2 = 1.78×10^{-30} kg
Time	t	s	ms, μs, ns, ps	1 s = 10^3 ms = 10^6 μs = 10^9 ns = 10^{12} ps
Current	I	A	mA, μA, nA, pA	1 A = 10^3 mA = 10^6 μA = 10^9 nA
Temperature	T	K		T (in K) = T (in °C) + 273.16
Mass density	ρ	kg/m^3	g/cm^3	1 kg/m^3 = 10^{-3} g/cm^3
Current density	j	A/m^2		
Velocity	υ	m/s		
Acceleration	a	m/s^2		
Frequency	ν	Hz		1 Hz = 1 s^{-1}
Electric charge	q	C	e	1 e = 1.602×10^{-19} C
Force	F	N		1 N = 1 kg · m · s^{-2}
Pressure	P	Pa	760 torr = 101.3 kPa	1 Pa = 1 N/m^2 = 7.5×10^{-3} torr
Momentum	p	N · s		1 N · s = 1 kg · m · s^{-1}
Energy	E	J	eV, keV, MeV	1 eV = 1.602×10^{-19} J = 10^{-3} keV
Power	P	W		1 W = 1 J/s = 1 V · A

1.1.3. Classification of radiation

Radiation, the transport of energy by electromagnetic waves or atomic particles, can be classified into two main categories depending on its ability to ionize matter. The ionization potential of atoms, i.e. the minimum energy required to ionize an atom, ranges from a few electronvolts for alkali elements to 24.6 eV for helium which is in the group of noble gases. Ionization potentials for all other atoms are between the two extremes.

— Non-ionizing radiation cannot ionize matter because its energy per quantum is below the ionization potential of atoms. Near ultraviolet radiation, visible light, infrared photons, microwaves and radio waves are examples of non-ionizing radiation.
— Ionizing radiation can ionize matter either directly or indirectly because its quantum energy exceeds the ionization potential of atoms. X rays, γ rays, energetic neutrons, electrons, protons and heavier particles are examples of ionizing radiation.

1.1.4. Classification of ionizing radiation

Ionizing radiation is radiation that carries enough energy per quantum to remove an electron from an atom or a molecule, thus introducing a reactive and potentially damaging ion into the environment of the irradiated medium. Ionizing radiation can be categorized into two types: (i) directly ionizing radiation and (ii) indirectly ionizing radiation. Both directly and indirectly ionizing radiation can traverse human tissue, thereby enabling the use of ionizing radiation in medicine for both imaging and therapeutic procedures.

— Directly ionizing radiation consists of charged particles, such as electrons, protons, α particles and heavy ions. It deposits energy in the medium through direct Coulomb interactions between the charged particle and orbital electrons of atoms in the absorber.
— Indirectly ionizing radiation consists of uncharged (neutral) particles which deposit energy in the absorber through a two-step process. In the first step, the neutral particle releases or produces a charged particle in the absorber which, in the second step, deposits at least part of its kinetic energy in the absorber through Coulomb interactions with orbital electrons of the absorber in the manner discussed above for directly ionizing charged particles.

1.1.5. Classification of indirectly ionizing photon radiation

Indirectly ionizing photon radiation consists of three main categories: (i) ultraviolet, (ii) X ray and (iii) γ ray. Ultraviolet photons are of limited use in medicine. Radiation used in imaging and/or treatment of disease consists mostly of photons of higher energy, such as X rays and γ rays. The commonly accepted difference between the two is based on the radiation's origin. The term 'γ ray' is reserved for photon radiation that is emitted by the nucleus or from other particle decays. The term 'X ray', on the other hand, refers to radiation emitted by electrons, either orbital electrons or accelerated electrons (e.g. bremsstrahlung type radiation).

With regard to their origin, the photons of the indirectly ionizing radiation type fall into four categories: characteristic (fluorescence) X rays, bremsstrahlung X rays, photons resulting from nuclear transitions and annihilation quanta.

1.1.6. Characteristic X rays

Orbital electrons have a natural tendency to configure themselves in such a manner that they inhabit a minimal energy state for the atom. When a vacancy is opened within an inner shell, as a result of an ionization or excitation process, an outer shell electron will make a transition to fill the vacancy, usually within a nanosecond for solid materials. The energy liberated in this transition may be released in the form of a characteristic (fluorescence) photon of energy equal to the difference between the binding energies of the initial and final vacancies. Since different elements have different binding energies for their electronic shells, the energy of the photon released in this process will be characteristic of the particular atom. Rather than being emitted as a characteristic photon, the transition energy may also be transferred to an orbital electron that is then emitted with kinetic energy that is equal to the transition energy less the electron binding energy. The emitted orbital electron is called an Auger electron.

1.1.7. Bremsstrahlung

The word 'bremsstrahlung' can be translated from its original German term as 'braking radiation', and is a name aptly assigned to the phenomenon. When light charged particles (electrons and positrons) are slowed down or 'negatively' accelerated (decelerated) by interactions with other charged particles in matter (e.g. by atomic nuclei), the kinetic energy that they lose is converted to electromagnetic radiation, referred to as bremsstrahlung radiation. The energy spectrum of bremsstrahlung is non-discrete (i.e. continuous) and ranges between zero and the kinetic energy of the initial charged particle. Bremsstrahlung plays a central role in modern imaging and therapeutic equipment, since it can be used to produce X rays on demand from an electrical energy source. The power emitted in the form of bremsstrahlung photons is proportional to the square of the particle's charge and the square of the particle's acceleration.

1.1.8. Gamma rays

When a nuclear reaction or spontaneous nuclear decay occurs, the process may leave the product (daughter) nucleus in an excited state. The nucleus can then make a transition to a more stable state by emitting a γ ray photon and the process is referred to as γ decay. The energy of the photon emitted in γ decay is characteristic of the nuclear energy transition, but the recoil of the emitting atom produces a spectrum centred on the characteristic energy. Gamma rays typically have energies above 100 keV and wavelengths less than 0.1 Å.

1.1.9. Annihilation quanta

When a parent nucleus undergoes β plus decay or a high energy photon interacts with the electric field of either the nucleus or the orbital electron, an energetic positron may be produced. In moving through an absorber medium, the positron loses most of its kinetic energy as a result of Coulomb interactions with absorber atoms. These interactions result in collision loss when the interaction is with an orbital electron of the absorber atom and in radiation loss (bremsstrahlung) when the interaction is with the nucleus of the absorber atom. Generally, after the positron loses all of its kinetic energy through collision and radiation losses, it will undergo a final collision with an available orbital electron (due to the Coulomb attractive force between the positively charged positron and a local negatively charged electron) in a process called positron annihilation. During annihilation, the positron and electron disappear and are replaced by two oppositely directed annihilation quanta, each with an energy of 0.511 MeV. This process satisfies a number of conservation laws: conservation of electric charge, conservation of linear momentum, conservation of angular momentum and conservation of total energy.

A percentage of positron annihilations occur before the positron expends all of its kinetic energy and the process is then referred to as in-flight annihilation. The two quanta emitted in in-flight annihilation are not of identical energies and do not necessarily move in absolute opposite directions.

1.1.10. Radiation quantities and units

Accurate measurement of radiation is very important in all medical uses of radiation, be it for diagnosis or treatment of disease. In diagnostic imaging procedures, image quality must be optimized, so as to obtain the best possible image with the lowest possible radiation dose to the patient to minimize the risk of morbidity. In radiotherapy, the prescribed dose must be delivered accurately and precisely to maximize the tumour control probability (TCP) and to minimize the normal tissue complication probability (NTCP). In both instances, the risk of morbidity includes acute radiation effects (radiation injury) as well as late radiation-induced effects, such as induction of cancer and genetic damage.

Several quantities and units were introduced for the purpose of quantifying radiation and the most important of these are listed in Table 1.2. Also listed are the definitions for the various quantities and the relationships between the old units and the SI units for these quantities. The definitions of radiation related physical quantities are as follows:

— *Exposure X* is related to the ability of photons to ionize air. Its unit, roentgen (R), is defined as a charge of 2.58×10^{-4} coulombs produced per kilogram of air.
— *Kerma K* (acronym for kinetic energy released in matter) is defined for indirectly ionizing radiation (photons and neutrons) as energy transferred to charged particles per unit mass of the absorber.
— *Dose* (also referred to as absorbed dose) is defined as energy absorbed per unit mass of medium. Its SI unit, gray (Gy), is defined as 1 joule of energy absorbed per kilogram of medium.
— *Equivalent dose* H_T is defined as the dose multiplied by a radiation weighting factor w_R. When different types of radiation are present, H_T is defined as the sum of all of the individual weighted contributions. The SI unit of equivalent dose is the sievert (Sv).
— *Effective dose E* of radiation is defined as the equivalent dose H_T multiplied by a tissue weighting factor w_T. The SI unit of effective dose is also the sievert (Sv).
— *Activity* $\mathcal{A}$ of a radioactive substance is defined as the number of nuclear decays per time. Its SI unit, becquerel (Bq), corresponds to one decay per second.

TABLE 1.2. RADIATION QUANTITIES, UNITS AND CONVERSION BETWEEN OLD AND SI UNITS

Quantity	Definition	SI unit	Old unit	Conversion
Exposure X	$X = \dfrac{\Delta Q}{\Delta m_{air}}$	$2.58 \times \dfrac{10^{-4}\,\mathrm{C}}{\mathrm{kg\ air}}$	$1\ \mathrm{R} = \dfrac{1\ \mathrm{esu}}{\mathrm{cm}^3\ \mathrm{air_{STP}}}$	$1\ \mathrm{R} = 2.58 \times \dfrac{10^{-4}\,\mathrm{C}}{\mathrm{kg\ air}}$
Kerma K	$K = \dfrac{\Delta E_{tr}}{\Delta m}$	$1\ \mathrm{Gy} = 1\dfrac{\mathrm{J}}{\mathrm{kg}}$	—	—
Dose D	$D = \dfrac{\Delta E_{ab}}{\Delta m}$	$1\ \mathrm{Gy} = 1\dfrac{\mathrm{J}}{\mathrm{kg}}$	$1\ \mathrm{rad} = 100\ \dfrac{\mathrm{erg}}{\mathrm{g}}$	1 Gy = 100 rad
Equivalent dose H_T	$H_T = D w_R$	1 Sv	1 rem	1 Sv = 100 rem
Effective dose E	$E = H_T w_T$	1 Sv	1 rem	1 Sv = 100 rem
Activity $\mathcal{A}$	$\mathcal{A} = \lambda N$	$1\ \mathrm{Bq} = 1\ \mathrm{s}^{-1}$	$1\ \mathrm{Ci} = 3.7 \times 10^{10}\ \mathrm{s}^{-1}$	$1\ \mathrm{Bq} = \dfrac{1\ \mathrm{Ci}}{3.7 \times 10^{10}}$

1.2. BASIC DEFINITIONS FOR ATOMIC STRUCTURE

The constituent particles forming an atom are protons, neutrons and electrons. Protons and neutrons are known as nucleons and form the nucleus of the atom. Protons have a positive charge, neutrons are neutral and electrons have a negative charge mirroring that of a proton. In comparison to electrons, protons and neutrons have a relatively large mass exceeding the electron mass by a factor of almost 2000 (note: $m_p/m_e = 1836$).

The following general definitions apply to atomic structure:

— Atomic number Z is the number of protons and number of electrons in an atom.
— Atomic mass number A is the number of nucleons in an atom, i.e. the number of protons Z plus the number of neutrons N in an atom: $A = Z + N$.
— Atomic mass m_a is the mass of a specific isotope expressed in atomic mass units u, where 1 u is equal to one twelfth of the mass of the ^{12}C atom (unbound, at rest and in the ground state) or 931.5 MeV/c^2. The atomic mass is smaller than the sum of the individual masses of the constituent particles because of the intrinsic energy associated with binding the particles (nucleons) within the nucleus. On the other hand, the atomic mass is larger than the nuclear mass M because the atomic mass includes the mass contribution of Z orbital

electrons while the nuclear mass M does not. The binding energy of orbital electrons to the nucleus is ignored in the definition of the atomic mass.

While for ^{12}C the atomic mass is exactly 12 u, for all other atoms m_a does not exactly match the atomic mass number A. However, for all atomic entities, A (an integer) and m_a are very similar to one another and often the same symbol (A) is used for the designation of both. The mass in grams equal to the average atomic mass of a chemical element is referred to as the mole (mol) of the element and contains exactly 6.022×10^{23} atoms. This number is referred to as the Avogadro constant N_A of entities per mole. The atomic mass number of all elements is, thus, defined such that A grams of every element contain exactly N_A atoms. For example, the atomic mass of natural cobalt is 58.9332 u. Thus, one mole of natural cobalt has a mass of 58.9332 g and by definition contains 6.022×10^{23} entities (cobalt atoms) per mole of cobalt.

The number of atoms N_a per mass of an element is given as:

$$\frac{N_a}{m} = \frac{N_A}{A} \tag{1.1}$$

The number of electrons per volume of an element is:

$$Z\frac{N_a}{V} = \rho Z\frac{N_a}{m} = \rho Z\frac{N_A}{A} \tag{1.2}$$

The number of electrons per mass of an element is:

$$Z\frac{N_a}{m} = Z\frac{N_A}{A} \tag{1.3}$$

It should be noted that $Z/A \approx 0.5$ for all elements with one notable exception of hydrogen for which $Z/A = 1$. Actually, Z/A slowly decreases from 0.5 for low Z elements to 0.4 for high Z elements. For example, Z/A for ^4He is 0.5, for ^{60}Co is 0.45 and for ^{235}U is 0.39.

If it is assumed that the mass of a molecule is equal to the sum of the masses of the atoms that make up the molecule, then, for any molecular compound, there are N_A molecules per mole of the compound where the mole in grams is defined as the sum of the atomic mass numbers of the atoms making up the molecule. For example, 1 mole of water (H_2O) is 18 g of water and 1 mole of carbon dioxide (CO_2) is 44 g of carbon dioxide. Thus, 18 g of water or 44 g of carbon dioxide contain exactly N_A molecules (or 3 N_A atoms, since each molecule of water and carbon dioxide contains three atoms).

1.2.1. Rutherford model of the atom

At the beginning of the 20th century, the structure of the atom was not well known. Scientific pioneers such as Dalton, Mendeleev and Thomson, among others, were developing a common theory through their endeavours. Often noted as a significant contribution to the modern understanding of the atom, is the work performed by Rutherford and his colleagues Geiger and Marsden in 1909. Through observation of the behaviour of positively charged α particles traversing a thin gold foil, Rutherford concluded that the positive charge and most of the mass of the atom are concentrated in the atomic nucleus (diameter of a few femtometres) and negative electrons are spread over the periphery of the atom (diameter of a few ångströms). This work was significant because it introduced a new specialty of physics (nuclear physics) and demonstrated that the atom is not simply a single particle, but instead is made up of smaller subatomic particles, organized in an atom with well defined characteristics.

1.2.2. Bohr model of the hydrogen atom

Bohr expanded the Rutherford atomic model in 1913 using a set of four postulates that combine classical, non-relativistic mechanics with the concept of angular momentum quantization. The Bohr model of the atom can be said to resemble a 'planetary model' in that the protons and neutrons occupy a dense central region called the nucleus and the electrons orbit the nucleus as planets orbit the sun. The Bohr model introduces the concept that the angular momenta of orbital electrons revolving around the nucleus in allowed orbits, radii of the allowed electronic orbits (shells), velocities of orbital electrons in allowed orbits and binding energies of orbital electrons in allowed orbits within the atom, are restricted to certain discrete states. This means that angular momenta, radii, velocities and binding energies of orbital electrons are quantized.

While scientific theory was later expanded to include the necessary principles of quantum mechanics in our understanding of the atom, the Bohr model is elegant and provides a simplistic, yet practical, view of the atom that is still used for teaching atomic principles, and successfully deals with one-electron entities, such as the hydrogen atom, the singly ionized helium atom and the doubly ionized lithium atom.

1.3. BASIC DEFINITIONS FOR NUCLEAR STRUCTURE

According to the Rutherford–Bohr atomic model, most of the atomic mass is concentrated in the atomic nucleus consisting of Z protons and $(A - Z)$ neutrons, where Z is the atomic number and A the atomic mass number of a given nucleus. In nuclear physics, the convention is to designate a nucleus X as ${}^{A}_{Z}X$, where A is its atomic mass number and Z its atomic number; for example, the ^{60}Co nucleus is identified as ${}^{60}_{27}$Co and the ^{226}Ra nucleus as ${}^{226}_{88}$Ra. The atomic number Z is often omitted in references to an atom because the atom is already identified by its 1–3 letter symbol. In ion physics, the convention is to designate ions with + or – superscripts. For example, ${}^{4}_{2}He^{+}$ stands for a singly ionized helium atom and ${}^{4}_{2}He^{2+}$ stands for a doubly ionized helium atom, also known as the α particle. With regard to relative values of atomic number Z and atomic mass number A of nuclei, the following conventions apply:

— An element may be composed of atoms that all have the same number of protons, i.e. have the same atomic number Z, but have a different number of neutrons (have different atomic mass numbers A). Such atoms of identical Z but differing A are called isotopes of a given element.
— The term 'isotope' is often misused to designate nuclear species. For example, ^{60}Co, ^{137}Cs and ^{226}Ra are not isotopes, since they do not belong to the same element. Rather than isotopes, they should be referred to as nuclides. On the other hand, it is correct to state that deuterium (with a nucleus called deuteron) and tritium (with a nucleus called triton) are heavy isotopes of hydrogen or that ^{59}Co and ^{60}Co are isotopes of cobalt. Thus, the term 'radionuclide' should be used to designate radioactive species; however, the term 'radioisotope' is often used for this purpose.
— A nuclide is an atomic species characterized by its nuclear composition (A, Z and the arrangement of nucleons within the nucleus). The term 'nuclide' refers to all atomic forms of all elements. The term 'isotope' is narrower and only refers to various atomic forms of a given chemical element.

In addition to being classified into isotopic groups (common atomic number Z), nuclides are also classified into groups with a common atomic mass number A (isobars) and a common number of neutrons (isotones). For example, ^{60}Co and ^{60}Ni are isobars with 60 nucleons each (A = 60), and ${}^{67}_{31}$Ga, ${}^{67}_{32}$Ge and ${}^{67}_{33}$As are isobars with atomic mass number 67, while ^{3}H (tritium) and ^{4}He are isotones with two neutrons each ($A - Z = 2$), and ${}^{12}_{6}$C, ${}^{13}_{7}$N and ${}^{14}_{8}$O are isotones with six neutrons each.

A tool for remembering these definitions is as follows: isoto**p**es have the same number of **p**rotons Z; isoto**n**es have the same number of **n**eutrons, $A - Z$; isob**a**rs have the same mass number $\mathbf{A}$.

If a nucleus exists in an excited state for some time, it is said to be in an isomeric (metastable) state. Isomers are, thus, nuclear species that have a common atomic number Z and a common atomic mass number A. For example, ^{99m}Tc is an isomeric state of ^{99}Tc and ^{60m}Co is an isomeric state of ^{60}Co.

1.3.1. Nuclear radius

The radius R of a nucleus with atomic mass number A is estimated from the following expression:

$$R = R_0 \sqrt[3]{A} \tag{1.4}$$

where R_0 is the nuclear radius constant equal to 1.25 fm. Since the range of A in nature is from 1 to about 250, nuclear radius ranges from about 1 fm for a proton to about 8 fm for heavy nuclei.

1.3.2. Nuclear binding energy

The sum of the masses of the individual components of a nucleus that contains Z protons and $(A - Z)$ neutrons is larger than the actual mass of the nucleus. This difference in mass is called the mass defect (deficit) Δm and its energy equivalent Δmc^2 is called the total binding energy E_B of the nucleus. The total binding energy E_B of a nucleus can, thus, be defined as the energy liberated when Z protons and $(A - Z)$ neutrons are brought together to form the nucleus.

The binding energy per nucleon (E_B/A) in a nucleus (i.e. the total binding energy of a nucleus divided by the number of nucleons in the given nucleus) varies with the number of nucleons A and is of the order of ~8 MeV/nucleon.

A plot of the binding energy per nucleon E_B/A in megaelectronvolts per nucleon against the atomic mass number in the range from 1 to 250 is given in Fig. 1.1 and shows a rapid rise in E_B/A at small atomic mass numbers, a broad maximum of about 8.7 MeV/nucleon around $A \approx 60$ and a gradual decrease in E_B/A at large A. The larger the binding energy per nucleon (E_B/A) of an atom, the larger is the stability of the atom. Thus, the most stable nuclei in nature are the ones with $A \approx 60$ (iron, cobalt, nickel). Nuclei of light elements (small A) are generally less stable than nuclei with $A \approx 60$, and the heaviest nuclei (large A) are also less stable than nuclei with $A \approx 60$.

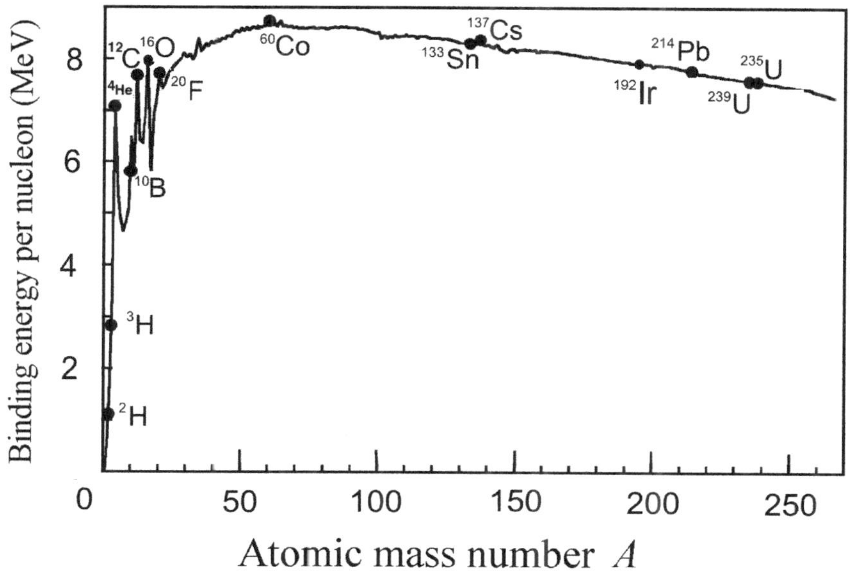

FIG. 1.1. Binding energy per nucleon in megaelectronvolts per nucleon against atomic mass number A. Data are from the National Institute of Science and Technology (NIST).

1.3.3. Nuclear fusion and fission

The peculiar shape of the E_B/A versus A curve (Fig. 1.1) suggests two methods for converting mass into energy: (i) fusion of nuclei at low A and (ii) fission of nuclei at large A:

— Fusion of two nuclei of very small mass, e.g. ${}^{2}_{1}\text{H} + {}^{3}_{1}\text{H} \rightarrow {}^{4}_{2}\text{He} + \text{n}$, will create a more massive nucleus and release a certain amount of energy. Experiments using controlled nuclear fusion for production of energy have so far not been successful in generating a net energy gain, i.e. the amount of energy consumed is still larger than the amount created. However, fusion remains an active field of research and it is reasonable to expect that in the future controlled fusion will play an important role in the production of electrical power.

— Fission attained by bombardment of certain elements of large mass (such as ^{235}U) by thermal neutrons in a nuclear reactor will create two lower mass and more stable nuclei, and transform some mass into kinetic energy of the two product nuclei. Hahn, Strassman, Meitner and Frisch described fission in 1939, and, in 1942, Fermi and colleagues at the University of Chicago carried out the first controlled chain reaction based on nuclear fission.

Since then, fission reactors have become an important means of production of electrical power.

1.3.4. Two-particle collisions and nuclear reactions

A common problem in nuclear physics and radiation dosimetry is the collision of two particles in which a projectile with mass m_1, velocity υ_1 and kinetic energy $(E_K)_1$ strikes a stationary target with mass m_2 and velocity $\upsilon_2 = 0$. The probability or cross-section for a particular collision as well as the collision outcome depends on the physical properties of the projectile (mass, charge, velocity, kinetic energy) and the stationary target (mass, charge).

As shown schematically in Fig. 1.2, the collision between the projectile and the target in the most general case results in an intermediate compound that subsequently decays into two reaction products: one of mass m_3 ejected with velocity υ_3 at an angle θ to the incident projectile direction, and the other of mass m_4 ejected with velocity υ_4 at an angle ϕ to the incident projectile direction.

Two-particle collisions are classified into three categories: (a) elastic scattering, (b) inelastic collisions and (c) nuclear reactions:

(a) Elastic scattering is a special case of a two-particle collision in which the products after the collision are identical to the products before collision, i.e. $m_3 = m_1$ and $m_4 = m_2$, and the total kinetic energy and momentum before the collision are equal to the total kinetic energy and momentum, respectively, after the collision.

(b) In inelastic scattering of a projectile m_1 on the target m_2, similarly to elastic scattering, the reaction products after collision are identical to the initial products, i.e. $m_3 = m_1$ and $m_4 = m_2$; however, the incident projectile transfers a portion of its kinetic energy to the target in the form of not only kinetic energy but also intrinsic excitation energy E^*.

(c) During a nuclear reaction, a collision between a projectile m_1 and a target m_2 takes place and will result in the formation of two reaction products m_3 and m_4, with the products having new atomic numbers. This process is shown schematically in Fig. 1.2. In any nuclear reaction, a number of physical quantities must be conserved, most notably charge, linear momentum and mass–energy. In addition, the sum of atomic numbers Z and the sum of atomic mass numbers A before and after the collision must also be conserved.

The Q value of a nuclear reaction is defined as the difference between the total rest energy before the reaction $(m_1c^2 + m_2c^2)$ and the total rest energy after the reaction $(m_3c^2 + m_4c^2)$ or:

$$Q = (m_1c^2 + m_2c^2) - (m_3c^2 + m_4c^2) \tag{1.5}$$

Each two-particle collision possesses a characteristic Q value that can be either positive, zero or negative. For $Q > 0$, the collision is termed 'exothermic' (also called exoergic) and results in a release of energy; for $Q = 0$, the collision is termed 'elastic' and for $Q < 0$, the collision is termed 'endothermic' (also called endoergic), and to take place, it requires an energy transfer from the projectile to the target. An exothermic reaction can occur spontaneously, while an endothermic reaction cannot take place unless the projectile has kinetic energy exceeding the threshold energy $(E_K)_{thr}$ given as:

$$(E_K)_{thr} = \frac{(m_3c^2 + m_4c^2)^2 - (m_1c^2 + m_2c^2)^2}{2m_2c^2} \approx -Q\left(1 + \frac{m_1}{m_2}\right) \tag{1.6}$$

where m_1c^2, m_2c^2, m_3c^2 and m_4c^2 are the rest energies of the projectile m_1, target m_2 and reaction products m_3 and m_4, respectively.

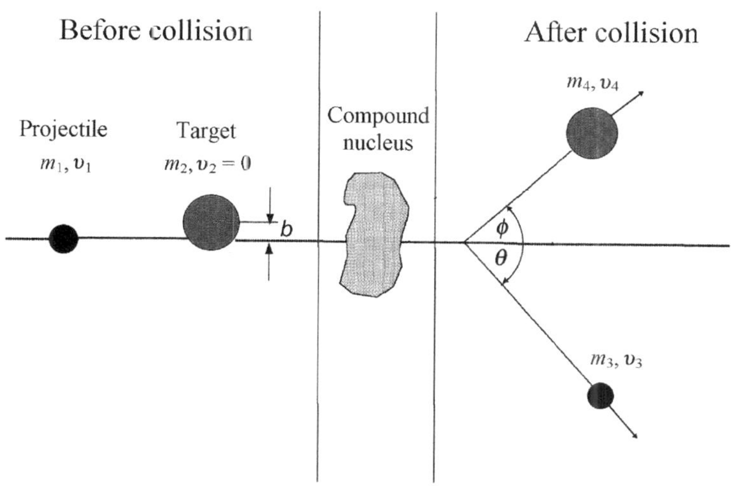

FIG. 1.2. Schematic representation of a two-particle collision of a projectile (incident particle) of mass m_1 and velocity υ_1 striking a stationary target with mass m_2 and velocity $\upsilon_2 = 0$. An intermediate compound entity is formed temporarily that subsequently decays into two reaction products of mass m_3 and m_4.

1.4. RADIOACTIVITY

Radioactivity, also known as radioactive decay, nuclear decay, nuclear disintegration and nuclear transformation, is a spontaneous process by which an unstable parent nucleus emits a particle or electromagnetic radiation and transforms into a more stable daughter nucleus that may or may not be stable. The unstable daughter nucleus will decay further in a decay series until a stable nuclear configuration is reached. Radioactive decay is usually accompanied by emission of energetic particles or γ ray photons or both.

All radioactive decay processes are governed by the same general formalism that is based on the definition of the activity $\mathcal{A}(t)$ and on a characteristic parameter for each radioactive decay process, the radioactive decay constant λ with dimensions of reciprocal time, usually in s^{-1}. The main characteristics of radioactive decay are as follows:

- The radioactive decay constant λ multiplied by a time interval that is much smaller than $1/\lambda$ represents the probability that any particular atom of a radioactive substance containing a large number $N(t)$ of identical radioactive atoms will decay (disintegrate) in that time interval. An assumption is made that λ is independent of the physical environment of a given atom.
- The activity $\mathcal{A}(t)$ of a radioactive substance containing a large number $N(t)$ of identical radioactive atoms represents the total number of decays (disintegrations) per unit time and is defined as a product between $N(t)$ and λ, i.e.:

$$\mathcal{A}(t) = \lambda N(t) \tag{1.7}$$

The SI unit of activity is the becquerel (Bq) given as 1 Bq = 1 s^{-1}. The becquerel and hertz both correspond to s^{-1}, but hertz refers to the frequency of periodic motion, while becquerel refers to activity.

The old unit of activity, the curie (Ci), was initially defined as the activity of 1 g of ^{226}Ra; 1 Ci $\cong 3.7 \times 10^{10}$ s^{-1}.

Subsequently, the activity of 1 g of ^{226}Ra was determined to be 3.665×10^{10} s^{-1}; however, the definition of the activity unit curie (Ci) was kept as 1 Ci = 3.7×10^{10} s^{-1}. Since the unit of activity the becquerel is 1 s^{-1}, the SI unit becquerel (Bq) and the old unit curie (Ci) are related as follows: 1 Ci = 3.7×10^{10} Bq and, consequently, 1 Bq = $(3.7 \times 10^{10})^{-1}$ Ci = 2.703×10^{-11} Ci.

Specific activity a is defined as activity $\mathcal{A}$ per unit mass m, i.e.:

$$a = \frac{\mathcal{A}}{m} = \frac{\lambda N}{m} = \frac{\lambda N_A}{A} \tag{1.8}$$

where N_A is Avogadro's number.

Specific activity a of a radioactive atom depends on the decay constant λ and on the atomic mass number A of the radioactive atom. The units of specific activity are Bq/kg (SI unit) and Ci/g (old unit).

1.4.1. Decay of radioactive parent into a stable or unstable daughter

The simplest form of radioactive decay involves a radioactive parent nucleus P decaying with decay constant λ_P into a stable or unstable daughter nucleus D:

$$P \xrightarrow{\lambda_P} D \tag{1.9}$$

The rate of depletion of the number of radioactive parent nuclei $N_P(t)$ is equal to activity $\mathcal{A}_P(t)$ at time t defined as the product $\lambda N(t)$ in Eq. (1.7). We, thus, have the following expression:

$$\frac{dN_P(t)}{dt} = -\mathcal{A}_P(t) = -\lambda_P N_P(t) \tag{1.10}$$

The fundamental differential equation in Eq. (1.10) for $N_P(t)$ can be rewritten in general integral form:

$$\int_{N_P(0)}^{N_P(t)} \frac{dN_P(t)}{N_P} = -\int_0^t \lambda_P \, dt \tag{1.11}$$

where $N_P(0)$ is the initial condition represented by the number of radioactive nuclei at time $t = 0$.

Assuming that λ_P is constant, Eq. (1.11) can be solved to obtain:

$$\ln\frac{N_P(t)}{N_P(0)} = -\lambda_P t \quad (1.12)$$

or

$$N_P(t) = N_P(0)e^{-\lambda_P t} \quad (1.13)$$

Based on the definition of activity given in Eq. (1.7), the activity of parent nuclei P at time t can be expressed as follows:

$$\mathcal{A}_P(t) = \lambda_P N_P(t) = \lambda_P N_P(0)e^{-\lambda_P t} = \mathcal{A}_P(0)e^{-\lambda_P t} \quad (1.14)$$

where $\mathcal{A}_P(0) = \lambda_P N_P(0)$ is the initial activity of the radioactive substance.

The decay law of Eq. (1.14) applies to all radioactive nuclides irrespective of their mode of decay; however, the decay constant λ_P is different for each parent radioactive nuclide P and is the most important defining characteristic of a radioactive nuclide.

Two special time periods called half-life $(T_{1/2})_P$ and mean or average life τ_P are used to characterize a given radioactive parent substance P. The half-life $(T_{1/2})_P$ of a radioactive substance P is the time during which the number of radioactive nuclei of the substance decays to half of the initial value $N_P(0)$ present at time $t = 0$. It can also be stated that in the time of one half-life the activity $\mathcal{A}_P(t)$ of a radioactive substance decreases to one half of its initial value $\mathcal{A}_P(0) = \lambda_P N_P(0)$:

$$N_P[t = (T_{1/2})_P] = \frac{1}{2}N_P(0) = N_P(0)e^{-\lambda_P (T_{1/2})_P} \quad (1.15)$$

and

$$\mathcal{A}_P[t = (T_{1/2})_P] = \frac{1}{2}\mathcal{A}_P(0) = \mathcal{A}_P(0)e^{-\lambda_P (T_{1/2})_P} \quad (1.16)$$

From Eqs (1.15) and (1.16), it is noted that $e^{-\lambda_P (T_{1/2})_P}$ must equal 1/2, resulting in the following relationship between the decay constant λ_P and half-life $(T_{1/2})_P$:

$$\lambda_P = \frac{\ln 2}{(T_{1/2})_P} = \frac{0.693}{(T_{1/2})_P} \quad (1.17)$$

Mean (average) life τ_P of a radioactive parent P is defined as the time required for the number N_P of radioactive atoms or its activity $\mathcal{A}_P$ to fall to 1/e = 0.368 (or 36.8%) of the initial number of nuclei $N_P(0)$ or of the initial activity $\mathcal{A}_P(0)$, respectively. Thus, the following expressions describe the mean half-life:

$$N_P(t=\tau_P)=\frac{1}{e}N_P(0)=0.368N_P(0)=N_P(0)e^{-\lambda_P\tau_P} \tag{1.18}$$

and

$$\mathcal{A}_P(t=\tau_P)=\frac{1}{e}\mathcal{A}_P(0)=0.368\mathcal{A}_P(0)=\mathcal{A}_P(0)e^{-\lambda_P\tau_P} \tag{1.19}$$

From Eqs (1.18) and (1.19), it is noted that $e^{-\lambda_P\tau_P}$ must be equal to $1/e = e^{-1} = 0.368$, resulting in $\lambda_P\tau_P = 1$ and $\tau_P = 1/\lambda_P$. We now get the following relationship between mean life τ_P and half-life $(T_{1/2})_P$ using Eq. (1.17) and $\tau_P = 1/\lambda_P$:

$$\lambda_P=\frac{\ln 2}{(T_{1/2})_P}=\frac{1}{\tau_P} \tag{1.20}$$

and

$$\tau_P=\frac{(T_{1/2})_P}{\ln 2}=1.44(T_{1/2})_P \tag{1.21}$$

A typical example of a radioactive decay for initial condition $\mathcal{A}_P(t=0)=\mathcal{A}_P(0)$ is shown in Fig. 1.3 with a plot of parent activity $\mathcal{A}_P(t)$ against time t given in Eq. (1.14).

1.4.2. Radioactive series decay

The radioactive decay of parent P into stable daughter D, discussed in Section 1.4.1, is the simplest known radioactive decay process; however, the decay of a radioactive parent P with decay constant λ_P into a radioactive (unstable) daughter D which in turn decays with decay constant λ_D into a stable or unstable grand-daughter G, i.e. ($P \xrightarrow{\lambda_P} D \xrightarrow{\lambda_D} G$), is much more common and results in a radioactive decay series for which the last decay product is stable.

The parent P in the decay series follows a straightforward radioactive decay described by Eq. (1.16) for the rate of change of the number of parent

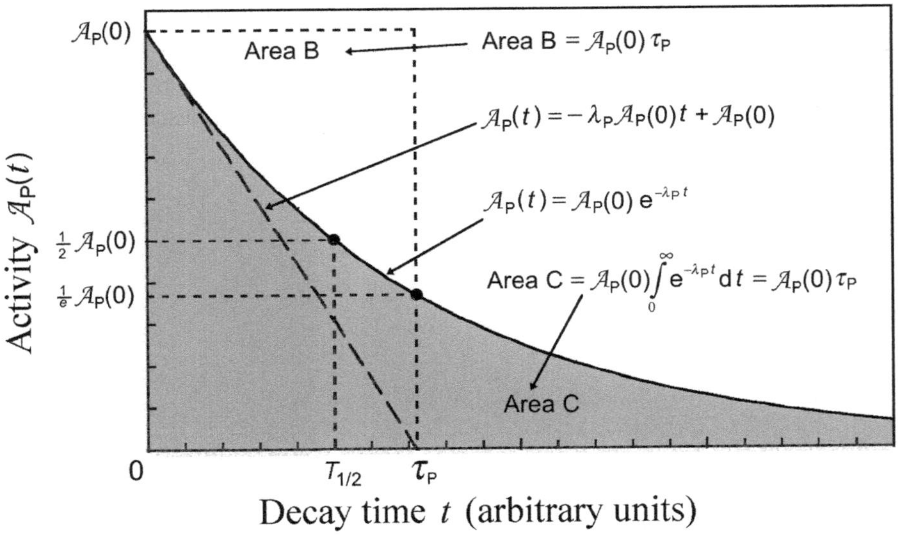

FIG. 1.3. Activity $\mathcal{A}_P(t)$ plotted against time t for a simple decay of a radioactive parent P into a stable or unstable daughter D. The concepts of half-life $(T_{1/2})_P$ and mean life τ_P are also illustrated. The area under the exponential decay curve from $t = 0$ to $t = \infty$ is equal to the product $\mathcal{A}_P(0)\tau_P$ where $\mathcal{A}_P(0)$ is the initial activity of the parent P. The slope of the tangent to the decay curve at $t = 0$ is equal to $\lambda_P\mathcal{A}_P(0)$ and this tangent crosses the abscissa axis at $t = \tau_P$.

nuclei $dN_P(t)/dt$. The rate of change of the number of daughter nuclei $dN_D(t)/dt$, however, is more complicated and consists of two components, one being the supply of new daughter nuclei D through the decay of P given as $\lambda_P N_P(t)$ and the other being the loss of daughter nuclei D from the decay of D to G given as $-\lambda_D N_D(t)$, resulting in the following expression for $dN_D(t)/dt$:

$$\frac{dN_D(t)}{dt} = \lambda_P N_P(t) - \lambda_D N_D(t) = \lambda_P N_P(0) e^{-\lambda_P t} - \lambda_D N_D(t) \tag{1.22}$$

With the initial conditions for time $t = 0$ assuming that (i) the initial number of parent nuclei P is $N_P(t = 0) = N_P(0)$, and (ii) there are no daughter D nuclei present, i.e. $N_D(t = 0) = 0$, the solution of the differential equation in Eq. (1.22) reads as follows:

$$N_D(t) = N_P(0)\frac{\lambda_P}{\lambda_D - \lambda_P}\left[e^{-\lambda_P t} - e^{-\lambda_D t}\right] \tag{1.23}$$

Recognizing that the activity of the daughter $\mathcal{A}_D(t)$ is $\lambda_D N_D(t)$, the daughter activity $\mathcal{A}_D(t)$ is written as:

$$\mathcal{A}_D(t) = N_P(0)\frac{\lambda_D \lambda_P}{\lambda_D - \lambda_P}\left[e^{-\lambda_P t} - e^{-\lambda_D t}\right] = \mathcal{A}_P(0)\frac{\lambda_D}{\lambda_D - \lambda_P}\left[e^{-\lambda_P t} - e^{-\lambda_D t}\right]$$
$$= \mathcal{A}_P(0)\frac{1}{1 - \dfrac{\lambda_P}{\lambda_D}}\left[e^{-\lambda_P t} - e^{-\lambda_D t}\right] = \mathcal{A}_P(t)\frac{\lambda_D}{\lambda_D - \lambda_P}\left[1 - e^{-(\lambda_D - \lambda_P)t}\right] \tag{1.24}$$

where

$\mathcal{A}_D(t)$ is the activity at time t of the daughter nuclei equal to $\lambda_D N_D(t)$;
$\mathcal{A}_P(0)$ is the initial activity of the parent nuclei present at time $t = 0$;

and $\mathcal{A}_P(t)$ is the activity at time t of the parent nuclei equal to $\lambda_P N_P(t)$.

While for initial conditions $\mathcal{A}_P(t = 0) = \mathcal{A}_P(0)$ and $\mathcal{A}_D(t = 0) = 0$, the parent P activity $\mathcal{A}_P(t)$ follows the exponential decay law of Eq. (1.14) shown in Fig. 1.3, the daughter D activity $\mathcal{A}_D(t)$ starts at 0, then initially rises with time t, reaches a maximum at a characteristic time $t = (t_{max})_D$, and then diminishes to reach 0 at $t = \infty$. The characteristic time $(t_{max})_D$ is given as follows:

$$(t_{max})_D = \frac{\ln\dfrac{\lambda_P}{\lambda_D}}{\lambda_P - \lambda_D} \tag{1.25}$$

1.4.3. Equilibrium in parent–daughter activities

In many parent P → daughter D → grand-daughter G relationships, after a certain time t the parent and daughter activities reach a constant ratio independent of a further increase in time t. This condition is referred to as radioactive equilibrium and can be analysed by examining the behaviour of the activity ratio $\mathcal{A}_D(t)/\mathcal{A}_P(t)$ obtained from Eq. (1.24) as:

$$\frac{\mathcal{A}_D(t)}{\mathcal{A}_P(t)} = \frac{\lambda_D}{\lambda_D - \lambda_P}\left[1 - e^{-(\lambda_D - \lambda_P)t}\right] = \frac{1}{1 - \dfrac{\lambda_P}{\lambda_D}}\left[1 - e^{-(\lambda_D - \lambda_P)t}\right] \tag{1.26}$$

Three possibilities merit special consideration:

(a) The half-life of the daughter exceeds that of the parent: $(T_{1/2})_D > (T_{1/2})_P$ resulting in $\lambda_D < \lambda_P$. The activity ratio $\mathcal{A}_D(t)/\mathcal{A}_P(t)$ of Eq. (1.26) is written as:

$$\frac{\mathcal{A}_D(t)}{\mathcal{A}_P(t)} = \frac{\lambda_D}{\lambda_P - \lambda_D}\left[e^{(\lambda_P - \lambda_D)t} - 1\right] \tag{1.27}$$

The ratio $\mathcal{A}_D(t)/\mathcal{A}_P(t)$ increases exponentially with time t, indicating that no equilibrium between the parent activity $\mathcal{A}_P(t)$ and daughter activity $\mathcal{A}_D(t)$ will be reached.

(b) The half-life of the daughter is shorter than that of the parent: $(T_{1/2})_D < (T_{1/2})_P$ or $\lambda_D > \lambda_P$.
The activity ratio $\mathcal{A}_D(t)/\mathcal{A}_P(t)$ at large t becomes a constant equal to $\lambda_D/(\lambda_D - \lambda_P)$ and is then independent of t and larger than unity, implying transient equilibrium, i.e.:

$$\frac{\mathcal{A}_D(t)}{\mathcal{A}_P(t)} = \frac{\lambda_D}{\lambda_D - \lambda_P} = \text{const} > 1 \tag{1.28}$$

(c) The half-life of the daughter is much shorter than that of the parent: $(T_{1/2})_D \ll (T_{1/2})_P$ or $\lambda_D \gg \lambda_P$.
For relatively large time $t \gg t_{max}$, the activity ratio $\mathcal{A}_D(t)/\mathcal{A}_P(t)$ of Eq. (1.28) simplifies to:

$$\frac{\mathcal{A}_D(t)}{\mathcal{A}_P(t)} \approx 1 \tag{1.29}$$

The activity of the daughter $\mathcal{A}_D(t)$ very closely approximates that of its parent $\mathcal{A}_P(t)$, i.e. $\mathcal{A}_D(t) \approx \mathcal{A}_P(t)$, and they decay together at the rate of the parent. This special case of transient equilibrium in which the daughter and parent activities are essentially identical is called secular equilibrium.

1.4.4. Production of radionuclides (nuclear activation)

In 1896, Henri Becquerel discovered natural radioactivity, and in 1934 Frédèric Joliot and Irène Curie-Joliot discovered artificial radioactivity. Most natural radionuclides are produced through one of four radioactive decay chains, each chain fed by a long lived and heavy parent radionuclide. The vast majority of currently known radionuclides, however, are human-made and artificially produced through a process of nuclear activation which uses bombardment of a stable nuclide with a suitable energetic particle or high energy photons to induce a nuclear transformation. Various particles or electromagnetic radiation generated by a variety of machines are used for this purpose, most notably neutrons from nuclear reactors for neutron activation, protons from cyclotrons or synchrotrons for proton activation, and X rays from high energy linear accelerators for nuclear photoactivation.

Neutron activation is important in production of radionuclides used for external beam radiotherapy, brachytherapy, therapeutic nuclear medicine and nuclear medicine imaging also referred to as molecular imaging; proton activation is important in production of positron emitters used in positron emission tomography (PET) imaging; and nuclear photoactivation is important from a radiation protection point of view when components of high energy radiotherapy machines become activated during patient treatment and pose a potential radiation risk to staff using the equipment.

A more in depth discussion of radionuclide production can be found in Chapter 4.

1.4.5. Modes of radioactive decay

Nucleons are bound together to form the nucleus by the strong nuclear force that, in comparison to the proton–proton Coulomb repulsive force, is at least two orders of magnitude larger but of extremely short range (only a few femtometres). To bind the nucleons into a stable nucleus, a delicate equilibrium between the number of protons and the number of neutrons must exist. For light (low A) nuclear species, a stable nucleus is formed by an equal number of protons and neutrons ($Z = N$). Above the nucleon number $A \approx 40$, more neutrons than protons must constitute the nucleus to form a stable configuration in order to overcome the Coulomb repulsion among the charged protons.

If the optimal equilibrium between protons and neutrons does not exist, the nucleus is unstable (radioactive) and decays with a specific decay constant λ into a more stable configuration that may also be unstable and decay further, forming a decay chain that eventually ends with a stable nuclide.

Radioactive nuclides, either naturally occurring or artificially produced by nuclear activation or nuclear reactions, are unstable and strive to reach more stable nuclear configurations through various processes of spontaneous radioactive decay that involve transformation to a more stable nuclide and emission of energetic particles. General aspects of spontaneous radioactive decay may be discussed using the formalism based on the definitions of activity $\mathcal{A}$ and decay constant λ without regard for the actual microscopic processes that underlie the radioactive disintegrations.

A closer look at radioactive decay processes shows that they are divided into six categories, consisting of three main categories of importance to medical use of radionuclides and three categories of less importance. The main categories are: (i) alpha (α) decay, (ii) beta (β) decay encompassing three related decay processes (beta minus, beta plus and electron capture) and (iii) gamma (γ) decay encompassing two competing decay processes (pure γ decay and internal conversion). The three less important radioactive decay categories are: (i) spontaneous fission, (ii) proton emission decay and (iii) neutron emission decay.

Nuclides with an excess number of neutrons are referred to as neutron-rich; nuclides with an excess number of protons are referred to as proton-rich. The following features are notable:

— For a slight proton–neutron imbalance in the nucleus, radionuclides decay by β decay characterized by transformation of a proton into a neutron in β^+ decay, and transformation of a neutron into a proton in β^- decay.
— For a large proton–neutron imbalance in the nucleus, the radionuclides decay by emission of nucleons: α particles in α decay, protons in proton emission decay and neutrons in neutron emission decay.
— For very large atomic mass number nuclides ($A > 230$), spontaneous fission, which competes with α decay, is also possible.

Excited nuclei decay to their ground state through γ decay. Most of these transformations occur immediately upon production of the excited state by either α or β decay; however, a few exhibit delayed decays that are governed by their own decay constants and are referred to as metastable states (e.g. ^{99m}Tc).

Nuclear transformations are usually accompanied by emission of energetic particles (charged particles, neutral particles, photons, etc.). The particles released in the various decay modes are as follows:

— Alpha particles in α decay;
— Electrons in β^- decay;
— Positrons in β^+ decay;
— Neutrinos in β^+ decay;
— Antineutrinos in β^- decay;
— Gamma rays in γ decay;
— Atomic orbital electrons in internal conversion;
— Neutrons in spontaneous fission and in neutron emission decay;
— Heavier nuclei in spontaneous fission;
— Protons in proton emission decay.

In each nuclear transformation, a number of physical quantities must be conserved. The most important of these quantities are: (i) total energy, (ii) momentum, (iii) charge, (iv) atomic number and (v) atomic mass number (number of nucleons).

The total energy of particles released by the transformation process is equal to the net decrease in the rest energy of the neutral atom, from parent P to daughter D. The disintegration (decay) energy, often referred to as the Q value for the radioactive decay, is defined as follows:

$$Q = \{M(\mathrm{P}) - [M(\mathrm{D}) + m]\}c^2 \tag{1.30}$$

where M(P), M(D) and m are the nuclear rest masses (in unified atomic mass units u) of the parent, daughter and emitted particles, respectively.

For radioactive decay to be energetically possible, the Q value must be greater than zero. This means that spontaneous radioactive decay processes release energy and are called exoergic or exothermic. For $Q > 0$, the energy equivalent of the Q value is shared as kinetic energy between the particles emitted in the decay process and the daughter product. Since the daughter generally has a much larger mass than the other emitted particles, the kinetic energy acquired by the daughter is usually negligibly small.

1.4.6. Alpha decay

In α decay, a radioactive parent nucleus P decays into a more stable daughter nucleus D by ejecting an energetic α particle. Since the α particle is a ^{4}He nucleus ($^4_2\mathrm{He}^{2+}$), in α decay the parent's atomic number Z decreases by two and its atomic mass number A decreases by four:

$$^A_Z\mathrm{P} \rightarrow {}^{A-4}_{Z-2}\mathrm{D} + {}^4_2\mathrm{He}^{2+} = {}^{A-4}_{Z-2}\mathrm{D} + \alpha \tag{1.31}$$

Naturally occurring α particles have kinetic energies between 4 and 9 MeV; their range in air is between 1 and 10 cm, and their range in tissue is between 10 and 100 μm.

Typical examples of α decay are the decay of ^{226}Ra with a half-life of 1602 a into ^{222}Rn which is also radioactive and decays with a half-life of 3.82 d into ^{218}Po:

$$^{226}_{88}\text{Ra} \xrightarrow[T_{1/2}=1602\text{ a}]{} {}^{222}_{86}\text{Rn} + \alpha$$

and

$$^{222}_{86}\text{Rn} \xrightarrow[T_{1/2}=3.82\text{ d}]{} {}^{218}_{84}\text{Po} + \alpha \qquad (1.32)$$

1.4.7. Beta minus decay

In beta minus (β^-) decay, a neutron-rich parent nucleus P transforms a neutron into a proton and ejects an electron e^- and an electronic antineutrino $\bar{\nu}_e$. Thus, in β^- decay, the atomic number of the daughter increases by one, i.e. $Z_D = Z_P + 1$, the atomic mass number remains constant, i.e. $A_D = A_P$, and the general relationship for β^- decay is given as:

$$^{A}_{Z}\text{P} \rightarrow {}^{A}_{Z+1}\text{D} + \text{e}^- + \bar{\nu}_\text{e} \qquad (1.33)$$

A typical example of β^- decay is the decay of ^{60}Co with a half-life of 5.26 a into an excited state of ^{60}Ni ($^{60}_{28}\text{Ni}^*$). Excited states of ^{60}Ni progress to the ground state of ^{60}Ni instantaneously (within 10^{-12} s) through emission of γ rays in γ decay:

$$^{60}_{27}\text{Co} \xrightarrow[T_{1/2}=5.26\text{ a}]{} {}^{60}_{28}\text{Ni}^* + \text{e}^- + \bar{\nu}_\text{e} \qquad (1.34)$$

1.4.8. Beta plus decay

In beta plus (β^+) decay, a proton-rich parent nucleus P transforms a proton into a neutron and ejects a positron e^+ and an electronic neutrino ν_e. Thus, in β^+ decay, the atomic number of the daughter decreases by one, i.e. $Z_D = Z_P - 1$, the atomic mass number, just as in β^- decay, remains constant, i.e. $A_D = A_P$, and the general relationship for β^+ decay is written as:

$$^{A}_{Z}\text{P} \rightarrow {}^{A}_{Z-1}\text{D} + \text{e}^+ + \nu_\text{e} \qquad (1.35)$$

Radionuclides undergoing β^+ decay are often called positron emitters and are used in medicine for functional imaging with the special imaging technique

PET. The most common tracer for PET studies is fluorodeoxyglucose (FDG) labelled with ^{18}F which serves as a good example of β^+ decay:

$$ {}^{18}_{9}\mathrm{F} \xrightarrow[T_{1/2}=110\ \mathrm{min}]{} {}^{18}_{8}\mathrm{O} + \mathrm{e}^+ + \nu_\mathrm{e} \tag{1.36} $$

1.4.9. Electron capture

Electron capture radioactive decay may occur when an atomic electron ventures inside the nuclear volume, is captured by a proton, triggers a proton to neutron transformation, and an ejection of an electronic neutrino ν_e. In electron capture, as in β^+ decay, the atomic number of the daughter decreases by one, i.e. $Z_\mathrm{D} = Z_\mathrm{P} - 1$, and its atomic mass number, just as in β^- and β^+ decay, remains constant, i.e. $A_\mathrm{D} = A_\mathrm{P}$. The general relationship for electron capture decay is written as:

$$ {}^{A}_{Z}\mathrm{P} + \mathrm{e}^- \xrightarrow[T_{1/2}=60\ \mathrm{d}]{} {}_{Z-1}^{\ \ A}\mathrm{D} + \nu_\mathrm{e} \tag{1.37} $$

A simple example of electron capture decay is the decay of ^{125}I with a half-life of 60 d into an excited state of ^{125}Te which then decays to the ground state of ^{125}Te through γ decay and internal conversion:

$$ {}^{125}_{53}\mathrm{I} + \mathrm{e}^- \xrightarrow[T_{1/2}=60\ \mathrm{d}]{} {}^{125}_{52}\mathrm{Te}^* + \nu_\mathrm{e} \tag{1.38} $$

1.4.10. Gamma decay and internal conversion

Alpha decay as well as the three β decay modes (β^-, β^+ and electron capture) may produce a daughter nucleus in an excited state without expending the full amount of the decay energy available. The daughter nucleus will then reach its ground state, either instantaneously or with some time delay (isomeric metastable state), through one of the following two processes:

(a) By emitting the excitation energy in the form of one or more γ photons in a decay process referred to as γ decay.

(b) By transferring the excitation energy to one of its associated atomic orbital electrons (usually a K shell electron) in a process called internal conversion. The vacancy left behind by the ejected orbital electron is filled by a transition from a higher atomic shell, resulting in characteristic X rays and/or Auger electrons.

In most radioactive α or β decays, the daughter nucleus de-excitation occurs instantaneously (i.e. within 10^{-12} s), so that the emitted γ rays are referred to as if they were produced by the parent nucleus (e.g. ^{60}Co γ rays). The γ rays produced from isomeric transitions are attributed to the isomeric daughter product (e.g. ^{99m}Tc γ rays).

The γ decay process and the internal conversion process may be represented, respectively, as follows:

$$ {}^{A}_{Z}X^{*} \rightarrow {}^{A}_{Z}X + \gamma \tag{1.39}$$

and

$$ {}^{A}_{Z}X^{*} \rightarrow {}^{A}_{Z}X^{+} + e^{-} \rightarrow {}^{A}_{Z}X \tag{1.40}$$

where

${}^{A}_{Z}X^{*}$ stands for an excited state of the nucleus ${}^{A}_{Z}X$;

and ${}^{A}_{Z}X^{+}$ is the singly ionized state of atom ${}^{A}_{Z}X$ following internal conversion decay.

An example of γ decay is the transition of an excited ${}^{60}_{28}Ni^{*}$ nucleus, resulting from the β^{-} decay of ${}^{60}_{27}Co$, into stable ${}^{60}_{28}Ni$ through an emission of two γ rays with energies of 1.17 and 1.33 MeV. An example of internal conversion decay is the decay of excited ${}^{125}_{52}Te^{*}$ which results from an electron capture decay of ^{125}I into stable ${}^{125}_{52}Te$ through emission of 35 keV γ rays (7 %) and internal conversion electrons (93%).

1.4.11. Characteristic (fluorescence) X rays and Auger electrons

A large number of radionuclides used in nuclear medicine (e.g. ^{99m}Tc, ^{123}I, ^{201}Tl, ^{64}Cu) decay by electron capture and/or internal conversion. Both processes leave the atom with a vacancy in an inner atomic shell, most commonly the K shell. The vacancy in the inner shell is filled by an electron from a higher level atomic shell and the binding energy difference between the two shells is either emitted as a characteristic X ray (fluorescence photon) or transferred to a higher shell orbital electron which is then emitted from the atom as an Auger electron with a kinetic energy equal to the transferred energy minus the binding energy of the emitted Auger electron. Emission of characteristic photons and Auger electrons is discussed further in Section 1.6.4.

1.5. ELECTRON INTERACTIONS WITH MATTER

As an energetic charged particle, such as an electron or positron, traverses an absorbing medium, it experiences a large number of Coulomb interactions with the nuclei and orbital electrons of absorber atoms before its kinetic energy is expended. In each interaction, the charged particle's path may be altered (elastic or inelastic scattering) and it may lose some of its kinetic energy that will be transferred to the medium or to photons. Charged particle interactions with orbital electrons of the absorber result in collision (ionization loss); interactions with nuclei of the absorber result in radiation loss. Each of these possible interactions between the charged particle and absorber atom is characterized by a specific cross-section (probability) σ for the particular interaction. The energy loss of the charged particle propagating through an absorber depends on the properties of the particle, such as its mass, charge, velocity and energy, as well as on the properties of the absorber, such as its density and atomic number.

Stopping power is a parameter used to describe the gradual loss of energy of the charged particle as it penetrates into an absorbing medium. Two classes of stopping power are known: collision stopping power s_{col} results from charged particle interaction with orbital electrons of the absorber and radiation stopping power s_{rad} results from charged particle interaction with nuclei of the absorber. The total stopping power s_{tot} is the sum of the collision stopping power and the radiation stopping power.

1.5.1. Electron–orbital interactions

Coulomb interactions between the incident electron or positron and orbital electrons of an absorber result in ionizations and excitations of absorber atoms. Ionization is described as ejection of an orbital electron from the absorber atom, thereby producing an ion. Excitation, on the other hand, is defined as the transfer of an orbital electron of the absorber atom from an allowed orbit to a higher allowed orbit (shell), thereby producing an excited atom. Atomic excitations and ionizations result in collision energy loss and are characterized by collision (also known as ionization) stopping powers.

1.5.2. Electron–nucleus interactions

Coulomb interactions between the incident electron or positron and nuclei of the absorber atom result in particle scattering. The majority of these scattering events are elastic and result in no energy loss. However, when the scattering is inelastic, the incident charged particle loses part of its kinetic energy through production of X ray photons referred to as bremsstrahlung radiation. This energy loss is characterized by radiation stopping powers and is governed by the Larmor relationship which states that the rate of energy loss is proportional to the square of the particle's acceleration and the square of the particle's charge.

1.6. PHOTON INTERACTIONS WITH MATTER

1.6.1. Exponential absorption of photon beam in absorber

The most important parameter used in characterization of X ray or γ ray penetration into absorbing media is the linear attenuation coefficient μ. This coefficient depends on the energy $h\nu$ of the photon and the atomic number Z of the absorber, and may be described as the probability per unit path length that a photon will have an interaction with the absorber. The attenuation coefficient μ is determined experimentally by aiming a narrowly collimated monoenergetic photon beam $h\nu$ onto a suitable radiation detector and placing an absorber material of varying thickness x between the photon source and the detector. The absorber decreases the detector signal intensity from $I(x)$ which is measured with no absorber in the beam ($x = 0$) to $I(x)$ measured with an absorber of thickness $x > 0$ in the beam.

$\mathrm{d}I(x)/\mathrm{d}x$, the rate of change in beam intensity $I(x)$ transmitted through an absorber of thickness x, is equal to the product of the attenuation coefficient μ and the beam intensity $I(x)$ at thickness x (see Eq. (1.41)). Alternatively, it can be said that an absorber of thickness $\mathrm{d}x$ reduces the beam intensity by $\mathrm{d}I$ and the fractional reduction in intensity $-\mathrm{d}I/I$ is equal to the product of the attenuation coefficient μ and the absorber layer thickness $\mathrm{d}x$ (see Eq. (1.42)). The following expressions are obtained, respectively:

$$\frac{\mathrm{d}I(x)}{\mathrm{d}x} = -\mu I(x) \tag{1.41}$$

and

$$-\frac{\mathrm{d}I}{I} = \mu\, \mathrm{d}x \tag{1.42}$$

where the negative sign is used to indicate a decrease in signal $I(x)$ with an increase in absorber thickness x.

It should be noted that Eqs (1.41) and (1.42) can be considered identical.

The form of Eq. (1.41) is identical to the form of Eq. (1.10) that deals with simple radioactive decay; however, it must be noted that in radioactive decay the product $\lambda N(t)$ is defined as activity $\mathcal{A}(t)$, while in photon beam attenuation the product $\mu I(x)$ does not have a special name and symbol.

Integration of Eq. (1.42) over absorber thickness x from 0 to x and over intensity $I(x)$ from the initial intensity $I(0)$ (no absorber) to intensity $I(x)$ at absorber thickness x, gives:

$$\int_{I(0)}^{I(x)} \frac{\mathrm{d}I}{I} = -\int_{0}^{x} \mu \,\mathrm{d}x \tag{1.43}$$

resulting in:

$$I(x) = I(0)\mathrm{e}^{-\mu x} \tag{1.44}$$

where it is assumed that in a homogeneous absorber the attenuation coefficient μ is uniform and independent of absorber thickness x.

1.6.2. Characteristic absorber thicknesses

Equation (1.44) represents the standard expression for the exponential attenuation of a monoenergetic narrow photon beam. A typical exponential plot of intensity $I(x)$ against absorber thickness x of Eq. (1.44) is shown in Fig. 1.4 for a monoenergetic and narrow photon beam. The figure also defines three special absorber thicknesses used for characterization of photon beams: half-value layer (HVL), mean free path (MFP) and tenth-value layer (TVL):

— HVL (or $x_{1/2}$) is defined as the thickness of a homogeneous absorber that attenuates the narrow beam intensity $I(0)$ to one half (50%) of the original intensity, i.e. $I(x_{1/2}) = 0.5I(0)$. The relationship between the HVL $x_{1/2}$ and the attenuation coefficient μ is determined from the basic definition of the HVL as follows:

$$I(x_{1/2}) = 0.5I(0) = I(0)\mathrm{e}^{-\mu x_{1/2}} \tag{1.45}$$

resulting in:

$$\frac{1}{2} = \mathrm{e}^{-\mu x_{1/2}} \quad \text{or} \quad \mu x_{1/2} = \ln 2 = 0.693$$

and

$$\text{HVL} = x_{1/2} = \frac{\ln 2}{\mu} \tag{1.46}$$

— MFP (or $\bar{x}$) or relaxation length is the thickness of a homogeneous absorber that attenuates the beam intensity $I(0)$ to $1/\text{e} = 0.368$ (36.8%) of its original intensity, i.e. $I(\bar{x}) = 0.368I(0)$. The photon MFP is the average distance a photon of energy $h\nu$ travels through a given absorber before undergoing an interaction. The relationship between the MFP $\bar{x}$ and the attenuation coefficient μ is determined from the basic definition of the MFP as follows:

$$I(\bar{x}) = \frac{1}{\text{e}} I(0) = 0.368 I(0) = I(0)\text{e}^{-\mu \bar{x}} \tag{1.47}$$

resulting in:

$$\frac{1}{\text{e}} = \text{e}^{-\mu \bar{x}} \qquad \text{or} \qquad \mu \bar{x} = 1$$

and

$$\text{MFP} = \bar{x} = \frac{1}{\mu} \tag{1.48}$$

— TVL (or $x_{1/10}$) is the thickness of a homogeneous absorber that attenuates the beam intensity $I(0)$ to one tenth (10%) of its original intensity, i.e. $I(x_{1/10}) = 0.1I(0)$. The relationship between the TVL $x_{1/10}$ and the attenuation coefficient μ is determined from the basic definition of the TVL as follows:

$$I(x_{1/10}) = 0.1 I(0) = I(0)\text{e}^{-\mu x_{1/10}} \tag{1.49}$$

resulting in:

$$\frac{1}{10} = \text{e}^{-\mu x_{1/10}} \qquad \text{or} \qquad \mu x_{1/10} = \ln 10 = 2.303$$

and

$$\text{TVL} = x_{1/10} = \frac{\ln 10}{\mu} \tag{1.50}$$

From Eqs (1.46), (1.48) and (1.50), the linear attenuation coefficient μ may be expressed in terms of $x_{1/2}$, $\bar{x}$ and $x_{1/10}$, respectively, as follows:

$$\mu = \frac{\ln 2}{x_{1/2}} = \frac{1}{\bar{x}} = \frac{\ln 10}{x_{1/10}} \tag{1.51}$$

resulting in the following relationships among the characteristic thicknesses:

$$x_{1/2} = (\ln 2)\bar{x} = \frac{\ln 2}{\ln 10} x_{1/10} = 0.301 x_{1/10} \tag{1.52}$$

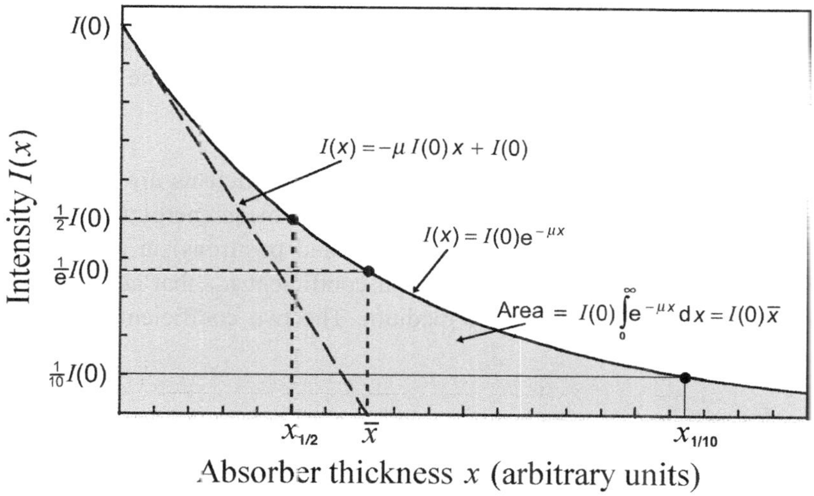

FIG. 1.4. Intensity I(x) against absorber thickness x for a monoenergetic photon beam. Half-value layer $x_{1/2}$, mean free path $\bar{x}$ and tenth-value layer $x_{1/10}$ are also illustrated. The area under the exponential attenuation curve from $x = 0$ to $x = \infty$ is equal to the product $I(0)\bar{x}$ where I(0) is the initial intensity of the monoenergetic photon beam. The slope of the tangent to the attenuation curve at $x = 0$ is equal to $\mu I(0)$ and this tangent crosses the abscissa (x) axis at $x = \bar{x}$.

1.6.3. Attenuation coefficients

In addition to the linear attenuation coefficient μ, three other related attenuation coefficients are in use for describing photon beam attenuation characteristics in absorbers: mass attenuation coefficient μ_m, atomic attenuation coefficient ${}_a\mu$ and electronic attenuation coefficient ${}_e\mu$. The attenuation coefficients are related as follows:

$$\mu = \rho \, \mu_{m} = n^{\square} \, {}_{a}\mu = Z \, n^{\square} \, {}_{e}\mu \tag{1.53}$$

where

ρ is the mass density of the absorber;

$n^{\square}$ is the number of atoms N_a per volume V of the absorber, i.e. $n^{\square} = N_a / V$, and $N_a / V = \rho N_a / m = \rho N_A / A$ with m the mass of the absorber, N_A Avogadro's number of atoms per mole and A the atomic mass of the absorber in grams per mole;

Z is the atomic number of the absorber;

and $Z\, n^{\square}$ is the number of electrons per unit volume of absorber, i.e. $Z\, n^{\square} = \rho Z N_A / A$.

In radiation dosimetry, two energy-related coefficients are in use: (i) the energy transfer coefficient μ_{tr} that accounts for the mean energy transferred $\bar{E}_{tr}$ from photons to charged particles (electrons and positrons) in a photon–atom interaction; and (ii) the energy absorption coefficient μ_{ab} that accounts for the mean energy absorbed $\bar{E}_{ab}$ in the medium. The two coefficients are given as follows:

$$\mu_{tr} = \mu \frac{\bar{E}_{tr}}{h\nu} \tag{1.54}$$

and

$$\mu_{ab} = \mu \frac{\bar{E}_{ab}}{h\nu} \tag{1.55}$$

The light charged particles (electrons and positrons) released or produced in the absorbing medium through various photon interactions will either:

(a) Deposit their energy to the medium through Coulomb interactions with orbital electrons of the absorbing medium (collision loss also referred to as ionization loss); or
(b) Radiate their kinetic energy away in the form of photons through Coulomb interactions with the nuclei of the absorbing medium (radiation loss).

Typical examples of the mass attenuation coefficient μ/ρ are shown in Fig. 1.5 with plots of μ/ρ against photon energy $h\nu$ (in solid dark curves) for carbon and lead in the energy range from 0.001 to 1000 MeV. Carbon with $Z = 6$ is an example of a low Z absorber; lead with $Z = 82$ is an example of a high Z absorber. Comparing the two absorbers, it can be noted that at intermediate photon energies (around 1 MeV), carbon and lead have a similar μ/ρ of about 0.1 cm^2/g. On the other hand, at low photon energies, the μ/ρ of lead significantly exceeds the μ/ρ of carbon, and at energies above 10 MeV, the μ/ρ of carbon is essentially flat while the μ/ρ of lead increases with increasing energy.

1.6.4. Photon interactions on the microscopic scale

The general trends in μ/ρ depicted in Fig. 1.5 reflect the elaborate dependence of μ/ρ on the energy $h\nu$ of the photon and the atomic number Z of the absorber. In penetrating an absorbing medium, photons may experience various interactions with the atoms of the medium. On a microscopic scale, these interactions involve either the nuclei of the absorbing medium or the orbital electrons of the absorbing medium:

(a) Photon interactions with the nucleus of the absorber atoms may be direct photon–nucleus interactions (photonuclear reaction) or interactions between the photon and the electrostatic field of the nucleus (nuclear pair production);
(b) Photon interactions with orbital electrons of absorber atoms are characterized as interactions between the photon and either a loosely bound electron (Compton effect, triplet production) or a tightly bound electron (photoelectric effect, Rayleigh scattering).

A loosely bound electron is an electron whose binding energy E_B is much smaller in comparison with the photon energy $h\nu$, i.e. $E_B \ll h\nu$. An interaction between a photon and a loosely bound electron is considered to be an interaction between a photon and a ‘free’ (i.e. unbound) electron.

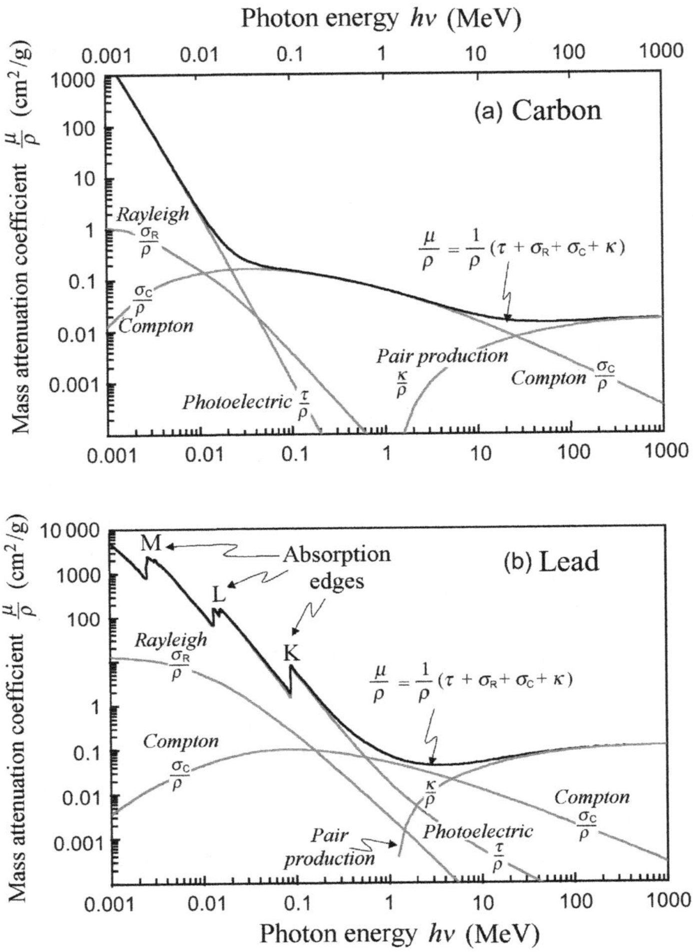

FIG. 1.5. Mass attenuation coefficient μ/ρ against photon energy hν in the range from 1 keV to 1000 MeV for carbon (a) and lead (b). In addition to the total coefficients μ/ρ, the individual coefficients for the photoelectric effect, Rayleigh scattering, Compton scattering and pair production (including triplet production) are also shown. Data are from the National Institute of Science and Technology (NIST).

A tightly bound electron is an electron whose binding energy E_B is comparable to, larger than or slightly smaller than the photon energy $h\nu$. For a photon interaction to occur with a tightly bound electron, the binding energy E_B of the electron must be of the order of, but slightly smaller than, the photon energy, i.e. $E_B \leq h\nu$. An interaction between a photon and a tightly bound electron is considered an interaction between a photon and the atom as a whole.

As far as the photon fate after the interaction with an atom is concerned, there are two possible outcomes: (i) the photon disappears and is absorbed completely (photoelectric effect, nuclear pair production, triplet production, photonuclear reaction) and (ii) the photon is scattered and changes its direction but keeps its energy (Rayleigh scattering) or loses part of its energy (Compton effect).

The most important photon interactions with atoms of the absorber are: the Compton effect, photoelectric effect, nuclear pair production, electronic pair production (triplet production) and photonuclear reactions. In some of these interactions, energetic electrons are released from absorber atoms (photoelectric effect, Compton effect, triplet production) and electronic vacancies are left in absorber atoms; in other interactions, a portion of the incident photon energy is used to produce free electrons and positrons. All of these light charged particles move through the absorber and either deposit their kinetic energy in the absorber (dose) or transform part of it back into radiation through production of bremsstrahlung radiation.

The fate of electronic vacancies produced in photon interactions with absorber atoms is the same as the fate of vacancies produced in electron capture and internal conversion. As alluded to in Section 1.4.11, an electron from a higher atomic shell of the absorber atom fills the electronic vacancy in a lower shell and the transition energy is emitted either in the form of a characteristic X ray (also called a fluorescence photon) or an Auger electron and this process continues until the vacancy migrates to the outer shell of the absorber atom. A free electron from the environment will eventually fill the outer shell vacancy and the absorber ion will revert to a neutral atom in the ground state.

A vacancy produced in an inner shell of an absorber atom migrates to the outer shell and the migration is accompanied by emission of a series of characteristic photons and/or Auger electrons. The phenomenon of emission of Auger electrons from an excited atom is called the Auger effect. Since each Auger transition converts an initial single electron vacancy into two vacancies, a cascade of low energy Auger electrons is emitted from the atom. These low energy electrons have a very short range in tissue but may produce ionization densities comparable to those produced in an α particle track.

The branching between a characteristic photon and an Auger electron is governed by the fluorescence yield ω which, as shown in Fig. 1.6, for a given electronic shell, gives the number of fluorescence photons emitted per vacancy in the shell. The fluorescence yield ω can also be defined as the probability of emission of a fluorescence photon for a given shell vacancy. Consequently, as also shown in Fig. 1.6, $(1 - \omega)$ gives the probability of emission of an Auger electron for a given shell vacancy.

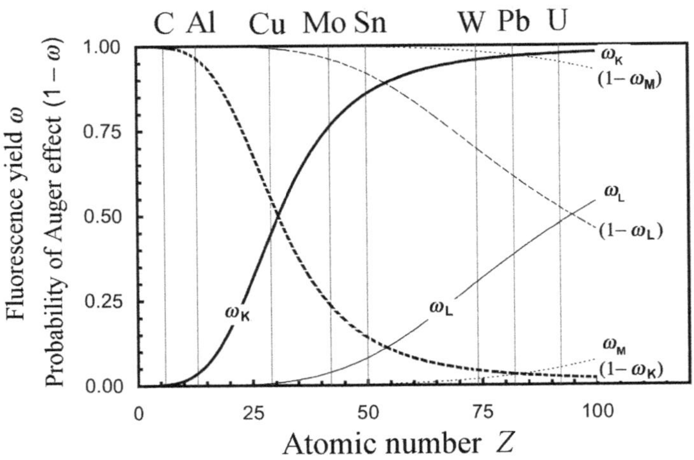

FIG. 1.6. Fluorescence yields ω_K, ω_L and ω_M, against atomic number Z of the absorber. Also shown are probabilities for the Auger effect, given as (1 – ω). Data are from the National Institute of Science and Technology (NIST).

1.6.5. Photoelectric effect

In the photoelectric effect (sometimes called the 'photoeffect'), the photon interacts with a tightly bound orbital electron of an absorber atom, the photon disappears and the orbital electron is ejected from the atom as a so-called photoelectron, with a kinetic energy E_K given as:

$$E_K = h\nu - E_B \tag{1.56}$$

where

$h\nu$ is the incident photon energy;

and E_B is the binding energy of the ejected photoelectron.

A general diagram of the photoelectric effect is provided (see Fig. 1.9(a)).

For the photoelectric effect to happen, the photon energy $h\nu$ must exceed the binding energy E_B of the orbital electron to be ejected and, moreover, the closer $h\nu$ is to E_B, the higher the probability of the photoelectric effect happening. The photoelectric mass attenuation coefficient τ/ρ is plotted in Fig. 1.5 for carbon and lead as one of the grey curves representing the components of the total μ/ρ attenuation coefficient. The sharp discontinuities in the energy $h\nu$ are called absorption edges and occur when $h\nu$ becomes equal to the binding energy E_B of a given atomic shell. For example, the K absorption edge occurs at $h\nu = 88$ keV in lead, since the K shell binding energy E_B in lead is 88 keV. Absorption edges for carbon occur at $h\nu < 1$ keV and, thus, do not appear in Fig. 1.5(a).

As far as the photoelectric attenuation coefficient dependence on photon energy $h\nu$ and absorber atomic number Z is concerned, the photoelectric atomic attenuation coefficient ${}_a\tau$ goes approximately as $Z^5/(h\nu)^3$, while the photoelectric mass attenuation coefficient τ/ρ goes approximately as $Z^4/(h\nu)^3$.

As evident from Fig. 1.5, the photoelectric attenuation coefficient τ/ρ is the major contributor to the total attenuation coefficient μ/ρ at relatively low photon energies where $h\nu$ is of the order of the K shell binding energy and less than 0.1 MeV. At higher photon energies, first the Compton effect and then pair production become the major contributors to the photon attenuation in the absorber.

1.6.6. Rayleigh (coherent) scattering

In Rayleigh scattering (also called 'coherent scattering'), the photon interacts with the full complement of tightly bound atomic orbital electrons of an absorber atom. The event is considered elastic in the sense that the photon loses essentially none of its energy $h\nu$ but is scattered through a relatively small scattering angle θ. A general diagram of Rayleigh scattering is given (see Fig. 1.9(b)).

Since no energy transfer occurs from photons to charged particles, Rayleigh scattering plays no role in the energy transfer attenuation coefficient and energy absorption coefficient; however, it contributes to the total attenuation coefficient μ/ρ through the elastic scattering process. The Rayleigh atomic attenuation coefficient ${}_a\sigma_R$ is proportional to $Z^2/(h\nu)^2$ and the Rayleigh mass attenuation coefficient σ_R/ρ is proportional to $Z/(h\nu)^2$.

As a result of no energy transfer from photons to charged particles in the absorber, Rayleigh scattering is of no importance in radiation dosimetry. As far as photon attenuation is concerned, however, the relative importance of Rayleigh scattering in comparison to other photon interactions in tissue and tissue equivalent materials amounts to only a few per cent of the total μ/ρ but it should not be neglected.

1.6.7. Compton effect (incoherent scattering)

The Compton effect (also called 'incoherent scattering' or 'Compton scattering') is described as an interaction between a photon and a free as well as stationary electron. Of course, the interacting electron is not free, rather it is bound to a nucleus of an absorbing atom, but the photon energy $h\nu$ is much larger than the binding energy E_B of the electron ($E_B \ll h\nu$), so that the electron is said to be loosely bound or essentially 'free and stationary'.

In the Compton effect, the photon loses part of its energy to the recoil (Compton) electron and is scattered as a photon $h\nu'$ through a scattering angle θ, as shown schematically in Fig. 1.7. In the diagram, the interacting electron is at the origin of the Cartesian coordinate system and the incident photon is oriented in the positive direction along the abscissa (x) axis. The scattering angle θ is the angle between the direction of the scattered photon $h\nu'$ and the positive abscissa axis while the recoil angle ϕ is the angle between the direction of the recoil electron and the positive abscissa axis. A general diagram of the Compton effect is given (see Fig. 1.9(c)).

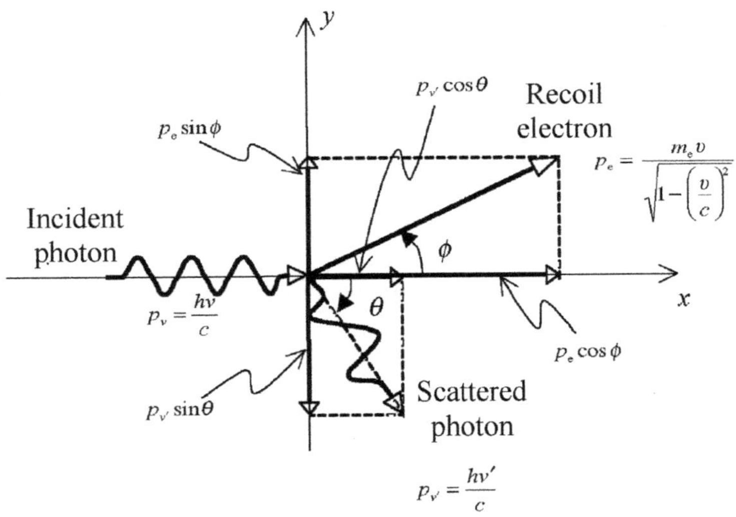

FIG. 1.7. Schematic diagram of the Compton effect in which an incident photon of energy $h\nu = 1$ MeV interacts with a 'free and stationary' electron. A photon with energy $h\nu' = 0.505$ MeV is produced and scattered with a scattering angle $\theta = 60°$.

Considerations of conservation of energy and momentum result in the following three equations for the Compton effect:

(a) Conservation of energy:

$$h\nu + m_e c^2 = h\nu' + m_e c^2 + E_K \qquad \text{or} \qquad h\nu = h\nu' + E_K \tag{1.57}$$

(b) Conservation of momentum on the abscissa (x) axis:

$$p_\nu = \frac{h\nu'}{c}\cos\theta + \frac{m_e \upsilon}{\sqrt{1-\frac{\upsilon^2}{c^2}}}\cos\phi \tag{1.58}$$

(c) Conservation of momentum on the ordinate (y) axis:

$$0 = -\frac{h\nu'}{c}\sin\theta + \frac{m_e \upsilon}{\sqrt{1-\frac{\upsilon^2}{c^2}}}\sin\phi \tag{1.59}$$

where

m_ec^2 is the rest energy of the electron (0.511 MeV);
E_K is the kinetic energy of the recoil (Compton) electron;
υ is the velocity of the recoil (Compton) electron;

and c is the speed of light in a vacuum (3×10^8 m/s).

From equations describing conservation of energy (Eq. (1.57)) and conservation of momentum (Eqs (1.58) and (1.59)), the basic Compton equation (also referred to as the Compton wavelength-shift equation) can be derived and is expressed as follows:

$$\lambda' - \lambda = \Delta\lambda = \frac{h}{m_e c}(1-\cos\theta) = \lambda_C(1-\cos\theta) \tag{1.60}$$

where

λ is the wavelength of the incident photon (c/ν);
λ' is the wavelength of the scattered photon (c/ν');
$\Delta\lambda$ is the wavelength shift in Compton effect ($\lambda' - \lambda$);

and λ_C, defined as $\lambda_C = h/(m_e c) = 2\pi\hbar c/(m_e c^2) = 0.024$ Å, is the so-called Compton wavelength of the electron.

From the Compton equation (Eq. (1.60)), it is easy to show that the scattered photon energy $h\nu'$ and the recoil electron kinetic energy E_K depend on the incident photon energy $h\nu$ as well as on the scattering angle θ and are, respectively, given as:

$$h\nu'(h\nu,\theta) = h\nu \frac{1}{1+\varepsilon(1-\cos\theta)} \tag{1.61}$$

and

$$E_{\mathrm{K}}^{\mathrm{C}}(h\nu,\theta) = h\nu - h\nu' = h\nu - h\nu \frac{1}{1+\varepsilon(1-\cos\theta)} = h\nu \frac{\varepsilon(1-\cos\theta)}{1+\varepsilon(1-\cos\theta)} \tag{1.62}$$

where ε is the incident photon energy $h\nu$ normalized to electron rest energy $m_{\mathrm{e}}c^2$, i.e. $\varepsilon = h\nu/(m_{\mathrm{e}}c^2)$.

Using Eq. (1.61), it is easy to show that energies of forward-scattered photons (θ = 0), side-scattered photons ($\theta = \pi/2$) and backscattered photons ($\theta = \pi$) are in general given as follows:

$$h\nu'\big|_{\theta=0} = h\nu \tag{1.63}$$

$$h\nu'\big|_{\theta=\frac{\pi}{2}} = \frac{h\nu}{1+\varepsilon} \tag{1.64}$$

and

$$h\nu'\big|_{\theta=\pi} = \frac{h\nu}{1+2\varepsilon} \tag{1.65}$$

For very large incident photon energies ($h\nu \to \infty$), they are given as:

$$h\nu'\big|_{\theta=0} = h\nu \tag{1.66}$$

$$h\nu'\big|_{\theta=\frac{\pi}{2}} = m_{\mathrm{e}}c^2 \tag{1.67}$$

and

$$h\nu'\big|_{\theta=\pi} = \frac{m_{\mathrm{e}}c^2}{2} \tag{1.68}$$

From the conservation of momentum equations (Eqs (1.58) and (1.59)), the following expression for the relationship between the scattering angle θ and recoil electron angle ϕ can be derived:

$$\cot\phi = (1+\varepsilon)\tan\frac{\theta}{2} \tag{1.69}$$

and

$$\tan\phi = \frac{1}{1+\varepsilon}\cot\frac{\theta}{2} \tag{1.70}$$

Since the range of θ is from 0 (forward-scattering) through $\pi/2$ (side-scattering) to π (backscattering), it is noted that the corresponding range of ϕ is from $\phi = \pi/2$ at $\theta = 0$ through to $\phi = (1 + \varepsilon)^{-1}$ for $\theta = \pi/2$ to $\phi = 0$ at $\theta = \pi$.

The Compton electronic attenuation coefficient ${}_{\mathrm{e}}\sigma_{\mathrm{C}}$ steadily decreases with increasing $h\nu$ from a theoretical value of 0.665×10^{-24} cm^2/electron (known as the Thomson cross-section) at low photon energies to 0.21×10^{-24} cm^2/electron at $h\nu$ = 1 MeV, 0.51×10^{-24} cm^2/electron at $h\nu$ = 10 MeV, and 0.008×10^{-24} cm^2/electron at $h\nu$ = 100 MeV.

Since Compton interaction is a photon interaction with a free electron, the Compton atomic attenuation coefficient ${}_{\mathrm{a}}\sigma_{\mathrm{C}}$ depends linearly on the absorber atomic number Z, while the electronic coefficient ${}_{\mathrm{e}}\sigma_{\mathrm{C}}$ and the mass coefficient σ_{C}/ρ are essentially independent of Z. This independence of Z can be observed in Fig. 1.5, showing that σ_{C}/ρ for carbon (Z = 6) and lead (Z = 82) at intermediate photon energies (~1 MeV), where Compton effect predominates, are equal to about 0.1 cm^2/electron irrespective of Z.

Equation (1.62) gives the energy transferred from the incident photon to the recoil electron in the Compton effect as a function of the scattering angle θ. The maximum energy transfer to recoil electron occurs when the photon is backscattered ($\theta = \pi$) and the Compton maximum energy transfer fraction $(f_{\mathrm{C}})_{\max}$ is then given as:

$$(f_{\mathrm{C}})_{\max} = \frac{(E_{\mathrm{K}}^{\mathrm{C}})_{\max}}{h\nu} = \frac{2\varepsilon}{1+2\varepsilon} \tag{1.71}$$

The mean energy transfer in the Compton effect $\bar{f}_{\mathrm{C}}$ is determined by normalizing (to incident photon energy $h\nu$) the mean energy transferred to the Compton electron $\bar{E}_{\mathrm{K}}^{\mathrm{C}}$. This quantity is very important in radiation dosimetry and is plotted against incident photon energy $h\nu$ in the Compton graph presented in Fig. 1.8. The figure shows that the fractional energy transfer to recoil electrons

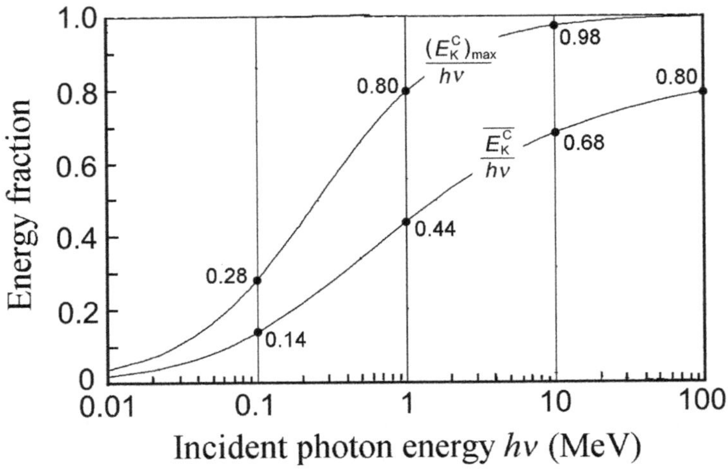

FIG. 1.8. Maximum and mean fractions of incident photon energy hν transferred to the recoil electron in the Compton effect. Data are from the National Institute of Science and Technology (NIST).

is quite low at low photon energies ($\bar{f}_C = 0.02$ at $h\nu = 0.01$ MeV) and then slowly rises through $\bar{f}_C = 0.44$ at $h\nu = 1$ MeV to reach $\bar{f}_C = 0.80$ at $h\nu = 100$ MeV and approaches one asymptotically at very high incident photon energies.

1.6.8. Pair production

When the incident photon energy $h\nu$ exceeds $2m_ec^2 = 1.022$ MeV, with m_ec^2 being the rest energy of the electron and positron, the production of an electron–positron pair in conjunction with a complete absorption of the incident photon by the absorber atom becomes energetically possible. For the effect to occur, three quantities must be conserved: energy, charge and momentum. To conserve the linear momentum simultaneously with total energy and charge, the effect cannot occur in free space; it can only occur in the Coulomb electric field of a collision partner (atomic nucleus or orbital electron) that can take up a suitable fraction of the momentum carried by the photon. Two types of pair production are known:

— If the collision partner is an atomic nucleus of the absorber, the pair production event is called nuclear pair production and is characterized by a photon energy threshold slightly larger than two electron rest masses ($2m_ec^2 = 1.022$ MeV).

— Less probable, but nonetheless possible, is pair production in the Coulomb field of an orbital electron of an absorber atom. The event is called electronic pair production or triplet production and its threshold photon energy is $4m_ec^2 = 2.044$ MeV.

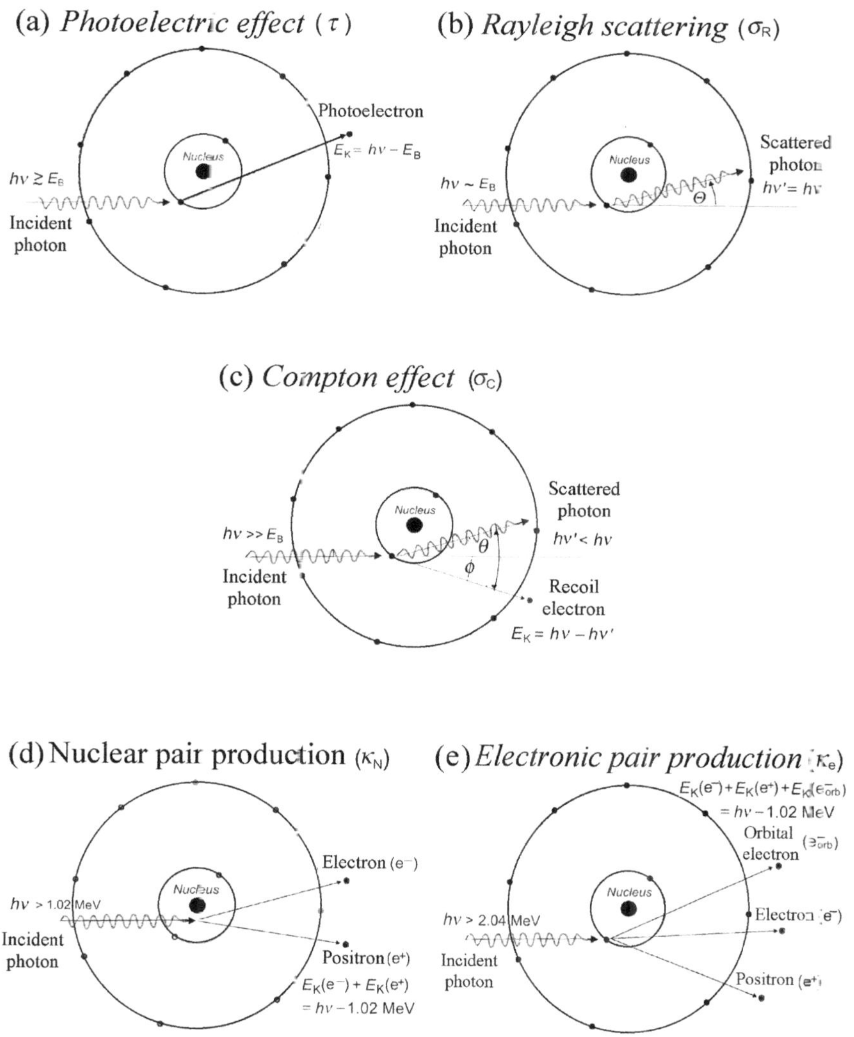

FIG. 1.9. Schematic diagrams of the most important modes of photon interaction with atoms of an absorber: (a) photoelectric effect; (b) Rayleigh scattering; (c) Compton effect; (d) nuclear pair production; and (e) electronic pair production (triplet production).

The two pair production attenuation coefficients, despite having different origins, are usually dealt with together as one parameter referred to as pair production. The component that the nuclear pair production contributes usually exceeds 90%. Nuclear pair production and electronic pair/triplet production are shown schematically in Figs 1.9(d) and (e), respectively.

The probability of pair production is zero for photon energy below the threshold value and increases rapidly with photon energy above the threshold. The pair production atomic attenuation coefficient ${}_{a}\kappa$ and the pair production mass attenuation coefficient κ/ρ vary approximately as Z^2 and Z, respectively, where Z is the atomic number of the absorber.

1.6.9. Relative predominance of individual effects

As is evident from the discussion above, photons have several options for interaction with absorber atoms. Five of the most important photon interactions are shown schematically in Fig. 1.9. Nuclear and electronic pair production are usually combined and treated under the header 'pair production'.

The probability for a photon to undergo any one of the various interaction phenomena with an absorber depends on the energy $h\nu$ of the photon and the atomic number Z of the absorber. In general, the photoelectric effect predominates at low photon energies, the Compton effect at intermediate energies and pair production at high photon energies.

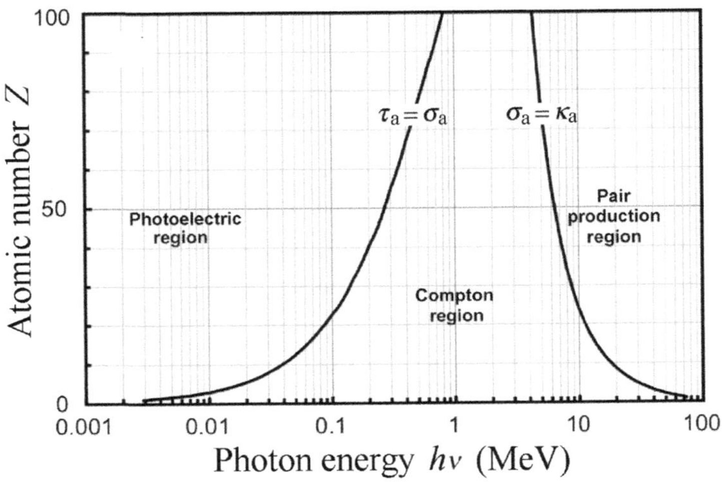

FIG. 1.10. Representation of the relative predominance of the three main processes of photon interaction with an absorber atom: the photoelectric effect, Compton effect and pair production.

Figure 1.10 shows the regions of relative predominance of the three most important individual effects with $h\nu$ and Z as parameters. The two curves display the points on the ($h\nu$, Z) diagram for which $\sigma_C = \tau$ at low photon energies and for which $\sigma_C = \kappa$ for high photon energies and, thus, delineate regions of photoelectric effect predominance at low photon energies, Compton effect predominance at intermediate photon energies and pair production predominance at high photon energies. Figure 1.10 also indicates how the regions of predominance are affected by the absorber atomic number. For example, a 100 keV photon will interact with a lead absorber ($Z = 82$) predominantly through the photoelectric effect and with soft tissue ($Z_{eff} \approx 7.5$) predominantly through the Compton effect. A 10 MeV photon, on the other hand, will interact with lead predominantly through pair production and with tissue predominantly through the Compton effect.

1.6.10. Macroscopic attenuation coefficients

For a given photon energy $h\nu$ and absorber atomic number Z, the macroscopic attenuation coefficient μ and energy transfer coefficient μ_{tr} are given as a sum of coefficients for individual photon interactions discussed above (photoelectric, Rayleigh, Compton and pair production):

$$\mu = \rho \frac{N_A}{A}({}_a\tau + {}_a\sigma_R + {}_a\sigma_C + {}_a\kappa) \tag{1.72}$$

and

$$\mu_{tr} = \rho \frac{N_A}{A}\left[{}_a\tau_{tr} + ({}_a\sigma_C)_{tr} + {}_a\kappa_{tr}\right] = \rho \frac{N_A}{A}\left[{}_a\tau \bar{f}_{PE} + {}_a\sigma_C \bar{f}_C + {}_a\kappa \bar{f}_{PP}\right] \tag{1.73}$$

where all parameters are defined in sections dealing with the individual microscopic effects.

It should be noted that in Rayleigh scattering there is no energy transfer to charged particles.

The energy absorption coefficient μ_{ab} (often designated μ_{en} in the literature) is derived from μ_{tr} of Eq. (1.73) as follows:

$$\mu_{ab} = \mu_{en} = \mu_{tr}(1 - \bar{g}) \tag{1.74}$$

where $\bar{g}$ is the mean radiation fraction accounting for the fraction of the mean energy transferred from photons to charged particles and subsequently lost by charged particles through radiation losses. These losses consist of two components: the predominant bremsstrahlung loss and the small, yet not always negligible, in-flight annihilation loss.

1.6.11. Effects following photon interactions with absorber and summary of photon interactions

In the photoelectric effect, Compton effect and triplet production, vacancies are produced in atomic shells of absorber atoms through the ejection of orbital electrons from atomic shells. For the diagnostic range and megavoltage range of photons used for diagnosis and treatment of disease with radiation, the shell vacancies occur mainly in inner atomic shells and are followed by characteristic radiation or Auger electrons, the probability of the former given by fluorescence yield ω (see Fig. 1.6).

Pair production and triplet production are followed by the annihilation of the positron with a 'free' electron producing two annihilation quanta, most commonly with an energy of 0.511 MeV each and emitted at 180° from each other to satisfy conservation of energy, momentum and charge.

BIBLIOGRAPHY

ATTIX, F.H., Introduction to Radiological Physics and Radiation Dosimetry, Wiley, New York (1986).

CHERRY, S.R., SORENSON, J.A., PHELPS, M.E., Physics in Nuclear Medicine, 3rd edn, Saunders, Philadelphia, PA (2003).

EVANS, R.D., The Atomic Nucleus, Krieger Publishing, Malabar, FL (1955).

HENDEE, W., RITENOUR, E.R., Medical Imaging Physics, 4th edn, Wiley, New York (2002).

JOHNS, H.E., CUNNINGHAM, J.R., The Physics of Radiology, 3rd edn, Thomas, Springfield, IL (1984).

KHAN, F., The Physics of Radiation Therapy, 4th edn, Lippincott, Williams and Wilkins, Baltimore, MD (2009).

KRANE, K., Modern Physics, 3rd edn, Wiley, New York (2012).

PODGORSAK, E.B., Radiation Physics for Medical Physicists, 2nd edn, Springer, Heidelberg, New York (2010).

ROHLF, J.W., Modern Physics from α to Z^0, Wiley, New York (1994).

CHAPTER 2

BASIC RADIOBIOLOGY

R.G. DALE
Department of Surgery and Cancer,
Faculty of Medicine,
Imperial College London,
London, United Kingdom

J. WONDERGEM*
Division of Human Health,
International Atomic Energy Agency,
Vienna

2.1. INTRODUCTION

Radiobiology is the study (both qualitative and quantitative) of the actions of ionizing radiations on living matter. Since radiation has the ability to cause changes in cells which may later cause them to become malignant, or bring about other detrimental functional changes in irradiated tissues and organs, consideration of the associated radiobiology is important in all diagnostic applications of radiation. Additionally, since radiation can lead directly to cell death, consideration of the radiobiological aspects of cell killing is essential in all types of radiation therapy.

2.2. RADIATION EFFECTS AND TIMESCALES

At the microscopic level, incident rays or particles may interact with orbital electrons within the cellular atoms and molecules to cause excitation or ionization. Excitation involves raising a bound electron to a higher energy state, but without the electron having sufficient energy to leave the host atom. With ionization, the electron receives sufficient energy to be ejected from its orbit and to leave the host atom. Ionizing radiations (of which there are several types) are, thus, defined through their ability to induce this electron ejection process, and

---

* Present address: Department of Radiology, Leiden University Medical Centre, Leiden, Netherlands.

the irradiation of cellular material with such radiation gives rise to the production of a flux of energetic secondary particles (electrons). These secondary particles, energetic and unbound, are capable of migrating away from the site of their production and, through a series of interactions with other atoms and molecules, give up their energy to the surrounding medium as they do so.

This energy absorption process gives rise to radicals and other chemical species and it is the ensuing chemical interactions involving these which are the true causatives of radiation damage. Although the chemical changes may appear to operate over a short timescale (~10^{-5} s), this period is nonetheless a factor of ~10^{18} longer than the time taken for the original particle to traverse the cell nucleus. Thus, on the microscopic scale, there is a relatively long period during which chemical damage is inflicted (Table 2.1).

It is important to note that, irrespective of the nature of the primary radiation (which may be composed of particles and/or electromagnetic waves), the mechanism by which energy is transferred from the primary radiation beam to biological targets is always via the secondary electrons which are produced. The initial ionization events (which occur near-instantaneously at the microscopic level) are the precursors to a chain of subsequent events which may eventually lead to the clinical (macroscopic) manifestation of radiation damage.

Expression of cell death in individual lethally damaged cells occurs later, usually at the point at which the cell next attempts to enter mitosis. Gross (macroscopic and clinically observable) radiation effects are a result of the wholesale functional impairment that follows from lethal damage being inflicted to large numbers of cells or critical substructures. The timescale of the whole process may extend to months or years. Thus, in clinical studies, any deleterious health effects associated with a radiation procedure may not be seen until long after the diagnostic test or treatment has been completed (Table 2.1).

TABLE 2.1. THE TIMESCALES OF RADIATION EFFECTS

Action	Approximate timescale
Initial ionizing event	10^{-18} s
Transit of secondary electrons	10^{-15} s
Production of ion radicals	10^{-10} s
Production of free radicals	10^{-9} s
Chemical changes	10^{-5} s
Individual cell death	Hours–months
Gross biological effects	Hours–years

2.3. BIOLOGICAL PROPERTIES OF IONIZING RADIATION

2.3.1. Types of ionizing radiation

In nuclear medicine, there are four types of radiation which play a relevant role in tumour and normal tissue effects: gamma (γ) radiation, beta (β) radiation, alpha (α) particles and Auger electrons.

2.3.1.1. Gamma radiation

Gamma radiation is an electromagnetic radiation of high energy (usually above 25 keV) and is produced by subatomic particle interactions. Electromagnetic radiation is often considered to be made up of a stream of wave-like particle bundles (photons) which move at the speed of light and whose interaction properties are governed mainly by their associated wavelength. Although the collective ionization behaviour of large numbers of photons can be predicted with great accuracy, individual photon interactions occur at random and, in passing through any type of matter, a photon may interact one or more times, or never. In each interaction (which will normally involve a photoelectric event, a Compton event or a pair production event), secondary particles are produced, usually electrons (which are directly ionizing) or another photon of reduced energy which itself can undergo further interactions. The electrons undergo many ionizing events relatively close to the site of their creation and, therefore, contribute mostly to the locally absorbed dose. Any secondary photons which may be created carry energy further away from the initial interaction site and, following subsequent electron-producing interactions, are responsible for the dose deposition occurring at sites which are more distant from the original interaction.

2.3.1.2. Beta radiation

Beta radiation is electrons emitted as a consequence of β radionuclide decay. A β decay process can occur whenever there is a relative excess of neutrons (β^-) or protons (β^+). One of the excess neutrons is converted into a proton, with the subsequent excess energy being released and shared between an emitted electron and an anti-neutrino. Many radionuclides exhibit β decay and, in all cases, the emitted particle follows a spectrum of possible energies rather than being emitted with a fixed, discrete energy. In general, the average β energy is around one third of the maximum energy. Most β emitting radionuclides also emit γ photons as a consequence of the initial β decay, leaving the daughter nucleus in an excited, metastable state. Since β particles are electrons, once ejected from the host atom, they behave exactly as do the electrons created following the passage of a γ ray, giving up their energy (usually of the order of several hundred kiloelectronvolts) to other atoms and molecules through a series of collisions.

For radionuclides which emit both β particles and γ photons, it is usually the particulate radiation which delivers the greatest fraction of the radiation dose to the organ which has taken up the activity. For example, about 90% of the dose delivered to the thyroid gland by ^{131}I arises from the β component. On the other hand, the γ emissions contribute more significantly to the overall whole body dose.

2.3.1.3. Alpha particles

Alpha radiation is emitted when heavy, unstable nuclides undergo decay. Alpha particles consist of a helium nucleus (two protons combined with two neutrons) emitted in the process of nuclear decay. The α particles possess approximately 7000 times the mass of a β particle and twice the electronic charge, and give up their energy over a very short range (<100 μm). Alpha particles usually possess energies in the megaelectronvolt range, and because they lose this energy in such a short range are biologically very efficacious, i.e. they possess a high linear energy transfer (LET; see Section 2.6.3) and are associated with high relative biological effectiveness (RBE; see Section 2.6.4).

2.3.1.4. Auger electrons

Radionuclides which decay by electron capture or internal conversion leave the atom in a highly excited state with a vacancy in one of the inner shell electron orbitals. This vacancy is rapidly filled by either a fluorescent transition (characteristic X ray) or non-radiative (Auger) transition, in which the energy gained by the electron transition to the deeper orbital is used to eject another electron from the same atom. Auger electrons are very short range, low energy particles that are often emitted in cascades, a consequence of the inner shell atomic vacancy that traverses up through the atom to the outermost orbital, ejecting additional electrons at each step. This cluster of very low energy electrons can produce ionization densities comparable to those produced by an α particle track. Thus, radionuclides which decay by electron capture and/or internal conversion can exhibit high LET-like behaviour close (within 2 nm) to the site of the decay.

2.4. MOLECULAR EFFECTS OF RADIATION AND THEIR MODIFIERS

Radiation induced damage to biological targets may result from direct or indirect action of radiation (Fig. 2.1):

- Direct action involves ionization or excitation (via Coulomb interactions) of the atoms in the biological target. This gives rise to a chain of events which eventually leads to the observable (macroscopic) damage. In normally oxygenated mammalian cells, the direct effect accounts for about one third of the damage for low LET radiations such as electrons and photons.
- Indirect action involves radiation effects on atoms or molecules which are not constituent parts of the biological target. Since cells exist in a rich aqueous environment, the majority of indirect actions involve the ionization or excitation of water molecules. The free radicals subsequently created may then migrate and damage the adjacent biological targets. Indirect action is the main cause of radiation damage and, in normoxic cells, accounts for about two thirds of the damage.

Indirect action is predominant with low LET radiation, e.g. X and γ rays, while direct action is predominant with high LET radiation, e.g. α particles and neutrons.

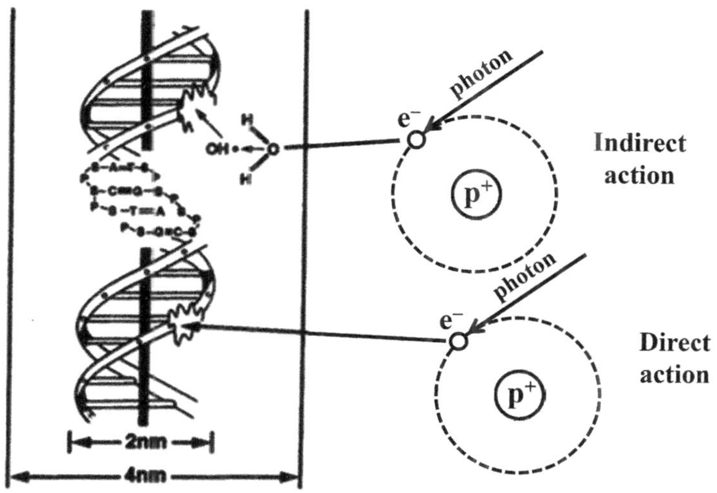

FIG. 2.1. Illustration of the difference between direct and indirect damage to cellular DNA.

2.4.1. Role of oxygen

Radiation effects may be influenced by several factors, especially the presence or absence of oxygen. The free radicals (denoted by a dot placed to the right of the atomic symbol) produced as a result of direct or indirect effects are very reactive and seek to interact with other molecules which can share or donate electrons. Molecular oxygen (O_2) has two unpaired electrons and readily reacts with free radicals, causing an increased likelihood that DNA (deoxyribonucleic acid) will be damaged by the indirect process. Important reactions via which oxygen can increase biological damage are:

$$R^{\bullet} + O_2 \rightarrow RO_2^{\bullet} \text{ (highly toxic)}$$
$$H^{\bullet} + O_2 \rightarrow HO_2^{\bullet}$$
$$HO_2^{\bullet} + HO_2^{\bullet} = H_2O_2 \text{ (highly toxic)} + O_2$$

where R represents an organic molecule.

The oxygen enhancement ratio (OER) is given by the dose in hypoxia (total absence of oxygen) divided by the dose in air required to achieve an equivalent biological effect. For low LET radiation, such as γ rays, the OER has a value of ~3. For high LET radiation, such as α particles, the OER decreases to almost 1.0.

2.4.2. Bystander effects

Bystander effects occur when a cell which has not been traversed by a charged particle is damaged as a result of radiation interactions occurring in neighbouring cells. The discovery of the bystander effect poses a challenge to the traditional view that all radiation damage stems from direct interactions of charged particles with critical cellular targets. For this reason, it still remains controversial in radiobiology. A possible explanation is that irradiated cells may send out a stress signal to nearby cells, which may elicit a response, e.g. the initiation of apoptosis, in those cells. The overall relevance of the bystander effect is presently difficult to gauge. It is probably most significant in radiation protection considerations involving low doses since it amplifies the overall radiation effect in situations where not all of the cells in a tissue are subjected to particle transversal, i.e. the overall radiation risk to that tissue is higher than would be expected from consideration of the gross response exhibited by those cells which have been directly traversed by charged particles.

2.5. DNA DAMAGE AND REPAIR

2.5.1. DNA damage

DNA damage is the primary cause of cell death caused by radiation. Radiation exposure produces a wide range of lesions in DNA such as single strand breaks (SSBs), double strand breaks (DSBs), base damage, protein–DNA cross-links and protein–protein cross-links (see Fig. 2.1). The number of DNA lesions generated by irradiation is large, but there are a number of mechanisms for DNA repair. As a result, the percentage of lesions causing cell death is very small. The numbers of lesions induced in the DNA of a cell by a dose of 1–2 Gy are approximately: base damages: >1000; SSBs: ~1000; DSBs: ~40. DSBs play a critical role in cell killing, carcinogenesis and hereditary effects. There are experimental data showing that the initially produced DSBs correlate with radiosensitivity and survival at lower dose, and that unrepaired or misrepaired DSBs also correlate with survival after higher doses. Furthermore, there is experimental evidence for a causal link between the generation of DSBs and the induction of chromosomal translocations with carcinogenic potential.

2.5.2. DNA repair

DNA repair mechanisms are important for the recovery of cells from radiation and other damaging agents. There are multiple enzymatic mechanisms for detecting and repairing radiation induced DNA damage. DNA repair mechanisms, such as base excision repair, mismatch repair and nucleotide excision repair, respond to damage such as base oxidation, alkylation and strand intercalation. Excision repair consists of cleavage of the damaged DNA strand by enzymes that cleave the polynucleotide chain on either side of the damage, and enzymes which cleave the end of a polynucleotide chain allowing removal of a short segment containing the damaged region. DNA polymerase can then fill in the resulting gap using the opposite undamaged strand as a template. For DSBs, there are two primary repair pathways, non-homologous end joining (NHEJ) and homologous recombination. NHEJ repair operates on blunt ended DNA fragments. This process involves the repair proteins recognizing lesion termini, cleaning up the broken ends of the DNA molecule, and the final ligation of the broken ends. DSB repair by homologous recombination utilizes sequence homology with an undamaged copy of the broken region and, hence, can only operate in late S/G2-phases of the cell cycle. Undamaged DNA from both strands is used as a template to repair the damage. In contrast to NHEJ, the repair process of homologous recombination is error-free. Repair by NHEJ operates throughout the cell cycle but dominates in G1/S-phases. The process is error-prone because it does not rely on sequence homology. Unrepaired or misrepaired damage to DNA will lead to mutations and/or chromosome damage in the exposed cell. Mutations might lead to cancer or hereditary effects (when germ cells are exposed), whereas severe chromosome damage often leads to cell death.

2.6. CELLULAR EFFECTS OF RADIATION

2.6.1. Concept of cell death

Radiation doses of the order of several grays may lead to cell loss. Cells are generally regarded as having been 'killed' by radiation if they have lost reproductive integrity, even if they have physically survived. Loss of reproductive integrity can occur by apoptosis, necrosis, mitotic catastrophe or by induced senescence. Although all but the last of these mechanisms ultimately results in physical loss of the cell, this may take a significant time to occur.

Apoptosis or programmed cell death can occur naturally or result from insult to the cell environment. Apoptosis occurs in particular cell types after low doses of irradiation, e.g. lymphocytes, serous salivary gland cells, and certain cells in the stem cell zone in testis and intestinal crypts.

Necrosis is a form of cell death associated with loss of cellular membrane activity. Cellular necrosis generally occurs after high radiation doses.

Reproductive cell death is a result of mitotic catastrophe (cells attempt to divide without proper repair of DNA damage) which can occur in the first few cell divisions after irradiation, and it occurs with increasing frequency after increasing doses.

Ionizing radiation may also lead to senescence. Senescent cells are metabolically active but have lost the ability to divide.

2.6.2. Cell survival curves

A quantitative understanding of many aspects of biological responses to radiation may be made by consideration of the behaviour of the underlying cell survival (dose response) characteristics. Although the practical determination of cell survival curves is potentially fraught with experimental and interpretational difficulties and is best performed by persons who are experts in such procedures, an appreciation of the structure and meaning of such curves, even in a purely schematic context, can be very helpful in understanding the role played by the various factors which influence radiation response.

Figure 2.2 shows the typical shape of a cell survival curve for mammalian tissue. Physical radiation dose is plotted on the linear horizontal axis while fractional cell survival is plotted on the logarithmic vertical axis. Each of the individual points on the graph represents the fractional survival of cells resulting from delivery of single acute doses of the specified radiation, which in this case is assumed to be γ radiation. (In the context of the subject, an acute dose of radiation may be taken to mean one which is delivered at high dose rate, i.e. the radiation delivery is near instantaneous.) Mammalian cell survival curves plotted in this way are associated with two main characteristics: a finite initial slope (at zero dose) and a gradually increasing slope as dose is increased.

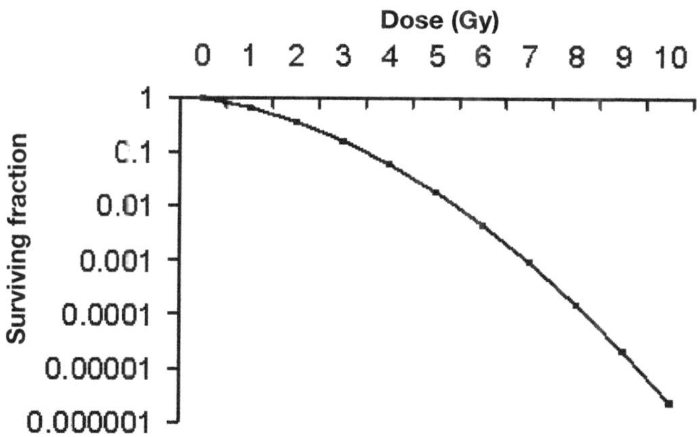

FIG. 2.2. A radiation cell survival curve plots the fraction of plated cells retaining colony forming ability (cell surviving fraction) versus radiation absorbed dose.

2.6.3. Dose deposition characteristics: linear energy transfer

As noted above, the energy transfer to the absorbing medium (whether that be animate or inanimate material) is via secondary electrons created by the passage of the primary ionizing particle or ray. LET is a measure of the linear rate at which radiation is absorbed in the absorbing medium by the secondary particles and is defined by the International Commission on Radiation Units and Measurements (ICRU) as being the quotient dE/dl, where dE is the average energy locally imparted to the medium by a charged particle of specified energy in traversing a distance dl. The unit usually employed for LET is kiloelectronvolt per micrometre and some representative values are listed in Table 2.2.

TABLE 2.2. THE LINEAR ENERGY TRANSFER OF DIFFERENT RADIATIONS

Radiation type	Linear energy transfer (keV/μm)
^{60}Co γ rays	0.2
250 kVp X rays	2.0
10 MeV protons	4.7
2.5 MeV α particles	166
1 MeV electrons	0.25
10 keV electrons	2.3
1 keV electrons	12.3

For radiobiological studies in particular, the concept of LET is problematic since it relates to an average linear rate of energy deposition but, at the microscopic level (i.e. at dimensions comparable with the critical cellular targets), the energy deposited per unit length along different parts of a single track may vary dramatically. In particular, as charged particles lose energy in their passage through a medium via the result of collision and ionizing processes, the LET rises steeply to its highest value towards the very end of their range. The change in LET value along the track length is one reason why average LET values correlate poorly with observed (i.e. macroscopic) biological effects. For these reasons, the directly measured RBE is of much greater use as an indicator of the differing biological efficacies of various radiation types.

2.6.4. Determination of relative biological effectiveness

For a given biological end point, the RBE of the high LET radiation is defined as the ratio of the isoeffective doses for the reference (low LET) and the high LET radiation (Fig. 2.3). The reference radiation is usually ^{60}Co γ rays or high energy (250 kVp) X rays.

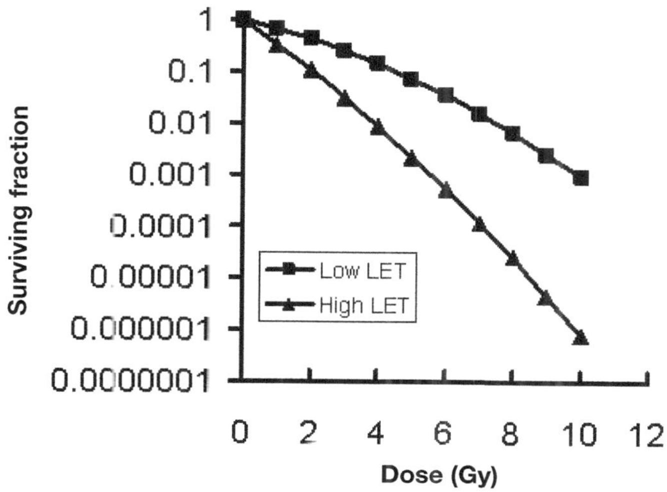

FIG. 2.3. The relative biological effectiveness of a radiation is defined as the ratio of the dose required to produce the same reduction in cell survival as a reference low linear energy transfer (LET) radiation.

If the respective low and high LET isoeffective doses are d_{L} and d_{H}, then:

$$\mathrm{RBE} = \frac{d_{\mathrm{L}}}{d_{\mathrm{H}}} \tag{2.1}$$

If the basic cell survival curves are described in terms of the linear–quadratic (LQ) model, then the surviving fraction S as a function of acute doses at low and high LET is respectively given as:

$$S_{\mathrm{L}} = \exp\left(-\alpha_{\mathrm{L}} d_{\mathrm{L}} - \beta_{\mathrm{L}} d_{\mathrm{L}}^2\right) \tag{2.2}$$

$$S_{\mathrm{H}} = \exp\left(-\alpha_{\mathrm{H}} d_{\mathrm{H}} - \beta_{\mathrm{H}} d_{\mathrm{H}}^2\right) \tag{2.3}$$

where the suffixes L and H again respectively refer to the low and high LET instances.

Figure 2.4 shows an example of how the RBEs determined at any particular end point (cell surviving fraction) vary with changing dose for a given radiation fraction size for a low LET radiation. The maximum RBE (RBE_{max}) occurs at zero dose and, in terms of microdosimetric theory, corresponds to the ratio between the respective high and low LET linear radiosensitivity constants, α_H and α_L, i.e.:

$$\mathrm{RBE}_{\max} = \frac{\alpha_H}{\alpha_L} \tag{2.4}$$

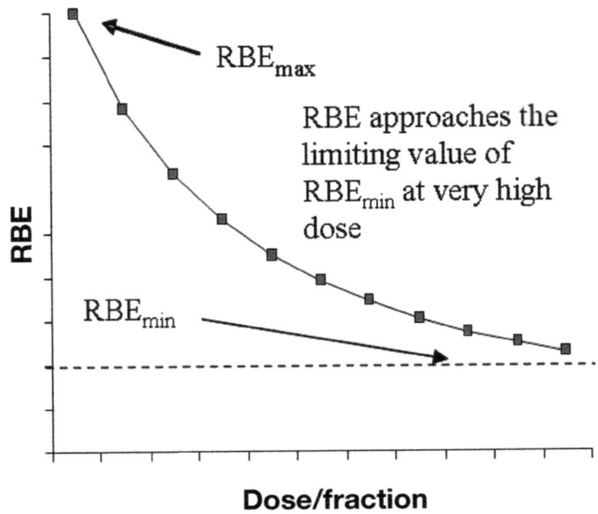

FIG. 2.4. Relative biological effectiveness (RBE) as a function of the radiation dose per fraction.

If the quadratic radiosensitivity coefficients (β_H and β_L) are unchanged with changing LET (i.e. $\beta_H = \beta_L$), then, at high doses, the RBE tends to unity. However, this constancy of β, assumed by the theory of Kellerer and Rossi, has been challenged and, if β does change with LET, then RBE will tend asymptotically to an alternative minimum value ($\mathrm{RBE}_{\min}$) given by:

$$\mathrm{RBE}_{\min} = \sqrt{\frac{\beta_H}{\beta_L}} \tag{2.5}$$

and the 'working' RBE at any given dose per fraction is given as:

$$\mathrm{RBE} = \frac{(\alpha/\beta)_L \mathrm{RBE}_{\max} + \sqrt{(\alpha/\beta)_L^2 \mathrm{RBE}_{\max}^2 + 4d_L \mathrm{RBE}_{\min}^2 \left[(\alpha/\beta)_L + d_L\right]}}{2\left[(\alpha/\beta)_L + d_L\right]} \tag{2.6}$$

when expressed in terms of the low LET dose per fraction d_L or:

$$\mathrm{RBE} = \frac{-(\alpha/\beta)_L + \sqrt{(\alpha/\beta)_L^2 + 4d_H\left[(\alpha/\beta)_L\,\mathrm{RBE}_{max} + \mathrm{RBE}_{min}^2 d_H\right]}}{2d_H} \tag{2.7}$$

when expressed in terms of the high LET dose per fraction d_H.

Figure 2.4 was derived using $\mathrm{RBE}_{max} = 5$, $\mathrm{RBE}_{min} = 1$ and $(\alpha/\beta)_L = 3$ Gy, but the general trend of a steadily falling RBE with increasing dose per fraction is independent of the chosen values. Clearly, the assumption of a fixed value of RBE, if applied to all fraction sizes, could lead to gross clinical errors and Eqs (2.6) and (2.7) make the point that determination of RBEs in a clinical setting is potentially complex and will depend on accurate knowledge of RBE_{max} and (if it is not unity) RBE_{min}. Although there is not yet clear evidence over whether or not there is a consistent trend for RBE_{min} to be non-unity, the possibility is nevertheless important as it may hold very significant implications.

Figure 2.4 also shows schematically how the rate of change of RBE with changing dose per fraction is influenced by the existence of a non-unity RBE_{min} parameter. Even for a fixed value of RBE_{max}, the potential uncertainty in the RBE values at the fraction sizes likely to be used clinically might themselves be very large if RBE_{min} is erroneously assumed to be unity. These uncertainties would be compounded if there were an additional linkage between RBE_{max} and the tissue α/β value.

As is seen from Eqs (2.6) and (2.7), the RBE value at any particular dose fraction size will also be governed by the low LET α/β ratio (a tissue dependent parameter which provides a measure of how tissues respond to changes in dose fractionation) and the dose fraction size (a purely physical parameter) at the point under consideration. Finally, and as has been shown through the earlier clinical experience with neutron therapy, the RBE_{max} value may itself be tissue dependent, likely being higher for the dose-limiting normal tissues than for tumours. This tendency is borne out by experimental evidence using a variety of ion species as well as by theoretical microdosimetric studies. This potentially deleterious effect may be offset by the fact that, in carbon-, helium- and argon-ion beams, LET (and, hence, RBE) will vary along the track in such a way that it is low at the entry point (adjacent to normal tissues) and highest at the Bragg peak located in the tumour. However, although this might be beneficial, it does mean that the local RBE is more spatially variable than is indicated by Eq. (2.6).

Owing to the difficulties in setting reference doses at which clinical inter-comparisons could be made more straightforward, Wambersie proposed that a distinction be made between the ‘reference’ RBE and the ‘clinical’

RBE. Thus, the reference RBE might be that determined at 2 Gy fractions on a biological system end point representative, for example, of the overall late tolerance of normal tissues. As more clinical experience of using the particular radiation becomes available, a more practical 'clinical' RBE evolves, this being the reference RBE empirically weighted by collective clinical experience and by volume effects related to the beam characteristics, geometry or technical conditions.

2.6.5. The dose rate effect and the concept of repeat treatments

When mammalian cells are irradiated, it is helpful to visualize their subsequent death as resulting from either of two possible processes. In the first process, the critical nuclear target (DNA) is subjected to a large deposition of energy which physically breaks both strands of the double helix structure and disrupts the code sufficiently to disallow any opportunity of repair. This process can be thought of as a single-hit process and the total amount of DNA damage created this way is directly proportional to the dose delivered.

In the second process, an ionizing event occurs and releases only sufficient energy to disrupt the coding carried by one strand of the DNA. Following this event, and if the irradiation continues, two outcomes are possible: either the broken strand will restore itself to its original state (no lethality) or, prior to full repair taking place, a second, independent radiation event may occur in the same location and damage the opposite strand of the DNA, a complementary action between the two damaged strands then leading to cell lethality in what is called a two-hit process. Since this route depends on there being two independent events, each having a probability proportional to dose, the number of damaged DNA targets created this way is proportional to dose × dose, i.e. dose^2. Once created, the radiation damage due to these two possible routes is indistinguishable (i.e. both processes are lethal). From this simplified description, it is clear that the observed radiation response characterized in the cell survival curve will consist of two components: one linear with dose and the other quadratic, i.e. proportional to dose^2. This phenomenological description qualitatively explains the shape of a radiation survival curve, with a finite initial slope at low dose followed by an increasingly downward curvature as dose increases.

However, the amount of damage created in the second process is dependent on the ability of the second break to be induced before the first break has repaired itself and, thus, is dependent on the dose rate.

Figure 2.5 shows a range of response curves in which the doses are delivered at four different dose rates, the individual doses taking proportionately longer to deliver as dose rate is reduced. This graph illustrates that reducing the dose rate causes the overall shape of the response curve to become less 'curvy' than in the acute case, but that the initial slope remains unchanged. When the doses are all delivered at a very low dose rate, as is the case for most radionuclide therapies, the response is essentially a straight line, when the curves are plotted on a log-linear scale, as is common practice for radiation survival curves. This means that the low dose response is purely exponential.

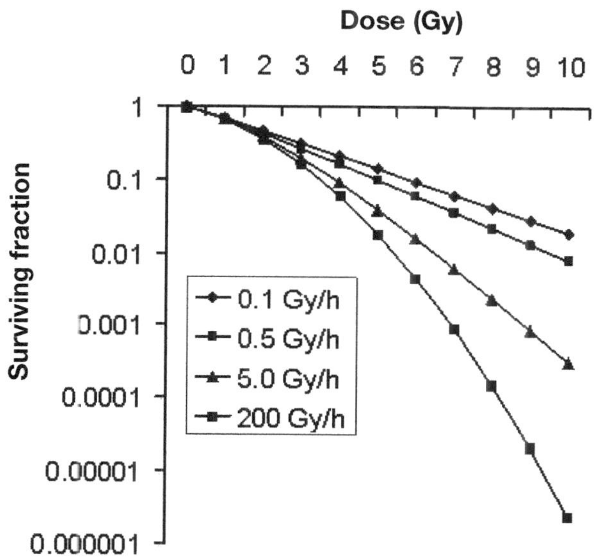

FIG. 2.5. Surviving fraction as a function of dose for different dose rates. It is important to note that most radionuclide therapies are delivered at low dose rate in the range of 0.1–0.5 Gy/h, when the survival curve is almost linear.

2.6.6. The basic linear–quadratic model

The basic equation describing the shape of the cell survival curves shown in Fig. 2.2 is referred to as the LQ model, which has a biophysical origin. Cell survival following delivery of an acute dose d is given as:

$$S = \exp\left(-\alpha d - \beta d^2\right) \tag{2.8}$$

where α (in units of Gy^{-1}) and β (in units of Gy^{-2}) are the respective linear and quadratic sensitivity coefficients.

If the treatment is repeated in N well spaced fractions, then the net survival is S_N, where:

$$S_N = S^N = \exp\left(-N\alpha d - N\beta d^2\right) \tag{2.9}$$

Taking natural logarithms on both sides of Eq. (2.9) and dividing throughout by α leads to:

$$\frac{\ln S_N}{\alpha} = -Nd - \frac{Nd^2}{(\alpha / \beta)} \tag{2.10}$$

2.6.7. Modification to the linear–quadratic model for radionuclide therapies

Targeted radionuclide therapy normally involves irradiation of the tumour/normal tissues at a dose rate which is not constant but which reduces as treatment proceeds, as a consequence of the combination of radionuclide decay and biological clearance of the radiopharmaceutical. To allow for this, a more extensive formulation of the LQ model is required.

2.6.8. Quantitative intercomparison of different treatment types

In many aspects of LQ modelling, a term called the 'biological effective dose' (BED) is employed to assess and inter-compare different treatment types. BED is defined as:

$$\text{BED} = -\frac{\ln S_N}{\alpha} = Nd\left[1 + \frac{d}{(\alpha / \beta)}\right] \tag{2.11}$$

Although the parameters α and β are rarely known in detail for individual tumours or tissues, values of the ratio α/β (in units of grays) are becoming increasingly known from clinical and experimental data. In general, α/β is systematically higher (5–20 Gy) for tumours than for critical, late-responding normal tissues (2–5 Gy) and it is this difference which provides the BED concept with much of its practical usefulness.

For non-acute treatments (those in which the dose delivery is protracted over a long time period on account of a lower dose rate), the BED is re-written as:

$$\mathrm{BED} = Nd\left[1 + \frac{d\,g(t)}{(\alpha/\beta)}\right] \tag{2.12}$$

where $g(t)$ is a function of the time t taken for delivery:

$$g(t) = \frac{2}{\mu}\left[1 - \frac{1 - \exp(-\mu t)}{\mu t}\right] \tag{2.13}$$

and where μ is the mono-exponential time constant relating to the repair of sublethal damage. μ is related to the tissue repair half-time ($T_{1/2}$) via:

$$\mu = \frac{0.693}{T_{1/2}} \tag{2.14}$$

For a treatment delivery at constant dose rate R, the delivered dose d is related to treatment time t via $d = Rt$, thus:

$$\mathrm{BED} = Rt\left[1 + \frac{2R}{\mu(\alpha/\beta)}\left\{1 - \frac{1 - \exp(-\mu t)}{\mu t}\right\}\right] \tag{2.15}$$

When $t > 12$ h, Eq. (12.15) simplifies to:

$$\mathrm{BED} = Rt\left[1 + \frac{2R}{\mu(\alpha/\beta)}\right] \tag{2.16}$$

2.6.9. Cellular recovery processes

At lower doses and dose rates, cellular recovery may play an important role in the fixation of the radiation damage. There are three broad types of cellular radiation damage:

(a) Lethal damage in which the cellular DNA is irreversibly damaged to such an extent that the cell dies or loses its proliferative capacity;
(b) Sublethal damage in which partially damaged DNA is left with sufficient capacity to restore itself over a period of a few hours, provided there is no further damage during the repair period;

(c) Potentially lethal damage in which repair of what would normally be a lethal event is made possible by manipulation of the post-irradiation cellular environment.

2.6.10. Consequence of radionuclide heterogeneity

The effectiveness per unit dose of a radiopharmaceutical depends on the heterogeneity of the radionuclide distribution. Global non-uniformity of a source distribution, which results in pockets of cells (tumour or normal tissue) receiving less than the average dose will almost always result in a greater fraction of cell survivors, than if all cells receive a uniform dose. The one possible exception would be if a radiopharmaceutical would selectively localize at sensitive target cells, within an organ, that are key for organ regeneration or function, e.g. crypt cells in the colon. The cellular response also depends on the microdosimetry, especially if the radiopharmaceutical in question selectively localizes on the cell surface or internalizes within a certain cohort of cells within a tumour/normal organ. Radiolabels that selectively localize on the surface of cells or are internalized may exhibit geometric enhancement factors that modulate a response. The reader is referred to ICRU Report 67 on absorbed dose specification in nuclear medicine for more details.

2.7. GROSS RADIATION EFFECTS ON TUMOURS AND TISSUES/ORGANS

2.7.1. Classification of radiation damage (early versus late)

Cells which are lethally affected by radiation may continue to function for some time after the infliction of the damage, only dying when attempting to undergo subsequent cell division (mitosis). Clinically observed radiation effects in whole tissues or organs reflect the damage inflicted to large numbers of constituent cells and, thus, appear on a timescale which is governed largely by the underlying proliferation rates of those cells. Such observable effects are classified as being either late or early, depending on the speed at which they manifest themselves following irradiation. Late effects appear months or years after irradiation and appear in structures which proliferate very slowly, e.g. kidney. Early (or acute) effects appear within days, weeks or months of irradiation and are associated with fast-proliferating epithelial tissues, e.g. bone marrow, mucosa, intestinal tract, etc.

In most types of radiotherapy, it is the late effects which are considered to be most critical and which generally limit the total dose which may be delivered to the tumour. If the radiation tolerance of the late-responding tissues is exceeded, then the subsequent reactions, depending on the tissues in which they arise, may seriously affect mobility and/or quality of life, and may even be life threatening. Such problems arise long after the completion of treatment and are, thus, impossible to correct. These are the serious considerations which are at the heart of the therapeutic index concept discussed below (see Section 2.7.3). Acute reactions in radiotherapy, although they may be unpleasant, are usually transient and easier to control by adjustment of the treatment dose delivery pattern and/or simple medication. In radionuclide therapies, it is in most instances possible to circumvent acute radiation toxicities once they begin to occur, such as by accelerating clearance of the radiopharmaceutical. Chronic toxicities, such as to the kidney, usually occur at times which are long relative to the lifetime of the radionuclide. Hence, considerable importance should be attributed to the administration of safe activities of therapeutic radionuclides that do not exceed any dose limiting constraints.

2.7.2. Determinants of tumour response

Irrespective of the mechanism used to achieve tumour targeting, the potential advantage of radionuclide therapy over other forms of radiation therapy is its ability to deliver dose to both the local disease and to occult tumour deposits.

In nuclear medicine, the primary determinants of treatment effectiveness are:

- The tumour specificity of the radionuclide carrier.
- The homogeneity of uptake of the carrier within the targeted tumour(s).
- The intrinsic RBE (see Section 2.6.4) of the radiation used for the therapy: this is determined primarily by the nature of the radionuclide emissions (e.g. α particles, β particles, low energy γ rays, Auger electrons, etc.).
- The range of the particles, as determined by their energies.
- The total dose delivered.
- The responsiveness of the targeted tumour cells to radiation. This will be determined by radiobiological properties such as cellular radiosensitivity and the variations of sensitivity within the cell cycle, the oxygen status of the cells (fully oxic, partially oxic or hypoxic), the ability of the cells to recover from sublethal radiation damage and the degree to which tumour growth (repopulation) may occur during therapy.

These factors are complementary and interactive, and should not be considered in isolation from each other. Thus, for example, significant non-uniformity of uptake within the tumour may result in dose 'cold spots', but the detrimental potential of these might be offset by the selection of a radionuclide which emits particles of sufficient range to produce a cross-fire effect within the cold spots from those adjacent cells which are properly targeted. The significance of cold spot and cross-fire effects is further dependent on the size of the tumour deposit under consideration.

2.7.3. The concept of therapeutic index in radiation therapy and radionuclide therapy

The therapeutic index of a particular radiation treatment (often referred to in older publications as the 'therapeutic ratio') is a measure of the resultant damage to the tumour vis a vis the damage to critical normal structures. Treatments with a high therapeutic index will demonstrate good tumour control and low normal tissue morbidity; treatments with a low therapeutic index will be associated with a low tumour control and/or high morbidity. There have been several attempts to provide quantitative definitions of therapeutic index, but it is usually sufficient to consider therapeutic index as being a qualitative concept — any new treatment which, relative to an existing treatment, improves tumour control and/or reduces morbidity is said to be associated with an improved therapeutic index.

In conventional (external beam) radiotherapy, the normal tissues at risk will be those immediately adjacent to the tumour being treated. Doses to the normal tissues (along with the risk of toxicity) may be reduced by attention to a combination of physical and radiobiological factors. In targeted radionuclide therapy, the tumour may be single and discrete (as is the case in most external beam therapy) or may consist of distributed masses or metastatic deposits at several locations within the body. The normal tissues at risk may themselves be widely distributed but, more particularly, may be a reflection of the particular uptake pattern of the targeting compound being used for the therapy.

2.7.4. Long term concerns: stochastic and deterministic effects

The radiation detriment which results from radiation exposure may be classified as being either stochastic or deterministic in nature. Stochastic effects (e.g. hereditary damage, cancer induction) are those for which the likelihood of them occurring is dose related, but the severity of the resultant condition is not related to the dose received. Deterministic effects (e.g. cataract induction, general radiation syndromes, bone marrow ablation, etc.) manifest themselves with a severity which is dose related. In general, it is predominantly stochastic effects which need to be considered as potential side effects from diagnostic uses of radionuclides, although deterministic damage may result if the embryo or fetus is irradiated. For radionuclide therapy applications, the concerns relate to both stochastic and deterministic effects.

2.8. SPECIAL RADIOBIOLOGICAL CONSIDERATIONS IN TARGETED RADIONUCLIDE THERAPY

2.8.1. Radionuclide targeting

Tumour targeted radiotherapy is a very promising approach for the treatment of wide-spread metastasis and disseminated tumour cells. This technique aims to deliver therapeutic irradiation doses to the tumour while sparing normal tissues by targeting a structure that is abundant in tumour cells, but rare in normal tissues. This can be done by using antibodies labelled with a therapeutic relevant radionuclide acting against a specific tumour target. Radiolabelled antibody therapy has already become common in the treatment of non-Hodgkin's lymphoma, e.g. ^{131}I-tositumomab (Bexxar®) and ^{90}Y-ibritumomab tiuxetan (Zevalin®), and exhibits great potential for being extended to other diseases. A good example is epidermal growth factor (EGF) labelled with ^{125}I which will bind EGF receptors. EGF receptors are overexpressed on tumour cells in many malignancies such as highly malignant gliomas. At present, several other radiolabelled antibodies are being used in experimental models and in clinical trials to study their feasibility in other types of cancer.

2.8.2. Whole body irradiation

Conventional external beam radiotherapy involves controlled irradiation of a carefully delineated target volume. Normal structures adjacent to the tumour will likely receive a dose, in some cases a moderately high dose, but the volumes involved are relatively small. The rest of the body receives only a minimal dose, mostly arising from radiation scattered within the patient from the target volume and from a small amount of leakage radiation emanating from the treatment machine outside the body.

Targeted radionuclide therapies are most commonly administered intravenously and, thus, can give rise to substantial whole body doses and, in particular, doses to the radiation sensitive bone marrow. Once the untargeted activity is removed from the blood, it may give rise to substantial doses in normal structures, especially the kidneys. Furthermore, the activity taken up by the kidneys and targeted tumour deposits may (if γ ray emissions are involved) continue to irradiate the rest of the body.

2.8.3. Critical normal tissues for radiation and radionuclide therapies

Since the radiation doses used in radionuclide therapies are much higher than the doses used for diagnosis, (prolonged) retention of the pharmaceuticals within the blood circulation and, hence, increased accumulation of radionuclides in non-tumour cells, might lead to unwanted toxicities. The bone marrow, kidney and liver are regarded as the main critical organs for systemic radionuclide therapy. Other organs at risk are the intestinal tract and the lungs. The bone marrow is very sensitive towards ionizing radiation. Exposure of the bone marrow with high doses of radiation will lead to a rapid depression of white blood cells followed a few weeks later by platelet depression, and in a later stage (approximately one month after exposure) also by depression of the red blood cells. In general, these patients could suffer from infections, bleeding and anaemia. Radiation damage to the gastrointestinal tract is characterized by de-population of the intestinal mucosa (usually between 3 and 10 days) leading to prolonged diarrhoea, dehydration, loss of weight, etc. The kidneys, liver and lungs will show radiation induced damage several months after exposure. In kidneys, a reduction of proximal tubule cells is observed. These pathological changes finally lead to nephropathy. In the liver, hepatocytes are the radiosensitive targets. Since the lifespan of the cells is about a year, deterioration of liver function will become apparent between 3 and 9 months after exposure. In lungs, pulmonary damage is observed in two waves: an acute wave of pneumonitis and later fibrosis.

The determinants of normal tissue response from radionuclide studies is a large subject due to the diversity of radiopharmaceuticals with differing pharmacokinetics and biodistribution, and the widely differing responses and tolerances of the critical normal tissues. A principal determinant of the type of toxicity depends on the radionuclide employed. For example, isotopes of iodine localize in the thyroid (unless blocked), salivary glands, stomach and bladder. Strontium, yttrium, samarium, fluorine, radium, etc. concentrate in bone. Several radiometals, such as bismuth, can accumulate in the kidney. If these radionuclides are tightly conjugated to a targeting molecule, the biodistribution and clearance are determined by that molecule. For high molecular weight targeting agents, such as an antibody injected intravenously, the slow plasma clearance results in marrow toxicity being the principal dose limiting organ. For smaller radiolabelled peptides, renal toxicity becomes of concern. When studying a new radiopharmaceutical or molecular imaging agent, it is always important to perform a detailed study of the biodistribution at trace doses, to ensure the absence of radionuclide sequestration within potentially sensitive tissue, such as the retina of the eye or the germ cells of the testes.

A review of normal tissue toxicities resulting from radionuclide therapies is given by Meredith et al. (2008).

2.8.4. Imaging the radiobiology of tumours

The development of molecular imaging using positron emission tomography (PET) has given rise to new radiotracers which have the potential to assess several features of radiobiological relevance for therapy planning. One tracer that is becoming more widely available for PET imaging is fluorothymidine. This radiotracer exhibits the property of becoming selectively entrapped within cells that are progressing through S-phase (DNA replication) of the cell cycle, thus providing a signal which should be proportional to cell proliferation, and minimizing the signal from cells in G_0 or in cell cycle arrest. The ability to selectively identify only replicating cells separate from all tumour cells present within the computed tomography-determined tumour volume may present an excellent opportunity for more accurate measures of the initial viable tumour burden as well as evaluating tumour response. Complementary to measuring tumour response is the measurement of therapeutic efficacy through radiotracers that selectively target cell death. Radiotracers are under development with the ability to selectively bind to receptors expressed on cells undergoing programmed cell death, e.g. radiolabelled annexin V. Another area of active research is in the field of hypoxia imaging. Cells within a tumour microenvironmental region of low partial oxygen pressure, i.e. hypoxia, are known to exhibit a great radio-resistance to both radiation and chemotherapy relative to those under normoxic conditions. A number of PET radiotracers are under evaluation for imaging tumour hypoxia with PET, including fluoromisonidazole (^{18}F-FMISO), fluoroazomycin arabinoside (^{18}F-FAZA) and copper-diacetyl-bis(N4-methylthiosemicarbazone) (^{64}Cu-ATSM). The ability to measure the radiobiological attributes of a tumour prior to therapy may provide invaluable information concerning the relative resistance/aggressiveness of tumours, leading to improved management of these patients.

2.8.5. Choice of radionuclide to maximize therapeutic index

The choice of the optimum radionuclide to maximize the therapeutic index depends on a number of factors. First, the range of the emitted particles from the radionuclide should depend on the type of tumour being treated. For leukaemia or micrometastatic deposits, consisting of individual or small clusters of tumour cells, there is a distinct advantage of using radionuclides which emit very short range particles. Since α particles have ranges of <100 μm in tissue, α particle emitters would have an advantage, if the targeting molecule were able to reach all tumour cells. However, α particle emitting radionuclides are not widely available and are extremely expensive. In addition, the short range of α particles can be a disadvantage for bulk tumours. For these reasons, almost all therapeutic radionuclides utilized in the clinic today consist of medium (^{131}I) or long range (^{90}Y, ^{186}Re) β emitters. These radionuclides are advantageous when treating solid tumours for which target receptor (antigen) expression may be heterogeneous, or with non-uniform delivery, due to the greater cross-fire range of their β emissions (ranging up to a 1 cm range in unit density tissue).

A second important consideration is the choice of radionuclide half-life. If the half-life is too short, then the radiolabelled tumour targeting agent may have insufficient time to reach its target, resulting in a minimal therapeutic index. Increasing the half-life will increase the therapeutic index, but render the patient radioactive for a longer period of time, resulting in prolonged confinement, greater expense and radiation risks to staff and family. Pure β emitting radionuclides such as ^{90}Y and ^{32}P have advantages in that they minimize the exposure to personnel assisting the patient. The half-life of the radionuclide should ideally match the biological uptake and retention kinetics of the tumour-targeting carrier used. For large protein carriers such as antibodies, radionuclides with half-lives of several days are required to optimize the therapeutic index. For smaller molecular targeting agents such as peptides, short lived radionuclides may be better suited to minimize radioactive waste.

Thirdly, it is necessary to consider radiochemistry, ease and stability of the radiolabelled end product. All of these factors need to be taken into consideration in order to produce the optimum therapeutic targeting compound for use in clinical therapeutic applications.

BIBLIOGRAPHY

DALE, R.G., JONES, B. (Eds), Radiobiological Modelling in Radiation Oncology, The British Institute of Radiology, London (2007).

HALL, E.J., GIACCIA, A.J., Radiobiology for the Radiologist, 6th edn, Lippincott, Williams and Wilkins, Philadelphia, PA (2006).

INTERNATIONAL COMMISSION ON RADIATION UNITS, Absorbed-dose Specification in Nuclear Medicine, Rep. 67, Nuclear Technology Publishing, Ashford, United Kingdom (2002).

MEREDITH, R., WESSELS, B., KNOX, S., Risks to normal tissue from radionuclide therapy, Semin. Nucl. Med. **38** (2008) 347–357.

CHAPTER 3

RADIATION PROTECTION

S.T. CARLSSON
Department of Diagnostic Radiology,
Uddevalla Hospital,
Uddevalla, Sweden

J.C. LE HERON
Division of Radiation, Transport and Waste Safety,
International Atomic Energy Agency,
Vienna

3.1. INTRODUCTION

Medical exposure is the largest human-made source of radiation exposure, accounting for more than 95% of radiation exposure. Furthermore, the use of radiation in medicine continues to increase worldwide — more machines are accessible to more people, the continual development of new technologies and new techniques adds to the range of procedures available in the practice of medicine, and the role of imaging is becoming increasingly important in day to day clinical practice. The introduction of hybrid imaging technologies, such as positron emission tomography/computed tomography (PET/CT) and single photon emission computed tomography (SPECT)/CT, means that the boundaries between traditional nuclear medicine procedures and X ray technologies are becoming blurred. Worldwide, the total number of nuclear medicine examinations is estimated to be about 35 million per year.

In Chapter 2, basic radiation biology and radiation effects were described, demonstrating the need for a system of radiation protection. Such a system allows the many beneficial uses of radiation to be utilized, but at the same time ensures detrimental radiation effects are either prevented or minimized. This can be achieved by having the objectives of preventing the occurrence of deterministic effects and of limiting the probability of the stochastic effects to a level that is considered acceptable. In a nuclear medicine facility, consideration needs to be given to the patient, the staff involved in performing the nuclear medicine procedures, members of the public and other staff that may be in the nuclear medicine facility, carers and comforters of patients undergoing procedures, and persons who may be undergoing a nuclear medicine procedure as part of a biomedical research project.

This chapter discusses how the objectives stated above are achieved through a system of radiation protection, and how such a system should be applied practically in a hospital in general and in nuclear medicine specifically.

3.2. BASIC PRINCIPLES OF RADIATION PROTECTION

The means for achieving the objectives of radiation protection have evolved over many years to the point where, for some time, there has been a reasonably consistent approach throughout the world — namely the 'system of radiological protection', as espoused by the International Commission on Radiological Protection (ICRP). The following will briefly describe this system, specifically as it applies to nuclear medicine.

3.2.1. The International Commission on Radiological Protection system of radiological protection

The principles of radiation protection and safety upon which the IAEA safety standards are based are those developed by the ICRP. The detailed formulation of these principles can be found in ICRP publications and they cannot easily be paraphrased without losing their essence. However, a brief, although simplified, summary of the principles is given in this section.

The ICRP recommends a system of radiological protection to cover all possible exposure situations. There are many terms associated with the system and some of these will now be introduced.

The ICRP in its Publication 103 [3.1] divides all possible situations of where exposure can occur into three types — planned exposure situations, emergency exposure situations and existing exposure situations. For the practice of nuclear medicine, only the first situation is relevant. The use of radiation in nuclear medicine is a planned exposure situation — it needs to be under regulatory control, with an appropriate authorization in place from the regulatory body before operation can commence. Misadministration, spills and other such incidents or accidents can give rise to what is called potential exposure, but these remain part of the planned exposure situation as their occurrence is considered in the granting of an authorization. It should be noted that the ICRP has used the term 'practice' to describe a planned exposure situation such as the operation of a nuclear medicine facility.

The ICRP then puts exposure of individuals into three categories — medical exposure, occupational exposure and public exposure:

- — Medical exposure refers primarily to exposure incurred by patients for the purpose of medical diagnosis or treatment. It also refers to exposures incurred by individuals helping in the support and comfort of patients undergoing diagnosis or treatment, and by volunteers in a programme of biomedical research involving their exposure.
- — Occupational exposure is the exposure of workers incurred in the course of their work.
- — Public exposure is exposure incurred by members of the public from all exposure situations, but excluding any occupational or medical exposure.

All three need to be considered in the nuclear medicine facility.

An individual person may be subject to one or more of these categories of exposure, and for radiation protection purposes such exposures are dealt with separately.

The ICRP system has three fundamental principles of radiological protection, namely:

- — The principle of justification: Any decision that alters the radiation exposure situationshould do more good than harm.
- — The principle of optimization of protection: The likelihood of incurring exposures, the number of people exposed and the magnitude of their individual doses should all be kept as low as reasonably achievable (ALARA), taking into account economic and societal factors.
- — The principle of limitation of doses: The total dose to any individual from regulated sources in planned exposure situations other than medical exposure of patients should not exceed the appropriate limits recommended by the ICRP. Recommended dose limits are given in Table 3.1.

In a nuclear medicine facility, occupational and public exposures are subject to all three principles, whereas medical exposure is subject to the first two only. More detail on the application of the ICRP system for radiological protection as it applies to a nuclear medicine facility is given in the remainder of this chapter.

TABLE 3.1. RECOMMENDED DOSE LIMITS IN PLANNED EXPOSURE SITUATIONS[a]

Type of limit	Occupational	Public
Effective dose	20 mSv per year, averaged over defined periods of 5 years[b]	1 mSv in a year[c]
Annual equivalent dose in:		
Lens of the eye[d]	20 mSv	15 mSv
Skin[e, f]	500 mSv	50 mSv
Hands and feet	500 mSv	—

[a] Limits on effective dose are for the sum of the relevant effective doses from external exposure in the specified time period and the committed effective dose from intakes of radionuclides in the same period. For adults, the committed effective dose is computed for a 50 year period after intake, whereas for children it is computed for the period up to reaching 70 years of age.

[b] With the further provision that the effective dose should not exceed 50 mSv in any single year. Additional restrictions apply to the occupational exposure of pregnant women.

[c] In special circumstances, a higher value of effective dose could be allowed in a single year, provided that the average over 5 years does not exceed 1 mSv/a.

[d] In 2011, the ICRP recommended that the occupational dose limit be lowered from the previous 150 mSv/a to 20 mSv/a, averaged over 5 years, and with no more than 50 mSv in any single year.

[e] The limitation on effective dose provides sufficient protection for the skin against stochastic effects.

[f] Averaged over a 1 cm^2 area of skin regardless of the area exposed.

3.2.2. Safety standards

Safety standards are based on knowledge of radiation effects and on the principles of protection described above. In this respect, the development of safety standards by the IAEA follows a well established approach. The United Nations Scientific Committee on the Effects of Atomic Radiation (UNSCEAR), a body set up by the United Nations in 1955, compiles, assesses and disseminates information on the health effects of radiation and on levels of radiation exposure due to different sources; this information was taken into account in developing the standards. Following a decision made in 1960, the IAEA safety standards are based on the recommendations of the ICRP, which also take account of the scientific information provided by UNSCEAR.

Purely scientific considerations, however, are only part of the basis for decisions on protection and safety, and the safety standards implicitly encourage decision makers to make value judgements about the relative importance of different kinds of risks and about the balancing of risks and benefits. General acceptance of risk is a matter of consensus and, therefore, international safety standards should provide a desirable international consensus for the purpose of protection.

For these reasons, international consensus is integral to the IAEA safety standards, which are prepared with the wide participation of and approval by its Member States and relevant international organizations. The current version of what is commonly called the Basic Safety Standards (BSS) is entitled Radiation Protection and Safety of Radiation Sources: International Basic Safety Standards (2014) [3.2]. The BSS are jointly sponsored by the European Commission, the Food and Agriculture Organization of the United Nations, the IAEA, the International Labour Organization, the OECD Nuclear Energy Agency, the Pan American Health Organization (PAHO), the United Nations Environment Programme and the World Health Organization (WHO).

The BSS comprises five sections: Introduction, General requirements for protection and safety, Planned exposure situations, Emergency exposure situations and Existing exposure situations, as well as four schedules. The purpose of the BSS is to establish basic requirements for protection against exposure to ionizing radiation and for the safety of radiation sources that may deliver such exposure. The requirements of the BSS underpin the implementation of radiation protection in a nuclear medicine facility, supplemented by the relevant IAEA Safety Guides and Safety Reports.

3.2.3. Radiation protection quantities and units

The basic dosimetry quantity for use in radiation protection is the mean organ or tissue dose D_T given by:

$$D_T = \varepsilon_T / m_T \tag{3.1}$$

where

m_T is the mass of the organ or tissue T;

and ε_T is the total energy imparted by radiation to that tissue or organ.

The International System of Units (SI) unit of mean organ dose is joules per kilogram (J/kg) which is termed gray (Gy).

Owing to the fact that different types of ionizing radiation will have different effectiveness in damaging human tissue at the same dose, and the fact that the probability of stochastic effects will depend on the tissue irradiated, it is necessary to introduce quantities to account for these factors. Those quantities are equivalent dose and effective dose. Since they are not directly measurable, the International Commission on Radiation Units and Measurements (ICRU) has defined a set of operational quantities for radiation protection purposes (area monitoring and personal monitoring): the ambient dose equivalent, directional dose equivalent and personal dose equivalent.

Regarding internal exposure from radionuclides, the equivalent dose and the effective dose are not only dependent on the physical properties of the radiation but also on the biological turnover and retention of the radionuclide. This is taken into account in the committed dose quantities (equivalent and effective).

3.2.3.1. Equivalent dose

It is a well known fact in radiobiology that densely ionizing radiation such as α particles and neutrons will cause greater harm to a tissue or organ than γ rays and electrons at the same mean absorbed dose. This is because the dense ionization events will result in a higher probability of irreversible damage to the chromosomes and a lower chance of tissue repair. To account for this, the organ dose is multiplied with a radiation weighting factor in order to get a quantity that more closely reflects the biological effect on the irradiated tissue or organ. This quantity is called the equivalent dose and is defined as:

$$H_{\mathrm{T}} = w_{\mathrm{R}} D_{\mathrm{T,R}} \tag{3.2}$$

where

$D_{\mathrm{T,R}}$ is the mean tissue or organ dose delivered by type R radiation;

and w_{R} is the radiation weighting factor.

For X rays, γ rays and electrons, $w_{\mathrm{R}} = 1$; for α particles, $w_{\mathrm{R}} = 20$. The SI unit of equivalent dose is joules per kilogram (J/kg), which is termed sievert (Sv). In a situation of exposure from different types of radiation, the total equivalent dose is the sum of the equivalent dose from each type of radiation.

3.2.3.2. Effective dose

The relationship between the probability of stochastic effects and equivalent dose is found to depend on the organ or tissue irradiated. To account for this, tissue weighting factors w_T are introduced. They should represent the relative contribution of an organ or tissue T to the total detriment due to the stochastic effects resulting from a uniform irradiation of the whole body. The total tissue weighted equivalent dose is called effective dose and is defined as:

$$E = \sum w_T H_T \tag{3.3}$$

where H_T is the equivalent dose in organ or tissue T.

The sum is performed over all organs and tissues of the human body considered to be sensitive to the induction of stochastic effects. Recommended tissue weighting factors are found in ICRP Publication 103 [3.1]. Despite depending on the sex and age of the person, for the purposes of radiation protection, the values for tissue weighting factors are taken as constants and are applicable to the average population.

The use of effective dose has many advantages in practical radiation protection. Very different exposure situations (e.g. internal and external exposure by different types of radiation) can be combined and result in a single value, the effective dose.

3.2.3.3. Committed dose

When radionuclides are taken into the body, the resulting dose is received throughout the period of time during which they remain in the body. The total dose delivered during this period of time is referred to as the committed dose and is calculated as a specified time integral of the rate of receipt of the dose. The committed equivalent dose is defined as:

$$H_T(\tau) = \int_{t_0}^{t_0+\tau} \dot{H}_T(t)\,\mathrm{d}t \tag{3.4}$$

where

t_0 is the time of intake;

and τ is the integration time.

For workers and adult members of the general public, τ is taken to be 50 years while for children 70 years is regarded as appropriate.

The committed effective dose is given by:

$$E(\tau) = \sum_{T} w_T H_T(\tau) \tag{3.5}$$

3.2.3.4. Operational quantities

For all types of external radiation, the operational quantities for area monitoring are defined on the basis of a dose equivalent value at a point in the ICRU sphere. It is a sphere of tissue-equivalent material (30 cm in diameter with a density of 1 g/cm^3 and a mass composition of: 76.2% oxygen, 11.1% carbon, 10.1% hydrogen and 2.6% nitrogen). For radiation monitoring, it adequately approximates the human body in regards to the scattering and attenuation of the radiation fields under consideration.

The operational quantities for area monitoring defined in the ICRU sphere should retain their character of a point quantity. This is achieved by introducing the terms 'expanded' and 'aligned' radiation field in the definition of these quantities. An expanded radiation field is a hypothetical field in which the spectral and the angular fluence have the same values at all points of a sufficiently large volume equal to the values in the actual field at the point of interest. The expansion of the radiation field ensures that the whole ICRU sphere is thought to be exposed to a homogeneous radiation field with the same fluence, energy distribution and direction distribution as at the point of interest of the real radiation field. If all radiation is aligned in the expanded radiation field so that it is opposed to a radius vector $\boldsymbol{\Omega}$ specified for the ICRU sphere, the aligned and expanded radiation field is obtained. In this hypothetical field, the ICRU sphere is homogeneously irradiated from one direction, and the fluence of the field is the integral of the angular differential fluence at the point of interest in the real radiation field over all directions. In the expanded and aligned radiation field, the value of the dose equivalent at any point in the ICRU sphere is independent of the direction distribution of the radiation in the real radiation field.

For area monitoring, the operational quantity for penetrating radiation is the ambient dose equivalent, $H^*(10)$:

— The ambient dose equivalent at a point in a radiation field is the dose equivalent that would be produced by the corresponding expanded and aligned field in the ICRU sphere at a depth of 10 mm on the radius vector opposing the direction of the aligned field.

For area monitoring, the operational quantity for low-penetrating radiation is the directional dose equivalent $H'(0.07, \boldsymbol{\Omega})$:

— The directional dose equivalent at a point in a radiation field is the dose equivalent that would be produced by the corresponding expanded field in the ICRU sphere at a depth of 0.07 mm in a specified direction $\boldsymbol{\Omega}$.

The personal dose equivalent $H_p(d)$ is defined as:

— The equivalent dose at a depth d in soft tissue below a specified point on the body.

The relevant depth is $d = 10$ mm for penetrating radiations (photon energies above 15 keV), while depths $d = 0.07$ mm and $d = 3$ mm are used for weakly penetrating radiations (photon energies below 15 keV) in skin and the lens of the eye, respectively.

3.3. IMPLEMENTATION OF RADIATION PROTECTION IN A NUCLEAR MEDICINE FACILITY

3.3.1. General aspects

Implementation of radiation protection in a nuclear medicine facility must fit in with, and be complementary to, the systems for implementing medical practice in the facility. Radiation protection must not be seen as something imposed from 'outside' and separate to the real business of providing medical services and patient care. Most countries have their own radiation protection legislation and regulatory framework, typically requiring any facility or person wishing to provide or perform nuclear medicine procedures to have an appropriate authorization from the radiation protection regulatory body. The requirements to be fulfilled in order to be granted such an authorization will vary from country to country, but in general, compliance with the requirements of the BSS would be expected.

To achieve a high standard of radiation protection, the most important thing is to establish a safety based attitude in every individual, such that protection and accident prevention are regarded as a natural part of daily duties. This objective is basically achieved by education and training, and encouraging a questioning and learning attitude, but also by a positive and cooperative attitude from the national authorities and the employer in supporting radiation protection with sufficient resources, both in terms of personnel and money. A feeling of responsibility can only be achieved if the people involved regard the rules and regulations as necessary, and are a support to and not a hindrance in their daily work. Every individual should also know their responsibilities through formal assignment of duties.

3.3.2. Responsibilities

3.3.2.1. Licensee and employer

The licensee of a nuclear medicine facility, through the authorization issued by the regulatory body, has the prime responsibility for applying the relevant national regulations and meeting the conditions of the licence. The licensee may appoint other people to carry out actions and tasks related to these responsibilities, but the licensee retains overall responsibility. In particular, the nuclear medicine physician, the medical physicist, the nuclear medicine technologist, the radiopharmacist and the radiation protection officer (RPO) all have key roles and responsibilities in implementing radiation protection in a nuclear medicine facility, and these are discussed in more detail below.

The BSS need to be consulted for details on all of the requirements for radiation protection that are assigned to licensees. Employers are also assigned many responsibilities, in cooperation with the licensee, for occupational radiation protection. Key responsibilities for the licensee include ensuring that the necessary personnel (nuclear medicine physicians, medical physicists, nuclear medicine technologists, radiopharmacists and an RPO) are appointed, and that the individuals have the necessary education, training and competence to perform their respective duties. Clear responsibilities for personnel must be assigned; a radiation protection programme (RPP) must be established and the necessary resources provided; a comprehensive quality assurance (QA) programme must be established; and education and training of personnel supported.

3.3.2.2. Nuclear medicine specialist

The general medical and health care of the patient is, of course, the responsibility of the individual physician treating the patient. However, when the patient presents in the nuclear medicine facility, the nuclear medicine specialist has the particular responsibility for the overall radiation protection of the patient. This means responsibility for the justification of a given nuclear medicine procedure for the patient, in conjunction with the referring medical practitioner, and responsibility for ensuring the optimization of protection in the performance of the examination or treatment.

3.3.2.3. *Nuclear medicine technologist*

The technologist has a key position, and their skill and care to a large extent determine the optimization of the patient's exposure.

3.3.2.4. *Radiation protection officer*

It is highly recommended that the licensee appoint a person to oversee and implement radiation protection matters in the hospital. This person is called an RPO or radiation safety officer. The RPO should have a good theoretical and practical knowledge of the properties and hazards of ionizing radiation, as well as protection. In addition, the RPO should possess necessary knowledge of all the appropriate legislation and codes of practice relating to the uses of ionizing radiation in the relevant medical area, e.g. nuclear medicine. The RPO, unless also a qualified medical physicist in nuclear medicine, has no responsibilities for radiation protection in medical exposure.

3.3.2.5. *Medical physicist*

The medical physicist is a person who by education and training is competent to practise independently in one or more of the subfields in medical physics. For instance, a medical physicist in nuclear medicine should have a comprehensive knowledge of the imaging equipment used, including performance specifications, physical limitations of the equipment, calibration, quality control and image quality. The medical physicist should also be qualified in handling radiation protection matters associated with nuclear medicine, and has particular responsibilities for radiation protection in medical exposure, including the requirements pertaining to imaging (for diagnostic procedures), calibration, dosimetry and QA. Whenever possible, a medical physicist should serve as the RPO (see above). Other important tasks for the medical physicist are to be responsible for QA and for the local continuing education in radiation protection of the nuclear medicine staff and other health professionals.

3.3.2.6. *Other personnel*

Other personnel that may have responsibilities in radiation protection in nuclear medicine include radiopharmacists and other staff that may have been trained to perform special tasks, such as contamination tests or some quality control tests.

3.3.3. Radiation protection programme

The BSS require a licensee (and employer where appropriate) to develop, implement and document a protection and safety programme commensurate with the nature and extent of the risks of the practice to ensure compliance with radiation protection standards. Such a programme is often called an RPP and each nuclear medicine facility should have one. The RPP for a nuclear medicine facility is quite complex as it needs to cover all relevant aspects of protection of the worker, the patient and the general public. The details of such an RPP can be found in Ref. [3.3].

For an RPP to be effective, the licensee needs to provide for its implementation, including the resources necessary to comply with the programme and arrangements to facilitate cooperation between all relevant parties.

3.3.4. Radiation protection committee

An effective way to supervise compliance with the RPP is the formation of a committee for radiation protection. Since a representative of the management is usually a member of the radiation protection committee, communication with the representative may be the most appropriate. The members of the radiation protection committee should include an administrator representing the management, the chief nuclear medicine physician, a medical physicist, the RPO, a nuclear medicine technologist, possibly a nurse for patients undergoing therapy with radiopharmaceuticals, and a maintenance engineer.

3.3.5. Education and training

According to the BSS, provision must be made to ensure that all personnel on whom protection and safety depend are appropriately trained and qualified so that they understand their responsibilities and perform their duties with appropriate judgement and according to defined procedures. Such personnel clearly include the nuclear medicine physician (or other medical specialist wishing to perform nuclear medicine procedures), nuclear medicine technologist, medical physicist, radiopharmacist and the RPO. However, there are additional staff that may also need appropriate training, such as nurses working with radioactive patients and maintenance staff. Details about appropriate levels of training are given in Ref. [3.3].

3.4. FACILITY DESIGN

It is an important task for the medical physicist to be actively involved in the planning and design of the nuclear medicine facility. Factors that are to be considered are:

— Safety of sources;
— Optimization of protection for staff and the general public;
— Preventing uncontrolled spread of contamination;
— Maintaining low background where most needed;
— Fulfilment of national requirements regarding pharmaceutical work.

3.4.1. Location and general layout

The location of the nuclear medicine facility within the hospital or clinic is not critical, but a few factors need to be considered. It should be readily accessible, especially for outpatients, who constitute the majority of the patients. The facility should also be located away from radiotherapy sources and other strong sources of ionizing radiation such as a cyclotron, which can interfere with the measuring equipment. Isolation wards for patients treated with radionuclides should be located outside of the nuclear medicine facility.

The general layout of the nuclear medicine facility should take into account a possible separation of the work areas and the patient areas. It is also essential to reduce uncontrolled spread of contamination. This will be achieved by locating rooms for preparation of radiopharmaceuticals as far away as possible from rooms for measurements and patient waiting areas. Another important factor is to reduce the transport of unsealed sources within the facility. The general layout is from a low activity area close to the entrance to high activity areas at the opposite end. More details regarding floor planning and additional topics can be found in the IAEA's Nuclear Medicine Resources Manual [3.4]. It should be borne in mind that the design of facilities is an important tool in the optimization of protection of workers and the general public. This is further discussed in Section 3.6.2.

3.4.2. General building requirements

The design of the facility should take into consideration the type of work to be performed and the radionuclides (and their activity) intended to be used. The ICRP's concept of categorization of hazard can be used in order to determine the special needs concerning ventilation and plumbing, and the materials used in walls, floors and work-benches. The different rooms in the facility will be categorized as low, medium or high hazard areas.

Of special concern in a nuclear medicine facility is the risk of contamination, and if contamination occurs, the ability to contain it and clean it up. Therefore, the floors and work-benches should generally be finished in an impermeable material which is washable and resistant to chemical change, with all joints sealed. The floor cover should be curved to the wall. The walls should also be easily cleaned. Chairs and beds used in high hazard areas should be easily decontaminated. However, some attention has to be given to the comfort of the patients, for instance in the waiting areas.

Rooms in which unsealed sources, especially radioactive aerosols or gases, may be produced or handled should have an appropriate ventilation system that includes a fume hood, laminar air flow cabinet or glove box. It should be noted that this might also be necessary in the examination room depending on the radiopharmaceutical used in ventilation scintigraphy. Details regarding fume hoods, etc. are given in Chapter 9.

If the regulatory body allows the release of aqueous waste to the sewer, a dedicated sink needs to be used, and this needs to be easily decontaminated. Local rules for the discharge shall be available.

A separate bathroom for the exclusive use by injected patients is recommended. A sign requesting patients to always sit down, flush the toilet well and wash their hands should be displayed to lower the risk of contamination of the floor and to ensure adequate dilution of excreted radioactive materials. The bathroom should include a sink as a normal hygiene measure and should be finished in materials that are easily decontaminated. Local rules should be available for cleaning the toilet. The patient toilet facilities should not be used by hospital staff as it is likely that the floor, toilet seat and taps will frequently be contaminated.

Drain-pipes from the nuclear medicine facility should go as directly as possible to the main building sewer. It should be noted that some countries require that drain-pipes from a nuclear medicine facility and especially from isolation wards for patients undergoing radionuclide therapy end up in a delay tank.

3.4.3. Source security and storage

The licensee needs to establish a security system to prevent theft, loss, unauthorized use or damage to sources. It should be included in all steps from ordering and delivery of the sources to disposal of spent sources. Only authorized personnel are permitted to order radionuclides. Routines for delivery and unpacking shipments should be available, as well as routines for safe handling and storage of sources. Records of all sources should be kept. The user is always responsible for the security of sources and, in principle, it should be possible to identify where an individual source is located or how it has been used, even if it has left the facility in a patient. The regulatory body should promptly be informed in cases of lost or stolen sources.

When a radioactive source is not in use, it should always be stored. In a nuclear medicine facility, the sources are generally stored in the room where preparation of radiopharmaceuticals is undertaken. Storage of sources is further discussed in Chapter 9.

It is necessary to consider the possible consequences of an accidental fire and to take steps to minimize the risk of this. Careful selection of non-flammable construction materials when building the storage facility will greatly reduce this hazard. The storage facility should not be used to hold any highly flammable or highly reactive materials. Liaison with the local firefighting authority is necessary and their advice should be sought regarding provision of firefighting equipment in the vicinity of the radioactive waste store.

3.4.4. Structural shielding

Structural shielding should be considered in a busy facility where large activities are handled and where many patients are waiting and examined. In a PET/CT facility, structural shielding is always necessary and the final design will generally be determined by the PET application because of the high activities used and because of the high energy of the annihilation radiation. Careful calculations should be performed to ensure the need and construction of the barrier. Such calculations should include not only walls but also the floor and ceiling, and must be made by a qualified medical physicist. Radiation surveys should always be performed to ensure the correctness of the calculations.

The correct design of protective barriers is of the utmost importance not only from a protection but also from an economic point of view. If the basic calculations are wrong, it will become very expensive to correct the mistakes later when the whole construction is completed. It is, therefore, very important that a qualified expert, such as a medical physicist, be consulted in the planning stage.

3.4.5. Classification of workplaces

With regard to occupational exposure, the BSS require the classification of workplaces as controlled areas or as supervised areas.

In a controlled area, individuals follow specific protective measures to control radiation exposures. It will be necessary to designate an area as controlled if it is difficult to predict doses to individual workers or if individual doses may be subject to wide variations. The controlled area must be delineated and it is convenient to use existing structural boundaries, which should already be considered at the planning stage of a facility.

A supervised area is any area for which occupational exposure conditions are predictable and stable. They are kept under review even though specific additional protective measures and safety provisions are not normally needed.

In a nuclear medicine facility, the rooms for preparation, storage (including radioactive waste) and injection of the radiopharmaceuticals will be controlled areas. Owing to the potential risk of contamination, the imaging rooms and waiting areas for injected patients might also be classified as controlled areas. The area housing a patient to whom therapeutic amounts of activity have been given will also be a controlled area. In the case of pure β emitters, such as ^{90}Y, ^{89}Sr or ^{32}P, which are not excreted from the body, the area may not need to be classified as a controlled area.

3.4.6. Workplace monitoring

Workplace monitoring means checking the facility for the presence of radiation or radioactive contamination. The two basic types of workplace monitoring are exposure monitoring and contamination monitoring. Exposure monitoring (sometimes called 'area monitoring' or 'radiation surveying') consists of measuring radiation levels (in microsieverts per hour) at various points using an exposure meter or survey meter. Contamination monitoring is the search for extraneous radioactive material deposited on surfaces.

Routine workplace monitoring should be performed at predefined places in the facility as defined by the RPO. It is an advantage if one member of staff is appointed to take the measurements. The staff member should be well trained in handling the instrument. The results should be recorded and investigated if they exceed the investigation levels predefined by the RPO.

More details regarding workplace monitoring are given in Chapters 9 and 20.

3.4.7. Radioactive waste

The radioactive waste in a nuclear medicine facility comprises many different types of waste. It may be of high activity, such as a technetium generator, or of low activity, such as from biomedical procedures or research. In addition, it may have a long or short half-life and it may be in a solid, liquid or gaseous form. Radioactive waste needs to be safely managed because it is potentially hazardous to human health and the environment. Through good practices in the use of radionuclides, the amount of waste can be significantly reduced but not eliminated. It is important that safe waste management, in full compliance with all relevant regulations, is considered and planned for at the early stages of any projects involving radioactive materials. It is the responsibility of the licensee to provide safe management of the radioactive waste. It should be supervised by the RPO and local rules should be available.

Containers to allow segregation of different types of radioactive waste should be available in areas where the waste is generated. The containers must be suitable for the purpose (volume, shielding, being leakproof, etc.). Each type of waste should be kept in separate containers that are properly labelled to supply information about the radionuclide, physical form, activity and external dose rate.

A room for interim storage of radioactive waste should be available. The room should be locked, properly marked and, if necessary, ventilated. Flammable waste should be placed separately. It is essential that all waste be properly packed in order to avoid leakage during storage. Biological waste should be refrigerated or put in a freezer. Records should be kept, so that the origin of the waste can be identified.

The final disposal of the radioactive waste produced in the nuclear medicine facility includes several options: storage for decay and disposal as cleared waste into the sewage system (aqueous waste), through incineration or transfer to a landfill site (solid waste), or transfer of sources to the vendor or to a special waste disposal facility outside of the hospital.

For many of the wastes generated in hospitals, storage for decay is a useful option because the radionuclides generally have short half-lives. This can be done in the hospital and may include some treatment of the wastes to ensure safe storage. Other types of waste containing radionuclides with longer half-lives must be transferred to a special waste treatment, storage and disposal facility outside of the hospital. One option is to return the source to the vendor. This is an attractive option for radionuclide generators and might also be useful for sealed sources used in a quality control programme. The option of returning the source should be provided for in the purchase process.

For diagnostic patients there is generally no need for collection of excreta. Ordinary toilets can be used. For therapy patients, there are different policies in different countries, either to use separate toilets equipped with delay tanks or an active treatment system, or to allow the excreta to be released directly into the sewage system. This is further discussed in Chapter 20.

3.5. OCCUPATIONAL EXPOSURE

Detailed requirements for protection against occupational exposure are given in Section 3 of the BSS, and recommendations on how to meet these requirements are given in IAEA Safety Guides [3.5–3.7]. All of these safety standards are applicable to nuclear medicine practice, and in addition Ref. [3.3] provides further specific advice. A summary of the most relevant issues for nuclear medicine is given in this section.

3.5.1. Sources of exposure

Exposure of workers may arise from unsealed sources either through external irradiation of the body or through entry of radioactive substances into the body. The main precautions required in dealing with external irradiation depend on the physical characteristics of the emitted radiation and the activity as reflected by the specific dose rate constant as well as the half-life of the radionuclide. When a radionuclide enters the body, the internal exposure will depend on factors such as the physical and chemical properties of the radionuclide, the activity and the biokinetics.

Every type of work performed in a nuclear medicine facility will make a contribution to the external exposure of the worker: unpacking radioactive material, activity measurements, storage of sources, preparation of radiopharmaceuticals, administration of radiopharmaceuticals, patient handling and examination, care of the radioactive patient and handling of radioactive waste. Generally, the yearly effective dose to staff working full time in nuclear medicine with optimized protection should be well below 5 mSv.

Among the different tasks involved, the highest effective dose is received from the patient at injection and imaging. The dose rate close to the patient can be quite high, for instance, 300 μSv/h at 0.5 m from a patient who has received 350 MBq of ^{18}F.

High equivalent dose to the fingers can be received in preparation and administration of radiopharmaceuticals, even if proper shielding is used. Injecting eight patients per day with 400 MBq of ^{99m}Tc per patient has been reported to give a mean and maximum equivalent dose to the fingers of 80 and 330 mSv/a, respectively, even if syringe shields are used. Without shielding, the maximum equivalent dose will be about 2500 mSv/a.

Higher risk of internal exposure due to contamination is associated with radioactive spills, animal experiments, emergency surgery of a therapy patient and autopsy of a therapy patient. However, traces of the radionuclides used in a nuclear facility can be found almost everywhere, especially on door handles, taps, some specific equipment and in the patient's toilet. Some procedures, such as ventilation scans, might also cause contamination of both personnel and equipment. Whole body measurements of workers have revealed an equilibrium internal contamination of up to 10 kBq of ^{99m}Tc, which will result in an effective dose of ~0.05 mSv/a. Although this is a small fraction of the external exposure, every precaution must be taken to avoid contamination of the facility.

Of special concern is contamination of the skin, since this can result in extremely high local equivalent doses. For instance, 1 kBq of ^{18}F will result in an initial equivalent dose rate to the skin of 0.8 mSv/h. The activity on the hands after elution, preparation and administration of ^{99m}Tc radiopharmaceuticals has been reported to be 0.02–200 kBq, which results in an initial skin dose of 0.005–50 mSv/h.

3.5.2. Justification, optimization and dose limitation

Nuclear medicine workers have no personal benefit from exposure. Therefore, justification of occupational exposure must be included in justification of the nuclear medicine practice itself. The risks in radiation work should not be greater than for any other similar work. The upper limit of a tolerable risk for the individual is determined by the dose limits (see Table 3.1). However, through optimized protection, the incurred effective dose should be further reduced. Besides facility and equipment design, shielding of sources, handling of sources as well as personal protective equipment are important in the optimization of occupational radiation protection. Optimization is also achieved through education and training, resulting in awareness and involvement in radiation protection.

From the examples above, it should be clear that the dose limits for workers can be exceeded if the necessary protective precautions are not taken. Radiation protection measures must be applied in each step of the work with radiopharmaceuticals in the nuclear medicine facility, including work with the patient.

The principal parties responsible for occupational exposure are licensees and employers, and they should ensure that the exposure is limited and that protection is optimized. The worker also has responsibilities and must follow the rules and procedures as well as using the devices for monitoring and the protective equipment and tools provided, and in all aspects cooperate with the employer in order to improve the protection standard in the workplace.

3.5.3. Conditions for pregnant workers and young persons

It is generally accepted that the unborn child should be afforded the same protection level as a member of the general public, meaning that a dose limit of 1 mSv should be applied once pregnancy is declared. Good operational procedures should ensure that the radiation doses received by staff working in nuclear medicine facilities are well below any occupational dose limits. Therefore, there is generally no need for a pregnant member of staff to change her duties based on the expected dose to the embryo or fetus. However, removal of pregnant women from work in laboratories where large quantities of radionuclides are prepared and administered, and from nursing teams responsible for patients who have been treated with radionuclides should be considered. These staff members could receive a dose to the embryo or fetus comparable with the public dose limit over the period of the pregnancy. Since all doses should be reduced whenever possible, some supervisors will consider it prudent to reassign pregnant staff to non-radiation duties if this is possible. Many nuclear medicine facility managers would also accept requests from women to be reassigned to other duties for reasons beyond radiation protection. Previous personal monitoring results can help guide any decisions, noting that the dose to the fetus from external radiation is not likely to exceed 25% of the personal dosimeter measurement.

According to the BSS, no person under the age of 16 years is to be subjected to occupational exposure, and no person under the age of 18 years is to be allowed to work in a controlled area unless supervised and then only for the purpose of training.

3.5.4. Protective clothing

Suitable personal protective clothing should be provided for the use of all persons employed in work in controlled areas. The protective clothes should be adequate to prevent any contamination of the body of the worker for whom it is provided and should include gloves, laboratory coats, safety glasses and shoes or overshoes, as well as caps and masks for aseptic work.

A question frequently asked is whether lead aprons are useful for nuclear medicine work. Wearing a lead apron at all times will reduce the effective dose by a factor of about two. It is, therefore, a matter of judgement whether this dose reduction compensates for the effort of wearing an apron. In some hospitals, lead aprons are used in the case of prolonged injections and high activity.

3.5.5. Safe working procedures

The safety of the work in nuclear medicine is based on facility design as well as on the use of protective clothing and the use of protective equipment and tools as discussed above. These measures together with working procedures aimed to minimize external exposure, risk of contamination and spread of contamination, will optimize protection of workers. Work with unsealed sources should always be supported by written local rules.

In order to minimize the risk of contamination in handling radiopharmaceuticals, clean operation conditions and good laboratory practice should be adopted, and protective clothing used. The work area should be kept tidy and free from articles not required for work. It should be monitored periodically and be cleaned often enough to ensure minimal contamination. No food or drink, cosmetic or smoking materials, crockery or cutlery should be brought into an area where unsealed radioactive substances are used. They should not be stored in a refrigerator used for unsealed radioactive substances. Handkerchiefs should never be used in these areas.

All manipulation for preparation, dispensing and administration of radioactive materials should be carried out in such a way that the spread of contamination is minimized. That includes preparing and dispensing radiopharmaceuticals over a drip tray covered with absorbing paper as well as using absorbing compresses at administrations. Any spills of radioactive material should be immediately covered with absorbent material to prevent the spread of material. If the spill cannot be cleaned up immediately, it must be marked to warn other personnel of its location. Decontamination of the area must begin as soon as possible.

When wearing gloves which may be contaminated, unnecessary contact with all other objects should be avoided. Gloves should be removed and disposed of in the radioactive waste bin as soon as work with radioactive substances is finished.

After finishing work with the potential for contamination, the protective clothing should be removed and placed in an appropriate container. Hands should be washed and monitored.

In order to minimize external exposure, the three fundamental measures of protection should be applied: time, distance and shielding. As far as possible, the time of exposure should be as short as possible. Of course, this is important in work where high exposure rates can be expected, such as in the preparation of radiopharmaceuticals. However, limiting exposure time should not compromise the quality of work or the use of other protective measures.

Direct handling of vials, syringes or other sources which produce a significant radiation field is not recommended. Forceps or tongs should be used to reduce the radiation exposure by increasing the distance between the source and the hands. Properly designed vial and syringe shields must be used wherever practicable. In cases where unshielded sources are handled or the exposure time is prolonged, the work should be performed behind a properly designed lead glass shield or similar type of protective barrier.

Radioactive waste should not be stored in the work area but transferred to a separate radioactive waste storage room as soon as possible.

A patient undergoing a nuclear medicine imaging study is a source of radiation exposure and contamination. Contact with these patients by nursing staff presents little hazard, as the radiation dose rate is quite low, and accumulated dose to any single individual would not be significant. However, for nuclear medicine staff that spend a great deal of time in the immediate vicinity of these patients, the accumulated radiation dose can be significant. These workers should, whenever possible, maximize their distance from the patient and spend as little time as possible in close proximity to the patient.

In summary, the following protective approaches can reduce external exposure significantly:

- — For preparation and dispensing of radiopharmaceuticals, working behind a lead glass bench shield, and using shielded vials and syringes;
- — For administration of radiopharmaceuticals to patients, using lead aprons in the case of prolonged injection and high activity, and using a syringe shield;
- — During examinations, when the distance to the patient is short, using a movable transparent shield.

3.5.6. Personal monitoring

The licensee and employer have the joint responsibility to ensure that appropriate personal monitoring is provided to staff. This normally means that the RPO would specify which workers need to be monitored routinely, the type of monitoring device to be used and the body position where the monitor should be worn, bearing in mind that some countries may have specific regulatory requirements on these issues. Further, the regulatory body is likely to have specified the monitoring period and the time frame for reporting monitoring results.

Staff to be monitored in a nuclear medicine facility should include all those who work routinely with radionuclides or with the patients who have received administrations of radiopharmaceuticals. This will include nursing staff who either work routinely in nuclear medicine or nurse patients who have received radionuclide therapy and staff dealing with excreta from radionuclide therapy. Monitoring would not normally be extended to those that come into occasional contact with nuclear medicine patients.

There are several types of external personal dosimetry systems and the system to use is dependent on national or local conditions. In many countries, the service is centralized to the regulatory body or provided through third party personal dosimetry providers. Occasionally, some large hospitals have their own personal dosimetry service. In all cases, the dosimetry provider must be approved by the regulatory body.

Finger monitoring should be carried out occasionally on staff that regularly prepare and administer radioactive substances to patients, and also when setting up an operation which requires the routine handling of large quantities of radionuclides. After handling unsealed radionuclides, the hands should be monitored. It may, therefore, be convenient to mount a suitable contamination monitor near the sink where hands are washed. Care should be taken to ensure that the monitor itself does not become contaminated. In high background areas, it will be necessary to shield the detector, and it may be convenient to have a foot or elbow operated switch to activate the monitor.

Monitoring for internal contamination is rarely necessary in nuclear medicine on radiation protection grounds but it may be useful in providing reassurance to staff. The circumstances in which internal monitoring becomes advisable are those where staff use significant quantities of ^{131}I for thyroid therapy. They should be included in a programme of thyroid uptake measurements.

In other circumstances where it is necessary to assess the intake of γ emitting radionuclides (e.g. after a serious incident), the use of a whole body counter may be appropriate. Such equipment should be available at national referral centres. The possible use of an uncollimated gamma camera should also be considered.

Sometimes, a more detailed monitoring survey may be indicated if staff doses have increased (or it is anticipated that they may do so in the future) as a result of either the introduction of new examinations or procedures, or a change in the nuclear medicine facility's equipment. The RPO should decide who should be monitored and at which monitoring sites.

Individual monitoring results must be analysed and records must be kept. It is vital that the individual monitoring results are regularly assessed and the cause of unusually high dosimeter readings should be investigated by the RPO, with ensuing corrective actions where appropriate. The administrative arrangements, the scope and nature of the individual monitoring records, and the length of time for which records have to be kept may differ among countries.

3.5.7. Monitoring of the workplace

The BSS require licensees to develop programmes for monitoring the workplace. Such programmes are described in Section 3.4.6 and in Chapters 9 and 20.

3.5.8. Health surveillance

According to the BSS, the licensee needs to make arrangements for appropriate health surveillance in accordance with the rules established by the national regulatory body. The primary purpose of health surveillance is to assess the initial and continuing fitness of employees for their intended tasks. The health surveillance programme should be based on the general principles of occupational health.

No specific health surveillance related to exposure to ionizing radiation is necessary for staff involved in nuclear medicine procedures. Only in the case of overexposed workers at doses much higher than the dose limits would special investigations involving biological dosimetry and further extended diagnosis and medical treatment be necessary.

Counselling should be available to workers such as women who are or may be pregnant, individual workers who have or may have been exposed substantially in excess of dose limits and workers who may be worried about their radiation exposure.

3.5.9. Local rules and supervision

According to the BSS, employers and licensees must, in consultation with the workers or through their representatives:

— Establish written local rules and procedures necessary to ensure adequate levels of protection and safety for workers and other persons;
— Include in the local rules and procedures the values of any relevant investigation level or authorized level, and the procedure to be followed in the event that any such value is exceeded;
— Make the local rules and procedures, the protective measures and safety provisions known to those workers to whom they apply and to other persons who may be affected by them;
— Ensure that any work involving occupational exposure be adequately supervised and take all reasonable steps to ensure that the rules, procedures, protective measures and safety provisions be observed.

These local rules should include all working procedures involving unsealed sources in the facility such as:

— Ordering radionuclides;
— Unpacking and checking the shipment;
— Storage of radionuclides;
— General rules for work in controlled and supervised areas;
— Preparation of radiopharmaceuticals;
— Personal and workplace monitoring;
— In-house transport of radionuclides;
— Management of radioactive waste;
— Administration of radiopharmaceuticals to the patients;
— Protection issues in patient examinations and treatments;
— Routine cleaning of facilities;
— Decontamination procedures;
— Care of radioactive patients.

It is the responsibilty of the licensee of the nuclear medicine facility to ensure that local rules are established, maintained and continually reviewed. The RPO would have significant involvement in this process.

3.6. PUBLIC EXPOSURE

3.6.1. Justification, optimization and dose limitation

According to the BSS, public exposure is exposure incurred by members of the public from radiation sources, excluding any occupational or medical exposure.

The three ICRP principles described in Section 3.2.1 apply to public exposure arising from the practice of nuclear medicine. Just as for occupational exposure, the justification of public exposure is based on the justification of the practice of nuclear medicine. The exposure of the general public is ultimately restricted by the application of dose limits (see Table 3.1), but in the first instance the application of the principle of optimization of protection ensures that public doses will be ALARA.

The licensee is responsible for controlling public exposure arising from a nuclear medicine facility. The presence of members of the public in or near the nuclear medicine facility needs to be considered when designing the shielding and flow of persons in the facility.

The sources of exposure of the general public are primarily the same as for workers. Hence, the use of structural shielding and the control of sources, waste and contamination are fundamental to controlling exposure of the public. There are, however, some additional situations that need special consideration. These include the release of patients examined or treated with radiopharmaceuticals.

3.6.2. Design considerations

The general layout of the nuclear medicine facility should take into account the protection of members of the public. The areas for storage and preparation of radiopharmaceuticals must be well separated from public areas such as waiting rooms. The movement of radionuclides must be minimized. For example, the room for preparation and dispensing of radiopharmaceuticals and the room for administration should be adjacent and connected by a pass through. Areas where significant activities of radionuclides are present must be appropriately shielded. Access must be restricted so that members of the public are not allowed into controlled areas. Radioactive waste must be stored in a secure location away from areas accessible to the public. Since a patient still waiting for administration of the radiopharmaceutical is regarded as a member of the public, separate waiting rooms and toilets for injected and not injected patients should be considered in order to minimize both external exposure and the spread of contamination.

3.6.3. Exposure from patients

Every precaution must be taken to ensure that the doses received by individuals who come close to a patient or who spend some time in neighbouring rooms remain below the dose limit for the public and below any applicable dose constraint. For almost all diagnostic procedures, the maximum dose that could be received by another person due to external exposure from the patient is a fraction of the annual public dose limit and it should not normally be necessary to issue any special radiation protection advice to the patient. One exception is restrictions on breast-feeding a baby, which will be further discussed in Section 3.7.2.4. Another exception is an intensive use of positron emitters which may require structural shielding based on the exposure of the public as discussed above (Section 3.4.4). For patients who have undergone radionuclide therapy, specific advice should be given regarding restrictions on their contact with other people. This is discussed separately in Chapter 20.

3.6.4. Transport of sources

One possible source of exposure of the general public is transport of sources. It is performed both inside and outside the nuclear medicine facility. Inside the facility, the transport includes distribution of the radioactive sources from the storage area to where it will be used. Such transport should be limited as far as possible by the facility design. The transport that takes place should be performed according to optimized radiation protection conditions as given by local rules.

The transport of radioactive sources to and from the nuclear medicine facility should follow the internationally accepted IAEA Regulations for the Safe Transport of Radioactive Material [3.8]. These Regulations include basic rules for the transport itself and regulations about the shape and labelling of packages.

In general, the package is built in several parts. It should be mechanically safe and reduce the effect of potential fire and water damage. The package should be labelled with a sign. There are three different labels: I–White, II–Yellow and III–Yellow. In all cases, the radionuclide and its activity should be specified. The label gives some indication of the dose rate D at the surface of the package:

— Category I–White $D \leq 0.005$ mSv/h
— Category II–Yellow $0.005 < D \leq 0.5$ mSv/h
— Category III–Yellow $0.5 < D \leq 2$ mSv/h

A more exact figure of the radiation around the package is given by the transport index which is the maximum dose rate (mSv/h) at a distance 1 m from the surface of the package multiplied by a hundred.

3.7. MEDICAL EXPOSURE

The detailed requirements given in Section 3 of the BSS are applicable to medical exposure in nuclear medicine facilities. Furthermore, Ref. [3.9] describes strategies to involve organizations outside the regulatory framework, such as professional bodies (nuclear medicine physicians, medical physicists, nuclear medicine technologists, radiopharmacists), whose cooperation is essential to ensure compliance with the BSS requirements for medical exposures. Examples that may illustrate this point include the adoption of protocols for calibration of unsealed sources and for QA and for reporting accidental medical exposure. Reference [3.3] provides further specific advice. A summary of the most relevant issues for nuclear medicine is given in this section.

3.7.1. Justification of medical exposure

The BSS state that:

> "Medical exposures shall be justified by weighing the expected diagnostic or therapeutic benefits…that they yield against the radiation detriment that they might cause, with account taken of the benefits and the risks of available alternative techniques that do not involve medical exposure."

The principle of justification of medical exposure should not only be applied to nuclear medicine practice in general but also on a case by case basis, meaning that any examination should be based upon a correct assessment of the indications for the examination, the actual clinical situation, the expected diagnostic and therapeutic yields, and the way in which the results are likely to influence the diagnosis and the medical care of the patient. The nuclear medicine specialist has the ultimate responsibility for the control of all aspects of the conduct and extent of nuclear medicine examinations, including the justification of the given procedure for a patient. The nuclear medicine specialist should advise and make decisions on the appropriateness of examinations and determine the techniques to be used. In justifying a given diagnostic nuclear medicine procedure, relevant international or national guidelines should be taken into account.

Any nuclear medicine procedure that occurs as part of a biomedical research project (typically as a tool to quantify changes in a given parameter under investigation) is considered justified if the project has been approved by an ethics committee.

3.7.2. Optimization of protection

The principle of optimization of protection is applied to nuclear medicine procedures that have been justified, and can be summarized as follows. For diagnostic nuclear medicine procedures, the patient exposure should be the minimum necessary to achieve the clinical purpose of the procedure, taking into account relevant norms of acceptable image quality established by appropriate professional bodies and relevant diagnostic reference levels (DRLs).

For therapeutic nuclear medicine procedures, the appropriate radiopharmaceutical and activity are selected and administered so that the activity is primarily localized in the organ(s) of interest, while the activity in the rest of the body is kept ALARA.

The implementation of optimization of protection for patients in nuclear medicine is quite complex and includes equipment design, choice of radiopharmaceutical and activity, procedure considerations, DRLs, calibration, clinical dosimetry and QA, as well as special considerations for children, pregnant women and lactating women. This is further discussed in the following sections.

3.7.2.1. Administered activity and radiopharmaceuticals

For diagnostic procedures, it is necessary for the nuclear medicine specialist in cooperation with the medical physicist to determine the optimum activity to administer in a certain type of examination, taking the relevant DRL (see below) into account. For any given procedure used on an individual patient, the optimum activity will depend on the body build and weight of the patient, the patient's metabolic characteristics and clinical condition, the type of equipment used, the type of study (static, dynamic, tomographic) and the examination time.

For a given type of imaging equipment, the diagnostic value of the information obtained from an examination will vary with the amount of administered activity. There is a threshold of administered activity below which no useful information can be expected. Above this level, the diagnostic quality will increase steeply with increasing activity. Once an acceptable image quality has been reached, a further increase of the administered activity will only increase the absorbed dose and not the value of the diagnostic information.

It should also be noted that limiting the administered activity below the optimum, even for well intentioned reasons, will usually lead to a poor quality of the result which may cause serious diagnostic errors. It is very important to avoid failure to obtain the required diagnostic information; failure would result in unnecessary (and, therefore, unjustified) irradiation and may also necessitate repetition of the test.

If more than one radiopharmaceutical can be used for a procedure, consideration should be given to the physical, chemical and biological properties for each radiopharmaceutical, so as to minimize the absorbed dose and other risks to the patient while at the same time providing the desired diagnostic information. Other factors affecting the choice include availability, shelf life, instrumentation and relative cost. It is also important that the radiopharmaceuticals used are received from approved manufacturers and distributors, and are produced according to national and international requirements. This is a requirement also for in-house production of radiopharmaceuticals for PET studies.

The activity administered to a patient should always be determined and recorded. Knowing the administered activity makes it possible to estimate the absorbed dose to different organs as well as the effective dose to the patient. Substantial reduction in absorbed dose from radiopharmaceuticals can be achieved by simple measures such as hydration of the patient, use of thyroid blocking agents and laxatives.

3.7.2.2. Optimization of protection in procedures

The nuclear medicine procedure starts with the request for an examination or treatment. The request should be written and contain basic information about the patient's condition. This information should help the nuclear medicine specialist to decide about the most appropriate method to use and to decide how urgent the examination is. The patient should then be scheduled for the examination or treatment and be informed about when and where it will take place. Some basic information about the procedure should also be given, especially if it requires some preparation of the patient, such as fasting. These initial measures require an efficient and reliable administrative system. In parallel to these routines, the nuclear medicine facility has to ensure that the radiopharmaceutical to be used is available at the time of the scheduled procedure.

When the patient appears in the nuclear medicine facility, they should be correctly identified using the normal hospital or clinic routines. The patient should be informed about the procedure and have the opportunity to ask questions about it. A fully informed and motivated patient is the basis for a successful examination or treatment. Before the administration of the radiopharmaceutical, the patient should be interviewed about possible pregnancy, small children at home, breast-feeding and other relevant questions which might have implications for the procedure. Before administration, the technologist or doctor should check the request and ensure that the right examination or treatment is scheduled and that the right radiopharmaceutical and the right activity are dispensed. If everything is in order, the administration can proceed. The administered activity should always be recorded for each patient.

While most adults can maintain a required position without restraint or sedation during nuclear medicine examinations, it may be necessary to immobilize or sedate children, so that the examination can be completed successfully. Increasing the administered activity to reduce the examination time is an alternative that can be used in elderly patients with pain.

Optimization of protection in an examination means that equipment should be operated within the conditions established in the technical specifications, thus ensuring that it will operate satisfactorily at all times, in terms of both the tasks to be accomplished and radiation safety. More details are given in Chapters 8 and 15. Particular procedural considerations for children, pregnant women and lactating women are given in the following subsections.

Optimization of protection in radionuclide therapy means that a correctly calculated and measured activity should be administered to the patient in order to achieve the prescribed absorbed dose in the organ(s) of interest, while the radioactivity in the rest of the body is kept as low as reasonably achievable. Optimization also means using routines to avoid accidental exposures of the patient, the staff and members of the general public. Radionuclide therapy is further discussed in Chapter 20.

The availability of a written manual of all procedures carried out by the facility is highly desirable. The manual should regularly be revised as part of a QA programme.

3.7.2.3. Pregnant women

Special consideration should be given to pregnant women exposed to ionizing radiation due to the larger probability of inducing radiation effects in individuals exposed in utero compared to exposed adults. As a basic rule, it is recommended that diagnostic and therapeutic nuclear medicine procedures of women likely to be pregnant be avoided unless there are strong clinical indications.

In order to avoid unintentional irradiation of the unborn child, a female of childbearing age should be evaluated regarding possible pregnancy or a missed period. This should be done when interviewing and informing the woman prior to the examination or treatment. It is also common to place a poster in the waiting area requesting a woman to notify the staff if she is or thinks she is pregnant. If the patient is found not to be pregnant without any doubt, the examination or treatment can be performed as planned. If pregnancy is confirmed, careful consideration should be given to other methods of diagnosis or to the postponement of the examination until after delivery. If, after consultation between the referring physician and the nuclear medicine specialist, these options are not feasible, then the examination should be performed, but the process of optimization of protection needs to also consider protection of the embryo/fetus.

In order to reduce the fetal dose, it may sometimes be possible to reduce the administered activity and acquire images for longer times, but great care must be taken not to compromise the quality of the result. After the administration of radiopharmaceuticals, frequent voiding should be ensured to minimize exposure from the bladder. This contribution to the fetal dose can be further reduced by administering the radiopharmaceutical when the bladder is partially filled, rather than immediately after voiding.

Of special concern is also the use of CT in PET/CT or SPECT/CT examinations. Routine diagnostic CT examinations of the pelvic region with and without contrast injection can lead to a dose of 50 mSv to the uterus which is assumed to be equivalent to the fetal dose in early pregnancy. It is important to use low dose CT protocols and to reduce the scanning area to a minimum when PET/CT or SPECT/CT scanning is indicated in a pregnant patient.

Pregnant women should not be subject to therapy with a radioactive substance unless the application is life-saving. Following treatment with a therapeutic activity of a radionuclide, female patients should be advised to avoid pregnancy for an appropriate period. More details are given in Ref. [3.3].

If the fetal dose is suspected to be high (e.g. >10 mSv), it should be carefully determined by a qualified medical physicist and the pregnant woman should be informed about the possible risks. The same procedure should be applied in the case of an inadvertent exposure, which can be incurred by a woman who later was found to have been pregnant at the time of the exposure or in emergency situations.

Exposure of a pregnant patient at a time when the pregnancy was not known often leads to her apprehension because of concern about the possible effects on the fetus. It may lead to a discussion regarding termination of pregnancy due to the radiation risks. Many misunderstandings and lack of knowledge, also among physicians, have probably resulted in unnecessary termination of pregnancies. It is generally considered that for a fetal dose of less than 100 mGy, as in most diagnostic procedures, termination of pregnancy is not justified from the point of radiation risks. At higher doses, individual circumstances should be taken into account. This is an ethical issue and the national authorities should give guidance.

3.7.2.4. Lactating women

When nuclear medicine examinations are requested for women who are breast-feeding, they present a potential radiation hazard to the baby. This is due to uptake of some radiopharmaceuticals in breast tissue followed by excretion into the breast milk. The dose to the baby depends on various factors such as the radiopharmaceutical, the amount of milk and the time between the administration of the radiopharmaceutical to the mother and the feeding of the child. The mother also represents a source of external exposure and contamination when feeding or cuddling the baby. The dose will depend on the time the child is held, the distance from the mother's body and personal hygiene. Some restrictions on breast-feeding and advice to the mother are necessary in order to minimize the exposure of the baby to an acceptable level. The baby is a member of the public and a typical constraint on the dose from a single source of exposure (in this case, per episode) is 0.3 mSv.

Before a nuclear medicine examination or therapy with radionuclides, the woman should be asked, orally or in writing, whether she is breast-feeding a child. A notice requesting the patient to inform the staff about breast-feeding should also be prominently displayed in the waiting area. If the answer is yes, consideration should be given as to whether the examination or treatment could reasonably be delayed until she has ceased breast-feeding. If not, advice about restriction of breast-feeding dependent on the diagnostic or therapeutic procedure should be given to the patient.

It is the responsibility of the nuclear medicine specialist in cooperation with the medical physicist to establish local rules regarding breast-feeding and close contact between the mother and the child after a nuclear medicine examination or treatment. The rules should be based on recommendations given by international and national authorities as well as professional organizations. Some guidance is found in Ref. [3.3].

3.7.2.5. Children

Optimization of protection for an examination of a child is basically an optimization of the administered activity. There are several approaches to the problem of how to calculate the administered activity for children. It should be the minimum consistent with obtaining a diagnostic result. As this is the same principle which is applied to adult doses, the normal activity administered to adults should be used as a guide, bearing in mind that the average adult body weight is 70 kg. For children or young persons, body weight should always be measured and the adult administered activity should then be scaled down. Opinions differ as to how the scaling should be achieved. Simply reducing the activity in proportion to body weight may, in some types of investigation, result in inadequate image quality. Another method is based on the principle of scaling in proportion to body surface area. This approach should give the same image count density as that for an adult patient, although the effective dose is higher. As a general guide, activities less than 10% of the normal adult activity should not be administered.

In hybrid imaging, the CT protocol should be optimized by reducing the tube current–time product (mAs) and tube potential (kV) without compromising the diagnostic quality of the images. Careful selection of slice width and pitch as well as scanning area should also be done. It is important that individual protocols based on the size of the child are used. The principles behind such protocols should be worked out by the medical physicist and the responsible specialist.

Since the examination times in nuclear medicine examinations are quite long, there may be problems in keeping the child still during the examination. Even small body motions can severely interfere with the quality of the examination and make it useless. There are several methods of mechanical support to fasten the child. Drawing the child's attention to something else such as a television programme can also be useful for older children. Sometimes, even sedation or general anaesthesia may be necessary.

3.7.2.6. Calibration

The licensee of a nuclear medicine facility needs to ensure that a dose calibrator or activity meter is available for measuring activity in syringes or vials. The validity of measurements should be ensured by regular quality control of the instrument, including periodic reassessment of its calibration, traceable to secondary standards.

3.7.2.7. Clinical (patient) dosimetry

The licensee of a nuclear medicine facility should ensure that appropriate clinical dosimetry by a medical physicist is performed and documented. For diagnostic nuclear medicine, this should include representative typical patient doses for common procedures. For therapeutic nuclear medicine, this needs to be for each individual patient, and includes absorbed doses to relevant organs or tissues.

3.7.2.8. Diagnostic reference levels

Many investigations have shown a large spread of administered activities for a certain type of diagnostic nuclear medicine examination between different hospitals within a country, even if the equipment used is similar in performance. Even though no dose limits are applied to medical exposure, the process of optimization should result in about the same administered activity for the same type of examination and for the same size of patient.

The concept of a DRL provides a tool for the optimization of protection in medical exposure. In the case of nuclear medicine, the DRL is given as administered activity for a certain type of examination and for a normal sized patient. DRLs are aimed to assist in the optimization of protection by helping to avoid unnecessarily high activities to the patient or too low activities to provide useful diagnostic information. DRLs are normally set at the national level as a result of consultation between the health authority, relevant professional bodies and the radiation protection regulatory body.

3.7.2.9. Quality assurance for medical exposures

The BSS require the licensee of the nuclear medicine facility to have a comprehensive programme of QA for medical exposures. The programme needs to have the active participation of the medical physicists, nuclear medicine specialists, nuclear medicine technologists and radiopharmacists, and needs to take into account principles established by international organizations, such as the WHO and PAHO, and relevant professional bodies.

The programme of QA for medical exposures should be complementary to and part of the wider programme of QA for radiation protection — the latter also including occupational and public exposure. In turn, this programme needs to be part of and harmonized with the nuclear medicine facility's quality management system. Section 3.9 discusses the wider QA programme, while the remainder of this subsection deals with some aspects of the programme as it applies to medical exposures.

The programme of QA for medical exposures should include:

- — Measurements by, or under the oversight of, a medical physicist of the physical parameters of medical radiological equipment at the time of acceptance and commissioning prior to clinical use on patients, periodically thereafter, and after any major maintenance that could affect patient protection;
- — Implementation of corrective actions if measured values of the physical parameters are outside established tolerance limits;
- — Verification of the appropriate physical and clinical factors used in patient diagnosis or treatment;
- — Records of relevant procedures and results;
- — Periodic checks of the appropriate calibration and conditions of operation of dosimetry and monitoring of equipment.

In addition, the licensee needs to ensure that there are regular and independent audits of the programme of QA for medical exposures, their frequency depending on the complexity of the nuclear medicine procedures performed and the risks involved.

The above indicates, among other actions, the need for quality control tests on the equipment. More details regarding quality control of equipment used in diagnosis will be found in other chapters of this book.

3.7.3. Helping in the care, support or comfort of patients

Certain patients, such as children, the elderly or the infirm, may have difficulty during a nuclear medicine procedure. Occasionally, people knowingly and voluntarily (other than in their employment or occupation) may volunteer to help in the care, support or comfort of patients. In such circumstances, the dose to these persons (excluding children and infants) should be constrained so that it is unlikely that it will exceed 5 mSv during the period of a patient's diagnostic examination or treatment. The dose to children visiting patients who have ingested radioactive materials should be similarly constrained to less than 1 mSv. Special concern should be given to members of the family of a patient who has received radionuclide therapy. This is further discussed in Chapter 20.

Sometimes, a nurse escorting a patient to the nuclear medicine facility is asked to provide assistance during a procedure. Any resultant exposure should be regarded as occupational, and the nurse should have received education and training on this role.

3.7.4. Biomedical research

The exposure of humans for biomedical research is deemed not to be justified unless it is in accordance with the provisions of the Helsinki Declaration [3.10] and follows the guidelines for its application prepared by the Council for International Organizations of Medical Sciences [3.11]. It is also subject to the approval of an ethics committee.

The use of radioactive trace substances is common in biomedical research. Diagnostic nuclear medicine procedures may be part of a biomedical research project, typically as a means for quantifying changes in a given parameter under investigation or assessing the efficacy of a treatment under investigation. An exposure as part of biomedical research is treated on the same basis as a medical exposure and, therefore, is not subject to dose limits. However, in all investigations involving exposure of humans, a careful estimation of the radiation dose to the volunteer should be made. The associated risk should then be weighed against the benefit for the patient or society. Recommendations are given by the ICRP. The BSS require the use of dose constraints, on a case by case basis, in the process of optimization.

3.7.5. Local rules

The management of patients in the nuclear medicine facility should be supported by written local rules covering all procedures that may affect medical exposure. These local rules should be signed by the responsible person and known to every member of the staff and should include:

— Routines for patient identification and information;
— Prescribed radiopharmaceutical and activity for adults and children for different types of examination, including methods used to adjust the activity to the single patient and routes of administration;
— Management of patients that are pregnant or might be pregnant;
— Management of breast-feeding patients;
— Routines for safe preparation and administration of radiopharmaceuticals including activity measurements;
— Procedures in case of misadministration of the radiopharmaceutical;
— Detailed procedure manuals for every type of examination including handling of equipment.

3.8. POTENTIAL EXPOSURE

3.8.1. Safety assessment and accident prevention

Unintended and accidental exposure may occur due to equipment failure, human error or a combination of both. Although such events can be identified by a careful safety assessment, their details and the time of occurrence cannot be predicted. These exposures are called potential exposures. It is the responsibility of the licensee to take measures in order to prevent such events as far as possible and, in case they occur, mitigate their consequences.

According to the BSS, the licensee needs to conduct a safety assessment applied to all stages of the design and operation of the nuclear medicine facility, and present the report to the regulatory body if required. The safety assessment needs to include, as appropriate, a systematic critical review of identification of possible events leading to unintended or accidental exposure. In practice, this means that all procedures in which unsealed sources are involved in the work should be listed and for every procedure it should be asked what can go wrong. Some examples are given in Table 3.2.

TABLE 3.2. EXAMPLES OF WHAT CAN POTENTIALLY GO WRONG IN A NUCLEAR MEDICINE FACILITY

Procedure and involvement	What can go wrong?
Patients involved	
Request and scheduling	Wrong patient scheduled
Identification at arrival	Wrong patient identified
Information	Missed pregnancy or breast-feeding
Administration of radiopharmaceutical	Misadministration (wrong patient, wrong activity, wrong radiopharmaceutical)
Waiting	Contamination of waiting area (vomiting, incontinence)
Examination	Inconclusive due to contamination, equipment and/or software failure
Workers involved	
Ordering of sources	Unauthorized ordering
Receipt and unpacking of shipments	Damage to package, contamination
Storage of sources	Unshielded sources, high dose rates, loss of sources
Preparation and administration of radiopharmaceutical	High doses recorded, contamination of workers and facilities
Handling of radioactive waste	Contamination of workers and facilities
General public involved	
Storage of sources	Loss of sources
Handling of sources	Contamination of facility
Radioactive waste	Loss of sources, contamination of facilities
Radioactive patient	Escape of hospitalized patient, medical emergency, death of patient

Undertaking a safety assessment requires using one's imagination to try to define an event that could result in a potential exposure, even if the event has never occurred before. For instance, what should be done if a patient who just received 15 GBq of ^{131}I escapes from the isolation ward and the hospital and is seriously injured in a road accident?

If an unintended or accidental medical exposure occurs, the licensee is required to determine the patient doses involved, identify any corrective actions needed to prevent recurrence and implement the corrective measures. There may be a requirement to report the event to the regulatory body.

A well established RPP is fundamental in accident prevention together with a high level of safety culture in the organization and among the people working in a nuclear medicine facility. The content of an RPP as well as the importance of well established working procedures in order to protect patients, workers and the general public have been discussed in the sections above. It should be stressed that documentation of the procedures used in the facility is also important in accident prevention. Other important factors are a well working QA programme and a programme for continuing education and training which includes not only the normal practices, but also accidental situations and lessons learned from accidents.

3.8.2. Emergency plans

According to the BSS, the licensee needs to prepare emergency procedures on the basis of events identified by the safety assessment. The procedures should be clear, concise and unambiguous, and need to be posted visibly in places where their need is anticipated. An emergency plan needs to, as a minimum, list and describe:

- — Predictable incidents and accidents, and measures to deal with them;
- — The persons responsible for taking actions, with full contact details;
- — The responsibilities of individual personnel in emergency procedures (nuclear medicine physicians, medical physicists, nuclear medicine technologists, etc.);
- — Equipment and tools necessary to carry out the emergency procedures;
- — Training and periodic drills;
- — The recording and reporting system;
- — Immediate measures to avoid unnecessary radiation doses to patients, staff and the public;
- — Measures to prevent access of persons to the affected area;
- — Measures to prevent spread of contamination.

The most likely accident in a nuclear medicine facility is contamination of workers, patients, equipment and facilities. It can range from small to very large spillages of radioactivity, for example, serious damage to the technetium generator or spillage of several gigabecquerels of ^{131}I. The procedures of cleaning up a small amount of contamination should be known and practised by every technologist in the facility. The cleaning procedures in cases of more severe contamination should always be supervised by the RPO. Local rules should be established that define serious contamination based on radionuclide, activity and whether it is contamination of a person or equipment and facilities. It is recommended that the facility have an emergency kit readily available in case of contamination. Such a kit should contain:

— Protective clothing, e.g. overshoes, gloves;
— Decontamination materials for the affected areas, including absorbent materials for wiping up spills;
— Decontamination materials for persons;
— Warning notices;
— Portable monitoring equipment (in working order and regularly checked);
— Bags for waste, tape, labels, pencils.

Several severe accidents in medical exposures in nuclear medicine have been reported and are solely associated with radionuclide therapy and especially when using ^{131}I in treatment of thyroid diseases. Several incidents with misadministration of radiopharmaceuticals in diagnostic nuclear medicine have also been reported. These include examination of the wrong patient or administration of the wrong radiopharmaceutical or the wrong activity. The most common incident is to administer the wrong radiopharmaceutical. Even if this does not cause severe injury to the patient, it is a non-justified exposure with increased radiation risks. It will also lead to a delayed diagnosis, increased cost and increased workload because the examination will have to be repeated. Last but not least, it will cause reduced confidence in the practice of nuclear medicine.

Other accidents and incidents that also involve the general public include the possible death of a patient containing radionuclides. In diagnostic nuclear medicine, such an incident can generally be left without specific measures. However, in radionuclide therapy, emergency plans have to be available on how to handle the cadaver. Since this is a sensitive issue, depending on ethical and religious rules and traditions, advice should be available from the national authorities.

3.8.3. Reporting and lessons learned

In the event of an incident or accident, the licensee has the responsibility to ensure that a comprehensive investigation takes place and a report is produced that includes the following information:

— A description of the incident by all persons involved;

— Methods used to estimate the radiation dose received by those involved in the incident and implications of those methods for possible subsequent litigation;
— Methods used to analyse the incident and to derive risk estimates from the data;
— The subsequent medical consequences for those exposed;
— The particulars of any subsequent legal proceedings that may ensue;
— Conclusions drawn from the evaluation of the incident and recommendations on how to prevent a recurrence of such an accident.

In the case of a misadministration or an accident in radionuclide therapy, the responsible nuclear medicine specialist should be promptly informed. They should then inform the referring physician and the patient. The medical physicist should make dose calculations and the staff involved in the accident should independently describe their view of the accident. Conclusions regarding any deficits in the procedures should be drawn and necessary changes implemented. Finally, the licensee may need to submit the report to the regulatory body.

In order to avoid future accidents, it is important to learn from previous ones. The initiating event and the contributing factors can always be identified. This information provides material that should be used to prevent future accidents. This is achieved by informing all members of staff about the accident or incident, which means that it is very important to have an efficient reporting system and a programme for local education and training that also includes potential exposures.

3.9. QUALITY ASSURANCE

3.9.1. General considerations

The International Organization for Standardization defines QA as all planned and systematic actions needed to provide confidence that a structure, system or component will perform satisfactorily in service. Satisfactory performance in nuclear medicine implies the optimum quality of the entire process. Since an examination or therapy is justified only if the procedure benefits the patient, QA in the whole process of nuclear medicine is an important aspect of radiation protection.

The BSS require the licensee of the nuclear medicine facility to have established a QA programme that provides adequate assurance that the specified requirements relating to protection and safety are satisfied, and that provides quality control mechanisms and procedures for reviewing and assessing the overall effectiveness of protection and safety measures.

It is a common and growing practice that hospitals or clinics implement a quality management system for all of the medical care received in diagnosis and treatment, i.e. covering the overall nuclear medicine practice. The QA programme envisaged by the BSS should be part of the wider facility quality management system. In the hospital or clinic, it is common to include QA as part of the RPP or, conversely, to include the RPP as part of a more general QA programme for the hospital or clinic. Regardless of its organization, it is important that radiation protection is an integral part of a system of quality management. The remainder of this section considers aspects of QA applied to a nuclear medicine facility that are covered in the BSS. Specific details with respect to medical exposure are covered in Section 3.7.2.9.

An effective QA programme requires a strong commitment from the nuclear medicine facility's management to provide the necessary resources of time, personnel and budget. It is recommended that the nuclear medicine facility establish a group that actively works with QA issues. Such a QA committee should have a representative from management, a nuclear medicine physician, a medical physicist, a nuclear medicine technologist and an engineer as members. The QA committee should meet regularly and review the different components of the programme.

The QA programme should cover the entire process from the initial decision to adopt a particular procedure through to the interpretation and recording of results, and should include ongoing auditing, both internal and external, as a systematic control methodology. The maintenance of records is an important part of QA. One important aspect of any QA programme is continuous quality improvement. This implies a commitment of the staff to strive for continuous improvement in the use of unsealed sources in diagnosis and therapy, based on new information learned from their QA programme and new techniques developed by the nuclear medicine community at large. Feedback from operational experience and lessons learned from accidents or near misses can help identify potential problems and correct deficiencies, and should, therefore, be used systematically, as part of the continuous quality improvement.

A QA programme should cover all aspects of diagnosis and therapy, including:

— The prescription of the procedure by the medical practitioner and its documentation (supervising if there is any error or contraindication);
— Appointments and patient information;
— Clinical dosimetry;
— Optimization of examination protocol;

— Record keeping and report writing;
— Quality control of radiopharmaceuticals and radionuclide generators;
— Acceptance and commissioning;
— Quality control of equipment and software;
— Waste management procedures;
— Training and continuing education of staff;
— Clinical audit;
— General outcome of the nuclear medicine service.

Further details on the general components of a QA programme and the associated quality control tests are given in Ref. [3.3]. The WHO has also published guidelines on QA in nuclear medicine [3.12], covering the organization of services, the training of personnel, the selection of procedures, quality control requirements for instrumentation and radiopharmaceuticals, as well as the interpretation and evaluation of results. The IAEA has several other relevant publications on QA for various aspects of nuclear medicine (see the Bibliography for details).

3.9.2. Audit

The QA programme should be assessed on a regular basis either as an external or internal audit or review. Audits of activities within the QA programme should be scheduled on the basis of the status and importance of the activity. Management should establish a process for such assessments to identify and correct administrative and management problems that may prevent achievement of the objectives. Audits and reviews should be conducted by persons who are technically competent to evaluate the processes and procedures being assessed, but do not have any direct responsibility for those activities. These may be staff from other work areas within the organization (internal audit), or an independent assessment by other organizations (external audit). External audits are generally a requirement for an accredited practice.

The quality audit should be performed in accordance with written procedures and checklists. It should include medical, technical and procedural checks, with the objective to enhance the effectiveness and efficiency of the QA programme. Any major changes in the QA programme should initiate an audit. The result should be documented and necessary correction initiated and followed up.

REFERENCES

[3.1] INTERNATIONAL COMMISSION ON RADIOLOGICAL PROTECTION, Recommendations of the ICRP, Publication 103, Elsevier (2008).

[3.2] EUROPEAN COMMISSION, FOOD AND AGRICULTURE ORGANIZATION OF THE UNITED NATIONS, INTERNATIONAL ATOMIC ENERGY AGENCY, INTERNATIONAL LABOUR ORGANIZATION, OECD NUCLEAR ENERGY AGENCY, PAN AMERICAN HEALTH ORGANIZATION, UNITED NATIONS ENVIRONMENT PROGRAMME, WORLD HEALTH ORGANIZATION, Radiation Protection and Safety of Radiation Sources: International Basic Safety Standards, IAEA Safety Standards Series No. GSR Part 3, IAEA, Vienna (2014).

[3.3] INTERNATIONAL ATOMIC ENERGY AGENCY, Applying Radiation Safety Standards in Nuclear Medicine, Safety Reports Series No. 40, IAEA, Vienna (2005).

[3.4] INTERNATIONAL ATOMIC ENERGY AGENCY, Nuclear Medicine Resources Manual, IAEA, Vienna (2006).

[3.5] INTERNATIONAL ATOMIC ENERGY AGENCY, Occupational Radiation Protection, IAEA Safety Standards Series No. RS-G-1.1, IAEA, Vienna (1999).

[3.6] INTERNATIONAL ATOMIC ENERGY AGENCY, Assessment of Occupational Exposure Due to Intakes of Radionuclides, IAEA Safety Standards Series No. RS-G-1.2, IAEA, Vienna (1999).

[3.7] INTERNATIONAL ATOMIC ENERGY AGENCY, Assessment of Occupational Exposure Due to External Sources of Radiation, IAEA Safety Standards Series No. RS-G-1.3, IAEA, Vienna (1999).

[3.8] INTERNATIONAL ATOMIC ENERGY AGENCY, Management of Waste from the Use of Radioactive Material in Medicine, Industry, Agriculture, Research and Education, IAEA Safety Standards Series No. WS-G-2.7, IAEA, Vienna (2005).

[3.9] INTERNATIONAL ATOMIC ENERGY AGENCY, Regulations for the Safe Transport of Radioactive Material, 2012 Edition, IAEA Safety Standards Series No. SSR-6, IAEA, Vienna (2012).

[3.10] WORLD MEDICAL ASSOCIATION, 18th World Medical Assembly, Helsinki, 1974, as amended by the 59th World Medical Assembly, Seoul (2008).

[3.11] COUNCIL FOR INTERNATIONAL ORGANIZATIONS OF MEDICAL SCIENCES, WORLD HEALTH ORGANIZATION, International Ethical Guidelines for Biomedical Research Involving Human Subjects, CIOMS, Geneva (2002).

[3.12] WORLD HEALTH ORGANIZATION, Quality Assurance in Nuclear Medicine, WHO, Geneva (1982).

BIBLIOGRAPHY

EUROPEAN COMMISSION, European Guidelines on Quality Criteria for Computed Tomography, Rep. EUR 16262 EN, Office for Official Publications of the European Communities, Brussels (1999).

INTERNATIONAL ATOMIC ENERGY AGENCY (IAEA, Vienna)

Quality Control of Nuclear Medicine Instruments 1991, IAEA-TECDOC-602 (1991).

Radiological Protection for Medical Exposure to Ionizing Radiation, IAEA Safety Standards Series No. RS-G-1.5 (2002).

IAEA Quality Control Atlas for Scintillation Camera Systems (2003).

Quality Assurance for Radioactivity Measurement in Nuclear Medicine, Technical Reports Series No. 454 (2006).

Radiation Protection in Newer Medical Imaging Techniques: PET/CT, Safety Reports Series No. 58 (2008).

Quality Assurance for PET and PET/CT Systems, IAEA Human Health Series No. 1 (2009).

Quality Assurance for SPECT Systems, IAEA Human Health Series No. 6 (2009).

Quality Management Audits in Nuclear Medicine Practices (2009).

Radiation Protection of Patients (RPoP),
https://rpop.iaea/RPOP/RPOP/Content/index.htm

INTERNATIONAL COMMISSION ON RADIOLOGICAL PROTECTION

Radiological Protection of the Worker in Medicine and Dentistry, Publication 57, Pergamon Press, Oxford and New York (1989).

Radiological Protection in Biomedical Research, Publication 62, Pergamon Press, Oxford and New York (1991).

Radiological Protection in Medicine, Publication 105, Elsevier (2008).

Pregnancy and Medical Radiation, Publication 84, Pergamon Press, Oxford and New York (2000).

INTERNATIONAL COMMISSION ON RADIATION UNITS AND MEASUREMENTS, Quantities and Units in Radiation Protection Dosimetry, ICRU Rep. 51, Bethesda MD (1993).

MADSEN, M.T., et al., AAPM Task Group 108: PET and PET/CT shielding requirements, Med. Phys. **33** (2006) 1.

SMITH, A.H., HART, G.C. (Eds), INSTITUTE OF PHYSICAL SCIENCES IN MEDICINE, Quality Standards in Nuclear Medicine, IPSM Rep. No. 65, York (1992).

CHAPTER 4

RADIONUCLIDE PRODUCTION

H.O. LUNDQVIST
Department of Radiology, Oncology and Radiation Science,
Uppsala University,
Uppsala, Sweden

4.1. THE ORIGINS OF DIFFERENT NUCLEI

All matter in the universe has its origin in an event called the 'big bang', a cosmic explosion releasing an enormous amount of energy about 14 billion years ago. Scientists believe that particles such as protons and neutrons, which form the building blocks of nuclei, were condensed as free particles during the first seconds. With the decreasing temperature of the expanding universe, the formation of particle combinations such as deuterium (heavy hydrogen) and helium occurred. For several hundred million years, the universe was plasma composed of hydrogen, deuterium, helium ions and free electrons. As the temperature continued to decrease, the electrons were able to attach to ions, forming neutral atoms and converting the plasma into a large cloud of hydrogen and helium gas. Locally, this neutral gas slowly condensed under the force of gravity to form the first stars. As the temperature and the density in the stars increased, the probability of nuclear fusion resulting in the production of heavier elements increased, culminating in all of the elements in the periodic table that we know today. As the stars aged, consuming their hydrogen fuel, they eventually exploded, spreading their contents of heavy materials around the universe. Owing to gravity, other stars formed with planets around them, composed of these heavy elements. Four and a half billion years have passed since the planet Earth was formed. In that time, most of the atomic nuclei consisting of unstable proton–neutron combinations have undergone transformation (radioactive decay) to more stable (non-radioactive) combinations. However, some with very long half-lives remain: ^{40}K, ^{204}Pb, ^{232}Th and the naturally occurring isotopes of uranium.

The discovery of these radioactive atoms was first made by Henri Becquerel in 1896. The chemical purification and elucidation of some of the properties of radioactive substances was further investigated by Marie Skłodowska-Curie and her husband Pierre Curie. Since some of these long lived radionuclides generated more short lived radionuclides, a new scientific tool had been discovered that was later found to have profound implications in what today is known as nuclear medicine. George de Hevesy was a pioneer in demonstrating the practical uses of the new radioactive elements. He and his colleagues used a radioactive isotope of lead, ^{210}Pb, as a tracer (or indicator) when they studied the solubility of sparingly soluble lead salts. De Hevesy was also the first to apply the radioactive tracer technique in biology when he investigated lead uptake in plants (1923) using ^{212}Pb. Only one year later, Blumengarten and Weiss carried out the first clinical study, when they injected ^{212}Bi into one arm of a patient and measured the arrival time in the other arm. From this study, they concluded that the arrival time was prolonged in patients with heart disease.

4.1.1. Induced radioactivity

In the beginning, nature was the supplier of the radioactive nuclides used. Isotopes of uranium and thorium generated a variety of radioactive heavy elements such as lead, but radioactive isotopes of light elements were not known. Marie Curie's daughter Irène, together with her husband Frédéric Joliot took the next step. Alpha emitting sources had long been used to bombard different elements, for example, by Ernest Rutherford who studied the deflection of α particles in gold foils. The large deflections observed in this work led to the conclusion that the atom consisted of a tiny nucleus of protons with orbiting electrons (similar to planets around the sun). However, Joliot–Curie also showed that the α particles induced radioactivity in the bombarded foil (in their case, aluminium foil). The induced radioactivity had a half-life of about 3 min. They identified the emitted radiation to be from ^{30}P created in the nuclear reaction $^{27}Al(\alpha, n)^{30}P$.

They also concluded that:

> "These elements and similar ones may possibly be formed in different nuclear reactions with other bombarding particles: protons, deuterons and neutrons. For example, ^{13}N could perhaps be formed by capture of a deuteron in ^{12}C, followed by the emission of a neutron."

The same year, this was proved to be true by Ernest Lawrence in Berkeley, California and Enrico Fermi in Rome. Lawrence had built a cyclotron capable of accelerating deuterons up to about 3 MeV. He soon reported the production of ^{13}N with a half-life of 10 min. Thereafter, the cyclotron was used to produce several other biologically important radionuclides such as ^{11}C, ^{32}P and ^{22}Na. Fermi realized that the neutron was advantageous for radionuclide production. Since it has no charge, it could easily enter into the nucleus and induce a nuclear reaction. He immediately made a strong neutron source by sealing up ^{232}Ra gas with beryllium powder in a glass vial. The α particle emitted from ^{232}Ra caused a nuclear reaction in beryllium and a neutron was emitted, ^{9}Be(α, n)^{12}C.

Fermi and his research group started a systematic search by irradiating all available elements in the periodic system with fast and slow neutrons to study the creation of induced radioactivity. From hydrogen to oxygen, no radioactivity was observed in their targets, but in the ninth element, fluorine, their hopes were fulfilled. In the following weeks, they bombarded some 60 elements and found induced radioactivity in 40 of them. They also observed that the lighter elements were usually transmuted into radionuclides of a different chemical element, whereas heavier elements appeared to yield radioisotopes of the same element as the target.

These new discoveries excited the scientific community. From having been a rather limited technique, the radioactivity tracer principle could suddenly be applied in a variety of fields, especially in life sciences. De Hevesy immediately started to study the uptake and elimination of ^{32}P phosphate in various tissues of rats and demonstrated, for the first time, the kinetics of vital elements in living creatures. Iodine-128 was soon after applied in the diagnosis of thyroid disease.

This was the start of the radiotracer technology in biology and medicine as we know it today.

One early cyclotron produced nuclide of special importance was ^{11}C since carbon is fundamental in life sciences. Carbon-11 had a half-life of only 20 min but by setting up a chemical laboratory close to the cyclotron, organic compounds labelled with ^{11}C were obtained in large amounts. Photosynthesis was studied using $^{11}CO_2$ and the fixation of carbon monoxide in humans by inhaling ^{11}CO. However, 20 min was a short half-life and the use of ^{11}C was limited to the most rapid biochemical reactions. It must be remembered that the radio-detectors used at that time were primitive and that the chemical, synthetic and analytical tools were not adapted to such short times. A search to find a more long lived isotope of carbon resulted in the discovery in 1939 of ^{14}C produced in the nuclear reaction ^{13}C(d, p)^{14}C.

Unfortunately, ^{14}C produced this way was of limited use since the radionuclide could not be separated from the target. However, during the bombardments, a bottle of ammonium nitrate solution had been standing close to the target. By pure chance, it was discovered that this bottle also contained ^{14}C, which had been produced in the reaction ^{14}N(n, p)^{14}C.

The deuterons used in the bombardment consist of one proton and one neutron with a binding energy of about 2 MeV. When high energy deuterons hit a target, it is likely that the binding between the particles breaks and that a free neutron is created in what is called a ‘stripping reaction’. The bottle with ammonium nitrate had, thus, unintentionally been neutron irradiated. Since no carbon was present in the bottle (except small amounts from solved airborne carbon dioxide), the ^{14}C produced this way was of high specific radioactivity. It was also very easy to separate from the target. In the nuclear reaction, a 'hot' carbon atom was created, which formed $^{14}CO_2$ in the solution. By simply bubbling air through the bottle, the ^{14}C was released from the target.

The same year, tritium was discovered by deuteron irradiation of water. One of the pioneers 'Martin Kamen' stated:

> "Within a few months, after the scientific world had somewhat ruefully concluded that development of tracer techniques would be seriously handicapped because useful radioactive tracers for carbon, hydrogen, oxygen and nitrogen did not exist, ^{14}C and ^{3}H were discovered".

Before the second world war, the cyclotron was the main producer of radionuclides since the neutron sources at that time were very weak. However, with the development of the nuclear reactor, that situation changed. Suddenly, a strong neutron source was available, which could easily produce almost unlimited amounts of radioactive nuclides including biologically important elements, such as ^{3}H, ^{14}C, ^{32}P and ^{35}S, and clinically interesting radionuclides, such as ^{60}Co (for external radiotherapy) and ^{131}I, for nuclear medicine. After the war, a new industry was born which could deliver a variety of radiolabelled compounds for research and clinical use at a reasonable price.

However, accelerator produced nuclides have a special character, which makes them differ from reactor produced nuclides. Today, their popularity is increasing again. Generally, reactor produced radionuclides are most suitable for laboratory work, whereas accelerator produced radionuclides are more useful clinically. Some of the most used radionuclides in nuclear medicine, such as ^{111}In, ^{123}I and ^{201}Tl, and the short lived radionuclides, ^{11}C, ^{13}N, ^{15}O and ^{18}F, used for positron emission tomography (PET), are all cyclotron produced.

4.1.2. Nuclide chart and line of nuclear stability

During the late 19th century, chemists learned to organize chemical knowledge into the periodic system. Radioactivity, when it was discovered, conflicted with that system. Suddenly, various samples, apparently with the same chemical behaviour, were found to have different physical qualities such as half-life, emitted type of radiation and energy. The concept of 'isotopes' or elements occupying the 'same place' in the periodic system (from the Greek 'ίσος τόπος' (isos topos) meaning 'same place') was introduced by Soddy 1913, but a complete explanation had to await the discovery of the neutron by Chadwick in 1932.

The periodic system was organized according to the number of protons (atom number) in the nucleus, which is equal to the number of electrons to balance the atomic charge. The nuclide chart consists of a plot with the number of neutrons in the nucleus on the x axis and the number of protons on the y axis (Fig. 4.1).

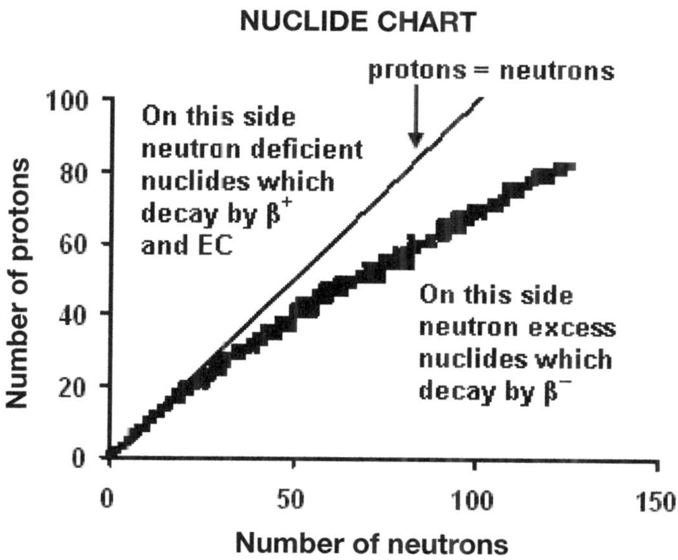

FIG. 4.1. Chart of nuclides. The black dots represent 279 naturally existing combinations of protons and neutrons (stable or almost stable nuclides). There are about 2300 proton/neutron combinations that are unstable around this stable line.

Figure 4.2 shows a limited part of the nuclide chart. The formal notation for an isotope is ${}^{A}_{Z}X$, where X is the element name (e.g. C for carbon), A is the mass number ($A = Z + N$), Z is the number of protons in the nucleus (atom number) and N the number of neutrons in the nucleus.

The expression above is overdetermined. If the element name X is known, so is the number of protons in the nucleus, Z. Therefore, the simplified notation ${}^{A}X$ is commonly used.

Some relations of the numbers of protons and neutrons have special names such as:

— Isotopes: the number of protons is constant (Z = constant).
— Isotones: the number of neutrons is constant (N = constant).
— Isobars: The mass number is constant (A = constant).

Of these expressions, only the isotope concept is generally used. It is important to understand that whenever the expression 'isotope' is used, it must always be related to a specific element or group of elements, for example, isotopes of carbon (e.g. ^{11}C, ^{12}C, ^{13}C and ^{14}C).

Number of protons																	
9									^{17}F	^{18}F	^{19}F	^{20}F	^{21}F	^{22}F	^{23}F	^{24}F	^{25}F
8						^{13}O	^{14}O	^{15}O	^{16}O	^{17}O	^{18}O	^{19}O	^{20}O	^{21}O	^{22}O	^{23}O	^{24}O
7						^{12}N	^{13}N	^{14}N	^{15}N	^{16}N	^{17}N	^{18}N	^{19}N	^{20}N	^{21}N	^{22}N	^{23}N
6				^{9}C	^{10}C	^{11}C	^{12}C	^{13}C	^{14}C	^{15}C	^{16}C	^{17}C	^{18}C	^{19}C	^{20}C		
5				^{8}B		^{10}B	^{11}B	^{12}B	^{13}B	^{14}B	^{15}B		^{17}B				
4				^{7}Be	^{8}Be	^{9}Be	^{10}Be	^{11}Be	^{12}Be		^{14}Be						
3				^{6}Li	^{7}Li	^{8}Li	^{9}Li		^{11}Li								
2		^{3}He	^{4}He		^{6}He		^{8}He										
1	^{1}H	^{2}H	^{3}H														
0		n															
	0	1	2	3	4	5	6	7	8	9	10	11	12	13	14	15	16

Number of neutrons

FIG. 4.2. A part of the nuclide chart where the lightest elements are shown. The darkened fields represent stable nuclei. Nuclides to the left of the stable ones are radionuclides deficient in neutrons and those to the right, rich in neutrons.

In the nuclide chart (Fig. 4.1), the stable nuclides fall along a monotonically increasing line called the stability line. The stability of the nucleus is determined by competing forces: the 'strong force' that binds the nucleons (protons and neutrons) together and the Coulomb force that repulses particles of like charge, e.g. protons. The interplay between the forces is illustrated in Fig. 4.3.

For best stability, the nucleus has an equal number of protons and neutrons. This is a quantum mechanic feature of bound particles and in Fig. 4.1 this is illustrated by a straight line. It is also seen that the stability line follows the straight line for the light elements but that there is considerable deviation (neutron excess) for the heavier elements. The explanation is the large Coulomb force in the heavy elements which have many protons in close proximity. By diluting the charge by non-charged neutrons, the distance between the charges increases and the Coulomb force decreases.

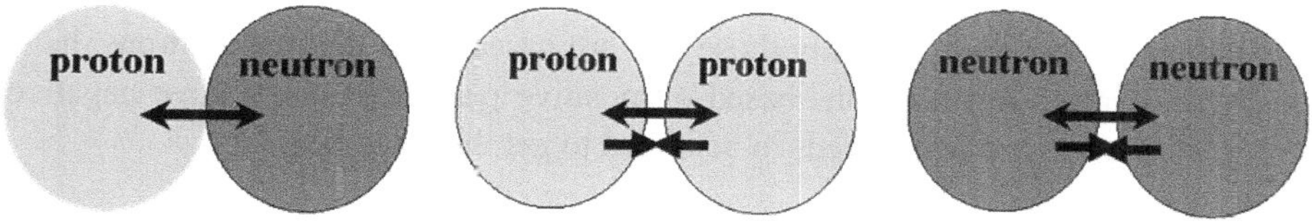

FIG. 4.3. Between the proton and a neutron, there is a nuclear force that amounts to 2.225 MeV. The nucleons form a stable combination called deuterium, an isotope of hydrogen. In a system of two protons, the nuclear force is equally strong to a neutron–proton, but the repulsive Coulomb forces are stronger. Thus, this system cannot exist. The nuclear force between two neutrons is equally strong and there is no Coulomb force. Nevertheless, this system cannot exist due to other repulsive forces, a consequence of the rules of pairing quarks.

4.1.3. Binding energy, Q-value, reaction threshold and nuclear reaction formalism

There are no barriers and no repulsive forces between a free proton and neutron, and they can fuse at low kinetic energies to form a deuterium nucleus, which has a weight somewhat smaller than the sum of the free neutron and proton weights. This mass difference can be converted into energy using Albert Einstein's formula $E = mc^2$ and is found to be 2.2 MeV. This is also the energy released as a γ photon in the reaction. To separate the two nucleons in the deuterium nucleus, at least 2.2 MeV have to be added. The energy gained or lost in a nuclear reaction is called the Q-value. In a somewhat more complex reaction, $^{14}N(p, \alpha)^{11}C$, the Q-value is calculated as the difference between the summation of the mass of the particles before the reaction (p, ^{14}N) from the mass of the particles after the reaction (α, ^{11}C). It should be noted that it is the mass of the nucleus and not the atomic mass that is used. Using a Q-value calculator[1], the Q-value for the reaction $^{14}N(p, \alpha)^{11}C$ is –2923.056 keV. This means that the proton, when it reaches the ^{14}N nucleus, has to have a kinetic energy of at least 2.93 MeV in order to make the reaction possible.

However, before it hits the nucleus, the proton has to overcome the barrier created by the repulsive Coulomb force between the proton and the positive ^{14}N nucleus. During the passage, the proton loses some energy and the starting value, called the threshold value, must then exceed the Q-value. The same calculator gives the threshold value of 3.14 MeV for the ^{11}C production reaction.

The reaction energy (the 'Q-value') is positive for exothermal reactions (spontaneous reactions) and negative for endothermal reactions. Since all radioactive decays are spontaneous, they need to have positive Q-values. Some reactions used

[1] For example, http://nucleardata.nuclear.lu.se/database/masses/

to produce radionuclides, mainly those that are based upon thermal neutrons, have positive Q-values but reactions based on positive particles usually have negative Q-values, e.g. extra energy needs to be added to get the reaction going.

4.1.4. Types of nuclear reaction, reaction channels and cross-section

As seen in Fig. 4.1, the radionuclides to the right of the stability line have an excess of neutrons compared to the stable elements and they are preferentially produced by irradiating a stable nuclide with neutrons. The radionuclides to the left are neutron deficient or have an excess of charge and, hence, they are mainly produced by irradiating stable elements by a charged particle, e.g. p or d. Although these are the main principles, there are exceptions.

Usually, the irradiating particles have a large kinetic energy that is transferred to the target nucleus to enable a nuclear reaction (the exception being thermal neutrons that can start a reaction by thermal diffusion). Figure 4.4 shows schematically an incoming beam incident upon the target, where it may be scattered and absorbed. It can transfer its energy totally or partly to the target nucleus and can interact with parts of or the whole of the target nucleus. Since the produced activity should be high, the target is also usually thick.

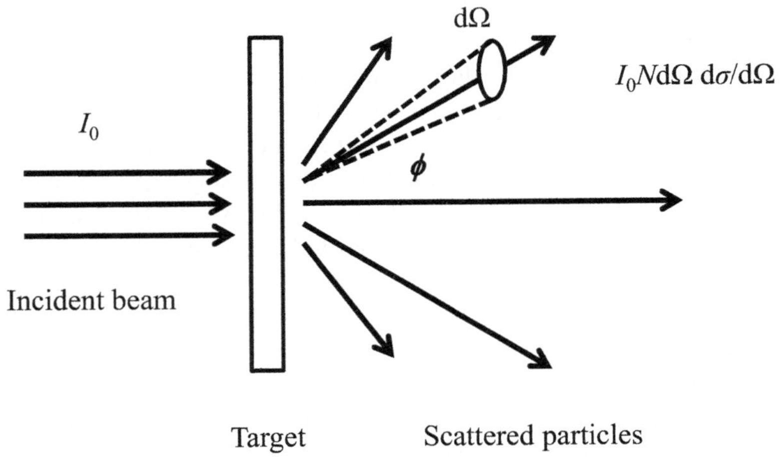

FIG. 4.4. Target irradiation. A nuclear physicist is usually interested in the particles coming out, their energy and angular distribution, but the radiochemist is mainly interested in the transformed nuclides in the target.

In radionuclide production, the nuclear reaction always involves a change in the number of protons or neutrons. Reactions that result in a change in the number of protons are preferable because the product becomes a different element, facilitating chemical separation from the target, compared to an (n, γ) reaction, where the product and target are the same.

Neutrons can penetrate the target at down to thermal energies. Charged particles need to overcome the Coulomb barrier to penetrate the nucleus (Fig. 4.5).

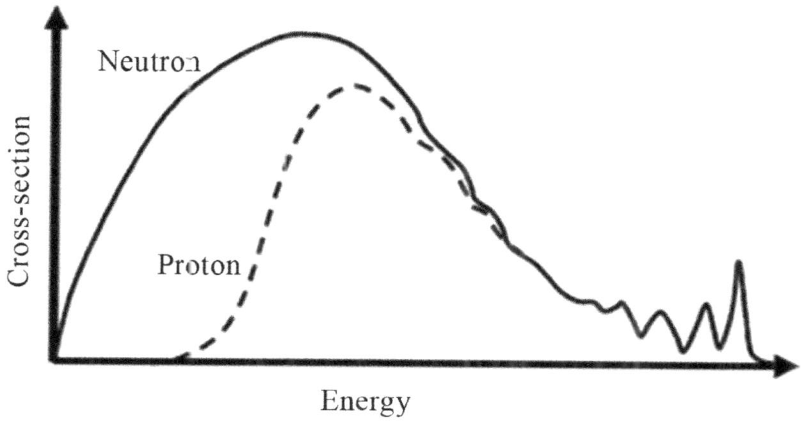

FIG. 4.5. General cross-sectional behaviour for nuclear reactions as a function of the incident particle energy. Since the proton has to overcome the Coulomb barrier, there is a threshold that is not present for the neutron. Even very low energy neutrons can penetrate into the nucleus to cause a nuclear reaction.

The parameter cross-section σ is the probability of a certain nuclear reaction happening and is expressed as a surface. It is the probability that a particle will interact per unit surface area of target. The geometrical cross-section of a uranium nucleus is roughly 10^{-28} m^2, and this area has also been taken to define the unit for cross-section barn (b). This is not an International System of Units unit but is commonly used to describe reaction probabilities in atomic and nuclear physics.

For fast particle reactions, the probability is usually less than the geometrical cross-section area of the nucleus, with probabilities in the range of millibarns. However, the probability of a hit is a combination of the area of both the nucleus and the incoming particle. The Heisenberg uncertainty principle states that the position and the momentum of particles cannot be simultaneously known to arbitrarily high precision. This implies that particles of well defined but low energy, such as thermal neutrons, will have a large uncertainty in their position. One may also say that they are increasing in size and nuclear reactions involving thermal neutrons may have very large cross-sections, sometimes of the order of several thousand barns.

The general equation for a nuclear reaction is:

a + A → b + B + Q

where a is the incoming particle and A is the target nucleus in the ground state (the entrance channel). Depending on the energy and the particle involved, several nuclear reactions may happen, each with its own probability (cross-section). Each nuclear reaction creates an outgoing channel, where b is the outgoing particle or particles and B is the rest nucleus. Q is the reaction energy and can be both negative and positive.

A common notation of a nuclear reaction is A(a, b)B. If the incoming particle is absorbed, there is a capture process type (n, α) and in a reaction of type (p, n) charge exchange occurs. If many particles are expelled, the reaction can be referred to as (p, 3n). Each such reaction is called a reaction channel and is characterized by an energy threshold (an energy that makes the nuclear reaction possible, opens up the channel) and a probability (cross-section) varying with the incoming particle energy. A schematic illustration of different reaction channels opened in proton irradiation is given in Fig. 4.6.

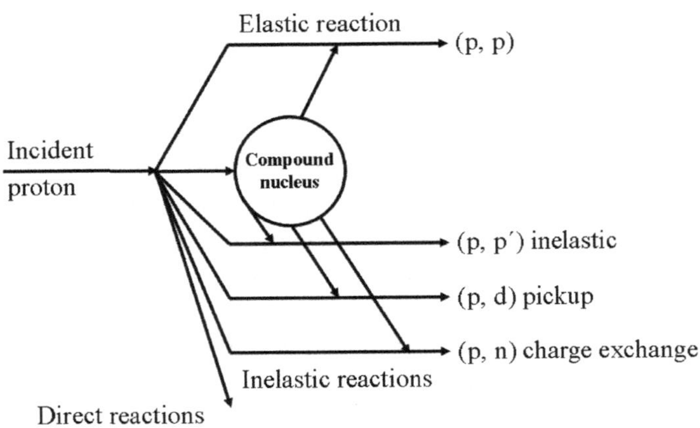

FIG. 4.6. A schematic figure showing some reaction channels upon proton irradiation.

Different reaction mechanisms can operate in the same reaction channel. Here, two ways are differentiated:

— The formation of a compound nucleus;
— Direct reactions.

The compound nucleus has a large probability to be formed in a central hit of the nucleus and is preferable at low energies close to the energy threshold of the reaction channel. Here, the incoming particle is absorbed and an excited compound nucleus formed. This compound nucleus will rapidly (~10^{-19} s) undergo decay (fragment) with the isotropic emission of neutrons and γ rays. Direct reactions preferentially occur at the edge of the nucleus or at high energies. The incoming energy is directly transferred to a nucleon (knock-on reaction) giving two outgoing particles. The outgoing particles usually have high energy and are emitted in about the same direction as the incoming particle.

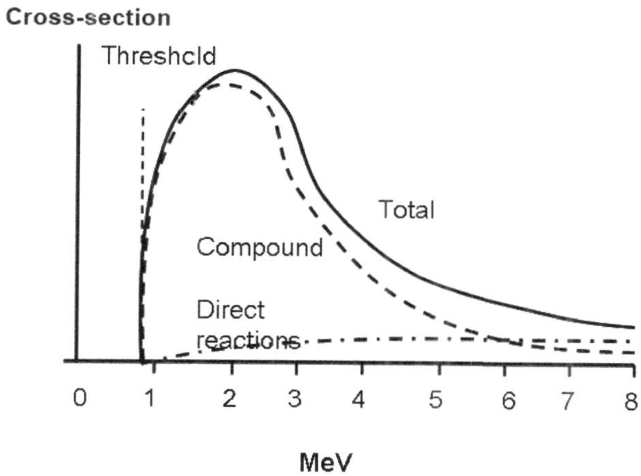

FIG. 4.7. A schematic view of particle energy variations of a cross-section for direct nuclear reactions and for forming a compound nucleus.

The production of radionuclides is due to a mixture of these two reaction types. Their probability varies with energy in different ways. The direct reactions are heavily associated with the geometrical size of the nucleus, and the cross-section is usually small and fairly constant with energy. The highest probability of forming a compound nucleus is just above the reaction threshold as seen in Fig. 4.7.

4.2. REACTOR PRODUCTION

There are two major ways to produce radionuclides: using reactors (neutrons) or particle accelerators (protons, deuterons, α particles or heavy ions). Since the target is a stable nuclide, either a neutron-rich radionuclide (reactor produced) or a neutron deficient radionuclide (accelerator produced) is generally obtained.

4.2.1. Principle of operation and neutron spectrum

A nuclear reactor is a facility in which a fissile atomic nucleus such as ^{235}U, ^{239}Pu or ^{241}P absorbs a low energy neutron and undergoes nuclear fission. In the process, fast neutrons are produced with energies from about 10 MeV and below (the fission neutron spectrum). The neutrons are slowed down in a moderator (usually water) and the slowed down neutrons start new fissions. By regulating this nuclear chain reaction, there will be a steady state of neutron production with a typical neutron flux of the order of 10^{14} neutrons · cm^{-2} · s^{-1}.

Since neutrons have no charge and are, thus, unaffected by the Coulomb barrier, even thermal neutrons (0.025 eV) can enter the target nucleus and cause a nuclear reaction. However, some nuclear reactions, depending upon the cross-section, require fast neutrons (energy < 10 MeV).

A reactor produces a neutron cloud in which the target is placed so that it will be isotropically irradiated. Placing the target in different positions exposes it to neutrons of different energy. Usually, the reactor facility has a pneumatic system for placing targets at predefined positions. One has to consider the heat that is generated in the reactor core, since the temperature at some irradiation positions may easily reach 200°C. The reactor is characterized by the energy spectrum, the flux (neutrons · cm^{-2} · s^{-1}) and the temperature at the irradiation position.

Most reactors in the world are for energy production and, for safety reasons, cannot be used for radionuclide production. Usually, only national research reactors are flexible enough for use in radioisotope production.

4.2.2. Thermal and fast neutron reactions

The most typical neutron reaction is the (n, γ) reaction in which a thermal neutron is captured by the target nucleus forming a compound nucleus. The decay energy is emitted as a prompt γ ray. A typical example is the reaction $^{59}Co(n, \gamma)^{60}Co$ that produces an important radionuclide used in external therapy. However, since the produced radionuclide is of the same element as the target, the specific activity a, i.e. the radioactivity per mass of the sample, is low. This type of nuclear reaction is of little interest when labelling radiopharmaceuticals. In light elements, other nuclear reactions resulting from thermal neutron irradiation are possible, such as (n, p). Table 4.1 lists possible production reactions for some biologically important radionuclides.

TABLE 4.1. TYPICAL NUCLEAR REACTIONS IN A REACTOR FOR RADIONUCLIDE PRODUCTION

Type of neutrons	Nuclear reaction	Half-life $T_{1/2}$	Cross-section σ (mb)
Thermal	$^{59}Co(n, \gamma)^{60}Co$	5.3 a	2000
	$^{14}N(n, p)^{14}C$	5730 a	1.75
	$^{33}S(n, p)^{33}P$	25 d	0.015
	$^{35}Cl(n, \alpha)^{32}P$	24 d	0.05
Fast	$^{24}Mg(n, p)^{24}Na$	15 h	1.2
	$^{35}Cl(n, \alpha)^{32}P$	14 d	6.1

Nuclear reactions with thermal neutrons are attractive for many reasons. The yields are high due to large cross-sections and the high thermal neutron fluxes available in the reactor. In some cases, the yields are sufficiently high to use these reactions as the source of charged secondary particles, e.g. $^{6}Li(n, \alpha)^{3}H$ for the production of high energy ^{3}H ions, which can then be used for the production of ^{18}F by $^{16}O(^{3}H, n)^{18}F$. The target used is $^{6}LiOH$, in which the produced ^{3}H ions will be in close contact with the target ^{16}O. A drawback of this production is that when the target is dissolved the solution is heavily contaminated with ^{3}H water that might be difficult to remove. Today, with an increasing number of hospital based accelerators, there is little need of neutron produced ^{18}F.

Another reactor produced neutron deficient radionuclide is ^{125}I:

$$^{124}Xe(n, \gamma)^{125}Xe\ (T_{1/2} = 17\ h) \rightarrow {}^{125}I\ (T_{1/2} = 60\ d)$$

This is currently the common way of producing high quality ^{125}I. A drawback is that ^{124}Xe has a natural abundance of 0.1%. To increase the production yield, one needs to work with expensive enriched targets. However, these can be reused several times. This is an example of a generator system where the mother is shorter lived than the daughter. Although there is no need to make a separation between the mother and daughter, the target, after irradiation, has to be stored for some days to allow the decay of ^{125}Xe to be complete. The expensive target gas ^{124}Xe is carefully removed and the ^{125}I is washed out from the walls of the target capsule.

Many reactor produced radionuclides emit high energy β particles that contribute to the absorbed dose (but not the imaging signal) to patients, which is a drawback in diagnostic procedures. However, a few β emitting isotopes result in daughter nuclei that emit γ rays with long de-excitation times (metastable excited levels), instead of the more common prompt (10^{-14} s) γ emission. Such radioisotopes are suitable for nuclear medicine imaging, since they principally yield γ radiation, with some electron emission, a consequence of internal conversion. The most commonly used radionuclide in nuclear medicine, ^{99m}Tc, is of this type. The 'm' after the atomic mass signifies that this is the metastable version of the radionuclide.

In radionuclide therapy, in contrast to diagnostic applications, the emission of high energy β radiation is desirable. Most radionuclides for radiotherapy are, therefore, reactor produced. Examples include ^{90}Y, ^{131}I and ^{177}Lu. A case of interest to study is ^{177}Lu, which can be produced in two different ways using thermal neutrons. The most common production route is still the (n, γ) reaction on ^{176}Lu, which opposes two conventional wisdoms in practical radionuclide production for biomolecular labelling:

(a) Not to use a production that yields the same product element as the target since it will negatively affect the labelling ability due to the low specific radioactivity;
(b) Not to use a target that is radioactive.

However, ^{176}Lu is a natural radioactive isotope of lutetium with an abundance of 2.59%. Figure 4.8 shows how ^{177}Lu needs to be separated from the dominant ^{175}Lu to decrease the mass of the final product. This method of production works because the high cross-section (2020 b) of ^{176}Lu results in a high fraction of the target atoms being converted to ^{177}Lu, yielding an acceptable specific radioactivity of the final product.

On the right of Fig. 4.8, an indirect way to produce ^{177}Lu from ^{176}Yb is also shown. This method of production utilizes a generator nuclide ^{177}Yb, produced by an (n, γ) reaction, which then decays to ^{177}Lu. In principle, by chemically separating lutetium from ytterbium, one would obtain the highest possible specific radioactivity. However, the chemical separation between two lanthanides is not trivial and, thus, it is difficult to obtain ^{177}Lu without substantial contamination of the target material Yb that may compete in the labelling procedure. Furthermore, the cross-section for this reaction is almost a thousandfold lower, resulting in a much lower product yield.

Reactions involving fast neutrons usually have cross-sections that are of the order of millibarns, which, coupled with the much lower neutron flux at higher energy relative to thermal neutron fluxes, leads to lower yields. However, there are some important radionuclides, e.g. ^{32}P that have to be produced this way. Figure 4.9 gives the details of this production.

	^{175}Lu	^{176}Lu	^{177}Lu	^{178}Lu
$T_{1/2}$ Abundance (%) σ(mb)	Stable 97.41	3.78 10^{10} a 2.59 2020	6.734 d	28.4 min
$T_{1/2}$ Abundance (%) σ(mb)				

^{175}Lu	^{176}Lu	^{177}Lu	^{178}Lu
Stable 97.41	3.78 10^{10} a 2.59 2020	6.734 d	28.4 min
^{174}Yb Stable 31.8	^{175}Yb 4.185 d	^{176}Yb Stable 12.7 2.85	^{177}Yb 1.911 h 31.8

FIG. 4.8. Production of ^{177}Lu from ^{176}Lu (left) and from ^{176}Yb (right).

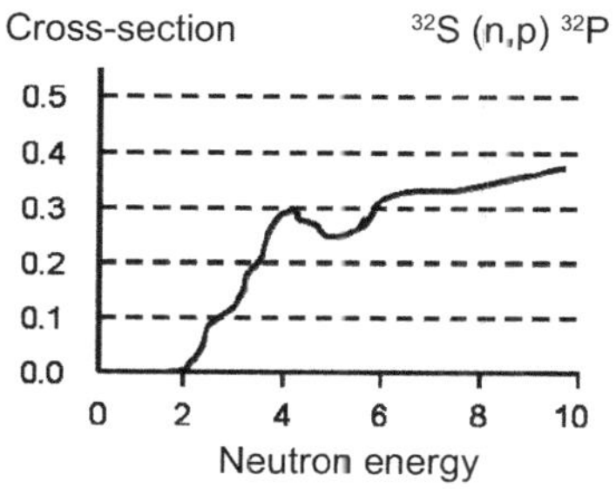

	^{32}S	^{33}S	^{34}S
$T_{1/2}$ Abundance (%) σ(mb)	Stable 95.02 varying	Stable 0.75 varying	Stable 4.21 varying
	^{31}P	^{32}P	^{33}P
$T_{1/2}$ Abundance (%) σ(mb)	Stable 100	14.26 d	25.4 d

FIG. 4.9. Data for the production of ^{32}P in the nuclear reaction ^{32}S(n, p)^{32}P. The reaction threshold is 0.51 MeV. From the cross-section data, it can be seen that there is no substantial yield until an energy of about 2 MeV. The yield is an integration of the cross-section data and the neutron energy spectrum. A practical cross-section can be calculated to about 60 mb

4.2.3. Nuclear fission, fission products

Uranium-235 is not only used as fuel in a nuclear reactor but it can also be used as a target to produce radionuclides. Uranium-235 irradiated with thermal neutrons undergoes fission with a cross-section of 586 b. The fission process results in the production of two fragments of ^{235}U nucleus plus a number of free neutrons. The sum of the fragments' mass will be close to the mass of ^{235}U, but they will vary according to Fig. 4.10.

The masses of the ^{99}Mo and ^{134}Sn produced in the reaction:

$$^{235}\mathrm{U} + \mathrm{n} \rightarrow {}^{236}\mathrm{U} \rightarrow {}^{99}\mathrm{Mo} + {}^{134}\mathrm{Sn} + 3\mathrm{n}$$

are marked in Fig. 4.10. Some medically important radionuclides are produced by fission, such as ^{90}Y (therapy) and ^{99m}Tc (diagnostic). They are not produced directly but by a generator system:

$$^{90}\mathrm{Sr}\ (28.5\ \mathrm{a}) \rightarrow {}^{90}\mathrm{Y}\ (2\ 3\ \mathrm{d})\ \text{and}\ {}^{99}\mathrm{Mo}\ (2.7\ \mathrm{d}) \rightarrow {}^{99m}\mathrm{Tc}\ (6\ \mathrm{h})$$

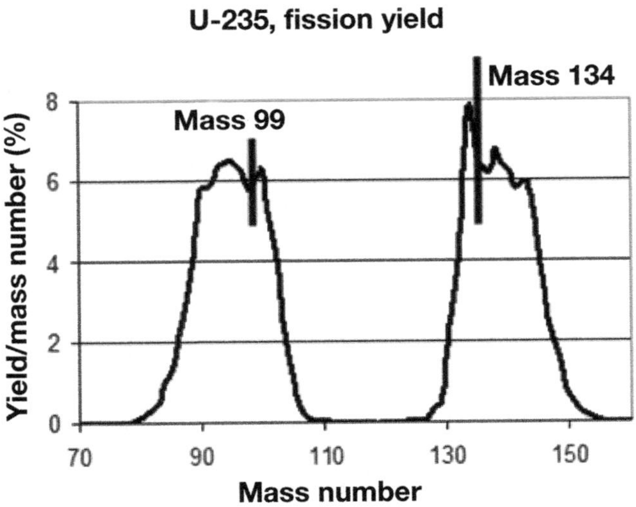

FIG. 4.10. The yield of fission fragments as a function of mass.

The primary radionuclides produced are then ^{90}Sr and ^{99}Mo, or more precisely the mass numbers 90 and 99.

Another important fission produced radionuclide in nuclear medicine, both for diagnostics and therapy, is ^{131}I. The practical fission cross-section for this production is the fission cross-section of ^{235}U multiplied by the fraction of fragments having a mass of 131 or $586 \times 0.029 = 17$ b. The probability of producing a mass of 131 is 2.9% per fission. Iodine-131 is the only radionuclide with a mass of 131 that has a half-life of more than 1 h, meaning that all of the others will soon have decayed to ^{131}I.

4.3. ACCELERATOR PRODUCTION

Charged particles, unlike neutrons, are unable to diffuse into the nucleus, but need to have sufficient kinetic energy to overcome the Coulomb barrier. However, charged particles are readily accelerated to kinetic energies that open up more reaction channels than fast neutrons in a reactor. An example is seen in Fig. 4.11 that also illustrates alternative opportunities with: p, d, ^{3}He and ^{4}He or α, to produce practical and economic nuclear reactions.

$$^{127}\mathrm{I(p,\,5n)}^{123}\mathrm{Xe} \rightarrow {}^{123}\mathrm{I}$$

$$^{124}\mathrm{Xe(p,\,np)}^{123}\mathrm{Xe} \rightarrow {}^{123}\mathrm{I}$$

$$^{123}\mathrm{Te(p,\,n)}^{123}\mathrm{I}$$

$$^{122}\mathrm{Te(d,\,n)}^{123}\mathrm{I}$$

$$^{124}\mathrm{Te(p,\,2n)}^{123}\mathrm{I}$$

$$^{121}\mathrm{Sb}(^{4}\mathrm{He,\,2n})^{123}\mathrm{I}$$

$$^{121}\mathrm{Sb}(^{3}\mathrm{He,\,n})^{123}\mathrm{I}$$

$$^{123}\mathrm{Sb}(^{3}\mathrm{He,\,3n})^{123}\mathrm{I}$$

FIG. 4.11. Various nuclear reactions that produce ^{123}I. All of the reactions have been tried and can be performed at relatively low particle energies. The ^{123}Xe produced in the first two reactions decays to ^{123}I with a half-life of about 2 h. In the first reaction, the ^{123}Xe is separated from the target and then decays, while in the second reaction, the ^{123}I is washed out of the target after decay.

An accelerator in particle physics can be huge, as in the European Organization for Nuclear Research (CERN), with a diameter of more than 4 km. Accelerators for radionuclide production are much smaller, as they need to accelerate particles to much lower energies. The first reaction in Fig. 4.11, where five neutrons are expelled, is the most energy demanding, as it requires a proton energy of about 5 × 10 MeV = 50 MeV (the rule of thumb is that about 10 MeV are required per expelled particle). All of the other reactions require 20 MeV or less (see Table 4.2).

Another advantage with accelerator production is that it is usually easy to find a nuclear reaction where the product is a different element from the target. Since different elements can be separated chemically, the product can usually be of high specific radioactivity, which is important when labelling biomolecules.

A technical difference between reactor and accelerator irradiation is that in the reactor the particles come from all directions but in the accelerator the particles have a particular direction. The number of charged particles is often smaller and is usually measured as an electric current in microamperes (1 μA = 6×10^{12} protons/s but 3×10^{12} alpha/s because of the two charges of the α particle).

TABLE 4.2. CHARACTERIZATION OF ACCELERATORS FOR RADIONUCLIDE PRODUCTION

Proton energy (MeV)	Accelerated particles	Used for
<10	Mainly single particle, p or d	PET
10–20	Usually p and d	PET
30–40	p and d, ^{3}He and ^{4}He may be available	PET, commercial production
40–500	Usually p only	Often placed in national centres and have several users

A drawback in accelerator production is that charged particles are stopped more efficiently than neutrons; for example, 16 MeV protons are stopped in 0.6 mm Cu. A typical production beam current of 100 μA hitting a typical target area of 2 cm^2 will then put 1.6 kW in a volume of 0.1 cm^3, which will evaporate most materials if not efficiently cooled. In addition, the acceleration of the beam occurs in a vacuum but the target irradiation is at atmospheric pressure or in gas targets at 10–20 times over pressure. To separate the vacuum from the target, the beam has to penetrate foils that will absorb some particle energy and they will also become strongly activated.

4.3.1. Cyclotron, principle of operation, negative and positive ions

There are several types of accelerator, all of which can, in principle, be used for radionuclide production. The dominant one for radionuclide production is currently the cyclotron that was invented by Lawrence in the early 1930s. Cyclotrons were first installed in hospitals in the 1960s, but during the past two decades, hospital based small cyclotrons yielding 10–20 MeV protons have become fairly common, especially with the rise of PET.

A cyclotron is composed of four systems:

(a) A resistive magnet that can create a magnetic field of 1–2 T;
(b) A vacuum system down to 10^{-5} Pa;
(c) A high frequency system (about 40 MHz) providing a voltage with a peak value of about 40 kV, although these figures can vary considerably for different systems;
(d) An ion source that can ionize hydrogen to create free protons as well as deuterium and α particles.

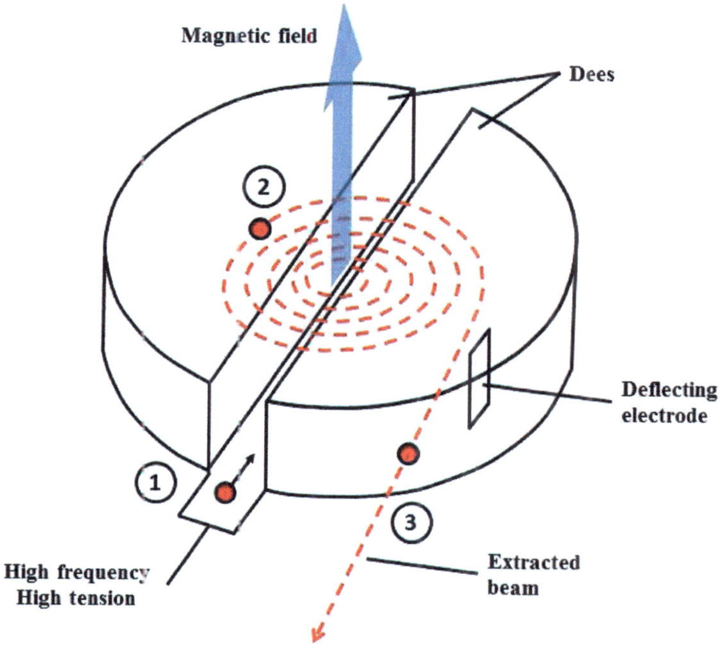

FIG. 4.12. The cyclotron principle. A negative ion is injected into the gap between D-shaped magnets (Dees) (1). An alternating electric field is applied across the gap, which causes the charge to accelerate. The magnetic force on a moving charge forces it to bend into a semicircular path of ever increasing radius (2). The applied electric field is reversed in direction each time the charged particle reaches the gap, so that it is continuously accelerated, until finally being ejected (3).

The inside of a cyclotron is shown in Fig. 4.12. The ion source is usually placed inside the vacuum and in the centre (internal), but, in larger machines, can be external. The ions are then injected from the outside through a central hole in the magnet. The main idea of the ion source is to have a slow flow of gas that is made into plasma by an arc discharge. The desired ion species are extracted through a collimator and accelerated in a static electric field. There are several types of ion source with different operating characteristics. In modern accelerators, negative ions, protons or deuterium with two orbit electrons are usually used. These facilitate extraction of the beam.

The ions leave the ion source with some velocity. Since the vacuum chamber is in a magnetic field, the ions move in a circular orbit. Inside the vacuum chamber, there are two electrodes, historically called the 'Dees' since the first ones have the shape of the letter D. These electrodes are hollow, which enables the ions to move freely inside the electrodes. There is a gap between the electrodes called the acceleration gap. If a voltage is applied between the electrodes, the ions will experience the potential gradient when traversing the gap between the electrodes. If the voltage polarity is switched at the correct rate, the ions will be continuously accelerated when crossing the gap, thus resulting in an increase in the ions' energy and velocity. As their velocity increases, the ions will move into a circular orbit of increasing radius. The time taken for the ions to return to the gap is independent of their radius in accelerators <30 MeV. For the cyclotron to operate correctly, it is necessary for the frequency of the electric field across the Dees to be the same as the frequency of the circulating ions, so that the polarity changes upon each traversal of the ions across the Dees.

In commercial accelerators, with high beam currents of several milliamperes, it is usual to have an internal target for radionuclide production located inside the chamber. In accelerators with lower beam currents <100 μA, such as those dedicated for PET hospital facilities, it is more common to extract the beam onto an external target system. The modes of extraction depend upon whether positive or negative ions are accelerated. Extraction of positive ions is made by using a deflector that applies a static electric field which acts upon the particles when in the outer orbits. Some beam current is invariably lost in the process and the deflector often becomes quite radioactive.

Modern proton/deuterium accelerators usually accelerate negative ions that are more easily extracted. In these systems, a thin carbon foil is used that will strip away the two orbit electrons. As a consequence, the particles suddenly change from negative to positive charge and are effectively bent out of the magnetic field with an almost 100% extraction efficiency and with little activation.

The extracted beam can either be transported further in a beam optical transport system or will hit a production target directly. The target is usually separated from the vacuum by metallic foils that are strong enough to withstand the pressure difference and the heat from the beam energy, as it is transferred and absorbed by the foils. The reason why two foils are used is that the heat produced by the beam passage has to be removed, which is facilitated by a flow of helium gas between the foils. Helium is preferred as the cooling medium since no induced activity will be produced in this gas.

4.3.2. Commercial production (low and high energy)

If the proton energy is >30 MeV, the particles tend to be relativistic, i.e. their mass and their cycle time in orbit increase. A constant frequency of the accelerating electric field would cause the ions to come out of phase. This can be compensated for either by increasing the magnetic field as a function of the cyclotron radius (isochronic cyclotrons) or by decreasing the radiofrequency during acceleration (synchrocyclotrons). Such accelerators tend to be more complex and expensive and, for this reason, 30 MeV is a typical energy for commercial accelerators that need to have large beam currents and to be both reliable and cost effective.

Commercial accelerators usually run beam currents of several milliamperes. Since it is technically difficult to extract such high beam currents due to heating problems in the separating foils, most commercial accelerators use internal targets, i.e. targets that are placed inside the cyclotron vacuum as shown schematically in Fig. 4.13.

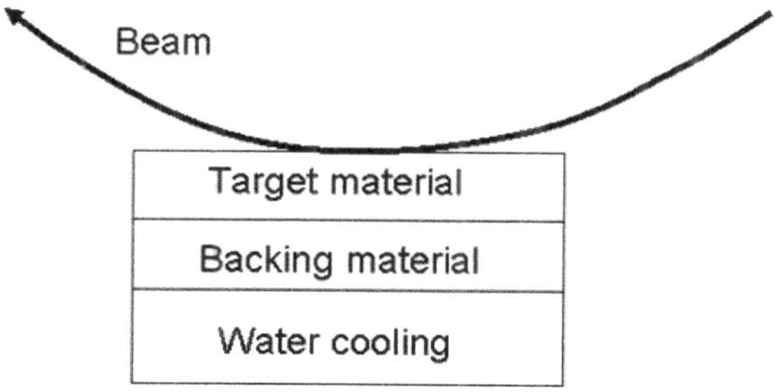

FIG. 4.13. Schematic image of an internal target. The target material is usually thin (a few tenths of a micrometre) and evaporated on a thicker backing plate. The target ensemble is water cooled on the back. An advantage is that the beam is spread out over a large area which facilitates cooling.

Many patients in nuclear medicine undergo single photon emission computed tomography (SPECT) investigations. Besides reactor produced ^{99m}Tc, commercial cyclotrons commonly produce ^{67}Ga, ^{111}In, ^{123}I and ^{201}Tl. In addition, some PET radionuclides, such as ^{124}I, are becoming commercially available. Increasing demand for the ^{68}Ge/^{68}Ga generator has also led to commercial production of the cyclotron produced mother nuclide ^{68}Ge. Only a few radionuclides of medical interest require production energies above 30 MeV. A limited number of high energy accelerators with high beam currents, usually at national physics laboratories, have the capacity for the production of, for example, ^{52}Fe and ^{61}Cu and other isotopes used for research activities.

4.3.3. In-house low energy production (PET)

Commercial accelerators dedicated to PET radioisotope production are limited both in energy (<20 MeV) and in beam current (<100 μA). Many production routes utilize gases or water as target materials and, therefore, external targets are to be preferred. Owing to the relatively low beam current, extraction is not a problem. Since internal targets need to be taken in and out of the cyclotron vacuum, they are not usually implemented in PET cyclotrons.

The importance of choosing the right reaction and target material is crucial and is illustrated by the production of ^{18}F. There are several nuclear reactions that can be applied (Table 4.3).

TABLE 4.3. DIFFERENT NUCLEAR REACTIONS FOR THE PRODUCTION OF ^{18}F

$^{20}Ne(d, \alpha)^{18}F$	The nascent ^{16}F will be highly reactive. In the noble gas Ne, it will diffuse and stick to the target walls; difficult to extract
$^{21}Ne(p, \alpha)^{18}F$	Same as above; in addition, the abundance of ^{21}Ne is low (0.27%) and needs enrichment
$^{19}F(p, d)^{18}F$	The product and target are the same element; poor specific radioactivity
$^{16}O(\alpha, d)^{18}F$	Cheap target but accelerators that can accelerate α particles to 35 MeV are expensive and not common
$^{16}O(d, \gamma)^{18}F$	Small cross-section and no practical yields can be obtained
$^{18}O(p, n)^{18}F$	Expensive enriched target material but the proton energy is low (low cost accelerator), which makes this the nuclear reaction of choice

Not only the nuclear reaction is important, but also the chemical composition of the target. To irradiate ^{18}O as a gas would be the purest target (only target nuclide present) but handling a highly enriched gas in addition to the hot-atom chemistry is complicated. Still, for some applications, this might be the best choice. To irradiate ^{18}O as an oxide and a solid target is possible but the process following irradiation to dissolve the target and to chemically separate ^{18}F is complex, has a low yield and other elements in the oxide could potentially contribute unwanted radioactivity. Enriched ^{18}O water is a target of choice as ^{18}O is the dominant nucleus and hydrogen does not contribute to any unwanted radioactivity. There is usually no need for target separation as water containing ^{18}F can often be directly used in the labelling chemistry. The target water can also, after being diluted with saline, be injected directly into patients, e.g. ^{18}F-fluoride for PET bone scans. Water targets will produce ^{18}F-fluoride for use in stereospecific nucleophilic substitutions. An alternative production route is neon gas production, $^{20}Ne(d, \alpha)^{18}F$. Adding $^{19}F_2$ gas to the neon as a carrier yields $^{18}F^{19}F$ that can be used for electrophilic substitution. Adding carrier lowers the specific radioactivity of the labelled product.

A problem is the heat generated when the beam is stopped in a few millilitres of target water. High pressure targets that force the water to remain in the liquid phase can overcome some of these problems but production is usually limited to beam currents <40 μA. Gas and solid targets are advantageous as they can withstand higher beam currents.

There are also several options for the production of ^{11}C. These include: $^{10}B(d, n)^{11}C$, $^{11}B(p, n)^{11}C$ and $^{14}N(p, \alpha)^{11}C$. The reactions on boron are made as solid target irradiations while the reaction on nitrogen is a gas target application.

The routine production routes of common positron emitters associated with PET are summarized in Table 4.4.

TABLE 4.4. COMMONLY USED RADIONUCLIDES IN PET

Radionuclide	Nuclear reaction	Yield (GBq)
^{15}O	$^{14}N(d, n)^{15}O$ gas target	15
^{13}N	$^{16}O(p, \alpha)^{13}N$ liquid target	5
^{11}C	$^{14}N(p, \alpha)^{11}C$ gas target	40
^{18}F	$^{18}O(p, n)^{18}F$ liquid target	100

Oxygen-15 is produced by deuteron bombardment of natural nitrogen through the $^{14}N(d, n)^{15}O$ nuclear reaction. An alternative is the $^{15}N(p, n)^{15}O$ reaction if a deuterium beam is not available. In this case, the target needs to be enriched. In the nitrogen target, ^{15}O-labelled molecular oxygen is produced directly. Direct production of ^{11}C-labelled carbon dioxide is possible by mixing the target gas with 5% natural carbon dioxide as a carrier. Water labelled with ^{15}O is preferably made by processing ^{15}O-labelled molecular oxygen.

Carbon-11 is produced by proton bombardment of natural nitrogen. By adding a small amount of oxygen to the target gas (<0.5%), carbon dioxide ($^{11}CO_2$) will be produced. Adding 5% hydrogen to the target will produce methane ($^{11}CH_4$).

Liquid targets are today by far the most popular and widely used for the production of ^{13}N. The reaction of protons on natural water produces nitrate and nitrite ions, which can be converted to ammonia by reduction. Water targets can also be used to form ammonia directly with the addition of a reducing agent, e.g. ethanol or hydrogen.

4.3.4. Targetry, optimizing the production regarding yield and impurities, yield calculations

When the nucleus is hit by an energetic particle, a complex interplay between physical and statistical laws determines the result. Important parameters are the entrance particle energy, the target thickness and the reaction channel cross-sections for the particle energies in the target. Computer codes such as ALICE and TALYS are available to calculate the size and the energy dependence of the cross-section for a certain reaction channel but they are not easy to apply; hence, caution should be exercised when interpreting the results from such codes. However, a rough estimation of the irradiating particle energy can be obtained using a well known rule of thumb in radionuclide production (illustrated in Fig. 4.14).

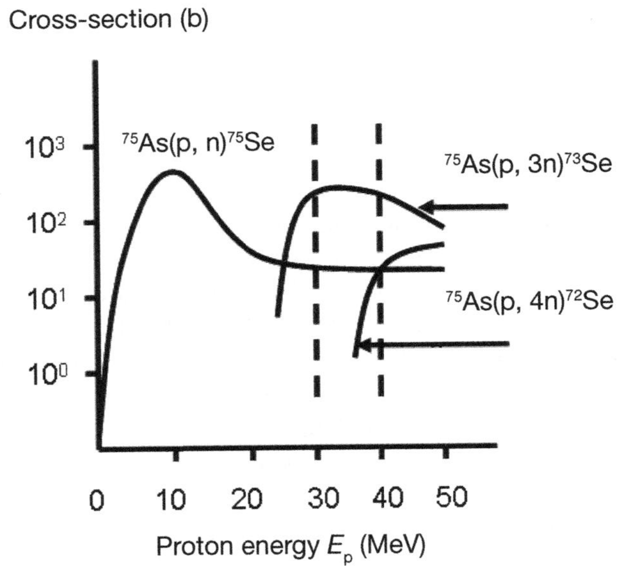

FIG. 4.14. Excitation functions of ^{75}As(p, xn)^{72,73,75}Se reactions. The optimal energy for the production of ^{73}Se is to use a proton energy of 40 MeV that is degraded to 30 MeV in the target.

The maximum cross-sections are found at about 10, 30 and 40 MeV for the (p, n), (p, 3n) and (p, 4n) reactions, respectively. Thus, it takes about 10 MeV to expel a nucleon, i.e. a proton of 50 MeV can cover radionuclide productions that involve the emission of about five nucleons. At low energy, there is a disturbing production of ^{75}Se and if excessively high proton energy is used, another unwanted radionuclide impurity is produced, namely ^{72}Se. The latter impurity can be avoided completely by restricting the proton energy to an energy lower than the threshold for the (p, 4n) reaction. The impurity that results from the (p, n) reaction cannot be avoided but can be minimized by using a target thickness that avoids the lower proton energies (having the highest (p, n) cross-sections).

Figure 4.14 highlights the fact that the chosen production parameters are a compromise. A proton range of 40–30 MeV uses the (p, 3n) cross-section well. Some ^{72}Se contamination is acceptable in order to increase the yield of ^{73}Se. An important factor is the half-life of ^{75}Se ($T_{1/2}$ = 120 d), ^{73}Se ($T_{1/2}$ = 7.1 h) and ^{72}Se ($T_{1/2}$ = 8.5 d). Sometimes, it is possible to wait for the decay of the radioactive contaminants. Although not the case here, sometimes a long half-life contaminant is not a serious disadvantage. If the product half-life is long, then there may be little product decay over the target irradiation time compared to short lived radionuclides.

The practical set-up when undertaking radionuclide production is as follows. A suitable As target is made and irradiated with 40 MeV protons. The thickness of the target is such that it decreases the proton energy to 30 MeV. This then gives a radioactivity yield of the desired radionuclide at the end of bombardment, which is mainly dependent upon the beam current and the irradiation time. The yield is usually expressed in gigabecquerels per microampere hours (GBq/µA · h), i.e. the produced radioactivity per time integrated beam current. If possible, it is endeavoured to keep the radioactivity of the contaminants at low levels (<1%). However, from the end of bombardment, the ratio of the product relative to any long lived radio-contaminants begins to decrease.

4.4. RADIONUCLIDE GENERATORS

Whenever a radionuclide (parent) decays to another radioactive nuclide (daughter), this is called a radionuclide generator. Most natural radioactivity is produced in generator systems starting with uranium isotopes and ^{232}Th, and involves about fifty radioactive daughters. Several radionuclides used in nuclear medicine are produced by generator systems such as the ^{99}Mo production of ^{99m}Tc, which subsequently decays to ^{99}Tc. The extremely long half-life of ^{99}Tc ($T_{1/2}$ = 2.1 × 10^5 a) means that ^{99m}Tc can be safely used as a clinical isotope without any radiological concerns. In other nuclides, the creation of a radioactive nuclide may be more important, e.g. the positron emitter ^{52}Fe ($T_{1/2}$ = 8 h) decays to ^{52}Mn ($T_{1/2}$ = 21 min) which is also a positron emitter. Furthermore, radionuclides used in therapy may themselves be generators such as ^{211}At ($T_{1/2}$ = 7 h) decaying to ^{211}Po ($T_{1/2}$ = 0.5 s) or ^{223}Ra, which generates a series of relatively short lived radioactive daughters in situ.

When talking about generators in nuclear medicine, a special case is usually considered in which a long lived mother generates a short lived daughter, which after labelling is administrated to the patient. Generally, this is a practical way to deliver short lived radionuclides to hospitals which otherwise, for logistical reasons, would not have been possible. The half-life should be sufficiently long so that the radionuclide can be delivered to hospitals, and provide the radioactive product for a number of patients over days or weeks. A typical example is the ^{99}Mo/^{99m}Tc generator (Fig. 4.15), which produces the most used radionuclide in nuclear medicine. The half-life of the parent (2.7 d) is adequate for transport and delivery, and the daughter has a suitable half-life (6 h) for patient investigations. The generator is used for about two to three half-lives of the parent (1 week) after which time it is renewed.

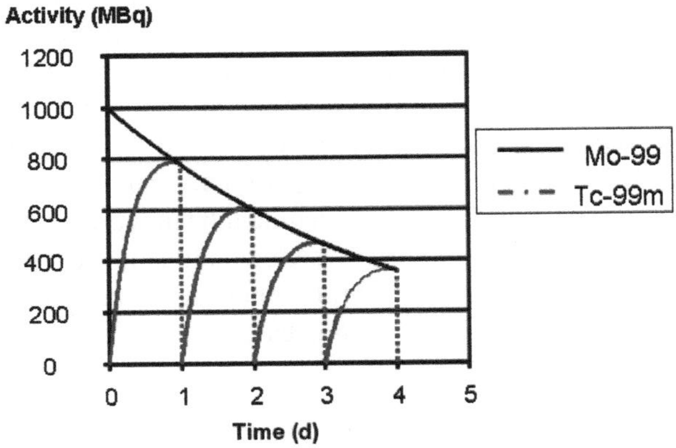

FIG. 4.15. Elution of a ^{99}Mo/^{99m}Tc generator. The generator has a nominal activity of 1000 MBq on day 0 (Monday). It is eluted daily, five times a week, yielding 1000, 780, 600, 470 and 360 MBq.

4.4.1. Principles of generators

Generator systems require that the parent is a reactor or accelerator produced by the methods described above and that the daughter radionuclide of interest can easily be separated from the parent. The ^{99}Mo/^{99m}Tc generator exhibits these characteristics. Most commercial generators use column chromatography, in which ^{99}Mo is adsorbed onto alumina. Eluting the immobilized ^{99}Mo on the column with physiological saline elutes the soluble ^{99m}Tc in a few millilitres of liquid. In fact, most generators in nuclear medicine use ion exchange columns in much the same way due to its simplicity of handling.

In generator systems, the daughter radionuclide is formed at the rate at which the parent decays, $\lambda_P \times N_P$. It also decays at the same rate, $\lambda_D \times N_D$, as the parent, once a state of transient equilibrium has been reached. The equations that describe the relationship between parent and daughter are provided in Chapter 1.

Another generator of increasing importance is ^{68}Ge, which has a half-life of 271 d that produces a short lived positron emitter ^{68}Ga ($T_{1/2}$ = 68 min). This is produced as a +3 ion that can be tagged, using a chelating agent such as DOTA, to small peptides, e.g. ^{68}Ga-DOTATOC. Owing to the long half-life of the mother, the generator can be operated for up to two years and can be eluted every 5 h. One problem with such a long lived generator is keeping it sterile, and furthermore, the ion exchange material is exposed to high radiation doses that may reduce the elution efficiency and the quality of the product.

The ^{90}Sr/^{90}Y generator is used to produce the therapeutic radionuclide ^{90}Y. This generator is not distributed to hospitals but is operated in special laboratories on account of radiation protection considerations associated with the long lived parent. The daughter, ^{90}Y, has a half-life of 2.3 d which is adequate for transport of the eluted ^{90}Y to distant hospitals.

^{81}Rb (4.5 h)/^{81m}Kr (13.5 s) for ventilation studies and ^{82}Sr (25.5 d)/^{82}Rb (75 s) for cardiac PET studies are examples of other generators with special requirements due to the extremely short half-life of the eluted product. Recently, generator systems producing α emitters for therapy have become available, e.g. ^{225}Ac (10 d)/^{213}Bi (45.6 min).

4.5. RADIOCHEMISTRY OF IRRADIATED TARGETS

During target irradiation, a few atoms of the wanted radionuclide are produced within the bulk target material. The energy released in a nuclear reaction is large relative to the electron binding energies and the radionuclide is, therefore, usually 'born' almost naked with no or few orbit electrons. This 'hot atom' will undergo chemical reactions depending on the target composition. In a gas or liquid target, these hot atom reactions may even cause the activity to be lost in covalent bonds to the target holder material. During irradiation, the target is also heated and its structure and composition may change. A pressed powder target may be sintered and become more ceramic, which makes it more difficult to dissolve. The target may melt and the radioactivity may diffuse in the target and even possibly evaporate. In designing a separation method, all of these factors have to be considered. Fast, efficient and safe methods are required to separate the few picograms of radioactive product from the bulk target material which is present in gram quantities.

Separation of the radionuclide already starts in the target as demonstrated in the production of $^{11}CO_2$. Carbon-11 is produced in a (p, α) reaction on nitrogen gas. To enable the production of CO_2, some trace amounts of oxygen gas (0.1–0.5%) are added. However, at low beam currents, mainly CO will be formed, since the target will not be heated. At high beam currents, the CO will be oxidized to the chemical form CO_2. The separation, made by letting the target through a liquid nitrogen trap, is simple and efficient. By adding hydrogen gas instead, the product will be CH_4.

The skill in hot-atom chemistry is to obtain a suitable chemical form of the radioactive product, especially when working with gas and liquid targets. Solid targets are usually dissolved and chemically processed to obtain the wanted chemical form for separation.

4.5.1. Carrier-free, carrier-added systems

The concept of specific activity a, i.e. the activity per mass of a preparation, is essential in radiopharmacy. If 100% of the product contains radioactive atoms, often called the theoretical a, then the relationship between the activity $\mathcal{A}$ in becquerels and the number of radioactive atoms N is given by $N = \mathcal{A}/\lambda$, where λ is the decay constant (1/s). The decay constant can be calculated from the half-life $T_{1/2}$ in seconds as $\lambda = \ln(2)/T_{1/2}$.

The specific activity a expressed as activity per number of radioactive atoms is then $\mathcal{A}/N = \lambda = \ln(2)/T_{1/2}$. For a short lived radionuclide, a will be relatively large compared to a long lived isotope. For example, a for ^{11}C ($T_{1/2} = 20$ min) is 1.5×10^8 times larger than for ^{14}C ($T_{1/2} = 5730$ a).

The specific activity a expressed in this way is a theoretical value that is rarely obtained in practical work. When producing ^{11}C, the target gas and target holder will contain stable carbon that will dilute the radioactive carbon as well as compete in the labelling process afterwards. A more empirical way to define a is to divide the activity by the total mass of the element under consideration. This value for ^{11}C will usually be a few thousand times lower than the theoretical value, while the production of ^{14}C can come closer to the theoretical a.

In the labelling process, a is usually expressed as the activity per number of molecules (a sum of labelled and unlabelled molecules). Instead of using the number of atoms or molecules, it is common to use the mole concept by dividing N by Avogadro's number ($N_A = 6.022 \times 10^{23}$). A common unit for a is then gigabecquerels per micromole.

If the radioactive atoms are produced and separated from the target without any stable isotopes, the process is said to be 'carrier-free'. If stable isotopes are introduced as being a contaminant in the target or in the separation procedure, the process is said to have 'no carrier added', i.e. no stable isotope is deliberately added. Both of these processes usually give a high final *a*. However, it may be necessary to use a target of the same element or it may be necessary to add extra mass of the same element in order for the separation process to work. In this case, carrier is added deliberately and *a* will usually be low.

It should be noted that a carrier does not necessarily need to be of the same element. When labelling a radiopharmaceutical with a chelator and metal ions, any ion fitting into the chelator will compete. An example is labelling a peptide with ^{111}In, when the activity will usually be delivered as $InCl_3$ in a weak acid. By sampling the activity with a stainless steel needle, Fe ions will be released and will probably completely ruin the labelling process by outnumbering the ^{111}In atoms.

4.5.2. Separation methods, solvent extraction, ion exchange, thermal diffusion

After irradiation, the small amount of desired radioactivity (of the order of nanomoles) usually needs to be separated from the bulk of the target in a suitable form for the following labelling process and at high *a*. The separation time should be related to the half-life of the radionuclide and should take at most one half-life. Solid targets usually have to be dissolved, which is simple for salts such as NaI but more complicated for, for example, Ni foils where boiling aqua regia may have to be used. To speed up this process, the Ni foil can be replaced by a pressed target of Ni powder that will increase the metal surface and will speed up the dissolving process.

In general, two principles are used: liquid extraction and ion exchange. In liquid extraction, usually two liquids that do not mix are used, e.g. water and an organic solvent. The target element and the produced activity of another element should have different relative solubility in the liquids. The two liquids and the dissolved target are mixed by shaking, after which two phases are formed. The phase with a high concentration of the wanted radioactive product is sampled and is usually separated again one or more times to reduce the target mass in that fraction. The relative solubility can be optimized by varying the pH or by adding a complexing agent.

In the ion exchange mechanism, an ion in the liquid phase (usually an aqueous phase) is transferred to a solid phase (organic or ceramic material). To maintain the charge balance, a counter ion is released from the solid phase. This ion may be a hydrogen ion. In the ion exchange mechanism, the distribution ratio is often a function of the pH. Furthermore, complexing agents can be used to modify the distribution ratio. The dissolved target is adjusted to obtain the right pH and other separation conditions, and is then put on to a column containing the ion exchange material. The optimal separation conditions would be that the small mass of desired radioactivity but not the bulk target material sticks to the column. The column can then be small, and after washing and change of pH, the desired activity can be eluted in a small volume. Under other conditions, large amounts of ion exchange material have to be used to prevent saturation of binding sites and leakage of the target material. This also means that large liquid volumes have to be used, implying poorer separation. The two techniques are often performed together by using liquid extraction to reduce the target mass, after which ion exchange is used to make the final separation.

Occasionally, thermal separation techniques may be applied, which have the advantage that they do not destroy the target (important when expensive enriched targets are used) and that they lend themselves to automation. As an example of such dry methods, the thermal separation of ^{76}Br ($T_{1/2}$ = 16 h) is described. The target is $Cu_2{}^{76}Se$, a selenium compound that can withstand some heat. The nuclear reaction used is ^{76}Se(p, n)^{76}Br.

The process is as follows:

(a) The target is placed in a tube and heated, under a stream of argon gas, to evaporate the ^{76}Br activity by dry distillation (Fig. 4.16);

(b) A temperature gradient is applied to separate the deposition areas of ^{76}Br and traces of co-evaporated selenide in the tube by thermal chromatography;

(c) The ^{76}Br activity deposited on the tube wall is dissolved in small amounts of buffer or water.

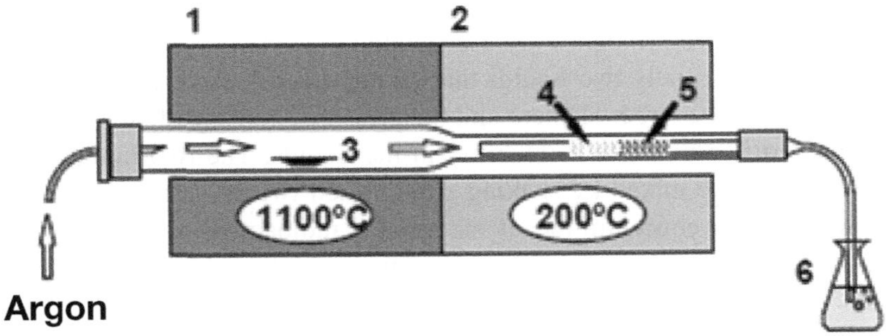

FIG. 4.16. A schematic description of the ^{76}Br separation equipment: (1) furnace, (2) auxiliary furnace, (3) irradiated target, (4) deposition area of selenium, (5) deposition area of ^{76}Br, (6) gas trap.

Separation yields of 60–70% are achieved by this method, with a separation time of about 1 h. Since dry distillation permits the extraction of radiobromine without destroying the target, the Cu_2Se targets are reusable. Considering the rather expensive ^{76}Se-enriched target material, this is a practical prerequisite for this type of production. The chemical form of the ^{76}Br activity after separation, analysed by ion exchange high performance liquid chromatography and thin-layer chromatography, was almost exclusively found to be bromide.

4.5.3. Radiation protection considerations and hot-box facilities

Besides the desired activity, the irradiated target usually contains a number of other radionuclides of varying elements, half-lives and γ energies. The presence of such contaminants needs to be taken into account when planning radiopharmaceutical labelling. An example is the production of ^{35}S using the reaction ^{35}Cl(n, p)^{35}S. At first glance, NaCl would be a suitable target due to the low atomic weight of sodium, a single isotope (^{23}Na) and a salt that is easy to dissolve. However, ^{23}Na has a huge thermal neutron cross-section for producing ^{24}Na, which has a half-life of 15 h and abundant γ energies up to 2.75 MeV. This target would be extremely hot, demanding lead protection more than 30 cm thick. If instead KCl were used, the emitted γ radiation energy would be substantially lower and decay times shorter.

After irradiation, the target is usually stored before processing to allow any short lived radionuclides to decay. Depending on the half-life, this ‘cooling period’ can be from minutes to months, but should not exceed one half-life of the desired radionuclide. The place used for this depends on the source activity and the energy and abundance of the γ emissions. Separation of fairly pure β and γ emitters may require just some distance and some plastic shielding, and can be performed in a standard fume hood, while targets with a high γ emission need significant lead shielding.

Handling reactor or accelerator produced radioactivity of the order of several hundred gigabecquerels requires adequate radiation protection, usually in the form of lead shields, hot-boxes, lead shielded fume hoods and laminar air flow benches. Typical lead thicknesses required by common radionuclides are indicated in Table 4.5.

The radioactive target and the radionuclide separation is often the first step in labelling a radiopharmaceutical. The hot-box then has to fulfil the requirements both to protect the operator from the radiation and to protect the pharmaceutical from the surroundings. The first step usually requires a negative pressure hood to prevent eventual airborne radioactivity to leak out into the laboratory, while the second step requires a high positive pressure to be applied across the pharmaceutical to avoid contact with less pure air from the laboratory.

TABLE 4.5. DOSE RATES AND LEAD SHIELDING REQUIRED FOR DIFFERENT RADIONUCLIDES, DETERMINED BY THE GAMMA RADIATION ABUNDANCE AND ENERGY[a]

	Dose rate (mSv/h) at 1 m per TBq			
	^{99m}Tc	^{111}In	^{18}F	^{124}I
	18	81	135	117
	Thickness of lead shield (cm) giving 1 μSv/h			
TBq	^{99m}Tc	^{111}In	^{18}F	^{124}I
0.1	0.28	1.0	5.8	20
1.0	0.36	1.3	7.1	22
10.0	0.43	1.6	8.5	27

[a] Calculations made with RadProCalculator (http://www.radprocalculator.com/).

These contradictory conditions are usually handled by having a box in the box, i.e. the pharmaceutical is processed in a closed facility at over pressure placed in the hot-box having low pressure. The classical hot-box design, with manipulators to manually process the radioactivity remotely, as seen in Fig. 4.17, is gradually being replaced by lead protected chambers housing an automatic chemistry system or a chemical robot making the pharmaceutical computer controlled.

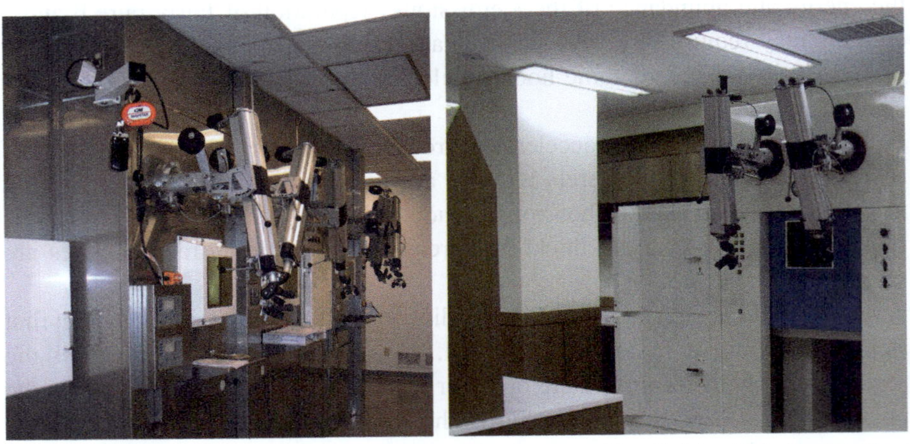

FIG. 4.17. Examples of modern hot-box designs (courtesy of Von Gahlen Nederland B.V.).

CHAPTER 5

STATISTICS FOR RADIATION MEASUREMENT

M.G. LÖTTER
Department of Medical Physics,
University of the Free State,
Bloemfontein, South Africa

5.1. SOURCES OF ERROR IN NUCLEAR MEDICINE MEASUREMENT

Measurement errors are of three general types: (i) blunders, (ii) systematic errors or accuracy of measurements, and (iii) random errors or precision of measurements.

Blunders produce grossly inaccurate results and experienced observers easily detect their occurrence. Examples in radiation counting or measurements include the incorrect setting of the energy window, counting heavily contaminated samples, using contaminated detectors for imaging or counting, obtaining measurements of high activities, resulting in count rates that lead to excessive dead time effects, and selecting the wrong patient orientation during imaging. Although some blunders can be detected as outliers or by duplicate samples and measurements, blunders should be avoided by careful, meticulous and dedicated work. This is especially important where results will determine the diagnosis or treatment of patients.

Systematic errors produce results that differ consistently from the correct results by some fixed amount. The same result may be obtained in repeated measurements, but overestimating or underestimating the true value. Systematic errors are said to influence the accuracy of measurements. Measurement results having systematic errors will be inaccurate or biased. Examples of a systematic error are:

- When an incorrectly calibrated ionization chamber is used for measurement of radiation dose.
- When during thyroid uptake studies with ^{123}I the count rate of the reference standard results in dead time losses. The percentage of thyroid uptake will be overestimated.
- When in sample counting the geometry of samples and the position within the detector are not the same as in the reference sample.

— When during blood volume measurements the tracer leaks out of the blood compartment. The theory of the method assumes that the tracer will stay in the blood compartment. The leaking of the tracer will consistently overestimate the measured blood volume.
— When in calculation of the ventricular ejection fraction during gated blood pool studies the selected background counts underestimate the true ventricular background counts, the ejection fraction will be consistently underestimated.

Measurement results affected by systematic errors are not always easy to detect, since the measurements may not be too different from the expected results. Systematic errors can be detected by using reference standards. For example, radionuclide standards calibrated at a reference laboratory should be used to calibrate source calibrators to determine correction factors for each radionuclide used for patient treatment and diagnosis.

Measurement results affected by systematic errors can differ from the true value by a constant value and/or by a fraction. Using 'golden standard' reference values, a regression curve can be calculated. The regression curve can be used to convert systematic errors to a more accurate value. For example, if the ejection fraction is determined by a radionuclide gated study, it can be correlated with the 'golden standard' values.

Random errors are variations in results from one measurement to the next, arising from actual random variation of the measured quantity itself, as well as physical limitations of the measurement system.

Random error affects the reproducibility, precision or uncertainty in the measurement. Random errors are always present when radiation measurements are performed because the measured quantity, namely the radionuclide decay, is a random varying quantity. The random error during radiation measurements introduced by the measured quantity, that is the radionuclide decay, is demonstrated in Fig. 5.1. Figure 5.1 shows the energy spectrum of a ^{57}Co source in a scattering medium and measured with a scintillation detector probe. The energy spectrum represented by square markers is the measured energy spectrum with random noise due to radionuclide decay. The solid line spectrum represents the energy spectrum without random noise. The variation around the solid line of the data points, represented by markers, is a result of random error introduced by radionuclide decay.

The influence of the random error of the measurement system introduced by the scintillation detector is also demonstrated in Fig. 5.1. Cobalt-57 emits photons of 122 keV and with a perfect detection system all of the counts are expected at 122 keV. The measurements are, however, spread around 122 keV as a result of the random error introduced by the scintillation detector during the detection of each γ photon. When a γ photon is detected with the scintillation detector, the number of charge carriers generated will vary randomly. The varying number of charge carriers will cause varying pulse heights at the output of the detector and this variation determines the spread around the true photon energy of 122 keV. The width of the photopeak determines the energy resolution of the detection system.

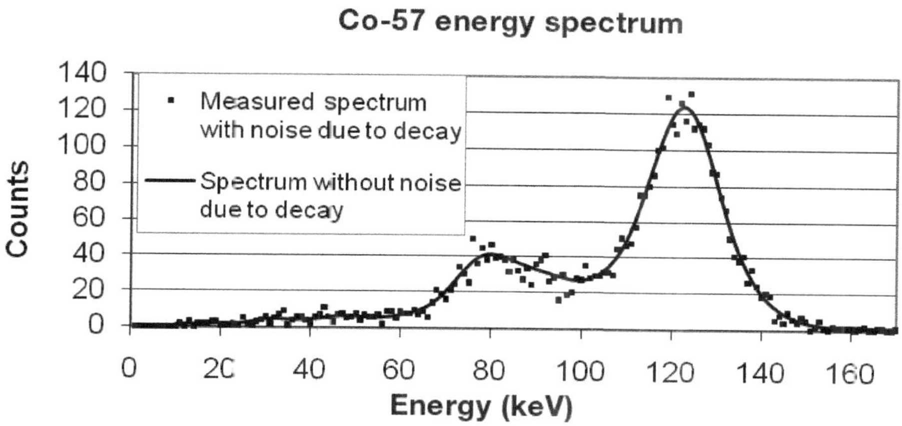

FIG. 5.1. Energy spectrum of a ^{57}Co source in a scattering medium obtained with a scintillation detector.

Random errors also play a significant role in radionuclide imaging. Here, the random error as a result of the measured quantity, namely radionuclide decay, will significantly influence the visual quality of the image. This is because the number of counts acquired in each pixel is subject to random error. It is shown that the relative random error decreases as the number of counts per pixel increases. The visual effect of the random error as a result of the measured quantity is demonstrated in Fig. 5.2. Technetium-99m planar bone scans (acquired on a 256 × 256 matrix) were acquired with a scintillation camera. Image acquisition was terminated at a total count of 21, 87 and 748 kcounts. When the total number of counts per image are increased, the counts per pixel increase and the random error decreases, resulting in improved visual image quality. As the accumulated counts are increased, the ability to visualize anatomical structures and, more importantly, tumour volumes, significantly increases. The random error introduced by the measuring system or imaging device, such as a scintillation camera, also influences image quality. This is as a result of the energy resolution and intrinsic spatial resolution of imaging devices that are influenced by random errors during the detection of each γ photon. The energy resolution of the system will determine the ability of the system to reject lower energy scattered γ photons and improve image contrast.

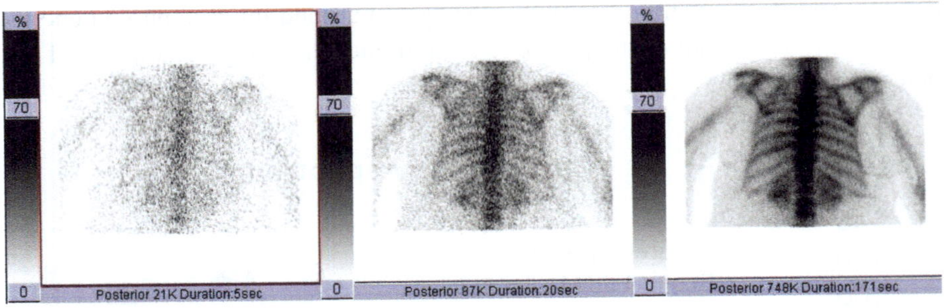

FIG. 5.2. The influence of random error as a result of radionuclide decay or counting statistics is demonstrated for imaging. Technetium-99m posterior planar bone images (256 × 256) using a scintillation camera were acquired to total counts of 21, 87 and 748 kcounts.

It is possible for a measurement to be precise (small random error) but inaccurate (large systematic error), or vice versa. For example, for the calculation of the ejection fraction during gated cardiac studies, the selection of the background region of interest (ROI) will be exactly reproducible when a software algorithm is used. However, if the algorithm is such that the selected ROI does not reflect the true ventricular background, the measurement will be precise but inaccurate. Conversely, individual radiation counts of a radioactive sample may be imprecise because of the random error, but the average value of a number of measurements will be accurate, representing the true counts acquired.

Random errors are always present and play a significant role in radiation counting and imaging. It is, therefore, important to analyse the random errors to determine the associated uncertainty. This is done using methods of statistical analysis. The remainder of the chapter describes methods of analysis.

The analysis of radiation measurements and imaging forms a subgroup of general statistical analysis. In this chapter, the focus is on statistical analysis for radiation counting and imaging measurements, although some methods described will be applicable to a wider class of experimental data as described in Sections 5.2, 5.3 and 5.5.

5.2. CHARACTERIZATION OF DATA

5.2.1. Measures of central tendency and variability

5.2.1.1. Dataset as a list

Two measurements of the central tendency of a set of measurements are the mean (average) and median. It is assumed that there is a list of N independent measurements of the same physical quantity:

$x_1, x_2, x_3, \ldots x_i \ldots\ldots x_N$

It is supposed that the dataset is obtained from a long lived radioactive sample counted repeatedly under the same conditions with a properly operating counting system. As the disintegration rate of the radioactive sample undergoes random variations from one moment to the next, the number of counts recorded in successive measurements is not the same as the result of random errors in the measurement.

The experimental mean $\bar{x}_e$ of the set of measurements is defined as:

$$\bar{x}_e = \frac{x_1 + x_2 + \ldots + x_N}{N} \tag{5.1}$$

$$= \frac{\sum_{i=1}^{N} x_i}{N} \tag{5.2}$$

The following procedure is followed to obtain the median. The list of measurements must first be sorted by size. The median is the middlemost measurement if the number of measurements is odd and is the average of the two middlemost measurements if the number of measurements is even. For example, to obtain the median of five measurements, 7, 13, 6, 10 and 14, they are first sorted by size: 6, 7, 10, 13 and 14. The median is 10. The advantage of the median over the mean is that the median is less affected by outliers. An outlier is a blunder and is much greater or much less than the others.

The measures of variability, random error and precision of a list of measurements are the variance, standard deviation and fractional standard deviation, respectively.

The variance σ_e^2 is determined from a set of measurements by subtracting the mean from each measurement, squaring the difference, summing the squares and dividing by one less than the number of measurements:

$$\sigma_e^2 = \frac{(x_1 - \overline{x}_e)^2 + (x_2 - \overline{x}_e)^2 + \dots\dots + (x_N - \overline{x}_e)^2}{N-1}$$
$$= \frac{1}{N-1}\sum_{i=1}^{N}(x_i - \overline{x}_e)^2 \quad (5.3)$$

where N is the total number of measurements and $\overline{x}_e$ is the experimental mean.

The standard deviation σ_e is the square root of the variance:

$$\sigma_e = \sqrt{\sigma_e^{\ 2}} \quad (5.4)$$

The fractional standard deviation σ_{eF} (fractional error or coefficient of variation) is the standard deviation divided by the mean:

$$\sigma_{eF} = \frac{\sigma_e}{\overline{x}_e} \quad (5.5)$$

The fractional standard deviation is an important measure to evaluate variability in measurements of radioactivity. The inverse of the fractional standard deviation $1/\sigma_{eF}$ in imaging is referred to as the signal to noise ratio.

5.2.1.2. Dataset as a relative frequency distribution function

It is often convenient to represent the dataset by a relative frequency distribution function $F(x)$. The value of $F(x)$ is the relative frequency with which the number appears in the collection of data in each bin. By definition:

$$F(x) = \frac{\text{number of occurrences of the value } x \text{ in each bin}}{\text{number of measurements } (N)} \quad (5.6)$$

The distribution is normalized, that is:

$$\sum_{x=0}^{\infty} F(x) = 1 \quad (5.7)$$

As long as the specific sequence of numbers is not important, the complete data distribution function represents all of the information in the original dataset in list format.

Figure 5.3 illustrates a demonstration of the application of the relative frequency distribution. The scintillation counter measurements appear noisy due to the random error as a result of the measured quantity (the radionuclide decay) (Fig. 5.3(a)). The measurements fluctuate randomly above and below the mean of 90 counts. A histogram (red bars) of the relative frequency distribution of the fluctuations in the measurements can be constructed by plotting the relative frequency of the measured counts (Fig. 5.3(b)). The x axis represents the range of possible counts that were measured in each of the bins with a six count interval. The y axis represents the relative frequencies with which the particular count values occur. The most common value, that is 26% of the measurements, is near the mean of 90 counts. The values of the other measurements are substantially higher or lower than the mean. The measured frequency distribution histogram agrees well with the expected calculated normal distribution (blue curve).

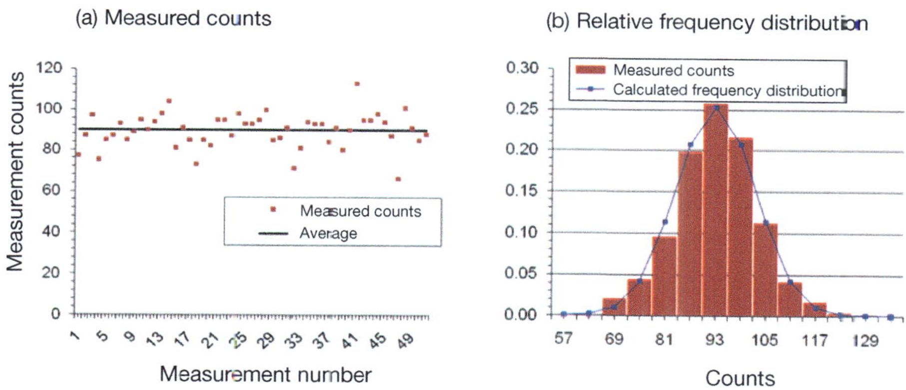

FIG. 5.3. One thousand measurements were made with a scintillation counter. (a) The graph shows the variations observed for the first 50 measurements. (b) The graph (red bars) shows the histogram of the relative frequency distribution for the measurements as well as the expected calculated frequency distribution.

The relative frequency distribution is a useful tool to provide a quick visual summary of the distribution of measurement values and can be used to identify outliers such as blunders or the correct functioning of equipment.

Three measurements of the central tendency for a frequency distribution are the mean (average), median and mode:

- — The mode of a frequency distribution is defined as the most frequent value or the value at the maximum probability of the frequency distribution.
- — The median of a frequency distribution is the value at which the integral of the frequency distribution is 0.5; that is, half of the measurements will be smaller and half will be larger than the median.

The experimental mean $\overline{x}_e$ using the frequency distribution function can be calculated. The experimental mean is obtained by calculating the first moment of the frequency distribution function. The equation for calculating the mean can also be derived from the equation for calculating the mean for data in a list (Eq. (5.2)). The sum of measurements $\sum_{i=1}^{N} x_i$ is equal to the sum of the measurements in each 'bin' in the frequency distribution function. The sum of the measurements for each bin is obtained by multiplying the value of the bin i and the number of occurrences of the value x_i.

$$\begin{aligned} \overline{x}_e &= \frac{\sum_{i=1}^{N} x_i}{N} \\ &= \sum_{i=1}^{N} \frac{[\text{value of bin } (i)]\ [\text{number of occurences of the value } x_i]}{N} \\ &= \sum_{i=1}^{N} x_i F(x_i) \end{aligned} \tag{5.8}$$

The experimental sample variance can be calculated using the frequency distribution function:

$$\sigma_e^2 = \frac{N}{N-1} \sum_{i=1}^{N} (x_i - \overline{x}_e)^2 F(x_i) \tag{5.9}$$

The standard deviation and the fractional standard deviation are given by Eqs (5.4) and (5.5).

The frequency distribution provides information and insight on the precision of the experimental sample mean and of a single measurement. Figure 5.3 demonstrates the distribution of counting measurements around the true mean $(\bar{x}_t)$. The value of the true mean is not known but the experimental sample mean $(\bar{x}_e)$ can be used as an estimate of the true mean $(\bar{x}_t)$:

$$(\bar{x}_t) \approx (\bar{x}_e) \tag{5.10}$$

In routine practice, it is often impractical to obtain multiple measurements and one must be satisfied with only one measurement. This is especially the case during radionuclide imaging and nuclear measurements on patients. The frequency distribution of the measurements will determine the precision of a single measurement as an estimate of the true value. The probability that a single measurement will be close to the true mean depends on the relative width or dispersion of the frequency distribution curve. This is expressed by the variance σ^2 (Eq. (5.9)) or standard deviation σ of the distribution. The standard deviation σ is a number such that 68.3% of the measurement results fall within $\pm\sigma$ of the true mean $\bar{x}_t$.

Given the result of a given measurement x, it can be said that there is a 68.3% chance that the measurement is within the range $x \pm \sigma$. This is called the 68.3% confidence interval for the true mean $\bar{x}_t$. There is 68.3% confidence that $\bar{x}_t$ is in the range $x \pm \sigma$. Other confidence intervals can be defined in terms of the standard deviation σ. They are summarized in Table 5.1. The 50% confidence interval (0.675σ) is referred to as the probable error of the true mean $\bar{x}_t$.

TABLE 5.1. CONFIDENCE LEVELS IN RADIATION MEASUREMENTS

Range	Confidence level for true mean $\bar{x}_t$
$\bar{x}_t \pm 0.675\sigma$	50.0
$\bar{x}_t \pm 1.000\sigma$	68.3
$\bar{x}_t \pm 1.640\sigma$	90.0
$\bar{x}_t \pm 2.000\sigma$	95.0
$\bar{x}_t \pm 3.000\sigma$	99.7

5.3. STATISTICAL MODELS

Under certain conditions, the distribution function that will describe the results of many repetitions of a given measurement can be predicted. A measurement is defined as counting the number of successes x resulting from a given number of trials n. Each trial is assumed to be a binary process in that only two results are possible: the trial is either a success or not. It is further assumed that the probability of success p is constant for all trials.

To show how these conditions apply in real situations, Table 5.2 gives four separate examples. The third example gives the basis for counting nuclear radiation events. In this case, a trial consists of observing a given radioactive nucleus for a period of time t. The number of trials n is equivalent to the number of nuclei in the sample under observation, and the measurement consists of counting those nuclei that undergo decay. We identify the probability of success as p. For radioactive decay:

$$p = (1 - e^{-\lambda t}) \tag{5.11}$$

where λ is the decay constant of the radionuclide.

The fifth example demonstrates the uncertainty associated with the energy determination during scintillation counting. The light photons generated in the scintillator following interaction with an incoming γ ray will eject electrons at the photocathode of the photomultiplier tube (PMT). Typically, one electron ejected for every five light photons results in a probability of success of 1/5.

TABLE 5.2. EXAMPLES OF BINARY PROCESSES

Trial	Definition of success	Probability of success p
Tossing a coin	Heads	1/2
Rolling a die	A six	1/6
Observing a given radionuclide for time t	The nucleus decays during observation	$(1 - e^{-\lambda t})$
Observing a given γ ray over a distance x in an attenuating medium	The γ ray interacts with the medium during observation	$(1 - e^{-\mu x})$
Observing light photons generated in a scintillator	An electron is ejected from the photocathode	1/5

5.3.1. Conditions when binomial, Poisson and normal distributions are applicable

Three statistical models are used: the binomial distribution, the Poisson distribution and the Gaussian or normal distribution. Figure 5.4 shows the distribution for the three models. The distributions were generated by using a Microsoft Office Excel spreadsheet.

5.3.1.1. *Binomial distribution*

This is the most general model and is widely applicable to all constant p processes (Fig. 5.4). Binomial distribution is rarely used in nuclear decay applications. One example in which the binomial distribution must be used is when a radionuclide with a very short half-life is counted with a high counting efficiency.

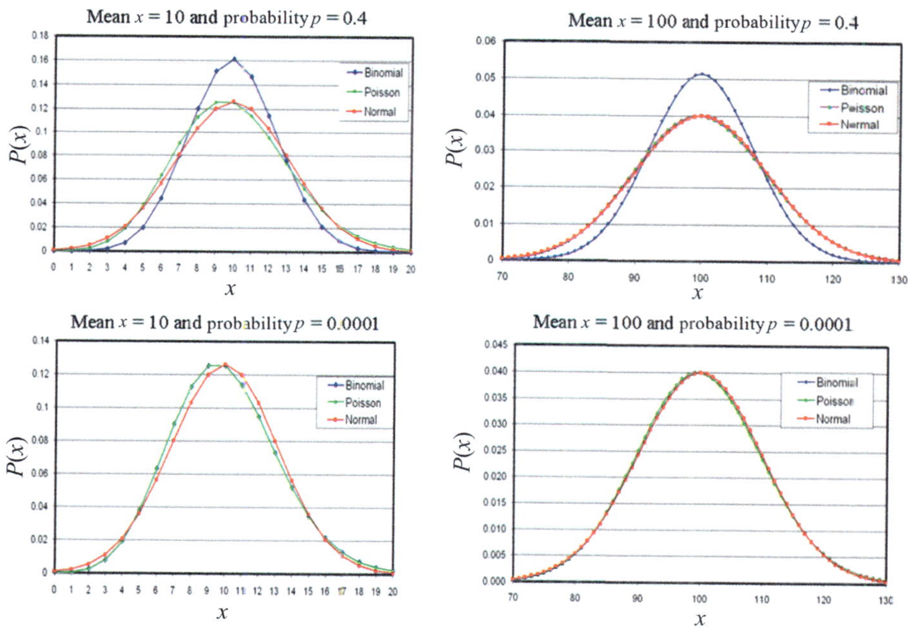

FIG. 5.4. Probability distribution models for successful event probability p = 0.4 and p = 0.0001 for $\bar{x}$ = 10 and $\bar{x}$ = 100, respectively.

5.3.1.2. *Poisson distribution*

The model is a direct mathematical simplification of the binomial distribution under conditions that the event probability of success p is small (Fig. 5.4). For nuclear counting, this condition implies that the chosen observation time is small compared to the half-life of the source, or that the detection efficiency is low. The Poisson distribution is an important distribution. When the success rate is low, the true experimental distribution is asymmetric and a Poisson distribution must then be used since the normal distribution is always symmetrical. This is demonstrated in Fig. 5.4 for $p = 0.0001$ and $\bar{x} = 10$.

5.3.1.3. Gaussian or normal distribution

The third important distribution is the normal or Gaussian, which is a further simplification if the mean number of successes $\bar{x}$ is relatively large (>30). At this level of success, the experimental distribution will be symmetrical and can be represented by the Gaussian distribution (Fig. 5.4). The Gaussian model is widely applicable to many applications in counting statistics.

It should be emphasized that the distribution of all of the above models becomes identical for processes with a small individual success probability p and with a large enough number of trials such that the expected mean number of successes $\bar{x}$ is large. This is demonstrated in Fig. 5.4 for $p = 0.0001$ and $\bar{x} = 100$.

5.3.2. Binomial distribution

Binomial distribution is the most general of the statistical models discussed. If n is the number of trials for which each trial has a success probability p, then the predicted probability of counting exactly x successes is given by:

$$P(x) = \frac{n!}{(n-x)!x!} p^x (1-p)^{n-x} \tag{5.12}$$

$P(x)$ is the predicted probability distribution function, and is defined only for integer values of n and x. The value of $n!$ (n factorial) is the product of integers up to n, that is $1 \times 2 \times 3 \times \ldots\ldots\ldots \times n$. The values of $x!$ and $(n - x)!$ are similarly calculated.

The properties of the binomial distribution are as follows:

— The distribution is normalized:

$$\sum_{x=0}^{n} P(x) = 1 \tag{5.13}$$

— The mean value $\bar{x}$ of the distribution using Eq. (5.8) is given by:

$$\bar{x} = \sum_{x=0}^{n} xP(x) \tag{5.14}$$

If Eq. (5.12) is substituted for $P(x)$, the mean value $\bar{x}$ of the distribution is given by:

$$\bar{x} = pn \tag{5.15}$$

The sample variance for a set of experimental data has been defined by Eq. (5.9). By analogy, the predicted variance σ^2 is given by:

$$\sigma^2 = \sum_{x=0}^{\infty} (x - \bar{x})^2 P(x) \tag{5.16}$$

If Eq. (5.12) is substituted for $P(x)$, the predicted variance σ^2 of the distribution will be:

$$\sigma^2 = np(1-p) \tag{5.17}$$

If Eq. (5.15) is substituted for np:

$$\sigma^2 = \bar{x}(1-p) \tag{5.18}$$

The standard deviation σ is the square root of the predicted variance σ^2:

$$\sigma = \sqrt{np(1-p)} = \sqrt{\bar{x}(1-p)} \tag{5.19}$$

The fractional standard deviation σ_F is given by:

$$\sigma_F = \frac{\sqrt{np(1-p)}}{np} = \sqrt{\frac{(1-p)}{np}}$$

$$\sigma_F = \frac{\sqrt{\bar{x}(1-p)}}{\bar{x}} = \sqrt{\frac{(1-p)}{\bar{x}}} \tag{5.20}$$

Equation (5.19) predicts the amount of fluctuation inherent in a given binomial distribution in terms of the basic parameters, namely the number of trials n and the success probability p, where $\bar{x} = pn$.

5.3.2.1. Application example of binomial distribution

The operation of a scintillation detector (Section 6.4) is considered. It consists of a scintillation crystal mounted on a PMT in a light tight construction. Firstly, when a γ ray interacts with the crystal, it generates n light photons.

Secondly, the light photons then eject x electrons from the photomultiplier photocathode. Thirdly, these electrons are then multiplied to form a pulse that can be further processed. For each γ ray that interacts with the scintillator, the number of light photons n, electrons ejected x and multiplication vary statistically during the detection of the different γ rays. This variation determines the energy resolution of the system.

In this example, the second stage is illustrated, that is the ejection of electrons from the photocathode. The variation or the standard deviation and fractional standard deviation for the number of electrons x that are ejected can be calculated using the binomial distribution as is given by Eqs (5.19) and (5.20).

The typical values for a scintillation counter are as follows. It is assumed that the 142 keV γ rays emitted by ^{99m}Tc are being counted. It is further assumed that it uses 100 eV to generate a light photon in the scintillation crystal when a γ ray interacts with the crystal. Therefore, if all of the energy of a single 142 keV photon is absorbed, n = 142 000/100 = 1420 light photons will be emitted. It is assumed that these light photons fall on the photocathode of the PMT to generate x electrons for each γ ray absorbed. It is further assumed that five light photons are required to eject one electron.

For the binomial distribution, the probability of a light photon ejecting an electron is p = 1/5 and the number of trials n will be the number of light photons generated for each γ ray. This will be 1420. Equation (5.15) can be used to calculate the predicted mean number of electrons ejected for each γ ray:

$$\bar{x} = pn = \frac{1}{5} \times 1420 = 284 \text{ electrons}$$

The standard deviation (Eq. (5.19)) and relative standard deviation (Eq. (5.20)) can be calculated using the binomial distribution:

$$\sigma = \sqrt{\bar{x}(1-p)} = \sqrt{284(1-1/5)} = 15$$

and

$$\sigma_{\mathrm{F}} = \sqrt{\frac{(1-p)}{\bar{x}}} = \sqrt{\frac{(1-1/5)}{284}} = 0.053 \tag{5.21}$$

Therefore, the contribution to the overall standard deviation at the electron ejection stage at the photocathode is 5.3%. The variation in the number of electrons will influence the pulse height obtained for each γ ray. The variation in the pulse height during the detection of γ rays will determine the width of the photopeak (Fig. 5.1) and the energy resolution of the system (Sections 5.7.1 and 6.4).

5.3.3. Poisson distribution

Many binary processes can be characterized by a low probability of success for each individual trial. This includes nuclear counting and imaging applications in which large numbers of radionuclides make up the sample or number of trials, but a relatively small fraction of these give rise to recorded counts. Similarly, during imaging, many γ rays are emitted by the administered imaging radionuclide, for every one that interacts with the tissue. In addition, during nuclear counting, many γ rays strike the detector for every single recorded interaction.

Under these conditions, the approximation that the probability p is small ($p \ll 1$) will hold and some mathematical simplifications can be applied to the binomial distribution. The binomial distribution reduces to the form:

$$P(x) = \frac{(pn)^x \mathrm{e}^{-pn}}{x!} \tag{5.22}$$

The relation $pn = \bar{x}$ holds for this distribution as well as for the binomial distribution:

$$P(x) = \frac{(\bar{x})^x \mathrm{e}^{-\bar{x}}}{x!} \tag{5.23}$$

Equation (5.23) is the form of the Poisson distribution.

For the calculation of binomial distribution, two parameters are required: the number of trials n and the individual success probability p. It is noted from Eq. (5.23) that only one parameter, the mean value $\bar{x}$, is required. This is a very useful simplification because using only the mean value of the distribution, all other values of the Poisson distribution can be calculated. This is of great help for processes in which the mean value can be measured or estimated, but for which there is no information about either the individual probability or size of the sample. This is the case in nuclear counting and imaging.

The properties of the Poisson distribution are as follows. The Poisson distribution is a normalized frequency distribution (see Eqs (5.6) and (5.7)):

$$\sum_{x=0}^{n} P(x) = 1 \tag{5.24}$$

The mean value or first moment for the Poisson distribution is calculated by inserting the Poisson distribution (Eq. (5.22)) into the equation to calculate the mean for a frequency distribution (Eq. (5.8)):

$$\overline{x} = \sum_{x=0}^{\infty} xP(x) = pn \tag{5.25}$$

This is the same result as was obtained for the binomial distribution.

The predicted variance of the Poisson distribution differs from that of the binomial distribution and can be derived from Eqs (5.9) and (5.22):

$$\sigma^2 = \sum_{x=0}^{\infty} (x - \overline{x})^2 P(x) = pn \tag{5.26}$$

From the result of Eq. (5.26), the predicted variance is reduced to the important general equation:

$$\sigma^2 = \overline{x} \tag{5.27}$$

The predicted standard deviation is the square root of the predicted variance (Eq. (5.4)):

$$\sigma = \sqrt{\sigma^2} = \sqrt{\overline{x}} \tag{5.28}$$

The predicted standard deviation of any Poisson distribution is just the square root of the mean value that characterizes the same distribution.

The predicted fractional standard deviation σ_F (fractional error or coefficient of variation) is the standard deviation divided by the mean (Eq. (5.5)):

$$\sigma_F = \frac{\sigma}{\overline{x}} = \frac{1}{\sqrt{\overline{x}}} = \frac{1}{\sigma} \tag{5.29}$$

The fractional standard deviation is the inverse of the square root of the mean value of the distribution.

Equations (5.28) and (5.29) are important equations and frequently find application in nuclear detection and imaging.

5.3.4. Normal distribution

The Poisson distribution holds as a mathematical simplification to the binomial distribution within the limit $p < 1$. If, in addition, the mean value of the distribution is large (>30), additional simplification can generally be carried out which leads to a normal or Gaussian distribution:

$$P(x) = \frac{1}{\sqrt{2\pi\overline{x}}} e^{\left(\frac{(x-\overline{x})^2}{2x}\right)} \tag{5.30}$$

The distribution function is only defined for integer values of x.

Figure 5.4 for $\overline{x} = 100$ and $p = 0.0001$ demonstrates that for these values the normal distribution is identical to the Poisson and binomial distributions. The normal distribution is always symmetrical or 'bell-shaped' (Fig. 5.4). It shares the following properties with the Poisson distribution:

— It is normalized (see Section 5.2.1.2 and Eqs (5.6) and (5.7)):

$$\sum_{x=0}^{n} P(x) = 1 \tag{5.31}$$

— The distribution is characterized by a single parameter $\overline{x} = pn$.

— The predicted variance of the normal distribution is given by the mean of x:

$$\sigma^2 = \overline{x} \tag{5.32}$$

— The predicted standard deviation is the square root of the predicted variance (Eq. (5.4)):

$$\sigma = \sqrt{\sigma^2} = \sqrt{\overline{x}} \tag{5.33}$$

— The predicted fractional standard deviation σ_F (fractional error or coefficient of variation) is the standard deviation divided by the mean (Eq. (5.5)):

$$\sigma_F = \frac{\sigma}{\overline{x}} = \frac{1}{\sqrt{\overline{x}}} = \frac{1}{\sigma} \tag{5.34}$$

The fractional standard deviation is the inverse of the square root of the mean value of the distribution.

5.3.4.1. Continuous normal distribution: confidence intervals

In experiments where the sample size is small, there are only a few discrete outcomes. As the sample size increase, so does the number of possible sample outcomes. As the sample size approaches infinity, there is, in effect, a continuous distribution of outcomes. In addition, some random variables, such as height and weight, are essentially continuous and have continuous distributions. In these situations, the probability of a single event is not small as was assumed for the discrete Poisson and normal distributions, and the equation $\sigma = \sqrt{\bar{x}}$ does not apply. The continuous normal distribution is given by:

$$P(x) = \frac{1}{\sigma\sqrt{2\pi}} e^{-\frac{1}{2}\left(\frac{x-\bar{x}}{\sigma}\right)^2} \tag{5.35}$$

The properties (Fig. 5.5) of the continuous normal distribution are:

- It is a continuous, symmetrical curve with both tails extending to infinity.
- All three measures of central tendency, mean, median and mode, are identical.
- It is described by two parameters: the arithmetic mean $\bar{x}$ and the standard deviation σ.

The mean $\bar{x}$ determines the location of the centre of the curve and the standard deviation σ represents the spread around the mean.

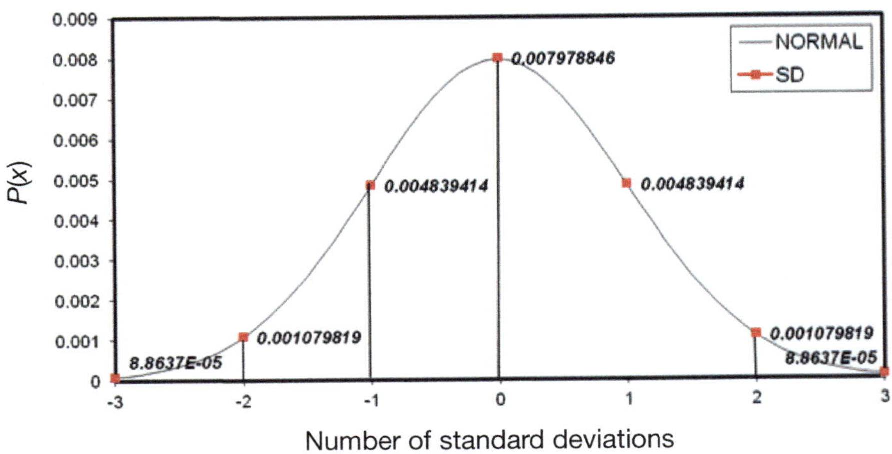

FIG. 5.5. The continuous normal distribution indicating the probability levels at different standard deviations (SDs) from the mean.

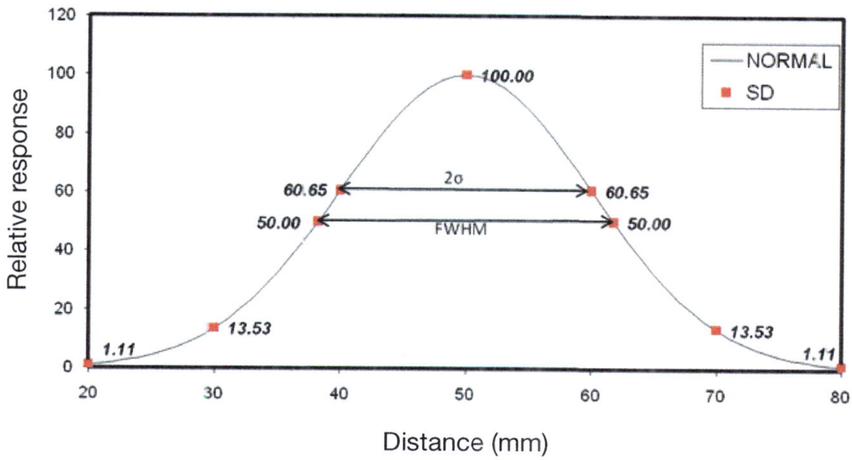

FIG. 5.6. Line source response curve obtained from a scintillation camera fitted to a normal distribution model. Image resolution is measured as the distance of the full width at half maximum (FWHM) of the percentage response. The standard deviation (SD) σ is the half width at a percentage response of 60.65%.

All continuous normal distributions have the property that between the mean and one standard deviation 68% is included on either side, between the mean and two standard deviations 95%, and between the mean and three standard deviations 99.7% of the total area under the curve.

5.3.4.2. Continuous normal distribution: applications in medical physics

The normal distribution is often used in radionuclide measurements and imaging to fit to experimental data. In this case, the equation is modified as follows:

$$P(x) = 100\mathrm{e}^{-\frac{1}{2}\left(\frac{x-\bar{x}}{\sigma}\right)^2} \tag{5.36}$$

where the maximum value of the distribution at $\bar{x}$ is normalized to 100.

The spatial resolution of imaging devices such as scintillation cameras and positron emission tomography equipment is determined as the full width at half maximum (FWHM) response of a normal distribution fitted to a point or line spread function (Fig. 5.6). The FWHM of the imaging device used in Fig. 5.6

was 23.6 mm. The relation for a normal distribution between the FWHM and standard deviation σ can be derived by setting $P(x) = 50$ and solving Eq. (5.36):

$$\text{FWHM} = 2.355\sigma \tag{5.37}$$

For the imaging system used in Fig. 5.6, the standard deviation $\sigma = 10$ mm. The value of the response $P(x)$ is 60.65% at a distance of $\sigma = \overline{x}$ (Eq. (5.36)). The value of the standard deviation σ can, therefore, also be obtained from the measured percentage response curve by finding the x value at a percentage response of 60.65% (Fig. 5.6).

In radionuclide energy spectroscopy, the photopeak distribution can be fitted to a normal distribution (Fig. 5.1). The energy resolution of scintillation detectors is expressed as the FWHM of the photopeak distribution divided by the photopeak energy E. The energy spectrum in medical physics applications is measured in kiloelectronvolts or megaelectronvolts. The fractional energy resolution R_E is:

$$R_E = \frac{\text{FWHM}}{E} = \frac{2.355\sigma}{E} \tag{5.38}$$

5.4. ESTIMATION OF THE PRECISION OF A SINGLE MEASUREMENT IN SAMPLE COUNTING AND IMAGING

5.4.1. Assumption

A valuable application of counting statistics applies to the case in which only a single measurement of a particular quantity is available and the uncertainty associated with that measurement is required. The square root of the sample variance σ should be a measure of the deviation of any one measurement from the true mean value and will serve as an index of the degree of precision that should be associated with a measurement from that set.

As only a single measurement is available, the sample variance cannot be calculated directly using Eqs (5.3) or (5.9) and must be estimated by analogy with an appropriate statistical model. The appropriate theoretical distribution can be matched to the available data if the measurement has been drawn from a population whose theoretical distribution function is predicted by either a Poisson or Gaussian distribution. As the value of the single measurement x is the only information available, it is assumed that the mean of the distribution is equal to the single measurement:

$$\overline{x} \approx x \tag{5.39}$$

Having obtained an assumed value for $\overline{x}$, the entire predicted probability distribution function $P(x)$ is defined for all values of x.

The expected sample variance s^2 can be expressed in terms of the variance of the selected statistical model:

$$s^2 = \sigma^2 = \overline{x} \approx x \tag{5.40}$$

Therefore, the best estimate of the deviation σ from the true mean, which should typify a single measurement x, is given by:

$$\sqrt{s} = \sigma \approx \sqrt{x} \tag{5.41}$$

To illustrate the application of Eq. (5.41), it is assumed that the probability distribution function is Gaussian with a large value for the measurement x. The range of values $x \pm \sigma$ or $x \pm \sqrt{x}$ will contain the true mean with 68% probability.

If it is assumed that there is a single measurement $x = 100$, then:

$$\sigma \approx \sqrt{x} = \sqrt{100} = 10$$

In Table 5.3, the various options available in quoting the uncertainty to be associated with the single measurement are shown. The conventional choice is to quote the measurement x plus or minus the standard deviation σ or 100 ± 10. This interval is expected to contain the true mean $\overline{x}$ with a probability of 68%. The probability that the true mean is included in the range can be increased by expanding the interval associated with the measurement as is shown in Table 5.3. For example, to achieve a 99% probability that the true mean is included, the interval must be expanded by 2.58σ. In the example, the range is then 100 ± 25.8.

When errors are reported, the associated probability level should be stated in the report under methods.

TABLE 5.3. EXAMPLES OF ERROR INTERVALS FOR A SINGLE MEASUREMENT $x = 100$

Interval (relative σ)	Interval (values)	Probability that the true mean $\bar{x}$ is included (%)
$x \pm 0.67\sigma$	93.3–106.7	50
$x \pm 1.00\sigma$	90.0–110.0	68
$x \pm 1.64\sigma$	83.6–116.4	90
$x \pm 2.00\sigma$	80.0–120.0	95
$x \pm 2.58\sigma$	74.2–125.8	99
$x \pm 3.00\sigma$	70.0–130.0	99.7

5.4.2. The importance of the fractional σ_F as an indicator of the precision of a single measurement in sample counting and imaging

The relation between the precision and a single counting measurement x is given by Eq. (5.40). The precision, expressed as the standard deviation σ, will increase proportionally to the square root of the measurement x. Thus, if the value of the single measurement x increases, the standard deviation will also increase. The increase in the standard deviation will be smaller than that of the measurement x. The relation between the standard deviation and the single measurement is best demonstrated by calculating the relative or fractional standard deviation σ_F:

$$\sigma_F = \frac{\sigma}{x} = \frac{\sqrt{x}}{x} = \frac{1}{\sqrt{x}} \tag{5.42}$$

Thus, the recorded number of counts or the value of the single measurement x completely determines the relative precision. The relative precision decreases as the number of counts increases. Therefore, to achieve a required relative precision, a minimum number of counts must be accumulated.

The following example illustrates the important relation between the relative precision and the number of counts recorded. If 100 counts are recorded, the relative standard deviation is 10%. If 10 000 counts are recorded, the relative standard deviation reduces to 1%. This example demonstrates the importance of acquiring enough counts to meet the required precision.

It is easier to achieve the required precision when samples in counting tubes are measured than when in vivo measurements on patients are performed. The single measurement from a high count rate radioactive sample in a counting tube will be obtained in a short time. However, if a low activity sample is measured, the measurement time will have to be increased to achieve the desired precision. The desired precision can be conveniently obtained by using automatic sample counters. These counters can be set to stop counting after a preset time or preset counts have been reached. By choosing the preset count option, the desired precision can be achieved for each sample.

The acquisition time of in vivo measurements using collimated detector probes, such as thyroid iodine uptake studies or imaging studies, can often not be increased to achieve the desired precision as a result of patient movement. In these single measurements, high sensitivity radiation detectors or a higher, but acceptable radioactive dose, can be selected.

The precision of a single measurement is very important during radionuclide imaging. If the number of counts acquired in a picture element or pixel is low, a low precision is obtained. There will then be a wide range of fluctuations between adjacent pixels. As a result of the poor quality of the images, it would only be possible to identify large defect volumes or defects with a high contrast. To detect a defect, the measured counts from the defect must lie outside the range of the background measurement plus or minus two standard deviations ($x \pm 2\sigma$). During imaging, the number of counts measured in a target volume will be determined by the acquisition time, activity within the target volume and the sensitivity of the measuring equipment. The sensitivity of imaging equipment can be increased by increasing the FWHM spatial resolution. There is a trade-off between single sample counting precision and the spatial resolution of the imaging device to obtain images that would provide the maximum diagnostic value during visual interpretation of the images by nuclear medicine physicians.

Counting statistics are also very important during image quantification such as measuring renal function, left ventricular ejection fraction and tumour uptake. During quantification, the accumulated counts by an organ or within a target volume have to be accurately determined. In quantification studies, the background activity, attenuation and scatter contributions have to be corrected. These procedures further reduce the precision of quantification.

5.4.3. Caution on the use of the estimate of the precision of a single measurement in sample counting and imaging

All conclusions are based on the measurement of a counted number of success (number of heads in coin tossing). In nuclear measurements or imaging, the estimate of the precision of a single measurement by using $\sigma = \sqrt{x}$ can only be applied if x represents a counted number of success, that is the number of events recorded in a given observation time.

The estimate of the precision of a single measurement by using $\sigma = \sqrt{x}$ cannot be used if x is not a directly measured count. For example, the association does not apply to:

— Counting rates;
— Sums or differences of counts;
— Averages of independent counts;
— Pixel counts following tomographic image reconstruction;
— Any derived quantity.

In these cases, the quantity is calculated as a function of the number of counts recorded. The error to be associated with that quantity must be calculated according to the error propagation methods outlined in the next section.

5.5. PROPAGATION OF ERROR

The preceding section described methods for estimating random error or the precision of a single measurement during nuclear measurements or imaging. Most procedures in nuclear medicine involve multiple nuclear measurements and imaging procedures for the calculation of results such as thyroid iodine uptake, ejection fraction, renal clearance, blood volume or red cell survival time, on which clinical diagnosis is based. Similarly, internal dosimetry is performed using nuclear measurements and imaging data. To estimate the corresponding precision in the derived quantity, how the error associated with the initial measurements propagates through the calculations that were performed to arrive at the required result has to be followed. This is done by applying the error of propagation formulas. The variables used in the calculation of errors must be independent to avoid effects of correlation. It is assumed that the error in nuclear measurements arises only from random fluctuations in the decay rate and is statistically independent of other errors.

The error of propagation formulas applies to measurements that are obtained from a continuous distribution as well as to Poisson and discrete normal distributions. The measurements from continuous distributions will be represented by x_1, x_2, x_3... with variances of $\sigma(x_1)^2, \sigma(x_2)^2, \sigma(x_3)^2$... These equations can be used to estimate precision in measurements such as height and weight.

Discrete nuclear measurements with Poisson or normal distribution are represented by N_1, N_2, N_3... with variances of $\sigma(N_1)^2, \sigma(N_2)^2, \sigma(N_3)^2$... or N_1, N_2, N_3...

5.5.1. Sums and differences

The product x_s of the sums or difference of a series of measurements with a continuous normal distribution is given by:

$$x_s = x_1 \pm x_2 \pm x_3 \ldots \tag{5.43}$$

The variance of x_s is given by:

$$\sigma(x_1 \pm x_2 \pm x_3 \ldots)^2 = \sigma(x_1)^2 + \sigma(x_2)^2 + \sigma(x_3)^2 \ldots \tag{5.44}$$

The standard deviation is given by:

$$\sigma(x_1 \pm x_2 \pm x_3 \ldots) = \sqrt{\sigma(x_1)^2 + \sigma(x_2)^2 + \sigma(x_3)^2 \ldots} \tag{5.45}$$

The fractional standard deviation is given by:

$$\sigma_F(x_1 \pm x_2 \pm x_3 \ldots) = \frac{\sqrt{\sigma(x_1)^2 + \sigma(x_2)^2 + \sigma(x_3)^2 \ldots}}{x_1 \pm x_2 \pm x_3 \ldots} \tag{5.46}$$

For counting measurements or measurements with a Poisson or discreet normal distribution, the variance is given by:

$$\sigma(N_1 \pm N_2 \pm N_3 \ldots)^2 = N_1 \pm N_2 \pm N_3 \ldots \tag{5.47}$$

The standard deviation is given by:

$$\sigma(N_1 \pm N_2 \pm N_3 \ldots) = \sqrt{N_1 + N_2 + N_3 \ldots} \tag{5.48}$$

The fractional standard deviation is given by:

$$\sigma_F(N_1 \pm N_2 \pm N_3 \ldots) = \frac{\sqrt{N_1 + N_2 + N_3 \ldots}}{N_1 \pm N_2 \pm N_3 \ldots} \tag{5.49}$$

These equations apply to mixed combinations of sums and differences.

TABLE 5.4. UNCERTAINTY AFTER SUMMING AND SUBTRACTING COUNTS

	$N_2 \ll N_1$			$N_2 \approx N_1$		
	N	σ	σ_F	N	σ	σ_F
N_1	500	22.4	0.0447	500	22.4	0.0447
N_2	10	3.2	0.3162	450	21.2	0.0471
$N_1 - N_2$	490	22.6	0.0461	50	30.8	0.6164
$N_1 + N_2$	510	22.6	0.0443	950	30.8	0.0324

The influence on the standard deviation and fractional standard deviation of summing and subtracting values N_1 and N_2 is demonstrated in Table 5.4. The following conclusions can be drawn:

— The standard deviation σ for $N_1 - N_2$ and $N_1 + N_2$ is the same for the same values of N_1 and N_2, but the fractional standard deviation σ_F is different;
— The fractional standard deviation for differences is large when the differences between the values are small.

This is the reason why it is important to limit the background to a value as low as possible in counting procedures. In imaging, when scatter or background correction is performed by subtraction, image quality deteriorates as a result of the increased uncertainty in the pixel values.

5.5.2. Multiplication and division by a constant

We define:

$$x_M = Ax \quad (5.50)$$

where A is a constant.

Then:

$$\sigma_M = A\sigma_x \quad (5.51)$$

and

$$\sigma_{\mathrm{F}} = \frac{A\sigma_x}{Ax} = \frac{\sigma_x}{x} \tag{5.52}$$

For counting measurements or measurements with a Poisson or discreet normal distribution, the following applies:

$$x_{\mathrm{M}} = AN \tag{5.53}$$

Then:

$$\sigma_{\mathrm{M}} = A\sqrt{N} \tag{5.54}$$

and

$$\sigma_{\mathrm{F}} = \frac{1}{\sqrt{N}} \tag{5.55}$$

Similarly, if:

$$x_{\mathrm{D}} = \frac{x}{B} \tag{5.56}$$

where B is also a constant:

$$\sigma_{\mathrm{M}} = \frac{\sigma_x}{B} \tag{5.57}$$

and

$$\sigma_{\mathrm{F}} = \frac{\sigma_x}{B}\frac{B}{x} = \frac{\sigma_x}{x} \tag{5.58}$$

For counting measurements or measurements with a Poisson or discreet normal distribution, the following apply:

$$x_{\mathrm{D}} = \frac{N}{B} \tag{5.59}$$

$$\sigma_{\mathrm{M}} = \frac{\sqrt{N}}{B} \tag{5.60}$$

and

$$\sigma_F = \frac{1}{\sqrt{N}} \tag{5.61}$$

It should be noted that multiplying (Eqs (5.52) and (5.55)) or dividing (Eqs (5.58) and (5.61)) a value by a constant does not change the fractional standard deviation.

5.5.3. Products and ratios

The uncertainty in the product or ratio of a series of measurements x_1, x_2, x_3... is expressed in terms of the fractional uncertainties in the individual results, $\sigma_F(x_1)$, $\sigma_F(x_2)$, $\sigma_F(x_3)$...

The product x_P of the products or ratios of a series of measurements with a continuous normal distribution is given by:

$$x_P = x_1 \overset{\times}{\div} x_2 \overset{\times}{\div} x_3 \overset{\times}{\div} \dots\dots\dots\dots\dots \tag{5.62}$$

The notation $\overset{\times}{\div}$ means $x_1 \times x_2$ or $x_1 \div x_2$. These equations apply to mixed combinations of sums and differences.

The fractional variance of x_P is given by:

$$\sigma_F(x_1 \overset{\times}{\div} x_2 \overset{\times}{\div} x_3 \overset{\times}{\div} \dots\dots)^2 = \sigma_F(x_1)^2 + \sigma_F(x_2)^2 + \sigma_F(x_3)^2 \dots\dots \tag{5.63}$$

The fractional standard deviation is given by:

$$\sigma_F(x_1 \overset{\times}{\div} x_2 \overset{\times}{\div} x_3 \overset{\times}{\div} \dots\dots) = \sqrt{\sigma_F(x_1)^2 + \sigma_F(x_2)^2 + \sigma_F(x_3)^2 \dots\dots} \tag{5.64}$$

The standard deviation is given by:

$$\sigma(x_1 \overset{\times}{\div} x_2 \overset{\times}{\div} x_3 \overset{\times}{\div} \dots) = \sqrt{\sigma_F(x_1)^2 + \sigma_F(x_2)^2 + \sigma_F(x_3)^2 \dots} \times (x_1 \overset{\times}{\div} x_2 \overset{\times}{\div} x_3 \overset{\times}{\div} \dots) \tag{5.65}$$

For counting measurements or measurements with a Poisson or discreet normal distribution, the product or ratio is given by:

$$N_P = N_1 \overset{\times}{\div} N_2 \overset{\times}{\div} N_3 \overset{\times}{\div} \dots\dots \tag{5.66}$$

The fractional variance of N_P is given by:

$$\sigma_F(N_1 \overset{\times}{\div} N_2 \overset{\times}{\div} N_3 \overset{\times}{\div} \dots\dots)^2 = \frac{1}{N_1} + \frac{1}{N_2} + \frac{1}{N_3} + \dots\dots \tag{5.67}$$

The fractional standard deviation is given by:

$$\sigma_F(N_1 \overset{\times}{\div} N_2 \overset{\times}{\div} N_3 \overset{\times}{\div} \ldots\ldots) = \sqrt{\frac{1}{N_1} + \frac{1}{N_2} + \frac{1}{N_3} + \ldots\ldots} \tag{5.68}$$

The standard deviation is given by:

$$\sigma(N_1 \overset{\times}{\div} N_2 \overset{\times}{\div} N_3 \overset{\times}{\div} \ldots\ldots) = \sqrt{\frac{1}{N_1} + \frac{1}{N_2} + \frac{1}{N_3} + \ldots\ldots} (N_1 \overset{\times}{\div} N_2 \overset{\times}{\div} N_3 \overset{\times}{\div} \ldots\ldots) \tag{5.69}$$

5.6. APPLICATIONS OF STATISTICAL ANALYSIS

5.6.1. Multiple independent counts

5.6.1.1. Sum of multiple independent counts

If it is supposed that there are n repeated counts from the same source for equal counting times and the results of the measurements are N_1, N_2, N_3....... N_n and their sum is N_s, then:

$$N_s = N_1 + N_2 + N_3 \ldots \tag{5.70}$$

According to the propagation of error for sums and Eq. (5.48):

$$\sigma_{N_s} = \sqrt{N_1 + N_2 + N_3 \ldots} = \sqrt{N_s} \tag{5.71}$$

The results show that the standard deviation for the sum of all counts is the same as if the measurement had been carried out by performing a single count, extending over the period represented by all of the counts.

5.6.1.2. Mean value of multiple independent counts

If the mean value $\bar{N}$ of the n independent counts referred to in the previous section is calculated, then:

$$\bar{N} = \frac{N_s}{n} \tag{5.72}$$

Equation (5.72) is an example of dividing an error-associated quantity $\bar{N}$ by a constant n. Equation (5.51), therefore, applies and the standard deviation of the mean or standard error is given by:

$$\sigma_{\bar{N}} = \frac{\sigma_{N_s}}{n} = \frac{\sqrt{N_s}}{n} = \frac{\sqrt{n\bar{N}}}{n} = \sqrt{\frac{\bar{N}}{n}} \tag{5.73}$$

It should be noted that the standard deviation for a single measurement N_i (Eq. (5.41)) is $\sigma_{N_i} = \sqrt{N_i}$.

A typical count will not differ greatly from the mean $N_i \approx \bar{N}$. Thus, the mean value based on n independent counts will have an expected error that is smaller by a factor of $\sqrt{n}$ compared with any single measurement on which the mean is based. To improve the statistical precision of a given measurement by a factor of two, the counting time must, therefore, be increased four times.

5.6.2. Standard deviation and relative standard deviation for counting rates

If N counts are accumulated over time t, then the counting rate R is given by:

$$R = \frac{N}{t} \tag{5.74}$$

In the above equation, it is assumed that the time t is measured with a very small uncertainty, so that t can be considered a constant. The calculation of the uncertainty associated with the counting rate is an application of the propagation of errors, multiplying by a constant (Eq. (5.60)):

$$\sigma_R = \frac{\sigma_x}{t} = \frac{\sqrt{N}}{t} = \sqrt{\frac{R}{t}} \tag{5.75}$$

The fractional standard deviation is calculated using Eq. (5.61):

$$\sigma_F = \frac{\sigma_x}{tR} = \frac{\sqrt{N}}{tR} = \sqrt{\frac{1}{tR}} \tag{5.76}$$

The above equations illustrate the calculation of uncertainties if calculations are required to obtain a value, and the equation for a single value (Section 5.3) cannot be applied. The following example illustrates the use of the equations.

5.6.2.1. Example: comparison of error of count rates and counts accumulated

The activity of two samples is measured. Sample 1 is counted with a counter that is set to stop when a count of 10 000 is reached. It takes 100 s to reach 10 000 counts. Sample 2 is counted using an automatic sample changer. The activity of the sample is given as 10 000 counts per second (cps) and the sample was counted for 100 s.

Calculating the counting error associated with the measurements of samples 1 and 2:

Sample 1:

The counts acquired: $N = 10\,000$ counts
Standard deviation (Eq. (5.41)): $\sigma_N = 100$ counts
Fractional standard deviation (Eq. (5.42)): $\sigma_{\mathrm{F}} = 0.01 = 1\%$

Sample 2:

The count rate: 10 000 cps
Standard deviation (Eq. (5.75)): $\sigma_R = \sqrt{\dfrac{10\,000}{100}} = 10$ cps

Fractional standard deviation (Eq. (5.76)):

$$\sigma_{\mathrm{F}} = \sqrt{\frac{1}{10\,000 \times 100}} = 0.001 = 0.1\%$$

Although the counts acquired for sample 1 and the count rate of sample 2 were numerically the same, the uncertainties associated with the measurements were very different. When calculations on counts are performed, it must be determined whether the value is a single value or whether it is a value that has been obtained by calculation.

5.6.3. Effects of background counts

Background counts are those counts that do not originate from the sample or target volume or are unwanted counts such as scatter. The background counts during sample counting consist of electronic noise, detection of cosmic rays, natural radioactivity in the detector, and down scatter radioactivity from non-target radionuclides in the sample. During in vivo measurements, such as measurement of thyroid iodine uptake or left ventricular ejection fraction, radiation from non-target tissue will also contribute to background. Scattered radiation from target as well as non-target tissue will influence quantification and will be included in the background. To obtain the true net counts, the background is subtracted from the gross counts accumulated. The uncertainty of the true target counts can be calculated using Eqs (5.48) and (5.49), and the uncertainty of true count rates can be calculated using Eqs (5.75) and (5.76).

If the background count is N_b, and the gross counts of the sample and background is N_g, then the net sample count N_s is:

$$N_s = N_g - N_b \tag{5.77}$$

The standard deviation for N_s counts is given by Eq. (5.48):

$$\sigma(N_s) = \sqrt{N_g + N_b} \tag{5.78}$$

The fractional standard deviation for N_s counts is given by Eq. (5.49):

$$\sigma_F(N_s) = \frac{\sqrt{N_g + N_b}}{N_g - N_b} \tag{5.79}$$

If the background count rate is R_b, acquired in time t_b, and the gross count rate of the sample and background is R_g, acquired in time t_g, then the net sample count rate R_s is:

$$R_s = R_g - R_b \tag{5.80}$$

The standard deviation for a count rate R_s is given by Eqs (5.45) and (5.75):

$$\sigma(R_s) = \sqrt{\frac{R_g}{t_g} + \frac{R_b}{t_b}} \tag{5.81}$$

The fractional standard deviation for a count rate R_s is given by Eqs (5.46) and (5.76):

$$\sigma_F(R_s) = \frac{\sqrt{\frac{R_g}{t_g} + \frac{R_b}{t_b}}}{R_g - R_b} \tag{5.82}$$

If the same counting time t is used for both sample and background measurement:

$$\sigma(R_s) = \frac{\sqrt{R_g + R_b}}{\sqrt{t}} \tag{5.83}$$

and

$$\sigma_F(R_s) = \frac{\sqrt{R_g + R_b}}{\sqrt{t}(R_g - R_b)} \tag{5.84}$$

5.6.3.1. Example: error in net target counts following background correction

The following example illustrates the application to determine the uncertainty in the measurement of target volume counts following background correction. A planar image of the liver is acquired for the detection of tumours. Two equal sized ROIs, ROI1 and ROI2, were selected to cover the areas of the two potential tumours. The gross counts N_g in ROI1 were 484 counts (Table 5.5) and in ROI2 484 counts. The background counts N_b selected over normal tissue of the same area as for the gross counts were 441 and 169 counts. How to calculate the uncertainties in the tumor volume net counts is presented.

The difference and error associated with the difference (Eq. (5.77) – Eq. (5.79)) when $N_g \approx N_b$ are:

$$N_g - N_b = 484 - 441 = 43 \text{ counts}$$

$$\sigma(N_g - N_b) = \sqrt{484 + 441} = 30.4 \text{ counts}$$

$$\sigma_F(N_g - N_b) = \frac{\sqrt{484 + 441}}{484 - 441} = 0.7073$$

$$\sigma_P(N_g - N_b) = 70.7\%$$

The influence on the standard deviation and fractional standard deviation of background correction for $N_g \approx N_b$ and $N_g \gg N_b$ is demonstrated in Table 5.5. The following conclusion can be drawn: the fractional σ_F and percentage σ_P standard deviations significantly increase when the background increases relative to the net counts.

This is the reason why it is important in measurements of radioactivity to acquire as many counts as possible to decrease the uncertainty in detection of target volume radioactivity. The following example illustrates the application to determine the uncertainty in the measurement of target volume count rate following background correction.

TABLE 5.5. CALCULATION OF UNCERTAINTIES IN COUNTS AS A RESULT OF BACKGROUND CORRECTION

$N_g \approx N_b$					$N_g \gg N_b$				
Source	Counts	σ counts	σ_F	σ_P (%)	Source	Counts	σ counts	σ_F	σ_P (%)
N_g	484	22.0	0.0455	4.5	N_g	484	22.0	0.0455	4.5
N_b	441	21.0	0.0476	4.8	N_b	169	13.0	0.0769	7.7
N_s	43	30.4	0.7073	70.7	N_s	315	25.6	0.0811	8.1
$3\sigma(N_s)$	91	Counts	Not significant		$3\sigma(N_s)$	77	Counts	Significant	

5.6.3.2. Example: error in net target count rate following background correction

A planar image of the liver is acquired for the detection of tumours. Two equal sized ROIs, ROI1 and ROI2, were selected to cover the areas of the two potential tumours. The gross count rate R_g in ROI1 was 484 counts per minute (cpm) (Table 5.6) and in ROI2 484 cpm. The background count rates R_b selected over normal tissue of the same area as for the gross counts were 441 and 169 cpm. The acquisition time of the image was 2 min. How to calculate the uncertainties in the tumor volume net counts is presented.

The difference and error associated with the difference (Eq. (5.80) – Eq. (5.82)) when $R_g \approx R_b$ are:

$$R_g - R_b = 484 - 441 = 43 \text{ cpm}$$

$$\sigma(R_g - R_b) = \sqrt{\frac{484}{2} + \frac{441}{2}} = 21.5 \text{ cpm}$$

$$\sigma_F(R_g - R_b) = \frac{\sqrt{\frac{484}{2} + \frac{441}{2}}}{484 - 441} = 0.5001$$

$$\sigma_P(R_g - R_b) = 50.0\%$$

The influence on the standard deviation and fractional standard deviation of background correction for $R_g \approx R_b$ and $R_g \gg R_g$ is demonstrated in Table 5.6.

Again, it is shown that the fractional standard deviation σ_F significantly increases when the background count rate increases relative to the net target count rate.

TABLE 5.6. CALCULATION OF UNCERTAINTIES IN COUNT RATES AS A RESULT OF BACKGROUND CORRECTION

$R_g \approx R_b$					$R_g \gg R_b$				
Source	Count rate (cpm)	σ (cpm)	σ_F	σ_P (%)	Source	Count rate (cpm)	σ_F (cpm)	σ_F	σ_P (%)
R_g	484	15.6	0.0321	3.2	R_g	484	15.6	0.0321	3.2
R_b	441	14.8	0.0337	3.4	R_b	169	9.2	0.0544	5.4
R_s	43	21.5	0.5001	50.0	R_s	315	18.1	0.0574	5.7
t	2	Minutes			t	2	Minutes		
$3\sigma\,(R_s)$	65	Not significant			$3\sigma\,(R_s)$	54	Significant		

5.6.4. Significance of differences between counting measurements

If N_1 and N_2 counts are measured in two counting measurements, the difference $(N_1 - N_2)$ between the measured counts may be a result of random variations in the counting rate or may be as a result of an actual difference. The statistical significance of the difference is evaluated by comparing it to the expected random error expressed as the standard deviation σ_d of the difference. If $(N_1 - N_2) > 2\sigma(N_1 - N_2)$, there is a 5% chance that the difference is caused by random error (see Table 5.3). If:

$$N_1 - N_2 > 3\sigma(N_1 - N_2) \tag{5.85}$$

there is a 0.3% chance that the difference is caused by random error and this difference is considered significant.

The examples in the previous section to determine whether tumours were present following a liver scan illustrate the application to determine the significance of the difference between two counts (Table 5.5). The net counts and uncertainty over two tumour areas were calculated. Do the counts over the tumour areas significantly differ from the normal background area?

For the difference for $N_g \approx N_b$ (Table 5.5) to be significant, Eq. (5.85) must apply.

The difference of 43 cpm was less than the norm of $3\sigma(N_1 - N_2)$ and the difference is, therefore, not significant. It can be concluded with a smaller than 0.3% chance that there is not a tumour present.

An example when $N_g \gg N_b$ is also given in Table 5.5. In this case, the 315 cpm counts difference was larger than $3\sigma(N_1 - N_2)$ of 77 cpm. The difference in this case is significant. It can be concluded with a smaller than 0.3% chance that there is a tumour present.

The significance of differences between the counting rates of samples can also be calculated. Two counting rates, R_1 and R_2, are acquired using counting times t_1 and t_2.

The uncertainty associated with the difference is given by applying Eqs (5.45) and (5.75):

$$\sigma(R_1 - R_2) = \sqrt{\frac{R_1}{t_1} + \frac{R_2}{t_2}} \tag{5.86}$$

For the difference $R_1 - R_2$ to be significant:

$$R_1 - R_2 > 3\sigma(R_1 - R_2) \tag{5.87}$$

The examples in the previous section (Table 5.6) to determine whether tumours were present following a liver scan illustrate an application to determine the significance of the difference between two count rates. The net count rate and uncertainty over two tumour areas were calculated. Do the count rates over the tumour areas significantly differ from the normal background area?

For the difference for $R_g \approx R_b$ (Table 5.6) to be significant, Eq. (5.87) must apply.

The difference count rate of 43 cpm was less than the 65 cpm which is the norm of $3\sigma(R_1 - R_2)$ and the difference is, therefore, not significant. It can be concluded with a smaller than 0.3% chance that there is not a tumour present.

An example when $R_g \gg R_b$ is also given in Table 5.6. In this case, the difference of 315 cpm was larger than $3\sigma(R_1 - R_2)$ which was 54 cpm. The difference in this case is significant. It can be concluded with a smaller than 0.3% chance that there is a tumour present.

5.6.5. Minimum detectable counts, count rate and activity

According to Eq. (5.85), if the difference of two measurements is larger than three standard deviations, the difference is considered significant. Therefore, the minimum net counts N_m that can be detected with 0.3% confidence is given by:

$$N_m = N_1 - N_2 = 3\sigma(N_1 - N_2) \tag{5.88}$$

or

$$N_m = N_g - N_b = 3\sigma(N_g - N_b) \tag{5.89}$$

Solving this equation for N_g will give the minimum detectable gross counts N_m:

$$N_g = \frac{(2N_b + 9) + \sqrt{72N_b + 81}}{2} \tag{5. 90}$$

An approximation can be used by assuming that $N_g \approx N_b$ and:

$$N_g \approx N_b + 3\sqrt{2N_b} \tag{5.91}$$

The minimum detectable activity A_m can be calculated:

$$A_m = \frac{N_m}{tS} \tag{5.92}$$

where

S is the sensitivity of the detection system usually expressed as count rate per becquerel;

and t is the time that the background was counted.

5.6.5.1. Example: calculation of minimum activity that can be detected

A detector is to be used to detect ^{131}I in the thyroid of radiation workers. The background count was 441 counts measured over a period of 5 min. The acquisition time for the thyroid was also 5 min. The sensitivity of the counter was 0.1 counts · s^{-1} · Bq^{-1}. What is the minimum activity that can be detected?

From Eq. (5.90):

$$N_g = \frac{(2N_b + 9) + \sqrt{72N_b + 81}}{2} = \frac{(2 \times 441 + 9) + \sqrt{72 \times 441 + 81}}{2} = 535 \text{ counts}$$

It should be noted that $N_g - N_b = 94$ counts and $3\sigma(N_g - N_b) = 94$ as was specified in Eq. (5.85). The minimum detectable radioactivity is:

$$A_m = \frac{(535-441)}{5\times 60\times 0.1} = 3.124 \text{ Bq}$$

The minimum detectable net count rate R_m is given by Eq. (5.89):

$$R_m = R_g - R_b > 3\sigma(R_g - R_b) \quad (5.93)$$

Solving this equation for R_g gives the minimum detectable gross count rate R_m:

$$R_g = \frac{\left(2R_b + \frac{9}{t_g}\right) + \sqrt{\frac{36R_b}{t_g} + \frac{81}{t_g^2} + \frac{36R_b}{t_b}}}{2} \quad (5.94)$$

An approximation can be used by assuming that $R_g \approx R_b$ and from Eqs (5.86) and (5.87):

$$R_g \approx R_b + 3\sqrt{\frac{R_b}{t_g} + \frac{R_b}{t_b}} \quad (5.95)$$

The minimum detectable activity A_m can be calculated:

$$A_m = \frac{R_m}{S} \quad (5.96)$$

where S is the sensitivity of the detection system usually expressed as count rate per becquerel.

5.6.5.2. Example 2: calculation of minimum activity that can be detected

A detector is to be used to detect ^{131}I in the thyroid of radiation workers. The background count rate was 441 cpm measured over a period of 5 min and the thyroid count rate was measured over 1 min. The sensitivity of the counter was 0.1 counts · s^{-1} · Bq^{-1}. What is the minimum activity that can be detected?

From Eq. 5.94:

$$R_g = \frac{\left(2\times 441 + \frac{9}{1}\right) + \sqrt{36\times\frac{441}{1} + \frac{81}{1^2} + 36\times\frac{441}{5}}}{2} = 515 \text{ cpm}$$

It should be noted that $R_g - R_b = 74$ cpm and $3\sigma(R_g - R_b) = 74$ cpm as was specified in Eq. (5.93). The minimum detectable radioactivity is:

$$A_m = \frac{(515 - 441)}{0.1 \times 60} = 12.28 \text{ Bq}$$

5.6.6. Comparing counting systems

It was concluded in Section 5.3.1 that a large number of counts have smaller uncertainties expressed as the fractional standard deviation. In Section 5.6.3, it was shown that if background counts increase, the uncertainty of the net counts expressed as fractional standard deviation rapidly increases. Thus, it is desirable to use a counting system with a high sensitivity and low background. However, when the detector sensitivity is increased, the system will also be more sensitive to background. The trade-off between sensitivity and background can be analysed as follows.

It is considered that results from systems 1 and 2 are compared. The acquisition times for gross and background counts are acquired over the same time. From Eq. (5.79):

$$\sigma_{F1}(N_{S1}) = \frac{\sqrt{N_{g1} + N_{b1}}}{N_{g1} - N_{b1}}$$

and

$$\sigma_{F2}(N_{S2}) = \frac{\sqrt{N_{g2} + N_{b2}}}{N_{g2} - N_{b2}}$$

The fractional uncertainties for the net sample counts obtained with the two systems are, therefore:

$$\frac{\sigma_{F1}(N_{S1})}{\sigma_{F2}(N_{S2})} = \frac{\dfrac{\sqrt{N_{g1} + N_{b1}}}{N_{g1} - N_{b1}}}{\dfrac{\sqrt{N_{g2} + N_{b2}}}{N_{g2} - N_{b2}}} \tag{5.97}$$

If $\dfrac{\sigma_{F1}(N_{S1})}{\sigma_{F2}(N_{S2})} < 1$, then system 1 is statistically the preferred system. If $\dfrac{\sigma_{F1}(N_{S1})}{\sigma_{F2}(N_{S2})} > 1$, then system 2 is preferred.

Systems can be compared using the count rate and fractional standard deviation for the count rate R_S (Eq. (5.82)). To compare systems 1 and 2, the ratio of the fractional standard deviation is calculated:

$$\frac{\sigma_{F1}(R_{S1})}{\sigma_{F2}(R_{S2})}=\frac{\dfrac{\sqrt{\dfrac{R_{g1}}{t_{g1}}+\dfrac{R_{b1}}{t_{b1}}}}{R_{g1}-R_{b1}}}{\dfrac{\sqrt{\dfrac{R_{g2}}{t_{g2}}+\dfrac{R_{b2}}{t_{b2}}}}{R_{g2}-R_{b2}}} \tag{5.98}$$

Equation (5.98) can be used to compare different counting times in the same system for measuring fixed geometry samples. However, to obtain the best energy window selection in a system, or to compare two systems, the same counting time t should be used:

$$\frac{\sigma_{F1}(R_{S1})}{\sigma_{F2}(R_{S2})}=\frac{\dfrac{\sqrt{R_{g1}+R_{b1}}}{R_{g1}-R_{b1}}}{\dfrac{\sqrt{R_{g2}+R_{b2}}}{R_{g2}-R_{b2}}} \tag{5.99}$$

It should be noted that Eqs (5.98) and (5.99) are the same except that in Eq. (5.99) counts are substituted by counting rates.

Equation (5.99) can also be used in planar imaging. Different collimators can be evaluated by comparing counts from a target region to a non-target or background region. However, in imaging, spatial resolution is also important and must be considered.

5.6.7. Estimating required counting times

It is supposed that it is desired to determine the net sample or target counting rate R_s to within a certain fractional uncertainty $\sigma_F(R_s)$. It is supposed further that the approximate gross sample R_{ga} and background R_{ba} counting rates are known from preliminary measurements. If a counting time t is to be used for both the sample or target and the background counting measurements, then the time required to achieve the desired level of statistical reliability is given by Eq. (5.84):

$$t = \frac{R_{ga} + R_{ba}}{[\sigma_F^2(R_s)](R_{ga} - R_{ba})^2}$$

5.6.7.1. Example: calculation of required counting time

The counting time for a thyroid uptake study using a collimated detector is to be determined. The preliminary measurement of the gross thyroid count rate is R_{ga} = 900 cpm and background count rate R_{ba} = 100 cpm. What counting time is required to determine the net count rate to within 5%?

R_{sa} = 900 – 100 = 800 cpm

$$t = \frac{(900+100)}{(0.05)^2 \times (900-100)^2} = \frac{1000}{(0.05)^2 \times (800)^2} = 0.625 \text{ min}$$

The time for both the thyroid and background counts is 0.625 min, resulting in a total time of 1.25 min.

5.6.8. Calculating uncertainties in the measurement of plasma volume in patients

A plasma volume (PV) measurement is required on a patient and the uncertainty in the PV measurement is to be calculated. The PV is measured by using the dilution principle. A labelled plasma sample of a known volume is prepared for injection into the patient. A standard sample with the same activity and volume is also prepared for counting. The standard sample is diluted before a sample is counted. Ten minutes after injection of the sample, a blood sample is obtained, the plasma separated from the blood and the blood sample counted. The PV is calculated using the following equation:

$$\text{PV} = \frac{R_s}{R_p} VD \tag{5.100}$$

where

Net count rate per millilitre of standard sample $R_s = R_{s+b} - R_b$;
R_b is the count rate of background;
R_{s+b} is the gross count rate per millilitre of standard sample;
Net count rate per millilitre of plasma sample $R_p = R_{p+b} - R_b$;

R_{p+b} is gross count rate per millilitre of plasma sample;
V is volume of standard sample in millilitres with percentage uncertainty $\sigma_P(V)$;

and D is dilution of standard sample for counting with percentage uncertainty $\sigma_P(D)$.

TABLE 5.7. APPLICATION OF THE PROPAGATION OF ERRORS PRINCIPLE TO THE CALCULATION OF UNCERTAINTIES

Values			Uncertainty in values			
Symbol	Value	Unit	Symbol	Calculation	σ	$\sigma_F(\%)$
t	10	min				
R_{s+b}	3200	cpm	$\sigma(R_{s+b})$	$\sqrt{\frac{R_{s+b}}{t}}$	17.89	0.559
R_b	200	cpm	$\sigma(R_b)$	$\sqrt{\frac{R_b}{t}}$	4.472	2.236
R_s	3000	cpm	$\sigma(R_s)$	$\sqrt{(\sigma(R_{s+b}))^2+(\sigma(R_b))^2}$	18.44	0.615
R_{p+b}	1200	cpm	$\sigma(R_{p+b})$	$\sqrt{\frac{R_{p+b}}{t}}$	10.95	0.913
R_p	1000	cpm	$\sigma(R_p)$	$\sqrt{(\sigma(R_{p+b}))^2+(\sigma(R_b))^2}$	11.83	1.183
$\frac{R_s}{R_p}$	3		$\sigma(R_s/R_p)$	$\frac{R_s}{R_p}\sqrt{\left(\frac{\sigma(R_s)}{R_s}\right)^2+\left(\frac{\sigma(R_p)}{R_p}\right)^2}$	0.040	1.333
V	5	mL	$\sigma(V)$		0.150	3.000
$\frac{R_s}{R_p}V$	15	mL	$\sigma\left(\frac{R_s}{R_p}V\right)$	$\left(\frac{R_s}{R_p}V\right)\sqrt{\left(\frac{\sigma(R_s/R_p)}{R_s/R_p}\right)^2+\left(\frac{\sigma(V)}{V}\right)^2}$	0.492	3.283
D	200		$\sigma(D)$		6.000	3.000
$\text{PV}=\frac{R_s}{R_p}VD$	3000	mL	$\sigma(\text{PV})$	$\text{PV}\sqrt{\left(\frac{\sigma((R_s/R_p)\text{V})}{(R_s/R_p)\text{V}}\right)^2+\left(\frac{\sigma(D)}{D}\right)^2}$	133	4.447

The following values were used and measured:

Counting time t = 10 min
$V \pm \sigma_P(V)$ = 5 ± 3% mL
$D \pm \sigma_P(D)$ = 200 ± 3%
R_{s+b} = 3200 cpm
R_{p+b} = 1200 cpm
R_b = 200 cpm

The uncertainties are calculated step by step by applying the propagation of errors principle (see Table 5.7)

The measured PV is, therefore, 3000 ± 133 mL or 3000 ± 4.447%. It should be noted that the uncertainty is expressed as one standard deviation. A spreadsheet can be used efficiently to do the calculations in the above table With a spreadsheet, the influence in changing the counting time or uncertainties in the measurement of the dilution and volume of the standard can be investigated. These spreadsheets are ideally suited for calculations of uncertainties in routine clinical investigations.

5.7. APPLICATION OF STATISTICAL ANALYSIS: DETECTOR PERFORMANCE

5.7.1. Energy resolution of scintillation detectors

We have directed our attention in the previous sections to determine the uncertainty associated with the number of counts measured in a radioactive sample or number of counts in an image pixel. Poisson statistics also play an important role in other aspects of the detection of radiation. A statistical process determines the energy resolution of a detector or the uncertainty associated with the energy measurement of a detected photon. This is the reason why the energy resolution of a solid state detector is significantly better than that of a scintillation detector. The type of detector and the energy of the detected photons determine the energy resolution or uncertainty in the energy of a detected photon. The energy resolution for a detector system and a specific radionuclide does not change from sample to sample. This is different from counting statistics where the uncertainty is determined by the number of counts accumulated during a measurement. Therefore, even for the same sample and same detector system, the uncertainty can change if measurements are repeated following the decay of the nuclide.

Another important consequence of statistics is that in scintillation cameras the location of the position of incoming photons is based on the pulses detected by the detectors. Therefore, the statistics of the detector system limits the spatial resolution that can be achieved with an imaging device. A clear understanding of the statistics associated with the detector when detecting a photon is, therefore, important.

In this section, we will investigate the statistical processes in scintillation detectors, since they are widely used in nuclear medicine for sample counting and imaging.

The operation of scintillation detectors can be considered a three stage process:

(a) The number x of light photons produced in the scintillator by the detected γ ray;
(b) The fraction p of the light photons that will eject electrons from the photocathode of the PMT;
(c) The multiplication M of these electrons multiplied at successive dynodes before being collected at the anode.

The average number N_e of electrons produced at the anode is given by:

$$N_e = xpM \tag{5.101}$$

The fractional variance σ_F^2 in the electron number N for a three stage cascade process is given by Eq. (5.63):

$$\sigma_F^2(N_e) = \sigma_F^2(x) + \sigma_F^2(px) + \sigma_F^2(pxM) \tag{5.102}$$

It can be shown that for dynodes with identical multiplication $\sigma_F^2(M) = \frac{1}{M-1}$. It is assumed that the production of light photons follows a Poisson distribution and, therefore, $\sigma_F^2(x) = \frac{1}{x}$. The fractional variance of the production of electrons from light photons at the photocathode is given by Eq. (5.20) as $\sigma_F^2(p) = \frac{1-p}{p}$:

$$\sigma_F^2(N_e) = \frac{1}{x} + \frac{1}{x}\frac{1-p}{p} + \frac{1}{xp}\frac{1}{(M-1)} \tag{5.103}$$

The fractional energy resolution R_E of detectors is expressed as the FWHM divided by the mean photon energy (Section 5.3.4.1). From Eq. (5.38):

$$R_E = \frac{\text{FWHM}}{E} = \frac{2.355\sigma(E)}{E} = \frac{2.355\sigma(N_e)}{N_e} \tag{5.104}$$

From Eqs (5.103) and (5.104):

$$R_E = 2.355\sqrt{\frac{1}{x\left(1 + \frac{1-p}{p} + \frac{1}{p(M-1)}\right)}} \tag{5.105}$$

5.7.2. Intervals between successive events

The time intervals separating random events are of interest in nuclear measurements. Such an application is the calculation and measurement of the paralysable dead time of counting systems.

If r is the average rate at which events are occurring, it follows that $r\,\mathrm{d}t$ is the differential probability that an event will take place in the differential time increment $\mathrm{d}t$. For a radiation detector with unity efficiency, the time interval for counting a single radionuclide is given by:

$$r = \left|\frac{\mathrm{d}N}{\mathrm{d}t}\right| = \lambda N$$

where

N is the number of radioactive nuclei;

and λ is their decay constant.

In order to derive a distribution function to describe the time interval between adjacent random events, it is first assumed that an event has occurred at time $t = 0$. What is the differential probability that the next event will take place within a differential time $\mathrm{d}t$ after a time interval t?

Two independent processes must take place: no events may occur within the time interval from 0 to t, but an event must take place in the next differential time increment $\mathrm{d}t$. The overall probability will then be given by the product of the probabilities characterizing the two processes, or:

Probability of next event taking place in dt after delay of t	= Probability of number of events during time from 0 to t	× Probability of event during dt	
$P_1(t)\,\mathrm{d}t$	$= P(0)$	$\times\, r\,\mathrm{d}t$	(5.106)

The first factor on the right hand side follows directly from the earlier discussion of the Poisson distribution. We seek the possibility that no events will be recorded over an interval of length t for which the average number of recorded events should be rt. From Eq. (5.23):

$$P(0) = \frac{(rt)^0 \mathrm{e}^{-rt}}{0!}$$

$$P(0) = \mathrm{e}^{-rt} \tag{5.107}$$

Substituting Eq. (5.107) into Eq. (5.106):

$$P_1(t)\,\mathrm{d}t = r\mathrm{e}^{-rt}\,\mathrm{d}t \tag{5.108}$$

$P_1(t)$ is now the distribution function for intervals between adjacent random events. Figure 5.7 shows the simple exponential shape of the distribution.

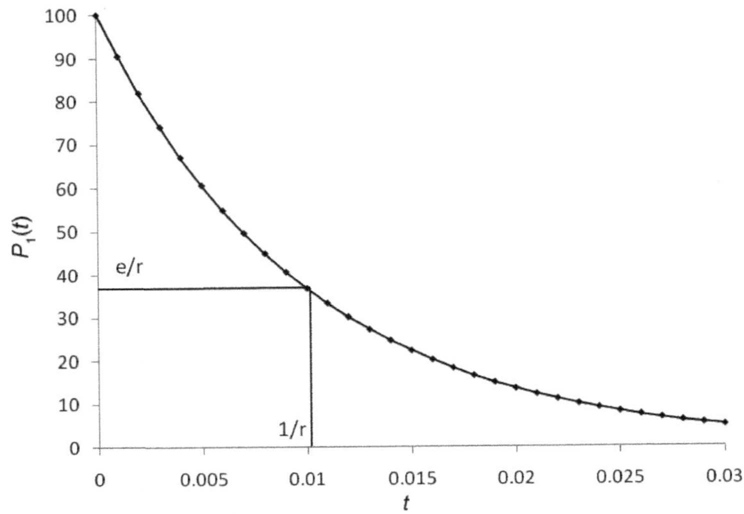

FIG. 5.7. Distribution for intervals between adjacent random events.

It should be noted that the most probable distribution is zero. The average interval length is calculated by applying Eq. (5.8):

$$\bar{t} = \frac{\int_0^\infty tP_1(t)\,\mathrm{d}t}{\int_0^\infty P_1(t)\,\mathrm{d}t} = \frac{\int_0^\infty t\mathrm{e}^{-rt}\,\mathrm{d}t}{\int_0^\infty \mathrm{e}^{-rt}\,\mathrm{d}t} = \frac{1}{r} \tag{5.109}$$

5.7.3. Paralysable dead time

In the paralysable dead time model, a fixed dead time τ follows each event during the live period of the detector. However, events that occur during the dead period, although not recorded, still create another fixed dead time τ on the system following the lost event. The recorded rate of events m is identical to the rate of occurrences of time intervals between true events, which exceed τ. The probability of intervals larger than τ can be obtained by integrating Eq. (5.108)

$$P_2(t)\,\mathrm{d}t = \int_{\tau}^{\infty} P_1(t)\,\mathrm{d}t = \mathrm{e}^{-r\tau} \tag{5.110}$$

The rate occurrence m of such intervals is obtained by multiplying Eq. (5.110) by the true rate r:

$$m = r\mathrm{e}^{-r\tau} \tag{5.111}$$

There is no explicit solution for r; it must be solved iteratively to calculate r from measurements of m and τ. This can be done using a spreadsheet.

BIBLIOGRAPHY

BUSHBERG, J.T., SEIBERT, J.A., LEIDHOLDT, E.M., BOONE, J.M., The Essential Physics of Medical Imaging, Lippincott Williams and Wilkins, London (2002).

CHERRY, S.R., SORENSON, J.A., PHELPS, M.E., Physics in Nuclear Medicine, Saunders, Los Angeles, CA (2003).

DELANEY, C.F.G., FINCH, E.C., Radiation Detectors, Clarendon Press, Oxford (1992).

KNOLL, G.F., Radiation Detection and Measurement, John Wiley and Sons, New York (1989).

NATIONAL ELECTRICAL MANUFACTURERS ASSOCIATION, Standards Publication NU 1-2007, Performance Measurements of Gamma Cameras (2007).

CHAPTER 6

BASIC RADIATION DETECTORS

C.W.E. VAN EIJK
Faculty of Applied Sciences,
Delft University of Technology,
Delft, Netherlands

6.1. INTRODUCTION

6.1.1. Radiation detectors — complexity and relevance

Radiation detectors are of paramount importance in nuclear medicine. The detectors provide a wide range of information including the radiation dose of a laboratory worker and the positron emission tomography (PET) image of a patient. Consequently, detectors with strongly differing specifications are used. In this chapter, general aspects of detectors are discussed.

6.1.2. Interaction mechanisms, signal formation and detector type

A radiation detector is a sensor that upon interaction with radiation produces a signal that can preferably be processed electronically to give the requested information. The interaction mechanisms for X rays and γ rays are the photoelectric effect, Compton scattering and pair formation, where the relative importance depends on the radiation energy and the interaction medium. These processes result in the production of energetic electrons which eventually transfer their energy to the interaction medium by ionization and excitation. Charged particles, such as α particles, transfer their energy directly by ionization and excitation. In all cases, the ionization results either in the production of charge carriers, viz. electrons and ions in a gaseous detection medium, and electrons and holes in a semiconductor detector material, or in the emission of light quanta in a scintillator. These processes represent the three major groups of radiation detectors, i.e. gas filled, semiconductor and scintillation detectors. In the former two cases, a signal, charge or current is obtained from the detector as a consequence of the motion of charge in the applied electric field (Figs 6.1(a) and (b)). In the scintillation detector, light emission is observed by means of a light sensor that produces observable charge or current (Fig. 6.1(c)). A detailed discussion is presented in Sections 6.2–6.4.

6.1.3. Counting, current, integrating mode

In radiology and radiotherapy, radiation detectors are operated in current mode. The intensities are too high for individual counting of events. In nuclear medicine, on the contrary, counting mode is primarily used. Observing individual events has the advantage that energy and arrival time information are obtained, which would be lost in current mode. In the case of a personal dosimeter, the detector is used in integrating mode. The dose is, for example, measured monthly. Furthermore, instead of real time observation, the information is extracted at a much later time after the actual interaction.

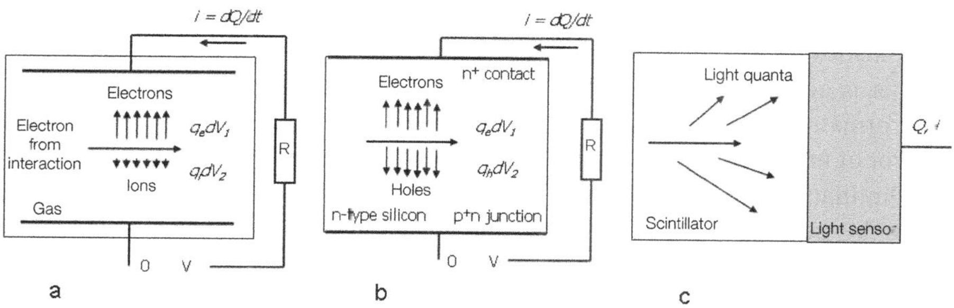

FIG. 6.1. Principle of operation of (a) a gas filled detector, i.e. an ionization chamber; (b) a semiconductor detector, i.e. a silicon detector; and (c) a scintillation detector. The former two detectors are capacitors. The motion of charge results in an observable signal. The light of a scintillation detector is usually detected by a photomultiplier tube.

6.1.4. Detector requirements

The quality of a radiation detector is expressed in terms of sensitivity, energy, time and position resolution, and the counting rate a detector can handle. Obviously, other aspects such as cost, machinability and reliability are also very important. The latter will not be discussed in this chapter.

6.1.4.1. Sensitivity

In radiation detection, the sensitivity depends on (i) the solid angle subtended by the detector and (ii) the efficiency of the detector for interaction with the radiation. The first point will be obvious and is not discussed further. In nuclear medicine, relevant X ray and γ ray energies are in the range of ~30–511 keV. The detection efficiency is governed by the photoelectric effect and Compton scattering only. The attenuation length (in centimetres) of the former is proportional to ρZ_{eff}^{3-4}, where ρ is the density and Z_{eff} is the effective atomic number of the compound. Compton scattering is almost independent of Z; it is just proportional to ρ. The density of a gas filled detector is three orders of magnitude smaller than that of a solid state detector. Thus, solid state detectors are very important in nuclear medicine. At 511 keV, even the highest possible ρ and Z_{eff} are needed. Gas filled detectors are used in dosimetry.

6.1.4.2. Energy, time and position resolution

Energy, time and position resolution depend on a number of factors. These are different depending on the physical property considered and the type of detector; yet, there is one aspect in common. Resolution is strongly coupled to the statistics of the number of information carriers. For radiation energy E, this number is given by $N = E/W$ in which W is the mean energy needed to produce an information carrier. Typical W values are shown in Table 6.1. As the smallest number of information carriers in the process of signal formation is determinative, for scintillation the effect of the light sensor is also shown. From the W values, it can be seen that semiconductor detectors produce the largest number of information carriers and inorganic scintillators coupled to a photomultiplier tube (PMT) the smallest. If a γ ray energy spectrum is measured, the observed energy resolution is defined as the width of a line at half height (FWHM: full width at half maximum) ΔE divided by its energy E. With $N = E/W$, and ΔN being the corresponding FWHM:

$$\frac{\Delta E}{E}=\frac{\Delta N}{N}=2.35\sqrt{\frac{FW}{E}} \qquad (6.1)$$

where

ΔN is 2.35σ for a Gaussian distribution;
σ^2 is FN for the variance;

and F is the Fano factor. For gas filled detectors, F = 0.05–0.20, for semiconductors $F \approx 0.12$. For a scintillator, $F = 1$.

Using the corresponding F and W values, it can be seen from Eq. (6.1) that the energy resolution of a semiconductor is ~16 times higher than that of an inorganic scintillator PMT. In this discussion, other contributions to the energy resolution were neglected, viz. from electronic noise in the case of the semiconductor detector and from scintillator and PMT related effects in the other case. Nevertheless, the large difference, by an order of magnitude, is characteristic of the energy resolutions.

TABLE 6.1. MEAN ENERGIES W TO PRODUCE INFORMATION CARRIERS

Detector type	W (eV)
Gas filled (electron–ion)	30
Semiconductor (electron–hole)	3
Inorganic scintillator (light quantum)	25
Inorganic scintillator + photomultiplier tube (electron)	100
Inorganic scintillator + silicon diode (electron–hole pair)	35

In nuclear medicine, time resolution is mainly of importance for PET. Time resolution depends primarily on two factors, the rise time and the height of the signal pulses. The effect of the former can be understood by considering that it is easier to measure the position of a pulse on a timescale with an accuracy of 100 ps if the rise time is 1 ns, than if it is 10 ns. The pulse height is important because there is noise as well. The higher the pulse relative to the noise, the easier it is to determine its position. In addition, time jitter due to pulse height (energy) variation will become less important. If time resolution is the issue, the fast response and fast rise time of inorganic scintillators and the fast response of the light sensors make the scintillator the preferred detector.

Position resolution can be obtained easiest by pixelating the detector at a pitch corresponding to the requested resolution. In nuclear medicine, position resolution is an issue in γ ray detection in the gamma camera and in single photon emission computed tomography (SPECT) and PET detection systems. In the latter, pixelated scintillators are used and the position resolution of a detector is determined by the pitch. More recently, studies have been published on the use of monolithic scintillator blocks in PET. Light detection occurs by means of pixelated sensors. In principle, this is analogous to the gamma camera. A relatively broad light distribution is measured using pixels that are smaller in size to define the centre of the distribution, thus obtaining a position resolution that is even better than the pixel size.

6.1.4.3. *Counting rate and dead time*

An achievable counting rate depends on (i) the response time of a detector, i.e. the time it takes to transport the charge carriers to form the signal or to emit the scintillation light, and (ii) the time needed to process the signals and to handle the data. For a better understanding, the concept of dead time is introduced. It is the minimum time separation between interactions (true events) at which these are counted separately. Non-paralysable and paralysable dead time are considered. In the former case, if within a period of time τ after a true event a second true event occurs, it cannot be observed. If the second event occurs at a time $t > \tau$, it will be counted. The dead period is of fixed length τ. Defining true event rate T (number per unit time) and counting rate R, the fraction of time the system is dead is given by $R\tau$ and the rate of loss of events is $TR\tau$. Considering that the latter is also $T - R$, the non-paralysable case can be derived:

$$R = \frac{T}{1 + T\tau} \tag{6.2}$$

If in the paralysable model a second event occurs at $t > \tau$ after the first event, it will be counted. If a second event occurs at $t < \tau$ after the first event, it will not be counted. However, in the paralysable case, if $t < \tau$, the second event will extend the dead time with a period τ from the moment of its interaction. If a third event occurs at $t > \tau$ after the first event but within a period of time τ after the second event, it will not be counted either. It will add another period of τ. The dead time is not of fixed length. It can become much larger than the basic period τ and in this case it is referred to as 'extendable' dead time. Only if an event occurs at time $> \tau$ after the previous event will it be counted. In this case, the counting rate is the rate of occurrences of time intervals $>\tau$ between events, for which the following can be derived:

$$R = T\mathrm{e}^{-T\tau} \tag{6.3}$$

Figure 6.2 demonstrates the relation between R and T for the two cases above and for the case of $\tau = 0$, i.e. $R = T$.

6.2. GAS FILLED DETECTORS

6.2.1. Basic principles

The mode of operation of a gas filled detector depends strongly on the applied voltage. In Fig. 6.3(a), the signal amplitude is shown as a function of the voltage V. If upon interaction with radiation an energetic electron ploughs through the gas, the secondary electrons produced will tend to drift to the anode and the ions to the cathode (see Fig. 6.1(a)). If the voltage is relatively low, the electric

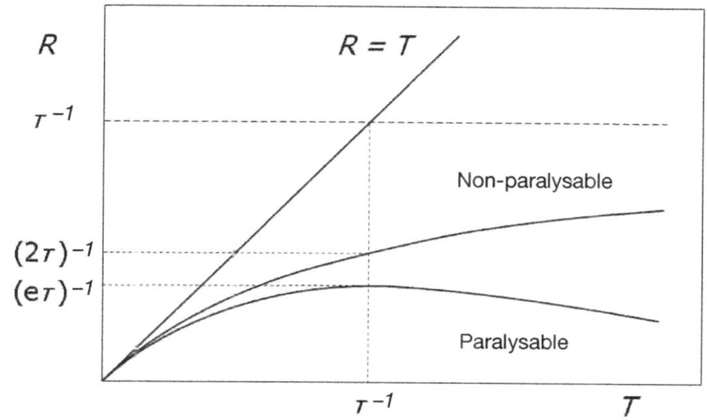

FIG. 6.2. Counting rate R as a function of true event rate T in the absence of dead time (R = T), in the non-paralysable case and in the paralysable case.

field E is too weak to efficiently separate the negative and positive charges. A number of them will recombine. The full signal is not observed — this is in the recombination region. Increasing the voltage, more and more electrons and ions escape from recombination. The region of full ionization is now reached. For heavier charged particles and at higher rates, this will happen at a higher voltage. The signal will become constant over a wide voltage range. Typical operating voltages of an ionization chamber are in the range of 500–1000 V.

For the discussion of operation at stronger electric fields, cylindrical detector geometry with a thin anode wire in the centre and a metal cylinder as cathode (see Fig. 6.3(b)) is introduced. The electric field $E(r)$ is proportional to the applied voltage V and inversely proportional to the radius r. At a certain voltage V_T, the threshold voltage, the electric field near the anode wire is so strong that a drifting electron will gain enough energy to ionize a gas atom in a collision. The proportional region is entered. If the voltage is further increased, the ionization zone will expand and an avalanche and significant gas amplification are obtained. At normal temperature and pressure, the threshold electric field $E_T \approx 10^6$ V/m. For parallel plate geometry with a depth of ~1 cm, this would imply that $V_T \approx 10$ kV, which is not practicable. Due to the r^{-1} dependence, in the cylindrical geometry, manageable voltages can be applied for proportional operation (1–3 kV). As long as the gas gain M is not too high ($M \approx 10^4$), it is independent of the deposited energy. This is referred to as the proportional region and proportional counter. If the voltage is further increased, space charge effects will start to reduce the effective electric field and, consequently, affect the gain. This process will start at a lower voltage for the higher primary ionization density events. The limited proportionality region is entered. With further increasing voltage, the pulse height will eventually become independent of the deposited energy. This is the Geiger–Müller region.

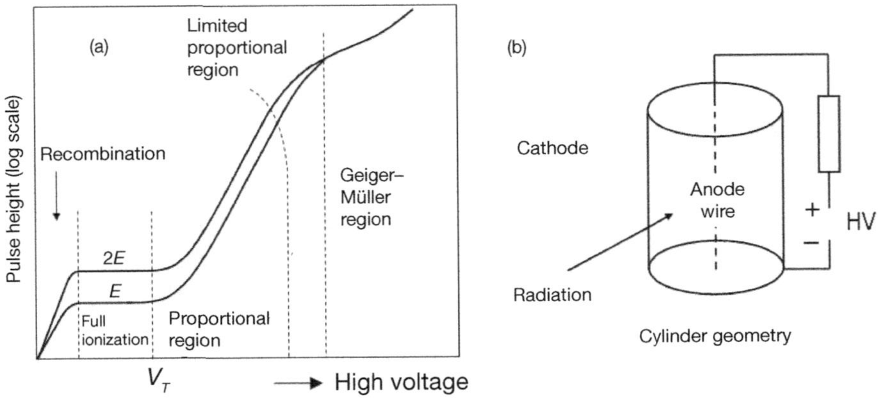

FIG. 6.3. (a) Pulse height as a function of applied high voltage for gas filled detectors; (b) cylindrical detector geometry.

Instead of one wire in a cylindrical geometry, many equidistant parallel anode wires at a pitch of 1–2 mm can be positioned in a plane inside a box with the walls as cathode planes. This multiwire proportional chamber (MWPC) is employed in autoradiography. The technique of photo-lithography made it possible to introduce micro-patterned detectors that operate analogously to the MWPC. Examples are the micro-strip gas chamber and the gas electron multiplier. Spatial resolutions are of the order of 0.1 mm.

6.3. SEMICONDUCTOR DETECTORS

6.3.1. Basic principles

As shown in Fig. 6.1(b), a semiconductor detector is a capacitor. If upon interaction with radiation, electrons are lifted from the valence band into the conduction band, the transport of the charge carriers in an applied electric field is observed. However, if a voltage difference is supplied to electrodes on opposite sides of a slab of semiconductor material, in general, too high a current will flow for practical use as a detector. At room temperature, electrons are lifted from the valence band into the conduction band by thermal excitation due to the small gap ($E_{gap} \approx 1$ eV). The resulting free electrons and holes cause the current. A solution is found in making a diode of the semiconductor, operated in reverse bias. Silicon is used as an example. The diode structure is realized by means of semiconductor-electronics technology. Silicon doped with electron-donor impurities, called n-type silicon, can be used to reduce the number of holes. Electrons are the majority charge carriers. Silicon with electron-acceptor impurities is called p-type silicon; the number of free electrons is strongly reduced. The majority charge carriers are the holes. When a piece of n-type material is brought into contact with a piece of p-type material, a junction diode is formed. At the junction, a space charge zone results, called a depletion region, due to diffusion of the majority charge carriers. When a positive voltage is applied on the n-type silicon side with respect to the p-type side, the diode is reverse-biased and the thickness of the depletion layer is increased. If the voltage is high enough, the silicon will be fully depleted. There are no free charge carriers left and there is virtually no current flowing. Only a small current will remain, the leakage or dark current.

To make a diode, n-type silicon is the starting material and a narrow zone is doped with impurities to make a p^+n junction, as indicated at the bottom of Fig. 6.1(b). The notation p^+ refers to a high doping concentration. For further reduction of the leakage current, high purity silicon and a blocking contact are used, i.e. an n^+ doping at the n-type side, also indicated in Fig. 6.1(b). If the leakage current is still problematic, the temperature can be decreased. The use of high purity semiconductor material is not only important for reducing the leakage current. Energy levels in the gap may trap charge carriers resulting from the interaction with radiation and the energy resolution of a detector would be reduced.

The above described approach is not the only way to make a detector. It is possible to start with p-type material and make an n^+p junction diode. Furthermore, it is possible to apply a combination of surface oxidation and deposition of a thin metal layer. Such contacts are called surface barrier contacts. If the thickness of a detector is <1 mm, it is even possible to use intrinsic silicon, symbol i, with p^+ and n^+ blocking contacts on opposite sides (p–i–n configuration). For thicker silicon detectors, yet another method is used. In slightly p-type intrinsic silicon, impurities are compensated for by introducing interstitial Li ions that act as electron donors. The Li ions can be drifted over distances of ~10 mm. Furthermore, if the bandgap of a semiconductor is large enough, just metal contacts will suffice.

Important parameters are the mobilities, μ_e and μ_h, and the lifetimes, τ_e and τ_h, of electrons and holes, respectively. The drift velocity $\upsilon_{e,h}$ in an electric field E is given by the product of the mobility and the field strength. Consequently, for a given detector size and electric field, the mobilities provide the drift times of the charge carriers and the signal formation times. From the mobilities and the lifetimes, information on the probability that the charge carriers will arrive at the collecting electrodes is obtained. The path length a charge carrier can travel in its lifetime is given by:

$$\upsilon_{e,h}\tau_{e,h} = \mu_{e,h}\tau_{e,h}E \tag{6.4}$$

If this is not significantly longer than the detector depth, charge carriers will be lost.

6.3.2. Semiconductor detectors

Some properties of semiconductor detector materials of relevance for nuclear medicine, viz. the density ρ, effective atomic number for photoelectric effect Z_{eff}, E_{gap} and W value, the mobilities $\mu_{e,h}$ and the products of the mobilities, and the lifetimes of the charge carriers, are presented in Table 6.2.

Silicon is primarily of interest for (position sensitive) detection of low energy X rays, β particles and light quanta. The latter are discussed in Section 6.4.2.2.

TABLE 6.2. PROPERTIES OF SEMICONDUCTOR DETECTOR MATERIALS

	ρ (g/cm³)	Z_{eff}	E_{gap} (eV)	W[a] (eV)	Mobility (cm²/Vs)		Mobility × lifetime (cm²/V)	
					μ_e	μ_h	$\mu_e\tau_e$	$\mu_h\tau_h$
Si (300 K)	2.3	14	1.12	3.6	1 350	480	>1	~1
Si (77 K)			1.16	3.8	21 000	11 000	>1	>1
Ge (77 K)	5.3	32	0.72	3.0	36 000	42 000	>1	>1
CdTe (300 K)	6.2	50	1.44	4.7	1 100	80	3×10^{-3}	2×10^{-4}
$Cd_{0.8}Zn_{0.2}Te$ (CZT-300 K)	~6	50	1.5–2.2	~5	1 350	120	4×10^{-3}	1×10^{-4}
HgI_2 (300 K)	6.4	69	2.13	4.2	70	4	5×10^{-3}	3×10^{-5}

[a] See Section 6.1.4.2.

For X ray detection in the range of ~300 eV to 60 keV, planar circular Li drifted p–i–n detectors — notated Si(Li) — are commercially available with a thickness up to 5 mm. Diameters are in the range of 4–20 mm. For typical field strengths of ~1000 V/cm, the drift times to the electrodes are on the order of tens of nanoseconds. Energy resolutions (FWHM) at 5.9 keV are ~130–220 eV if operated at 77 K. Position sensitive silicon detectors with a large variety of pixel structures are commercially available. Silicon detectors are also used in personal dosimeters.

Germanium, with its higher density and atomic number, is the basic material for high resolution γ ray spectroscopy. Detectors are made of high purity material. Large volume detectors are made of cylindrical crystals with their core removed (coaxial geometry). High purity n-type or p-type is used with the corresponding junction contacts on the outside and the blocking contacts on the inside. Germanium detectors are operated at 77 K. Cylindrical detectors up to a diameter of ~10 cm and a height of ~10 cm are commercially available. Drift times to the electrodes can be as large as ~100 ns. Typical energy resolutions are ~1 keV at 122 keV γ ray energy and ~2 keV at 1332 keV.

Cadmium telluride (CdTe) and cadmium zinc telluride (CZT) are of interest because their atomic number is significantly higher than that of germanium, and room temperature operation is possible due to the larger bandgap. High purity n-type or p-type material is used. The energy resolution is worse than that of Ge detectors, e.g. 2.5% FWHM at 662 keV. This is primarily due to the relatively short lifetime of the holes, resulting in incomplete charge collection. Electronic correction techniques are used and/or detectors with special electrode configurations (small pixels or grids) are made to observe the electron signal only. Detector dimensions are up to approximately 25 mm × 25 mm × 10 mm. Detectors of 25 mm × 25 mm × 5 mm with 16 pixels × 16 pixels are available. These detectors are used, for example, for innovation of SPECT.

In principle, HgI_2 (mercury iodide) is an attractive material for efficient γ ray detection because of the large density and high atomic number. Owing to the relatively large bandgap, room temperature operation is possible. However, the mobilities are low and charge collection, in particular of the holes, is poor. Consequently, application is limited to detector thicknesses ≤10 mm. Field strengths of 2500 V/cm are applied and analogous to CdTe and CZT, methods are used to observe the electron signal only. Detector areas are up to ~30 mm × 30 mm.

6.4. SCINTILLATION DETECTORS AND STORAGE PHOSPHORS

6.4.1. Basic principles

Scintillation of a material is the prompt emission of light upon interaction with radiation. In nuclear medicine, inorganic ionic crystals are most important. They combine high density and atomic number with a fast response and a high light yield, and large crystals can be grown. These crystals form the backbone for

X ray and γ ray detection. Another group is formed by organic scintillators, viz. crystals, plastics and liquids, which have a low density and atomic number, and are primarily of interest for counting of β particles. In some inorganic scintillator materials, metastable states (traps) are created that may live from milliseconds to months. These materials are called storage phosphors. Scintillators and storage phosphors are discussed later in this section. However, as light detection is of paramount importance, light sensors are introduced first.

6.4.2. Light sensors

6.4.2.1. Photomultiplier tubes

The schematic of a scintillation detector is shown in Fig. 6.4(a). A scintillation crystal is coupled to a PMT. The inside of the entrance window of the evacuated glass envelope is covered with a photocathode which converts photons into electrons. The photocathode consists of a thin layer of alkali materials with very low work functions, e.g. bialkali K_2CsSb, multialkali Na_2KSb:Cs or a negative electron affinity (NEA) material such as GaAs:Cs,O. The conversion efficiency of the photocathode η, called quantum efficiency, is strongly wavelength dependent (see Fig. 6.5). At 400 nm, η = 25–40%. The emitted electrons are focused onto the first dynode by means of an electrode structure. The applied voltage is in the range of 200–500 V, and the collection efficiency $\alpha \approx 95\%$. Typical dynode materials are BeO–Cu, Cs_3Sb and GaP:Cs. The latter is an NEA material. If an electron hits the dynode, electrons are released by secondary emission. These electrons are focused onto the next dynode and secondary electrons are emitted, etc. The number of dynodes n is in the range of 8–12. The signal is obtained from the last electrode, the anode. At an inter-dynode voltage of ~100 V, the multiplication factor per dynode $\delta \approx 5$. In general, a higher multiplication factor is applied for the first dynode, e.g. $\delta_1 \geq 10$, to improve the single-electron pulse resolution, and consequently the signal to noise ratio. Starting with N photons in the scintillator and assuming full light collection on the photocathode, the number of electrons N_{el} at the anode is given by:

$$N_{\text{el}} = \delta_1 \delta^{n-1} \alpha \eta N \tag{6.5}$$

Gains of 10^6–10^7 are obtained. A negative high voltage (1000–2000 V) is often used with the anode at ground potential and care must be taken of metal parts near the cathode. Furthermore, the detector housing should never be opened with the voltage on. Exposure to daylight would damage the photocathode permanently.

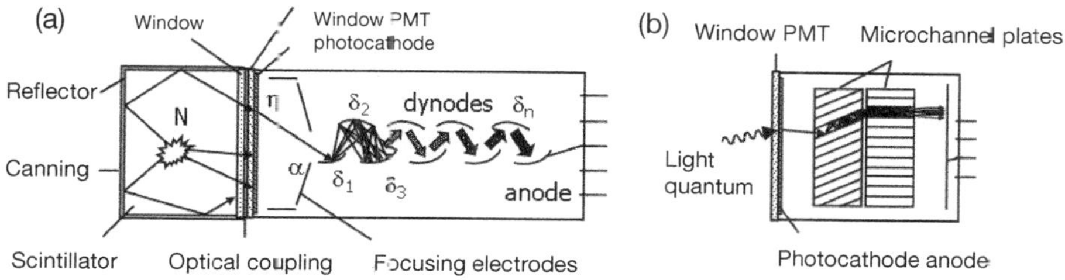

FIG. 6.4. (a) Schematic of a scintillation detector showing a scintillation crystal optically coupled to a photomultiplier tube (PMT); (b) schematic of a microchannel plate-photomultiplier tube.

PMTs are available with a large variety of specifications, including circular, square or hexagonal photocathodes. Cathode diameters are in the range of ~10 to ~150 mm. A ~50 mm diameter PMT has a length of ~150 mm including contact pins. Pixelated multi-anode PMTs exist as well. To optimize time resolution, special tubes are made with almost equal electron transit times to the anode, independent of the cathode position where an electron is emitted. Although the electron transit time is of the order of 30 ns, the transit time spread standard deviation is not more than ~250 ps, and the signal rise time ~1.5 ns.

A PMT aimed at ultra-fast timing is the microchannel plate (MCP)-PMT. For electron multiplication, it employs an MCP structure instead of a dynode configuration (see Fig. 6.4(b)). An MCP (thickness: ~1 mm) consists of a large number of closely packed hollow glass tubes (channel diameter: 5–50 μm). The inner surface of the tubes is covered with a secondary emission material, viz. PbO. The glass surfaces on the front and back side are covered with metal contacts. The MCP is placed in a vacuum, and a voltage of ~1000 V is applied between the contacts, positive on the back side. An electron that enters a glass tube on the front side will hit the wall and secondary emission will occur. The secondary electrons will be pulled to the back side by the electric field, hit the channel wall and produce secondaries, etc. Eventually, they will leave the tube at the back. Electron multiplication of $\sim 10^4$ can be obtained. In an MCP-PMT, two MCPs are used at a close distance. The glass tubes are at an angle, thus preventing ions from gaining too much energy. This structure of two MCPs is called a chevron. At voltages of ~3000 V, stable gains of the order of 10^6 are obtained. The advantage of the MCP-PMT is the short path length of the electrons, resulting in transit times of a few nanoseconds and transit time spreads of ~100 ps. MCP-PMTs are commercially available with circular (~10 mm diameter) and square photocathodes, the latter with multi-anode structures. The sensitivities range from 115 nm (MgF_2 window) to infrared.

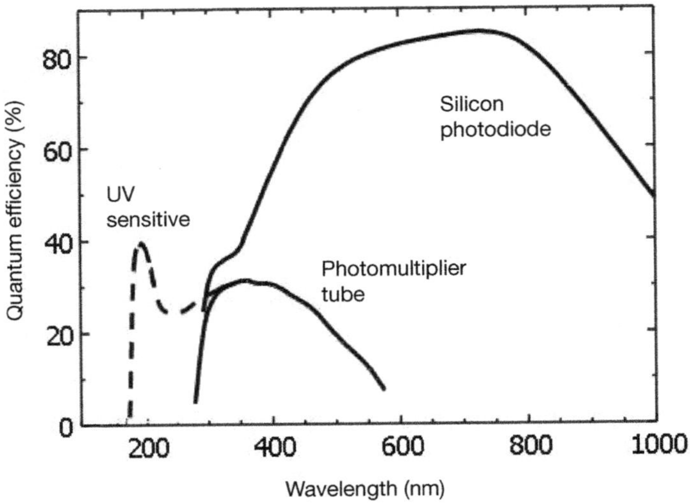

FIG. 6.5. Quantum efficiency as a function of scintillation wavelength for a blue sensitive photomultiplier tube (full line), a photomultiplier tube with sensitivity extended into the ultraviolet (dashed extension) and a silicon photodiode.

6.4.2.2. Silicon based photon sensors

Although PMTs are used on a large scale in nuclear medicine, the relatively large size, high voltages, small quantum efficiency and sensitivity to magnetic fields are a reason to prefer the use of silicon photodiodes in some applications. These diodes are usually of the p–i–n structure (PIN diodes). They have a thickness of ~2 mm including packaging, and are circular, rectangular or square, up to ~30 mm × 30 mm. Bias voltages are <150 V. The quantum efficiency of silicon diodes can be >80% at longer wavelengths (Fig. 6.5). The large capacitance of 20–300 pF, and leakage current, ~1–10 nA, are a disadvantage, resulting in a significant noise level that negatively affects energy resolution in spectroscopy.

An avalanche photodiode (APD) is the semiconductor analogue to the proportional counter. A high electric field is created in a small zone where a drifting electron can gain enough energy to produce an electron–hole (e–h) pair. An avalanche will result. The critical field for multiplication is ~10^7 V/m. The higher the voltage, the higher is the gain. Depending on the type, voltages are applied in the range of 50–1500 V. Gains are in the range of $M < 200$ to ~1000. The gain lifts the signal well above the noise as compared with the silicon diode. At a certain gain, the advantage is optimal. At very high electric fields, spontaneous charge multiplication will occur. The corresponding voltage is called the break-down voltage V_{br}. For gains of $M \approx 10^5 - 10^6$, an APD can be used at voltages $>V_{br}$, where it operates in Geiger mode. The pulses are equal in magnitude. Signal quenching techniques have to be used. Circular and square APDs are available with areas in the sub-square millimetre to ~1 cm^2 range. Various pixelated APDs are available, e.g. of 4 pixels × 8 pixels at a pitch of ~2.5 mm and a fill factor ≤40%.

In a hybrid photomultiplier tube (HPMT), the photoelectrons are accelerated in an electric field resulting from a voltage difference of ~10 kV, applied between the photocathode and a silicon diode which is placed inside the vacuum enclosure. The diode is relatively small, thus reducing the capacitance and, consequently, the noise level. As the production of 1 e–h pair will cost 3.6 eV, ~3000 e–h pairs are produced in the diode per impinging electron. Consequently, the signals from one or more photons can be observed well separated. Equipped with an APD, an overall gain of ~10^5 is possible. HPMTs have been made with pixelated diodes. Window diameters are up to ~70 mm.

The silicon photomultiplier (SiPM) is an array of tiny APDs that operate in Geiger mode. The dimensions are in the range of ~20 μm × 20 μm to 100 μm × 100 μm. Consequently, the number of APDs per square millimetre can vary from 2500 to 100. The fill factor varies from <30% for the smallest dimensions to ~80% for the largest. The signals of all of the APDs are summed. With gains of $M \approx 10^5 – 10^6$, the signal from a single photon can be easily observed. By setting a threshold above the one electron response, spontaneous Geiger pulses can be eliminated. The time spread of SiPM signals is very small, <100 ps. Excellent time resolutions have been reported. Arrays of 2 pixels × 2 pixels and 4 pixels × 4 pixels of 3 mm × 3 mm, each at a pitch of 4 mm, have been commercially produced. A 16 pixel × 16 pixel array of 50 mm × 50 mm has recently been introduced. Blue sensitive SiPMs have a photon detection efficiency of ~25% at 400 nm, including a 60% fill factor.

6.4.3. Scintillator materials

6.4.3.1. Inorganic scintillators

In an inorganic scintillator, the bandgap has to be relatively large to avoid thermal excitation and to allow scintillation photons to travel in the material without absorption ($E_{gap} \geq 4$ eV). Consequently, inorganic scintillators are based on ionic-crystal materials. Three steps for the production of scintillation photons are considered (Fig. 6.6): (i) interaction of radiation with the bulk material and thermalization of the resulting electrons and holes — on the energy scale, electrons end up at the bottom of the conduction band and holes at the top of the valence band; (ii) transport of these charge carriers to intrinsic or dopant luminescence centres; (iii) interaction with these centres, i.e. excitation, relaxation and scintillation. Using this model, the number of photons N_{ph} produced under absorption of a γ ray with energy E is:

$$N_{ph} = \frac{E}{\beta E_{gap}} SQ \tag{6.6}$$

The first term on the right is the number of e–h pairs at the bandgap edge. Typically, $\beta \approx 2.5$. S and Q are the efficiencies of steps (ii) and (iii).

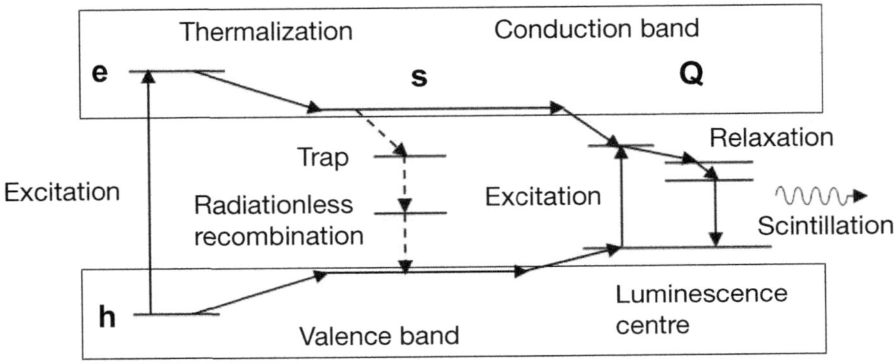

FIG. 6.6. Energy diagram showing the main process steps in an inorganic scintillator.

For the most relevant scintillators, the wavelength at emission maximum λ_{max}, light yield N_{ph}, best reported energy resolution at 662 keV R_{662} and the decay time of the scintillation pulse τ are presented in the last four columns of Table 6.3. In the first columns, some material properties are given, namely density, effective atomic number for photoelectric effect Z_{eff}, attenuation length at 511 keV, $1/\mu_{511}$, and the percentage of interaction by the photoelectric effect at 511 keV. These scintillators are commercially available. If hygroscopic, they are canned with reflective material (Fig. 6.4). Only BaF_2 and BGO have an intrinsic luminescence centre. The other scintillators have Tl^+ or Ce^{3+} ions as dopant luminescence centre. The cerium doped scintillators show a relatively fast response of the order of tens of nanoseconds due to the allowed 5d → 4f dipole transition of the Ce ion. The transitions of the Tl doped scintillators are forbidden and are, consequently, much slower. In general, mixed or co-doped crystals have advantages for crystal growing, response time, light yield or afterglow effects. Large variation is observed for light yields. This is mainly due to $S < 1$, i.e. there are traps of different kinds, resulting in loss of e–h pairs by non-radiative transitions. Using Eq. (6.6) and the proper values of E_{gap}, only $LaBr_3$:Ce appears to have $S \approx Q \approx 1$.

TABLE 6.3. SPECIFICATIONS OF SOME INORGANIC SCINTILLATORS

Scintillator	ρ (g/cm^3)	Z_{eff}	$1/\mu_{511}$ (mm)	Photoelectric effect (%)	λ_{max} (nm)	N_{ph} (photons/MeV)	R_{662} (%)	τ (ns)
NaI:Tl[a]	3.67	51	29	17	410	41 000	6.5	230
CsI:Tl	4.51	54	23	21	540	64 000	4.3	800, 10^4
BaF_2	4.88		23		220 310	1 500 10 000		0.8 600
$Bi_3Ge_4O_{12}$ (BGO)	7.1	75	10.4	40	480	8 900		300
$LaCl_3$:Ce[a]	3.86	49.5	28	15	350	49 000	3.3	25
$LaBr_3$:Ce[a]	5.07	46.9	22	13	380	67 000	2.8	16
$YAlO_3$:Ce (YAP)	5.5	33.6	21	4.2	350	21 000	4.4	25
$Lu_{0.8}Y_{0.2}Al$:Ce (LuYAP)	8.3	65	11	30	365	11 000		18
Gd_2SiO_5:Ce (GSO)	6.7	59	14.1	25	440	12 500	9	60
Lu_2SiO_5:Ce,Ca (LSO)	7.4	66	11.4	32	420	~36 000	7	36–43
$Lu_{1.8}Y_{0.2}SiO_5$:Ce (LYSO)	7.1		12		420	30 000	7	40

[a] Hygroscopic.

6.4.3.2. Organic scintillators — crystals, plastics and liquids

The scintillation mechanism of organic scintillators is based on molecular transitions. These are hardly affected by the physical state of the material. There are pure organic scintillator crystals such as anthracene, plastics such as polystyrene, and liquids such as xylene. Furthermore, there are solutions of organic scintillators in organic solid (plastic) and liquid solvents. Typical combinations are p-terphenyl in polysterene (plastic) and p-terphenyl in toluene. There are also systems with POPOP (para-phenylene-phenyloxazole) added for wavelength shifting. In general, organic scintillators luminesce at ~420 nm, have a light yield of ~10 000 photons/MeV of absorbed γ ray energy and the decay times are about 2 ns. The scintillators are usually specified by a commercial code.

6.4.3.3. Storage phosphors — thermoluminescence and optically stimulated luminescence

A storage phosphor is a material analogous to an inorganic scintillator. The difference is that a significant part of the interaction energy is stored in long-living traps. These are the memory bits of a storage phosphor. The lifetime must be long enough for the application considered. Readout is done either by thermal stimulation (heating) or by optical stimulation. An electron is lifted from the trap into the conduction band and transported to a luminescence centre. The intensity of the luminescence is recorded. These processes have been coined thermoluminescence and optically or photon stimulated luminescence. Storage phosphors have been used for dosimetry for more than fifty years (thermoluminescence dosimeter). In particular, LiF:Mg,Ti (commercial name TLD-100) is widely used. The sensitivity is in the range of ~50 μGy to ~1 Gy. A newer and more sensitive material is LiF:Mg,Cu,P (GR-200), with a sensitivity in the 0.2 μGy to 1 Gy range. Recently, an optically stimulated luminescent material has been introduced, Al_2O_3:C. The sensitivity is in the range of 0.3 μGy to 30 Gy. Storage phosphors are also used in radiography.

BIBLIOGRAPHY

INTERNATIONAL CONFERENCE ON INORGANIC SCINTILLATORS AND THEIR APPLICATIONS, SCINT 2007, IEEE Trans. Nucl. Sci. **55** (2008) 1029–1564.

— SCINT 2009, IEEE Trans. Nucl. Sci. **57** (2010) 1157–1520.

INTERNATIONAL WORKSHOP ON ROOM-TEMPERATURE SEMICONDUCTOR X- AND GAMMA-RAY DETECTORS (15th workshop), IEEE Trans. Nucl. Sci. **54** (2007) 761–880.

— (16th workshop), IEEE Trans. Nucl. Sci. **56** (2009) 1697–1884.

KNOLL, G.F., Radiation Detection and Measurement, 4th edn, John Wiley & Sons, New York (2010).

LEO, W.R., Techniques for Nuclear and Particle Physics Experiments, 2nd edn, Springer, Berlin (1994).

PROCEEDINGS OF NUCLEAR SCIENCE SYMPOSIUM AND MEDICAL IMAGING CONFERENCE (annually), IEEE Trans. Nucl. Sci. (recent volumes).

RODNYI, P.A., Physical Processes in Inorganic Scintillators, CRC Press, Boca Raton, FL (1997).

SCHLESINGER, T.E., JAMES, R.B. (Eds), Semiconductors for Room Temperature Nuclear Detector Applications, Academic Press, San Diego, CA (1995).

TAVERNIER, S., GEKTIN, A., GRINYOV, B., MOSES, W.M. (Eds), Radiation Detectors for Medical Applications, Springer, Dordrecht, Netherlands (2006).

CHAPTER 7

ELECTRONICS RELATED TO NUCLEAR MEDICINE IMAGING DEVICES

R.J. OTT
Joint Department of Physics,
Royal Marsden Hospital
and Institute of Cancer Research,
Surrey

R. STEPHENSON
Rutherford Appleton Laboratory,
Oxfordshire

United Kingdom

7.1. INTRODUCTION

Nuclear medicine imaging is generally based on the detection of X rays and γ rays emitted by radionuclides injected into a patient. In the previous chapter, the methods used to detect these photons were described, based most commonly on a scintillation counter although there are imaging devices that use either gas filled ionization detectors or semiconductors.

Whatever device is used, nuclear medicine images are produced from a very limited number of photons, due mainly to the level of radioactivity that can be safely injected into a patient. Hence, nuclear medicine images are usually made from many orders of magnitude fewer photons than X ray computed tomography (CT) images, for example. However, as the information produced is essentially functional in nature compared to the anatomical detail of CT, the apparently poorer image quality is overcome by the nature of the information produced.

The low levels of photons detected in nuclear medicine means that photon counting can be performed. Here each photon is detected and analysed individually, which is especially valuable, for example, in enabling scattered photons to be rejected. This is in contrast to X ray imaging where images are produced by integrating the flux entering the detectors. Photon counting, however, places a heavy burden on the electronics used for nuclear medicine imaging in terms of electronic noise and stability.

This chapter will discuss how the signals produced in the primary photon detection process can be converted into pulses providing spatial, energy and timing information, and how this information is used to produce both qualitative and quantitative images.

7.2. PRIMARY RADIATION DETECTION PROCESSES

As described in Chapter 6, the methods used for the detection of X ray and γ ray photons fall into three categories, namely the scintillation counter, gas filled detectors and semiconductors. Each of these techniques provides several detector types and requires different electronics to produce and utilize the signals.

7.2.1. Scintillation counters

Figure 7.1 shows a block diagram of a scintillation counter using a phosphor and photomultiplier combination, together with the basic electronics required to produce analogue and digital signals used to create an image. Table 6.3 shows that the phosphors used in nuclear medicine can produce 1500–67 000 optical photons per megaelectronvolt of energy deposited in the crystal and the light emission time can vary from less than 1 ns up to ~1 μs. Additionally, the amplification of the optical signal by a photomultiplier can vary by an order of magnitude or more depending on the photocathode quantum efficiency and the number of dynodes. From this, it can be seen that the pulses produced by the scintillation counter can vary substantially in both shape and amplitude, and that the electronic devices used to manipulate these signals must be flexible enough to account for these variations.

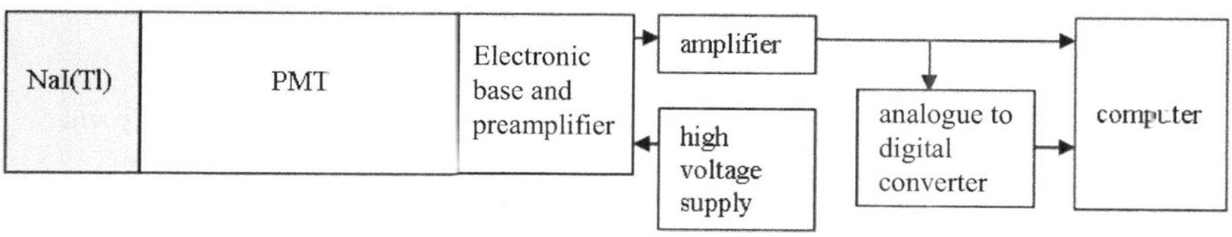

FIG. 7.1. Block diagram of a scintillation counter and associated electronics.

If the signal from the photomultiplier tube (PMT) anode is small, a preamplifier is needed prior to full amplification. This form of amplifier is usually incorporated into the PMT electronic base to minimize the noise generated prior to preamplification. Similar arguments apply to the use of solid state based light sensors such as photodiodes when coupled to phosphors.

Both PMTs and photodiodes require voltage supplies to produce signals — in the case of a PMT, this voltage supply can be 1–3 kV as each successive dynode typically requires 100–200 V to produce sufficient amplification of the electron signal. For a photodiode, the voltage required to totally deplete the device is usually a few tens of volts for a simple photodiode and more for an avalanche photodiode (APD).

7.2.2. Gas filled detection systems

Gas filled imaging systems convert the energy deposited by a γ ray photon directly into ion pairs. It takes 25–35 eV to produce a single ion pair, so the primary signal from ^{99m}Tc will be 4000–5000 electrons. This signal will be amplified in the gas detector using a high voltage (a few kilovolts) to produce an electron avalanche of typically 10^6–10^7 in a multiwire proportional chamber (MWPC) (Fig. 7.2). Typical dimensions of these devices for medical imaging are between 30 and 100 cm laterally by 10–20 cm in the direction of the γ ray.

These signals will clearly also need amplification if they are to be used as analogue and digital output pulses for image formation.

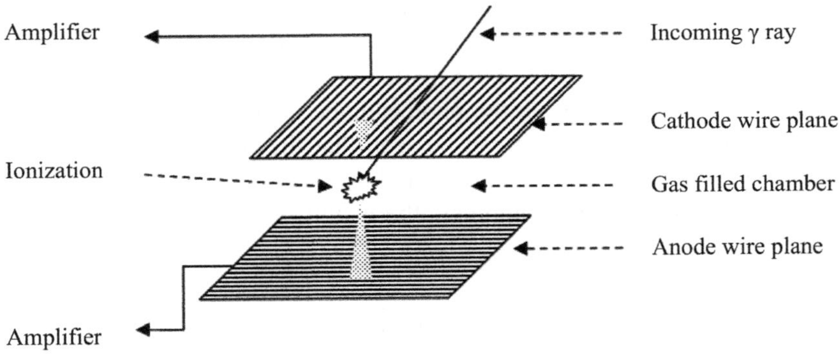

FIG. 7.2. Schematic of a two-plane multiwire proportional chamber detecting a γ ray.

7.2.3. Semiconductor detectors

A simple diode can be thought of as a solid state ionization chamber where the region between the p–n junction acts as a reservoir of electron–hole (e–h) pairs. Incoming radiation produces e–h pairs in the diode, the number of which is proportional to the energy deposited in the diode, and an array of diodes can function as a radiation imaging device. The energy needed to produce an e–h pair is ~3–5 eV (see Table 6.2) and so the size of the pulses produced varies less than for a scintillation counter. However, the signals will still require some form of amplification to produce useful analogue or digital information.

7.3. IMAGING DETECTORS

Having briefly discussed the production of signals by the three major ionizing radiation detection processes, it is necessary to understand how these methods are used to produce images in nuclear medicine. The two main imaging devices used are the gamma camera and the positron camera. For completeness, autoradiography imaging of tissue samples containing radiotracers is also described.

Generally, both gamma camera and positron camera systems use scintillation counters as the primary radiation detector because the stopping power for X rays and γ rays is good in the high density scintillating crystals used. However, there have been some examples of cameras using MWPCs and semiconductors, and a brief description is provided here.

7.3.1. The gamma camera

Invented by Hal Anger, the gamma camera is usually based on the use of a single large area (e.g. 50 cm × 40 cm of NaI(Tl)) phosphor coupled to up to a hundred PMTs. The camera (Fig. 7.3) can detect γ rays emitted by a radiotracer distributed in the body. The lead collimator placed in front of the scintillation counter selects the direction of the γ rays entering the device and allows an image of the biodistribution of the tracer to be made.

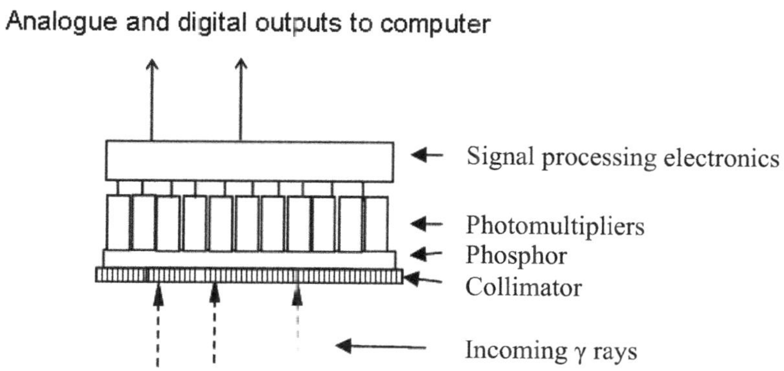

FIG. 7.3. Schematic of a gamma camera for detecting single photons.

The PMTs produce a signal that is proportional to the light generated in the crystal. Positional information can be obtained by comparing the size of the signals from different PMTs, whereas the energy information is related to the sum of the PMT signals. The accuracy of both the energy and the positional information depends on the stability of the signal production and also on the electronics used. The amplitude of analogue signals from the PMTs depends on both PMT and electronic stability, and the noise generated in the signal amplification process. The noise is kept as low as possible by amplification close to each PMT. The amplified/summed signal from the photomultiplier can be converted to a digital pulse train using an analogue to digital converter (ADC), where the number of pulses is proportional to the pulse height or charge in the signal. This signal is used to select events in which most of the energy of the γ ray is detected by the camera, a useful tool to reduce the effect on image production of γ rays scattered in the body prior to detection. It is important at this stage to include only those signals from PMTs that provide information above the intrinsic noise level of the electronics — this is done using some form of signal thresholding, such as a comparator.

Traditionally, the individual PMT pulses are digitized close to the PMTs and these signals are analysed using capacitor or resistor circuits to determine the positional information that is then sent to the computer system to form the image. The centroid of the energy/pulse height information provides a position that is closely related to the point at which the γ ray enters the crystal. Recent improvements for calculating this position based on the digital outputs from the PMTs uses nearest neighbour calculations based on a stored reference map of positional information. These methods provide more accurate estimates of the incident radiation entry point but are more demanding computationally.

Thus, the accuracy of the image production process is very much determined by the initial signal sizes and the subsequent amplification and digitization.

7.3.2. The positron camera

The positron camera is used to simultaneously detect the two annihilation photons produced by positron emitting tracers distributed in the body. The detectors are usually made of many thousands of small scintillating crystals coupled to up to a thousand PMTs. This means that there are many more amplifiers which may have to function at higher count rates than those used in a gamma camera. The detection of these two γ rays requires the addition of coincidence circuits used to select the pulses from a single annihilation event. Figure 7.4 illustrates the format of a positron camera based on multiple scintillating counters in which the signals can be read out to form a 3-D image.

The main difference in the electronics between the positron camera and the gamma camera is the large number of PMT channels involved and the count rates achieved — in both cases, factors of 10–20 are not unusual. In addition, the pulses from a positron camera must be carefully shaped to allow accurate timing information to be made in coincidence circuits to ensure that both annihilation photons from a single annihilation event are detected. Time jitter in the pulses will affect this process, allowing random photons from multiple nucleic decays to be included in the data acquisition. In addition, the recent introduction of so-called 'time of flight' cameras requires very accurate (sub-nanosecond) timing to be made between the two annihilation photons.

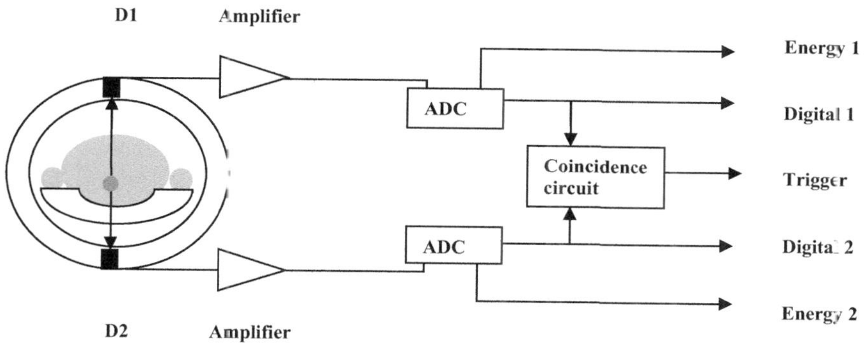

FIG. 7.4. Typical signal processing in a phosphor/photomultiplier tube based positron camera — pulses from two opposing detectors in the camera array are amplified, digitized, checked for coincidence and then used to provide positional and energy information for the event detected.

7.3.3. Multiwire proportional chamber based X ray and γ ray imagers

As shown in Fig. 7.2, an X or γ photon generates an ionization signal in the gas that is detected by the anode and cathode wire planes. The high voltages across the wire planes cause electron avalanches close to the nearest wires and these signals can be detected and amplified at either end of the wire.

In the PETRRA positron camera (Fig. 7.5), the initial γ ray detection is performed using blocks of barium fluoride. The vacuum ultraviolet produced in the crystal photo-ionizes the gas producing electrons that are subsequently amplified in the gas by a series of wire planes. The MWPC positional information is read out using delay lines coupled to the cathode wires. Signals are induced in the delay lines and detected at either end using amplifiers that produce signals with low time jitter. These signals are passed to constant fraction discriminators (CFDs) to produce fast timing signals and then to time digitizers. The time difference between the arrival of the signals at the two ends of the delay line is measured by the time to digital converters (TDCs) and provides the positional information — the accuracy of this information depends on the intrinsic properties of the delay lines and the spread of the signal at the wire planes, and in this system is ~4 mm. Pulses produced after the gas amplification region are used to provide the fast coincidence trigger to read the data into the computer — a timing resolution of ~2–3 ns is readily achievable.

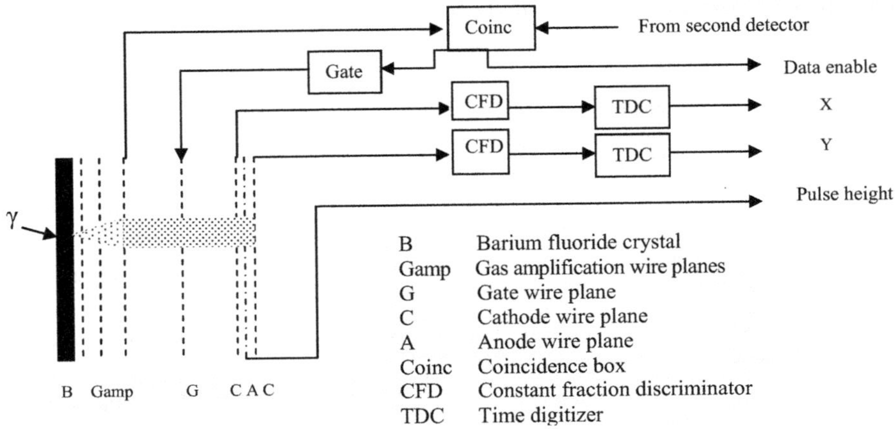

FIG. 7.5. Schematic of the pulse production and readout system for the PETRRA positron camera — wire planes are shown as dotted lines.

A further process to minimize the recording of single γ ray events in this system is the use of an electronic ‘gate’ in which the transport of ionization signals is controlled by an additional wire plane. This allows passage of the ionization to the anode/cathode part of the chamber only if two MWPC detectors have been triggered in fast coincidence. Anode signals can be used to measure the pulse height produced by each detected γ ray. Overall, the electronics of this camera has to manage count rates of several megahertz.

Similar detectors have been developed for imaging low energy γ rays, animal imaging and tissue autoradiography.

7.3.4. Semiconductor imagers

There have been several attempts to make nuclear medicine imaging devices using semiconductors as a γ ray camera. The need for a high Z material means that presently only germanium and cadmium zinc telluride (CZT) have potential as the primary γ ray detector. Germanium (in the form of GeLi) has been used as a 2-D strip detector where the signals from amplifiers at the end resistor chains could be used to determine the position of any interaction in the sensor. However, the modest stopping power of the material coupled with the need for a cryostat to reduce the intrinsic noise of the detector made this design impractical.

More practical systems based on room temperature operation of CZT have been developed by GE (the Alcyone system) and Spectrum Dynamics (the DSPECT system). In the case of the latter system designed specifically for cardiac imaging, ~1000 individual small CZT crystals are coupled to a tungsten collimator providing an intrinsic spatial resolution of 3.5–4.2 mm full width at half maximum and a sensitivity of approximately eight times that of a scintillator based camera — most of the increases in sensitivity are due to the collimator design.

Silicon photodiodes have been used as an alternative to PMTs for both gamma camera and positron camera designs. Here, APDs have been coupled to phosphors and because of their small size, a truly digital camera design is possible. In practice, due to the cost of APDs, only small systems have been developed. The recent development of silicon photomultipliers promises further improvements in nuclear medicine imaging.

7.3.5. The autoradiography imager

Autoradiography is based on the use of radioactive labels to determine the microscopic distribution of pharmaceuticals in tissues excised from humans or animals. A major use in humans is to detect areas of malignancy or tissue malfunction. In animals, the method is used to track the uptake of drugs, for instance. The pharmaceuticals are usually labelled with long lived radiotracers that have a short range β emission or low energy X ray or γ ray emission. Typical examples of tracers used are ^{3}H, ^{14}C, ^{32}P, ^{33}P and ^{125}I. Autoradiography imagers are required to detect the emissions with high efficiency as the levels of uptake in tissue samples are often very low. The gold standard for tissue radiography is film emulsion which produces a high resolution (μm) image of tissues, although these detectors have low efficiency for detecting the radiation involved. Images can take days to weeks to produce and this can be a severe limitation if diagnostic information is desired. Digital autoradiography systems based on the use of thin phosphors, gas filled detectors and silicon wafers can be 50–100 times more efficient although the spatial resolution is limited to typically a few tens of micrometres.

A phosphor based imager may use a very thin (50–100 μm) material such as GADOX or CsI(Tl) coupled to a high resolution sensor, such as a microchannel plate, a charge coupled device or a complementary metal oxide semiconductor APD (Fig. 7.6).

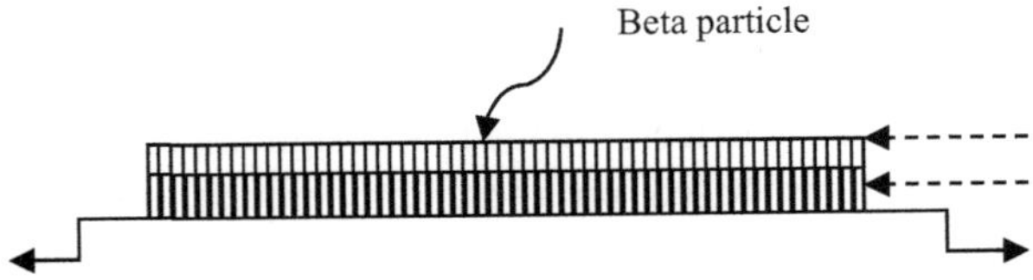

FIG. 7.6. Beta particle detection in an autoradiography system based on a segmented CsI(Tl) phosphor coupled to a complementary metal oxide semiconductor avalanche photodiode (CMOS APD).

The limitations of these devices are mostly the pixel size of the sensor and the noise in the sensor. Amplifiers with low noise are required and room temperature operation desirable. Such a device can have a resolution of <50 μm.

MWPC based autoradiography imagers have been built in which the sample is placed in intimate contact with the chamber gas — such a device will have a resolution of a few hundred micrometres. More recently, direct detection of β particles and X rays has been performed using charge coupled devices and complementary metal oxide semiconductor APDs. The advantage of such devices is that the spatial resolution can be improved further (down to a few micrometres) and ^{3}H can be imaged.

7.4. SIGNAL AMPLIFICATION

As discussed above, the primary signals from the radiation detectors are generally small and need to be amplified without the injection of high levels of noise into the signal readout system. A preamplifier is needed prior to the main amplification process if the signals from the detector are very small, for example, when a PMT has insufficient dynodes to provide a large output pulse. Preamplifiers are usually mounted immediately next to or as part of the output stage of the detector to minimize the noise produced prior to full amplification. The main amplifier can then be used to maximize and shape the signal (via current and/or voltage gain) without over-amplifying noise.

7.4.1. Typical amplifier

The output current from a PMT is directly proportional to the amount of light received from the phosphor. Although the PMT amplifies the electron signal produced at the photocathode by a large factor, the current produced at the anode is still very small. Amplifiers for PMTs are specially designed to transform this current into voltage which can be directly input into an analogue to digital converter or a comparitor. In order to achieve the optimum signal to noise ratio, the output current pulse is integrated in a capacitor, the resulting voltage forming the output signal. The capacitor is normally arranged in the feedback circuit of a wide bandwidth voltage amplifier chosen to have high input impedance and an extremely small input current. As data rates can be high (tens of kilcherz to megaherz), the operational frequency range of the amplifier must be able to cope with these rates. Ideally, the PMT anode is connected directly to the charge amplifier input, with a high value resistor providing a DC return path. Capacitative coupling can be a problem at low frequencies where the signal may be degraded. The charge amplifier integrates the current from the PMT, producing an output voltage pulse. Figure 7.7 illustrates a typical charge amplifier where the output voltage V_{out} is given by:

$$V_{\text{out}} = \frac{-1}{C} \int I(t)\, dt = \frac{-Q}{C} \tag{7.1}$$

The configuration has negative feedback that increases the effective input capacitance by a factor equal to the gain of the amplifier. This ensures that almost all of the current flows into the amplifier even though the PMT and wiring can have significant capacitance. The feedback also reduces the output impedance, so that the amplifier acts as a voltage source.

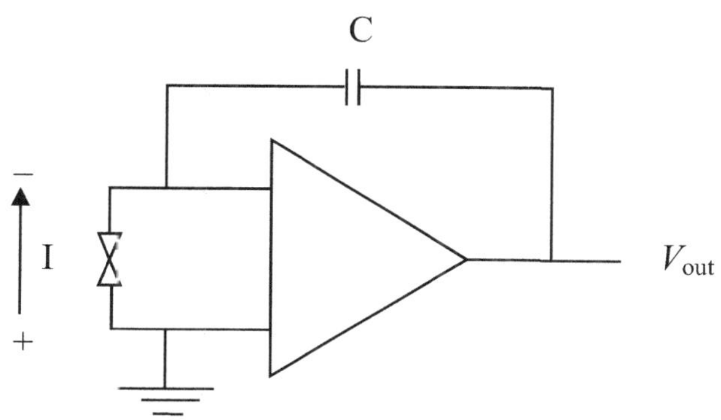

FIG. 7.7. A directly coupled charge amplifier producing a voltage output by integrating the current produced in the photomultiplier tube.

The shape of the output pulse is important for the measurement of both analogue and digital information, and is defined by the output stage of the amplifier (Fig. 7.8). The amplified signal is first passed through a CR (high pass) filter which improves the signal to noise ratio by attenuating the low frequencies, which contain a lot of noise and very little signal. The decay time of the pulse is also shortened by this filter.

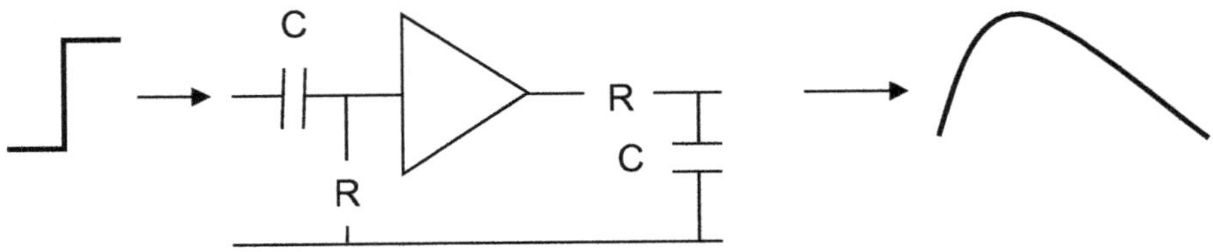

FIG 7.8. A CR–RC pulse shaping circuit.

Before the output of the amplifier, the pulse passes through an RC (low pass) filter which improves the signal to noise ratio by attenuating high frequencies, which contain excessive noise. The pulse rise time is lengthened by this filter. The combined effect produces a unipolar output pulse and with suitably chosen values, has an optimal signal to noise ratio.

7.4.2. Properties of amplifiers

The most important properties of an amplifier are gain, bandwidth, linearity, dynamic range, slew rate, rise time, ringing, overshoot, stability and noise:

— The gain of an amplifier is defined as the log ratio of the output power/voltage P_{out} to the input power/voltage P_{in} and is usually measured in decibels:

$$\text{gain [dB]} = 10\log(P_{out}/P_{in}) \tag{7.2}$$

Gain in charge amplifiers is often expressed in millivolts output per picocoulomb input charge.

— The bandwidth of an amplifier is defined as the range of frequencies that the amplifier operates and is often determined by frequencies at which the power output drops to half its normal value (the –3 dB point). This is an important feature of an amplifier attached to a high count rate detector as required in positron emission tomography (PET) imaging, for example.

— Amplifier linearity is limited when the gain of the amplifier is increased to saturation point, resulting in output pulse distortion. Clearly, this is important if the dynamic range of the pulses produced by the detector is large. Dynamic range is defined as the ratio of the smallest and largest useful output signals, with the former limited by the noise in the system and the latter by amplifier distortion.

- Rise time is often defined as the time taken for the output pulse to increase from 10–90% of its maximum and is a measure of the speed or frequency response of the amplifier.
- Slew rate is the maximum rate of change of the shape of the output signal for the whole range of input signals, usually expressed in volts per microsecond. This is very important if timing information is needed from the detector as a poor slew rate will distort the bigger signals, making them unsuitable for fast timing, as in PET. For PET applications, amplifier rise times of the order of a few nanoseconds are needed, with no shape distortion resulting from slew rate even on the biggest pulses.
- Ringing is a problem when an amplifier produces a pulse that either oscillates before reaching its maximum value or where the tail oscillates before reaching the baseline. This can be a serious problem if timing information is required or if the oscillations produce multiple triggers of the output electronics downstream of the amplifier.
- Stability is clearly an important parameter for an amplifier if the output signals are to be used for either analogue or digital purposes. It is essential that the amplifier output does not vary significantly for a given input signal as the processes used to determine positional, energy and timing information rely on the output for a given input being constant both in offset, amplitude and shape. Factors that affect stability are numerous but prime examples are variations in temperature, supply voltage and count rate as well as long term drift.
- Noise is a major impediment to the production of images using any of the devices discussed above. Examples include thermal noise caused by the thermal movement of charge carriers in resistors, shot noise caused by a random variation in the number of charge carriers and flicker or $1/f$ noise caused by the trapping or collisions of charge carriers in the structure of the silicon used in the electronics. These sources combine to produce a variation in the output signal of the combined detector/electronics system that can affect the quality of images produced. The root mean square noise of a system is defined as the square root of the absolute value of the sum of the squares of the noise variances.

For a system using PMTs, the dominant noise component is that associated with the number of photoelectrons produced at the photocathode as this is amplified by the gain of the PMT dynode chain and subsequent electronics. For a gas filled detector, the equivalent is the number of primary electrons produced at the first stage of the ionization process and for a silicon detector the important parameter is the initial number of e–h pairs produced.

7.5. SIGNAL PROCESSING

Once an amplified signal has been produced, it is then used to generate both analogue and digital information about the detected event. The analogue signal will relate to the energy deposited in the detector and is used, for example, to minimize the number of scattered γ rays accepted into the image production process. The digital signal is used to produce spatial and timing information.

7.5.1. Analogue signal utilization

The analogue information is generated by sending the pulse from the amplifier into a single or multichannel pulse height analyser. In a gamma camera, several 'energy windows' are available, whereby the pulse height or charge is compared with preset values that correspond to the known energies of the γ rays being detected. In the simplest case for imaging a single energy γ ray emission, two thresholds can be set to reject pulses that are above or below these values (Fig. 7.9).

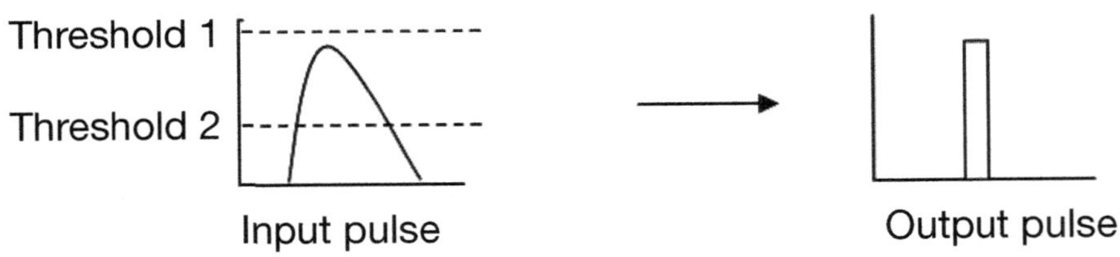

FIG. 7.9. A single channel pulse height analyser — an output pulse is produced when the input pulse is between the two thresholds. This system also functions as a single channel analogue to digital converter.

When a radiotracer that emits several different energy γ rays is being used, multiple thresholds can sort the information into several channels or images.

7.5.2. Signal digitization

Analogue signals are converted into digital signals that are subsequently used to provide spatial and temporal information about each detected event.

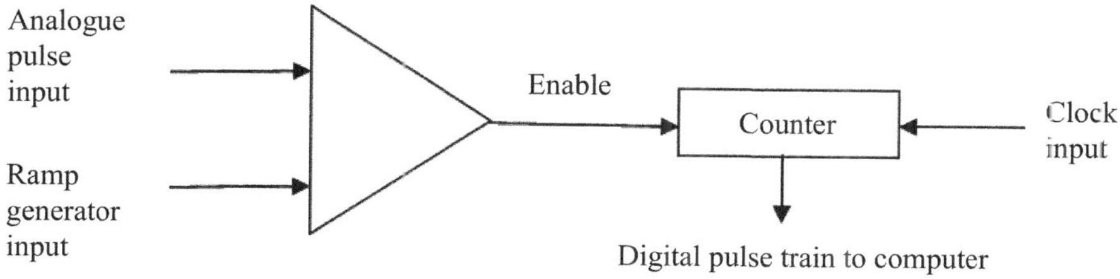

FIG. 7.10. Ramp-based single slope converter system for digitizing analogue pulses.

This is done using an ADC. The simplest method of digitizing an analogue signal is by using a single slope converter (Fig. 7.10). For this, a ramp signal is generated and at the same time a clock producing digital output pulses is started. When the ramp signal exceeds the input pulse, the clock is stopped and the

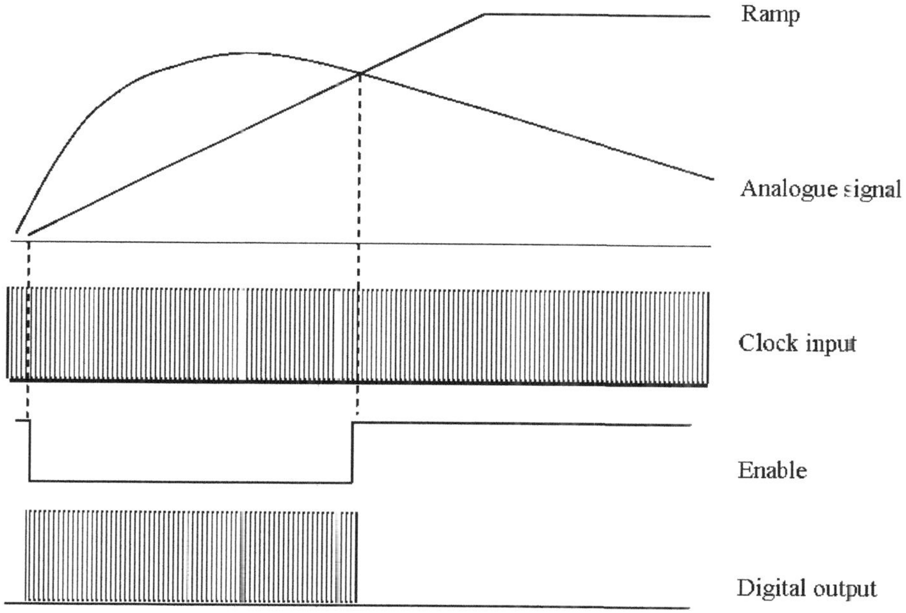

FIG. 7.11. Pulse sequence from the system illustrated in Fig. 7.10.

number of pulses generated corresponds to the amplitude of the signal. The faster the clock, the higher the accuracy of the digitization achieved. This is a relatively simple and low cost solution but is slow as the time taken to digitize the pulse is 2^N clock cycles. The pulse sequence producing the digital output is shown in Fig. 7.11. An important feature is that the analogue pulse shape must be constant to allow accurate conversion. It is clearly possible to have more than one ramp signal to provide several digitization regions if greater or lesser accuracy is needed in any region.

A faster method of analogue to digital conversion is possible by using a flash ADC. This is done using a large number of comparators (see Fig. 7.12), each having a different reference level. The output from each is the input into a logic box that produces the multiple bits of the digital signal. If an N bit output is needed, then $2^N - 1$ comparators are needed. The method is fast as conversion takes a single clock cycle but the system is complex and expensive and consumes a lot of power. Typically, between 8 and 12 bits may be needed for nuclear medicine imaging.

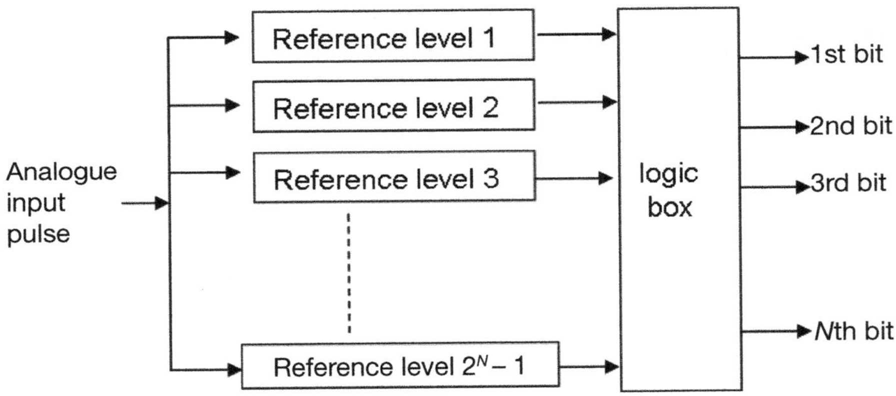

FIG. 7.12. Schematic of a FLASH analogue to digital converter producing N bits of digital data.

7.5.3. Production and use of timing information

In PET systems, the timing of events is very important as only pairs of annihilation photons from the same radioactive decay contribute positively to the image. As the single count rates in a PET scanner may be very high, fast timing is required for coincidence imaging and time of flight measurement. Coincidence timing systems can be based on the timing taken from the front edge of two pulses, from a ‘zero crossing’ point of the differentiated pulses or by using a constant fraction method. The main problem with using a simple front edge trigger is that the variation in pulse height of the analogue signals produces a large variation in timing. Figure 7.13 shows how the two pulses from detectors in a PET system are used to generate a coincidence with CFDs.

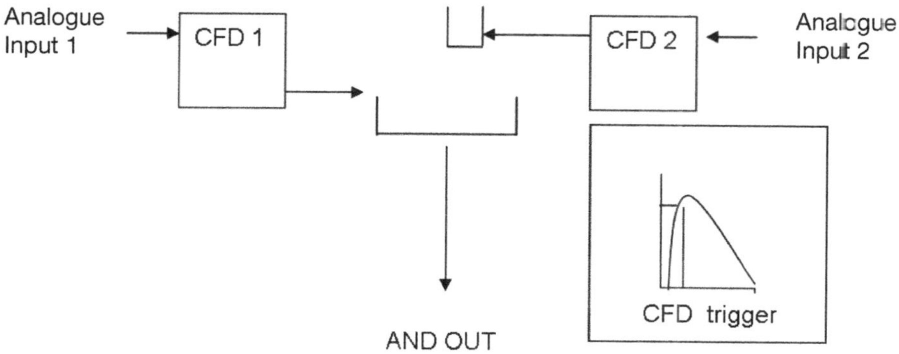

FIG. 7.13. The use of constant fraction discriminators (CFDs) to generate a fast coincidence output — the insert shows how the trigger point is set by a constant fraction of the pulse height.

The trigger points for the timing occur at a constant fraction of the shaped analogue signal, so that the timing is not affected by the different signal pulse heights. In this example, CFD1 generates a gate with a width set to more than twice the measured timing resolution of the detectors. If the pulse from CFD2 falls within this gate, a coincidence (AND) output is generated; otherwise, the event is rejected.

An alternative method of determining the timing from a pulse is to use the zero crossing technique (Fig. 7.14). In this method, the pulse is differentiated to produce a bipolar pulse — the timing is taken from the point where the pulse crosses a reference line that is usually tied to ground — hence, the 'zero crossing'. Again it is important that the pulse shapes are carefully controlled to minimize jitter in the timing information.

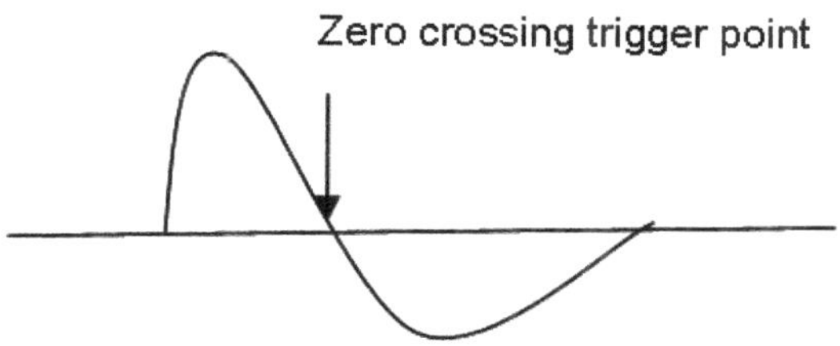

FIG. 7.14. A timing signal generated from the zero crossing point of a differentiated signal.

If the timing information is to be stored, then the two pulses from the CFDs can be input into a TDC. In this case, the first pulse starts and the second one stops a clock — the number of pulses generated is proportional to the time difference between the pulses. If a fast clock is used, excellent timing information is available for use in time of flight calculations, for example.

7.6. OTHER ELECTRONICS REQUIRED BY IMAGING SYSTEMS

7.6.1. Power supplies

Low voltage supplies are used to provide the power input for semiconductor systems where a few tens of volts are sufficient. In some cases, batteries may provide enough power but the need to maintain a constant current and voltage makes this a modest solution. Usually, a low voltage supply converts mains AC power, typically 240 V (or 110 V), into DC voltages of, for example, ±15 V and ±5 V to provide the line voltages for transistors and diodes. This is done by combining a transformer, which reduces the voltage, and a rectifier, typically a diode which allows only one half of the AC signal to pass — this is half-wave rectification (Fig. 7.15).

Full-wave rectification is achieved by using a diode bridge that allows both halves of the AC signal to be used, with one half being inverted. The oscillations are removed using a filter, usually capacitors. The smoothest DC output is provided by using a three phase AC input. The output is usually passed through a voltage regulator to stabilize the voltage and remove the last traces of ripple.

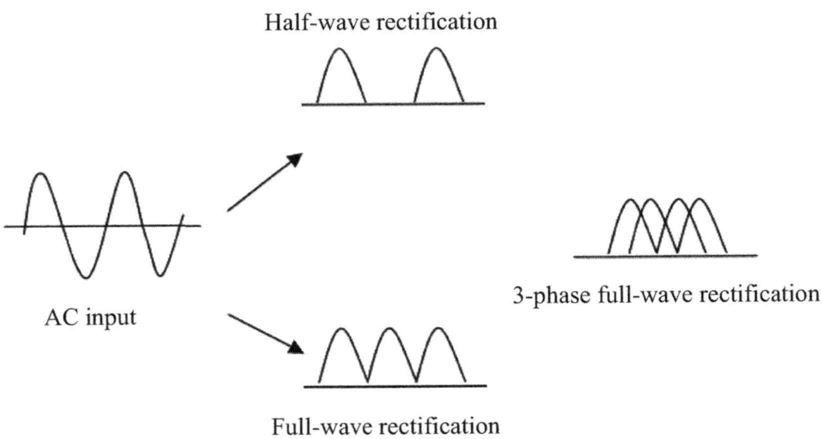

FIG. 7.15. Conversion of AC into DC using a transformer and rectifier system.

PMTs and MWPCs require power supplies that can provide voltages up to several kilovolts. For example, each pair of dynodes in a PMT usually has at least

100 V between them and even more may be used between the photocathode and the first dynode to maximize the early gain of the PMT. These power supplies usually have an oscillator and step up transformer operating at high frequency to provide the drive plus a voltage multiplier consisting of a stack of diodes and capacitors.

7.6.2. Uninterruptible power supplies

This form of support is needed for an imaging system to overcome loss of power during periods of mains supply interruption. In this case, the output power comes from the storage battery via some form of inverter. While the mains power is available, it charges the battery as well as providing power to the imaging device. If the mains supply is interrupted, the battery continues to provide support to ensure that the imaging system can continue to be used. The size of the battery support system depends directly on how long backup is needed or how long, for example, it takes the operator to save data and shut down the system. As in most imaging environments, the mains is replaced by a generator supply. The period of support is often short but usually several hours of supply is available from an uninterruptible power supply.

7.6.3. Oscilloscopes

In order to optimize the use of pulse generating equipment, an oscilloscope is essential. This type of device allows the pulses from the detectors to be displayed at various stages of generation prior to their use in image production. For example, the pulse sequences illustrated above can be displayed on an oscilloscope and this allows the equipment to be adjusted to provide the optimum analogue and digital pulse sequence, shape and size.

An oscilloscope allows the pulses to be displayed on a 2-D display, usually with the vertical axis representing voltage (pulse height) and the horizontal axis time. In addition to the amplitude of the pulses, the oscilloscope display can be used to analyse the frequency of the signals being studied and also to detect any pulse distortion such as oscillation or saturation. In an advanced form, the oscilloscope can function as a spectrum analyser over a wide range of pulse frequencies.

The original oscilloscopes were based on a cathode ray tube to display the pulses but more modern systems use liquid crystal displays connected to ADCs and other signal processing electronics. To the user, the oscilloscope will present as a box with a display screen, input connectors and various controls. The input from equipment can be done either directly using connecting cables/ sockets or through probes, often into a high impedance (e.g. 1 MΩ) or, for high frequency signals, 50 Ω. The trace on the oscilloscope screen is adjusted by various controls. The timebase control can adjust the horizontal display between, for example, 10 ns up to seconds and the pulse height control from millivolts to volts. Other controls include a beam finder, spot brightness and focus, graticule control (to provide a visual measurement grid), pulse polarity and trigger level controls, horizontal and vertical extent and position controls, selection of trigger source (particularly useful for pulse coincidence display) and sweep controls to provide single, multiple and delayed sweeps, for example. An example of an oscilloscope that can be used for examining pulses from imaging equipment is shown in Fig. 7.16.

FIG. 7.16. The front panel of the Tektronix 465 oscilloscope.

More recently, it has been possible to install oscilloscope software onto a computer to provide a low cost solution for pulse display.

7.7. SUMMARY

The information provided above gives a general overview of the equipment and electronics used in nuclear medicine imaging. In some cases, manufacturers have developed special multichannel electronics readout systems tailored to the detectors. These systems include the individual electronics elements described above in a compact design that increases speed and accuracy. Such systems are usually specific to the device involved.

BIBLIOGRAPHY

HOROWITZ, P., HILL, W., The Art of Electronics, Cambridge University Press (1982).

INIEWSKI, K. (Ed.), Medical Imaging: Principles, Detectors, and Electronics, John Wiley and Sons, Hoboken, NJ (2009).

TURCHETTA, R., Electronics signal processing for medical imaging, Phys. Med. Imaging Appl. **240** (2007) 273–276.

WEBB, S., The Physics of Medical Imaging, Hilger (1988).

CHAPTER 8

GENERIC PERFORMANCE MEASURES

M.E. DAUBE-WITHERSPOON
Department of Radiology,
University of Pennsylvania,
Philadelphia, Pennsylvania,
United States of America

8.1. INTRINSIC AND EXTRINSIC MEASURES

8.1.1. Generic nuclear medicine imagers

The generic nuclear medicine imager, whether a gamma camera, single photon emission computed tomography (SPECT) system or positron emission tomography (PET) scanner, comprises several main components: a detection system, a form of collimation to select γ rays at specific angles, electronics and a computing system to create the map of the radiotracer distribution. This section discusses these components in more detail.

The first stage of a generic nuclear medicine imager is the detection of the γ rays emitted by the radionuclide. In the case of PET, the radiation of interest are the 511 keV annihilation photons that result from the interaction of the positron emitted by the radionuclide with an electron in the tissue. For general nuclear medicine and SPECT, there is one or sometimes more than one γ ray of interest, with energies in the range of <100 to >400 keV.

The γ rays are detected when they interact and deposit energy in the crystal(s) of the imaging system. There are two main types of detector: crystals that give off light that can be converted to an electrical signal when the γ ray interacts ('scintillators') and semiconductors, crystals that generate an electrical signal directly when the γ ray deposits energy in the crystal. Scintillation detectors include NaI(Tl), bismuth germanate (BGO) and lutetium oxyorthosilicate (LSO); semiconductor detectors used in nuclear medicine imagers include cadmium zinc telluride (CZT). Radiation detectors are described in more detail in Chapter 6.

When a γ ray interacts in a scintillation crystal, it deposits some or all of its energy. This energy is re-emitted in the form of light with a wavelength dependent on the crystal material but not on the energy of the γ ray. The more energy deposited in the crystal, the greater the intensity of the light emitted. Scintillation crystals are coupled to photomultiplier tubes (PMTs), which serve to convert the scintillation light into an electrical signal. If scintillation light strikes the photocathode of the PMT, electrons are emitted from the photocathode by the photoelectron effect. The number of photoelectrons emitted depends on the intensity of the scintillation light and, therefore, the energy deposited in the crystal. The energy required to produce a single photoelectron is ~1000 eV, so only a few hundred to a thousand electrons are produced for each γ ray that interacts, well below the number needed to produce a measurable current. The PMT contains approximately ten stages that serve to increase the number of electrons by secondary emission of electrons from these dynodes. The signal at the output of the PMT is a measurable current, the amplitude of which is still proportional to the energy deposited in the crystal.

Semiconductor detectors operate differently: the γ ray still deposits some or all of its energy in the crystal through photoelectric absorption or, more likely, Compton scattering interactions. However, that energy is not re-emitted as scintillation light: instead, it creates electron–hole (e–h) pairs that are then collected by application of an electric field to create a measurable signal. The energy required to create an e–h pair is ~3 eV, so many more charge carriers are created in semiconductor detectors than in scintillators (see Chapter 6).

While the exact implementations vary from system to system, the electronics of nuclear medicine imagers have several common functions: they determine the location of interaction of the γ ray in the detector, calculate the energy deposited in the crystal and ascertain whether that energy falls within a prescribed range of desirable energies, and for PET systems, measure the times that the two annihilation photons interacted and evaluate whether the difference in those times falls within a desired timing window to have both come from the same annihilation event (i.e. from the same positron decay). If an event is determined to be valid, its position (and sometimes the energy and timing information) is sent to the computer to be stored along with the information for the many other valid events.

In order to create an image of the distribution of radiotracer, the measured locations of interaction of the γ rays must be converted to a 2-D or 3-D map through image reconstruction. For 2-D planar imaging with a stationary gamma camera, this can be as simple as displaying the number of events at each detector position. For PET or SPECT imaging, where measurements are made for many views around the subject, the data must be combined through a reconstruction algorithm. These techniques range from analytical methods such as filtered back projection to iterative algorithms where estimates of the distribution are calculated and refined based on a model of the imaging system. Not all events accepted are actually useful events with accurate position, energy and timing information. To obtain quantitative images (i.e. images whose counts are directly related to the amount of activity at each location), corrections must be applied for these unwanted events as part of the reconstruction process.

Performance measures aim to test one or more of the components, including both hardware and software, of a nuclear medicine imager.

8.1.2. Intrinsic performance

There are two general classes of measurements of scanner performance: intrinsic and extrinsic. Intrinsic measurements reflect the performance of a sub-part of the imager under ideal conditions. For example, measurements made on a gamma camera without a collimator will describe the best possible performance of the detector without the degrading effects of a collimator, although the collimator is essential for clinical imaging. For a PET scanner, intrinsic performance is often determined for a pair of detectors, rather than the entire system. Intrinsic measurements are useful because they reflect the best possible performance and can help isolate the source of any performance degradations observed clinically. However, these measures are typically performed under non-clinical conditions and will not reflect the performance of the nuclear medicine imager for patient studies. Intrinsic measures also tend to be measurements of an isolated characteristic of the system, rather than its impact on imaging studies. They reflect the limits of performance achievable by the detection system and electronics without collimators or image reconstruction.

8.1.3. Extrinsic performance

Extrinsic, or system, performance measures are made on the complete nuclear medicine imager under conditions that are more clinically realistic, although even these measures may not show the full clinical performance of the system. On a gamma camera, extrinsic measurements are made with the collimator in place; for SPECT and PET systems, the performance is often measured on the reconstructed image. The extrinsic performance of a system gives an indication of how well all of the components of the imager work together to yield the final image. As most extrinsic performance measurements attempt to isolate a single aspect of imaging performance (e.g. spatial resolution, count rate performance, sensitivity), the conditions of these measurements generally do not match the conditions encountered in patient imaging studies. However, the results of extrinsic performance measurements are generally good indicators of clinical performance or may provide useful information about system optimization for clinical studies.

8.2. ENERGY RESOLUTION

8.2.1. Energy spectrum

The amplitude of the signal from the detector depends on the energy deposited in the crystal. If the number of measured events with a given amplitude is plotted as a function of the amplitude (Fig. 8.1), the result is an energy spectrum. The shape of the energy spectrum depends on the radiotracer and γ rays emitted through its decay and the characteristics of the detector material, but all energy spectra have common features. There is one (or more than one) peak, called the photopeak, where the γ ray deposited all of its energy in the detector through one or more interactions. There is also a broad, lower energy region that reflects incomplete deposition of the γ ray's energy in the detector and/or Compton scattering of the γ rays in the body with the subsequent loss of energy before detection. Even in the absence of scattering material (i.e. for a point source in air), the photopeak is not a sharp peak but is blurred. This broadening, which depends on the properties of the detector, is due to statistical fluctuations in the detection of photons and conversion of the energy deposited in the crystal into an electrical signal. This effect is larger for scintillation detectors than for semiconductors. With scintillation detectors, there are several steps in the conversion process that are subject to statistical fluctuations, including the conversion of the γ ray's energy into scintillation light, collection of the scintillation light and conversion into photoelectrons at the PMT's photocathode, and multiplication of those photoelectrons at each dynode in the PMT. For semiconductor detectors, statistical uncertainty is introduced in the number of e–h pairs created when the γ ray deposits its energy and in the collection of these pairs.

The goal of nuclear medicine imaging is to map the distribution of radiotracers, so only γ rays that do not interact in the tissue before reaching the detectors are useful; any γ rays that scatter in the body first change their direction and do not provide an accurate measurement of the original radionuclide's location. Unscattered photons are those γ rays with energies in the photopeak. Nuclear medicine imagers accept events whose energies lie in a 'window' around the photopeak energy in order to reduce the contribution of lower energy, scattered γ rays. For PET scanners, a typical energy window is 440–650 keV for LSO detectors; for gamma cameras based on NaI(Tl) detectors, it is 15% of the photopeak energy (e.g. 129.5–150.5 keV for 140 keV γ rays from ^{99m}Tc, and 68–82 keV for the characteristic X rays from ^{201}Tl with a 20% window).

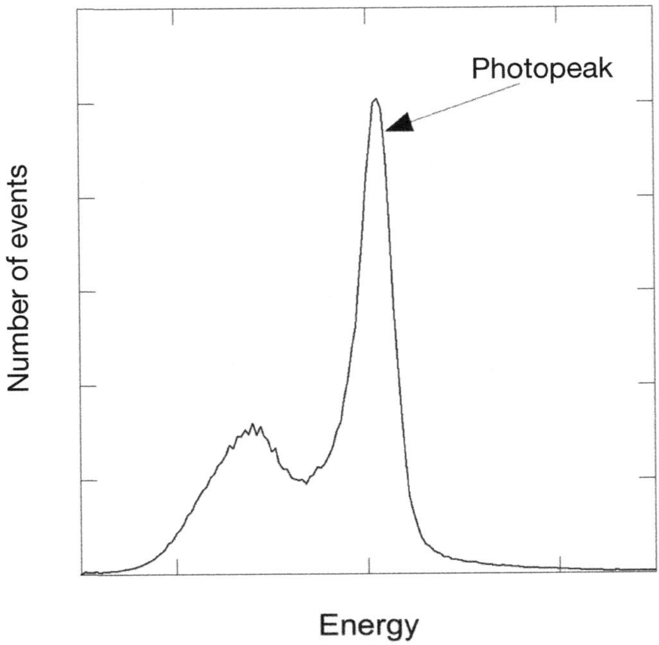

FIG. 8.1. An example of an energy spectrum, defined as the number of measured events with a given amplitude plotted as a function of the amplitude, where the amplitude depends directly on the energy deposited in the crystal.

8.2.2. Intrinsic measurement — energy resolution

The intrinsic ability of a detector to distinguish γ rays of different energies is reflected in its energy resolution. The energy resolution of a detector is defined as the full width of the photopeak at one half of its maximum amplitude, divided by the energy of the photopeak, and is typically expressed as a percentage. A smaller energy resolution value means that the detector is better able to distinguish between two γ rays whose energies are close to each other. The energy resolution depends on the energy of the γ ray approximately as $(\alpha + \beta E)^{1/2}/E$ and, therefore, the energy of the γ ray source must be specified when quoting the energy resolution of a system. The energy resolution worsens at lower energies because fewer photoelectrons are detected (in scintillation detectors) or e–h pairs are created (for semiconductors), so the statistical fluctuations in the measured signal are greater. In addition, the energy resolution of a complete imaging system is typically worse than that of small individual detectors due to slight differences in operating characteristics between detectors.

Only γ rays that have not scattered in the body will provide accurate information about the radiotracer distribution. Accordingly, the energy window is optimal if it includes as many photopeak events as possible, since they are more likely not to have interacted with the tissue, and as few lower energy events as possible, since they are more likely to be the result of one or more Compton scatter interactions in the tissue. As the energy resolution worsens, however, it is necessary to accept more low energy events because the photopeak includes lower energy γ rays. For example, for detection of 511 keV annihilation photons, the lower energy threshold for BGO (15–20% energy resolution) was typically set to 350–380 keV, while that for LSO (12% energy resolution) is 440–460 keV and for $LaBr_3$ (6–7% energy resolution) the lower energy threshold can be set as high as 480–490 keV without loss of unscattered γ rays.

8.2.3. Impact of energy resolution on extrinsic imager performance

The energy resolution is an intrinsic measure of detector performance; it defines the minimum width of the energy window for a given radiotracer. The energy window in turn affects the amount of scattered photons accepted. The ratio of scattered events to total measured events, the 'scatter fraction', is an extrinsic performance characteristic that is of concern, especially for quantitative imaging. In PET systems, for example, the clinical image is assumed to be linearly related to the activity uptake; because scatter adds a smoothly varying background to the image, it degrades the quantitative accuracy of the image and adds to the image noise, even when accurately estimated and subtracted.

There are two major types of scattered event, those where the initial γ ray scattered in the body and those where the γ ray was not completely absorbed in the detector but instead scattered, losing some but not all of its energy. In both cases, the measured energy of the γ ray is lower than the energy of the original photon because some energy is given up to the electron, and the measured position may no longer be related to the original source of the γ ray because the scattered photon does not travel along the same direction as the original γ ray. For typical patient sizes, scattering in the body is much more significant than detector scattering.

The scatter fraction is an extrinsic performance measure that describes the sensitivity of a nuclear medicine imager to scattered events. The measurement involves imaging a line source in a uniformly filled phantom of a specified size at a low activity level, where scattered and unscattered events can be reasonably well differentiated. As the amount of scatter depends on the size and distribution of scattering material in the scanner, the measured scatter fraction cannot be used to infer the amount or distribution of scatter in patient images. However, it is a good indicator of the relative sensitivity of the system to scatter.

The scatter fraction is directly related to the energy resolution of the system in the sense that the energy resolution determines the energy window, in particular the lower energy threshold. This determines the imager's ability to exclude scattered events. However, good energy resolution does not lead to a low scatter fraction unless the energy window used is made appropriately narrow; a scanner with 7% energy resolution will accept approximately as much scatter as one with 12% energy resolution if both systems have the same lower energy threshold. For this reason, measurement of the scatter fraction is a more clinically relevant parameter than the energy resolution.

8.3. SPATIAL RESOLUTION

8.3.1. Spatial resolution blurring

The spatial resolution of a nuclear medicine imager characterizes the system's ability to resolve spatially separated sources of radioactivity. An individual point source of activity does not appear at a single pixel in the image; rather, it is blurred over several pixels, largely due to statistical fluctuations in the detection of the γ rays. Sources whose measured activity distributions overlap cannot be distinguished as distinct sources and instead appear as a single, broad, low contrast source.

In addition to blurring small structures and edges, resolution losses also lead to a decrease in the contrast measure in these structures and at boundaries of the activity distribution. Activity in small structures is blurred into the background and vice versa. Areas of increased or decreased uptake are less easily detected because of this loss of contrast (the 'partial volume effect').

In imagers composed of many small crystals, the spatial resolution of the system is limited by the size of the detector elements. In gamma cameras with a single, large crystal coupled to an array of PMTs, the spatial sampling of the crystal determines the best spatial resolution achievable. The smaller the crystal element or the more finely sampled the detector, the better an event can be localized and the better the spatial resolution will be.

For a given size and sampling, crystals of different materials will have different spatial resolutions. This is because γ rays do not interact at the surface of a crystal but penetrate the crystal before interacting. If a crystal has a low density and low atomic number Z, γ rays will travel further before interacting, compared with a high density, high Z material. The ability to stop γ rays is referred to as the material's stopping power; detectors with higher stopping powers will have more accurate spatial localization than those with low stopping power because there is less inter-crystal scatter. The effect of stopping power becomes more apparent when γ rays enter the crystal at an oblique angle to the face of the crystal (e.g. near the radial edge of a system comprising a ring of detectors). In that case, the γ rays can completely pass through the entrance crystal before interacting in a neighbouring crystal. The γ ray is then mis-positioned as though it had entered the neighbouring crystal or in some intermediate location, depending on the relative amounts of energy deposited by the two interactions.

Spatial resolution is also affected by the energy of the photon and, for scintillation detectors, the efficiency of collection of the scintillation light by the PMTs. The energy of the γ ray that is deposited in the crystal determines the amplitude of the measured signal, which in turn defines how accurately it can be localized in the detector. The spatial resolution measured in a given crystal with ^{99m}Tc (140 keV) is inferior compared to that which would be measured with a 511 keV photon.

As will be discussed later, the spatial resolution can also depend on the count rate or amount of activity in the scanner. As the count rate increases, there is an increased chance that two events will be detected at the same time in nearby locations in the detector. These events will pile up and appear as a single event at an intermediate location with a summed energy. This can lead to a loss of resolution with increasing activity.

8.3.2. General measures of spatial resolution

There are several ways to characterize the spatial resolution, whether of a detector or of a complete system. The point spread function (PSF) and line spread function (LSF) are the profiles of measured counts as a function of position across the point/line source. Rather than showing the complete profiles, however, it is more convenient to characterize them by simple measures. The full width at half maximum (FWHM) and full width at tenth maximum (FWTM) are useful to describe the widths of the profile although they do not give information about any asymmetry in the response. The equivalent width was defined as a way to combine the FWHM and FWTM into a single parameter and describe the shape of the profile in a simple way; it is defined as the width of a box function with a height equal to the maximum amplitude of the profile and an area equal to the total number of counts in the profile above 1/20 of its maximum amplitude. Reducing the PSF or LSF to a few parameters carries with it a loss of information about the spatial response of the imager; for example, LSFs or PSFs can have very different shapes and still have the same FWHM.

The modulation transfer function (MTF) is one way to more completely characterize the ability of a system to reproduce spatial frequencies. The MTF is calculated as the Fourier transform of the PSF and is a plot of the response of a system to different spatial frequencies. High spatial frequencies correspond to fine detail and sharp edges, while low spatial frequencies correspond to coarse detail. The better the response at high frequencies, the smaller the structures that can be resolved. A flat response across all spatial frequencies means that the system most accurately reproduces the object. As it is difficult to compare imaging performance based on the MTF, however, the FWHM and FWTM are used to characterize spatial resolution.

8.3.3. Intrinsic measurement — spatial resolution

The intrinsic spatial resolution is a measure of the resolution at the detector level (or detector pair level for PET) without any collimation. It defines the best possible resolution of the system, since later steps in the imaging hardware degrade the resolution from the detector resolution. On gamma cameras, the intrinsic resolution is determined using a bar phantom with narrow slits of activity across the detector. On PET systems, the intrinsic resolution is measured as a source is moved between a pair of detectors operating in coincidence. The FWHM and FWTM of profiles of detected counts as a function of position are taken as measures of the intrinsic spatial resolution. In both cases, the intrinsic spatial resolution sets a limit on the resolution but does not translate easily into a clinically useful value because other components of the imager impact the resolution in the image.

8.3.4. Extrinsic measurement — spatial resolution

The spatial resolution of a nuclear medicine imager depends on many factors other than just the detectors. The linear and angular sampling play a significant role: to preserve the intrinsic resolution, the imager should be sampled every 0.1 × FWHM. Under-sampling leads to small structures being missed in the image. For single-photon imagers, a collimator is used to limit the direction of γ rays incident on the detector. Collimators are designed for specific purposes (e.g. sensitivity or resolution) and/or specific radionuclides. As the hole size and spacing of a collimator will affect the spatial sampling, each collimator will lead to different system spatial resolution.

The reconstruction processing performed to create tomographic images in SPECT or PET also affects the image resolution. Reconstruction algorithms are generally chosen to preserve as much fine detail and edge information as possible, while keeping image noise sufficiently low so that it is not confused with actual structure. The parameters of reconstruction can, therefore, change with the imaging study and with the number of events measured.

The spatial resolution is not constant throughout the imaging field of view (FOV). For PET systems, the resolution does not vary significantly with location of the source between two detectors in a detector pair, but the system's radial resolution often degrades as the source is moved radially outwards from the centre of the scanner. For gamma cameras, the resolution degrades as the source is moved away from the detector face. For this reason, system spatial resolution measurements are performed with the source at different locations in the imaging FOV.

Extrinsic measures of spatial resolution are made under more clinically realistic conditions and include the effects of the collimator (for single photon imaging) and reconstruction processing. The extrinsic spatial resolution is typically measured with a small point or line source of activity of a sufficiently low amount such that effects seen at high count rates (i.e. mis-positioning of events) are negligible. Measurements of system spatial resolution can be performed in air or with scattering material added. A stationary source is positioned at specified locations throughout the nuclear medicine imager's FOV. The spatial resolution is determined from the images, including any reconstruction or processing steps, by drawing profiles through the source. No spatial smoothing or other post-processing is performed. In addition, any resolution modelling or resolution recovery techniques applied during clinical reconstruction are not used in the measurement of extrinsic resolution. The extrinsic spatial resolution is distinguished from the intrinsic resolution because it includes many effects not seen with the intrinsic resolution: collimator blurring, linear and angular sampling, reconstruction algorithm, spatial smoothing, and impact of electronics.

While the extrinsic resolution measurement reflects the resolution of the complete imaging system, the spatial resolution achieved in patient images is typically somewhat worse than the extrinsic spatial resolution. The spatial sampling is finer than occurs clinically because the pixel size is typically smaller than that used for patient studies in order to sample the PSF or LSF sufficiently. For imagers that reconstruct the data, the reconstruction algorithm in the performance measurement is often not the technique applied to clinical data; an analytical algorithm such as filtered back projection is generally specified for tomographic systems to standardize results between systems. Another key determinant of the clinical resolution is noise in the data that necessitates noise reduction through spatial averaging (smoothing), which blurs the image. For data with high statistics, a sharp reconstruction algorithm can be applied, and the resulting image has good spatial resolution. For more typical nuclear medicine studies, where the number of detected events is limited, some form of spatial smoothing is applied, with the resulting blurring of fine structures.

8.4. TEMPORAL RESOLUTION

8.4.1. Intrinsic measurement — temporal resolution

As the activity in the FOV increases, events arrive closer to each other in time until the imager cannot distinguish individual events. The timing resolution, or resolving time, is the time needed between successive interactions in the detector for the two events to be recorded separately. The timing resolution is largely limited by the decay time of the crystal. For scintillators, the decay time can be as high as 250–300 ns or as low as 20–40 ns, depending on the detector material. Typically, the scintillation light does not decay with a single time constant but with a combination of fast (nanosecond) and slow (microsecond) components. For semiconductor detectors, the decay time is much smaller. In addition to the detector decay time, the various components of the electronics can contribute to the loss of temporal resolution. The timing resolution is generally of less interest than its impact on the count rate performance of the system.

The timing of events is critical for PET, where two annihilation photons must be detected within a timing window to be recorded as a valid event. The timing window must be set wide enough to measure valid coincidence events but not so wide that many coincidences between uncorrelated annihilation photons ('random coincidences') are accepted. Coincidence timing electronics are carefully designed so that a detector's signal is processed as quickly as possible, rather than waiting for the entire scintillation light to be measured. This allows the coincidence timing window to be set to <10 ns, limited by the time of flight of the two annihilation photons across the imager's diameter. For a ring diameter of 90 cm, the minimum coincidence time window would be 6 ns.

Recent developments in PET technology allow for the difference in times of arrival ('time of flight') of the two annihilation photons to be measured. For time of flight systems, the coincidence time window is still 4–6 ns but the time of flight difference can be measured with a resolution of 300–600 ps. This time of flight information is used in reconstruction to localize the events. The timing resolution is measured with a low-activity source of activity by recording a histogram of the number of events as a function of time difference.

8.4.2. Dead time

The consequence of a finite timing resolution is a loss of counts measured at higher activities. When two photons arrive within the resolving time of the detector, the two photons are seen by the electronics as a single event. One or both of the events may be lost, and the events are also mis-positioned in space. The random nature of radioactive decay means that there is always a possibility that two events will arrive within the resolving time of the detector; this possibility increases as the activity in the imager increases.

There are two kinds of dead time: non-paralysable and paralysable (see also Chapter 6). Non-paralysable dead time arises when an event causes the system to be unresponsive for a period of time, so that any later events that arrive during that time are not recorded. For paralysable dead time, the second event is not only not recorded but also extends the period for which the electronics are unresponsive. At moderate count rates, paralysable and non-paralysable dead times are the same; it is only at high count rates that the two types of dead time differ (see Fig. 8.2). It can be seen that systems with non-paralysable dead time saturate at high count rates, while those with paralysable dead time peak and then record fewer events as the activity increases. This leads to an ambiguity in the measured count rate: the same observed count rate corresponds to two different activity levels. The system dead time performance of nuclear medicine scanners is typically intermediate between paralysable and non-paralysable dead time because some components have paralysable dead time while other components have non-paralysable dead time.

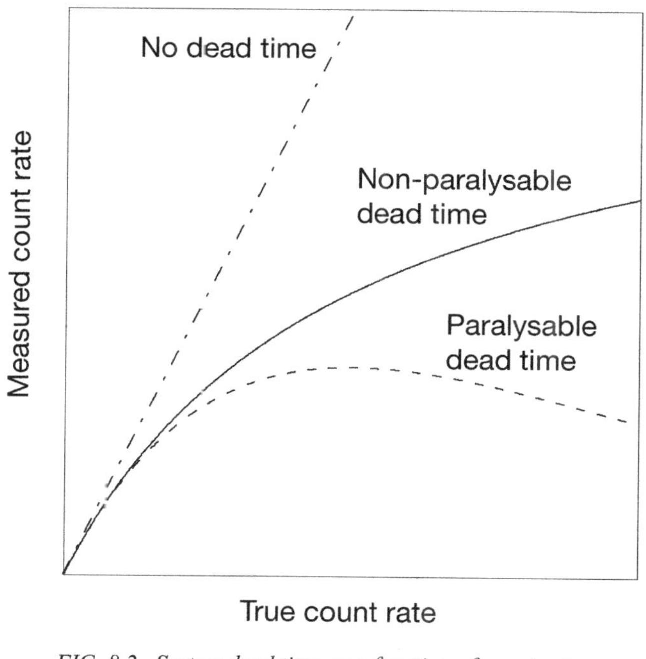

FIG. 8.2. System dead time as a function of count rate.

With increased dead time, additional activity injected in the patient does not lead to a comparable improvement in image quality or reduction in image noise. Dead time losses depend on the single event rate, coincidence count rate (for PET), and the analogue and digital design characteristics of the nuclear medicine imager. Dead time losses can depend on the activity distribution, especially for PET because of the different single photon and coincidence rate relationship with source distribution. They also depend on the radioisotope because dead time results from all γ rays that interact in the detector, not just the photons that fall within the energy window. For imaging studies with a large dynamic range (e.g. cardiac scans), count rate performance is critical.

To correct for event losses due to dead time, a correction based on a decaying source study is often applied to clinical data. The dead time correction will generally correct for the loss of counts, so that the number of counts in the image is independent of the count rate; it does not, however, compensate for the higher image noise that arises because fewer events are actually measured.

8.4.3. Count rate performance measures

The generic measurement of count rate performance involves determining the response of the nuclear medicine imager as a function of activity presented to the system. Typically, this requires starting with a high amount of activity and acquiring multiple images over time as the activity decays. The energy window is set at low activity levels and is not changed at higher activities to accommodate a shift in the photopeak due to pile-up effects. By comparing the observed events with the counts that would be expected after decay correction of events detected at low activities, the system dead time can be determined as a function of activity level. It is especially important to determine the maximum measurable count rate, since higher activities would result in no increase and perhaps a decrease in detected counts. While most count rate performance measures call for starting with a high activity and imaging as the activity decays, if too high an activity is used at the beginning of the measurement, the detector may show effects of saturation during later measurements at lower activities. Therefore, the amount of activity at the beginning of the study must be sufficient to measure the peak count rate but not be so high as to saturate the system for a significant period.

Intrinsic count rate performance measurements are performed with a source in air and without any detector collimation. This is typically performed only on gamma cameras. The system, or extrinsic, count rate performance is measured with the complete system, including any collimation or detector motion, and a distributed source with scattering material (e.g. a cylindrical phantom of specified dimensions or a source placed within scattering material). The scatter adds low energy photons that contribute to pile-up and dead time that are not present in the intrinsic measurement.

For PET, random coincidences also increase as the activity increases; whereas the true coincidence rate would increase linearly with activity in the absence of dead time losses, the random coincidence rate increases quadratically with activity, so that their impact becomes greater at higher count rates. The activity where the random rate equals the true event rate is of importance, in addition to the activity and count rate at which the true count rate saturates or peaks. A global measure of the impact of random coincidences and scatter on image quality is given in the noise equivalent count rate (NECR) defined as:

$$\text{NECR} = \frac{T^2}{T+S+kR} \tag{8.1}$$

where T, S and R are the true, scatter and random coincidence count rates, respectively, and k is a factor that is equal to one if a smooth estimate of random coincidences is used and two if a noisy estimate is used. This parameter does not include reconstruction effects or local image noise differences but can be useful in determining optimal activity ranges.

For systems that correct for dead time, it is important to apply dead time correction and to reconstruct the data in addition to looking at the count rates. The quantitative accuracy of the dead time correction is determined by looking at a large region of interest in decay-corrected, reconstructed images; the counts in the region of interest should be independent of activity level. It is also important to examine the images at high activities for artefacts that may arise due to spatially-varying mis-positioning effects or inaccuracies in various corrections with increased activity.

8.5. SENSITIVITY

8.5.1. Image noise and sensitivity

Images from nuclear medicine devices are typically noisy because the amount of activity that can be safely injected and/or the scan duration without patient discomfort or physiological changes in activity distribution is limited. The number of detected events for a given amount of activity in the imaging system's FOV is an important performance characteristic because a more efficient imager can achieve low image noise with lower injected activity than a less efficient system. Noise in the image can affect both visual (qualitative) image quality and quantitative accuracy, especially in areas of low uptake or low contrast. The relative response of a system to a given amount of activity is reflected in its sensitivity.

The sensitivity of a system is determined by many factors. The geometry of the imager, especially the solid angle of the detectors, as well as any collimation will determine how many photons reach the detectors. The stopping power and depth of the detectors will impact how many of these photons are detected. In addition, the radionuclide's energy, coupled with the imager's energy resolution and energy window, affect the number of accepted events. Finally, the number of counts measured in a given time for a fixed amount of activity depends on the source distribution and its position in the imager.

8.5.2. Extrinsic measure — sensitivity

All performance measurements of sensitivity are extrinsic; for single photon imaging, in particular, the collimator is a major source of loss of events, so it is more clinically interesting to know the sensitivity of the system with a particular collimator.

As noted above, the number of observed counts depends greatly on the activity distribution. For this reason, any measurement of sensitivity is performed under prescribed conditions that do not attempt to replicate patient activity distributions. The source configurations and definitions of sensitivity vary widely, however. For planar imaging, a shallow dish source without intervening scatter material is used, and the sensitivity is reported as a count rate per activity. For SPECT, a cylindrical phantom is filled uniformly with a known activity concentration, and the sensitivity is reported as a count rate per activity concentration. For whole body PET scanners, a line source that extends through the axial FOV is imaged with sequentially thicker sleeves of absorbing material, and the data are extrapolated to the count rate one would measure without any absorber; the sensitivity is then reported as a count rate per unit activity. Small animal PET systems use a point source in air centred in the scanner, and the count rate per activity, as well as the absolute sensitivity (in per cent) are reported. None of these sensitivity measurements can be used to predict the number of events that will be observed for patient studies; however, systems with higher sensitivity will generally record more events from a patient activity distribution than those with lower sensitivity.

8.6. IMAGE QUALITY

8.6.1. Image uniformity

The uniformity of response of a nuclear medicine imager across the FOV is important for both qualitative and quantitative image quality. All PMTs of a given type do not respond exactly the same way, and a correction for this difference in gain is applied before the image is formed. Collimators can also have defects that lead to non-uniformities in the image. For tomographic scanners, corrections for attenuation and unwanted events such as scatter can also affect the uniformity of the image.

Intrinsic uniformity is measured without a collimator by exposing the detector to a uniform activity distribution (e.g. from a distant, uncollimated point source). Intrinsic uniformity is measured at both low and high count rates, where mis-positioning effects become more pronounced. The extrinsic system uniformity is determined with a collimator in place (for single photon imaging), and images are processed or reconstructed as for clinical studies. In both cases, sufficient counts must be detected, so that image noise is low. Quantitative assessment of image uniformity includes variation of pixel counts in small regions across the FOV. However, because simple metrics of non-uniformity such as this do not provide a complete assessment of what the eye perceives in the image, a visual analysis is also important.

8.6.2. Resolution/noise trade-off

Most performance measurements are carried out under non-clinical conditions to isolate an aspect of the imager's performance. To include more of the effects seen in clinical data, some performance standards call for a measurement of image quality. The activity distribution is a series of small structures (e.g. spheres of varying diameters) in a background activity typical of the activity levels seen in patient studies. The activity is imaged for a clinically relevant time, so that the noise level in the data is comparable to that in typical patient studies. The data are processed in the same manner as clinical data. The resulting image, then, is a better representation of the resolution and noise seen clinically. Data analysis consists of such measures as sphere to background contrast recovery, noise in background areas and/or signal to noise ratio in the spheres. While still a simplistic and non-clinical distribution, the measurement gives a more relevant indication of clinical resolution/noise performance.

8.7. OTHER PERFORMANCE MEASURES

There are many other performance measures that reflect a given aspect of a nuclear medicine imager. For planar systems, the spatial linearity, or spatial distortion of the measured position of photons compared to the actual position, is important for good image quality. A number of nuclear medicine imaging systems incorporate anatomical (e.g. computed tomography or magnetic resonance imaging) imagers into the scanner, and the images from the different modalities must be registered spatially. Another area where spatial registration is necessary is in single photon systems where multiple energy windows are used, and the images acquired in the different windows must be overlaid to form the image. Quantitative linearity and calibration is an important measurement for systems such as PET scanners that aim to relate pixel values to activity concentrations.

CHAPTER 9

PHYSICS IN THE RADIOPHARMACY

R.C. SMART
Department of Nuclear Medicine,
St. George Hospital,
Sydney, Australia

9.1. THE MODERN RADIONUCLIDE CALIBRATOR

9.1.1. Construction of dose calibrators

Throughout the world, the instrument that is used in nuclear medicine to measure radioactivity is the calibrated re-entrant ionization chamber, commonly known as a radionuclide calibrator or dose calibrator. Commercial systems comprise a cylindrical well ionization chamber connected to a microprocessor-controlled electrometer providing calibrated measurements for a range of common radionuclides (Fig. 9.1). The chamber is usually constructed of aluminium filled with argon under pressure (typically 1–2 MPa or 10–20 atm). Dose calibrators with reduced gas pressure are available for positron emission tomography (PET) production facilities where very large activities may be measured.

FIG. 9.1. A typical dose calibrator (e.g. CRC 25R).

A well liner, made of low atomic number material (e.g. lucite (Perspex)) which can be removed for cleaning, prevents the ionization chamber from becoming accidentally contaminated. A sample holder is provided into which a vial or syringe can be placed to ensure that it is positioned optimally within the chamber. The dose calibrator may include a printer to document the activity measurements or an RS-232 serial communications port or USB port to interface the calibrator to radiopharmacy computerized management systems.

The chamber is typically shielded by the manufacturer with 6 mm of lead to ensure low background readings. Depending on the location of the dose calibrator, the user may require additional shielding, either to reduce background in the chamber or to protect the operator when measuring radionuclides of high energy and activity. However, this will alter the calibration factors due to backscattering of photons together with the emission of Pb K shell X rays arising from interactions within the lead shielding. If additional shielding is used, the dose calibrator should be recalibrated or correction factors determined to ensure that the activity readings remain correct.

As examples of commercial systems, the specifications of two widely used dose calibrators are given in Table 9.1.

TABLE 9.1. SPECIFICATIONS OF TWO COMMERCIAL DOSE CALIBRATORS

Specification	Capintec CRC-25R	Atomlab 200
Ionization chamber dimensions	26 cm deep × 6 cm diameter	26.7 cm deep × 7 cm diameter
Measurement range	Autoranging from 0.001 MBq to 250 GBq	Autoranging from 0.001 MBq to 399.9 GBq
Nuclide selection	8 pre-set, 5 user-defined (80 radionuclide calibrations in memory)	10 pre-set, 3 user-defined (94 radionuclide calibrations in manual)
Display units	Bq or Ci	Bq or Ci
Electrometer accuracy	<±2%	±1%
Response time	Within 2 s	1 s for activities >75 MBq
Repeatability	±1%	±0.3%

9.1.2. Calibration of dose calibrators

A dose calibrator can be calibrated in terms of activity by comparison with an appropriate activity standard that is directly traceable to a national primary standard. National primary standards are maintained by the relevant national metrology institute, such as the National Physical Laboratory (NPL) in the United Kingdom, the National Institute of Standards and Technology in the United States of America and the Australian Nuclear Science and Technology Organisation (ANSTO). Using the activity standard, a calibration factor for the ionization chamber can be determined for the specific radionuclide. The reciprocal of the calibration factor represents the efficiency ε_N of the ionization chamber for the radionuclide N.

The nuclide efficiency ε_N can be expressed as the sum of two components:

$$\varepsilon_N = \sum_i p_i(E_i) \cdot \varepsilon_i(E_i) \tag{9.1}$$

where

$p_i(E_i)$ is the emission probability per decay of photons of energy E_i;

and $\varepsilon_i(E_i)$ is the energy dependent photon efficiency of the ionization chamber.

Figure 9.2 illustrates a typical efficiency curve as a function of photon energy. Thin-walled aluminium chambers show a strong peak in efficiency at photon energies around 50 keV. This results from the rapid increase of the probability of photoelectric interactions in the filling gas with decreasing energy and the low energy cut-off with aluminium walls at about 20 keV.

Knowing the energy dependent photon efficiency curve for a specific ionization chamber will enable the nuclide efficiency for any radionuclide to be determined from the photon emission probability for each photon in its decay.

The 511 keV annihilation radiation will be measured when the activity of positron emitting radionuclides is to be assayed. A single calibration factor for all positron emitters cannot be used as the emission probability of the positrons must be taken into account. The probability (branching ratio) of positron emission for ^{11}C is 100% and for ^{18}F is 96.7%.

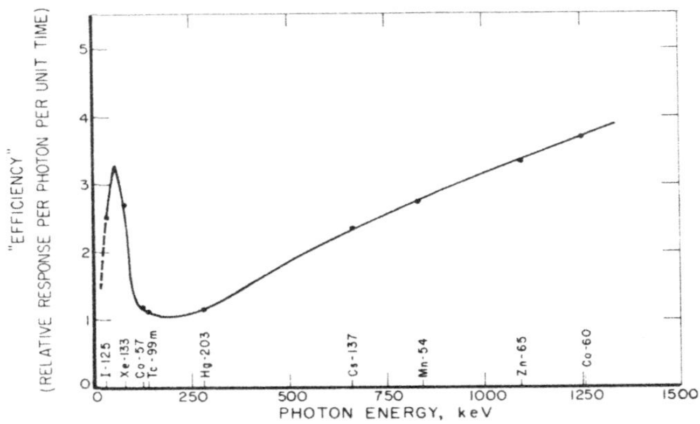

FIG. 9.2. Efficiency curve as a function of photon energy.

9.1.3. Uncertainty of activity measurements

The following sections describe the major sources of uncertainty in dose calibrator measurements.

9.1.3.1. Calibration factor

For medical radionuclides, such as ^{99m}Tc and ^{131}I, the uncertainty of national standards is typically in the range of 1–3%. However, when the standard is used to calibrate a medical dose calibrator, the uncertainty will be larger due to the inherent limit on instrument repeatability. Furthermore, the calibration factor will be for the particular vial size and thickness, and volume of solution, used for the national standard. The calibration factor for a different container (a syringe) and/or a different volume may vary from the established calibration by a significant amount (see Section 9.1.3.6).

9.1.3.2. Electronics

Electrometers measure the current output from the ionization chamber ranging from tens of femtoamperes up to microamperes — a dynamic range of 10^8, corresponding to activity levels from kilobecquerels to hundreds of gigabecquerels. Modern dose calibrators automatically adjust the range while older units required the operator to select the appropriate range. The potential for different linearity characteristics for each range may result in discontinuities when the range is changed. The effects of inherent inaccuracy, linearity and range changing are illustrated in Fig. 9.3. The linearity of the dose calibrator must be established over the full range of intended use when the unit is commissioned and verified as part of the quality control programme (see Section 9.2.1.2).

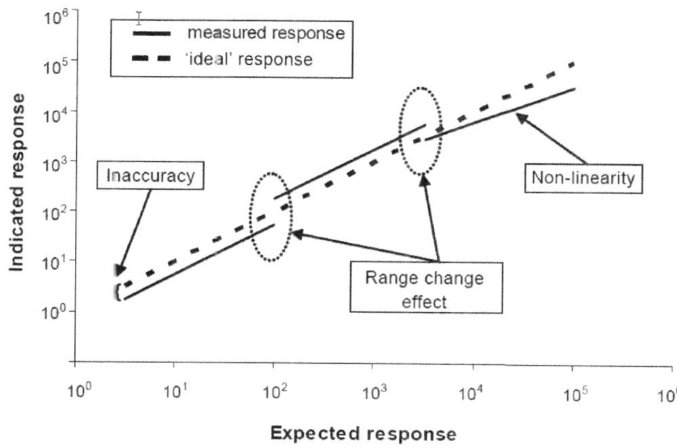

FIG. 9.3. Electrometer inaccuracies (courtesy of the National Physical Laboratory).

9.1.3.3. Statistical considerations

Repeated measurements on a single sample will not be identical because of the random nature of radioactive decay (see Chapter 5). If the measurement period remains constant, the precision of the measured activity will increase as the activity increases. Conversely, the precision will deteriorate for low activity sources. To compensate for this, many calibrators automatically adjust the measurement period depending on the activity level. This may vary from less than one second to tens of seconds for low activities (<1 MBq).

9.1.3.4. Ion recombination

As the activity of the source increases, the probability of recombination of the positive ions with electrons increases. At high source activities, this can become significant and lead to a reduction in the measured current. The effect of recombination is illustrated in Fig. 9.4. For most modern calibrators, the effects of recombination should be less than 1% when measuring 100 GBq of ^{99m}Tc.

9.1.3.5. Background radiation

When the source holder is empty, the dose calibrator will still record a non-zero reading due to background radiation. This will comprise natural background and background from sources within the radiopharmacy. It could also be due to contamination on either the source holder itself or the well liner. Most dose calibrators provide a background subtraction feature. An accurate measurement of the existing radiation level is made by the calibrator (usually integrating over several minutes to improve precision) which is then automatically subtracted from each subsequent reading. This may lead to erroneous results if the background radiation has changed since it was measured due to the presence of additional nearby sources or contamination. It is, therefore, essential to make regular checks of the background radiation level.

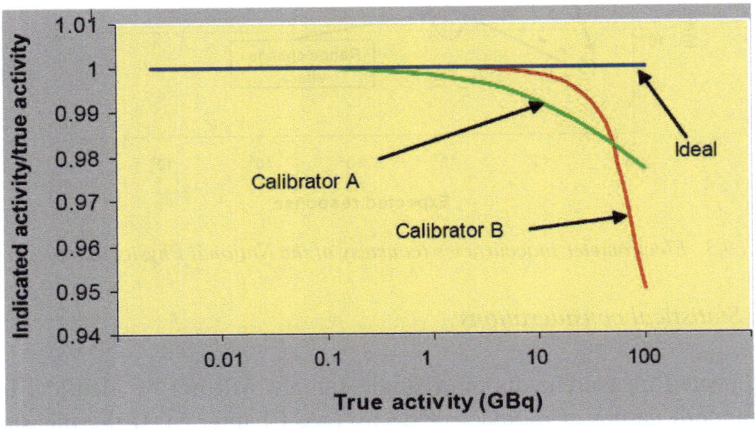

FIG. 9.4. Effects of recombination (courtesy of the National Physical Laboratory).

9.1.3.6. Source container and volume effects

Variations in the composition and thickness of the source container will give rise to corresponding variations in the measured activity. These effects will be most noticeable for low energy photon emitters and pure β emitters. Measurements made at NPL, United Kingdom (Table 9.2) have shown that variations in glass wall thicknesses, which were within the range of the vial manufacturing tolerances, could lead to errors of up to 7% for ^{125}I.

When the activity is drawn into a syringe, the source geometry will be different from that in a vial. Not only will the composition and thickness of the syringe wall be different from that of the vial, but the distribution of the source will also be different depending on the size of syringe used. This is clearly evident in Fig. 9.5, showing measurements at NPL for ^{111}In in three sizes of syringe (1, 2 and 5 mL) from two different manufacturers in comparison to those measured in a laboratory standard P6 vial. Also illustrated in Fig. 9.5 is the effect of changing the source volume without changing the activity. Self-absorption of the emitted

TABLE 9.2. REDUCTION IN DOSE CALIBRATOR RESPONSE DUE TO INCREASES IN GLASS WALL THICKNESS OF 0.08 AND 0.2 mm

Radionuclide	Reduction in response with increase in vial wall thickness of	
	0.08 mm	0.2 mm
^{125}I	3%	7%
^{123}I	0.6%	1.5%
^{111}In	0.2%	0.4%
^{131}I	0.1%	0.25%

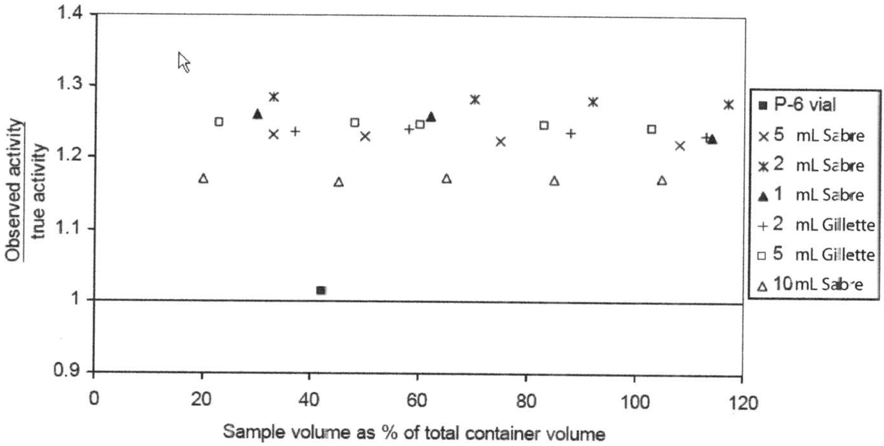

FIG. 9.5. The effects of geometry and sample size on dose calibrator readings, demonstrated for ^{111}In measured in varying syringes (reproduced from Ref. [9.1]).

radiation will change as the source volume changes. This will be particularly important for radionuclides with low energy components such as ^{123}I. For ^{99m}Tc, the correction will usually be less than 1% but should be confirmed for a new dose calibrator or when the supplier of the syringes changes.

9.1.3.7. Source position

The manufacturer's source holder is designed to keep the source at the area of maximum response on the vertical axis of the well. Variations in response due to changes in vertical height or horizontal position of a few millimetres are usually insignificant.

9.1.3.8. Source adsorption

Certain radiopharmaceuticals have been observed to adsorb to the surface of the container. For example, up to 30% of the activity of ^{201}Tl has been found to be adsorbed onto the glass of P6 vials. ^{99m}Tc-tetrofosmin has been shown to adsorb onto the surface of syringes, such that some types of syringe may retain as much as 19% of the activity. Of this, 6% adhered to the rubber plunger with the remainder attached to the plastic syringe barrel. Up to 15% of ^{99m}Tc-macroaggregate of albumin (MAA) may adhere to the syringe, although the amount on the rubber plunger is usually no more than 1%. The possibility of activity adsorption should be considered whenever the facility uses syringes from a different manufacturer.

9.1.4. Measuring pure β emitters

The detection efficiency of ionization chambers for β radiation is low as most, if not all, of the β particles are absorbed in the source solution (self-absorption), in the walls of the container or in the walls of the ionization chamber. The dose calibrator response from β particles will be almost entirely from bremsstrahlung radiation (see Section 1.1.7). In the energy region of interest for ionization chamber measurements, the bremsstrahlung photon spectrum is roughly the same shape as the β particle energy distribution. The average β particle energy is, therefore, a good parameter with which to characterize the ionization chamber response to the bremsstrahlung radiation.

Bremsstrahlung radiation flux is proportional to the square of the atomic number of the absorbing material. Thus, in argon-filled ionization chambers, significant activities are required in order to obtain a precise estimate of the activity. However, as substantial activities of radionuclides are required to be used therapeutically, reliable measurements are possible using pure β emitters used clinically such as ^{90}Y, ^{89}Sr and ^{32}P. However, geometry factors (see Section 9.1.3.6) will be even more important and the system must be calibrated for the specific containers and volumes to be used clinically. Manufacturers are now producing dose calibrators specifically for the measurement of β emitters. These use a sodium iodide detector instead of an ionization chamber, resulting in a significantly increased detection efficiency; however, as the manufacturers state in their product literature, measurements still require exacting attention to the sample container, the sample volume and activity concentration to achieve accurate results.

Most commercially available ionization chambers are provided with calibration factors for commonly used β emitters, although these will usually correspond to the activity within a vial rather than a syringe. The type of vial used in the calibration is often unspecified, so the user should verify the calibration in the vials normally used in the practice. Similarly, the calibration of the activity within the size of syringe to be used clinically should be established. Published results comparing the intrinsic efficiencies of dose calibrators from five different manufacturers found that all systems had a good calibration for ^{32}P, a reduction in efficiency of approximately 10–20% for ^{89}Sr, and a wide divergence in efficiency for ^{90}Y. For this radionuclide, the results obtained using the calibration factors supplied by the manufacturers ranged from 64 to 144% of the true value, re-emphasizing the need for the calibration to be confirmed within the nuclear medicine department.

Several β emitters used for radionuclide therapy include a γ ray component. These radionuclides include ^{131}I (364 keV, 81.5% abundance) and ^{186}Re (137 keV, 9.5% abundance). For these radionuclides, the ionization chamber efficiency is primarily determined by the γ contribution and the manufacturer's supplied calibrations will usually be accurate to within ±10%.

9.1.5. Problems arising from radionuclide contaminants

Unfortunately, it is often not possible for a solution of a radionuclide to be totally free of other radionuclides. The proportion of the total radioactivity that is present as a specific radionuclide is defined as the radionuclide purity. National and international pharmacopoeia specify the radionuclidic purity of a radiopharmaceutical. For example, the European Pharmacopoeia entry for ^{67}Ga-citrate injection requires that no more than 0.2% of the total radioactivity be due to ^{66}Ga. This requirement must be met at all times up to the expiry time of the product. The US Pharmacopoeia is less stringent, specifying that not less than 99% of the total radioactivity be present as ^{67}Ga at the time of calibration.

The presence of contaminants, even when less than 1% of the total activity, can have a marked effect on the ionization chamber current and, thus, on the measured activity. The British Pharmacopoeia specification for ^{201}Tl-thallous chloride requires that "Not more than 2.0 percent of the total radioactivity is due to thallium-202 and not less than 97.0 percent is due to thallium-201." Thallium-202 has a half-life of 12.2 d and the predominant photon energy is 440 keV. Another possible contaminant is ^{200}Tl which has a half-life of 1.09 d and prominent energies at 368 keV and 1.2 MeV. Both of these radionuclide contaminants will have a high efficiency in a dose calibrator. As the half-life of ^{202}Tl is significantly longer than that of ^{201}Tl, the relative proportion of ^{202}Tl to ^{201}Tl will increase over time. If the accuracy of a dose calibrator is to be checked with a ^{201}Tl source, the apparent accuracy could change depending on when the measurements are taken relative to the stated calibration date. The presence of these high energy contaminants will have an adverse effect on image quality due to increased septal penetration and will also lead to an increased radiation dose to the patient. The effective dose, in millisieverts per megabecquerel, for ^{200}Tl, ^{201}Tl and ^{202}Tl is 0.238, 0.149 and 0.608, respectively. It should be noted that these problems will be increased if the radiopharmaceutical is administered prior to the nominal calibration date, as the proportion of ^{200}Tl will be higher.

9.2. DOSE CALIBRATOR ACCEPTANCE TESTING AND QUALITY CONTROL

9.2.1. Acceptance tests

Acceptance tests for dose calibrators should include measurements of the accuracy, reproducibility, linearity and geometry response. These are required to ensure that the unit meets the manufacturer's specifications and to give baseline figures for subsequent quality control.

9.2.1.1. Accuracy and reproducibility

The accuracy is determined by comparing activity measurements using a traceable calibrated standard with the supplier's stated activity, corrected for radioactive decay. The accuracy is expressed in per cent deviation from the actual activity and should be measured for all radionuclides to be used routinely. It is recommended that measurements of a long lived source, for example ^{137}Cs, be recorded at the time of initial testing for each radionuclide setting to be used clinically for later quality control.

The reproducibility, or constancy, can be assessed by taking repeated measurements of the same source. If the sample holder is removed from the chamber between each measurement, the measured reproducibility will include any errors associated with possible variations in source position.

9.2.1.2. Linearity

There are several approaches to the measurement of the linearity response of a dose calibrator. Typically, a vial containing a high activity of ^{99m}Tc is measured repeatedly over a period of at least 5 d. During this time, a 100 GBq source will decay to 0.1 MBq. It is essential that the initial activity represents the highest activity that is likely to be used in clinical practice, which will usually be the first elution from a new Mo/Tc generator. A semi-log plot of the measurements, corrected for background, should follow the expected decay of the radionuclide. Any deviation from the expected line at high activities indicates saturation of response of the ionization chamber. Accurate background measurements, at the time of each assay, are essential as the background will become an increasing component of the reading as the source decays. Deviations from linearity at low activities are likely to be due to radionuclide impurities, such as ^{99}Mo in vials containing ^{99m}Tc.

Another approach that can be used to check the linearity requires a series of radioactive sources that cover the range of activities to be measured. The sources should all be prepared from the same stock solution and the dispensed volumes measured accurately by weighing the vials pre- and post-dispensing. The volume of liquid in each vial should be adjusted with a non-radioactive solution, so that the volume is identical in each vial, to eliminate any geometry dependency in the measurement. The measured activities are corrected for decay to the time of measurement of the first vial and plotted against the dispensed volumes to assess the calibrator linearity. The error in this method will be increased if there are any small variations in the vial wall thickness as the same vial is not used for all measurements.

Finally, linearity can be assessed by repeated measurements on a single vial using a series of graded attenuators appropriate for a specified test source to reduce the measured ionization current. These are typically a series of concentric cylinders that fit over the vial. The attenuation through each cylinder must be accurately known to use this method.

9.2.1.3. Geometry

The measured activity may vary with the position of the source within the ionization chamber, with the composition of the vial or syringe, or with the volume of liquid within the vial or syringe. Appropriate correction factors must be established for the containers and radionuclides to be used clinically, especially if radionuclides that have a substantial component of low energy photons, such as ^{123}I, are to be used. For each vial or syringe to be used clinically, a series of measurements should be undertaken in which the activity remains constant, but the volume is increased from 10 to 90% of the maximum volume by the addition of water or saline. Corrected for decay, a plot of activity against volume should be a straight horizontal line. Any deviations from this can be used to calculate the appropriate correction factor.

Similarly, vial to syringe correction factors can be determined by measuring the activity transferred from the vial to the syringe (original vial activity minus residual activity) and comparing this to the activity measured in the syringe itself.

Geometry dependencies should not change over time; however, if the practitioner changes the manufacturer of the syringes or obtains the radiopharmaceuticals in a different vial size, a new set of calibration factors should be determined.

9.2.2. Quality control

9.2.2.1. Background check

As noted in Section 9.1.3.5, when the source holder is empty, the dose calibrator will still record an 'activity' due to background radiation. This will come from natural background, from sources within the radiopharmacy and/or from contamination present on the source holder or well liner. It is a useful practice to keep a spare source holder and a spare well liner, so that if contamination is detected the contaminated item can be removed from service to be decontaminated, or left until the radioactivity has decayed.

At a minimum, the background should be determined each morning before the dose calibrator is used and recorded. The background subtraction feature, if available, can be used at that time to remove the measured background from subsequent measurements. The technologist should also confirm the absence of any additional background before all activity measurements during the day.

9.2.2.2. Check source reproducibility

A long lived check source should be used on a daily basis to confirm the constancy of the response of the dose calibrator. Sealed radioactive sources of ^{57}Co and ^{137}Cs, shaped to mimic a vial, are available commercially for this purpose. The check source should be measured on all radionuclide settings that are used clinically. Although the recorded activity of a ^{137}Cs source on the ^{99m}Tc setting will not be a correct measurement of its activity, a reading outside of that expected from previous results may indicate a faulty dose calibrator or a change in calibration factor, in this case of ^{99m}Tc.

9.3. STANDARDS APPLYING TO DOSE CALIBRATORS

The International Electrotechnical Commission (IEC) has published two standards [9.2, 9.3] and a technical report [9.4] relating to dose calibrators. IEC standards are often adopted by national standards organizations. Reference [9.3] is for manufacturers to use to ensure that the equipment performance is specified in a standardized way, while Ref. [9.4] is aimed at the users of dose calibrators.

There should also be national standards covering dose calibrators. The American National Standards Institute publication ANSI N42.13-2004 [9.5] is often referenced by US manufacturers. This specifies the minimum requirements in terms of accuracy and reproducibility for dose calibrators:

— "The accuracy of the instruments, at activity levels above 3.7 MBq shall be such that the measured activity of a standard source shall be within ±10% of the stated activity of that source" [9.5];

— "The reproducibility…shall be such that all of the results in a series of ten consecutive measurements on a source of greater than 100 μCi (3.7×10^6 Bq) in the same geometry shall be within ±5% of the average measured activity for that source" [9.5].

9.4. NATIONAL ACTIVITY INTERCOMPARISONS

National metrology institutes are responsible for the development and maintenance of standards, including activity standards. These institutes, often in collaboration with the relevant national professional body, have undertaken national comparisons of the accuracy of the dose calibrators used in clinical practice. Such comparisons have used, where possible, the clinical radionuclides ^{67}Ga, ^{123}I, ^{131}I, ^{99m}Tc and ^{201}Tl, and have been carried out in Argentina, Australia, Brazil, Cuba, the Czech Republic, Germany, India and the United Kingdom. In some countries, such as Cuba and the Czech Republic, participation in the comparison is mandatory, while in many other countries it is voluntary. The surveys can also be used to measure the reproducibility of the calibrators.

As an example, Table 9.3 shows the results from a survey undertaken in Australia in 2007.

TABLE 9.3. SUMMARY OF THE RESULTS OF THE DOSE CALIBRATOR SURVEY UNDERTAKEN IN AUSTRALIA IN 2007

Radionuclide	^{99m}Tc	^{131}I	^{67}Ga	^{201}Tl
No. of calibrators	167	164	116	162
Within ±5% error	86%	80%	84%	73%
Within ±10% error	98%	95%	97%	94%
Within ±10% reproducibility	100%	100%	100%	100%

These surveys also offer the opportunity for the calibration factor to be adjusted if a dose calibrator is found to be operating with an error of >10%.

9.5. DISPENSING RADIOPHARMACEUTICALS FOR INDIVIDUAL PATIENTS

9.5.1. Adjusting the activity for differences in patient size and weight

Protocols used in nuclear medicine practices should specify the usual activity of the radiopharmaceutical to be administered to a standard patient. In most western countries, the standard patient is taken to be one whose weight is in the range 70–80 kg. However, many patients fall outside of this range. If a fixed activity is used for all patients, this will lead to an unnecessarily high radiation exposure to an underweight patient and may lead to images of unacceptable quality or very long imaging times in obese patients.

There have been various approaches to determining the activity to be administered. These are usually designed to provide a constant count density in the image to maintain image quality or to provide a constant effective dose to the patient. For example, it has been shown that for myocardial perfusion scans using ^{99m}Tc-tetrofosmin, the activity should be increased by 150% for a 110 kg patient and by 200% for a 140 kg patient in order to maintain image quality without increasing imaging time.

It has been shown, using the radiation dose tables provided in International Commission on Radiological Protection (ICRP) publications 53, 80 and 106 [9.6–9.8], that the effective dose (mSv/MBq) can be expressed as a simple power function of body weight. Scaling factors for the activity, to give a constant effective dose can, therefore, be derived from the expression $(W/70)^a$, where W represents the weight of the person and the power factor a is specific for the radiopharmaceutical. Again, using ^{99m}Tc-tetrofosmin as an example, a is found to be –0.834. Although the dosimetry models are only available up to 70 kg, this power function can be extrapolated to derive scaling factors for patients whose weight exceeds 70 kg. Using this approach, the activity should be increased by 146% for a 110 kg patient and by 178% for a 140 kg patient. This approach is useful, but should be used with caution. The extrapolated activity would lead to comparable organ and tissue doses for a patient of large body build but not for a patient of similar weight due to large body fat deposits as the biodistribution of the radiopharmaceutical would not be the same in these two cases. Table 9.4 presents the a value for common radiopharmaceuticals.

9.5.2. Paediatric dosage charts

Children are approximately three times more radiosensitive than adults, so determining the appropriate activity to be administered for paediatric procedures is essential. In addition to the scaling factor to be applied to the adult activity, a

TABLE 9.4. THE POWER FACTOR *a* RELATING BODY WEIGHT TO A CONSTANT EFFECTIVE DOSE ACCORDING TO THE EXPRESSION $(W/70)^a$ FOR 14 COMMON RADIOPHARMACEUTICALS

Radiopharmaceutical	*a* value	Radiopharmaceutical	*a* value
^{99m}Tc-DMSA	–0.706	^{99m}Tc-IDA	–0.840
^{99m}Tc-DTPA	–0.801	^{99m}Tc-tetrafosmin	–0.834
^{99m}Tc-MAG3	–0.520	^{99m}Tc-red cells	–0.859
^{99m}Tc-HMPAO	–0.849	^{99m}Tc-white cells	–0.869
^{99m}Tc-MAA	–0.842	^{18}F-FDG	–0.782
^{99m}Tc-sestamibi	–0.871	^{67}Ga-citrate	–0.931
^{99m}Tc-phosphonates	–0.763	^{123}I or ^{131}I iodide	–1.11

minimum activity must be specified in order to ensure adequate image quality. In the past, the scaling factors were assessed using weight alone or body surface area obtained from both height and weight. These two methods can give rise to quite different scaling factors. For example, the scaling factor for a 20 kg child is 29% of the adult activity using weight alone, but 43% when based on body surface area.

Recently, the European Association of Nuclear Medicine (EANM) Dosimetry and Paediatric Committees have prepared a dosage card which recognizes that a single scaling factor is not optimal for all radiopharmaceuticals. They used the methodology presented in Section 9.5.1 and were able to establish that radiopharmaceuticals could be grouped into three classes (renal, thyroid and others), with different scaling factors for each class. A dosage card is available on the EANM web site that gives the minimum recommended activity and a weight dependent scaling factor for each radiopharmaceutical which was determined to give weight independent effective doses. This dosage card is reproduced here as Fig. 9.6. To assist in these calculations, an on-line dosage calculator is available on the EANM web site[1], in which the user specifies the child's weight and the radiopharmaceutical, and the recommended activity is displayed.

[1] http://www.eanm.org/publications/dosage_calculator.php?navId=285

9.5.3. Diagnostic reference levels in nuclear medicine

The recommendations of the ICRP specifically exclude medical exposures from its system of dose limits, as the patient is directly benefiting from the radiation exposure. However, in Publication 73 (1996) [9.9], the ICRP introduced the term 'diagnostic reference level' (DRL) for patients. DRLs are investigation levels and are based on an easily measured quantity, usually the entrance surface dose in the case of diagnostic radiology, or the administered activity in the case of nuclear medicine. DRLs are referred to by the IAEA as guidance levels in Safety Reports Series No. 40 [9.10], published in 2005. This publication contains a table of guidance levels reflecting the values used in the early 1990s, when single photon emission computed tomography procedures were far less common. A survey of the use of DRLs in eight European countries, published in 2007 [9.11], showed that their introduction in nuclear medicine varied considerably. For example, France had set DRLs for 10 nuclear medicine procedures, Germany for 17 procedures, Italy for 48 procedures, while the United Kingdom listed 96 procedures. In some countries, the DRLs were set at the activities for which marketing approval had been given, while in other countries the DRLs were determined for each procedure by calculating the 75th percentile of the spread of data values collected from a survey of participating practices. The latter approach has been widely used to set DRLs in radiology and has been used in other parts of the world, such as Australia and New Zealand, to establish DRLs in nuclear medicine.

EANM
European Association of Nuclear Medicine
BIOMEDICAL IMAGING AND THERAPY FOR PERSONALIZED HEALTHCARE

Dosage Card (Version 1.2.2014)

Multiple of Baseline Activity

Weight kg	Class A	Class B	Class C	Weight kg	Class A	Class B	Class C
3	1	1	1	32	3.77	7.29	14.00
4	1.12	1.14	1.33	34	3.88	7.72	15.00
6	1.47	1.71	2.00	36	4.00	8.00	16.00
8	1.71	2.14	3.00	38	4.18	8.43	17.00
10	1.94	2.71	3.67	40	4.29	8.86	18.00
12	2.18	3.14	4.67	42	4.41	9.14	19.00
14	2.35	3.57	5.67	44	4.53	9.57	20.00
16	2.53	4.00	6.33	46	4.65	10.00	21.00
18	2.71	4.43	7.33	48	4.77	10.29	22.00
20	2.88	4.86	8.33	50	4.88	10.71	23.00
22	3.06	5.29	9.33	52-54	5.00	11.29	24.67
24	3.18	5.71	10.00	56-58	5.24	12.00	26.67
26	3.35	6.14	11.00	60-62	5.47	12.71	28.67
28	3.47	6.43	12.00	64-66	5.65	13.43	31.00
30	3.65	6.86	13.00	68	5.77	14.00	32.33

$$A[MBq]_{\text{Administered}} = BaselineActivity \times Multiple$$

a) For a calculation of the administered activity, the baseline activity value has to be multiplied by the multiples given above for the recommended radiopharmaceutical class (see reverse).

b) If the resulting activity is smaller than the minimum recommended activity, the minimum activity should be administered.

c) The national diagnostic reference levels should not be exceeded!

Examples:

a) ^{18}F FDP-PET Brain, 50 kg:	activity to be administered [MBq] = 14.0 x10.71 [MBq] ≈ 150 MBq
b) ^{123}I mIBG, 3 kg:	activity to be administered [MBq] = 28.0 x1 [MBq] = 28 MBq < 37 MBq (Minimum Recommended Activity) **→ activity to be administered: 37 MBq**

This card is based upon the publication by Jacobs F, Thierens H, Piepsz A, Bacher K, Van de Wiele C, Ham H, Dierckx RA. Optimized tracer-dependent dosage cards to obtain weight-independent effective doses. Eur J Nucl Med Mol Imaging. 2005 May; 32(5):581-8.

This card summarizes the views of the Paediatric and Dosimetry Committees of the EANM and reflects recommendations for which the EANM cannot be held responsible.
The dosage recommendations should be taken in context of „good practice" of nuclear medicine and do not substitute for national and international legal or regulatory provisions.

Android App   iPhone App

EANM Executive Secretariat
Hollandstrasse 14/Mezzanine · 1020 Vienna, Austria
Phone: +43-1-2128030, fax: +43-1-21280309
office@eanm.org - www.eanm.org - fb/officialEANM

FIG. 9.6. European Association of Nuclear Medicine (EANM) paediatric dosage card (courtesy of EANM).

Recommended Amounts in MBq

Radiopharmaceutical	Class	Baseline Activity (for calculation purposes only) MBq	Minimum Recommended Activity[1] MBq
^{123}I (Thyroid)	C	0.6	3
^{123}I Amphetamine (Brain)	B	13.0	18
^{123}I HIPPURAN (Abnormal renal function)	B	5.3	10
^{123}I HIPPURAN (Normal renal function)	A	12.8	10
^{123}I mIBG	B	28.0	37
^{131}I mIBG	B	5.6	35
^{18}F FDG-PET torso	B	25.9	26
^{18}F FDG-PET brain	B	14.0	14
^{18}F Sodium fluoride	B	10.5	14
^{67}Ga Citrate	B	5.6	10
^{99m}Tc ALBUMIN (Cardiac)	B	56.0	80
^{99m}Tc COLLOID (Gastric Reflux)	B	2.8	10
^{99m}Tc COLLOID (Liver/Spleen)	B	5.6	15
^{99m}Tc COLLOID (Marrow)	B	21.0	20
^{99m}Tc DMSA	B	6.8	18.5
^{99m}Tc DTPA (Abnormal renal function)	B	14.0	20
^{99m}Tc DTPA (Normal renal function)	A	34.0	20
^{99m}Tc ECD (Brain perfusion)	B	32.0	110
^{99m}Tc HMPAO (Brain)	B	51.8	100
^{99m}Tc HMPAO (WBC)	B	35.0	40
^{99m}Tc IDA (Biliary)	B	10.5	20
^{99m}Tc MAA / Microspheres	B	5.6	10
^{99m}Tc MAG3	A	11.9	15
^{99m}Tc MDP	B	35.0	40
^{99m}Tc Pertechnetate (Cystography)	B	1.4	20
^{99m}Tc Pertechnetate (Ectopic Gastric Mucosa)	B	10.5	20
^{99m}Tc Pertechnetate (Cardiac First Pass)	B	35.0	80
^{99m}Tc Pertechnetate (Thyroid)	B	5.6	10
^{99m}Tc RBC (Blood Pool)	B	56.0	80
^{99m}Tc SestaMIBI/Tetrofosmin (Cancer seeking agent)	B	63.0	80
^{99m}Tc SestaMIBI/Tetrofosmin[2] (Cardiac rest scan 2-day protocol min)	B	42.0	80
^{99m}Tc SestaMIBI/Tetrofosmin[2] (Cardiac rest scan 2-day protocol max)	B	63.0	80
^{99m}Tc SestaMIBI/Tetrofosmin[2] (Cardiac stress scan 2-day protocol min)	B	42.0	80
^{99m}Tc SestaMIBI/Tetrofosmin[2] (Cardiac stress scan 2-day protocol max)	B	63.0	80
^{99m}Tc SestaMIBI/Tetrofosmin[2] (Cardiac rest scan 1-day protocol)	B	28.0	80
^{99m}Tc SestaMIBI/Tetrofosmin[2] (Cardiac stress scan 1-day protocol)	B	84.0	80
^{99m}Tc Spleen (Denatured RBC)	B	2.8	20
^{99}Tc TECHNEGAS (Lung ventilation)[3]	B	70.0	100

[1] The minimum recommended activities are calculated for commonly used gamma cameras or positron emission tomographs. Lower activities could be administered when using systems with higher counting efficiency.

[2] The minimum and maximum values correspond to the recommended administered activities in the EANM/ESC procedural guidelines (Hesse B, Tagil K, Cuocolo A, et al). EANM/ESC procedural guidelines for myocardial perfusion imaging in nuclear Cardiology. Eur J Nucl Med Mol Imaging. 2005 Jul;32(7):855-97.

[3] This is the activity load needed to prepare the Technegas device. The amount of inhaled activity will be lower.

FIG. 9.6. European Association of Nuclear Medicine (EANM) paediatric dosage card (courtesy of EANM) (cont.).

9.6. RADIATION SAFETY IN THE RADIOPHARMACY

9.6.1. Surface contamination limits

Surface contamination with radioactivity could lead to contamination of a radiation worker and/or external irradiation of the skin of the worker. Internal contamination could arise from inhalation and/or ingestion of the radionuclide. The surface contamination limits given in Table 9.5 were derived based on a committed effective dose limit of 20 mSv/a and the models for inhalation and ingestion given in ICRP publications 30, 60 and 61 [9.12–9.14]. For each radionuclide, the most restrictive pathway (inhalation, ingestion or external irradiation) was used

TABLE 9.5. DERIVED LIMITS FOR SURFACE CONTAMINATION

Nuclide	Surfaces in designated areas, including protective clothing (Bq/cm^2)	Interiors of glove boxes and fume cupboards (Bq/cm^2)	Non-designated areas including personal clothing (Bq/cm^2)
^{18}F	100	1 000	5
^{32}P	100	1 000	5
^{51}Cr	1 000	10 000	50
^{67}Ga	1 000	10 000	50
^{89}Sr	100	1 000	5
^{90}Y	100	1 000	5
^{99m}Tc	1 000	10 000	50
^{111}In	1 000	10 000	50
^{123}I	1 000	10 000	50
^{125}I	100	1 000	5
^{131}I	100	1 000	5
^{177}Lu	1 000	10 000	50
^{201}Tl	1 000	10 000	50

9.6.2. Wipe tests and daily surveys

Surveys of the radiopharmacy must be undertaken to ensure that these surface contamination limits are not exceeded and that the operator is not unnecessarily exposed to external radiation. Exposure could result from sources inadvertently left on a bench and from contamination on bench surfaces. Aerosolized droplets from a syringe during dispensing may go unnoticed, so it is essential that all staff are aware that the dispensing area may be contaminated and always wear protective gloves when working in this area. All radiopharmaceutical elution, preparation, assay and administration areas should be surveyed at the end of each working day.

Surveys should initially be undertaken with a survey meter to ensure that no unexpected exposed sources are present in the radiopharmacy. All surfaces should then be checked for contamination using a contamination monitor with a probe appropriate to the radionuclides used. The background radiation levels in the radiopharmacy, particularly in the dispensing area, are often higher than elsewhere in the nuclear medicine department, so quantifying any contamination found using a probe is difficult. If a low energy β emitter is being used, it will prove difficult or impossible to detect with an external probe. In these situations, a wipe test should be used. A minimum area of 100 cm^2 should be wiped and then the activity on the wipe can be assessed using a pancake probe, or more accurately in a well counter. For low energy β emitters, such as ^{3}H or ^{14}C, liquid scintillation counting must be used. When quantifying the surface contamination, it is generally assumed that a wipe test using a dry wipe will remove one tenth of the contamination while a wet wipe will remove one fifth of the contamination.

9.6.3. Monitoring of staff finger doses during dispensing

Systematic studies of the dose to the hands of staff working in radiopharmacies have shown that finger doses may approach or exceed the annual dose limit of 500 mSv to the extremities. The most exposed parts of the hands are likely to be the tips of the index and middle fingers, and the thumb of the dominant hand, with exposure for the index finger being highest. The ICRP has recommended that finger dose monitoring be undertaken for any person handling more than 2 GBq/d and regular monitoring should be carried out if doses to the most exposed part of the hand exceed 6 mSv/month.

Although the dose to the finger tip will be the highest, it is much more practical to wear a ring monitor at the base of the finger. A thermoluminescent dosimeter chip mounted in a plastic ring is usually the most convenient type of monitor. Such monitors are often available in a variety of sizes. The ring should fit tightly, so that it is not inadvertently removed when the gloves are taken off.

The ICRP recommends that the ring monitor be worn on the middle finger with the element positioned on the palm side, and that a factor of three should be applied to derive an estimate of the dose to the tip. If the dosimeter element is worn facing towards the back of the hand, a factor of six should be applied.

The dose to the fingers is critically dependent on the dispensing technique used and the skill of the operator. It is important that staff undertake extensive training in the dispensing technique with non-radioactive solutions prior to dispensing radiopharmaceuticals for the first time. This is particularly important with PET radiopharmaceuticals as the specific dose rate constant is much higher for positron emitters than for radionuclides used for single photon imaging.

9.7. PRODUCT CONTAINMENT ENCLOSURES

9.7.1. Fume cupboards

A fume cupboard is an enclosed workplace designed to prevent the spread of fumes to the operator and other persons. The 'fumes' can be in the form of gases, vapours, aerosols or particulate matter. The fume cupboard is designed to provide operator protection rather than protection for the product within the cabinet. A fume cupboard would, therefore, not be suitable as an area for cell labelling procedures as this requires that the blood remain sterile at all times. Fume cupboards usually include a transparent safety screen which can be adjusted either vertically (more commonly) or horizontally to vary the size of the working aperture into the cabinet. Some cupboards are available with a lead glass safety screen to minimize the need for additional radiation shielding. The most common type of fume cupboard is known as a variable exhaust air volume fume cupboard which maintains a constant velocity of air into the cabinet (the face velocity) irrespective of the sash position. Figure 9.7 shows a fume cupboard suitable for use with radioactive materials.

Fume cupboards are available which discharge the exhaust air directly, or after carbon filtration, to the atmosphere, usually above the building. Other cabinets, known as recirculating fume cabinets, rely on filtration or absorption to remove airborne contaminants released in the cabinet, so that the air may be safely discharged back into the laboratory. Recirculating fume cabinets are not normally applicable for use with radioactive materials.

Any installed fume cupboards must meet the requirements of the local appropriate standard and any air discharged to the atmosphere must meet the requirements of the appropriate regulatory authority. The standard will usually specify the minimum face velocity through the working aperture (e.g. 0.5 m/s). This should be checked on a regular basis and should be measured with the aperture fully open and in its minimum position. At the minimum position, the face velocity may need to be higher to retain a constant exhaust rate from the cabinet.

Before initial use, and as part of a regular quality control schedule, a smoke test should be performed. This is to provide visual evidence of fume containment within, or escape from, the fume cupboard. Smoke is released in and around the fume cupboard and the visual pattern of airflow is observed. The results of the smoke test must be documented and any reverse flows from the confines of the cupboard corrected before subsequent use.

FIG. 9.7. Fume cupboard suitable for use with radioactive materials.

9.7.2. Laminar flow cabinets

Laminar flow cabinets provide a non-turbulent airstream of near constant velocity, which has a substantially uniform flow cross-section and with a variation in velocity of not more than 20%. Laminar flow cabinets provide product protection while a fume cupboard is designed to provide operator protection. The air supplied to the cabinet is usually passed through a high efficiency particulate air filter, which is designed to remove 99.999% of particles greater than 0.3 μm in size. It must be remembered that the laminar flow of air (usually vertical) will be disturbed by the presence of any objects within the cabinet, including shielding and the arms of the operator. During use, the filtered air may escape from the front of the cabinet, when the airflow is disturbed, so operator protection cannot be ensured.

9.7.3. Isolator cabinets

Isolator cabinets provide both operator and product protection. Figure 9.8 shows an example of a blood cell labelling isolator. The product is manipulated through glove ports so that the interior of the cabinet is maintained totally sterile and full operator protection is provided. Airflow within the isolator is deliberately designed to be turbulent so that there are no dead spaces within the cabinet. The unit illustrated incorporates a centrifuge which can be controlled externally. A dose calibrator can be included within the isolator, so that the cell suspension does not need to be removed from the isolator for the activity to be measured. The isolator incorporates timed interlocks on the vacuum door seals to ensure that the product remains sterile.

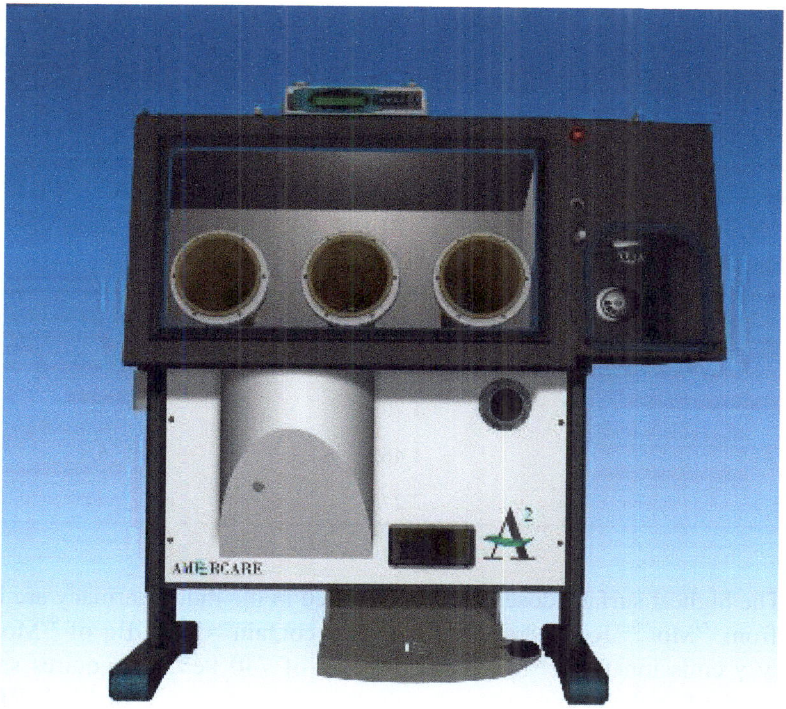

FIG. 9.8. Blood cell labelling isolator (courtesy of Amercare Ltd).

9.8. SHIELDING FOR RADIONUCLIDES

9.8.1. Shielding for γ, β and positron emitters

Shielding will be required in the walls of the radiopharmacy, in any containment enclosures, in a body shield to protect the operator at the dispensing station, and around individual vials and syringes containing radionuclides. Shielding of the walls of the radiopharmacy can be minimized by appropriate local shielding around the sources being handled. Shielding may be constructed from a variety of materials, including lead and concrete in walls, lead or tungsten in local shielding for γ emitting radionuclides, and aluminium or Perspex for pure β emitters. For positron emitters, the shielding will be primarily determined by the 511 keV annihilation photons rather than by the positrons themselves. A low atomic number material, such as aluminium or Perspex, is used for pure β emitters since this minimizes the production of bremsstrahlung radiation. As β radiation has a finite range in materials, determined by the maximum β energy, the thickness of the shielding needs to be greater than this range to ensure that all of the β particles are absorbed. Polymethyl methacrylate (Perspex or lucite) has a density of 1.19 g/cm^3, similar to the density of tissue and water, and is highly suitable for absorbing β particles. Table 9.6 gives the maximum β energy and the range in water for four pure β emitters used in nuclear medicine.

TABLE 9.6. MAXIMUM β ENERGY AND RANGE IN WATER FOR FOUR β EMITTERS USED CLINICALLY IN NUCLEAR MEDICINE. THE ELECTRON RANGE HAS BEEN DETERMINED USING THE CONTINUOUS SLOWING DOWN APPROXIMATION

Radionuclide	E_{max} (MeV)	Range in water (mm)
^{14}C	0.156	0.30
^{32}P	1.709	8.2
^{89}Sr	1.463	6.8
^{90}Y	2.274	11

The highest surface dose rates encountered in the radiopharmacy are likely to be from ^{99}Mo/^{99m}Tc generators which may contain >100 GBq of ^{99}Mo. The primary γ emission from ^{99}Mo has an energy of 740 keV, so requires several centimetres of lead shielding to reduce the dose rates to an acceptable level. The generator, as supplied, will already contain substantial shielding but additional shielding will usually be required. This may be available from the generator supplier specifically designed for their generator or it may be necessary to construct or purchase an additional shield. Figure 9.9 shows a generator supplied by ANSTO in Australia inside a dedicated lead ‘garage’. The body of the radiochemist is shielded by the garage doors while she is attaching the shielded elution vial prior to an elution of the generator. These shields are heavy (>20 kg), so it is important that the bench surfaces are strong enough to support the weight. The generator is itself quite heavy, so mechanical lifting devices may be necessary to prevent back injuries to staff when lifting the generator into position.

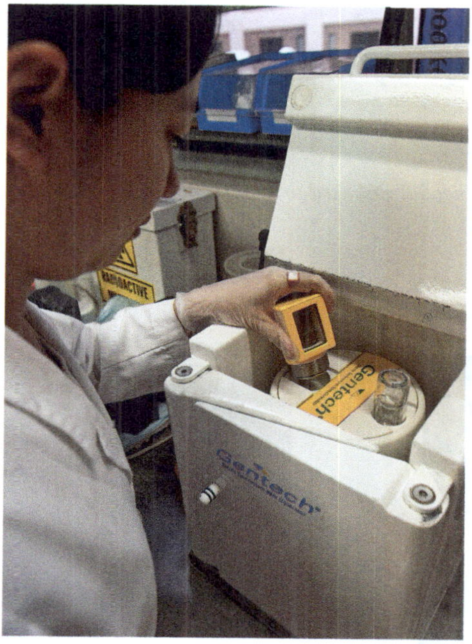

FIG. 9.9. Lead garage surrounding a $^{99}Mo/^{99m}Tc$ generator.

Vials of radiopharmaceuticals must be kept shielded. The shields are usually constructed so that only the rubber septum of the vial is accessible, thereby protecting the hands of the operator during dispensing (see Fig. 9.10). The vials themselves should never be held by the fingers once they contain radioactivity, instead long forceps should always be used (see Fig. 9.11).

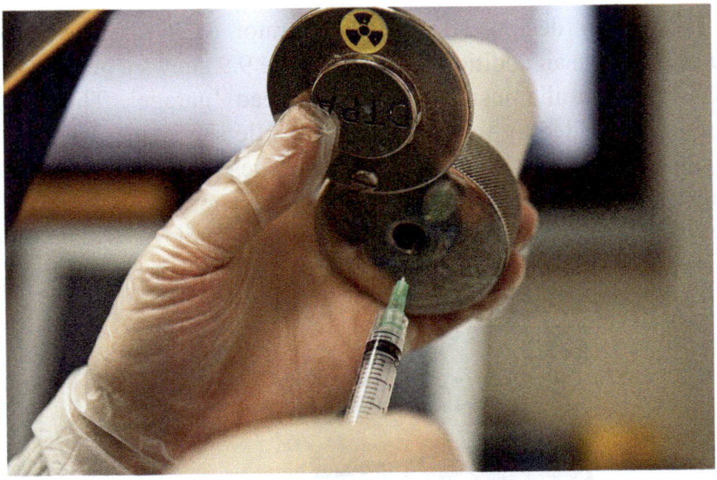

FIG. 9.10. Shielded vial used to hold reconstituted radiopharmaceuticals.

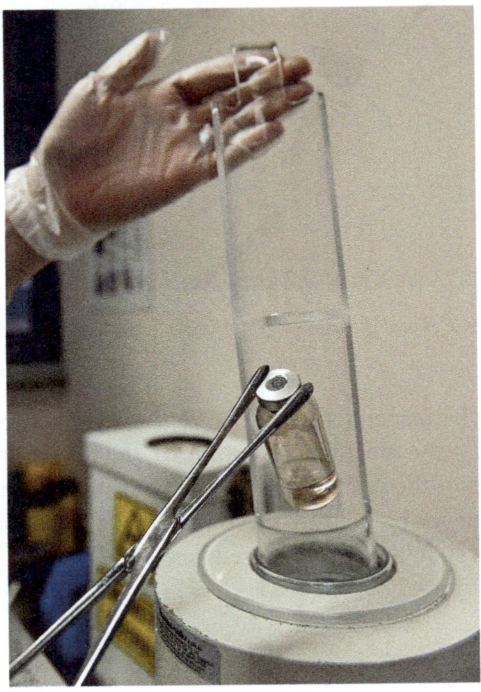

FIG. 9.11. Using long forceps to handle a vial containing radioactivity.

During radiopharmaceutical preparation, dispensing and administration to the patient, the activity is usually manipulated in syringes. The dose rates at the surface of the syringes may exceed 1 μSv · s^{-1} · MBq^{-1} depending on the volume of liquid and the size of the syringe. The plastic of the syringe provides little absorption of any high energy β particles, and for radionuclides used for therapy, the surface dose rates will be in excess of 10 mSv/s, so that the annual dose limit of 500 mSv to the extremities can easily be exceeded. Syringes should never be more than half filled, so that the syringe can be picked up near the plunger where the fingers are not over the activity. Syringe shields must be used whenever possible. These must be made of Perspex for the pure β emitters and of lead or tungsten for the γ emitters (see Fig. 9.12). A lead glass window is necessary to permit observation of the contents of the syringe. Syringe shields with a spring-loaded catch to hold the syringe in place are preferable to those using a screw, as the screw-thread wears quickly with use.

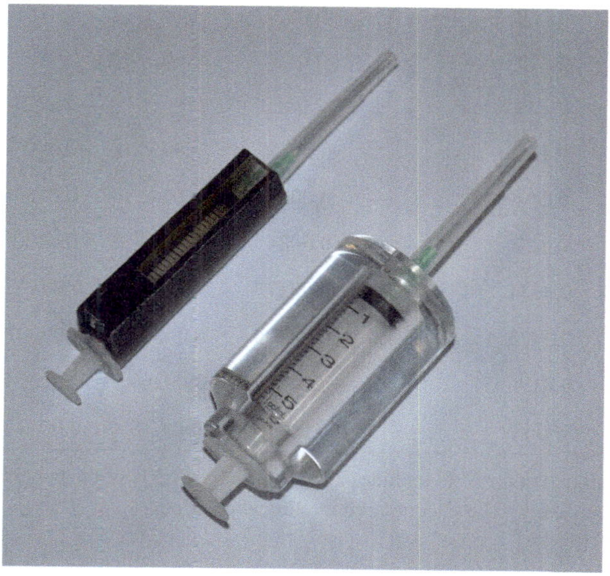

FIG. 9.12. A tungsten syringe shield for γ emitting radionuclides and a Perspex syringe shield for pure β emitters.

TABLE 9.7. MEASURED TRANSMISION FACTORS FOR LEAD

Thickness of lead (mm)	^{99}Mo	^{99m}Tc	^{67}Ga	^{131}I	^{201}Tl[a]	511 keV
0	1.00	1.00	1.00	1.00	1.00	1.00
1	0.876	0.105	0.455	0.769	0.136	0.891
2	0.776	0.00835	0.280	0.601	0.0709	0.787
3	0.694	6.52×10^{-4}	0.191	0.475	0.0557	0.690
4	0.623	5.09×10^{-5}	0.135	0.379	0.0485	0.602
5	0.561	3.97×10^{-6}	0.0983	0.306	0.0438	0.523
6	0.507	3.10×10^{-7}	0.0730	0.248	0.0430	0.452
7	0.458		0.0551	0.203	0.0422	0.390
8	0.414		0.0420	0.168	0.0415	0.336
9	0.375		0.0324	0.139	0.0408	0.289
10	0.340		0.0253	0.116	0.0291	0.249
12	0.279		0.0157	0.0829	0.0282	0.183
14	0.229		0.0102	0.0605	0.0273	0.135
16	0.188		0.00682	0.0451	0.0193	0.0990
18	0.154		0.00476	0.0342	0.0187	0.0728
20	0.127		0.00345	0.0263	0.0132	0.0535
25	0.0774		0.00177	0.0143	0.00893	0.0247
30	0.0473		0.00104	0.00805	0.00602	0.0114
40	0.0176		4.11×10^{-4}	0.00267	0.00274	0.00240
50	0.00659		1.71×10^{-4}	9.04×10^{-4}	0.00124	5.00×10^{-4}

[a] The transmission data for ^{201}Tl includes a contribution of 1.5% of the contaminant ^{200}Tl, the maximum level likely to be encountered in clinical practice.

9.8.2. Transmission factors for lead and concrete

Section 1.6 indicates that the attenuation of monoenergetic photons through materials such as lead or concrete will be exponential, characterized by the linear attenuation coefficient or the half-value layer (HVL). However, this is only correct for narrow beam geometries, using collimated beams of radiation, which are rarely encountered in practice. Furthermore, the attenuation of the radiation from radionuclides which emit more than one γ photon, such as ^{67}Ga and ^{131}I, cannot be expressed as a simple HVL.

Tables 9.7 and 9.8 give the measured broad beam transmission factors for lead and concrete for five radionuclides used in nuclear medicine and for 511 keV photons from positron emitters. This information can be used to calculate the required thickness of shielding around vials, for the body protection of the operator and for the walls of the radiopharmacy. The values for ^{201}Tl include a contribution from ^{200}Tl, a common contaminant, which has prominent energies at 368 keV and 1.2 MeV. A contribution of 1.5% of ^{200}Tl has been included which is the maximum likely value at the time of calibration.

TABLE 9.8. MEASURED TRANSMISSION FACTORS FOR CONCRETE (DENSITY: 2.35 g/cm^3)

Thickness of concrete (mm)	^{99}Mo	^{99m}Tc	^{67}Ga	^{131}I	^{201}Tl[a]	511 keV
0	1.00	1.00	1.00	1.00	1.00	1.00
10	0.845	0.779	0.884	0.916	0.759	0.958
20	0.718	0.607	0.769	0.825	0.581	0.909
30	0.614	0.473	0.661	0.735	0.449	0.852
40	0.527	0.368	0.564	0.649	0.349	0.789
50	0.454	0.287	0.477	0.570	0.274	0.722
60	0.393	0.224	0.402	0.498	0.217	0.653
70	0.341	0.174	0.338	0.434	0.173	0.584
80	0.296	0.136	0.282	0.377	0.139	0.518
90	0.253	0.106	0.236	0.327	0.112	0.456
100	0.225	0.0824	0.196	0.284	0.0912	0.399
120	0.172	0.0500	0.135	0.212	0.0612	0.301
140	0.132	0.0304	0.0928	0.158	0.0418	0.224
160	0.101	0.0184	0.0635	0.118	0.0290	0.166
180	0.0777	0.0112	0.0434	0.0879	0.0203	0.123
200	0.0598	0.00679	0.0296	0.0654	0.0143	0.0904
250	0.0312	0.00195	0.0113	0.0312	0.00607	0.0419

TABLE 9.8. MEASURED TRANSMISSION FACTORS FOR CONCRETE (DENSITY: 2.35 g/cm^3) (cont.)

Thickness of concrete (mm)	^{99}Mo	^{99m}Tc	^{67}Ga	^{131}I	^{201}Tl[a]	511 keV
300	0.0163	5.60×10^{-4}	0.00433	0.0149	0.00262	0.0194
400	0.00443	4.61×10^{-5}	6.30×10^{-4}	0.00339	4.95×10^{-4}	0.00417
500	0.00121	3.80×10^{-6}	9.16×10^{-5}	7.70×10^{-4}	9.39×10^{-5}	8.95×10^{-4}

[a] The transmission data for ^{201}Tl includes a contribution of 1.5% of the contaminant ^{200}Tl, the maximum level likely to be encountered in clinical practice.

9.9. DESIGNING A RADIOPHARMACY

Every radiopharmacy is unique and there is no one design that can be used in all situations. The requirements of a single camera practice using only ^{99m}Tc radiopharmaceuticals will be very different from a large teaching hospital with PET facilities and in-patient radionuclide therapy rooms. However, in addition to the general building requirements given in section 3.1.3 of Ref. [9.10], there are some general rules specific to a radiopharmacy that can be applied in most situations:

- The radiopharmacy should be located in an area that is not accessible to members of the public.
- There should be easy access from the radiopharmacy to the injection rooms and imaging rooms to minimize the distance that radioactive materials need to be transported.
- The radiopharmacy should not be adjacent to areas that require a low and constant radiation background such as a counting room.
- There should be an area within the radiopharmacy designated as a non-active area that is used for record keeping and/or computer entry.
- A refrigerator will be required for the storage of lyophilized radiopharmaceutical kits. A laboratory-grade unit is preferred to ensure that the temperature remains constant.
- A dedicated dispensing area with a body shield and lead glass viewing window will be required. This will normally be adjacent to the dose calibrator, so that the dispensed activity can be measured while the operator is still protected by the body shield. The thickness of the shield and window will depend on the radionuclide or radionuclides in use. PET radionuclides

will require substantial thickness and lead glass should be supplied as a single block rather than as a stack of thinner sheets.

- A storage area will be required for reconstituted radiopharmaceuticals, in shielded containers, together with radiopharmaceuticals purchased ready for dispensing such as ^{67}Ga-citrate and ^{201}Tl-chloride.
- The radiopharmacy must contain facilities for radioactive waste disposal. This will normally include separate shielded storage bins for short lived radionuclides such as ^{99m}Tc and for radionuclides with longer half-lives such as ^{131}I. In addition, there must be shielded containers for 'sharps', such as syringes with needles. A separate shielded storage bin may be required if a large number of bulky items, such as aerosol or Technegas kits, need to be stored.
- If a Mo/Tc generator is used, this should be positioned away from the dispensing area to minimize the dose received by the person dispensing the radiopharmaceuticals. Some countries require the generator to be housed inside a laminar flow cabinet. All local regulatory requirements must be taken into account when designing the radiopharmacy.
- If cell labelling procedures are to be performed, a dedicated area with a laminar flow cabinet or isolator will be required to ensure that the product remains sterile during the labelling procedure.
- A fume cupboard, together with an activated charcoal filter on the exhaust, will be required if radioiodination procedures are to be performed.
- Some radiopharmaceuticals require a heating step in their preparation. This is often performed using a temperature controlled heating block. This must be in a dedicated separately shielded area, particularly as several gigabecquerels of ^{99m}Tc are often involved. Similarly, the radiolabelling of blood samples may require local shielding of mixers and centrifuges.
- Wall, floor and ceiling surfaces should be smooth, impervious and durable, and free of externally mounted features such as pipes or ducts to facilitate any radioactive decontamination.
- Bench surfaces should be constructed of plastic laminate or resin composites or stainless steel, and benches must be able to safely withstand the weight of any required lead shielding.
- Hand washing facilities must be available which can be operated without the use of the operator's hands to prevent the spread of any contamination. An eye-wash should also be available.
- A contamination monitor must be available in a readily accessible location. A wall-mounted monitor to check for any hand contamination should be mounted near the exit from the radiopharmacy. A model which can be removed and used as a general contamination monitor is useful.

9.10. SECURITY OF THE RADIOPHARMACY

Until relatively recently, the safety of the staff when handling and storing radioactive materials was the sole concern when designing a radiopharmacy. The security of the radioactivity was often not specifically addressed. Unfortunately, it is now apparent that radioactive materials can be used for malicious purposes and the security of the radiopharmacy must now be considered.

The IAEA has categorized radioactive sources on a scale of 1 to 5, based on activity and nuclide, where category 1 is potentially the most hazardous. Sources categorized as 1, 2 or 3 are known as security enhanced sources. The security measures in place for safety purposes are considered adequate to ensure the physical security of category 4 and 5 sources. Legislation is now, or will be, in place in each jurisdiction to address the security of security enhanced sources. This currently only applies to sealed sources, and no sealed sources used in nuclear medicine are categorized as either 1, 2 or 3. However, the principles can be applied to unsealed sources. A Mo/Tc generator with an activity of greater than 300 GBq is a category 3 source.

Radioactive materials are at most risk of being stolen or lost when they are being transported to and from the facility. They will be in the appropriate transport container and, therefore, can be easily handled by someone with malicious intent. It is essential that all consignments of radioactive materials to the nuclear medicine facility are left in a secure area and not left, for example, on a loading dock. During working hours, all deliveries must be signed for by a designated staff member and the material safely unpacked and stored within the department. Some deliveries may occur outside of working hours. In this case, a dedicated secure area must be provided where the radioactive materials can be left. A key could be provided to the supplier for this area only, so that the radioactivity can be safely and securely delivered, but access to other parts of the facility is prevented. The supplier may need to be accompanied by the facility's security staff when delivering the shipment.

Whether secure access (such as electronic card access) to the radiopharmacy during working hours is required will depend on local requirements and the layout of the nuclear medicine department. It is essential that only trained nuclear medicine staff have access to the radiopharmacy. The need for controlled access needs to be balanced against the possibility of inadvertent contamination of the door or access mechanism by staff returning to the radiopharmacy.

9.11. RECORD KEEPING

The local regulations may specify the minimum records that must be kept at the facility, the form in which these must be kept (paper and/or electronic) and the time for which the records must be kept. Records can be generated as part of the quality assurance (QA) programme, for the receipt and subsequent administration of a radiopharmaceutical to a patient, and for waste disposal.

9.11.1. Quality control records

A key element of any QA programme is proper record keeping, so that any long term trends associated with a particular item of equipment or batch of radiopharmaceuticals can be identified and acted on before image quality and/or patient dose are compromised. Records should, at the very least, include details of:

— Acceptance testing of the dose calibrator;
— All constancy tests;
— Radiopharmaceutical testing.

Failures identified at acceptance or constancy testing and radiopharmaceutical testing, and the actions taken to remedy those failures, should be documented and these records kept for the lifetime of the equipment.

The following records should be kept for all generator elutions:

— Time of elution;
— Volume of eluate;
— Technetium-99m activity;
— Molybdenum-99 activity;
— Radionuclidic purity.

9.11.2. Records of receipt of radioactive materials

Complete records of the radionuclide, activity, chemical form, supplier, supplier's batch number and purchase date should be kept. On arrival, if a package containing radioactive material is suspected of being damaged, the package should be:

— Monitored for leakage with a wipe test;
— Checked with a survey meter for unexpectedly high external radiation levels.

If a package is damaged or suspected of being damaged, the supplier should be contacted immediately, and the details recorded.

9.11.3. Records of radiopharmaceutical preparation and dispensing

The preparation of radiopharmaceuticals needs to be performed in accordance with the manufacturer's requirements as specified in the product documentation, including any quality control such as thin-layer chromatography.

Records of each preparation should include the:

- — Name of the radiopharmaceutical;
- — Cold kit batch number;
- — Date of manufacture;
- — Batch number of final product;
- — Radiochemical purity results;
- — Expiry date.

A record for each patient dose dispensed must be kept with the:

- — Name of the patient;
- — Name of the radiopharmaceutical;
- — Measured radioactivity;
- — Time and date of measurement.

All unit patient doses (syringes, capsules or vials) supplied by a central radiopharmacy should identify the patient's name and the radionuclide and radiopharmaceutical form. These should be verified on arrival and the activity should be confirmed in a dose calibrator prior to administration to the patient, and recorded as above.

9.11.4. Radioactive waste records

Radioactive waste generated within a nuclear medicine facility usually consists of radionuclides with half-lives of less than one month. This waste will normally be stored on-site, be allowed to decay to background radiation levels and then be disposed of as normal waste or biologically contaminated waste (see Section 3.4.7). It is, therefore, not normally necessary to keep records of radioactive waste disposal from the facility, but it will be necessary to keep records of the waste in storage while it decays. In some circumstances, the waste will contain a single known radionuclide, such as ^{131}I from patients receiving radioiodine ablation therapy. In many cases, the waste will contain a mixture of short lived radionuclides. Each package of waste (bag, sharps container, wheeled bin) must be marked with the:

— Radionuclide, if known;
— Maximum dose rate at the surface of the container or at a fixed distance (e.g. 1 m);
— Date of storage.

The above information should be recorded, together with information identifying the location of the container within the store, and the likely release date (e.g. ten half-lives of the longest lived radionuclide in the container).

When the package is finally released for disposal, the record should be updated to record the dose rate at that time, which should be at background levels, the date of disposal, and the identification of the person authorizing its disposal.

Old sealed sources previously used for quality control or transmission scans, such as ^{137}Cs, ^{57}Co, ^{153}Gd and ^{68}Ge, should be kept in a secure store until the activity has decayed to a level permitted for disposal, or the source can be disposed of by a method approved by the regulatory authority.

REFERENCES

[9.1] TYLER, D.K., WOODS, M.J., Syringe calibration factors for the NPL Secondary Standard Radionuclide calibrator for selected medical radionuclides, Appl. Radiat. Isot. **59** (2003) 367–372.

[9.2] INTERNATIONAL ELECTROTECHNICAL COMMISSION, Calibration and Usage of Ionization Chamber Systems for Assay of Radionuclides, IEC 61145:1992, IEC (1992).

[9.3] INTERNATIONAL ELECTROTECHNICAL COMMISSION, Medical Electrical Equipment — Radionuclide Calibrators — Particular Methods for Describing Performance, IEC 61303:1994, IEC (1994).

[9.4] INTERNATIONAL ELECTROTECHNICAL COMMISSION, Nuclear Medicine Instrumentation — Routine Tests — Part 4: Radionuclide Calibrators, IEC/TR 61948-4:2006, IEC (2006).

[9.5] American National Standards Institute, Calibration and Usage of “Dose Calibrator” Ionization Chambers for the Assay of Radionuclides, ANSI N42.13-2004, ANSI (2004).

[9.6] INTERNATIONAL COMMISSION ON RADIOLOGICAL PROTECTION, Radiation Dose to Patients from Radiopharmaceuticals, Publication 53, Pergamon Press, Oxford and New York (1988).

[9.7] INTERNATIONAL COMMISSION ON RADIOLOGICAL PROTECTION, Radiation Dose to Patients from Radiopharmaceuticals (Addendum to ICRP Publication 53), Publication 80, Pergamon Press, Oxford and New York (1998).

[9.8] INTERNATIONAL COMMISSION ON RADIOLOGICAL PROTECTION, Radiation Dose to Patients from Radiopharmaceuticals — Addendum 3 to ICRP Publication 53, Publication 106, Elsevier (2008).

[9.9] INTERNATIONAL COMMISSION ON RADIOLOGICAL PROTECTION, Radiological Protection and Safety in Medicine, Publication 73, Pergamon Press, Oxford and New York (1996).

[9.10] INTERNATIONAL ATOMIC ENERGY, Applying Radiation Safety Standards in Nuclear Medicine, Safety Reports Series No. 40, IAEA, Vienna (2005).

[9.11] EUROPEAN ALARA NETWORK (2007), http://www.eu-alara.net/index.php?option=com_content&task=view&id=156&Itemid=53

[9.12] INTERNATIONAL COMMISSION ON RADIOLOGICAL PROTECTION, Limits for Intakes of Radionuclides by Workers, Publication 30, Pergamon Press, Oxford and New York (1979).

[9.13] INTERNATIONAL COMMISSION ON RADIOLOGICAL PROTECTION, 1990 Recommendations of the International Commission on Radiological Protection, Publication 60, Pergamon Press, Oxford and New York (1991).

[9.14] INTERNATIONAL COMMISSION ON RADIOLOGICAL PROTECTION, Annual Limits on Intake of Radionuclides by Workers Based on the 1990 Recommendations, Publication 61, Pergamon Press, Oxford and New York (1991).

BIBLIOGRAPHY

GADD, R., et al., Protocol for Establishing and Maintaining the Calibration of Medical Radionuclide Calibrators and their Quality Control, Measurement Good Practice Guide No. 93, National Physical Laboratory, UK (2006).

GROTH, M.J., Empirical dose rate and attenuation data for radionuclides in nuclear medicine, Australas. Phys. Eng. Sci. Med. **19** (1996) 160–167.

NATIONAL HEALTH AND MEDICAL RESEARCH COUNCIL (Australia), Recommended Limits on Radioactive Contamination on Surfaces in Laboratories, Radiation Health Series No. 38, NHMRC (1995).

SCHRADER, H., Activity Measurements with Ionization Chambers, Monographie Bureau International des Poids et Mesures No. 4 (1997).

CHAPTER 10

NON-IMAGING DETECTORS AND COUNTERS

P.B. ZANZONICO
Department of Medical Physics,
Memorial Sloan Kettering Cancer Center,
New York, United States of America

10.1. INTRODUCTION

Historically, nuclear medicine has been largely an imaging based specialty, employing such diverse and increasingly sophisticated modalities as rectilinear scanning, (planar) gamma camera imaging, single photon emission computed tomography (SPECT) and positron emission tomography (PET). Non-imaging radiation detection, however, remains an essential component of nuclear medicine. This chapter reviews the operating principles, performance, applications and quality control (QC) of the various non-imaging radiation detection and measurement devices used in nuclear medicine, including survey meters, dose calibrators, well counters, intra-operative probes and organ uptake probes. Related topics, including the basics of radiation detection, statistics of nuclear counting, electronics, generic instrumentation performance parameters and nuclear medicine imaging devices, are reviewed in depth in other chapters of this book.

10.2. OPERATING PRINCIPLES OF RADIATION DETECTORS

Radiation detectors encountered in nuclear medicine may generally be characterized as either scintillation or ionization detectors (Fig. 10.1). In scintillation detectors, visible light is produced as radiation excites atoms of a crystal or other scintillator and is converted to an electronic signal, or pulse, and amplified by a photomultiplier tube (PMT) and its high voltage (500–1500 V). In ionization detectors, free electrons produced as radiation ionizes a stopping material within a sensitive volume are electrostatically collected by a bias voltage (10–500 V) to produce an electron signal. In both scintillation and ionization detectors, the 'unprocessed' signal is then shaped and amplified. For some types of detector, the resulting pulses are sorted by their amplitude (or pulse height), which is related to the X ray or γ ray energy absorbed in the detector.

10.2.1. Ionization detectors

Detector materials in the most common ionization detectors are gaseous and such detectors are, therefore, often known as gas filled detectors; however, as discussed in the following, solid state ionization detectors also exist. The two most widely encountered gas ionization detectors in nuclear medicine are dose calibrators and Geiger counters. The principal difference between these detectors is the magnitude of the bias voltage between the anode and cathode, as indicated graphically in Fig. 10.2. When the bias voltage is less than 300 V, ion pairs (i.e. free electrons and positive ions) produced as radiation passes through the sensitive volume may recombine, thereby preventing at least some electrons from reaching the anode and yielding an artefactually low signal. The 0–300 V range is, therefore, called the recombination region.

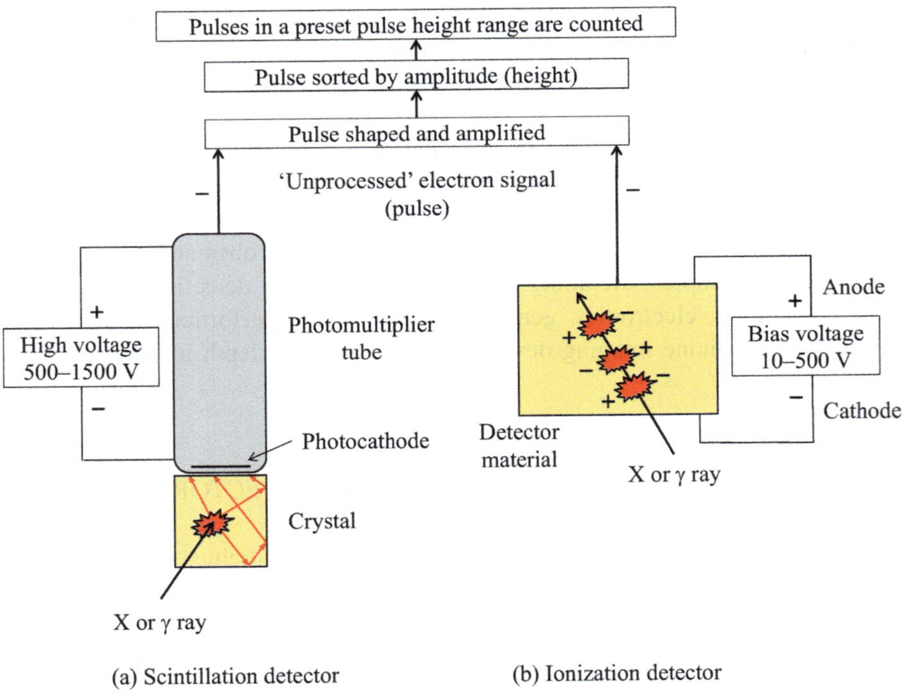

FIG. 10.1. Basic design and operating principles of (a) scintillation and (b) ionization detectors.

At a bias voltage of 300 V, all of the primary electrons (i.e. the electrons produced directly by ionization of the detector material by the incident radiation) are collected at the anode and the detector signal is, thereby, maximized. Since

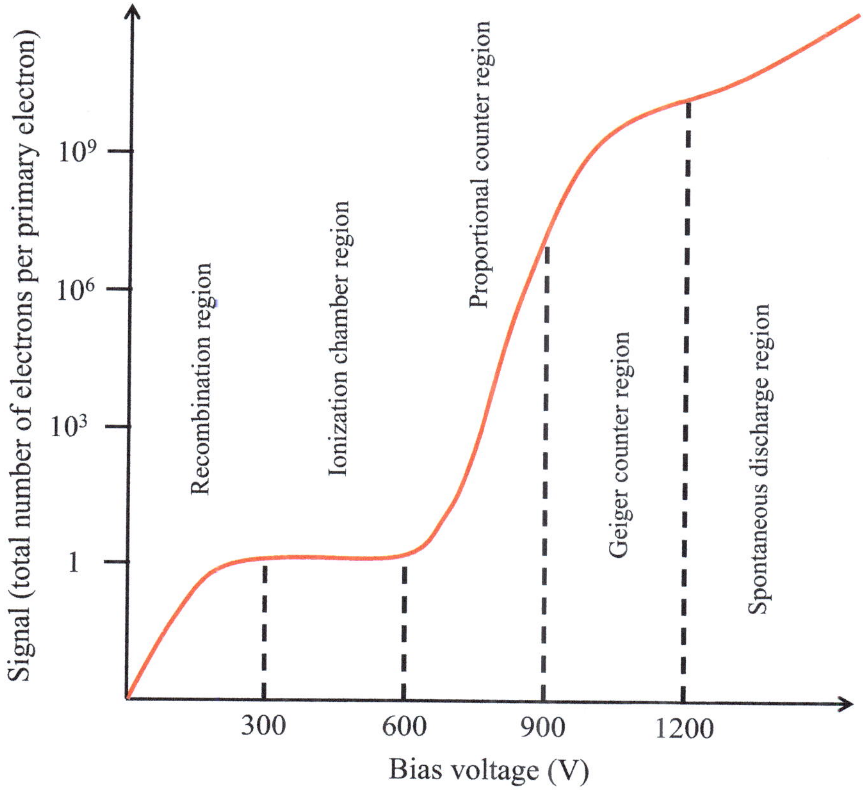

FIG. 10.2. The signal (expressed as the amplification factor, that is, the total number of electrons per primary electron produced in the detector material) as a function of the bias voltage for gas filled ionization detectors. The principal difference among such detectors is the magnitude of the bias voltage between the anode and cathode. The amplification factors and the voltages shown are approximate.

there are no additional primary electrons to collect, increasing the bias voltage further (up to 600 V) does not increase the signal. The 300–600 V range, where the overall signal is equivalent to the number of primary electrons and, therefore, proportional to the energy of the incident radiation, is called the ionization chamber region. At a bias voltage of 600–900 V, however, the large electrostatic force of attraction of the anode accelerates free electrons, as they travel towards the anode, to sufficiently high speeds to eject additional orbital electrons (i.e. secondary electrons) within the sensitive volume, contributing to an increasing overall signal — the higher the voltage, the more energetic the electrons and the more secondary electrons are added to the overall signal. The number of electrons comprising the overall signal is, thus, proportional to the primary number of electrons and the energy of the incident radiation, and the 600–900 V range is, therefore, called the proportional counter region. As the bias voltage is increased further, beyond 900 V (up to 1200 V), free electrons (primary and secondary) are accelerated to very high speeds and strike the anode with sufficient energy to eject additional electrons from the anode surface itself. These tertiary electrons are, in turn, accelerated back to the anode surface and eject even more electrons, effectively forming an electron 'cloud' over the anode surface and yielding a constant overall signal even with further increase in the bias voltage. The 900–1200 V range is called the Geiger counter (or Geiger–Müller) region. Importantly, the magnitude of the charge represented by this electron cloud is independent of the number of electrons initiating its formation. Therefore, in contrast to ionization chamber and proportional counter signals, the Geiger counter signal is independent of the energy of the incident radiation. Finally, beyond a bias voltage of 1200 V, atoms within the detector material are ionized even in the absence of ionizing radiation (i.e. undergo spontaneous ionization), producing an artefactual signal; the voltage range beyond 1200 V is known as the spontaneous discharge region.

Although the bias voltage is the principal difference among different types of gas filled ionization detectors, there may be other differences. The sensitive volume, for example, may or may not be sealed. Unsealed sensitive volumes contain only air at atmospheric (ambient) pressure. For detectors with unsealed volumes, the signal must be corrected by calculation for the difference between the temperature and pressure at which the detector was calibrated (usually standard temperature and pressure: 27°C and 760 mm Hg, respectively) and the ambient conditions at the time of an actual measurement. For detectors with sealed volumes, gases other than air (e.g. argon) may be used and the gas may be pressurized, providing higher stopping power, and, therefore, higher sensitivity, than detectors having a non-pressurized gas in the sensitive volume. In addition, different geometric arrangements of the anode and cathode, such as parallel plates (used in some ionization chambers), a wire along the axis of a cylinder (used in Geiger counters), etc., may be used.

The functional properties and, therefore, the applications of the various types of ionization detector — ionization chambers, proportional counters and Geiger counters — are largely dictated by their respective bias voltage dependent signal (Table 10.1). Ionization chambers are widely used in radiation therapy to calibrate the output of therapy units and in nuclear medicine as dose calibrators (i.e. devices used to assay radiopharmaceutical activities). The relatively low sensitivity of ionization chambers is not a major disadvantage for such applications, as the radiation intensities encountered are typically rather large. The stability of the response is an important advantage, however, as it allows the use of unconditioned AC electrical power (i.e. as provided by ordinary wall outlets). Proportional counters, because of their need for a stable bias voltage and, therefore, specialized power supplies, are restricted to research applications (e.g. in radiobiology) where both higher sensitivity and the capability of energy discrimination may be advantageous. Proportional counters often employ an unsealed, gas flow-through sensitive volume. Geiger counters, because of their high sensitivity and stability with respect to voltage (allowing the use of a portable power supply such as an ordinary battery), are widely used as survey meters to measure ambient radiation levels and to detect radioactive contamination. For such applications, sensitivity, and not energy discrimination, is critical. As with dose calibrators, Geiger counters have sealed sensitive volumes, avoiding the need for temperature–pressure corrections.

In addition to the more familiar gas filled ionization detectors, solid state ionization detectors are now available. Such detectors are based on a family of materials known as semiconductors. The pertinent difference among (crystalline) solids — conductors, insulators and semiconductors — is related to the widths of their respective electron 'forbidden' energy gaps. In a semiconductor, the highest energy levels occupied by electrons are completely filled but the forbidden gap is narrow enough (<2 eV) to allow radiative or even thermal excitation at room temperature, thereby allowing a small number of electrons to cross the gap and occupy energy levels among the otherwise empty upper energy levels. Such electrons are mobile and, thus, can be collected by a bias voltage, with the amplitude of the resulting signal being equivalent to the number of electrons produced by the radiation and, therefore, proportional to the radiation energy. Although many semiconductor materials have suitably large energy gaps (~2 eV), techniques must be available to produce crystals relatively free of structural defects. Defects (i.e. irregularities in the crystal lattice) can trap electrons produced by radiation and, thus, reduce the total charge collected, degrading the sensitivity and overall detector performance of semiconductors. Practical, reasonably economical crystal growing techniques have been developed for cadmium telluride (CdTe), cadmium zinc telluride (CZT) and mercuric iodide (HgI_2), and these detectors have been incorporated into commercial intra-operative gamma probes and, on a limited basis, small field of view gamma cameras.

TABLE 10.1. PROPERTIES OF GAS FILLED IONIZATION DETECTORS

	Ionization detector	Proportional counter	Geiger counter
Bias voltage operating range	300–600 V	600–900 V	900–1200 V
Response stable with respect to voltage?[a]	Yes	No	Yes
Sensitivity[b]	Low	Intermediate	High
Capable of energy discrimination?[c]	Yes	Yes	No
Applications	Dose calibrator	Research	Survey meter

[a] The stability with respect to the bias voltage corresponds to a constant signal over the respective detector's operating voltage range. In contrast to ionization detectors and Geiger counters, proportional counters are unstable with respect to the bias voltage and, thus, require specialized, highly stable voltage sources for constancy of response.

[b] The sensitivity of a detector is related to its amplification factor (see Fig. 10.2).

[c] If the total number of electrons comprising the signal is proportional to the number of electrons directly produced by the incident radiation and, therefore, proportional to its energy, as in ionization detectors and proportional counters, radiations of different energies can be discriminated (i.e. separated) on the basis of the signal amplitude.

10.2.2. Scintillation detectors

In scintillation detectors, radiation interacts with and deposits energy in a scintillator, most commonly, a crystalline solid such as thallium-doped sodium iodide (NaI(Tl)). The radiation energy thus deposited is converted to visible light. As the light is emitted isotropically (i.e. in all directions), the inner surface of the light-tight crystal housing is coated with a reflective material so that light emitted towards the sides and front of the crystal are reflected back towards a PMT (Fig. 10.3); this maximizes the amount of light collected and, therefore, the overall sensitivity of the detector. Interposed between the back of the crystal and the entrance window of the PMT is the light pipe, nowadays simply a thin layer of transparent optical gel. The light pipe optically couples the crystal to the PMT and, thus, maximizes the transmission (>90%) of the light signal from the crystal into the PMT. When struck by light from the crystal, the photocathode coated on the inner surface of the PMT emits electrons. Immediately beyond the photocathode (which is at ground, that is, 0 V) is the focusing grid, maintained at a relatively low positive voltage on the order of 10 V. As electrons pass through the focusing grid, they are attracted by a relatively large positive voltage, ~300 V, on the first of a series of small metallic elements called dynodes. The resulting high speed impact of each electron results in the ejection from the dynode surface of an average of three electrons. The electrons thus ejected are then attracted by the even larger positive voltage, ~400 V, on the second dynode. The impact of these electrons on the second dynode surface ejects an additional three electrons, on average, for each incident electron. Typically, a PMT has 10–12 such dynodes (or stages), each ~100 V more positive than the preceding dynode, resulting in an overall electron amplification factor of 3^{10}–3^{12} for the entire PMT. At the last anode, an output signal is generated. The irregularly shaped PMT output signal is then shaped by a preamplifier and further amplified into a logic pulse that can be further processed electronically. The resulting electrical pulses, whose amplitudes (or 'heights') are proportional to the number of electrons produced at the PMT photocathode are, therefore, also proportional to the energy of the incident radiation. These pulses can then be sorted according to their respective heights by an energy discriminator (also known as a pulse height analyser) and those pulses with a pulse height (i.e. energy) within the preset photopeak energy window (as indicated by the pair of dashed horizontal lines overlying the pulses in Fig. 10.3) are counted by a timer/scaler.

Advantageous features of scintillation detectors include:

- High electron density (determined by mass density ρ and effective atomic number Z_{eff});
- High light output;
- For certain applications such as PET, speed of light emission.

High mass density and effective atomic number maximize the crystal stopping power (i.e. linear attenuation coefficient μ) and, therefore, sensitivity. In addition, a higher atomic number crystal will have a higher proportion of photoelectric than Compton interactions, thus facilitating energy discrimination of photons which underwent scatter before entering the crystal. High light output reduces statistical uncertainty (noise) in the scintillation and associated electronic signal and, thus, improves energy resolution and scatter rejection. Other detector considerations include:

- Transparency of the crystal to its own scintillations (i.e. minimal self-absorption);
- Matching of the index of refraction η of the crystal to that of the photodetector (specifically, the entrance window ($\eta \approx 1.5$) of the PMT);
- Matching of the scintillation wavelength to the light response of the photodetector (the PMT photocathode, with maximum sensitivity in the 390–410 nm, or blue, wavelength range);
- Minimal hygroscopic behaviour.

To date, the most widely used scintillators in nuclear medicine include: NaI(Tl), bismuth germanate (BGO), cerium-doped lutetium oxyorthosilicate (LSO(Ce) or LSO) and cerium-doped gadolinium oxyorthosilicate (GSO(Ce) or GSO). NaI(Tl) is used in γ cameras/SPECT systems, well counters and organ uptake probes, and remains the most widely used scintillator in clinical practice; BGO, LSO and GSO are the scintillators of choice in PET scanners because of their higher stopping power for the 511 keV positron–negatron annihilation photons. Thallium- and sodium-doped caesium iodide (CsI(Tl) and CsI(Na), respectively) and cadmium tungstate as well as NaI(Tl), BGO and LSO have also been used in intra-operative probes.

10.3. RADIATION DETECTOR PERFORMANCE

Radiation detectors may be quantitatively characterized by many different performance parameters, particularly for those detectors such as γ cameras which localize (image) as well as count radiation. For non-imaging radiation detectors and counters, however, the most important performance parameters are sensitivity (or efficiency), energy resolution and count rate performance (or 'speed').

10.3.1. Sensitivity

Sensitivity (or efficiency) is the detected count rate per unit activity (e.g. in counts per minute per megabecquerel). As the count rate detected from a given activity is highly dependent on the source–detector geometry and intervening media, characterization of sensitivity can be ambiguous. There are two distinct components of overall sensitivity, geometric sensitivity and intrinsic sensitivity. Geometric sensitivity is the fraction of emitted radiations which intersect, or strike, the detector, that is, the fraction of the total solid angle subtended at the detector by the source. It is, therefore, directly proportional to the radiation-sensitive detector area and, for a point source, inversely proportional to the square of the source–detector distance. Intrinsic sensitivity is the fraction of radiation striking the detector which is stopped within the detector. Intrinsic sensitivity is directly related to the detector thickness, effective atomic number and mass density, and decreases with increasing photon energy, since higher energy photons are more penetrating and are more likely to pass through a detector without interacting.

Characteristic X rays and γ rays are emitted from radioactively decaying atoms with well defined discrete energies. Even in the absence of scatter, however, output pulses from absorption of these radiations will appear to originate over a range of energies, reflecting the relatively coarse energy resolution of the detector. For this reason, many radiation detectors employ some sort of energy-selective counting using an energy range, or window, such that radiations are only counted if their detected energies lie within that range (Figs. 10.3 and 10.4(a)). At least for scintillation detectors, a so-called '20% photopeak energy window', $E_\gamma \pm 10\%$ of E_γ, (e.g. 126–154 keV for the 140 keV γ ray of ^{99m}Tc) is employed, where E_γ is the photopeak energy of the X ray or γ ray being counted. For such energy-selective counting, overall sensitivity appears to increase as the photopeak energy window is widened. However, this results in acceptance of more scattered as well as primary (i.e. unscattered) radiations.

For each radionuclide and energy window (if applicable) for which a particular detector is used, the detector should be calibrated, that is, its sensitivity (e.g. in cpm/MBq) S determined, at installation and periodically thereafter:

$$S = \frac{R_g - R_b}{\mathcal{A}_0 e^{-\lambda \Delta t}} \tag{10.1}$$

where

- R_g is the gross (i.e. total) count rate (cpm) of the radionuclide source (RS),
- R_b is the background (BG), or blank, count rate (cpm);
- $\mathcal{A}_0$ is the activity (MBq) of the radionuclide source at calibration;
- λ is the physical decay constant (in month^{-1} or a^{-1}, depending on the half-life) of the calibration radionuclide;

and Δt is the time interval (in months or years, respectively, again depending on the half-life) between the calibration of the radionuclide and the current measurement.

As noted, sensitivity is highly dependent on the source–detector counting geometry (including the size and shape of the source and the source–detector distance), and the measured value, thus, applies exactly only for the geometry used for the measurement.

10.3.2. Energy resolution

Energy resolution quantifies the ability of a detector to separate, or discriminate, radiations of different energies. As illustrated in Fig. 10.4(b), energy resolution is generally given by the width of the bell shaped photopeak, specified as the full width at half maximum (FWHM = ΔE) height expressed as a percentage of the photopeak energy E_γ, FWHM $(\%) = \frac{\Delta E}{E_\gamma} 100\%$. It is related

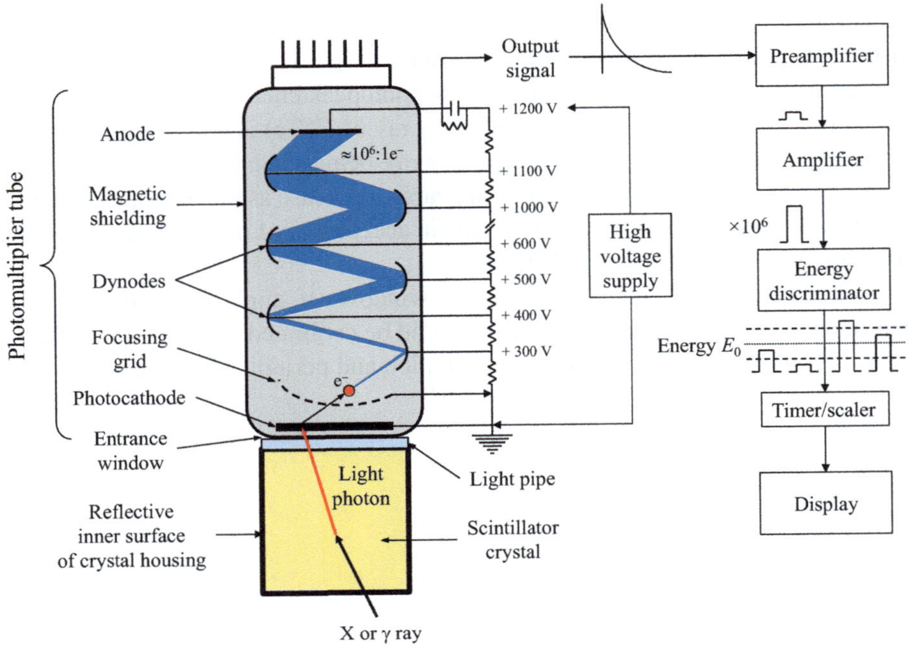

FIG. 10.3. The basic design and operating principle of photomultiplier tubes and scintillation detectors.

to the Poisson 'noise', or statistical uncertainty, inherent in the detection process. The importance of energy resolution lies in scatter rejection, particularly for imaging detectors. Radiation loses energy when undergoing Compton scatter within the source and the lower energy scattered radiations may, therefore, be discriminated from the primary radiations. However, the finite energy resolution of radiation detectors (i.e. the width of the photopeak in the energy spectrum) means that there will be overlap of scattered and primary radiations, as illustrated in Fig. 10.4(a). As energy resolution improves (i.e. the FWHM (%) decreases and the photopeak becomes narrower), the separation of unscattered and scattered radiations increases and more counts corresponding to scattered radiation may be eliminated, while discarding fewer counts corresponding to unscattered radiation.

10.3.3. Count rate performance ('speed')

Radiation detectors have a finite dead time or pulse resolving time τ — typically 5–10 μs for modern scintillation detectors — and associated count losses. The dead time is the length of time required for a counting system to

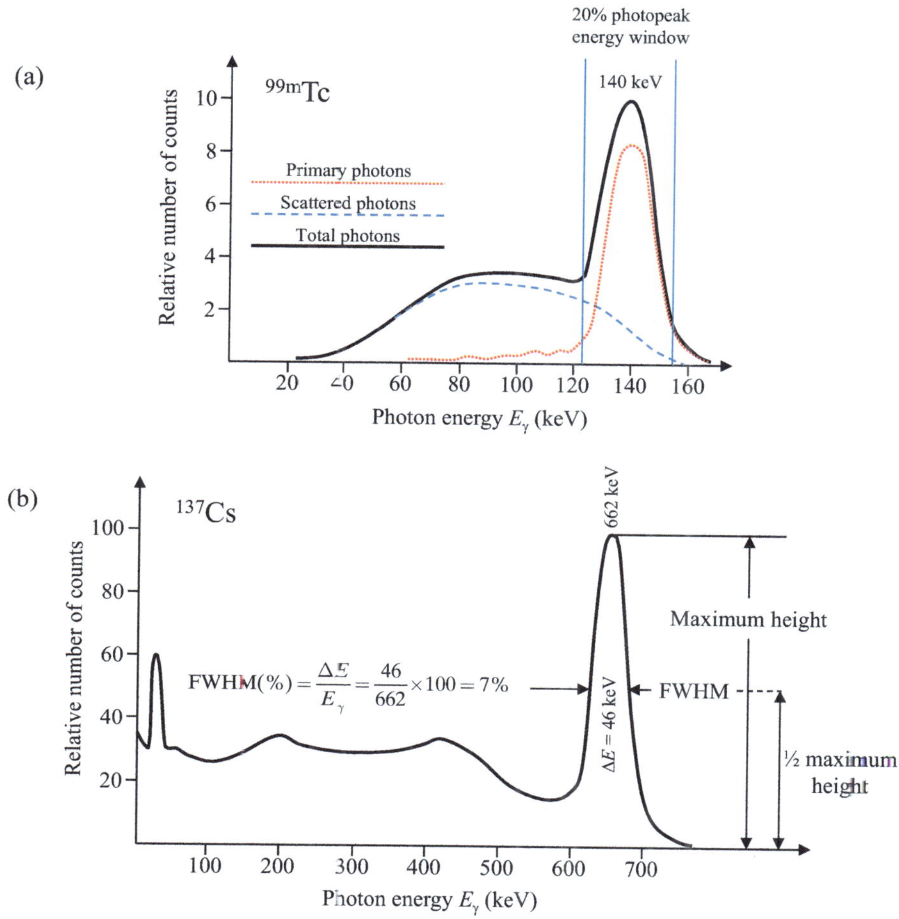

FIG. 10.4. (a) Energy spectrum for the 662 keV γ rays emitted by ^{137}Cs, illustrating the definition of energy resolution as the percentage full width at half maximum (FWHM) of the photopeak energy E_γ. (b) Energy spectrum for the 140 keV γ rays emitted by ^{99m}Tc, illustrating the contributions of primary (unscattered) and scattered radiation counts. In (a) and (b), the energy spectra were obtained with a thallium-doped sodium iodide (NaI(Tl)) scintillation detector.

record an event, during which additional events cannot be recorded. As a result, the measured count rate is lower than the actual count rate. Radiation detectors are characterized in terms of count rate performance as either non-paralysable or paralysable (Fig. 10.5). In non-paralysable systems, only radiation which is actually counted prevents the counting of subsequent radiation interacting with the detector during the dead time of that preceding radiation. In a paralysable detector, however, even radiation which is not counted (i.e. which interacts with the detector during the dead time of a previous event) prevents counting of subsequent incoming radiations during the time interval corresponding to its dead time. Geiger counters (with quenching gas) behave as non-paralysable systems but most detectors, including scintillation detector based systems, such as well counters, γ cameras and PET scanners, are paralysable. Modern scintillation detectors generally incorporate automated algorithms to yield count rates corrected for dead time count losses.

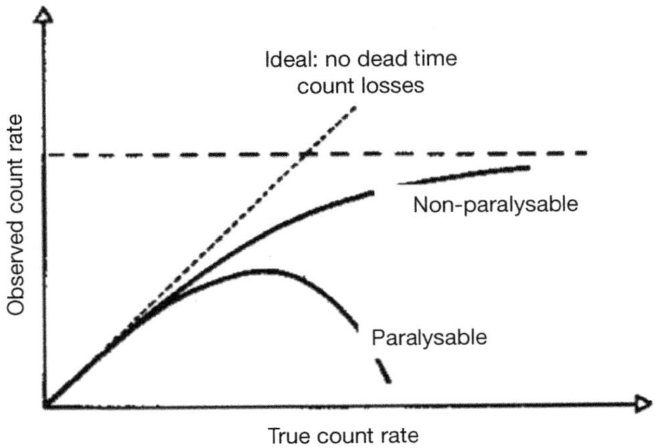

FIG. 10.5. The observed versus true count rates for paralysable and non-paralysable radiation detectors. For paralysable detectors, the observed count rate increases to a maximum value with increasing true count rate (e.g. with increasing activity) and then decreases as the true count rate is further increased. For non-paralysable detectors, the observed count rate also increases with increasing true count rate, asymptotically approaching a maximum value as the true count rate is further increased. In both cases, the maximum observed count rate is directly related to the detector's dead time τ.

10.4. DETECTION AND COUNTING DEVICES

10.4.1. Survey meters

Survey meters, an essential component of any radiation safety programme, are portable, battery operated, gas filled ionization detectors (or, to a much more limited extent, solid state scintillation detectors) used to monitor ambient radiation levels, that is, exposure rates (e.g. in coulombs per kilogram of air per hour ($C \cdot kg^{-1} \cdot h^{-1}$)) or count rates (e.g. in cpm). Among ionization detector survey meters, so-called 'cutie-pies' are relatively low sensitivity ionization chambers (i.e. are operated at a relatively low potential difference between the anode and cathode) and are designed for use where relatively high fluxes of X rays and γ rays are encountered. The more familiar Geiger counters are operated at a high potential difference (Fig. 10.2), providing a high electron amplification factor and, thus, high sensitivity. Geiger counters are, therefore, well suited for low level surveys, for example, checking for radioactive contamination. Both cutie-pies and Geiger counters are generally calibrated in terms of exposure rate. As an ionization chamber, the cutie-pie's electron signal depends on the energy of the detected X rays or γ rays and is, therefore, directly related to the exposure for all radionuclides. For Geiger counters, on the other hand, signal pulses have the same amplitude regardless of the energy of the incoming radiation. Thus, Geiger counter calibration results apply only to the particular radionuclide(s) used to calibrate the counter (see below). Solid state detectors employ a non-air-equivalent crystal as the detection medium and, thus, cannot measure exposure rates, only count rates.

10.4.2. Dose calibrator

The dose calibrator, used for assaying activities in radiopharmaceutical vials and syringes and in other small sources (e.g. brachytherapy sources), is a pressurized gas filled ionization chamber with a sealed sensitive volume configured in a 'well'-type geometry. While the intrinsic sensitivity of the dose calibrator, as that of other gas filled detectors, is relatively low, the well-type configuration of its sensitive volume provides high geometric efficiency[1], making the overall sensitivity entirely adequate for the relatively high radiopharmaceutical activities (of the order of 10–100 MBq) typically encountered in clinical nuclear medicine. Dose calibrators are equipped with isotope specific push-buttons and/or a potentiometer (with isotope-specific settings provided) to adjust for differences in energy dependent response and to thereby yield accurate readouts of activity (i.e. kBq or MBq) for any radioisotope.

10.4.3. Well counter

Well counters are used for high sensitivity counting of radioactive specimens such as blood or urine samples or 'wipes' from surveys of removable contamination (i.e. 'wipe testing'). Such counting results can be expressed in

[1] The solid angle subtended at the centre of a sphere by the total surface of the sphere is 4π steradians; a steradian is the unit of solid angle. A well-type detector configuration approximates a point source completely surrounded by a detector, yielding a per cent geometric efficiency of 100%, and is, therefore, referred to as a '4π' counting geometry.

terms of activity (e.g. MBq) using the measured isotope specific calibration factor (cpm/MBq) (see Eq. (10.1)). Such devices are generally comprised of a cylindrical scintillation crystal (most commonly, NaI(Tl)) with a circular bore (well) for the sample drilled part-way into the crystal and backed by a PMT and its associated electronics. An alternative design for well counters is the so-called 'through-hole' detection system in which the hole is drilled through the entire crystal. The through-hole design facilitates sample exchange, and because samples are centred lengthwise in the detector, yields a more constant response for different sample volumes as well as slightly higher sensitivity than the well counters. In both the well and through-hole designs, the crystal is surrounded by thick lead shielding to minimize the background due to ambient radiation.

Scintillation counters are often equipped with a multichannel analyser for energy (i.e. isotope) selective counting and an automatic sample changer for automated counting of multiple samples. Importantly, because of their high intrinsic and geometric efficiencies (resulting from the use of a thick crystal and a well-type detector configuration, respectively), well counters are extremely sensitive and, in fact, can reliably be used only for counting activities up to ~100 kBq; at higher activities, and even with dead time corrections applied, dead time counting losses may still become prohibitive and the measured counts inaccurate. Modern well counters often include an integrated computer which is used to create and manage counting protocols (i.e. to specify the isotope, energy window, counting interval, etc.), manage sample handling, and apply background, decay, dead time and other corrections, and, thus, yield dead time-corrected net count rate decay corrected to the start of the current counting session.

10.4.4. Intra-operative probes

Intra-operative probes (Fig. 10.6), small hand-held counting devices, are now widely used in the management of cancer, most commonly to more expeditiously identify and localize sentinel lymph nodes and, thereby, reduce the need for more extensive surgery as well as to identify and localize visually occult disease at surgery following systemic administration of a radiolabelled antibody or other tumour-avid radiotracer. Although intra-operative probes have been used almost exclusively for counting X rays and γ rays, beta (electron and positron) probes constructed with plastic scintillators have also been developed. In addition, small (~10 cm) field of view intra-operative γ cameras have recently become available. Intra-operative γ probes are available with either scintillation or semiconductor (ionization) detectors. Scintillation detector based probes have the advantages of relatively low cost and high sensitivity (mainly because of their greater thickness, ~10 mm versus only ~1 mm in ionization detectors), especially for medium to high energy photons.

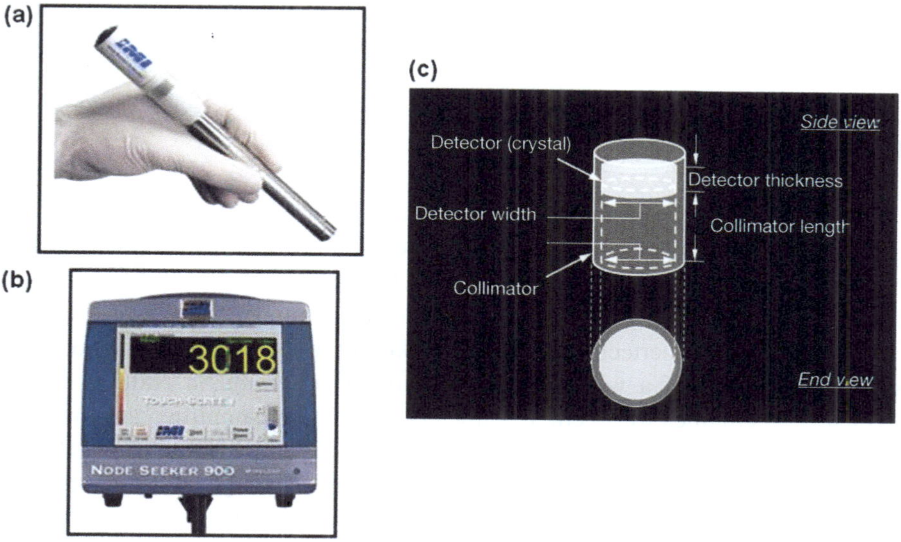

FIG. 10.6. A typical intra-operative probe (Node Seeker 900, Intra Medical Imaging LLC, Los Angeles, CA, United States of America). (a) Hand-held detector. (b) Control and display unit which not only displays the current count rate but also often emits an audible signal, the tone of which is related to the count rate, somewhat analogous to the audible signal produced by some Geiger counters. (c) A diagram of the detector and collimator assembly of a typical intra-operative probe, illustrating that the detector (crystal) is recessed from the collimator aperture. (Courtesy of Intra Medical Imaging LLC, Los Angeles, CA, USA.)

Disadvantages include bulkiness, and relatively poor energy resolution and scatter rejection relative to semiconductor based probes. In some scintillation–detector intra-operative probes, the light signal from the crystal is guided to a remote PMT through a flexible fibre-optic cable, allowing the probe assembly to be made relatively light and compact, and more like a surgical instrument. However, the significant loss of light in the long cable makes it more difficult to separate scatter from unscattered X rays and γ rays.

On the other hand, semiconductor based probes are compact and have excellent energy resolution and scatter rejection. To minimize structural imperfections which degrade energy resolution, semiconductor detectors are made relatively thin (only ~1 mm), but at the cost of lower intrinsic sensitivity. The main disadvantage of semiconductor detectors remains their limited thickness and resulting lower sensitivity, especially for medium to high energy X rays and γ rays. Nonetheless, while scintillation detectors can be made thicker and, therefore, more sensitive, semiconductor detectors produce more electrons per X ray and γ ray stopped, and, therefore, have a superior energy resolution. To date, the few clinical studies directly comparing scintillation and semiconductor intra-operative probes have not provided a clear choice between the two types of probe.

10.4.5. Organ uptake probe

Historically, organ uptake probes have been used almost exclusively for measuring thyroid uptakes and are, thus, generally known as 'thyroid' uptake probes.[2] Thyroid uptake (i.e. the decay-corrected per cent of administered activity in the thyroid) may be measured following oral administration of ^{131}I-iodide, ^{123}I-iodide or ^{99m}Tc-pertechnetate. The uptake probe is a radionuclide counting system consisting of a wide-aperture, diverging collimator, a NaI(Tl) crystal (typically ~5 cm thick by ~5 cm in diameter), a PMT, a preamplifier, an amplifier, an energy discriminator (i.e. an energy window) and a gantry (stand) (Figs 10.7(a) and (b)). Commercially available thyroid uptake probes are generally supplied as integrated, computerized systems with automated data acquisition and processing capabilities, yielding results directly in terms of per cent uptake.

Each determination of the thyroid uptake includes measurement of the thyroid (i.e. neck) count rate, the 'thigh' background count rate (measured over the patient's thigh and presumed to approximate the count contribution of extra-thyroidal neck activity), the standard count rate (often counted in a neck phantom simulating the thyroid/neck anatomy) and the ambient (i.e. 'room') background, with a 1–5 min counting interval for each measurement. Based on the foregoing measurements, and knowing the fraction of the administered activity which is in the standard, the thyroid uptake is calculated as follows:

$$\text{uptake (\%)} = \frac{C_{\text{neck}}/t_{\text{neck}} - C_{\text{thigh}}/t_{\text{thigh}}}{C_{\text{standard}}/t_{\text{standard}} - C_{\text{room}}/t_{\text{room}}} \times F \times 100\% \tag{10.2}$$

where

C is the total counts;
t is the measurement time;

and F is the fraction of administered activity in the standard.

---

2 At one time, organ uptake probes were also used to measure kidney time–activity data for the evaluation of renal function. In addition, organ uptake probes have been adapted to such well counter applications as counting of blood samples and wipes.

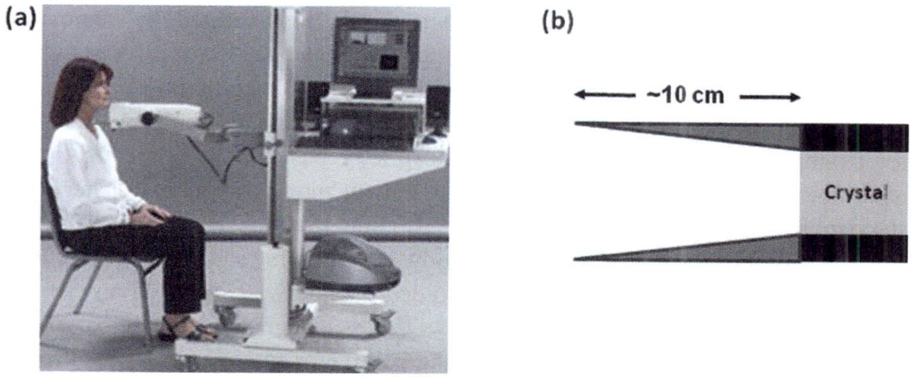

FIG. 10.7. (a) A typical organ ('thyroid') uptake probe system, including an integrated computer, set-up for a thyroid uptake measurement (AtomLab 950™ Thyroid Uptake System, Biodex Medical Systems, Shirley, NY, USA). The rather large neck to collimator aperture distance (typically of the order of 30 cm) should be noted. Although this reduces the overall sensitivity of the measurement of the neck count rate, it serves to minimize the effect of the exact size, shape and position of the thyroid, and the distribution of radioisotope within the gland. (b) A diagram (side view) of the open, or 'flat-field', diverging collimator typically used with thyroid uptake probes. (Courtesy of Biodex Medical Systems, Inc, Shirley, NY, USA.)

By including measurement of a standard activity with each uptake determination, corrections for radioactive decay and day to day variation in system sensitivity are automatic. This approach is sometimes known as the 'two-capsule' method, since one ^{131}I capsule is administered to the patient while a second, identical capsule serves as the standard and is counted with each uptake measurement. Alternatively, the patient capsule itself can be measured immediately before it is administered and then each subsequent uptake value for radioactive decay can be corrected from the time of measurement to the time of administration (by multiplying the right side of Eq. (10.2) by $e^{\lambda \Delta t}$ where λ is the physical decay constant of the administered isotope and Δt is the administration to measurement time interval). This is sometimes known as the 'one-capsule' method. For either the one- or two-capsule method, the fraction of administered activity in the standard is unity. Some centres administer radioiodine as a solution, which is more cost effective, rather than as a capsule. The standard is typically some dilution of the administered solution and the fraction of administered activity in the standard is, therefore, an independently determined value less than unity; for example, if the activity in the standard solution were 1/100th of the value in the administered solution, the value would equal 0.01.

Thyroid uptake measurements are now often performed by region of interest analysis of planar scintigraphic images of the neck and of a standard (i.e. phantom) acquired with a γ camera with parallel-hole collimation.

Organ uptake probes have also been used to measure total body activity, for example, as part of individualized dosimetry based radioiodine treatment of thyroid cancer. For this application, the patient may serve as his or her own standard by measuring the patient's total body count rate shortly (30–60 min) after administration of the radioisotope — to allow it to disperse somewhat throughout the body — but before the patient has voided or otherwise excreted any of the administered activity; in Eq. (10.3) below, this is designated time zero. Whole body measurements are performed with the collimator removed from the probe, the crystal oriented horizontally and at a height above the floor corresponding to the mid-height of the patient, either seated or standing, at a distance of ~3 m from the crystal. Further, anterior and posterior (i.e. conjugate-view) measurements are performed by having the patient facing towards and then away from the crystal for the respective measurements. The whole body activity (i.e. the per cent of administered activity in the body) is then calculated based on the geometric mean of the anterior and posterior count rates:

$$\text{Total body activity (\%)} = \frac{\left[\left(\dfrac{A}{t_{\text{A}}} - \dfrac{B}{t_{\text{B}}}\right) \times \left(\dfrac{P}{t_{\text{P}}} - \dfrac{B}{t_{\text{B}}}\right)\right]^{1/2}}{\left[\left(\dfrac{A(0)}{t_{\text{A}(0)}} - \dfrac{B(0)}{t_{\text{B}(0)}}\right) \times \left(\dfrac{P(0)}{t_{\text{P}(0)}} - \dfrac{B(0)}{t_{\text{B}(0)}}\right)\right]^{1/2}} \times 100\% \qquad (10.3)$$

where

A and P are the anterior and posterior total body counts, respectively;
B is the room (background) counts;
t_{A}, t_{P} and t_{B} are the counting intervals for anterior, posterior and room counts, respectively;

and (0) indicates the same quantities at time zero.

As above, the total body activity may be corrected for radioactive decay from the time of measurement to the time of administration (by multiplying the right side of Eq. (10.3) by $e^{\lambda \Delta t}$ where λ is the physical decay constant of the administered isotope and Δt is the administration to measurement time interval).

10.5. QUALITY CONTROL OF DETECTION AND COUNTING DEVICES

QC, which may be defined as an established set of ongoing measurements and analyses designed to ensure that the performance of a procedure or instrument is within a predefined acceptable range, is a critical component of routine nuclear medicine practice. The following is a brief review of routine QC procedures for non-imaging nuclear medicine instrumentation.

Documenting of QC procedures and organized, retrievable records of the results of such procedures are requirements of a sound, compliant QC programme. A written description of all QC procedures, including the acceptable (or tolerance) range of the results of each such procedure and the corrective action for an out of tolerance result, should be included in the facility's procedure manual. For each procedure, the written description should be signed and dated by the facility director, physicist or other responsible individual. For each QC test performed, the following data, as well as the initials or signature of the individual performing the test, should be recorded on a structured and suitably annotated form:

- — The test performed;
- — The date and time of the test;
- — The make, model and serial number of the device tested;
- — The make, model, serial number, activity at calibration and date of calibration of any reference source(s) used;
- — The result(s) of the test;
- — A notation indicating whether the test result was or was not acceptable (i.e. was or was not within the specified tolerance range).

Such records should be archived in chronological order in a secure but reasonably accessible location. It is generally helpful to track the results of QC tests longitudinally (e g. in the form of a graph of the numerical result versus the date of the test). In this way, long term trends in instrument performance, often imperceptible from one day to the next, may become apparent. Increasingly, of course, such records are maintained in electronic form. It is advisable, however, to also maintain records in hard copy form, both as a backup and for convenient review by regulators and other inspectors.

10.5.1. Reference sources

QC tests of nuclear medicine instrumentation are often performed not with the radionuclides that are used clinically but with longer lived surrogate radionuclides in the form of so-called reference sources. Such standards are commercially available in various activities and geometries, depending on the application. Importantly, in the USA, the certified activities of such reference sources must be traceable to the National Institute of Standards and Technology (NIST), formerly the National Bureau of Standards. NIST traceability helps ensure the accuracy of the calibrated activity. As such reference sources are long lived, a single standard may be used for months to years, avoiding the need to prepare sources on a daily or weekly basis and, thereby, avoiding possible inaccuracies in dispensing activity as well as the possibility of radioactive contamination. On the other hand, as with all sealed sources, reference sources must be checked for leakage of radioactivity (i.e. 'wipe-tested') periodically and an up to date inventory of such standards must be maintained. Reference sources are still radioactive at the end of their useful lifespan and must, therefore, be returned to the vendor or a third party or otherwise disposed of as radioactive waste.

A long lived radionuclide comprising a reference source must match, in terms of the frequency and energy of its X ray and γ ray emissions, the clinical radionuclide for which it acts as a surrogate in order to ensure that instrument responses to the clinical radionuclide and to its surrogate are comparable. Surrogate radionuclides commonly used in nuclear medicine and their physical properties and applications are summarized in Table 10.2.

10.5.2. Survey meter

QC tests of survey meters generally include a daily battery check, with a display indicating whether the voltage supplied by the battery is within the acceptable operating range. In order to confirm that the survey meter has not been contaminated (i.e. yields a reproducibly low exposure or count rate in the absence of radioactivity), the background exposure or count rate should be measured daily in an area remote from radioactive sources within the nuclear medicine facility, if such an area is reasonably accessible. In addition, survey meters should be checked daily for constancy of response by measuring the exposure or count rate of a long lived reference source in a reproducible measurement geometry. Aside from the short term decay of the reference source, the measured day to day exposure or count rates should agree within 10%; if not, the meter should be recalibrated.

Survey meters should be calibrated — that is, checked for accuracy — using suitable long lived reference sources at installation, annually and after any repair. If the source is 'small' (compared to the mean free path of its emitted X rays and γ rays within the material comprising the source) and the distance between the source and meter 'large' (compared to the dimensions of the source), then a point-source geometry is approximated and the expected dose rate $\dot{D}$ in air is given by the inverse square law:

$$\dot{D} = \frac{A_0 e^{-\lambda \Delta t} \Gamma_\delta}{d^2} \tag{10.4}$$

where

- A_0 is the activity of the reference source at calibration;
- λ is the physical decay constant of the radionuclide comprising the reference source;
- Δt is the time interval between the calibration of the reference source and the current measurement;
- Γ_δ is the air kerma rate constant (the subscript δ indicates that only photons with energies greater than δ, typically set at 20 keV, are included) of the radionuclide comprising the reference source;

and d is the distance between the reference source and the meter (Table 10.2).

The dose rates should be measured on each scale and, by appropriate adjustment of the source–meter distance, with two readings (~20% and ~80% of the maximum) on each scale. For all readings, the expected and measured dose rates should agree within 10%.

Many nuclear medicine facilities have their survey meters calibrated by the institutional radiation safety office or by a commercial calibration laboratory. In addition to a calibration report (typically, a one page document) specifying the reference source(s) used, the measurement procedure, and the measured and expected exposure rates, a dated sticker summarizing the calibration results should be affixed to the meter itself.

10.5.3. Dose calibrator

Among routine dose calibrator QC tests[3], constancy must be checked daily and accuracy and linearity at least quarterly; daily checks of accuracy are recommended. For the constancy test, an NIST-traceable reference source, such as ^{57}Co, ^{68}Ge or ^{137}Cs (Table 10.2), is placed in the dose calibrator and the

[3] At the installation of a dose calibrator, the geometry dependent response for ^{99m}Tc must be measured and volume dependent (2–25 mL) correction factors relative to the 'standard' volume (e.g. 10 mL) derived. This procedure is required periodically following installation.

TABLE 10.2. LONG LIVED RADIONUCLIDES COMPRISING REFERENCE SOURCES FOR INSTRUMENTATION QUALITY CONTROL

Radionuclide	Half-life	Physical decay constant λ	Photopeak energy E_γ and frequency of principal X ray or γ ray	Air kerma rate constant Γ_δ ($mGy \cdot m^2 \cdot h^{-1} \cdot GBq^{-1}$)[a]	Geometry and activity	Quality control application
^{57}Co	272 d	0.00254/d	122 keV (86%)	14.1	Test tube-size rod, ~37 kBq	Well counter constancy and accuracy
					Vial/small bottle, 185–370 MBq	Dose calibrator accuracy and constancy
^{68}Ge[b]	287 d	0.00241/d	511 keV (178%)	129	Test tube-size rod, 37 kBq	Well counter constancy and accuracy
					Vial/small bottle, 185–370 MBq	Dose calibrator accuracy and constancy
^{137}Cs	30 a	0.0231/a	662 keV (86%)	82.1	Test tube-size rod, 37 kBq	Well counter constancy and accuracy
					Vial/small bottle, 185–370 MBq	Dose calibrator accuracy and constancy

[a] The air kerma rate constant Γ_δ is equivalent to the older specific γ ray constant Γ.

[b] Germanium-68 in a sealed source is in secular equilibrium with its short lived, positron emitting daughter ^{68}Ga (half-life: 68 min).

activity reading on each scale recorded; day to day readings should agree within 10%. For the accuracy test (sometimes also known as the energy linearity test), at least two of the foregoing NIST-traceable reference sources are separately placed in the dose calibrator and the activity reading on each scale recorded. For each source, the measured activity on each scale and its current actual activity should agree within 10%.

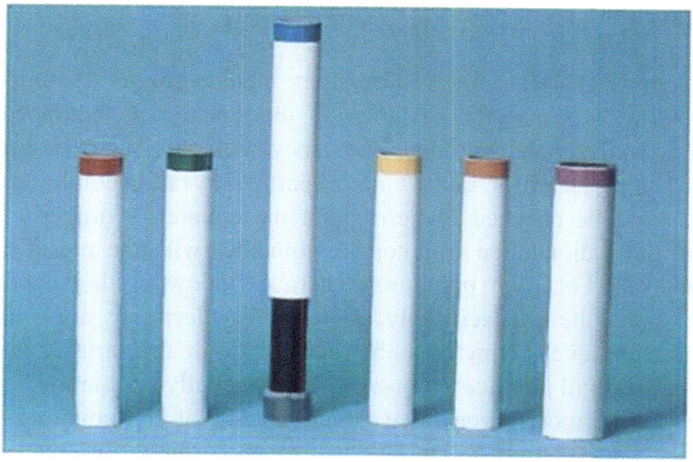

FIG. 10.8. Set of lead-lined plastic sleeves (Calicheck™ Dose Calibrator Linearity Test Kit, Calicheck, Cleveland, OH, USA) for evaluation of dose calibrator linearity by the shield method. The set is supplied with a 0.64 cm thick lead base, a colour coded unlined sleeve (to provide an activity measurement equivalent to the zero time point measurement of the decay method) and a six colour coded lead-lined sleeve providing attenuation factors nominally equivalent to decay over 6, 12, 20, 30, 40 and 50 h, respectively. (Courtesy of Calicheck, Cleveland, OH, USA.)

The quarterly check of linearity by the so-called ‘decay method’ begins with a high activity (~37 GBq), independently calibrated ^{99m}Tc source and its activity is assayed at 12 h intervals over three consecutive days. Over that time, equivalent to twelve half-lives of ^{99m}Tc, the activity decays to ~10 MBq. The measured activities are then plotted versus time on a semi-logarithmic graph and the best fit straight line drawn through the data points thus plotted (either ‘by eye’ or using a least squares curve-fitting algorithm). For each data point, the difference between the measured activity and the activity on the best fit straight line at that point should be less than 10%. An alternative approach to checking linearity is the ‘shield method’ in which lead sleeves of increasing thickness are placed in the dose calibrator with a ^{99m}Tc source (Fig. 10.8). By interposing increasing ‘decay-equivalent’ thicknesses (as specified by the manufacturer for the set of lead sleeves) between the source and the dose calibrator's sensitive volume, a decay-equivalent activity is measured for each sleeve. While the shield method is much faster than the decay method for checking linearity (taking minutes instead of days), an initial decay based calibration of the set of sleeves is recommended to accurately determine the actual decay equivalence of each shield.

10.5.4. Well counter

The routine QC tests for well counters include checks of the photopeak energy window (i.e. energy peaking) if the counter is equipped with a multichannel analyser, background, constancy and efficiency (or sensitivity). Prior to counting samples containing a particular radionuclide, the energy spectrum should be checked to verify that the counter is properly 'peaked', that is, that the radionuclide's photopeak coincides with the preset photopeak energy window[4]. For each photopeak energy window used, the background count rate should be checked daily. Importantly, electronic noise as well as ambient radiation levels, which may be relatively high and variable in a nuclear medicine facility, will produce a non-zero and potentially fluctuating background count rate. Furthermore, even trace contamination of the counting well will produce inaccurately high count rate values. Accordingly, a 'blank' (i.e. an empty counting tube or vial) should always be included to determine the current background count. To check constancy, at least one NIST-traceable reference source (Table 10.2) should likewise be counted each day; day to day net (i.e. gross minus background) count rates should agree within 10%.

In addition, as noted above, for each radionuclide for which a particular well counter is used, the counter should be calibrated — that is, its efficiency (sensitivity) (in cpm/kBq) determined — at installation, annually and after any repair (Eq. (10.1)).

10.5.5. Intra-operative probe

In addition to daily battery and background checks (as done for survey meters), QC tests of intra-operative probes should include a daily bias check for both the primary and any backup battery to verify that bias voltage (or high voltage) is within the acceptable range. As intra-operative probes may not provide a display of the energy spectrum, it may not be possible to visually check

---

[4] Isotope specific radionuclide counting or imaging with a scintillation detector is commonly performed using a 20% photopeak energy window, equivalent to an energy range of $E_\gamma \pm 10\%$ where E_γ is the X ray or γ ray energy of the radionuclide.

that the probe is properly peaked, that is, that the photopeak coincides with the preset photopeak energy window. The lower counts or count rates resulting from an inappropriate energy window may, therefore, go unnoticed. Thus, a long lived reference source or set of reference sources (such as ^{57}Co, ^{68}Ge and/or ^{137}Cs (Table 10.2)) should be available for daily checks of count rate constancy; a marked change (e.g. >±10%) in the net count rate from one day to the next may indicate an inappropriate energy window setting or some other technical problem. Ideally, the reference sources should each be incorporated into some sort of cap that fits reproducibly over the probe so that spurious differences in count rates due to variations in source–detector geometry are avoided.

10.5.6. Organ uptake probe

Aside from differences in counting geometry and sensitivity, uptake probes and well counters actually have very much in common and the QC procedures — checks of the photopeak energy window, background, constancy and efficiency — are, therefore, analogous. Importantly, however, efficiency should be measured more frequently — for each patient — than for a well counter, so that the probe net count rates can be reliably converted to thyroid uptakes for individual patients.

BIBLIOGRAPHY

CHERRY, S.R., SORRENSON, J.A., PHELPS, M.E., Physics in Nuclear Medicine, 3rd edn, Saunders, Philadelphia, PA (2003)

NINKOVIC, M.M., RAICEVIC, J.J., ANDROVIC, A., Air kerma rate constants for γ emitters used most often in practice, Radiat. Prot. Dosimetry **115** (2005) 247–250.

ZANZONICO, P., Routine quality control of clinical nuclear medicine instrumentation: A brief review, J. Nucl. Med. **49** (2008) 1114–1131.

ZANZONICO, P., HELLER, S., "Physics, instrumentation, and radiation protection", Clinical Nuclear Medicine (BIERSACK, H.J., FREEMAN, L.M., Eds), Springer Verlag, Berlin Heidelberg (2007) 1–33.

CHAPTER 11

NUCLEAR MEDICINE IMAGING DEVICES

M.A. LODGE, E.C. FREY
Russell H. Morgan Department of Radiology and Radiological Sciences,
Johns Hopkins University,
Baltimore, Maryland, United States of America

11.1. INTRODUCTION

Imaging forms an important part of nuclear medicine and a number of different imaging devices have been developed. This chapter describes the principles and technological characteristics of the main imaging devices used in nuclear medicine. The two major categories are gamma camera systems and positron emission tomography (PET) systems. The former are used to image γ rays emitted by any nuclide, while the latter exploit the directional correlation between annihilation photons emitted by positron decay. The first section of this chapter discusses the principal components of gamma cameras and how they are used to form 2-D planar images as well as 3-D tomographic images (single photon emission computed tomography (SPECT)). The second section describes related instrumentation that has been optimized for PET data acquisition. A major advance in nuclear medicine was achieved with the introduction of multi-modality imaging systems including SPECT/computed tomography (CT) and PET/CT. In these systems, the CT images can be used to provide an anatomical context for the functional nuclear medicine images and allow for attenuation compensation. The third section in this chapter provides a discussion of the principles of these devices.

11.2. GAMMA CAMERA SYSTEMS

11.2.1. Basic principles

The gamma camera, or Anger camera [11.1], is the traditional workhorse of nuclear medicine imaging and its components are illustrated in Fig. 11.1. Gamma camera systems are comprised of four basic elements: the collimator, which defines the lines of response (LORs); the radiation detector, which counts incident γ photons; the computer system, which uses data from the detector to create 2-D

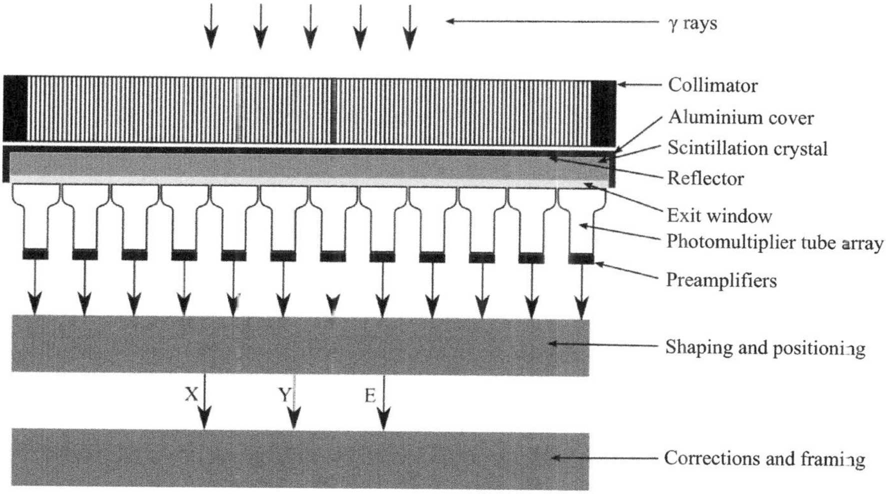

FIG. 11.1. Schematic diagram showing the major components of a gamma camera

histogram images of the number of counted photons; and the gantry system, which supports and moves the gamma camera and patient. The overall function of the system is to provide a projection image of the radioactivity distribution in the patient by forming an image of γ rays exiting the body. Forming an image means establishing a relationship between points on the image plane and positions in the object. This is sometimes referred to as an LOR: ideally, each position in the image provides information about the activity on a unique line through the object. In gamma cameras, single photons are imaged, in contrast to PET where pairs of photons are imaged in coincidence. Thus, in order to define an LOR, a lens is required, just as in optical imaging. However, the energies of γ rays are so high

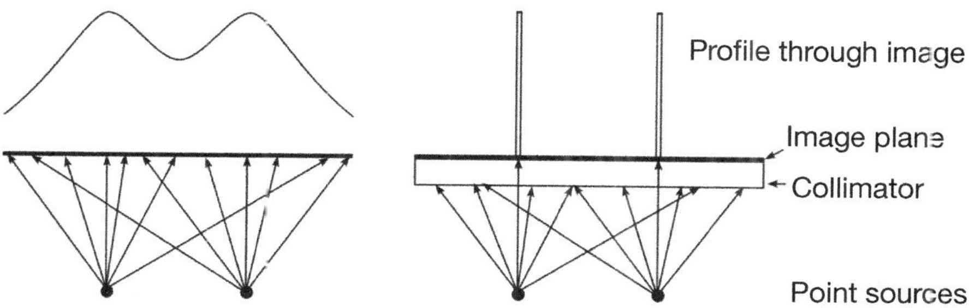

FIG. 11.2. The drawing on the left demonstrates the image of two point sources that would result without the collimator. It provides very little information about the origin of the photons and, thus, no information about the activity distribution in the patient. The drawing on the right illustrates the role of the collimator and how it defines lines of response through the patient. Points on the image plane are uniquely identified with a line in space.

that bending them is, for all practical purposes, impossible. Instead, collimators are used to act as a mechanical lens. The function of a collimator is, thus, to define LORs through the object. Figure 11.2 illustrates the function of and need for the collimator, and, thus, the basic principle of single photon imaging.

11.2.2. The Anger camera

11.2.2.1. Collimators

As mentioned above, the collimator functions as a mechanical lens: it defines LORs. The collimator accomplishes this by preventing photons emitted along directions that do not lie along the LOR from reaching the detector. Thus, collimators consist of a set of holes in a dense material with a high atomic number, typically lead. The holes are parallel to the LORs. Ideally, each point in the object would contribute to only one LOR. This requires the use of collimator holes that are very long and narrow. However, such holes would allow very few photons to pass through the collimator and be detected. Conversely, increasing the diameter or decreasing the length of the holes results in a much larger range of incident angles passing through the collimator. As illustrated in Fig. 11.3, this results in degraded resolution. As can be seen from this figure, each hole has a cone of response and the diameter of the cone of response is proportional to the distance from the face of the collimator.

As discussed above, changing the diameter of collimator holes changes the resolution and also the number of photons passing through the collimator. The noise in nuclear medicine images results from statistical variations in the number of photons counted in a given counting interval due to the random nature of radiation decay and interactions with the patient and camera. The noise is described by Poisson statistics, and the coefficient of variation (per cent noise) is inversely proportional to the square root of the number of counts. Thus, increasing the number of counts results in less noisy images. As a result, there is an inverse relationship between noise and spatial resolution for collimators: improving the resolution results in increased image noise and vice versa.

Another important characteristic of collimators is the opacity of collimator septa to incident γ rays. In an ideal collimator, the septa would block all incident radiation. However, in real collimators, some fraction of the radiation passes through or scatters in the septa and is detected. These phenomena are referred to as septal penetration and scatter. The amount of septal penetration and scatter depends on the energy of the incident photon, the thickness and composition of the septa, and the aspect ratio of the collimator holes. Since gamma cameras are used to image radionuclides with energies over a wide range, collimators are typically available that are appropriate for several energy ranges: low energy collimators are designed for isotopes emitting photons with energies lower than approximately 160 keV; medium energy collimators are designed for energies up to approximately 250 keV; and high energy collimators are designed for higher energies. It should be noted that in selecting the appropriate collimator for an isotope, it is important to consider not only the photon energies included in the image, but also higher energy photons that may not be included in the image. For example, in ^{123}I there are a number of low abundance high energy photons that can penetrate through or scatter in septa and corrupt the images. As a result, medium energy collimators are sometimes used for ^{123}I imaging, despite the main γ photopeak at 159 keV.

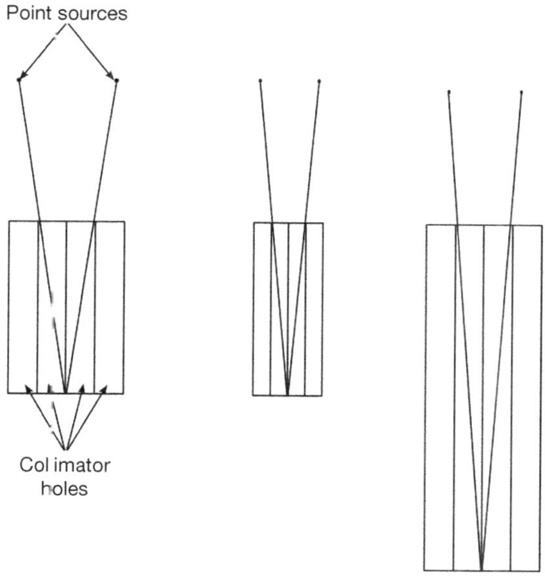

FIG. 11.3. Illustration of the concept of spatial resolution and how collimator hole length and diameter affect spatial resolution. The lines from the point source through the collimator indicate the furthest apart that two sources could be and still have photons detected at the same point on the image plane (assumed to be the back of the collimator). Thus, sources closer together than this would not be fully resolved (though they might be partially resolved). From this, we see that the resolution decreases as a function of distance. It should also be noted that the resolution improves proportionally with a reduction in the width of the collimator holes and improves (though not proportionally) with the hole length.

For multi-hole collimators, hole shape is an additional important factor in collimator design. The three most common hole shapes are shown in Fig. 11.4. Round holes have the advantage that the resolution is uniform in all directions in planes parallel to the face of the collimator. However, as discussed below, the sensitivity is relatively low because there is less open area for a given resolution and septal thickness. Hexagonal hole collimators are the most common design for gamma cameras using continuous crystals. They have the advantage of relatively direction independent response functions and higher sensitivity than a round hole collimator with the same resolution and septal thickness. Square hole collimators are especially appropriate for detectors that have pixelated crystals. Having squares holes that match the spacing and inter-gap distance of these detectors results in good sensitivity with these detectors. However, the resolution varies significantly depending on the direction, being worse by a factor of a $\sqrt{2}$ along the diagonal direction compared to parallel to the rows of holes.

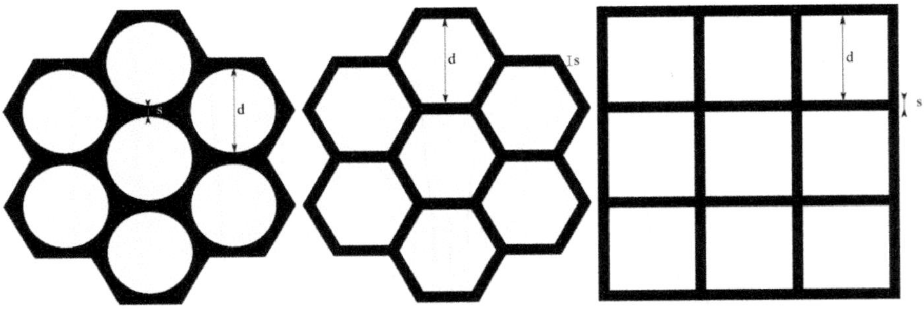

FIG. 11.4. Examples of the three major hole shapes used in multi-hole collimators. They are, from left to right: round, hexagonal and square holes. In all cases, black indicates septa and white open space. The diameter (often called flat-to-flat for square and hexagonal holes) is d and the septal thickness is s.

Multi-hole collimators typically have many thousands of holes. The uniformity of the image critically depends on the holes having uniform sizes and spacing. As a result, high quality fabrication is essential. The septa must be made of a high density, high Z material in order to stop the incident γ rays. Lead is the material of choice for most multi-hole collimators due to its relatively low cost, low melting temperature and malleability. Lead multi-hole collimators can be divided into two classes based on fabrication techniques: cast and foil collimators.

Fabrication of cast collimators involves the use of a mould that is filled with molten lead to form the collimator. The mould typically consists of two plates with holes at opposing positions that match each of the holes in the collimator. A set of pins is placed in the holes. Lead is then poured between the plates and the entire assembly is then carefully cooled. The plates and pins are removed leaving behind the collimator. This technology is especially well suited to making high and medium energy collimators as well as special purpose collimator geometries.

Foil collimators are created from thin lead foils. The foils are stamped and then glued together to build up the collimator layer by layer. Figure 11.5 shows a schematic of how two layers are stamped and glued to form the holes. It should be noted that in the stamping the septa that are glued must be thinner than the other walls in order to retain uniform septal thickness and, thus, maximize sensitivity. It is clear that precise stamping, alignment and gluing is essential to form a high quality collimator. The septa in foil collimators can be made thinner than in cast collimators. As a result, foil fabrication techniques are especially appropriate for low energy collimators. Understanding the fabrication technology can help in diagnosing problems with the collimator. For example, Fig. 11.6 shows an image with non-uniformities appearing as vertical stripes in the image of a sheet source. This was a foil collimator and the non-uniformities apparently originated from fabrication problems, resulting in some layers having different sensitivities compared to other layers.

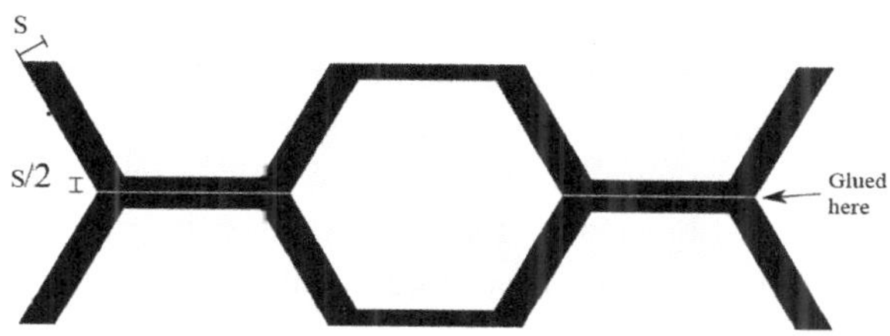

FIG. 11.5. Illustration of fabrication of foil collimator by gluing two stamped lead foils. It should be noted that the foils must be stamped so that the portions of the septa that are glued are half the thickness of the rest of the septa. Furthermore, careful alignment is essential to preserve the hole shapes.

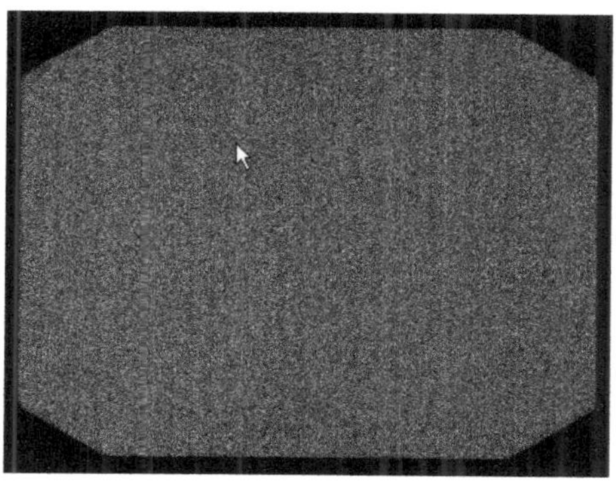

FIG. 11.6. Uniformity image of a defective foil collimator. The vertical stripes in the image result from non-uniform sensitivity of the collimator due to problems in the manufacturing process. The peppery texture is due to quantum noise and is visible because the grey level was expanded to demonstrate the non-uniformity artefacts.

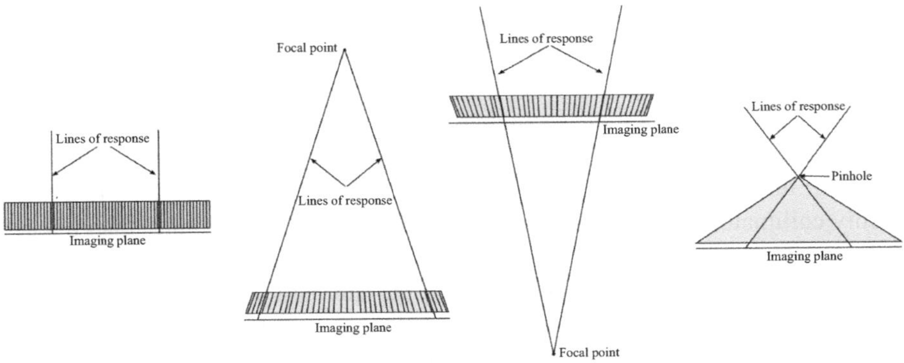

FIG. 11.7. Illustration of the four common collimator geometries: (left to right) parallel, converging, diverging and pinhole.

A final important characteristic of collimators is the hole geometry. There are four common geometries in nuclear medicine: parallel, converging, diverging and pinhole, as illustrated in Fig. 11.7. Parallel holes are the most commonly used collimators. The LORs are parallel, and there is, thus, a one to one relationship between the size of the object and image. In converging hole collimators, the LORs converge to a point (focal point) or line (focal line) in front of the collimator, and there is, thus, magnification of the image. These two collimator types are referred to as cone-beam and fan-beam collimators, respectively. These are useful for imaging small objects, such as the heart or brain, on a large camera as they provide an improved trade-off between spatial resolution and noise. In diverging hole collimators, the LORs converge to a point or line behind the collimator. This results in minification of the image. Diverging hole collimators are useful for imaging large objects on a small camera. However, they result in a poor resolution versus noise trade-off, and are, thus, infrequently used. Pinhole collimators use a single hole to define the LORs. In terms of geometry, they are similar to cone-beam collimators but with the focal point between the image plane and the object being imaged. As a result of this, the image is inverted compared to the object. In addition, the object can be either minified or magnified depending on whether the distance from the image plane to the focal point is less than or greater than the distance from the pinhole to the object plane. Pinhole collimators provide an improved resolution noise trade-off when objects are close to the pinhole. They are, thus, useful for imaging small objects such as the thyroid or small animals. Another advantage of pinholes is that there is only a single hole, referred to as an aperture, which determines the amount of collimator penetration and scatter. As a result, it is possible to fabricate the aperture from high density and high atomic number materials (e.g. tungsten, gold or depleted uranium), which can reduce collimator penetration and scatter. This makes these collimators appropriate for imaging radionuclides emitting high energy γ rays such as ^{131}I. In addition, pinhole collimators with changeable apertures can have different diameter pinholes. This allows selection of resolution/sensitivity parameters relatively easily.

The collimator properties can be most completely described by the collimator point source response function (PSRF), the noise-free image of a point source in air with unit activity using an ideal radiation detector, as a function of position in the object space. The shape of the collimator PSRF completely describes the resolution properties, and when normalized to unit total volume is referred to as the collimator point spread function (PSF).

Figure 11.8 shows some sample collimator PSFs for an ^{131}I point source imaged with a high energy general purpose collimator and a medium energy general purpose collimator. These PSFs are averaged over the position of the source with respect to the collimator and, thus, do not show the hole pattern. There are several things to note from this figure. First, using a properly designed collimator reduces septal scatter and penetration to very low levels, while they become significant for a collimator not designed to handle the high energies of ^{131}I. Second, the response function becomes wider as a function of distance, demonstrating the loss of resolution as a function of increasing source to collimator distance. Finally, there is evidence of the shape of the holes, which, in this case, were hexagonal. The shape can be barely discerned in the shape of the central portion of the response, which is due to photons passing through the collimator holes. The geometry of the collimator is more evident in the septal

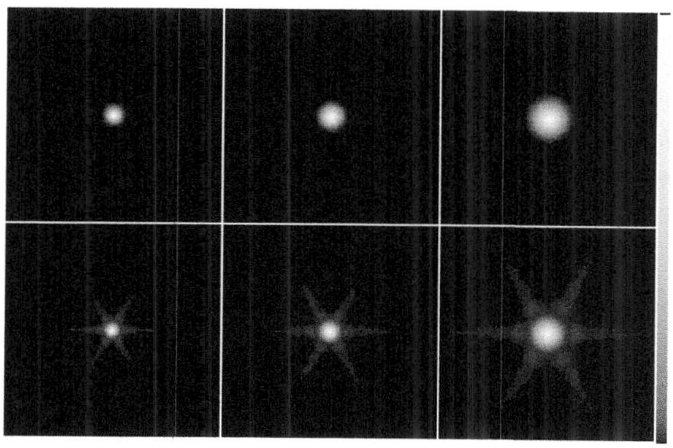

FIG. 11.8. Sample images of the point spread function for an ^{131}I point source at (left to right) 5, 10 and 20 cm from the face of a high energy general purpose collimator (top row) and a medium energy general purpose collimator (bottom row). The images are displayed on a logarithmic grey scale to emphasize the long tails of the point spread function resulting from septal penetration and scatter. The brightness of the image has been increased to emphasize the septal penetration and scatter artefacts.

penetration and scatter artefacts. In fact, the septal penetration is highest along angular directions where the path through the septa is thinnest, giving rise to the spoke-like artefacts in the directions perpendicular to the walls of the hexagonal holes.

Another useful way to describe and understand the resolution properties of the collimator is in terms of its frequency response. This can be described by the collimator modulation transfer function, which is the magnitude of the Fourier transform of the collimator PSF. Figure 11.9 shows some sample profiles through the collimator modulation transfer function. It should be noted that the collimator response does not pass high frequencies very well and, for some frequencies, the response is zero. This attenuation of high frequencies results in a loss of fine detail (i.e. spatial resolution) in the images. Finally, the cut-off frequency decreases with distance from the collimator and different collimator designs have different frequency responses.

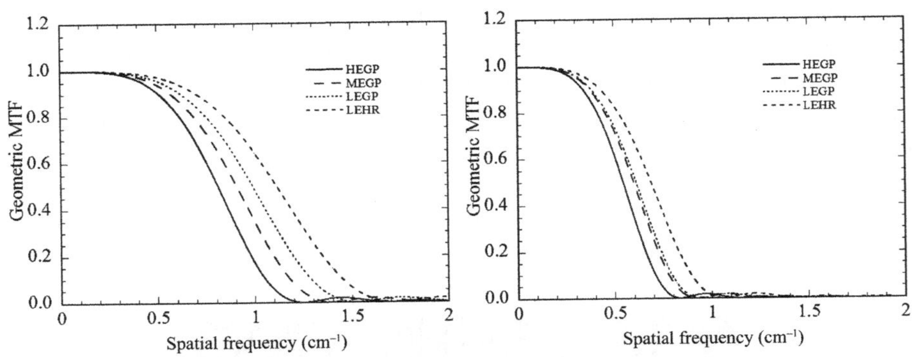

FIG. 11.9. Sample profile through the geometric modulation transfer functions (MTFs) for low, medium and high energy (HE, ME and LE, respectively) general purpose (GP) and high resolution (HR) collimators for a source 5 cm (left) and 20 cm (right) from the face of the collimator.

It is often desirable to summarize the collimator resolution in terms of one or two numbers rather than the entire response function. This is often done in terms of the width of the collimator PSRF at a certain fraction of its maximum value. For example, Fig. 11.10 shows a sample profile through a collimator PSF and the position of the full width at half maximum (FWHM) and the full width at tenth maximum (FWTM). To good approximation, the FWHM of a collimator is proportional to the distance from the face of the collimator. This holds for all distances except those very close to the collimator face, as illustrated in Fig. 11.11. The FWTM is useful for assessing the amount of septal penetration and scatter that are present. For a Gaussian response function, which is a good

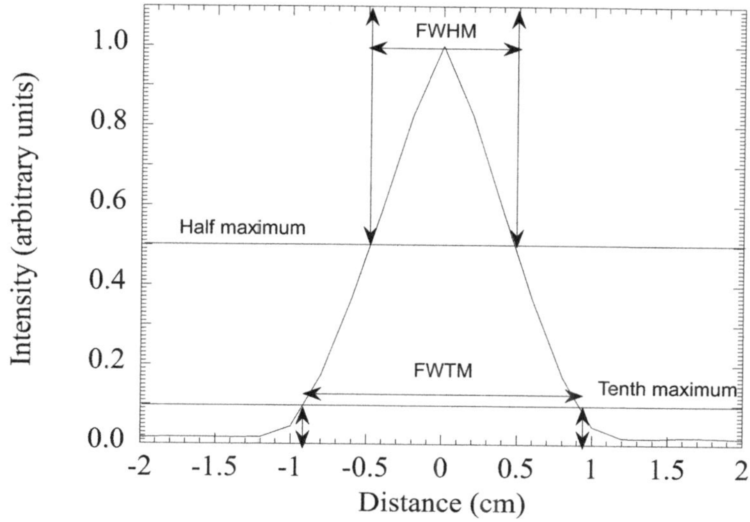

FIG. 11.10. Plot of the total collimator–detector point spread function for a medium energy general purpose collimator imaging ^{131}I, indicating the positions of the full width at half maximum (FWHM) and the full width at tenth maximum (FWTM).

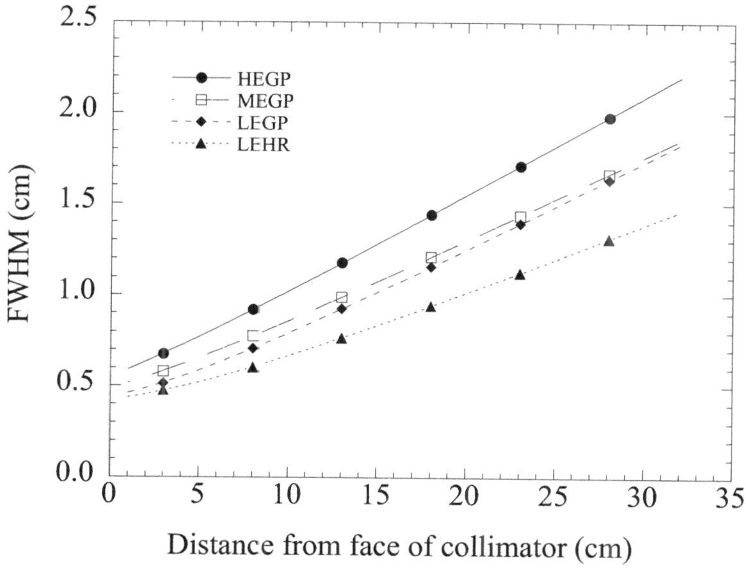

FIG. 11.11. Plot of the full width at half maximum (FWHM) of the geometric collimator–detector point source response function including a Gaussian intrinsic response with an FWHM of 4 mm as a function of distance from the face of the collimator for the same set of collimators described in Fig. 11.9.

approximation for the combination of a Gaussian intrinsic detector response with the geometric collimator response, the relationship between the FWHM and FWTM is given by FWTM/FWHM ≈ 1.86. Thus, if the FWTM is substantially larger than this factor times the FWHM, the response has been affected by factors other than the geometric response, such as septal penetration and scatter. It should be noted that the effects of septal penetration and scatter on the FWTM are less visible in a PSRF than they are in a line source response function.

The FWHM of the collimator resolution can be estimated from geometric arguments. Figure 11.12 shows a schematic that can be used to derive the resolution for a point source a distance Z from the face of the collimator. The collimator hole has a length L and a width d. The image plane (often taken to be the mean path of the primary photons in the crystal plus any physical gap) is a distance B behind the back face of the collimator. The photon passing through the collimator holes with the most oblique angle of incidence will have an incident angle defined by tan $\theta = d/L$. Thus, the extreme limits of the response function will be defined by this limit. If it is assumed that the geometric response function is triangular in shape, then the FWHM in Fig. 11.12 will be half of this distance. Using similar triangles, it can be shown that the FWHM is given by:

$$\text{FWHM} = \frac{d}{L}(Z + L + B) \tag{11.1}$$

Thus, it can be seen that, as described above, the FWHM is linearly related to the distance from the face of the collimator and is proportional to that distance when the distance is large compared to $L + B$.

The resolution of the collimator–detector system depends on both the resolution of the collimator and the intrinsic resolution of the gamma camera. For continuous-crystal gamma cameras, the intrinsic resolution can be modelled with a Gaussian function. In this case, the total response function for the collimator–detector is the convolution of the intrinsic resolution and the collimator response. If the collimator response is approximated by a Gaussian function, the FWHM is given by the Pythagorean sum of the intrinsic and collimator FWHMs:

$$\text{FWHM}_{\text{total}} = \sqrt{\text{FWHM}^2_{\text{collimator}} + \text{FWHM}^2_{\text{intrinsic}}} \tag{11.2}$$

Figure 11.11 shows a plot of the total geometric FWHM resolution for several collimators including the effects of a 4 mm Gaussian intrinsic resolution. The curves through the points represent a fit with a function of the above Pythagorean sum. It should be noted that except for distances close to the collimator, the resolution is linear with distance, indicating that the total resolution is dominated by the collimator resolution.

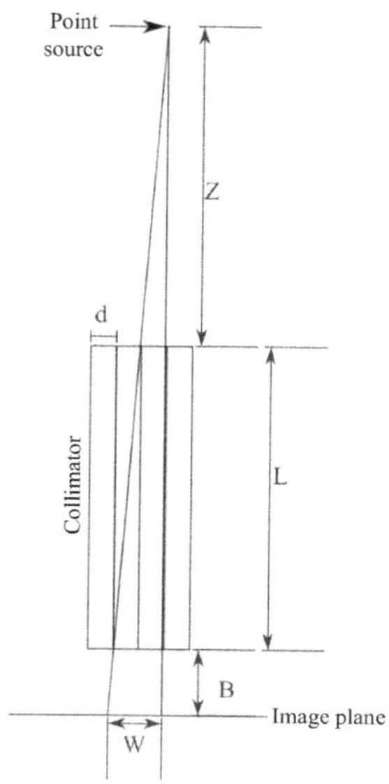

FIG. 11.12. Diagram illustrating the collimator geometry used to derive the expression for the full width at half maximum.

The integral of the collimator PSRF gives the sensitivity of the collimator. This is, in principle, a dimensionless quantity which gives the fraction of emitted photons that pass through the collimator, and is of the order of 10^{-3}–10^{-4} for typical nuclear medicine collimators. It is often useful to express the sensitivity in terms of counts per unit activity per unit time, for example counts per second per megabecquerel. For a parallel-hole collimator, the sensitivity is a function of two terms: the solid angle of the hole, which is a function of $(d/L)^2$, and the fraction of the active area of the collimator that is open area (hole) as compared to septa. The second term can be described in terms of the unit cell: the smallest geometric region that can be used to form the entire collimator by a set of simple translations. The sensitivity S of a parallel-hole collimator is given by:

$$S = \frac{a_{\text{open}}}{4\pi L^2} \frac{a_{\text{open}}}{a_{\text{total}}} \tag{11.3}$$

where

a_{open} is the open area of a collimator unit cell, i.e. the area of the hole itself;

and a_{total} is the total area of the cell including the part of the collimator septa lying in the unit cell.

From the above it can be seen that, for a parallel-hole collimator, the sensitivity is independent of the distance to the collimator face. This is because there is a balance between the decrease in sensitivity from a single hole and the increase in the number of holes through which photons can pass as a function of increasing distance from the collimator face. It should also be noted that a_{open} is proportional to d^2, which is proportional to the FWHM resolution. The term $a_{\text{open}}/a_{\text{total}}$ varies slowly as a function of d if $d \gg s$. Thus, the sensitivity is proportional to the square of the resolution:

$$S = k \times \text{FWHM}^2 \tag{11.4}$$

where the constant k depends only weakly on the FWHM. Since noise is directly related to the number of counts, there is a fundamental trade-off between resolution and noise. From the above, it is also evident that maximizing the ratio of $a_{\text{open}}/a_{\text{total}}$ is important in terms of reducing noise for a given resolution.

11.2.2.2. Scintillation crystals

The scintillation crystal in the gamma camera converts γ ray photons incident on the crystal into a number of visible light photons. The characteristics and principle of scintillation radiation sensors are described in more detail in Chapter 6. Ideally, the crystal would be dense and of a high Z material in order to stop all incoming γ rays with photoelectric events. It should have high light output to provide low quantum noise for energy and position estimation. The decay time of the light output needs to be fast enough to avoid a pile-up of pulses at the count rates experienced in nuclear medicine imaging procedures. The wavelength spectrum of the scintillation photons should be matched to the sensitivity of the photodetectors used to convert the scintillation signal into an electrical signal. In addition to these technical properties that directly affect image quality, there are a number of desirable material properties that influence the cost of the device. These include the cost of the raw material, the ease of growing large single crystals and the sensitivity to environmental factors such as humidity. Owing to its unique combination of desirable properties, the crystals used in gamma cameras based on photomultiplier tubes (PMTs) are typically made of NaI(Tl). Gamma cameras based on solid state photodetectors require a different light spectrum and typically CsI(Tl) is used in these devices. The scintillation properties of these materials are discussed in detail in Chapter 6.

As will be described below, the interaction position of the γ ray with the detector is estimated based on the distribution of the scintillation light to an array of PMTs. It is important that the distribution of light be as independent as possible of the depth of interaction in the crystal and depends in a predictable way on the lateral position. Further, absorption of scintillation photons by defects in the crystal is highly undesirable, as it will adversely affect the accuracy and precision of energy and position estimation. Thus, in order to make the lateral light distribution predictable, as even as possible and to minimize absorption, Anger cameras employ a large single crystal equal to the size of the field of view (FOV) of the camera. This can be as large as 60 cm × 40 cm in modern cameras. As NaI(Tl) is hygroscopic, it is important to hermetically seal the crystal in an enclosure. The back of the crystal must be optically coupled to the photodetector, so that the back part of the crystal enclosure consists of an optical glass exit window optically coupled to the crystal. The exit window lies between the crystal and the photodetector array. It serves several functions. First, it serves to hermetically seal the crystal. Second, the exit window allows scintillation photons to pass from the crystal into the photodetector array, and, thus, must be transparent in the emission range of the scintillator. To reduce internal reflections, it is desirable that the index of refraction be matched as closely as possible to that of the scintillator (n = 1.85 for NaI(Tl)) and the photodetector ($n \approx 1.5$ for borosilicate glasses used in the entrance windows of PMTs). The remainder of the enclosure should be light-tight to block out ambient light. The front face should be thin and of a low Z material — typically Al is used — to reduce the probability of incident γ ray absorption. Finally, to help collect incident light photons and improve energy resolution, the inside of the enclosure is a reflective layer. The use of specular versus diffusive reflectors affects the nature of the light response and has an impact on the variation in the light response on the photodetectors as a function of interaction depth, and, thus, impacts the precision of the position estimation.

One important parameter of the scintillation crystal related to camera performance is its thickness. The thickness is a trade-off between two characteristics: intrinsic resolution and sensitivity. The intrinsic resolution depends on the crystal thickness via the variation in the light distribution as a function of depth of interaction. Since the depth of interaction can vary over a wider range in a thicker crystal, there will be a larger variation in the light distribution and, thus, a larger uncertainty in the estimated lateral position of the interaction. In other words, thicker crystals generally have poorer intrinsic resolution. The functional relationship between the thickness and intrinsic resolution is complicated and depends on the details of the surface treatment of the scintillator, the photodetector array and the position estimation algorithm. For GE Millenium VG cameras, the FWHM intrinsic resolution for 140 keV photons using 0.953, 1.587 and 2.54 cm thick crystals is 3.5, 3.9 and 5.2 mm, respectively.

The intrinsic sensitivity of the crystal is related to crystal thickness by:

$$S_i = 1 - e^{-\mu t} \quad (11.5)$$

where

S_i is the intrinsic sensitivity;
μ is the linear attenuation coefficient of the crystal;

and t is the crystal thickness.

Since the linear attenuation coefficient decreases with energy, the intrinsic sensitivity also decreases with energy. Figure 11.13 shows a plot of the intrinsic sensitivity as a function of energy for several crystal thicknesses. For 140 keV, the sensitivity is ~92% for a 0.953 cm (3/8 in) thick crystal. This is the most common crystal thickness in commercial systems, though cameras with 5/8 and 1 in are available. These crystal thicknesses provide substantially improved sensitivity for radionuclides emitting medium and high energy photons such as ^{111}In and ^{131}I at the cost of a relatively minor reduction in intrinsic spatial resolution.

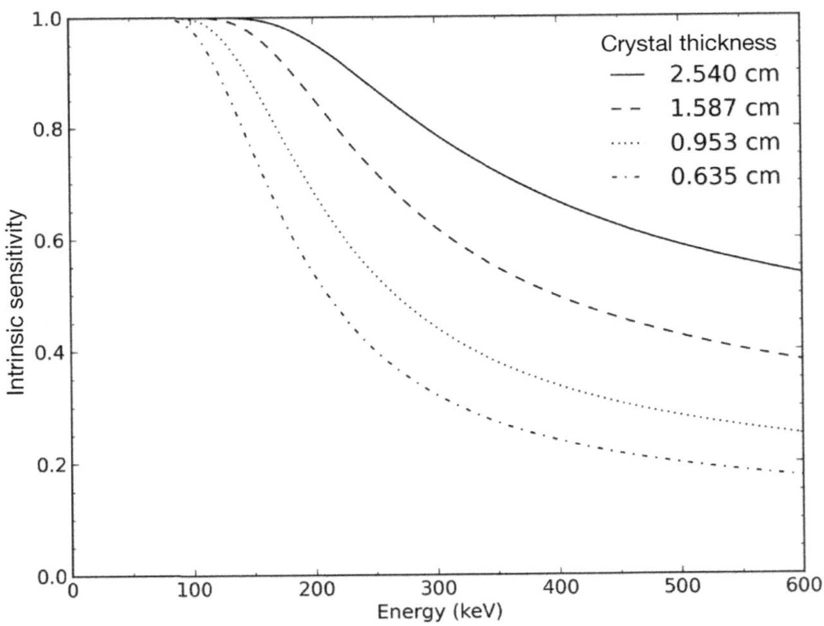

FIG. 11.13. Plot of the intrinsic sensitivity of a NaI scintillation crystal as a function of energy for several crystal thicknesses.

A final important property of the scintillation crystal is the light output. This is a characteristic of the scintillator material, and is the number of scintillation photons per unit energy deposited in the crystal by a γ photon. Thus, the total light is proportional to the energy deposited in the crystal, and can, therefore, be used to estimate the energy of the γ ray. The number of scintillation photons produced for a given event is a Poisson random variable. Thus, the larger the number of scintillation photons the smaller the coefficient of variation (standard deviation divided by the mean) of the mean number of photons, and, hence, the estimated photon energy. Thus, scintillators with high light output will provide higher energy resolution. In addition, as will be seen below, the light distribution over the photodetector array is used to estimate the interaction position. Since the light collected by each element in the array is also a Poisson random variable that is proportional to the light output, a larger light output will result in higher precision in the estimated position, and, thus, improved intrinsic spatial resolution. One reason that NaI(Tl) is used in gamma cameras is its high light output.

11.2.2.3. Photodetector array

The next element in the radiation detector is the photodetector array. This array measures the distribution of scintillation photons incident on the array and converts it into a set of pulses whose charge is proportional to the number of scintillation photons incident on each corresponding element in the array. As described below, the output of this array is used to compute the interaction position of the γ ray in the scintillator. In clinical gamma cameras, the photodetector array is comprised of a set of 30–90 PMTs arranged in a hexagonal close packed arrangement, as illustrated in Fig. 11.14. More details on the operation and characteristics of PMTs are provided in Chapter 6. In brief, PMTs have the advantage that they are very well understood, have a moderate cost, are relatively sensitive to low levels of scintillation light and have a very high gain. In some commercial designs, PMTs have been replaced by semiconductor detectors such as photodiodes. Generally, these devices are somewhat less sensitive and have a lower gain than PMTs, resulting in more noise in the charge signal and, thus, less precision in the energy and position estimated from the charge signal.

Since the position and energy are estimated from the set of charge signals from the elements in the photodetector array, it is highly desirable that the proportionality constants relating light intensity to charge be the same for all of the photodetectors. This can be ensured by choosing matching devices and by carefully controlling and matching the electronic gain. For PMTs, the gain is controlled by the bias voltage applied to the tubes. Since gain is also a function of temperature, the temperature of the photodetectors must be carefully controlled. The gains of PMTs are very sensitive to magnetic fields, even those as small as the Earth's magnetic field. Thus, the PMTs must be magnetically shielded using mu-metal. Finally, since the gains of tubes can drift over time, periodic recalibration is necessary.

One of the major advantages of the gamma camera is that the number of PMTs is much smaller than the number of pixels in images from the gamma camera. In other words, in contrast to semiconductor detectors where a separate set of electronics is required for each pixel, the gamma camera achieves a great reduction in cost and complexity by estimating the interaction position of the γ ray based on the output of the array of PMTs.

To understand the position estimation process, Fig. 11.15 is considered. This figure shows a cross-section through two PMTs, and the crystal and exit window. The number of photons collected by a PMT directly (i.e. without reflection) will be proportional to the solid angle subtended by it at the interaction point. As can be seen in the figure, the interaction position is offset to the right, and there is a smaller solid angle subtended by PMT 1 than by PMT 2. Thus, the signal from PMT 1 will be smaller than for PMT 2. If the interaction position moves to the left, so that it lies along the line separating the two PMTs, there will be an equal amount of light collected by each PMT. The relationship between the light collected by the two PMTs and the lateral interaction position can be used to estimate the interaction position, as will be described in more detail below. In addition, the total scintillation light collected by all of the PMTs is proportional to the energy deposited by the γ ray in the crystal. Thus, the total charge can be used to estimate the energy of the photon.

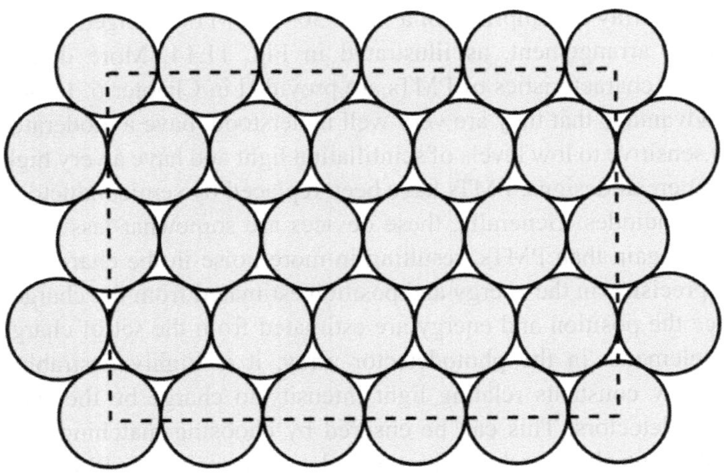

FIG. 11.14. Cross-section of a gamma camera at the back face of the entrance window showing the hexagonal close packed array of photomultiplier tubes. The dotted line indicates the approximate region where useful images can be obtained.

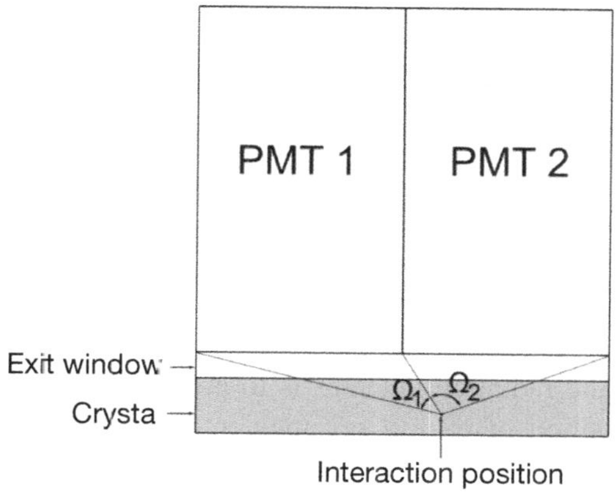

FIG. 11.15. Cross-section through two photomultiplier tubes (PMTs), the exit window and crystal in a gamma camera. The interaction position of a γ ray photon is indicated. The solid angles subtended by PMT 1 and 2 are Ω_1 and Ω_2, respectively.

11.2.2.4. Amplifiers and pulse shaping

The charge pulse from each PMT is very small and, thus, subject to noise. In addition, the scintillation photons are emitted randomly over a finite time (given by the scintillator's decay constant), making the charge pulse rather noisy. To make subsequent analysis of the pulse easier and more resistant to electrical noise, the pulse is amplified and shaped prior to processing to estimate the interaction position and photon energy. The components of this stage are a preamplifier and shaping amplifier.

The preamplifier integrates the charge pulse from the PMT to form a voltage pulse with height proportional to the input charge. The design of the preamplifier should be such that the voltage height is as independent as possible of the details of the charge pulse, such as decay and rise times. Preamplifiers are typically mounted directly to the PMT outputs in order to avoid corruption of the tiny charge pulses by electrical noise and interference.

Ideally, output pulses would have a very flat top to allow easy digitization of the pulse height and be very narrow to allow high pulse rates without pulse pile-up. However, the output pulse from the preamplifier typically has a relatively long decay time and is not very suitable for digitization and handling high pulse rates. As a result, the output of the preamplifier is fed into a shaping amplifier. Typically, shaping amplifiers use a combination of integration and differentiation stages to produce near Gaussian pulses. It should be noted that more recent commercial gamma cameras have used digital pulse processing methods to perform this function. This involves digitizing the output waveform from the preamplifier. This has a number of advantages including providing the ability to change the trade-off between energy resolution and count rate, depending on the requirements of the particular imaging procedure. In addition, this method also provides digital estimates of the pulse heights that can be used in digital position and energy estimation algorithms.

11.2.2.5. Position and energy estimation

The goal of the radiation detector is to provide an estimate of the energy and interaction position of each γ ray incident on the detector. The output of the photodetector array and amplifier system is a set of voltage signals for each photon. The sum of these voltages is proportional to the gamma camera energy and the position is a function of the set of voltage values. The position and energy estimation circuits estimate the γ ray energy and position from the set of voltage values from the photodetector array.

One way of doing this is to use a resistive network to divide the signals from the array elements among a set of four signals often referred to as X_+, X_-, Y_+ and Y_-, as illustrated in Fig. 11.16. The resistor values for each PMT are chosen so that the charge is divided in proportion to its position with respect to the centre of the array. For example, for a PMT in the centre horizontally, the resistances for X_+ and X_- would be equal. Similarly, for a PMT in the centre vertically, the resistances for Y_+ and Y_- would be equal.

Using the scheme described above, the energy E can be computed using:

$$E = X_+ + X_- + Y_+ + Y_- \tag{11.6}$$

However, one limitation of this method is that the total amount of light collected is dependent on position. For example, if the interaction is directly under a PMT, a larger fraction of the total light will be collected, resulting in a larger value of E than if the interaction is in the gap between PMTs. This means that the estimate of the energy will vary spatially. As discussed below, this has an impact on camera uniformity. As a result, the energy must be corrected based on the interaction position.

Under the assumption that the light collected by a PMT is proportional to the distance from its centre, and with the correct resistor values, the interaction position, defined by x and y can be computed using:

$$x = \frac{X_+ - X_-}{E} \quad \text{and} \quad y = \frac{Y_+ - Y_-}{E} \tag{11.7}$$

In early gamma cameras, the computations above were performed using analogue circuits. However, in recent cameras, the pulse heights (or the entire pulse) are digitized and the computations performed digitally.

Using the resistive summing and simple estimation approaches above results in a number of problems. First, light collected by phototubes is not linearly related to the distance from the interaction point. For example, the amount of light changes relatively little for PMTs at a large lateral distance from the interaction position. As a result, thresholding circuits are often added to exclude the signal from PMTs with small outputs (and, thus, far from the interaction point) from the position calculation. In addition, the distribution of light between two tubes changes more quickly when the interaction position lies between two tubes than it does when the interaction position is directly over a tube. This results in spatial non-linearities where images of line sources are bent towards the centre of PMTs. A final difficulty is that it is not possible to reliably estimate the position of photons interacting near the edge of the camera. In this case, almost all of the light will be collected by the nearest PMT and there will be little change in the relative amount of light as the interaction position moves.

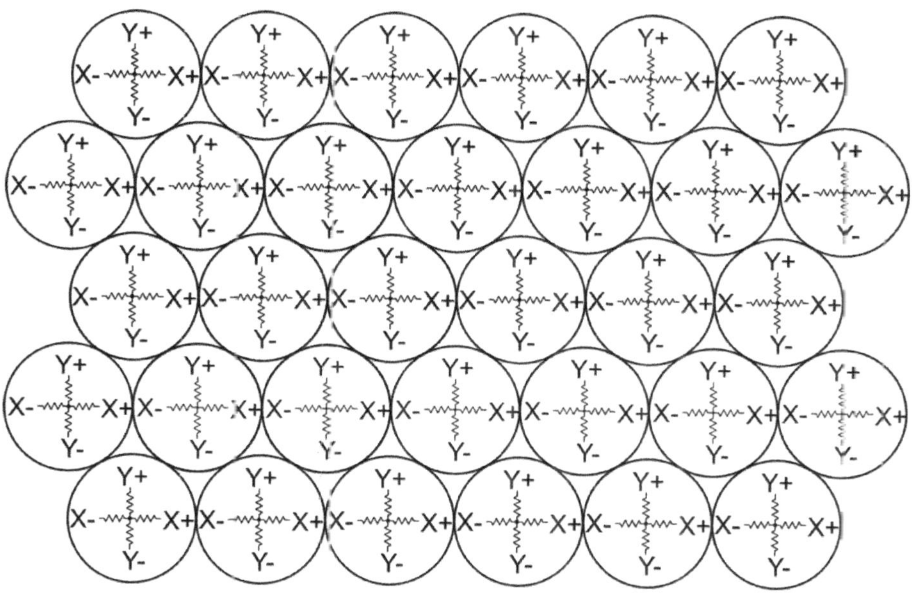

FIG. 11.16. Illustration of a resistive network used to implement position estimation. The output from each photomultiplier tube/preamplifier is divided by a resistive network with four outputs, X_+, X_-, Y_+ and Y_-. The corresponding signals from all of the photomultiplier tubes are connected to provide the summed signal.

As a result of the above difficulties, modern cameras use sophisticated correction and position estimation methods. The correction methods will be discussed in more detail below. Advanced position estimation methods involve digitizing the outputs of all of the PMTs and using them in the position estimation. In this case, position and energy estimation, and the various corrections are done by a digital signal processor or microprocessor, allowing a great deal more sophistication in the choice of algorithms. In some systems, these are done using statistical estimation techniques such as maximum-likelihood estimation.

As alluded to above, the values of the interaction position (x, y) and energy E are computed using equations similar to those shown above. The input variables X_+, X_-, Y_+ and Y_- are related to the charge signal from the PMT. These signals are proportional to the number of photoelectrons emitted from the photocathode. The emission of photoelectrons is the end result of a series of random processes that includes the number of scintillation photons produced, the number of these collected by the PMT, the number of photoelectrons produced for a given photon and the number of photoelectrons emitted which are focused onto the first dynode. The net result is that there is statistical variation in the values of the input variables for a given interaction position and energy. Thus, the values of position and energy are not exact, and are only estimates of the true quantities. Thus, there will be imprecision in the energy and position estimates resulting in finite intrinsic energy and spatial resolutions (see Chapter 8 for more details).

To a good approximation, both the energy and intrinsic spatial resolution can be characterized by a Gaussian function. The energy resolution results from variations in the total number of photoelectrons incident on the first dynode. The random variations can be approximated by a Poisson distribution and the variance in the energy resolution is, thus, approximately equal to the number of these photoelectrons. The approximate energy resolution of a gamma camera can, thus, be estimated as follows. A NaI(Tl) crystal produces, on average, 38 photons/keV. The quantum efficiency (fraction of incident scintillation photons that produce photoelectrons) of a PMT for the 415 nm emission of NaI(Tl) is approximately 12%. Thus, for a 140 keV photon, the number of photoelectrons collected is $140 \times 38 \times 0.12 = 638$ electrons. The FWHM is equal to approximately 2.35 times the standard deviation, so the fractional energy resolution is equal to $2.35 \times \sqrt{638}/638 = 9.3\%$. This is approximately equal to the energy resolution for a state of the art scintillation camera. It should also be noted from the above that the energy resolution is proportional to $E^{-0.5}$ and spatial variations in the collection efficiency will produce spatial variations in the energy resolution. Estimating the intrinsic spatial resolution is more difficult than estimating the energy resolution because of the more complicated estimation equation. However, typical intrinsic spatial resolutions are in the range of 3–5 mm, depending on the number of PMTs used and details of the estimation procedure.

11.2.2.6. Corrections

As mentioned above, the energy and position estimation are non-ideal, resulting in errors in energy and position estimates. These errors give rise to non-uniform sensitivity in the camera. Thus, to obtain clinically acceptable images, energy, spatial and uniformity corrections are needed. The need for these corrections is illustrated in Fig. 11.17. An image is shown resulting from a uniform distribution of γ rays on the camera with the collimator removed, often referred to as an intrinsic flood image. The substantial non-uniformity, the presence of edge packing artefacts near the edge of the FOV, and the visibility of the tube pattern should be noted.

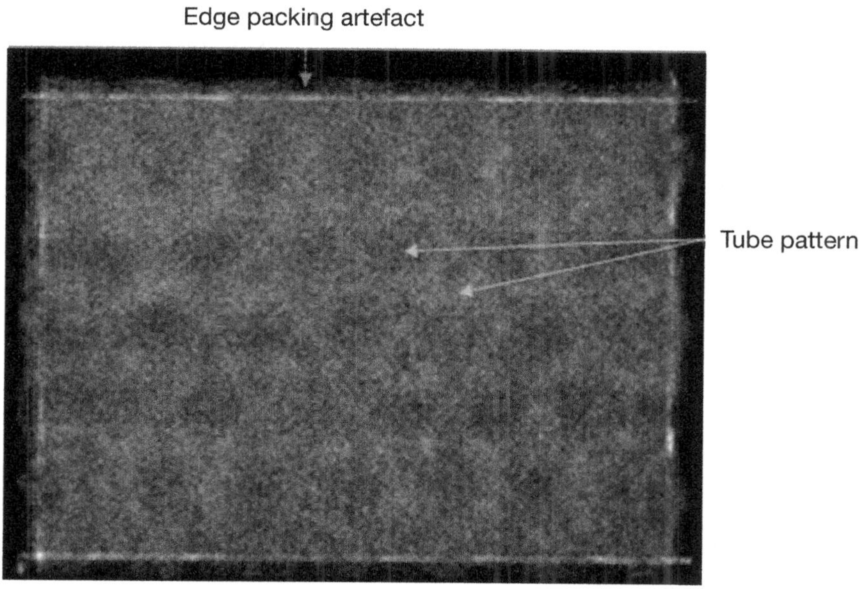

FIG. 11.17. Intrinsic flood image of gamma camera without energy, spatial or sensitivity corrections.

Energy corrections are needed because the estimated energy depends on spatial position. This behaviour can be understood in terms of variations of the fraction of the scintillation light collected as a function of interaction position. Since the energy is proportional to the light collected, differences in the fraction of light collected will result in a proportional change in the estimated energy. Figure 11.18 shows an example energy spectrum where the fraction of collected light was 2% lower or 2% higher than the average. This results in energy spectra shifted to lower or higher energy, respectively. Only photons falling within the acquisition energy window, in this case having a full width of 20% of the photopeak energy, centred at 140 keV, are counted. There are about 1.7% fewer photons counted in the two sample energy spectra than for the case of the average energy spectrum. Thus, the sensitivity is about 2% lower at these points. A typical energy correction algorithm measures the energy spectrum as a function of position in the image using a source or sources with known energies. A linear or higher order correction is then made to the estimated energy.

Spatial corrections are needed because of biases in estimated interaction positions. These biases result, as described above, from discrepancies in the formulas used to estimate the position and the actual behaviour. As mentioned, these usually result in lines being bent towards space between PMTs. Typically, separate corrections are made for the axial (y) and transaxial (x) directions. These corrections involve imaging a mask with a grid of holes or lines in combination with a flood source. This results in a series of bright spots or lines in the image that correspond to the known positions of the holes or lines in the mask. A function, typically a polynomial, is fit to the set of true points as a function of the set of measured points. This function can then be used to correct a measured position.

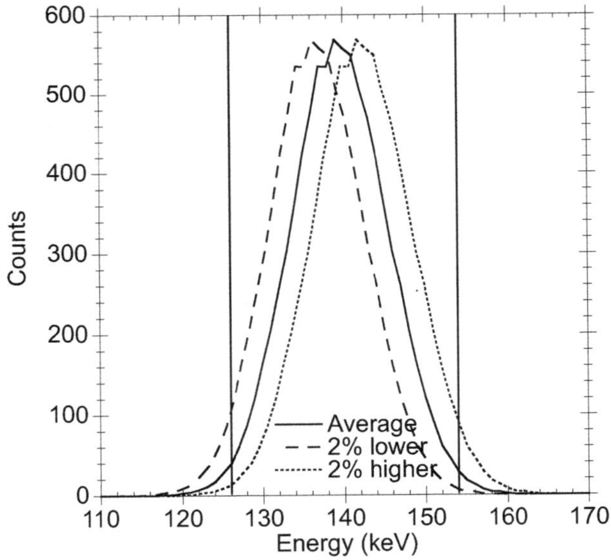

FIG. 11.18. Sample energy spectrum for 140 keV photons for the cases of average, 2% lower than average and 2% higher than average light collection efficiency. The variation in light collection efficiency results in a shift of the energy spectrum, which results in non-uniform sensitivity.

The final type of correction applied is a uniformity or sensitivity correction. The goal of this correction is to make images of a flood source as uniform as possible (see Fig. 11.19). There are two types of uniformity corrections: intrinsic, which corrects only for non-uniform sensitivity of the detector system (i.e. excluding the collimator), and extrinsic, which corrects for both detector and collimator non-uniformities. Uniformity corrections are made using a high-count flood image. Uniformity correction is implemented by, in essence, multiplying each pixel in acquired images by a factor equal to the average counts in the active portion of the flood image divided by the counts in the corresponding pixel in the flood image.

The number of counts in the flood image is critical in determining the ultimate uniformity of the image. This is especially important in SPECT where local non-uniformities can result in ring artefacts. To achieve this, the counts in the flood image should be such that the relative standard deviation (coefficient of variation) of the pixel counts resulting from Poisson counting statistics is less than the desired uniformity. For example, if uniformity correction to better than 1% is desired, the average number of counts per pixel in the uniformity flood should be greater than $1/0.01^2 = 10$ kcounts per pixel. The total number of counts in the flood, thus, depends on the number of active pixels in the image. Since non-uniformities are generally relatively low frequency, this restriction can be relaxed to some degree by the use of low pass filtering applied to the flood image.

Intrinsic flood images are usually acquired using a point (or syringe) source containing a small quantity of the isotope of interest. If the source is placed at a distance of more than five times the largest linear dimension of the camera

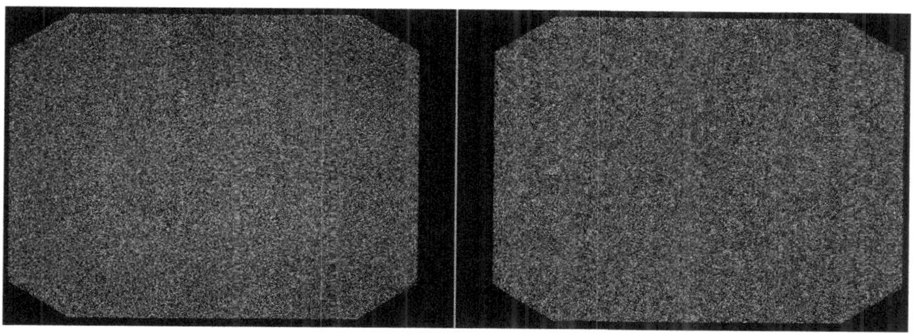

FIG. 11.19. Intrinsic flood images for a gamma camera having a poor (left) and good (right) set of corrections applied. It should be noted that the images are windowed so that the brightness represents a relatively small range of count values in order to amplify the differences. The peppery texture is due to quantum noise and is to be expected. The quantum noise is exaggerated because of the windowing used, and can be reduced by acquiring very high-count flood images.

FOV from the camera face, then the irradiation of the camera can be considered uniform. Since the uniformity of the camera will, in general, vary depending on the energy of the isotope and energy window used, this correction should ideally be made for each isotope and energy window used. The count rate for the acquisition should be within acceptable limits to avoid high count rate effects.

Extrinsic flood images are made using a flood or sheet source. Fillable flood sources have the advantage that they can be used for any isotope. However, great care must be made in filling the phantom to remove bubbles, mix the activity and maintain a constant source thickness. In addition, images must be obtained for each collimator used with a given isotope. As a result, ^{57}Co sheet sources are often used to obtain extrinsic flood images. These have the advantage of convenience but are, strictly speaking, valid for only a single isotope.

One way to take advantage of both approaches is to perform uniformity corrections using a combination of intrinsic flood images for each isotope used and an extrinsic flood image obtained for the collimator in question. This approach assumes that collimator uniformity is independent of energy and can, thus, be measured with, for example, a ^{57}Co sheet source. Not all equipment vendors support this approach. Some vendors assume that the energy and linearity corrections produce uniformity that is energy independent, and they, thus, recommend only the use of an extrinsic flood image for uniformity correction. Another approach is to first confirm the uniformity of all collimators via a sheet source flood image with intrinsic correction for ^{57}Co. Then, an extrinsic flood image for each isotope used is acquired and used in uniformity correction for that isotope, assuming that the collimator is sufficiently uniform. The best approach depends on the characteristics of the individual camera.

11.2.2.7. Image framing

The final step in generating gamma camera images is image framing. Image framing refers to building spatial histograms of the counts as a function of position and possibly other variables. This involves several steps, and is typically done either by microprocessors in the camera or in an acquisition computer. In this step, position is mapped to the elements in a 2-D matrix of pixels. The relationship between pixel index and physical position is linearly related to the ratio of the maximum dimension of the FOV of the camera to the number of pixels, the zoom factor and an image offset. The zoom factor allows enlarging the image so that an object of interest fills the image. This can be useful, for example, when imaging small objects. It results in a pixel size in the image that is a factor of 1/zoom factor as large as in the unzoomed (zoom factor equals one) image. It should be noted that even though the pixel size is decreased, the resolution of the image will not necessarily be improved as long as the original pixel size is smaller than the intrinsic resolution. For example, if the native pixel size is 3.2 and the intrinsic resolution is 4 mm, a zoom factor of two will result in a pixel size of 1.6 mm, but the intrinsic resolution will still be 4 mm. An image offset can be used to shift the image, so that an object of interest is in the centre of the acquired image.

In addition to adding counts to the appropriate pixel spatially, the framing algorithm performs a number of other important functions. The first is to reject photons that lie outside of the energy window of interest. This is done to reject scattered photons. Gamma cameras typically offer the ability to simultaneously frame images corresponding to more than one energy window. This can be useful for isotopes having multiple photopeaks, for acquiring data in energy windows used by scatter compensation algorithms or for acquiring simultaneous images of two or more radionuclides. Framing software typically enables the summation of photons from multiple, discontiguous energy windows into one image as well as simultaneously framing multiple images from different energy windows into different images. There is often a limited number of energy windows that can be framed into a single image, and a limit on the number of images that can be framed at one time. These limits may depend on the image size, especially if the framing is done by a microprocessor in the camera that has limited memory

A second important function provided by the framing system is the ability to obtain a sequence of dynamic images. This means that photons are recorded into a set of images depending on the time after the start of acquisition. For example, images could be acquired in a set of images with a frame duration of 10 s. Thus, for the first 10 s, photons are recorded into the first image; for the second 10 s, they are recorded into a second image; and so on. Thus, just as multiple images are obtained in the case of a multi-energy window acquisition, multiple images are obtained corresponding to a sequential set of time intervals. This is illustrated in Fig. 11.20, where seven dynamic frames are acquired for a time interval T. Dynamic framing is used for monitoring processes such as kidney function, gastric emptying, etc. The time frames are often not equal in duration as there may be more rapid uptake at early times and a later washout phase in which the change in activity with time is slower. The number or acquisition rate of dynamic frames is often limited due to constraints in framing memory and this limit can depend on the image size.

Gated acquisition is similar to dynamic acquisition in that photons are recorded into a set of images depending on the time they are detected. However, in gated acquisition, the time is relative to a physiological trigger, such as an electrocardiogram (ECG) signal that provides a signal at the beginning of each cardiac cycle. This is appropriate for processes that are periodic. The photons are counted into a set of frames, each of which corresponds to a subinterval of the time between the two triggers. For example, the bottom two illustrations in Fig. 11.20 should be considered. In both cases, the interval between gates is divided into four subintervals (in cardiac imaging, 8 or 16 frames are typically used, but 4 are illustrated in this example for simplicity). The photons arriving in each of the subintervals are counted in the corresponding frame. Thus, for cardiac imaging, the activity distribution during the first quarter of the cardiac cycle is imaged in frame 1, the second quarter in frame 2, etc. This is useful for assessing wall motion and thickening. Just as in dynamic and multiple window acquisitions, there may be limits on the size and number of frames due to limits in framing memory on the acquisition computer.

Since the gate is derived from a physiological signal, and physiological signals are not necessarily exactly periodic, the framing algorithm must handle the case of variable time intervals between gate signals. In cardiac imaging, this corresponds to variations in the heart rate. Figure 11.20 also illustrates two ways of dealing with this. In the first method, the frames correspond to fixed time intervals. What is typically done is to measure the average heart rate and to divide this by the number of frames per cycle. The photons arriving during each of these time intervals are then binned into the image corresponding to each time interval. Difficulty arises when there are variations in the length of the cardiac cycle. Figure 11.20 illustrates the case when there are beats that are 5% longer and 5% shorter than average. For fixed time interval framing, the actual interval for the fourth frame will be lengthened or shortened when the length of the cardiac cycle changes. This will result in motion blurring of the gated images. The alternative is to change the length of the subintervals for each beat based on the time between gates for that beat. This eliminates the problem with motion blurring described above. However, this also requires buffering of the events for the period between gates before they can be framed. One additional advantage of this approach is that bad beats (ones longer or shorter than the average beat by more than a certain fraction of the average beat length) can be discarded.

The final acquisition mode is list-mode acquisition. In list-mode acquisition, the energies and positions of incoming photons are simply saved to a file in the order in which they appear. Additional information is recorded in the form of events in the list-mode stream. These events include things such as physiological triggers, gantry or table motion, start and stop of acquisition, and timing marks which are injected at regular intervals. The advantage of list-mode data is that they contain all of the information obtained during the acquisition. As a result, it can be retrospectively reframed using different pixel sizes, energy windows, frame intervals, etc. However, the downside is that list-mode files are very large (typically eight or more bytes of data are stored for each photon). List-mode is often not made available for routine clinical use, but can be very useful for research use.

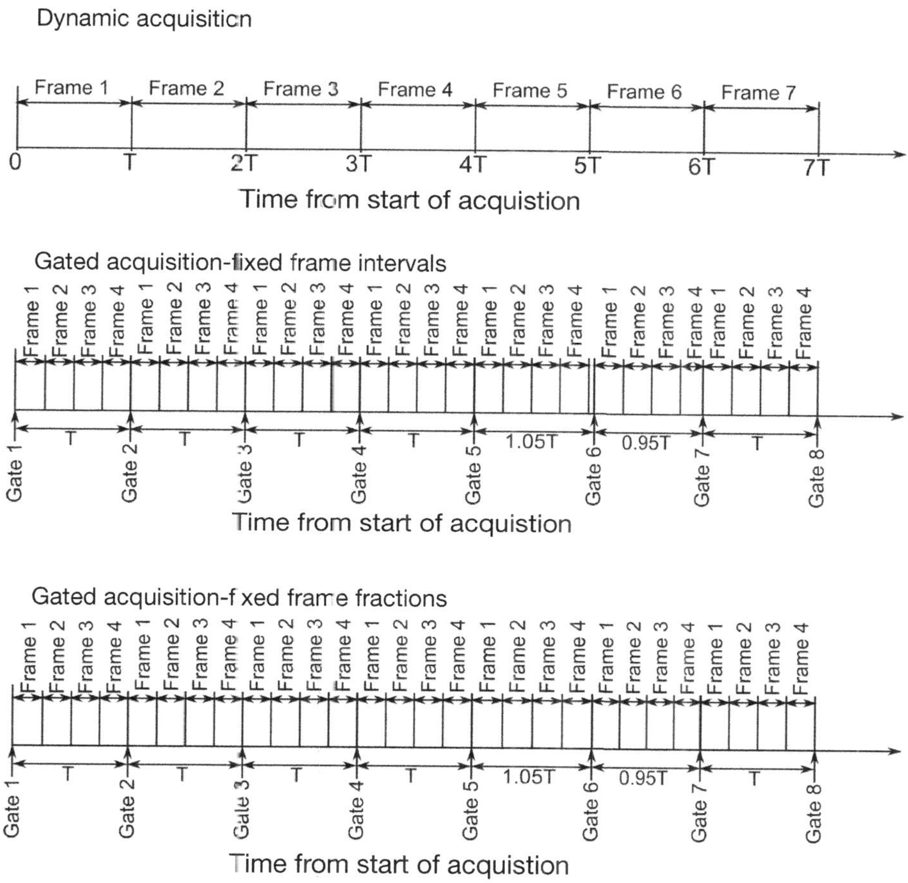

FIG. 11.20. Comparison of dynamic and gated acquisition modes. In all cases, time is along the horizontal axis. In dynamic acquisition, photons are framed into different images based on the time after acquisition. In this case, the interval between dynamic frames is T. For gated acquisition, photons are framed into a set of images based on the time in relation to the previous gate signal. The gate signal is derived from a physiological trigger such as the R wave in an electrocardiogram signal. For the two dynamic gating examples, the time interval between gate signals 1–5 and 7–8 are the same, but the interval between gates 5–6 and 6–7 are 5% longer and 5% shorter than the average interval. Two methods of dividing the time interval between gates, fixed intervals and fixed fractions are illustrated. In the fixed interval method, the photons are framed into frames with fixed widths. In this case, the extra time (5% of T) between gates 5–6 is not counted, and the interval into which photons are counted in frame 4 in the time interval between gates 6–7 is shortened by 20% (5% of T).

11.2.2.8. Camera housing

The camera housing contains the radiation detector and provides a mount for the collimators. The housing performs a number of important functions. First, it must provide radiation shielding so that photons can only enter the camera and be detected via the collimator. This shielding is made of lead and, since the crystal and PMT array is large, the housing must also be large and is rather heavy. In addition, the PMTs are very sensitive to magnetic fields. The housing, thus, includes mu-metal shields around the PMTs to screen magnetic fields. Without these, there can be variations in the sensitivity and uniformity due to changes in the camera position relative to ambient magnetic fields, including the magnetic field of the Earth. The PMTs and detection electronics are sensitive to variations in temperature and generate a non-negligible amount of heat. As a result, the housing must include a temperature control system, typically in the form of fans to circulate air and provide ventilation.

A final important function of the camera housing is to provide mounting for the collimators. High energy parallel and pinhole cameras can weigh more than 100 kg, so the mounting system must be sufficiently strong to securely support this much weight. The back face of the collimator (excluding pinhole collimators) must be in close proximity to the crystal in order to provide the highest possible resolution. Since collimators are often changed several times per day, the mounting system must provide for easy collimator removal and change. Most modern cameras include automatic or semi-automatic collimator exchange systems. Thus, the collimator retaining system often includes an automated locking system. Finally, modern cameras include touch sensors on the face of the collimators and often include proximity sensors. The touch sensors activate when the collimator face touches the patient or bed and disable camera motion. This is done for patient safety and in order to avoid injuring the patient or damaging the camera. A proximity sensor is sometimes also included. This typically consists of an array of light emitting diodes with an opposing array of photodiodes. They are mounted so that the light beam from the light emitting diodes is parallel to the face of the camera. Proximity to the patient can be detected when one of the light beams is interrupted by part of the patient. The proximity sensor can be used for automated sensing of the camera orbit that positions the cameras close to the patient at each camera position. Electrical connections between the housing and the collimator provide communication with the touch and proximity sensors as well as providing information about which collimator type is mounted.

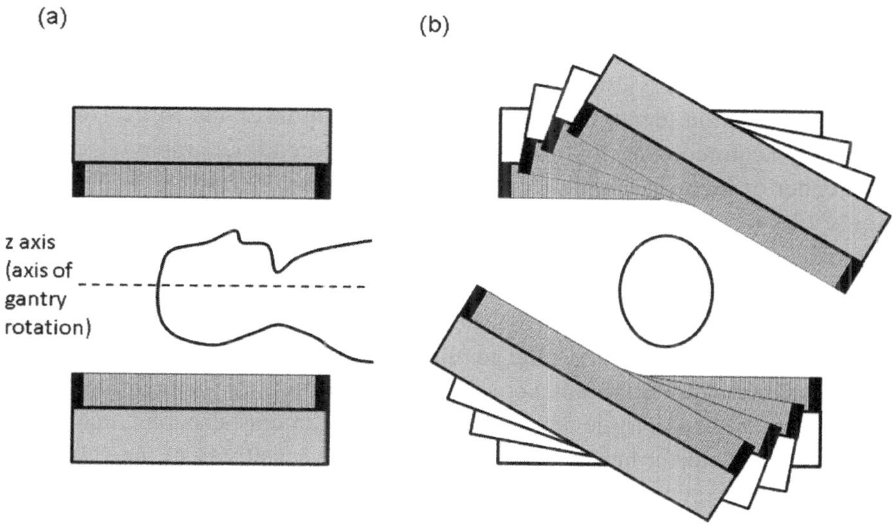

FIG. 11.21. (a) A cross-section of a dual head gamma camera capable of acquiring two views simultaneously. It should be noted in this example that the heads are oriented 180° apart, although a 90° degree configuration is also possible. SPECT data acquisition requires rotation of the gamma camera heads about the long axis of the patient as indicated. (b) A transverse slice with the position of four different camera orientations superimposed to illustrate the acquisition of multiple angular views.

11.2.3. SPECT

11.2.3.1. Gamma camera SPECT systems

In addition to the software requirements for image reconstruction, SPECT is associated with hardware requirements that are beyond those needed for planar imaging. Although various SPECT configurations have been developed, the most common implementation involves the use of a conventional gamma camera in conjunction with a gantry that allows rotation of the entire detector head about the patient [11.2]. The gantry rotation is about the long axis of the patient (see Fig. 11.21) and is typically performed in discrete steps (step and shoot), although continuous motion may also be supported. During rotation of the gantry, the patient bed typically does not move, so SPECT data acquisition is more similar to conventional CT than to spiral CT, in this respect. In this way, planar views of in vivo radioactivity distribution can be acquired at different angular orientations and these data can be used to form the projections that are required for image reconstruction by computed tomography.

In principle, rotation of the gamma camera about 180° allows for the acquisition of sufficient projections for tomographic reconstruction. However, in practice, opposing views acquired 180° apart differ due to various factors (photon attenuation, depth dependent collimator response) and SPECT data are commonly acquired over 360°. The theory of computed tomography determines the number of angular samples that are required, but for many SPECT studies, around 128 views may be acceptable. The time needed to acquire these multiple projections with adequate statistical quality is a practical problem for clinical SPECT where patient motion places a limit on the time available for data acquisition. In an effort to address this issue, a common approach in modern SPECT designs is to increase the number of detector heads, so that multiple views can be acquired simultaneously. Dual detector head systems currently predominate, although triple detector head gamma cameras also exist. Increasing the number of detector heads increases the effective sensitivity of the system for SPECT, at the expense of increasing cost. Dual head gamma cameras are often considered the preferred configuration for systems intended not just for SPECT but also for general purpose applications, including whole body studies where simultaneous acquisition of anterior and posterior planar images is required.

In addition to the rotational motion required for SPECT, flexibility is also required in the relative positioning of the detector heads. For general purpose SPECT with a dual head system, the two heads are typically oriented in an opposing fashion (sometimes referred to as H-mode) and 360° sampling is achieved by rotation of the gantry through 180°. In contrast, cardiac SPECT is often performed with the detectors oriented at 90° to each other (sometimes referred to as L-mode). In this mode, the gantry rotates through 90° and the two detectors acquire projections about 180° from the right anterior oblique position to the left posterior oblique position. Despite acquiring only 180° of data, this mode has advantages for cardiac applications as it can minimize the distance between the heart and the detectors, thus reducing attenuation and depth dependent losses in spatial resolution. Other approaches to minimizing the distance between the detectors and the patient during SPECT data acquisition involve further control of the rotational motion of the detector heads. For detectors rotating about a circular orbit, this involves adjusting the radius of rotation for individual studies, so as to minimize the source to collimator distance. Other options include detectors that rotate, not in a circular orbit, but in an elliptical orbit, or, alternatively, a variable rotational motion that contours to the outline of the body.

The flexibility of the motions that are available in modern SPECT systems makes it particularly important to ensure that the detectors are correctly aligned. This means that the specified angle of rotation is accurately achieved at all angles. The detector heads also need to be perfectly oriented parallel to the z axis of the system, such that each angular view is imaging the same volume. Furthermore, it is important that the centre of each angular projection is consistent with the centre of mechanical rotation. Errors due to these factors can potentially lead to a loss of spatial resolution and the introduction of image distortion or ring artefacts. In order to identify and correct these issues, an experimental centre of rotation procedure is employed. A small point source is placed in the FOV at an off-centre location. SPECT data acquisition is performed and deviations from the expected sinusoidal pattern are measured in the resulting sinograms.

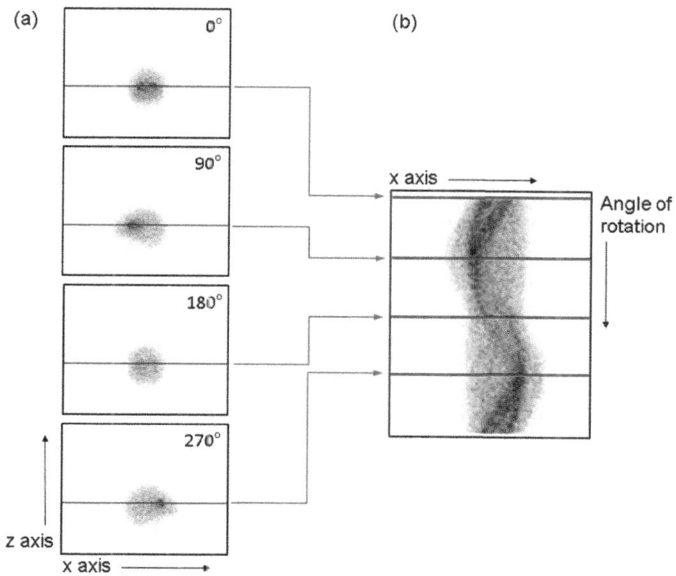

FIG. 11.22. (a) A series of planar views acquired at different angular orientations. A sample of four views has been extracted from a total of 128 views acquired about 360°. It should be noted that the z axis represents the axial position and is the axis of gantry rotation. (b) A sinogram corresponding to a particular axial location. The red lines in (b) indicate 1-D projections that have been extracted from the corresponding planar views shown in (a).

Although SPECT can be performed with a variety of collimator geometries, such as cone-beam or pinhole, much of the present discussion has assumed parallel-hole collimation. In this case, each planar view consists of multiple 1-D projections, each measured at different axial positions (Fig. 11.22). Each projection is defined by the holes of the collimator and approximates a series of parallel line integrals of the activity distribution in the FOV. In practice, these line integrals are substantially corrupted by the effects of photon attenuation, scatter and depth dependent collimator response. Each of these factors requires software correction and these corrections are described in the following paragraphs.

11.2.3.2. Attenuation correction

Standard tomographic reconstruction algorithms, such as those based on filtered back projection, assume that measured projections are line integrals through the object. However, in SPECT, the interaction of photons via photoelectric absorption and Compton scatter within the patient results in attenuated projections. The attenuated projections $P_\theta(t)$ can be described for the 2-D case by the equation:

$$P_\theta(t) = \int_0^\infty a(l\boldsymbol{n}_\theta + t\boldsymbol{m}_\theta) \mathrm{e}^{-\int_0^l \mu(l'\boldsymbol{n}_\theta + t\boldsymbol{m}_\theta)\,\mathrm{d}l'} \mathrm{d}l \tag{11.8}$$

where the geometry is illustrated in Fig 11.23. In this equation, the unit vectors $\boldsymbol{n}_\theta$ and $\boldsymbol{m}_\theta$ are as described in the legend of Fig. 11.23, t is the transaxial distance in the projection from the projected position of the origin and l is the distance along the projection line from the face of the detector. $a(\boldsymbol{x})$ is the activity distribution and gives the activity at point $\boldsymbol{x}$. It should be noted that the integral in the exponent represents the integral through the attenuation distribution $\mu(\boldsymbol{x})$ from the point $\boldsymbol{x} = l\boldsymbol{n}_\theta + t\boldsymbol{m}_\theta$. Thus, the exponential represents the attenuation of photons emitted at $\boldsymbol{x}$ as they travel back towards the detector.

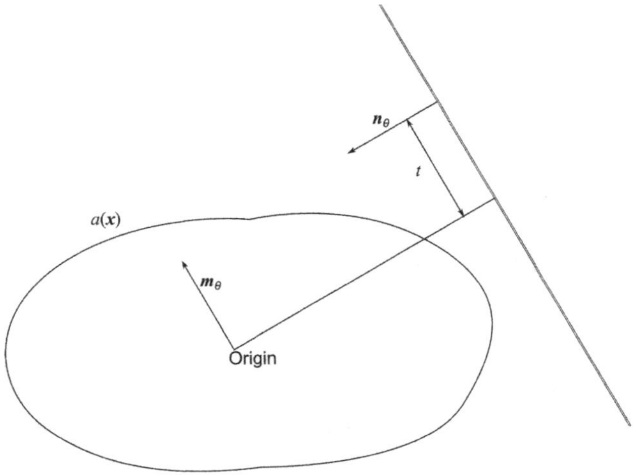

FIG. 11.23. Projection geometry used to describe the attenuated projection in Eq. (11.8). In this figure, the projection is at an angle θ. A parallel-hole collimator is assumed, and the unit vector $\boldsymbol{n}_\theta$ is perpendicular to the collimator and parallel to the projection rays. The unit vector $\boldsymbol{m}_\theta$ is parallel to the collimator face and perpendicular to $\boldsymbol{n}_\theta$. The variable t is the distance along the detector from the projected position of the origin.

As can be seen from the above equation, unlike PET, the attenuation is not constant for a projection ray, but instead varies along the ray. Using reconstruction methods that do not model this effect produces both artefacts and a loss of quantitative accuracy in the resulting images. The artefacts can include streak artefacts, resulting from highly attenuating objects such as bones, catheters or medical devices; shadows, due to higher attenuation between an object in some views than in others (e.g. breast or diaphragm artefacts in cardiac SPECT); and a generally reduced image intensity in the centre of the image.

The first requirement to compensate for attenuation is knowledge of the attenuation distribution in the patient. This is done by either assuming uniform attenuation inside the object and extracting information about the body outline from the emission data or using a direct transmission measurement. Assuming a uniform attenuation distribution in the patient is only valid in regions such as the head. Even in the head, bone and regions containing air, such as the sinuses, result in imperfect estimates of the attenuation distribution and lead to imperfect attenuation compensation. Myocardial perfusion imaging is an important application for SPECT and, since attenuation can produce artefacts that obscure actual perfusion defects, a number of commercial devices have been developed to allow measurement of the attenuation distribution in the body. All of these devices use transmission CT techniques to reconstruct the attenuation distribution inside the body. The devices that have been developed can be divided into two general classes: devices using radionuclide sources and devices based on X ray tube sources. In both cases, a source of X rays or γ radiation is aimed at the body and a detector on the opposite side of the body measures the transmitted intensity. The intensity $I_\theta(t)$ passing through the body for a source with incident intensity I_0, projection position t and projection view θ is given by:

$$I_\theta(t) = I_0(t)\mathrm{e}^{-\int_0^\infty \mu(l\mathbf{n}_\theta + t\mathbf{m}_\theta)\,\mathrm{d}l} \tag{11.9}$$

where the symbols and geometry are as in Fig. 11.23.

Acquiring sets of these transmission data for various angles allows reconstruction of the attenuation distribution. Tomographic reconstruction methods can be applied directly by noting that the negative of the log of the fraction of transmitted photons is a line integral through the attenuation distribution.

A number of transmission devices based on radionuclide sources have been developed and marketed (Fig. 11.24). All of these devices use the gamma camera to detect the transmission photons. The simplest of these designs is a sheet source of radioactivity. To avoid contaminating the projection data, either the transmission scan is acquired separately from the emission data or simultaneously using a radionuclide with a lower photopeak energy. Typically, ^{153}Gd is used as it has an energy lower than that of ^{99m}Tc and the transmission photons, thus, do not interfere with collection of emission data. To reduce patient dose and scatter in the transmission data, the source should be collimated. The disadvantages of sheet source designs are that they are expensive and high activities make them dangerous to handle. Using lower activity sources results in contamination of the transmission data.

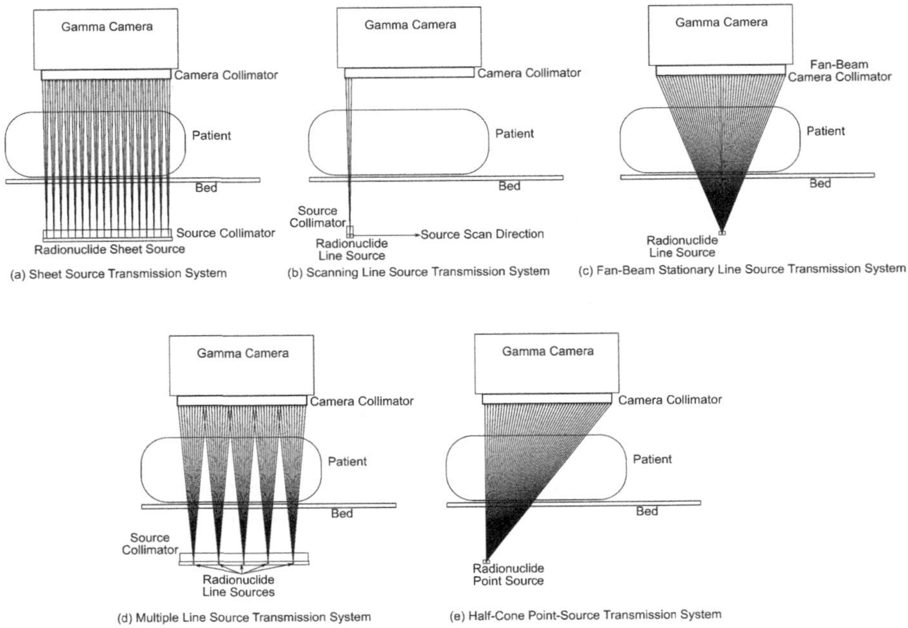

FIG. 11.24. Illustration of a number of proposed transmission scanning devices. It should be noted that in all cases the drawing represents a sagittal view of the system. However, for (d), the multiple line source system, the line sources are normally parallel to the long axis of the patient and, thus, perpendicular to the direction shown. Similarly, for (c), the fan-beam system, the fan beam is in the transaxial direction, the opposite of that pictured above. Furthermore, for the sheet source system, the source is continuous and the fan of rays is only shown to illustrate the limited range of directions of transmission source photons that pass through the source collimator.

As a result of these difficulties, line source transmission systems have been developed. The line sources are either scanned opposite the patient and parallel to the collimator or positioned at the focal spot of a fan-beam collimator. In the latter case, the projection geometry is different from the parallel-beam geometry described above, but the same basic principles apply. The fan-beam geometry has the advantage of having much higher sensitivity and, thus, of requiring lower activity sources. However, the fan-beam geometry results in magnification and, thus, the FOV is smaller than the size of the detector. It should be noted that in the fan-beam geometry the fan lies in the transaxial plane and, thus, the truncation is in the transaxial direction. For the scanning line source systems, an electronic window is used so that transmission data are acquired only in the region directly under the line source. To overcome the low sensitivity of parallel-beam geometry, high activities (1.85–18.5 GBq) are used. This reduces the contamination of the transmission data by emission activity. The scanning line source adds additional mechanical complexity to the system and, as a result, several other designs have been proposed.

In multiple line source systems, a small number of line sources are used. The line sources are placed close enough together so that the object is covered due to the finite acceptance angle of the camera collimator, in effect acting like a non-uniform sheet source. One advantage of this geometry is that weaker sources can be used for the sources near the edge of the camera, since these correspond to thinner regions of the body. This can help reduce the effects of high count rates. The final geometry proposed is the half-cone geometry shown above. In this design, a high energy ^{133}Ba point source is used. Many of the high energy photons penetrate through the collimator, creating a half-cone-beam geometry. This allows a parallel-hole collimator to be used both for transmission and emission imaging. The use of a point source simplifies shielding of the source.

In general, radionuclide transmission sources have a number of disadvantages. These include the fact that the source decays and must be replaced. There are also limits on transmission count rates imposed by the gamma camera, resulting in relatively noisy transmission images. In addition, if the count rate due to the emission activity within the patient is high, the transmission images can be degraded, resulting in inaccurate attenuation maps. Finally, with the sheet source, scanning line source and multiple line source geometries, the resolution of the transmission scan is limited by the combination of source and camera collimators. In general, these provide lower resolution transmission scans. However, one advantage of radionuclide based transmission systems is the potential to perform simultaneous imaging, thus eliminating the need for an additional transmission scan. Another advantage, especially when acquired simultaneously, is that registration of the emission and transmission images is guaranteed. Finally, the use of radionuclide sources with a small number of high energy photopeaks makes converting the transmission images into an attenuation map at the energy of the emission source easier than for X ray CT based systems, which use X ray tubes having continuous X ray energy spectra.

The second major kind of transmission scanning apparatus uses X ray sources. There are two major kinds: slow rotation and hybrid SPECT/CT systems. In the slow rotation devices, a low power X ray tube is attached to the gantry opposite a conventional X ray detector. The gantry is rotated around the patient, resulting in acquisition of X ray transmission data. In one commercially available slow-rotation device, the X ray detector has 1–4 rows of detectors with an effective axial length of 0.5–1.0 cm at the centre of rotation and can, thus, acquire transmission data for 1–4 slices per revolution. The patient is then translated on the bed to acquire data from additional slices. The resolution of these scanners can be as good as 1 mm transaxially but is 0.5–1 cm axially. Typical revolution times for this type of scanner are 20–30 s/rotation. Another commercial implementation uses a flat panel detector and cone-beam geometry. This provides very high resolution, though there is the potential for cone-beam artefacts in the reconstructed images. In hybrid SPECT/CT systems, a SPECT gantry and an independent diagnostic CT system are integrated. The CT images are acquired very rapidly using conventional CT technology either just before or just after acquisition of the SPECT data. The CT images acquired using these systems, as well as the acquisition protocols and options, are similar to those on diagnostic CT scanners.

X ray tube based methods for obtaining transmission images have a number of advantages and disadvantages. The major advantages are acquisition speed, the quality of the attenuation maps (high resolution and low noise) and the convenience of not needing to replace radionuclide sources. Among the disadvantages are that the image is not acquired simultaneously and is often acquired with the bed in a different position than used for the SPECT scan. There is, thus, the potential for mis-registration of the SPECT images and attenuation maps, resulting in degraded attenuation compensation. A second disadvantage is that the effects of motion, especially respiratory motion, during the attenuation scan are different to those during the emission scan. For example, in slow-rotation devices, streak artefacts in the diaphragm region caused by respiratory motion during transmission data acquisition are common. For hybrid devices, the CT scans are acquired very rapidly, so may freeze the patient at one particular phase of the respiratory cycle. Despite these potential disadvantages, the advantages of X ray tube based transmission scanning (including hybrid SPECT/CT systems) have meant that they have largely replaced devices based on radionuclide sources.

Using the above systems for attenuation compensation requires transforming the transmission CT image into an emission attenuation map. For radionuclide sources, this basically means translating the attenuation coefficients at the energy of the transmission source to that of the emission source. For emission and transmission source energies above 100 keV, this can be done by scaling the transmission CT image by the ratio of the attenuation coefficient in water at the emission energy divided by that at the transmission source energy. However, more accurate results can be obtained using the bilinear scaling method described below for use with X ray CT images.

For hybrid SPECT/CT systems where a conventional X ray CT image is transformed into an attenuation map, there are a number of additional considerations. First, transmission data are obtained using a polychromatic source. There can, thus, be substantial beam hardening. Fortunately, beam hardening and other corrections routinely applied in X ray CT scanners to produce images in Hounsfield units (HU) eliminate many of these concerns. In this case, the CT image in Hounsfield units can be transformed to the attenuation map via piecewise linear scaling, where, in effect, pixels with values less than 0 HU are treated as water with densities ranging from 0 to 1, pixel values between 0 and 1000 HU are treated as a mixture of bone and water, and pixel values greater than 1000 HU are treated as dense bone. Thus, for a pixel having a value h in Hounsfield units, the attenuation map value μ is given by:

$$\mu(h)=\begin{cases}\dfrac{1000+h}{1000}\mu_{\text{water}} & \text{for } h\leq 0\\ \mu_{\text{water}}+\dfrac{h}{h_{\text{bone}}}(\mu_{\text{bone}}-\mu_{\text{water}}) & \text{for } 0<h<h_{\text{bone}}\\ \dfrac{h}{h_{\text{bone}}}\mu_{\text{bone}} & \text{for } h>h_{\text{bone}}\end{cases} \tag{11.10}$$

where μ_{water} and μ_{bone} are the attenuation coefficients of water and bone, respectively, for the energy of the photopeak of the imaging radionuclide.

Once the attenuation map is obtained, attenuation correction can be implemented using analytical, approximate or statistical image reconstruction algorithms. Generally, analytical methods are not used due to their poor noise properties. Approximate methods include the Chang algorithm. This method is often used in regions of the body where the attenuation coefficient is assumed to be uniform and the actual attenuation distribution is not measured, but instead is approximated from the boundary of the object, which is usually assumed to be an ellipse. In the Chang method, an image reconstructed using filtered back projection (i.e. without attenuation compensation) is approximately compensated for attenuation. This approximate compensation is obtained for each voxel by dividing the uncorrected image signal by the average of the attenuation factors that correspond to each projection view. For a uniform attenuator with an assumed elliptical boundary, these attenuation factors can be calculated analytically. However, the Chang method is approximate and has poor noise properties.

Thus, for the best attenuation compensation, statistical iterative reconstruction methods should be used. Statistical iterative reconstruction methods are discussed in more detail in Chapter 13. These methods can be used to compensate for attenuation by incorporating a model of the attenuation process in the imaging matrix. Accurate attenuation compensation can be obtained with fast statistical iterative reconstruction methods, such as the ordered-subsets expectation-maximization (OSEM) algorithm, if an accurate attenuation map is available.

11.2.3.3. Scatter correction

In gamma camera imaging, a significant fraction of the detected photons are scattered in the body. This is due to the finite energy resolution of the gamma camera, which results in imperfect energy based scatter rejection. The scatter to primary ratio (SPR) depends on the radionuclide, energy window, energy resolution, source depth and the size of the object. For example, for ^{99m}Tc, the SPR is in the range of 0.2 for brain imaging and 0.6 for cardiac imaging. On the other hand, for ^{201}Tl cardiac imaging, the SPR can be greater than 1. Scatter results in loss of contrast, especially for cold objects in a warm background, and loss of quantitative accuracy.

Scatter correction requires estimating the scatter component of the projection data combined with a compensation method. Most frequently, the scatter component is estimated using data acquired in auxiliary energy windows. Perhaps the most common and flexible such method is the triple energy window (TEW) method. This method uses two scatter energy windows, one above and one below the photopeak window, as illustrated in Fig. 11.25. The scatter is estimated from the counts in the scatter windows using a trapezoidal approximation where counts in the scatter windows divided by their window widths are treated as the sides, and the scatter in the photopeak window is the area of the trapezoid. The estimated scatter counts in the photopeak window estimated using TEW s_{TEW} are given by:

$$s_{\mathrm{TEW}} = \left[\frac{c_{\mathrm{lower}}}{w_{\mathrm{lower}}} + \frac{c_{\mathrm{upper}}}{w_{\mathrm{upper}}}\right]\frac{w_{\mathrm{peak}}}{2} \tag{11.11}$$

where

c_{lower} and c_{upper} are the counts in the lower and upper scatter windows, respectively;

and w_{peak}, w_{lower} and w_{upper} are the widths of the photopeak, lower scatter and upper scatter windows, respectively.

It should be noted that in Fig. 11.25 the scatter windows are adjacent to the photopeak energy window, but this is not necessary, nor, in all cases, desirable. In fact, it is desirable to position the windows as close as possible to the photopeak window, while having only a small fraction of the photopeak photons detected in the scatter windows. The width of the scatter windows is a compromise between obtaining as accurate a scatter estimate as possible, which favours a narrow energy window, but having an estimate that is low in noise, which favours a wide energy window.

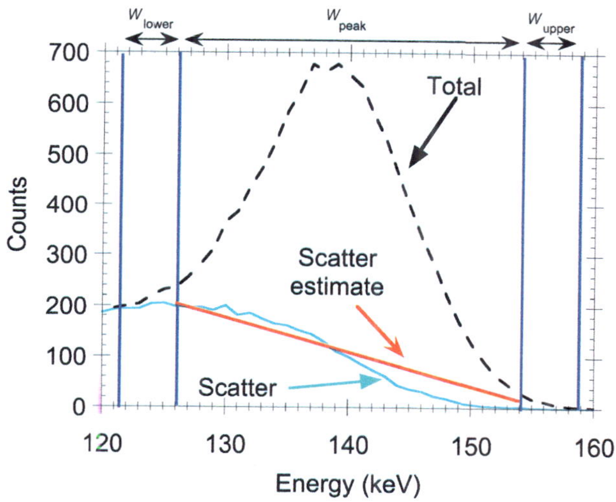

FIG. 11.25. Illustration of the use of a trapezoidal approximation to estimate the scatter in the photopeak energy window in the triple energy window method scatter compensation for ^{99m}Tc. It should be noted that in this example the windows are not necessarily optimally placed. In particular, the scatter windows are positioned such that there is a non-zero contribution from unscattered photons in the scatter energy windows. This is especially evident for the upper energy window. For the case of ^{99m}Tc, the counts in the upper window are often assumed to be zero.

Another method to estimate the scatter component in the projection data is via the use of scatter modelling techniques. These techniques use an estimate of the activity distribution and a mathematical algorithm to compute the scatter that would have been detected from the activity distribution. The mathematical techniques used range from accurate approximations to full Monte Carlo simulations.

As mentioned above, scatter correction is accomplished by combining scatter estimation and compensation methods. Methods of compensating for scatter include subtracting the scatter estimate from the projection data and, for

SPECT, including the projection data in the iterative reconstruction process. Owing to noise and errors in the scatter estimate, subtraction of a scatter estimate from measured projection data can lead to some negative pixel values. This can be a problem when the data are subsequently used with statistical iterative reconstruction algorithms. For scatter estimates obtained from energy windows, this effect can be reduced by low pass filtering the scatter estimate. Nevertheless, truncation of negatives is often used, though this can increase bias in the scatter compensation.

For SPECT, a better way to accomplish scatter compensation is to add the scatter estimate to the computed projection during the iterative reconstruction process. For example, algorithms such as OSEM and maximum-likelihood expectation-maximization (MLEM) involve back projecting the ratio of the measured and computed projections. In this case, the scatter estimate can simply be added to the computed projections. Another approach is to include scatter modelling in the projection matrix. Either of these approaches is superior to pre-subtraction of the scatter estimate.

11.2.3.4. Collimator response compensation

Images obtained with a gamma camera are degraded by the spatially varying collimator–detector response (CDR). For parallel-hole collimators, the CDR depends approximately linearly on the distance from the collimator face. The CDR has geometric, septal penetration and septal scatter components. These correspond, respectively, to photons passing through the collimator holes, photons passing through the septa without interacting, and photons scattering in the septa and resulting in a detected photon. The latter two effects tend to reduce image contrast, can produce star artefacts and introduce distance dependent sensitivity variations. Septal penetration and scatter can be reduced with the use of a properly designed collimator. However, for isotopes emitting photons with energies greater than ~300 keV, including isotopes having low abundance high energy photons not used for imaging, some level of these effects is almost unavoidable.

Since SPECT images contain information about the distance from the source to the collimator, it is possible to provide improved compensation for the CDR as compared to planar imaging. This can be accomplished using both analytical and iterative methods. However, analytical methods involve approximations and usually have suboptimal noise properties and are, generally, not commercially available. As a result, methods based on statistical iterative reconstruction have been developed and are commercially available. These methods model the CDR in the projection and back projection process. In SPECT, the matrix elements are not explicitly calculated and stored, but are implicitly calculated during the reconstruction using a projector and back projector algorithm. Thus, CDR compensation is accomplished by modelling the CDR in the projector and back projector. This is often implemented by rotating the image estimate, so that it is parallel to the collimator face at each projection view. In this orientation, the CDR is constant in planes parallel to the collimator face and can, thus, be modelled by convolution of the CDR for the corresponding distance. In order to do this, the distance from the plane to the face of the detector is needed. This is somewhat complicated by the use of non-circular orbits and, in this case, manufacturers need to store the orbit information (distance from the collimator face to the centre of rotation for each projection view) with the projection image. In addition, a way of estimating the CDR is needed. Analytical formulas exist for calculating the geometric component of the CDR. Alternatively, a Gaussian function fit to a set of measured point response functions measured in air can be used. Compensation for the full CDR, including septal penetration and scatter, requires a CDR that includes these effects. Analytical formulas do not exist, so either numerically calculated (e.g. using Monte Carlo simulations of the collimator–detector system) or measured CDRs are used. Various optimization and speed-up techniques have been implemented to reduce the time required for CDR modelling.

It should be noted that CDR compensation does not fully recover the loss of resolution of the collimator: the resolution remains limited and spatially varying and partial volume effects are still significant for small objects. In addition, CDR compensation results in correlated noise that can give a 'blobby' texture to the images (though the images do indeed seem qualitatively 'less noisy'), and results in ringing artefacts at sharp edges. Despite these limitations, CDR compensation has generally been shown to improve image quality for both detection and quantitative tasks and has been used as a way to allow reduced acquisition time.

11.3. PET SYSTEMS

11.3.1. Principle of annihilation coincidence detection

Radioactive decay via positron emission is at the heart of the PET image formation process [11.3]. Positrons are emitted from the nucleus during the radioactive decay of certain unstable, proton-rich isotopes. These isotopes achieve stability by a decay process that converts a proton to a neutron and is associated with the creation of a positron. A positron is the antimatter conjugate of an electron and has the same mass as an electron but positive charge. As with β decay, positrons are emitted from the nucleus with different energies. These energies have a continuous spectrum and a specific maximum value that is characteristic of the parent isotope. Once emitted from the nucleus, the positron propagates through the surrounding material and undergoes scattering interactions, changing its direction and losing kinetic energy (Fig. 11.26). Within a short distance, the positron comes to rest and combines with an electron from the surrounding matter. This distance is dependent on the energy of the positron, which is itself a function of the parent isotope and is typically on the order of a millimetre. The combination of a positron and an electron results in the annihilation of both particles and the creation of two photons, each with an energy of 511 keV, equivalent to the rest masses of the two original particles. Conservation of momentum, which is close to zero immediately before annihilation, ensures both photons are emitted almost exactly 180° apart. These characteristic photon emissions (known as annihilation radiation) — always 511 keV, always emitted simultaneously and almost exactly 180° apart — form the basis of PET and result in distinct advantages over single photon imaging in terms of defining the LOR.

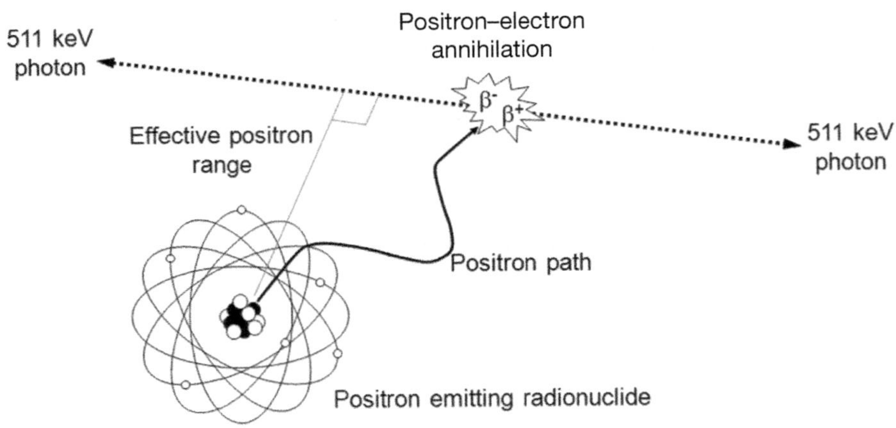

FIG. 11.26. Positrons emitted from a radioactive nucleus propagate through the surrounding material before eventually coming to rest a short distance from their site of emission. At this point, the positron annihilates with an electron, creating two 511 keV photons that are emitted approximately 180° apart. The perpendicular distance from the line defined by the two photons to the site of positron emission places a limit on the spatial resolution that can be achieved with PET systems.

The advantages of PET over SPECT, in terms of improved spatial resolution, statistical quality and quantitative accuracy, can be attributed to the fact that PET does not require a collimator and, therefore, eliminates the weakest link in the SPECT image formation process. Instead of physical collimation,

PET systems employ a form of detection that can be thought of as electronic collimation. If a positron source is surrounded by suitable detectors, both back to back photons from an individual positron decay can potentially be detected (Fig. 11.27). As both photons are emitted simultaneously, they will be detected at approximately the same time, allowing temporal acceptance criteria to be used to associate pairs of corresponding detection events. This mode of detection is referred to as coincidence detection and allows corresponding photon pairs to be distinguished from other unrelated, potentially numerous, photon detection events. As both photons that arise from positron decay are emitted almost exactly 180° apart, coincidence detection can be used to help localize the source of the photon emissions. In general, a line drawn between corresponding detectors can be assumed to intersect the point of photon emission, although information is usually not available about exactly where along that line the emission occurred. However, if a system of detectors is arranged at different positions around the source, multiple coincidence events can be recorded at different angular orientations. Over the course of an extended scanning period, a large number of coincidence events will be recorded and angular projections of the activity distribution can be estimated. These projections may then be used to reconstruct 3-D images using the methods of computed tomography.

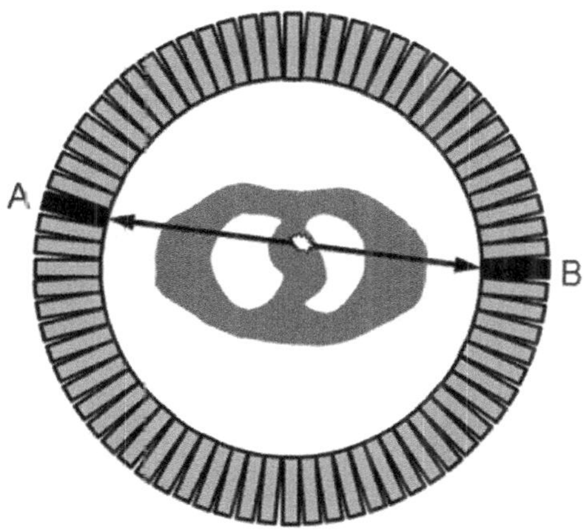

FIG. 11.27. The back to back photons that result from positron–electron annihilation can potentially be measured by detectors placed around the source. Coincidence detection involves the association of detection events occurring at two opposing detectors (A and B) based on the arrival times of the two photons. A line of response joining the two detectors is assumed to intersect the unknown location of the annihilation event. Coincidence detection obviates the need for a collimator and is sometimes referred to as electronic collimation.

11.3.2. Design considerations for PET systems

11.3.2.1. Spatial resolution

High spatial resolution is clearly an important design objective for PET imaging systems. As such, the trend in modern scanner systems has been to decrease the width of individual detectors and to increase the total number of detector elements surrounding the patient. The increased concentration of detector elements decreases the sampling interval and generally improves spatial resolution. Although modern designs involve detectors that are only a few millimetres wide, the need for high sensitivity means that they are often a few centimetres long. Problems can occur when photons are incident on one detector but penetrate through to an adjacent detector (Fig. 11.28(a)). The location within the detector (depth of interaction) is typically not measured and the detection event is assigned to a location at the face of the detector. This problem frequently occurs when detectors are arranged in a ring (or similar) configuration and gives rise to a loss of resolution at more peripheral locations. This resolution loss generally occurs in the radial direction as opposed to the tangential direction due to the angle of incidence of the photons on the detectors.

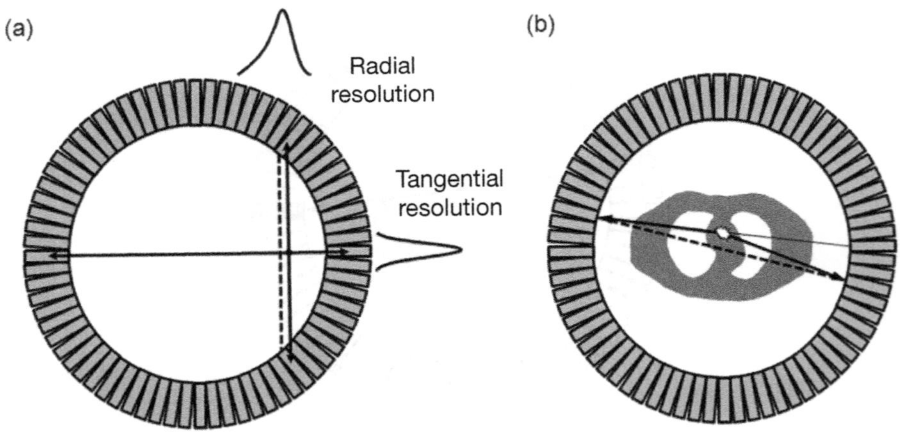

FIG. 11.28. (a) Photon penetration between adjacent detectors in a ring based system leads to mis-positioning of events. This primarily affects the radial component of spatial resolution which degrades with distance from the centre of the field of view. (b) Residual momentum of the positron and electron immediately before annihilation causes the two 511 keV photons to deviate slightly from the expected 180° angle. As a result, a line joining detection events does not intersect the exact point of annihilation. The extent of this non-collinearity is greatly exaggerated in the figure, but it does contribute to a loss of spatial resolution, especially for large diameter PET systems.

Another factor that influences spatial resolution is the distance between opposing detectors. This distance is relevant because of a small uncertainty in the relative angle of the 511 keV annihilation radiation (Fig. 11.28(b)). Although the basic assumption of coincidence detection is that annihilation radiation is emitted 180° apart, this is not strictly true. Positrons frequently annihilate before they have lost all momentum, and this residual momentum translates to a small deviation of about ±0.25° from the expected back to back emissions. This effect is referred to as non-collinearity and tends to degrade spatial resolution as detector separation increases. For PET systems with opposing detectors separated by only a few centimetres, such as those optimized for specific organs such as the brain or breast, this is not a major issue. However, for whole body systems, in which opposing detectors are typically separated by about 80 cm, the effect of non-collinear photons contributes a blurring with an FWHM of approximately 2 mm.

The distance travelled by a positron between its point of emission and annihilation is an additional factor that degrades the spatial resolution that can be achieved by PET systems. As previously discussed, this distance, or positron range, is dependent on the energy of the positron and also the type of material through which the positron is passing; a greater range is expected in tissue such as lung compared to soft tissue. It should be noted that software corrections have been implemented that model the effect of positron range and can potentially reduce the loss of resolution in reconstructed images.

11.3.2.2. Sensitivity

The best possible spatial resolution that can be obtained by a PET system is not always achieved in clinical practice due to statistical noise in the measured data (Fig. 11.29). In order to suppress this noise, clinical protocols generally employ low-pass filters or other similar image reconstruction methods, but the consequence is invariably a loss of spatial resolution. Improving the statistical

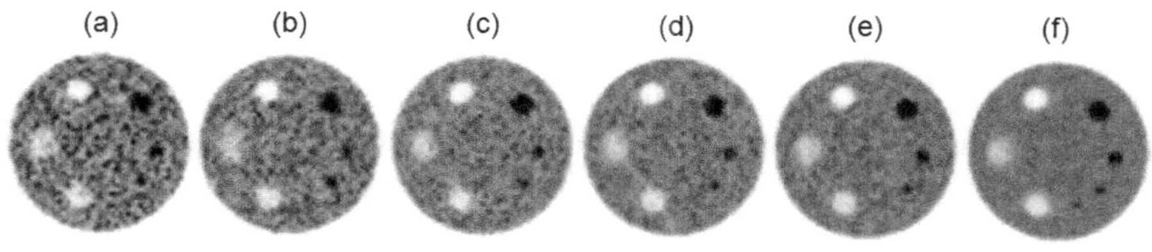

FIG. 11.29. Images of the same phantom, each showing different statistical quality. The images shown in (a), (b), (c), (d), (e) and (f) were acquired for 1, 2, 3, 4, 5 and 20 min, respectively. Increasing the acquisition time increases the total number of true coincidence events and reduces statistical variability in the image.

quality of the measured coincidence data not only reduces image noise but also allows the opportunity to reduce image smoothing and improve spatial resolution. The need to optimize this trade-off between statistical noise and spatial resolution influences both image reconstruction development and scanner design.

Noise in PET images is influenced by a number of factors, including the sensitivity of the detector system, the amount of radioactive tracer administered to the patient and the amount of time the patient can remain motionless for an imaging procedure. Limitations on the latter two factors mean that high sensitivity is an important objective for scanner design. Sensitivity is determined by the geometry of the detector arrangement and the absorption efficiency of the detectors themselves. Reducing the distance between opposing detectors increases the solid angle of acceptance and increases sensitivity. However, the requirement to accommodate all regions of the body imposes a minimum ring diameter for whole body systems. For such systems, extending the axial field of view (AFOV) of the detector system provides a mechanism for increasing sensitivity. Cost constraints have prevented the construction of PET systems that cover the entire length of the body. However, extended axial coverage can be achieved using systems with much smaller AFOVs by scanning sections of the body in a sequential fashion. For such systems, extending the AFOV of the scanner not only increases sensitivity but also reduces the number of bed translations required for whole body coverage. In addition to the geometry of the detector system, sensitivity is also determined by the absorption efficiency of the detectors. A high absorption efficiency for 511 keV photons is desirable in order to make best use of those photons that are incident upon the detectors. Absorption efficiency or stopping power of the detector material is, therefore, an important consideration for PET system design.

11.3.2.3. Quantitative accuracy

One of the strengths of PET is its capability to quantify physiological processes in vivo. A prerequisite for this kind of quantitative analysis is that the images accurately reflect the local activity concentration in the body. In order to ensure this kind of quantitative accuracy, it is important to minimize effects that corrupt the data and to correct residual corruption as necessary. Quantitative error can arise from many sources but is primarily due to random coincidence events, photon scatter within the body, photon attenuation within the body and detector dead time. Figure 11.30 illustrates some of these situations.

Figure 11.30(b) illustrates a type of unwanted coincidence event that can occur when photons from unrelated annihilation events are detected at approximately the same time. Two photons detected within a short time interval (coincidence timing window) will be associated with each other under the assumption that they originated from the same positron–electron annihilation. However, when there is a large amount of radioactivity within the FOV, it is very possible that photons from unrelated annihilation events will be detected within this time interval, leading to a spurious random coincidence. Figure 11.30(c) illustrates the case where a scattered photon is deflected from its original direction but still reaches the detectors. In this case, a coincidence event can potentially be recorded between the scattered and unscattered photons, leading to an inaccurate coincidence event whose LOR does not pass through the true location of the original annihilation. If not adequately corrected, such scattered coincidences contribute a spurious background to the image that is dependent upon the size and composition of the patient's body and is non-uniform across the FOV. In Fig. 11.30(d), one of the two annihilation photons has been attenuated within the body and only one photon is detected. No coincidence event is possible and the

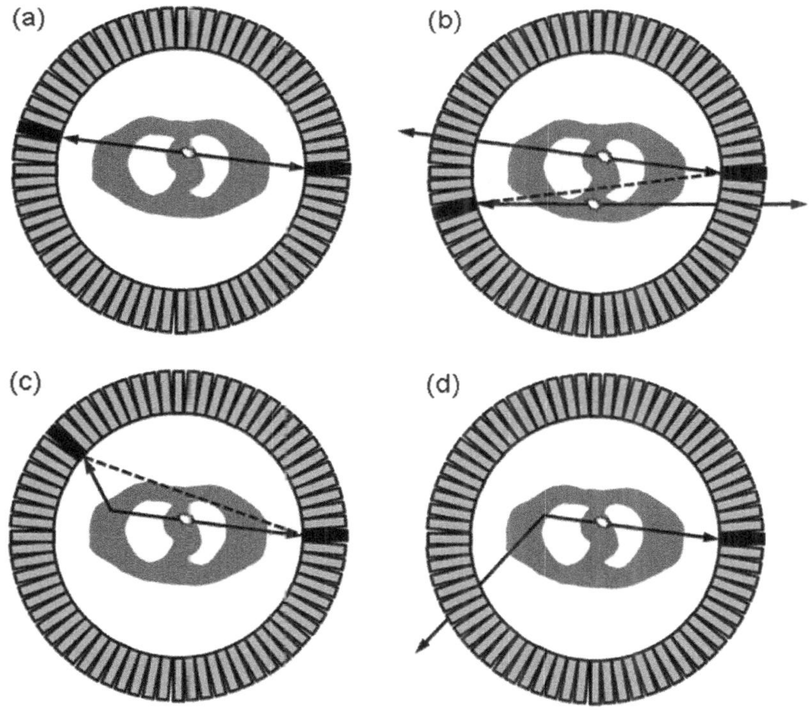

FIG. 11.30. (a) A true coincidence event can occur when both photons escape the body without interacting. (b) A random coincidence event occurs when two photons from unrelated annihilation events are detected at approximately the same time. (c) A scattered coincidence event can occur when either photon is scattered within the body but is still detected. (d) No coincidence event is recorded when one or both photons are attenuated, typically due to scatter out of the field.

scanner underestimates the signal that would be measured in the absence of attenuation. The degree of underestimation depends on the size and distribution of the patient's body and results in quantitative error and gross image artefacts that are non-uniform across the FOV. Attenuation can occur when one of the two photons undergoes photoelectric absorption within the body. However, a more likely occurrence is that one of the two photons is scattered within the body and is deflected out of the FOV, leaving only one photon to be detected. Another problem that can occur when large amounts of radioactivity are present is that true coincidence events can potentially be lost due to the limited count rate capability of the detection system. Each component of the system requires a finite amount of time to process each event and, if another photon is detected during this time, the second event will be lost. This dead time effect becomes significant at high count rates and leads to an underestimation of the local activity concentration in the images.

Software processing prior to (or during) image reconstruction can mitigate the above effects, but the accuracy of these corrections may not be reliable if the contamination overwhelms the signal from true coincidence events. PET systems are, therefore, designed to minimize the contribution of the various degrading factors described above. In terms of scanner design, very little can be done to reduce attenuation as photons that are absorbed within the body do not reach the detectors. However, scattered photons can potentially be rejected by the detection system if their energy falls outside a predetermined acceptance range. Annihilation radiation that is scattered within the body will emerge with energies less than 511 keV. The exact energy of the scattered photon will depend on the number of Compton scattering interactions that have occurred and the angle through which the photon was scattered. Energy discrimination can, therefore, be used to reject photons that have energies less than 511 keV and are assumed to have been scattered. This approach relies on the detection system having high energy resolution and, in practice, energy discrimination reduces but does not eliminate scattered coincidence events. The limited energy resolution of current PET systems means that, in order to avoid rejecting too many true (unscattered) coincidence events, the energy acceptance window is usually set to accept quite a broad range of energies around 511 keV. High energy resolution is, nevertheless, an important design objective, particularly for those scanner systems that detect a large proportion of scattered photons.

Decreasing the coincidence timing window decreases the number of random events as the shorter time interval reduces the likelihood of a coincidence occurring by chance between two unrelated photons. There is, however, only limited scope for reducing the duration of the coincidence time window as it is restricted by the timing resolution of the detector system and the fact that off-centre annihilations result in real time differences between detection events.

Optimization of the coincidence timing window for a particular scanner represents a compromise between wanting to reduce the number of random coincidence events without significantly reducing the number of true coincidences. Detector systems that are able to measure photon detection times with low variability (high timing resolution) are, therefore, desirable from the perspective of randoms reduction. Timing resolution also contributes to detector dead time as a shorter coincidence timing window reduces the likelihood of more than two photon detection events occurring. Other contributions to dead time include the time required by the detector to measure an individual photon event and the time spent processing coincidence events.

11.3.2.4. Other considerations

Spatial resolution, sensitivity and quantitative accuracy are the main physics issues influencing PET system design, but there are various other factors that need to be considered. One obvious issue is the overall cost of the system. This significantly influences design decisions as the detectors represent a significant fraction of the overall production cost. The choice of detector material, the thickness of the detectors, the diameter of the detector ring and the axial extent of the detectors all contribute to the total cost of the system. Another important design issue that affects the overall cost is integration of the PET system with a second modality in a combined scanner. In most cases, this means integration of the PET subsystem with a CT subsystem, although combined PET/magnetic resonance (MR) scanners exist and present more significant technical challenges. Combined PET/CT scanners have effectively replaced stand-alone PET for clinical applications and the optimal CT configuration included in a combined system is limited mainly by cost concerns. In practice, this means the number of slices to include in the multi-detector CT system. High performance multi-detector CT is important for systems intended for cardiac PET/CT applications, whereas lower slice capability may be adequate for oncological PET/CT. Other relevant design issues are related to the computer workstation used for acquiring and processing data. These include fast image reconstruction times, convenient integrated control of both PET and CT subsystems, and seamless integration with other institutional information technology systems such as clinical information systems and image archive systems.

11.3.3. Detector systems

11.3.3.1. Radiation detectors

Although different radiation detector designs have been used in PET, almost all current systems adopt an approach based on scintillation detectors. Scintillation detectors are inorganic crystals that emit scintillation light in the visible range when high energy photons are incident upon them (Fig. 11.31). This light is converted to an electrical signal by means of a photodetector, usually a PMT, coupled to the crystal material. Various scintillator materials have been used in PET, including thallium doped sodium iodide (NaI(Tl)), bismuth germanate (BGO) and cerium doped lutetium oxyorthosilicate (LSO). Table 11.1 shows some of the properties of the crystal materials that are relevant for PET applications. The properties of an ideal crystal for PET would include a high stopping power for 511 keV photons (high linear attenuation coefficient); short scintillation light decay time to reduce dead time and allow short coincidence time windows to reduce random coincidences; and high light output. High light output enables good energy resolution, which gives rise to improved scatter

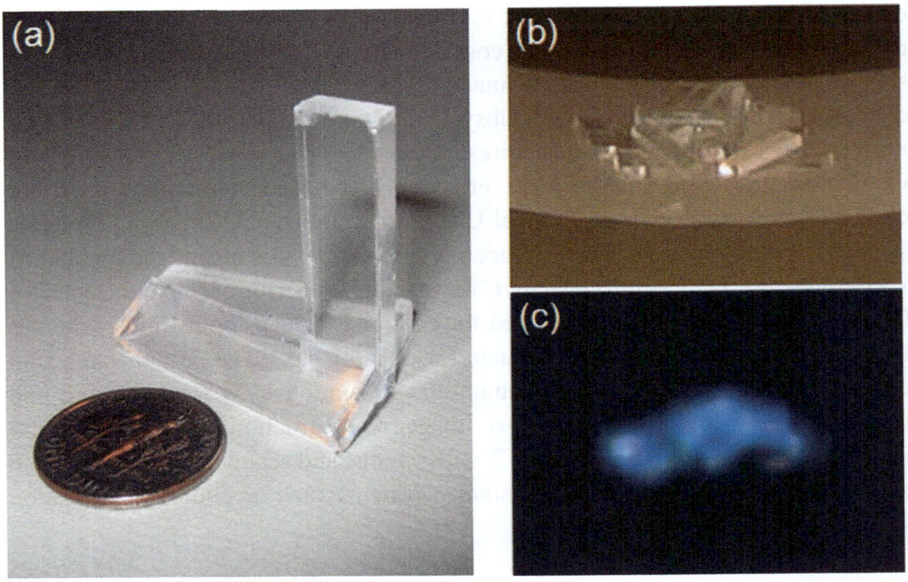

FIG. 11.31. Example of bismuth germanate crystals used for PET (a). Bismuth germanate samples photographed under room lighting (b) and in the presence of X ray irradiation and dimmed room lighting (c). The scintillation light seen in (c) is due to the interaction of radiation with the crystals, which causes electrons to become excited. When they return to their ground state, energy is emitted, partly in the form of visible light.

rejection. It also affords cost savings in the construction of a complete scanner system as the number of photodetectors required to resolve a given number of crystal elements can potentially be reduced.

TABLE 11.1. PROPERTIES OF SOME OF THE SCINTILLATORS USED IN PET. LINEAR ATTENUATION COEFFICIENTS AND ENERGY RESOLUTION ARE QUOTED FOR 511 keV

Property	NaI	BGO	LSO
Linear attenuation coefficient (cm^{-1})	0.34	0.95	0.87
Scintillation decay constant (ns)	230	300	40
Relative light output	100%	15%	75%
Energy resolution (%)	6.6	10.2	10.0

Note: BGO: bismuth germanate; LSO: lutetium oxyorthosilicate.

Although NaI(Tl) is ideal for lower energy single photon imaging, its relatively low linear attenuation coefficient for 511 keV photons makes it less attractive for PET applications. Sensitivity could potentially be increased by increasing the thickness of the crystals, which are typically 1–3 cm thick. However, the scope for substantially increasing crystal thickness is limited as it results in a loss of spatial resolution. This is because thicker crystals are prone to more significant depth of interaction problems as the apparent width of the detector increases for sources located off-centre. Thin crystals composed of a material with a high stopping power for 511 keV photons are, thus, desirable to ensure best possible sensitivity while maintaining spatial resolution. For this reason, BGO and, more recently, LSO have replaced NaI(Tl) as the scintillator of choice for PET.

BGO has the advantage of a high stopping power for 511 keV photons and has become an important scintillator for PET applications. However, it is not ideal in many respects as it has relatively poor energy resolution and a long crystal decay time. The poor energy resolution translates into a limited ability to reject scatter via energy discrimination. In addition, the long decay time translates into a greater dead time and increased number of random coincidences at high count rates. As such, BGO is well suited for scanner designs that minimize scatter and count rate via physical collimation, such as those with interplane septa (see Section 11.3.4.2). Attempts to increase sensitivity by removing the interplane septa typically result in a high scatter, high count rate environment for which BGO is not ideal.

Although LSO has a lower linear attenuation coefficient than BGO, its shorter crystal decay time and slightly improved energy resolution convey significant advantages as a PET scintillator. LSO has become the scintillator of choice for scanner designs that operate without interplane septa because its short decay time makes it well suited for high count rate applications. The fast decay time of LSO also enables a time of flight (TOF) data acquisition mode that will be discussed in Section 11.3.4.4. LSO has proved to be a successful crystal for PET detector applications despite the fact that the material contains around 2.6% ^{176}Lu, which is itself radioactive. A component of the emissions from ^{176}Lu is detected within the energy acceptance window and the dominant effect is to contribute random coincidences. In practice, the increased randoms rate is not a major problem for clinical studies, and the naturally occurring radiation has even been used for quality assurance. Lutetium-176 has a half-life of 3.8×10^{10} a and, thus, provides a long lived source of radiation that can be used to check consistency of detector response without the need for external sources. It should be noted that, for commercial reasons, some PET systems employ cerium doped lutetium yttrium oxyorthosilicate (LYSO(Ce)) which has substantially similar properties to LSO.

11.3.3.2. Detector arrangements

As discussed above, the interaction of 511 keV annihilation radiation with the scintillation crystals gives rise to optical light that can be detected by a suitable photodetector. A photodetector is a device that produces an electrical signal when stimulated by light of the sort emitted by a scintillation detector. For most PET applications, PMTs have been the preferred photodetector because their high gain results in an electrical output with a good signal to noise ratio. In addition, PMT output is proportional to the intensity of the incident light and, thus, proportional to the energy deposited in the crystal. This provides a mechanism for selective acceptance of only those detection events with energies within a specific range and can be used to reject scattered photons. In addition, PMTs provide high amplification with little degradation in the timing information that is essential for electronic collimation. Although PMTs are by far the most widely used photodetector for PET applications, they are somewhat bulky and highly sensitive to magnetic fields. For these reasons, they are often not used in combined PET/MR systems where space is limited and operation in high magnetic fields is a requirement. In these and some other applications, semiconductor based photodiodes are an alternative to PMTs. It should be noted that in these applications, the semiconductor device is used in conjunction with a scintillation detector and is used to detect scintillation light, not the annihilation photons. Avalanche photodiodes have much lower gains than PMTs but can be very small and have been shown to be effective in high magnetic field environments. Their low gain requires very low noise electronics and they are also sensitive to small temperature variations.

Space and cost constraints mean that individual scintillation crystals are not usually coupled directly to individual photodetectors in a one to one fashion. Instead, the most common arrangement is a block detector in which a group of crystal elements share a smaller number of PMTs (Fig. 11.32). The design of each block varies between manufacturers and scanner models but usually involves a matrix of crystal elements, a light guide and four PMTs. An example configuration might be an 8 × 8 array of closely packed 4.4 mm × 4.0 mm × 30 mm crystal elements, where the longest dimension is in the radial direction to maximize detection efficiency. The light guide allows light to be shared between four circular PMTs and the relative light distribution depends on the location of the crystal in which the photon interacted. The (x, y) position of the detection event is calculated from the outputs of the four PMTs using a weighted centroid algorithm, similar to the Anger logic of a gamma camera. Although individual crystals can be identified in this way, the response is not linear throughout the block due to differences in the locations of the different crystal elements relative to the PMTs. Experimentally determined look-up tables are used to relate the measured (x, y) position to a corresponding detector element, effectively performing a form of linearity correction. In this way, only four PMTs are needed to localize signals from a much greater number of crystal elements. The number of crystal elements divided by the number of PMTs in a PET system has been referred to as the encoding ratio. A high encoding ratio implies lower production costs and is, therefore, desirable.

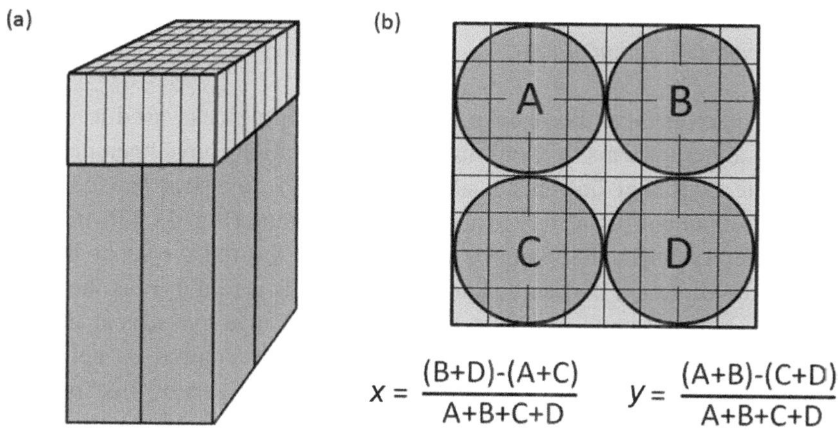

FIG. 11.32. (a) A PET detector block consisting of scintillator material coupled to an array of photomultiplier tubes. The scintillator is cut into an array of individual crystal elements. Four photomultiplier tubes are used to read out the signal from the 8 × 8 array of crystal elements. (b) The x and y position of each photon is determined from the signal measured by each of the four photomultiplier tubes labelled A–D, using the equations shown.

One of the advantages of the design described above is that each block operates independently of its surrounding blocks. This leads to good count rate performance as light is not transferred between blocks and the PMTs of one block are unaffected by detection events in an adjacent block. An alternative arrangement, referred to as quadrant sharing, increases the encoding ratio by locating the PMTs at the corners of adjacent blocks. This arrangement differs from the conventional block design in that each PMT can now be exposed to light from up to four different blocks. This can result in better spatial resolution and a higher encoding ratio but is also susceptible to greater dead time problems at high count rates.

Another alternative to the block design adopts an approach similar to that used in conventional gamma cameras. These Anger-logic designs involve detector modules that have a much larger surface area compared to conventional block detectors, e.g. 92 mm × 176 mm. Each module is comprised of many small crystal elements which are coupled, via a light guide, to an array of multiple PMTs. Light is spread over a larger area than in the block design and positional information is obtained using Anger-logic in the same way as a gamma camera. The PMTs used in this design are typically larger than those used in block detectors, increasing the encoding ratio. The larger area detector modules encourage more uniform light collection compared to block designs, which leads to more uniform energy resolution. However, a disadvantage of this design is that the broad light spread among many PMTs can lead to dead time problems at high count rates.

11.3.3.3. Scanner configurations

The detectors described above form the building blocks used to construct complete scanner systems. Various scanner configurations have been developed, although the dominant design consists of a ring of detectors that completely surrounds the patient (or research subject) in one plane (Fig. 11.33(a)). As with other scanner systems, this plane is referred to as the transverse or transaxial plane and the direction perpendicular to this plane is referred to as the axial or z direction. Several rings of detectors are arranged in a cylindrical geometry, allowing multiple transverse slices to be simultaneously acquired. As coincidence detection requires two opposing detectors, a full ring system of this sort allows coincidence data to be acquired at all angles around 180°. Although complete angular coverage is achieved in the transverse plane, there is much more limited coverage in the axial direction. Cost constraints and, to some extent, limited patient tolerance of extended tunnels mean that the detector rings usually extend for only a few centimetres in the axial direction. Human whole body systems typically have an AFOV of around 15–20 cm, although the trend in scanner design has been to increase the AFOV, thus increasing both sensitivity and the number of transverse slices that can be simultaneously acquired.

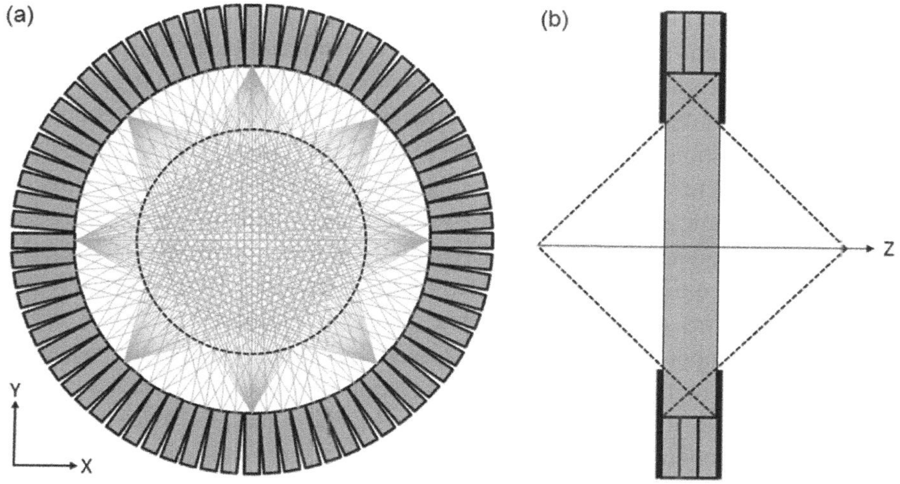

FIG. 11.33. (a) Full ring PET system shown in the transverse plane, indicating how each detector can form coincidence events with a specific number of detectors on the opposite side of the ring. For clarity, this fan-like arrangement of lines of response is shown for only eight detectors. The dashed line indicates how the imaging field of view is necessarily smaller than the detector ring diameter. (b) PET system shown in side elevation, indicating the limited detector coverage in the z direction. The shaded area indicates the coincidence field of view. The dashed lines indicate the singles field of view. End shields reduce, but do not eliminate, detection of single photons from outside of the coincidence field of view when operating in 3-D mode.

The diameter of the detector ring varies considerably between designs and this dimension reflects the intended research or clinical application. Small animal systems may have ring diameters of around 15 cm, brain oriented human systems around 47 cm and whole body human systems around 90 cm. Systems with ring diameters that can accommodate the whole body are clearly more flexible in terms of the range of studies that can be performed. However, smaller ring diameters have advantages in terms of increased sensitivity, owing to a greater solid angle of acceptance, and potentially better spatial resolution, owing to reduced photon non-collinearity effects. It should be noted that the spatial resolution advantage is complicated by greater depth of interaction problems as the detector ring diameter decreases and shorter crystals or depth of interaction measurement capability may be required. Furthermore, the effective imaging FOV is always smaller than the detector ring diameter because the acquisition of coincidence events between all possible detector pairs (such as those between nearby detectors in the ring) is not supported. In addition, PET systems have annular shields at the two ends of the detector ring that reduce the size of the patient port. These end shields are intended to decrease the contribution of single photons from outside the coincidence FOV (Fig. 11.33(b)). The coincidence FOV refers to the volume that the detector system surrounds, within which coincidence detection is possible. Single photons originating from outside the coincidence FOV cannot give rise to true coincidence events but may be recorded as randoms and can also contribute to detector dead time. Reducing the size of these end shields allows the patient port size to be increased but also leads to greater single photon contamination.

Unlike rotating camera SPECT systems, where different projections are acquired in a sequential fashion, full ring PET systems simultaneously acquire all projections required for tomographic image formation. This has an obvious advantage in terms of sensitivity, and it also enables short acquisition times, which can be important for dynamic studies. Full ring systems are, however, associated with high production costs and, for this reason, some early PET designs employed a partial ring approach. In these designs, two large area detectors were mounted on opposite sides of the patient and complete angular sampling was achieved by rotating the detectors around the z axis. Gaps in the detector ring led to reduced sensitivity and the partial ring design is now usually reserved for prototype systems. Another related approach to PET system design was to use dual head gamma cameras modified to operate in coincidence mode. The use of modified gamma cameras allowed for lower cost systems capable of both PET and SPECT. However, the poor performance of NaI based PET means that this approach has now been discontinued.

Current clinical systems have an AFOV that is adequate to cover most individual organs but in order to achieve coverage of the whole body, patient translation is required. Given the clinical importance of whole body oncology studies, the mechanism for translating the patient through the scanner has become an important component of modern PET systems. The patient bed or patient handling system has to be made of a low attenuation material but must still be able to support potentially very heavy patients. It must be capable of a long travel range, so as to allow whole body studies in a single pass without the need for patient repositioning. Precise motion control is also critical, particularly for PET/CT systems where accurate alignment of the two separately acquired modalities is essential. Advanced patient handling systems have been specifically developed for PET/CT to ensure that any deflection of the bed is identical for both the CT and PET acquisitions, thus ensuring accurate alignment irrespective of patient weight. Although most patient beds have a curved shape for improved patient comfort and better mechanical support, many manufacturers can also provide a flat pallet that is more compatible with radiation treatment positioning.

11.3.4. Data acquisition

11.3.4.1. Coincidence processing

The basis of coincidence detection is that pairs of related 511 keV annihilation photons can be associated together by the detector system based upon their times of measurement. Two photons detected within a short time interval are assumed to have arisen from the same positron–electron annihilation and a coincidence event is recorded. The time interval determining when events are considered to be coincident is denoted 2τ and is a system parameter that is not usually adjustable by the user. In order to minimize random coincidences, this interval should be kept as short as possible and for typical BGO based systems, 2τ may be around 12 ns. Shorter time windows are made possible by detector materials such as LSO that have faster scintillation decay times and, thus, better time resolution. Further reductions in the coincidence window are limited by differences in the arrival times of the two photons. For an electron–positron annihilation taking place at the edge of the transverse FOV, one photon would have to travel only a short distance, whereas the other photon might have to travel almost the diameter of the detector ring. Assuming a 90 cm ring diameter and a speed of light of 3×10^8 m/s, it can be seen that a maximum time difference of around 3 ns can be expected.

When a photon is incident upon a PET detector, an electrical pulse is generated (Fig. 11.34). A constant fraction discriminator then produces a digital logic pulse when the detector signal reaches a fixed fraction of the peak pulse height. This digital logic pulse is defined to have a duration τ and is fed into a coincidence module that determines whether it forms a coincidence event with any of the signals from other detectors in the ring. A coincidence event is indicated if there is an overlap in time between separate logic pulses from two different detectors. In other words, a coincidence would be identified if two detection events were recorded within a time interval no greater than τ. A photon detected at time t can, thus, form a coincidence with another photon detected within the interval $t + \tau$. It can alternatively form a coincidence with an earlier photon detected within $t - \tau$. For this reason, 2τ is often referred to as the coincidence time window. Consistent timing of signals from every detector in the system is clearly essential to ensure that true coincidences are effectively captured within the coincidence time window. Differences in performance of the various detector components and cable lengths can introduce variable time offsets. For this reason, time alignment corrections are performed to characterize and compensate for timing differences between different detectors in the ring.

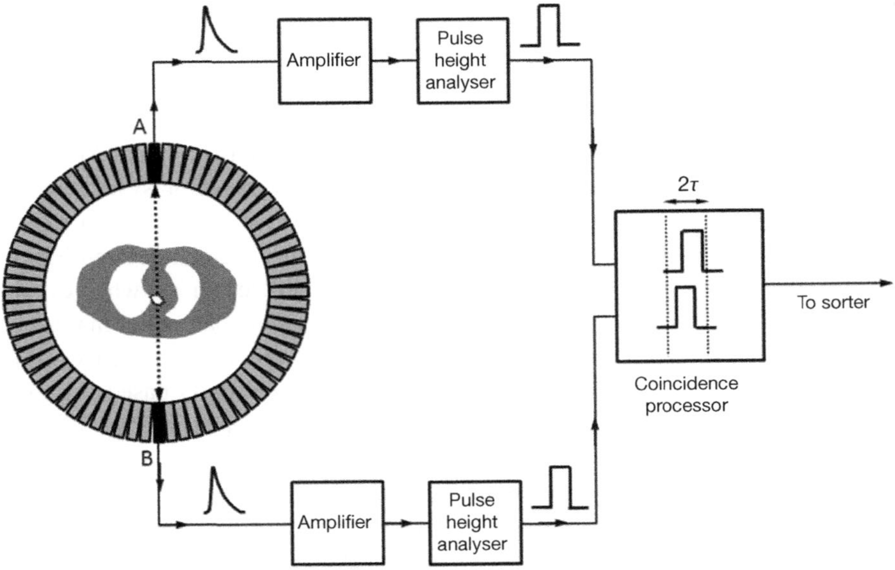

FIG. 11.34. Coincidence circuits allow two photon detection events to be associated with each other based upon their arrival times. Photons detected at A and B produce signals that are amplified and analysed to determine whether they meet the energy acceptance criteria. Those signals that fall within the energy acceptance window produce a logic pulse (width τ) that is passed to the coincidence processor. A coincidence event is indicated if both logic pulses fall within a specified interval (2τ).

Coincidence detection assumes that only two photons were detected, but with multiple disintegrations occurring concurrently, it is possible for three or more photons to be recorded by separate detectors within the coincidence time window. When this occurs, it is unclear which pair of detectors corresponds to a legitimate coincidence and multiple events of this sort are often discarded. This circumstance is most likely to occur when there is a large amount of activity in or around the FOV, and it contributes to count loss at high count rates. Another possible scenario is that only one photon is detected within the coincidence time window and no coincidence event will be recorded. These single photon detection events are a result of a number of reasons: the angle of photon emission was such that only one of the two annihilation photons was incident upon the detectors; one of the two annihilation photons was scattered out of the FOV; one of the two annihilation photons was absorbed within the body; and photons originating from outside the coincidence FOV. Although these single photons cannot form true coincidences, they are a major source of randoms.

The mechanism described above records what are known as prompt coincidence events which consist of true, random and scattered coincidences.

The relative proportion of each component depends on factors such as the count rate, the size of the attenuation distribution (patient size) and the acquisition geometry (2-D or 3-D). Only the trues component contributes useful information and the randoms and scattered coincidences need to be minimized. The randoms component is maintained as low as possible by setting the coincidence time window to the shortest duration consistent with the time resolution of the system. The scatter component is maintained as low as possible by energy discrimination. In addition to providing positional and timing information, the pulse produced by the detectors can be integrated over time to provide a measure of the energy deposited in the detector. In a block detector with four PMTs, the sum of the signals from each PMT is proportional to the total amount of scintillation light produced and, thus, the total energy deposited in the detector material. Under the assumption that the photon was completely absorbed in the detector, this signal provides a measure of the photon's energy and can be used to reject lower energy photons that have undergone Compton scattering within the patient. In practice, the energy resolution of most PET detector systems is such that the energy acceptance window must be set quite wide to avoid rejecting too many unscattered 511 keV photons. For BGO based systems, an energy acceptance range of 350–650 keV is typical. As small-angle scatter can result in only a small loss of energy, many of these scattered photons will be accepted within the energy window, despite the fact that they do not contribute useful information. Energy discrimination, therefore, reduces, but does not eliminate, scattered coincidence events and additional compensation is required.

11.3.4.2. Data acquisition geometries

Scanners consisting of multiple detector rings provide extended axial coverage and are advantageous for rapid acquisition of volumetric data. However, the presence of multiple detector rings raises issues concerning the optimum combinations of detectors that should be used to measure coincidence events. In a system with only one ring of detectors, the acquisition geometry is simple as each detector measures coincidence events with other detectors on the opposite side of the same ring. When additional detector rings are added to the system, it is possible to allow coincidence events to be recorded between detectors in different rings. This means including photons that were emitted in directions that are oblique to the transverse plane. Given that photons are emitted in all directions, increasing the maximum ring difference increases the angle of obliqueness that is accepted and, therefore, increases system sensitivity. The data acquisition geometry refers to the arrangement of detector pairs that are permitted to form coincidence events and, in practice, involves the presence or absence of interplane septa (Fig. 11.35). Data acquisition with septa in place is referred to as 2-D mode; data acquisition without any interplane septa is referred to as 3-D mode. The 2-D/3-D designation refers to the acquisition geometry rather than the resulting images as both modes produce similar volumetric images.

In 2-D acquisition mode, an array of septa is inserted between the detector rings. These septa are annular and are typically made of tungsten. The purpose of the septa is to physically absorb photons incident at large oblique angles relative to the transverse plane, allowing only those photons incident approximately orthogonal to the z axis of the scanner. These septa differ significantly from gamma camera parallel-hole collimators as in the PET case, no collimation is provided within the transverse planes. By physically rejecting almost all oblique photons from reaching the detectors, the count rate is substantially reduced, resulting in a low rate of random coincidences and low detector dead time. In addition, 2-D acquisition is associated with a low rate of scattered coincidence events since only photons emitted in and scattering within a transverse plane can pass through the septa. However, if a Compton interaction occurs, the likelihood is that the scattered photon will emerge at an oblique angle. Photons that undergo a Compton interaction are scattered through an angle that is distributed over 4π, so although the photon may be scattered in-plane, out-of-plane scatter is more likely. Photons scattered through oblique angles will be absorbed by the septa, effectively reducing the fraction of scattered coincidences that are measured.

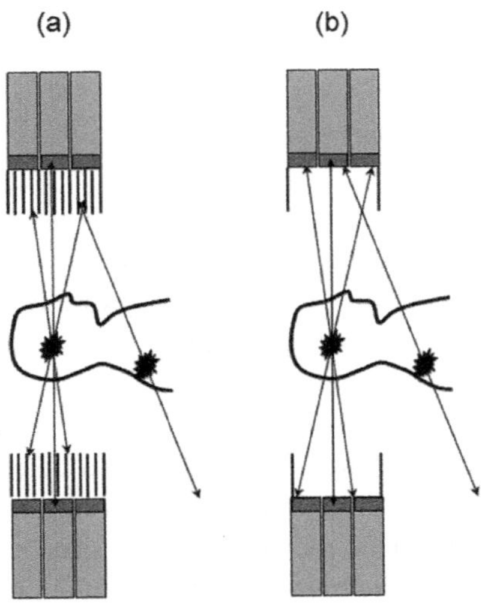

FIG. 11.35. (a) 2-D and (b) 3-D acquisition geometries. In 2-D mode, a series of annular septa are inserted in front of the detectors so as to absorb photons incident at oblique angles. In 3-D mode, these septa are removed, allowing oblique photons to reach the detectors. 3-D mode is associated with high sensitivity but also increased scatter and randoms fractions, the latter partly due to single photons from outside the coincidence field of view.

In 3-D acquisition mode, the septa are entirely removed from the FOV and there is no longer any physical collimation restricting the photons that are incident upon the detectors. Coincidence events can be recorded between detectors in different rings and potentially between all possible ring combinations. Photons emitted at oblique angles with respect to the transverse plane are no longer prevented from reaching the detectors, and system sensitivity is substantially increased compared to 2-D acquisition. Sensitivity gains by a factor of around five are typical, although the exact value depends on the scanner configuration and the source distribution. In 2-D mode, sensitivity varies slightly between adjacent slices but does not change greatly over the AFOV. In 3-D mode, the sensitivity variation in the axial direction is much greater and has a triangular profile with a peak at the central slice. The triangular axial sensitivity profile can be understood by considering a point source centrally located in the first slice at one of the extreme ends of the scanner. True coincidence events can only be recorded between detectors in the first ring. As the source is moved towards the central slice, coincidence events can be recorded between an increasing number of detector ring combinations, leading to an increase in sensitivity. As a consequence of the substantial sensitivity increase, 3-D acquisition is associated with higher detector count rates, leading to more randoms and greater dead time than corresponding acquisitions in 2-D mode. Furthermore, 3-D mode cannot take advantage of the scatter rejection afforded by interplane septa and, as a result, records a greatly increased proportion of scattered coincidence events.

The advantage of 3-D acquisition is its large increase in sensitivity compared to 2-D acquisition. This would be expected to result in images with improved statistical quality or, alternatively, comparable image quality with shorter scan times or reduced administered activity. In practice, evaluating the relative advantage of 3-D acquisition is complex as it is associated with substantial increases in both the randoms and scatter components. Both of these unwanted effects can be corrected using software techniques, but these corrections can themselves be noisy and potentially inaccurate. Furthermore, the relative contribution of randoms and scattered photons is patient specific. The randoms and scatter fractions are defined as the randoms or scatter count rate divided by the trues rate, and both increase with increasing patient size. Both randoms and scatter fractions are substantially higher in 3-D compared to 2-D mode. In 3-D mode, scatter fractions over 50% are common, whereas 15% is more typical for 2-D mode. Randoms fractions are more variable as they depend on the study, but randoms often exceed trues in 3-D mode.

A figure of merit that is sometimes useful when considering the performance of scanner systems is the noise equivalent count rate (NECR). The NECR is equivalent to the coincidence count rate that would have the same noise properties as the measured trues rate after correcting for randoms and scatter.

NECR is commonly used to characterize 3-D performance and, since the relative proportion of the different kinds of coincidence events is strongly dependent on object size, standardized phantoms have been developed. It is computed using:

$$\text{NECR} = \frac{T^2}{T + S + 2fR} \tag{11.12}$$

where

T, *S* and *R* are the true, scatter and random coincidence count rates, respectively;

and *f* is the fraction of the sinogram width that intersects the phantom.

For a given phantom, the NECR is a function of the activity in the FOV and is usually determined over a wide activity range as a radioactive phantom decays (Fig. 11.36). The reason for this count rate dependence is twofold: the randoms rate increases as the square of the single photon count rate (which is approximately proportional to the activity in the FOV) and the sensitivity of the scanner for trues decreases with increasing count rates as detector dead time becomes more significant.

An important factor when considering the relative performance of 2-D and 3-D acquisition modes is the characteristics of the detector material. In 2-D mode, the septa substantially reduce dead time, randoms and scatter, making the poor timing and energy resolution of BGO less of a limitation. BGO based systems are, thus, well suited to 2-D acquisition mode. However, for BGO, the sensitivity advantage of 3-D acquisition mode is substantially offset by the high randoms and scatter fractions that are encountered. For systems based on detectors such as LSO, the improved timing resolution can be used to reduce the coincidence time window and, thus, reduce the randoms fraction. The improved energy resolution also allows the lower level energy discriminator to be raised, resulting in a lower scatter fraction. LSO or similar fast detector materials are, thus, well suited to 3-D acquisition mode. The introduction of these detectors, along with improved reconstruction algorithms for 3-D data, means that 3-D acquisition mode now dominates. Many scanner systems no longer support 2-D mode as this allows the septa to be completely removed from the design, reducing cost and potentially increasing the patient port diameter.

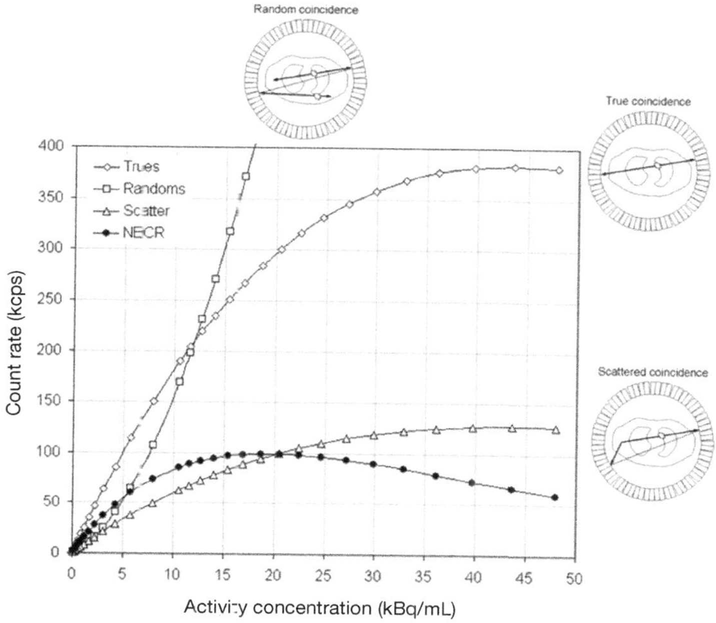

FIG. 11.36. The relative proportion of true, random and scattered coincidence events as a function of activity in the field of view. At low activities, the true coincidence count rate increases linearly with activity. However, at higher activities, detector dead time becomes increasingly significant. The trues rate increases less rapidly with increasing activity and can even decrease at very high activities. The randoms count rate increases with increasing activity as a greater number of photons are detected. The scatter count rate is assumed to be proportional to the trues rate. Scanner count rate performance can be characterized using the noise equivalent count rate (NECR), which is a function of the true, random and scatter coincidence count rates.

11.3.4.3. Data organization

The data recorded during a conventional PET acquisition are the total number of coincidence events measured between the various detector pairs. These data are typically binned into 2-D matrixes known as sinograms (Fig. 11.37). If a 2-D acquisition geometry is considered, each row of the sinogram represents a projection of the radionuclide distribution at a particular angle around the patient. These projections consist of coincidence events recorded between pairs of detectors, where each detector pair forms LORs that are approximately parallel to each other. The number of counts in each element of the projection is proportional to a line integral of the radionuclide distribution within the limitations imposed by the various physical effects such as scatter and attenuation. The sinogram is indexed along the y axis by angle and the x axis by distance. For a full ring system, angular sampling is usually evenly spaced over 180° but the sampling along each row is slightly non-linear. The separation of adjacent elements in the projection decreases towards the edges of the FOV owing to the ring geometry. Correction for this effect, known as arc correction, is required and is usually implemented during image reconstruction. Adjacent elements within a particular sinogram row would be expected to be associated with two parallel LORs joining detector pairs that are next to each other in the ring. In practice, improved sampling is achieved by also considering LORs that are offset by one detector. Despite the fact that these LORs are not exactly parallel to the others, these data are inserted into the sinogram rows as if they came from virtual detectors positioned in the gaps between the real detectors.

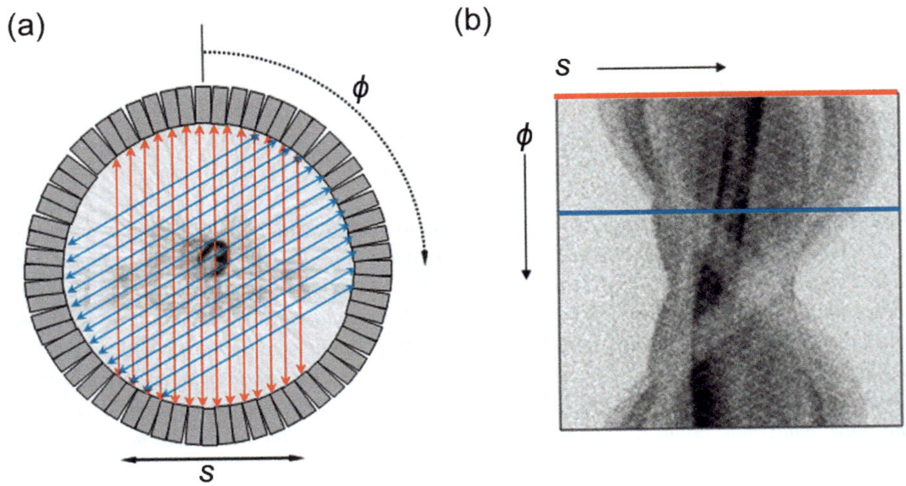

FIG. 11.37. Full ring PET scanners simultaneously measure multiple projections at different angles ϕ with respect to the patient. An example showing the orientation of two parallel projections is shown in (a). Projection data of this sort are typically stored in sinograms; an example is shown in (b). In a sinogram, each row represents a projection at a different angle ϕ. Each projection is made up of discrete elements that are indexed by s and contain the number of coincidence counts recorded along individual lines of response. The two example projections shown in (a) are also highlighted in sinogram (b).

The prior discussion of 2-D acquisition mode only considered coincidence events between detectors in a single ring, referred to as a direct plane. In practice, the interplane septa do not completely eliminate the possibility of coincidence events being detected between different nearby rings (Fig. 11.38). Inclusion of these slightly oblique events is advantageous as it increases sensitivity.

Coincidence events between detectors in immediately adjacent rings are combined into a sinogram that is considered to have been measured in a plane located between the two detector rings. This plane is referred to as a cross plane and is considered to be parallel to the direct planes, despite the fact that the contributing LORs are slightly oblique to these planes. As well as increasing sensitivity, inclusion of these cross planes increases axial sampling by producing $2N - 1$ slices from an N ring scanner. Sensitivity is further increased for cross planes by extending the ring difference between which coincidences are allowed from one to three or higher odd numbers. This principle is also applied to direct planes, resulting in coincidence events not just within the same ring but also between detectors with ring differences of two or higher even numbers. It should be noted that the data obtained from these different ring combinations are added together, so individual sinograms actually consist of LORs that were measured at slightly different oblique angles. The total number of ring combinations contributing to a direct plane plus those contributing to a cross plane is sometimes referred to as span. Within the limitations imposed by the septa, span can be increased, resulting in increased sensitivity and degraded spatial resolution in the axial direction.

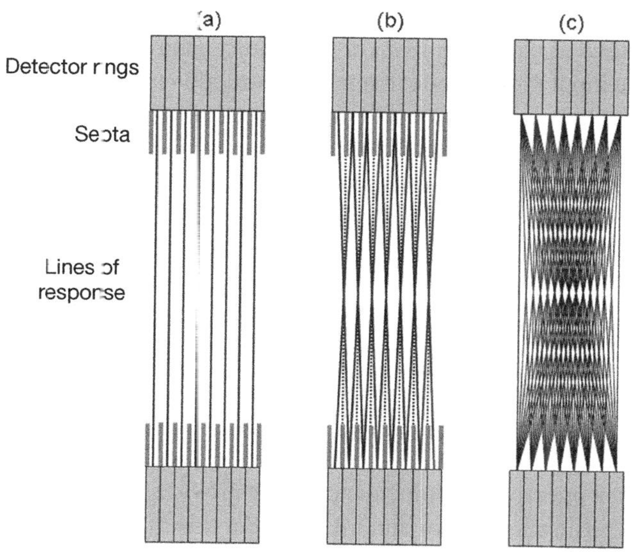

FIG. 11.38. Side elevation of an eight ring PET scanner in 2-D ((a) and (b)) and 3-D (c) acquisition modes. (a) Lines of response joining opposing detectors in the same ring forming direct planes. (b) Lines of response between detectors in adjacent rings. These lines of response are averaged to form cross planes (dotted line) that are assumed to be located at the mid-point between adjacent detectors. Both direct and cross planes are simultaneously acquired during 2-D acquisition. (c) 3-D acquisition in which each ring is permitted to form coincidence events with all other rings.

When the septa are removed, as is the case in 3-D acquisition mode, there is no longer any physical restriction on the detector rings that can be used to measure coincidence events. An N ring scanner could have a maximum ring difference of $N - 1$, resulting in up to N^2 possible sinograms. In 2-D mode, such a system would have a total of $2N - 1$ sinograms, so it can be seen that the total volume of data is substantially higher in 3-D mode. In order to reduce this volume for ease of manipulation, the maximum ring difference can be reduced. This has the effect of introducing a plateau on the axial sensitivity profile, converting it from a triangular to a trapezoidal form. Additionally, several possible ring combinations can be combined in a similar fashion to that indicated for 2-D acquisition mode. It should be noted that 3-D acquisition mode results in data that are redundant in the sense that only a subset of the sinograms (those in the transverse planes) are required for tomographic image reconstruction. The purpose of acquiring the additional oblique data is to increase sensitivity and reduce statistical noise in the resulting images.

In addition to the sinogram representation described above, some scanners also support list-mode acquisition. In this mode, coincidence events are not arranged into sinograms in real time but are recorded as a series of individual events. This stream of coincidence events is interspersed with time signals and potentially other signals from ECG or respiratory gating devices. These data can be used as the input to list-mode image reconstruction algorithms but may also be sorted into sinograms prior to image reconstruction. The advantage of acquiring in list-mode is that the sorting of the data into sinograms can be performed retrospectively. This provides a degree of flexibility that is very helpful when data are acquired in conjunction with physiological gating devices or when sequential images over time are of interest. For example, separate ECG gated and dynamic time series images can be obtained from the same cardiac list-mode acquisition. Furthermore, certain parameters can be retrospectively adjusted and do not have to match the parameters chosen at the time of acquisition.

11.3.4.4. Time of flight

Detectors operating in coincidence mode provide spatial information related to individual positron–electron annihilations but this information is not sufficient to determine the exact location of each event. A line joining the two detectors can be assumed to intersect the site of the annihilation but the exact position along this line cannot be determined. For this reason, PET systems measure signals from multiple events and the resulting projections are used to reconstruct images using computed tomography. However, it has long been appreciated that the difference in the detection times of the two annihilation photons provides a mechanism for precisely localizing the site of individual positron–electron annihilations (Fig. 11.39). Given that photons travel at the speed of light, essentially irrespective of the composition of the material through which they pass, the difference in the arrival times of the two photons can potentially be used to localize their original point of emission. This is clearly attractive because it means that each coincidence measurement provides significantly more information, promising substantial improvements in image statistical quality.

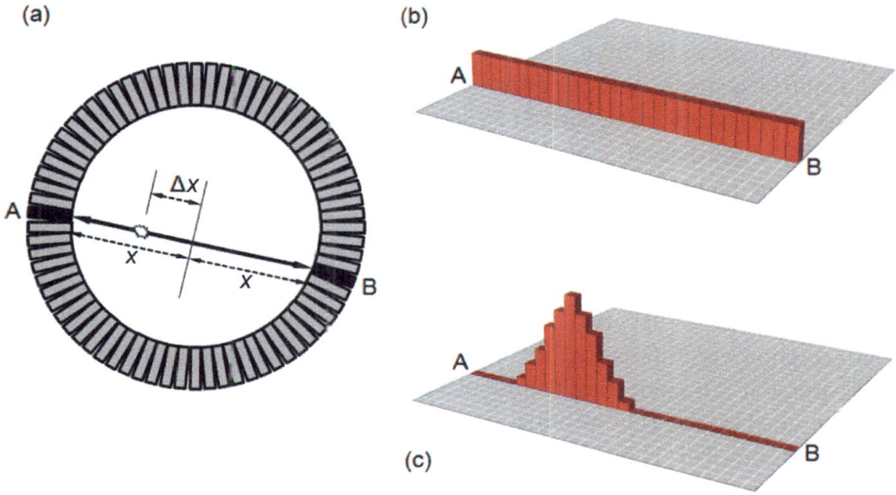

FIG. 11.39. (a) A coincidence event detected along a line of response between detectors A and B. The average time difference between the two detectors is given by $(x + \Delta x)/c - (x - \Delta x)/c = 2\Delta x/c$, where c is the speed of light. (b) With conventional PET, no information is available about the location of the annihilation event along the line of response. During reconstruction, the event is assigned with equal weight to all pixels between A and B. (c) With time of flight PET, the time difference between the signals recorded at detectors A and B is used to estimate the position of the annihilation event along the line of response. During reconstruction, events are weighted according to the detector time difference and a function that reflects the limited time resolution of the system.

Incorporating information derived from differences in the photon arrival times has been referred to as TOF mode and a number of PET systems have been developed that exploit this approach. A prerequisite for TOF PET systems is high timing resolution (Δt) as this determines the spatial uncertainty (Δx) with which the annihilation event can be localized. The two parameters are related by $\Delta x = c\Delta t/2$, where c is the speed of light (3×10^8 m/s). According to this equation, a timing resolution of 66 ps is required to achieve 1 cm depth resolution. Early TOF PET systems used detector materials that, although they had good timing resolution, suffered from a poor stopping power for 511 keV photons. The resulting low sensitivity of these devices could not be offset by the improved signal to noise ratio provided by the TOF information and interest in the method declined. Interest was subsequently rekindled with the introduction of LSO based systems, which have been able to combine timing resolutions of around 600 ps with high sensitivity. A timing resolution of 600 ps translates to a spatial uncertainty of 9 cm which, although clearly worse than the spatial resolution that can be achieved with conventional PET, does represent useful additional information.

In addition to the high performance required for conventional PET, TOF PET requires scanners optimized for high timing resolution. The additional TOF information has data management considerations because an extra dimension has been added to the dataset. TOF data may be acquired in list-mode and fed directly to a list-mode reconstruction algorithm that is optimized for TOF. Alternatively, the data may be reorganized into sinograms where the sinograms have an additional dimension reflecting a discrete number of time bins. Each coincidence event is assigned to a particular sinogram depending on the difference in the arrival times of the two photons. TOF sinograms also require dedicated reconstruction algorithms that incorporate the TOF information into the image reconstruction. An interesting feature of TOF PET is that the signal to noise ratio gain provided by the TOF information is greater for larger diameter distributions of radioactivity. This is related to the fact that the spatial uncertainty Δx becomes relatively less significant as the diameter increases. This has potential benefits for body imaging, particularly in large patients where high attenuation and scatter mean that image quality is usually poorest.

11.3.5. Data corrections

11.3.5.1. Normalization

Normalization refers to a software correction that is applied to the measured projection data in order to compensate for variations in the sensitivity of different LORs. Without such a correction, images display systematic variations in uniformity and pronounced artefacts that include spike and ring artefacts at the centre of the FOV (Fig. 11.40). It is somewhat analogous to the uniformity correction applied to gamma camera images. Sources of sensitivity variations include:

— Detector efficiency variations: The detection efficiency of a particular LOR depends on the efficiencies of the individual detectors involved. Individual detectors can have variable efficiency due to differences in PMT gain, differences in the performance of individual crystal elements and the position of the detector element within the larger detector array.

— Geometric effects: Geometric issues also influence the sensitivity of different LORs. Individual detector elements contribute to multiple LORs and measure photons that are incident over a range of angles. Photons incident normal to the face of the detector will have a shorter thickness of detector material in their path compared to those incident at more oblique angles. This results in a greater probability of detection for photons incident at more oblique angles. However, another geometric effect that exists in full ring block based systems occurs towards the edges of each projection where LORs are formed by detectors at larger oblique angles. Although oblique angles lead to a greater thickness of crystal, they also decrease the solid angle that the detectors present, reducing the sensitivity for those particular LORs.

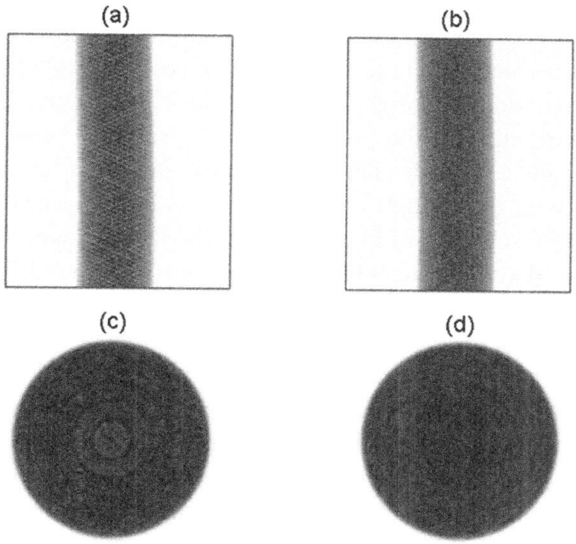

FIG. 11.40. Sinograms corresponding to a centrally located uniform cylinder before normalization (a) and after normalization (b). Transverse images are shown for reconstructions without normalization (c) and with normalization (d). The artefacts and non-uniformity seen in the image without normalization should be noted.

The acquisition geometry, either 2-D or 3-D mode, determines the way different detector pairs are combined and, thus, also influences sensitivity variations between different LORs. In addition, the presence of the interplane septa has a shadowing effect that reduces sensitivity. For this reason, separate

2-D and 3-D normalizations are required for systems capable of acquiring data in both modes. Normalization files are experimentally determined correction factors that are applied as multiplicative terms for each LOR. They are periodically updated to reflect the current state of the detector system and are applied to all subsequently acquired data. In order not to degrade the statistical quality of the patient data, normalization coefficients need to be measured with low statistical noise, and a variety of methods have been developed to achieve this.

11.3.5.2. Randoms correction

Randoms make up a potentially large component of all measured coincidence events and, if left uncorrected, will contribute to a loss of image contrast and quantitative accuracy. Random coincidences are generally smoothly distributed across the FOV but the magnitude of the randoms component depends on the count rate encountered during data acquisition. 3-D acquisition mode or studies involving large amounts of activity in or near the FOV are usually associated with high randoms fractions. Randoms correction is essential for all quantitative studies and is routinely implemented on almost all scanner systems.

One widely adopted correction method involves estimating the number of randoms contributing to the prompts (trues + scatter + randoms) using an additional coincidence circuit. This secondary coincidence circuit is acquired simultaneously with the prompt measurement but is only sensitive to random events. Preferential selection of randoms is achieved by delaying the logic pulse from one of the detectors such that it cannot form a true coincidence event between corresponding annihilation photons (Fig. 11.41). Although the time delay prevents measurement of true and scattered photons, it does not stop coincidence events being recorded by chance between unrelated 511 keV photons. As this secondary, or delayed, coincidence circuit is identical to the prompt circuit in all other respects, the number of counts in the delayed channel provides an estimate of the randoms in the prompt channel. In many implementations, the delayed data are automatically subtracted from the corresponding prompt data, providing an on-line randoms correction. Alternatively, the prompt and delayed data can be stored as separate sinograms for retrospective off-line subtraction, often as part of a statistical image reconstruction algorithm.

The delayed channel does not identify and remove individual random coincidences from the prompt measurement but, instead, estimates the average number of randoms that might be expected. As the delayed channel records counts in individual LORs over a limited scan duration, the randoms estimate is often noisy, leading to increased statistical uncertainty in the corrected data after randoms subtraction.

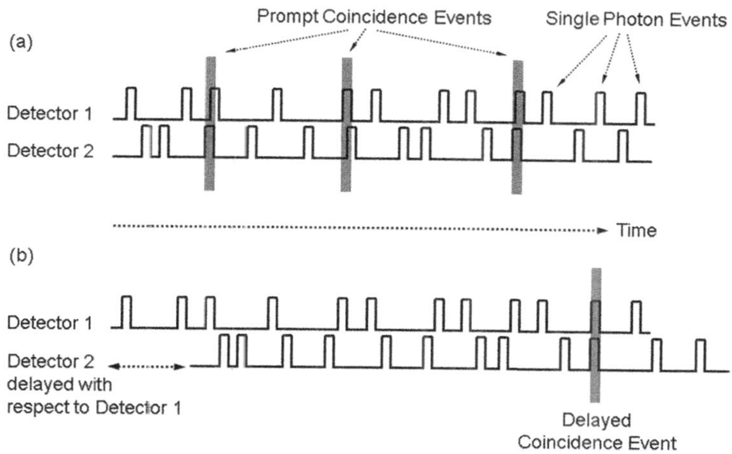

FIG. 11.41. Diagram illustrating the concept of how a delayed coincidence circuit can be used to estimate the number of random events in the prompt circuit. (a) Detection events from two opposing detectors, indicating three coincidence events in the prompt circuit. (b) Data from detector 2 delayed with respect to detector 1 and indicating one coincidence event in this delayed circuit. The temporal delay prevents true coincidence events from being recorded in the delayed circuit, but random coincidence events still occur with the same frequency as in the prompt circuit. If data are acquired with sufficient statistical quality, the total number of delayed coincidence events provides an estimate of the total number of randoms in the prompt circuit.

An alternative method of randoms correction is to estimate the randoms for each LOR using the singles rates at the corresponding detectors. If N_1 and N_2 are the singles rates at two opposing detectors, the rate of random coincidences between these detectors is given by $2\tau N_1 N_2$ where 2τ is the coincidence timing window. This approach has the advantage that singles count rates are substantially higher than coincidence count rates and, therefore, lead to randoms estimates with better statistical quality than the delayed channel method.

11.3.5.3. Attenuation correction

Despite their high energy, only a small fraction of the emitted 511 keV photon pairs escape the body without undergoing some form of interaction. Compton interaction is the most likely mechanism and, depending on the energy and direction of the scattered photon, may result in the detection of a scattered coincidence event. However, it is more likely that the scattered photon will not result in a coincidence event for a variety of reasons. The scattered photon may emerge from the Compton interaction along a path that is not incident upon the detectors. Alternatively, the scattered photon may undergo further Compton interactions, resulting in a lower energy and a greater likelihood of photoelectric absorption. Even if the scattered photon does reach the detectors, it may have lost so much energy that it does not meet the energy acceptance criteria of the scanner and will be rejected. Thus, as well as creating scattered coincidence events, Compton interactions lead to a much greater loss of true coincidence events. This underestimation of true counts is referred to as attenuation.

One of the advantages of PET over SPECT is the ease with which attenuation correction can be performed. This feature of coincidence detection can be understood by considering a point source located at depth x within a uniformly attenuating object with attenuation coefficient μ. Figure 11.42(a) shows an LOR passing through the point source and intersecting a thickness D of attenuating material. The probability of photon 1 escaping the object without undergoing any interactions p_1 is given by:

$$p_1 = \frac{I(x)}{I(0)} = \mathrm{e}^{-\mu x} \tag{11.13}$$

where

$I(x)$ is the beam intensity after passing through attenuating material of thickness x;

and $I(0)$ is the intensity in the absence of attenuation.

The probability that the corresponding photon 2 will also escape the object is given by p_2:

$$p_2 = \frac{I(D-x)}{I(0)} = \mathrm{e}^{-\mu(D-x)} \tag{11.14}$$

The probability of both photons escaping the body such that a coincidence event can occur is given by the product of p_1 and p_2:

$$p_1 \times p_2 = \mathrm{e}^{-\mu x} \times \mathrm{e}^{-\mu(D-x)} = \mathrm{e}^{-\mu D} \tag{11.15}$$

It can be seen that the probability of a coincidence event occurring decreases as the thickness of attenuating material increases. However, this probability is not dependent on the location of the source along a particular LOR. This result differs from the SPECT case where the attenuation experienced by a single photon source along a particular LOR is strongly dependent on the distance between the source location and the edge of the attenuating medium.

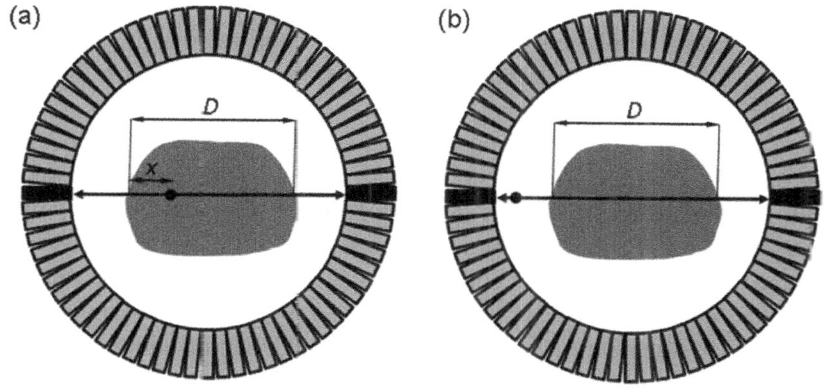

FIG. 11.42. When considering both back to back photons along a particular line of response, the attenuation experienced by a point source within the body (a) is independent of its location along the line and is given by $e^{-\mu D}$ for a uniformly attenuating object. In (b), the positron emitting transmission source is outside the patient but experiences the same attenuation as the internal source when considering coincidence events along the same line of response.

From Eq. (11.15), it can be seen that photon attenuation reduces the number of true coincidence events by a factor of $e^{-\mu D}$. In practice, the body is not composed of uniformly attenuating material, and the attenuation factor AF for a particular LOR is given by:

$$\mathrm{AF} = \mathrm{e}^{-\int \mu(x)\,\mathrm{d}x} \tag{11.16}$$

where $\mu(x)$ refers to the spatially-variant distribution of linear attenuation coefficients within the body and the integral is over the LOR joining opposing detectors.

As this attenuation factor is dependent only on the material along the LOR, it is identical to the attenuation experienced by an external source of 511 keV photons along the same line (Fig. 11.42(b)). Consider an external source of 511 keV photons that gives rise to a total of N_0 coincidence events along a particular LOR in the absence of any attenuation. When a patient is interposed, the number of coincidence events falls to N_x where:

$$N_x = N_0 \mathrm{e}^{-\int \mu(x)\,\mathrm{d}x} \tag{11.17}$$

From this equation, it can be seen that the attenuation factor can be obtained by dividing N_x by N_0. The attenuation correction factor is simply the reciprocal of the attenuation factor and is given by N_0/N_x. N_0 can be obtained from what is referred to as a blank scan, as it is acquired with nothing in the FOV. N_x can be obtained from what is referred to as a transmission scan, as it measures the coincidence events corresponding to the photons that pass through the patient while in the scanning position.

A variety of transmission systems have been developed for measuring patient specific attenuation correction factors. Early transmission systems used ^{68}Ge (271 d half-life, decays to the positron emitter ^{68}Ga) ring sources, although poor scatter rejection meant that the ring configuration was superseded by rotating rod sources. These sources, also consisting of ^{68}Ge, were oriented parallel to the z axis of the scanner and could be inserted close to the detectors at the periphery of the PET FOV during a transmission acquisition. Once in position, they would rotate around the patient in a similar fashion to the motion of an X ray tube in CT. Coincidence events were recorded in 2-D acquisition mode for all LORs as the source rotated continuously around the patient. The patient transmission data were used in conjunction with a similarly acquired blank scan to determine an attenuation correction sinogram that was then applied to the patient emission data, under the assumption that the patient did not move between scans. The rotating rod configuration had the advantage that it enabled transmission data to be acquired in conjunction with a spatial 'window' that tracked the current position of the rotating source. Coincidence events that were not collinear with the current position of the source, such as scattered coincidences, could be rejected, improving the quality of the transmission data. Rod windowing also helped reduce contamination of the transmission measurement by coincidence events not originating from the rotating sources but from a radiopharmaceutical within the patient. The ability to acquire transmission data in the presence of a positron emitting tracer within the body was of great practical significance as, without this capability, lengthy protocols were required involving transmission acquisition prior to tracer administration.

Rod windowing also provided the potential for simultaneous acquisition of emission and transmission data, although cross-contamination meant that separate emission and transmission acquisitions were usually preferred. The disadvantage of acquiring emission and transmission data in a sequential fashion was that scan times for both modes were necessarily lengthy in order to obtain data with sufficient statistical quality. Increasing the rod source activity was not an effective way of reducing transmission scan times as the source was located close to the detectors and dead time at the near-side detectors quickly became a limiting factor. As an alternative to ^{68}Ge, single photon emitters such as ^{137}Cs (30 a half-life, 662 keV) were used as a transmission source. Single photon transmission sources had the advantage that they could be shielded from the near-side detectors and could, thus, use much greater amounts of activity, leading to data with improved statistical quality and, in practice, shorter scan times.

The fact that the single photon emissions were at 662 keV as opposed to 511 keV and that the transmission data had a large scatter component was problematic but could be effectively suppressed using software segmentation algorithms.

With the introduction of PET/CT, the need for radionuclide transmission systems was eliminated as, with careful manipulation, the CT images can be used not just for anatomic localization but also for attenuation correction. CT based attenuation correction has a number of advantages, including the fact that the resulting attenuation correction factors have very low noise due to the high statistical quality of CT; rapid data acquisition, especially with high performance multi-detector CT systems; insensitivity to radioactivity within the body; and no requirement for periodic replacement of sources as is the case with ^{68}Ge based transmission systems. CT based attenuation correction is significantly different from radionuclide based transmission methods because the data are acquired on a separate, albeit well integrated, scanner system using X ray photons with energies that are very different from the 511 keV photons used in PET. Unlike monoenergetic PET photons, the photons used in CT consist of a spectrum of energies with a maximum value that is dependent on the peak X ray tube voltage (kVp). CT Hounsfield units reflect tissue linear attenuation coefficients that are higher than those applicable to PET photons as they are measured at an effective CT energy (~70 keV), which is substantially lower than 511 keV. An important step in the process of using CT images for attenuation correction is to scale the CT images to linear attenuation coefficients that are applicable to 511 keV photons. A number of slightly different approaches have been employed but usually involve multi-linear scaling of the CT Hounsfield units using functions specific for the X ray tube kVp setting (Fig. 11.43). The methods used are very similar to those described in Section 11.2.3.2 (Eq. (11.10)) for SPECT. After scaling, the CT images are filtered, so as to have a spatial resolution that is similar to that of the PET data and attenuation factors are calculated by integration in a manner indicated by Eq. (11.16). The integration, or forward projection, is performed over all directions measured by the PET system and, thus, provides attenuation correction factors for all LORs.

CT based attenuation correction has proved to be very effective although a number of potential problems require consideration. Patient motion, commonly motion of the arms or head, can cause the CT and PET images to be misregistered, leading to incorrect attenuation correction factors, which in turn cause image artefacts and quantitative error. Respiratory motion can also lead to similar problems as the CT and PET data are acquired over very different time intervals. CT data acquisition is extremely short and usually captures a particular phase in the respiratory cycle. In contrast, PET data are acquired over multiple breathing cycles and the resulting images represent an average position that will be somewhat blurred in regions where respiratory motion is significant.

In the area around the lung boundary where there is a sharp discontinuity in the body's attenuation properties, respiratory motion can lead to localized misregistration of the CT and PET images, and pronounced attenuation correction artefacts. Another consideration for CT based attenuation correction arises when the CT FOV is truncated such that parts of the body, usually the arms, are not captured on the CT image or are only partially included. This leads to under-correction for attenuation and corresponding artefacts in the PET images.

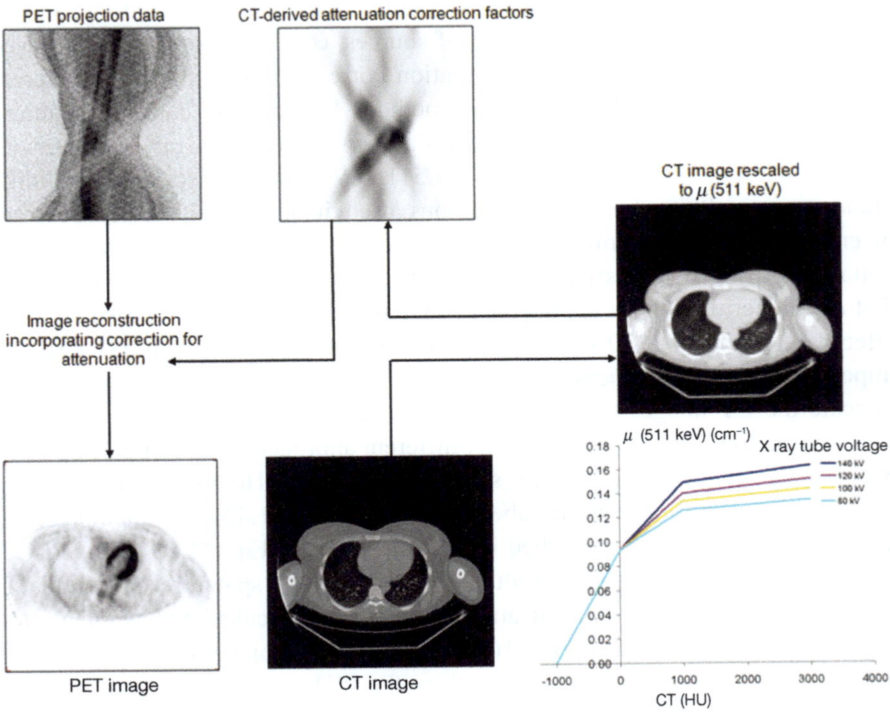

FIG. 11.43. For PET attenuation correction, CT images have to be rescaled from Hounsfield units (HU) to linear attenuation coefficients (μ) appropriate for 511 keV. Multi-linear scaling functions have been used. The rescaled CT images are then forward projected to produce attenuation correction factors that are applied to the PET emission data, prior to or during image reconstruction.

11.3.5.4. Scatter correction

Scatter correction is required because the limited energy resolution of PET systems means that scattered photons can only be partially rejected by energy discrimination. Uncorrected scatter forms a background in reconstructed images that reduces lesion contrast and degrades quantitative accuracy. This scatter background is a complex function of both the emission and attenuation distributions and is non-uniform across the FOV. In 2-D mode, physical collimation ensures that the scatter contribution is relatively low compared to 3-D mode and approximate corrections, based on scatter deconvolution, have been widely used. The form of the scatter distribution function can be measured experimentally using line sources at different positions in a water phantom. Analytical expressions derived from this scatter function can be determined and convolved with the projection data from individual patient studies to estimate the scatter distribution. This method assumes a uniform scattering medium and has limited accuracy in areas such as the thorax. It also cannot account for scatter between adjacent planes, which is significant for 3-D acquisition mode.

An alternative algorithm that has been applied to 3-D brain studies involves a tail fitting approach. In brain studies, the LORs that do not pass through the head are comprised of scatter that can be modelled by fitting a Gaussian function to the tails of each projection. This function can be interpolated to the projections that do pass through the head and used as an estimate of the scatter contribution along these LORs. This method provides a first order correction for scatter and has limited accuracy in areas of non-uniform attenuation or any area where the tails of the projections cannot be accurately measured.

More accurate scatter correction can be achieved in 3-D using a model based approach. This method makes use of the physics of Compton scattering to model the distribution of coincidence events for which one of the two photons experienced a single scattering interaction. The assumption that the scatter component in the measured data is dominated by these single scatter events has been shown to be reasonable. An initial estimate of the radionuclide distribution is first obtained from a reconstruction that does not include any scatter correction. This image is then used, in conjunction with an attenuation map derived from CT or transmission data, to estimate the scatter contribution to each LOR. Different implementations have been developed, but each makes use of the Klein–Nishina formula to determine the probability of a photon scattering through a certain angle and being detected by a certain detector. Determining these probabilities for all possible scattering positions and LORs is computationally demanding. However, good scatter estimates can be obtained by interpolating data estimated using a coarse grid of scattering locations and a subset of LORs. These data can then be interpolated for all LORs, resulting in an estimate of the scatter distribution that has been found to be highly accurate over a range of anatomical locations.

11.3.5.5. Dead time correction

Detector count rates vary between patients and can also vary within studies performed on the same patient. In order to achieve quantitatively accurate images, the true count rate should ideally increase linearly with increasing activity in the FOV. Although this is usually the case at low count rates, the PET scanner's response becomes increasingly non-linear at high count rates. Individual detector modules within a scanner require a finite period of time to process each detected photon. If a second photon is incident upon a detector while an earlier photon is still being processed, the secondary photon may be lost. The likelihood of this occurring increases at high count rates and results in an effective loss of sensitivity. This kind of count loss is referred to as dead time.

Detector dead time losses occur mostly in the detector front-end electronics. The signal produced by an individual photon is integrated for a fixed period of time in order to determine the position and energy of the incoming event. If a second photon is recorded during this integration time, the two signals will combine (pulse pile-up) in such a way as to become indistinguishable from each other. The resulting combined signal will be rejected if it exceeds the upper level energy discriminator. Alternatively, if it is recorded within the energy acceptance window, it may be treated as a single event with a position somewhere between the locations of the two individual photons. In this case, pulse pile-up contributes to a loss of spatial resolution as well as a loss of true counts. Other sources of dead time arise during coincidence event processing. When more than two events occur within the coincidence time window, it is impossible to determine the correct coincidence pair. In this circumstance, all detection events may be discarded, contributing to dead time losses, or alternatively, all possible coincidence events can be included, increasing the randoms component.

Dead time correction compensates for this loss of sensitivity. Corrections are usually based upon experimental measurements of the scanner's response to a decaying source of activity. After randoms correction, residual non-linearity in the scanner's response can be attributed to dead time. An analytical model of the scanner's count rate response can be determined from these experimental data and used for dead time correction of subsequent patient data. A global correction factor can be applied for a particular acquisition, assuming that dead time effects are similar for all detectors in the ring. Alternatively, different corrections can be applied to each detector block or group of blocks. Corrections can be determined based upon an estimate of the fraction of the acquisition period that each detector was busy processing events and unable to process other photons. Alternatively, the single photon rate at a particular detector can be used as input for a model of the scanner's dead time performance to estimate the magnitude of the dead time effect. Dead time correction only compensates for count losses and does not compensate for the event mis-positioning that can occur as a result of pulse pile-up.

11.3.5.6. Image calibration

The above corrections substantially eliminate the image artefacts and quantitative errors caused by the various physical effects that degrade PET data. As a result, the reconstructed images reflect the activity distribution within the FOV, within the limitations imposed by the system's limited spatial resolution. Furthermore, these reconstructed images can be used to quantify the in vivo activity concentration in a particular organ or tissue. Although this capability is not always fully exploited, the potential to accurately quantify images in terms of absolute activity concentration facilitates a range of potential applications.

After image reconstruction, including the application of the various physical corrections, PET images have arbitrary units, typically counts per voxel per second. Quantitative data can be extracted from the relevant parts of the image using region of interest techniques but cannot be readily compared with other related data such as measurements made with a radioactivity calibrator ('dose' calibrator). In order to convert the PET images into units of absolute activity concentration such as becquerels per millilitre, a calibration factor is required. This calibration factor is experimentally determined, usually using a uniform cylinder phantom. The cylinder is filled with a known volume of water, to which a known amount of radioactivity is added. After ensuring the radioactivity is uniformly distributed within the phantom, a fully corrected PET image is acquired. The calibration factor CF can be determined using:

$$\mathrm{CF} = \frac{\mathcal{A}}{V} \times \frac{p}{C} \tag{11.18}$$

where

$\mathcal{A}/V$ is the known activity concentration (Bq/mL) within the phantom;
C is the mean voxel data (counts · voxel^{-1} · s^{-1}) from a large region well within the cylinder part of the image;

and p is the positron fraction of the radionuclide used in the calibration experiment (typically ^{18}F, positron fraction 0.97).

The positron fraction is a property of the radionuclide and is the fraction of all disintegrations that give rise to the emission of a positron.

The above calibration assumes that the true activity within the phantom is accurately known. This can usually be achieved to an acceptable level of tolerance using an activity calibrator that has been calibrated for the isotope of interest using a long lived standard source that is traceable to a national metrology institute. In principle, a single calibration factor can be applied to subsequent studies performed with different isotopes as long as the positron fraction is known. Calibrated PET images can, thus, be determined by multiplying the raw image data by the calibration factor and dividing by the positron fraction for the particular isotope of interest.

11.4. SPECT/CT AND PET/CT SYSTEMS

11.4.1. CT uses in emission tomography

SPECT and PET typically provide very little anatomical information, making it difficult to precisely localize regions of abnormal tracer accumulation, particularly in oncology studies where disease can be widely disseminated. Indeed, it is often the case that the more specific the radiopharmaceutical, the less anatomical information is available to aid orientation. Relating radionuclide uptake to high resolution anatomic imaging (CT or MRI (magnetic resonance imaging)) greatly aids localization and characterization of disease but ideally requires the two images to be spatially registered. Retrospective software registration of images acquired separately on different scanner systems has proved to be effective in certain applications, notably for brain studies where rigid body assumptions are realistic. However, for most other applications, the rigid body assumption breaks down and the registration problem becomes much more difficult. Combined scanner systems, such as SPECT/CT and PET/CT, provide an alternative solution. The advantage of this hardware approach is that images from the two modalities are inherently registered with no need for further manipulation. Of course, this assumption can become unreliable if the patient moves during data acquisition, but, in general, combined scanner systems provide an accurate and convenient method for achieving image registration.

In addition to the substantial clinical benefit of registered anatomical and functional images, the coupling of CT with SPECT and PET systems provides an additional technical benefit. Although radionuclide sources have been used for attenuation correction, the availability of co-registered CT is particularly advantageous for this purpose. In the case of SPECT, the main advantages of CT based attenuation correction are greater accuracy and reliability compared to radionuclide sources, while in PET, the main advantage is an effective reduction in the overall duration of the scanning procedure owing to the speed with which

CT images can be acquired. In addition, there are a number of other subsidiary benefits to the introduction of CT to SPECT and PET systems. Radionuclide transmission sources and their associated motion mechanisms are somewhat cumbersome, particularly in the case of SPECT, and the addition of the CT allows this component to be removed from the design. Elimination of the transmission source from PET systems enabled the patient port size to be enlarged, allowing larger patients to be accommodated. Other benefits include use of the CT subsystem's localizing projection image to aid patient positioning, particularly for single bed-position PET scans. In addition, the potential of acquiring both a radionuclide study and diagnostic quality CT in the same scanning session has advantages in terms of convenience. In terms of quantitative image analysis, the availability of registered CT along with a radionuclide study can sometimes be useful for region of interest definition, particularly in research applications. Although giving rise to a number of significant benefits, it should be noted that the addition of CT to SPECT and PET instrumentation has led to an appreciable increase in the radiation dose received by patients undergoing radionuclide imaging procedures.

11.4.2. SPECT/CT

In many respects, SPECT might be expected to benefit more than PET from the addition of registered CT. SPECT has lower spatial resolution than PET. Many SPECT tracers are quite specific and often do not offer the kind of anatomical orientation that is provided by normal organ uptake with PET tracers such as fluorodeoxyglucose (FDG). Radionuclide transmission scanning is more awkward in SPECT compared to PET because gamma cameras need to be capable of multiple flexible modes of acquisition. Despite these considerations, the adoption of combined SPECT/CT instrumentation (Fig. 11.44) has been slower than that of PET/CT. Cost considerations no doubt contribute and there remains uncertainty about the level of CT performance that is required for a SPECT/CT system.

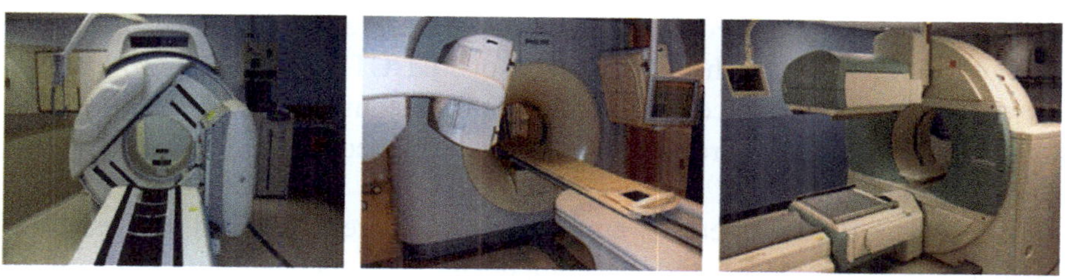

FIG. 11.44. Clinical SPECT/CT systems.

Early SPECT/CT designs, including successful commercial offerings, coupled SPECT with low performance CT. The SPECT subsystem was placed in front of the CT and data were acquired sequentially with the patient being translated between gantries by a common patient bed. Sequential as opposed to simultaneous data acquisition was required due to cross-talk between the CT and SPECT subsystems and was quickly established as the standard mode of operation for both SPECT/CT and PET/CT. Low power X ray tubes and slow rotation speeds meant that the CT component was by no means optimized for diagnostic quality imaging. The aim was to provide attenuation correction and a much needed anatomical context for the SPECT, while maintaining a relatively low cost.

The desire for improved CT image quality has led to the introduction of SPECT/CT systems that incorporate a high performance CT component with capabilities comparable to dedicated CT scanners. With this development, SPECT/CT now benefits from substantially improved CT image quality, faster data acquisition and a broader range of CT protocols. A number of different multi-detector CT slice configurations are available, as well as alternative designs including those based upon flat panel detectors and improvements of the original low cost, non-diagnostic CT. This broad range of CT capability may reflect a diversity of opinion about the role of combined SPECT/CT in clinical practice.

11.4.3. PET/CT

Concurrent technical developments in both PET and CT were exploited in the development of the combined PET/CT system [11.4]. In PET, new detector materials and approaches to image reconstruction increased spatial resolution and improved image statistical quality. In CT, the introduction of spiral scanning and multi-detector technology enabled extremely fast data acquisition over an extended FOV. These developments meant that the combined PET/CT system was well positioned to take advantage of the growing evidence that PET imaging with FDG could play a valuable role in oncology (Fig. 11.45). The addition of spatially registered anatomical information from the CT component of the combined scanner provided the impetus for widespread acceptance of PET in oncology and has driven the rapid growth of PET/CT instrumentation. PET/CT has now been rapidly accepted by the medical community, so much so that stand-alone PET systems are no longer being developed by the major commercial vendors.

Early PET/CT systems were not closely integrated and consisted of separately developed PET and CT components that operated independently of each other. Given that PET/CT acquisitions occur sequentially, as opposed to simultaneously, this approach was reasonable but it meant that multiple computer systems were required and the user interface could be awkward. In addition, the availability of the CT for attenuation correction meant that the PET transmission scanning system was somewhat redundant. Subsequent designs removed the PET transmission sources and moved the two subsystems towards greater integration. In some cases, this meant a more compact system with a continuous patient tunnel. In other cases, the PET and CT gantries were separated by a gap which allowed greater access to the patient during scanning. Removing the transmission scanning system also provided scope for increasing the size of the patient port, so as to accommodate larger patients. This was further achieved by removing the septa from the PET subsystem and decreasing the size of the end-shields used to reject out of FOV radiation. Although the PET and CT detectors remain separate subsystems, many software functions of a modern PET/CT system run on a common platform, including a common patient database containing both PET and CT data.

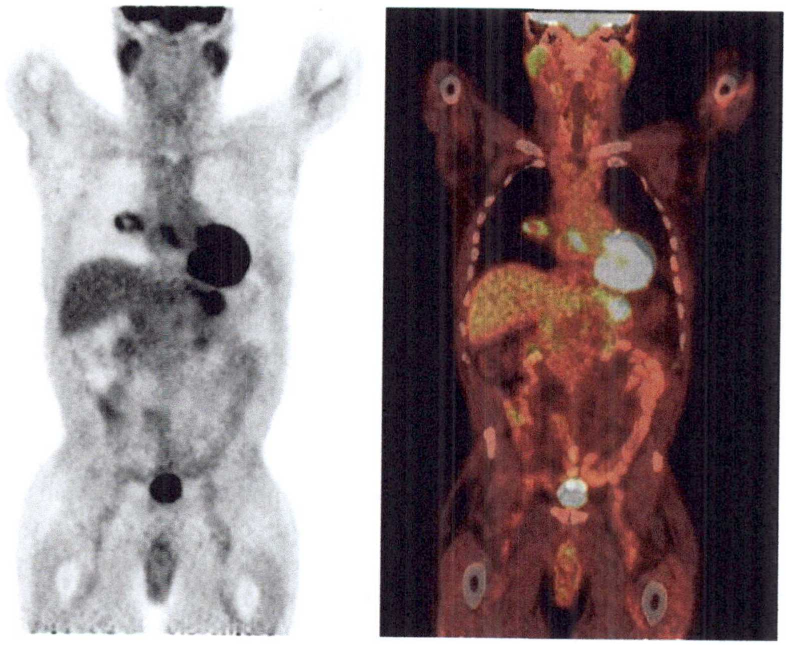

FIG. 11.45. Coronal images from a whole body fluorodeoxyglucose PET/CT study. (a) The PET data are shown in inverse grey scale. (b) The same PET data are shown in a colour scale, superimposed on the CT in grey scale.

The CT component of combined PET/CT systems continues to evolve as greater multi-detector capability becomes available. The initial PET/CT design involved single slice CT, whereas later systems have incorporated 4, 16, 64 or greater slice CT. Advanced CT detector capability allows for extended volume coverage in a single breath-hold or contrast phase and also facilitates a range of rapid diagnostic CT protocols, particularly those intended for cardiology. The level of CT capability required by a combined PET/CT system depends on the extent to which the system will be used for diagnostic quality CT. For many PET/CT oncology studies, state of the art CT is not necessary and low dose protocols are favoured. In these cases, the X ray tube current is reduced and intravenous contrast would typically not be administered. In addition, for whole body studies, data would not be acquired under breath-hold conditions in order to improve spatial alignment of the CT with the PET data that are acquired over multiple respiratory phases.

As far as CT based attenuation correction is concerned, the main advantage for PET is not so much improved accuracy, as ^{68}Ge transmission sources are perfectly adequate when obtained with sufficient statistical quality. The main advantage of CT based attenuation correction is the speed with which the data can be acquired. This is particularly important for whole body PET studies where transmission data are needed at each bed position, requiring extended scan durations to achieve adequate statistical quality. With multi-detector CT scanners, low noise images can be acquired over the whole body in only a few seconds. Replacing radionuclide transmission scans with CT had the effect of substantially reducing the overall duration of scanning procedures, particularly those requiring extended axial coverage. In turn, shorter scans increase patient comfort and reduce the likelihood of motion problems. With the shorter scan times, ‘arms up’ acquisition is better tolerated by patients, leading to reduced attenuation and improved image statistical quality. In addition to these methodological considerations, shorter scan durations allow for more patient studies in a given time period and, thus, more efficient utilization of the equipment.

Despite the relatively rapid scanning afforded by modern PET/CT systems, the PET component still typically requires many minutes of data acquisition and management of patient motion continues to be a problem. Motion can potentially cause misalignment of the PET and CT images, degradation of image spatial resolution and the introduction of characteristic artefacts. Although the CT images are acquired independently of the PET, the PET images involve CT based attenuation correction and are, thus, particularly susceptible to patient motion between the two scans. The types of motion usually encountered include gross movement, often of the arms or head, and respiratory motion. The former is hard to correct retrospectively, although external monitoring devices such as camera based systems can potentially provide solutions for head motion. External monitoring devices can also help reduce problems due to respiratory motion. Respiratory gated PET can reduce motion blurring by reconstructing images from data acquired only during particular phases of the respiratory cycle. This can be further refined by including a motion model into the PET reconstruction and incorporating CT data also acquired under respiratory control.

REFERENCES

[11.1] ANGER, H.O., Scintillation camera, Rev. Sci. Instrum. **29** (1958) 27–33.

[11.2] CHERRY, S.R., SORENSON, J.A., PHELPS, M.E., Physics in Nuclear Medicine, Saunders, Philadelphia, PA (2003).

[11.3] BAILEY, D.L., TOWNSEND, D.W., VALK, P.E., MAISEY, M.N., Positron Emission Tomography: Basic Sciences, Springer, London (2005).

[11.4] BEYER, T., et al., A combined PET/CT scanner for clinical oncology, J. Nucl. Med. **41** (2000) 1369–1379.

CHAPTER 12

COMPUTERS IN NUCLEAR MEDICINE

J.A. PARKER
Division of Nuclear Medicine and Department of Radiology,
Beth Israel Deaconess Medical Center,
Harvard Medical School,
Boston, Massachusetts,
United States of America

12.1. PHENOMENAL INCREASE IN COMPUTING CAPABILITIES

"I think there is a world market for about five computers" — this remark is attributed to Thomas J. Watson (Chairman of the Board of International Business Machines), 1943.

12.1.1. Moore's law

In 1965, Gordon Moore, a co-founder of Intel, said that new memory chips have twice the capacity of prior chips, and that new chips are released every 18 to 24 months. This statement has become known as Moore's law. Moore's law means that memory size increases exponentially. More generally, the exponential growth of computers has applied not only to memory size, but also to many computer capabilities, and since 1965, Moore's law has remained remarkably accurate. Further, this remarkable growth in capabilities has occurred with a steady decrease in price.

Anyone who has even a little appreciation of exponential growth realizes that exponential growth cannot continue indefinitely. However, the history of computers is littered with 'experts' who have prematurely declared the end of Moore's law. The quotation at the beginning of this section indicates that future growth of computers has often been underestimated.

12.1.2. Hardware versus 'peopleware'

The exponential growth of computer capabilities has a very important implication for the management of a nuclear medicine department. The growth in productivity of the staff of a department is slow, especially when compared to the growth in capabilities of a computer. This means that whatever decision was made in the past about the balance between staff and computers is now out of date. A good heuristic is: always apply more computer capacity and less people to a new task. Or stated more simply, hardware is 'cheap', at least with respect to what you learned in training or what you decided last time you considered the balance between hardware and 'peopleware'.

12.1.3. Future trends

In the near future, the increase in personal computer capability is likely to be due to an increase in the number of central processing units on a single processor chip (cores) and in the number of processing chips in a single computer. Multiple processing units have been a key feature of supercomputers for many years. Coordinating the large number of processors is often a bottleneck in the application of supercomputers for general purpose computing. Supercomputers have generally been applied to specific tasks, not to general purpose computing. The trend towards more cores in personal computers has also suffered from this bottleneck. Existing applications often run marginally faster on multicore computers than they do on a single core.

Multi-threaded programming, which has recently received more attention, ameliorates this bottleneck. Multi-threading is an efficient method of synchronizing subtasks that can be computed independently in parallel. Image processing, where different parts of the image can be processed independently, is well suited to multi-threading. Reworking just the intensive processing portions of the software as a multi-threaded application can greatly improve the overall speed on a multicore machine. In fact, several basic image processing packages are already multi-threaded. Thus, with relatively limited updating, nuclear medicine software should be able to take advantage of the trend towards multiple processors in a single computer.

An area where multiple processing units are currently used in personal computers is in graphical processing units (GPUs). GPUs, which perform the same operation on multiple parts of an image simultaneously, are classified as single instruction, multiple data processors. These units have traditionally been thought to be very difficult to program, but recently some programming tools are making them somewhat more readily accessible. They are still very difficult to program in comparison to multi-threading; however, the improved programming tools should make these processors more common for the most computing intensive data processing tasks. They are more likely to be used in the front-end computers within the imaging devices (see Chapter 11) than in workstations and servers that will be the focus of this chapter.

12.2. STORING IMAGES ON A COMPUTER

12.2.1. Number systems

12.2.1.1. Decimal, binary, hexadecimal and base 256

Everyone is familiar with the Arabic or decimal number system, which uses ten symbols, 0–9. Computer circuits are best at dealing with two values, fully on and fully off. The binary number system has two values, 0 and 1. These values can represent on/off, high/low or true/false. From right to left, the digits in a binary number represent ones, twos, fours, eights, sixteens, etc. Computers have become much better at performing conversions to decimal numbers, so that users generally do not have to worry about binary numbers; however, images are an exception. It is still useful to know something about binary numbers to understand image representation at a fundamental level.

Binary numbers are natural for computers, but they are very unnatural for humans. A compromise between humans and computers is the hexadecimal or base-16 number system. The decimal number, 1000, is represented by 3E8 in hexadecimal, which is much easier to remember than 1111101000 in binary. The hexadecimal number system uses 16 symbols: 0–9 are used for the first ten symbols and A–F are used for the last six symbols. From right to left, the digits in a hexadecimal number represent 1s, 16s, 256s, etc.

The value of a decimal digit is 10^n, where n is the position. The value of a binary digit is 2^n, and the value of a hexadecimal digit is 16^n. Four binary digits represent the same value as one hexadecimal digit: $2^4 = 16$. For this reason, one can easily convert back and forth between binary and hexadecimal, but not between binary and decimal.

Another interesting number system is used with the Internet (Table 12.1). It uses a base 256 number system, but rather than inventing a whole new set of symbols, it uses the corresponding decimal numbers as symbols. Since the decimal numbers from 0 to 255 vary in length, it needs to use periods to separate the digits. From right to left, the digits in an Internet address represent 1s, 256s, 256^2s, etc. The Internet number system is the most convenient compromise between the computer's preference for binary numbers and people's familiarity with the decimal system. An internet address can be considered to be a base 256 (2^8) number. The digits are separated by periods. Digit values are given by their decimal equivalents. One internet digit equals two hexadecimal digits or eight binary digits (see Table 12.1).

12.2.1.2. Kilo, mega, giga, tera

In the decimal system, kilo, mega, giga and tera are used to represent 1000, 1 000 000, 1 000 000 000 and 1 000 000 000 000, respectively. Using scientific notation, these numbers are 10^3, 10^6, 10^9 and 10^{12}, respectively. Twelve binary digits can be used to represent roughly the same range of numbers as three decimal digits: $2^{10} = 1024$ and $10^3 = 1000$. Therefore, these terms have been appropriated by the computer community to mean 2^{10}, 2^{20}, 2^{30} and 2^{40}.

TABLE 12.1. INTERNET ADDRESS NUMBER SYSTEM

	← 32 binary digits →							
Binary	0001	1000	0000	1001	1111	0011	0100	1110
Hexadecimal	1	8	0	9	F	3	4	E
Internet address		24.		9.		243.		78.

12.2.2. Data representation

Digital images are composed of individual picture elements, pixels, which represent a single point in the image. Each pixel is represented by a number or a series of numbers. There are several methods of representing each number.

12.2.2.1. Integer numbers

A group of binary digits (bits) can be interpreted in several different ways. The simplest is as a positive integer, numbers 0, 1, 2, 3, etc. The number of bits determines how large a number can be stored. Four bits, one hexadecimal digit, can represent 0–15. Eight bits can represent two hexadecimal digits, values 0–255. Eight bits are called a byte.

A byte is usually the smallest unit of storage that is used for a pixel. It is fairly limited, both in terms of the number of counts that can be collected in one pixel and in terms of how many colours can be specified by 1 byte. Given the architecture of modern computers, it makes more sense to use a number of bytes, e.g. 2 bytes (16 bits), 3 bytes (24 bits) and 4 bytes (32 bits). Table 12.2 shows the number of counts that can be represented using each of these formats. In nuclear medicine, a 2 byte image is sometimes called a word image. More generally, 'word' does not have a fixed meaning; it varies with the computer architecture. It is a less ambiguous way to describe larger integer values as 2 byte, 4 byte, etc.

A further complication is that pixels are sometimes represented by non-integral numbers of bytes. For example, computed tomography (CT) pixels are often 12 bits. Within computers, these pixels are often stored with 2 bytes, where the unused bits are set equal to zero. Some image formats even use a non-integral number of bytes per pixel.

TABLE 12.2. THE NUMBER OF COUNTS THAT CAN BE REPRESENTED IN DIFFERENT IMAGE FORMATS

	Bits	Range	Number of values
1 bit	1	0–1	2
1 byte	8	0–255	256
2 bytes	16	0–65 535	64 k
3 bytes	24	0–16 777 215	16 M
4 bytes	32	0–4 294 967 295	4 G
5 bytes	40	0–1 099 511 627 776	1 T

where

Symbol	Prefix	Power of 2	Value	Power of 10
k	kilo	2^{10}	1024	$\sim 10^3$
M	mega	2^{20}	1 048 567	$\sim 10^6$
G	giga	2^{30}	1 073 741 824	$\sim 10^9$
T	tera	2^{40}	1 099 511 627 776	$\sim 10^{12}$

12.2.2.2. Signed versus unsigned numbers

Positive integers are called unsigned integers. In ‘2s compliment’ arithmetic, negative integers, –1, –2, –3, etc., are represented by the largest, next largest, etc. binary numbers. The sign of the number is determined from the value of the highest order bit. If a signed number is displayed by mistake as an unsigned number, it is shown as if it had very high activity. Understanding the difference between signed and unsigned numbers makes it relatively easy to diagnose and know-how to solve this problem.

12.2.2.3. Floating point and complex numbers

Floating point representation is analogous to scientific notation where a number is represented by a mantissa times 10^n, where n is called the exponent. Instead of 10^n, computer representation uses 2^n. Using the Institute of Electrical and Electronics Engineers (IEEE) standard, both the number and the exponent are signed integers. For the single-precision, 32-bit IEEE standard, the magnitude of the largest number is about 3.4×10^{38} and the magnitude of the smallest non-zero number is about 1.2×10^{-38}. For the double-precision, 64-bit IEEE standard, the magnitude of the largest number is about 1.2×10^{308} and the magnitude of the smallest non-zero number is about 2.2×10^{-308}.

Floating point numbers are not generally used for raw nuclear medicine data; however, during processing, they can be quite useful both to represent fractions and because some processing involves large intermediary values.

Complex numbers consist of two parts, a real part and an imaginary part. They are written $x + \mathrm{i}y$ where x is the real part and $\mathrm{i}y$ is the imaginary part. The symbol i stands for $\sqrt{-1}$. In mathematics, the symbol j is often used instead of i. Each complex number is represented as two floating point numbers. Complex numbers only come up in nuclear medicine during processing. However, in magnetic resonance imaging (MRI), complex numbers are the most logical representation for the raw signals that come from the magnet.

12.2.2.4. Byte order

Memory is addressed by byte. When a number is stored using 4 bytes, then successive numbers start at addresses 0, 4, 8, etc. Confusion can arise because computer manufacturers did not adopt a standard way of assembling the bytes into a number. From least significant to most significant, some manufactures assembled 4-byte numbers using address 0 followed by address 1, e.g. 0123; other manufacturers assembled 4-byte numbers in the order 3210. The former is sometime called 'big-endian', the big address is at the end — the big values come first; the latter is sometimes called 'little-endian'.

12.2.3. Images and volumes

Images can be represented by 2-D functions. If x and y are taken to be horizontal and vertical, a 2-D function, $f(x, y)$, gives the value of the image at each point in the image. Ever increasingly, imaging equipment produces not a single image, but a 3-D volume of data, e.g. right to left, anterior to posterior and caudal to cephalic. A volume of data can be represented as a 3-D function, $f(x, y, z)$.

2-D, 3-D, 4-D, etc. are often used loosely in medicine, but it can be insightful to clearly understand the dimensions involved in an application. A relevant example is human vision. A single human eye can see in only two dimensions; the retina is a 2-D structure. From parallax as well as other physiological inputs, it is possible for people with binocular vision to perceive the depth of each point in the visual image. The most appropriate model of the perceived image is a multi-valued function with value intensity, hue, saturation and depth. In this model, the function is 2-D. This 2-D function represents a surface in a 3-D volume. Over time, the mind is able to construct a 3-D model from sequential 2-D surfaces.

On a more basic level, optics models an electromagnetic signal going through an aperture as a complex 2-D function. In addition to amplitude information, which can be sensed by the eye, the function also has phase information. Holography takes advantage of phase information, allowing the observer to 'see around' objects if there is no other object 'in the way'. However, basic physics limits the amount of information to a sparse (essentially 2-D) set of data contained within the 3-D space.

One of the challenges of data visualization is to facilitate the input of 3-D data using the 2-D channel afforded by the visual system. Often, one dimension is mapped into time using cine or a mouse to sequence through a stack of images. Rendering such as a reprojection or a maximum intensity projection can also help by providing an overview or by increasing the conspicuousness of the most important features in the data. Both of these visualization methods also use a sequence of images to overcome the limitations of the 2-D visual channel.

12.2.3.1. Continuous, discrete and digital functions

The real world is usually modelled as continuous in space and time. Continuous means that space or time can be divided into infinitesimally small increments. A function $f(x)$ is said to be a continuous function if both the independent variable x and the dependent variable f are represented by continuous values. The most natural model for the distribution of a radiopharmaceutical in the body is a continuous 3-D function.

An image that is divided into pixels is an example of a discrete function. The independent variables x and y, which can only take on the values at particular pixel locations, are digital. The dependent variable, intensity, is continuous. A function with independent variable(s) that are digital and dependent variable(s) that are continuous is called a discrete function.

A computer can represent only digital functions, where both the independent and dependent variable(s) are digital. Digital values provide a good model of continuous values if the coarseness of the digital representation is small with respect to the standard deviations of the continuous values. For this reason, digital images often provide verisimilar representations of the continuous world.

Nuclear medicine is intrinsically digital in the sense that nuclear medicine imaging equipment processes scintillation events individually. The original Anger scintillation cameras produced analogue horizontal and vertical position signals, but modern scintillation cameras process the signals digitally and the output is digital position signals. Most positron emission tomography (PET) cameras have discrete crystals, so that the lines of response are intrinsically digital.

12.2.3.2. Matrix representation

The surface of the gamma camera can be visualized as being divided into tiny squares. An element in a 2-D matrix can represent each of these squares. A 3-D matrix can represent a dynamic series of images or single photon emission computed tomography (SPECT) data collection. A 3-D matrix is equivalent to a 3-D digital function, $f[x, y, z]$. Both representations are equivalent to the computer program language representation, $f[z][y][x]$.

The lines of response in a PET camera without axial collimation are more complicated. Perhaps the easiest way to understand the data is to consider a ring of discrete detectors. The ring can be represented as a 2-D array — one axial and one tangential dimension. Any line of response (LOR) can be defined by two crystals, and each of those two crystals is defined by two dimensions. Thus, each LOR is defined by four dimensions. A 4-D matrix can represent the lines of response (not all crystal pairs are in coincidence, so the matrix is sparse, but that is a detail).

It is particularly simple for a computer to represent a matrix when the number of elements in each dimension is a power of two. In this case, the x, y and z values are aligned with the bits in the memory address. Early computer image dimensions were usually powers of two. Modern programming practice has considerably lessened the benefit of this simple addressing scheme. However, where hardware implementation is a large part of the task, there is still a strong tendency to use values that are a power of two.

12.3. IMAGE PROCESSING

This section will present an introduction to the general principles of image processing.

12.3.1. Spatial frequencies

12.3.1.1. Vision and hearing

Humans conceptualize images in terms of what they perceive. Thinking of images in terms of spatial frequencies is not natural. However, the relations between perception, $f(x, y)$, and a spatial frequency representation, $F(k_x, k_y)$, can be understood by appealing to an analogy to hearing. Sound waves striking the ear-drum can be represented by a function of time, $f(t)$. A pure tone is a sinusoidal variation in pressure as a function of time. Humans do not think of a sinusoid when they hear a pure tone; it is much more natural to think of a pure tone in terms of a musical note.

The cochlea transforms sinusoidal changes in pressure over time into firing of the single nerve in the auditory nerve that represents this tone. The pressure signal, $f(t)$, is transformed to frequencies. If ω represents the frequencies, the transformed signal can be represented by the function $F(\omega)$. It is recalled that angular frequency ω is 2π times frequency. In this simple case, $F(\omega)$ is an impulse at the frequency corresponding to the tone. The process that the cochlea performs can be represented mathematically by a Fourier transform. The 'time domain' representation, $f(t)$, of the pressure signal is transformed into a 'frequency domain' signal, $F(\omega)$, carried on the auditory nerve.

Humans hear tones; they do not feel the pressure waves. Thus, the frequency domain is more natural. Since sound waves are part of every one's early education, they are less natural than tones, but do not seem foreign. Spatial frequencies do seem foreign to most people. However, the relationship between $f(t)$ and $F(\omega)$ is exactly analogous to the relationship between $f(x, y)$ and $F(k_x, k_y)$. This analogy may help physicians and other non-mathematically oriented health professionals conceptualize the relationship between space and spatial frequencies.

12.3.1.2. Spatial sinusoids

A 'pure' spatial frequency corresponds to a sinusoidal function in the spatial domain. Sinusoidal means a sine, a cosine or a combination of the two. Figure 12.1 shows a number of different sinusoidal functions.

The spatial frequency variables k_x and k_y are often given in terms of the cycles per pixel. A signal that has successive maximums and minimums in adjacent pixels will have one complete cycle in two pixels or 0.5 cycles per pixel. The spatial frequency scale is often shown from 0 to 0.5 cycles per pixel, where 0 cycles per pixel represents a constant value in the image domain and 0.5 cycles per pixel represents an image which varies from maximum to minimum in 2 pixels.

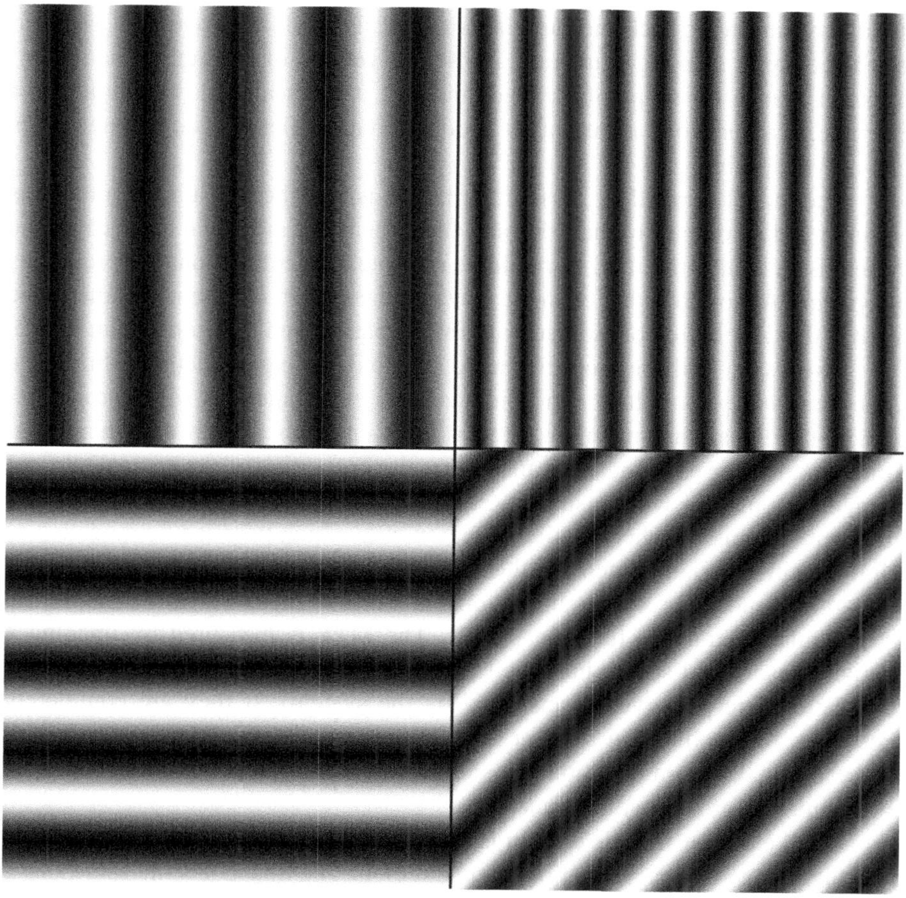

FIG. 12.1. The upper left quadrant is a sinusoid with a single frequency in the x direction; the upper right quadrant is a sinusoid with twice the frequency; the lower left quadrant is a sinusoid with a single frequency in the y direction; and the lower right quadrant is a single sinusoid which varies in both the x and y direction.

12.3.1.3. Eigenfunctions

Eigenfunctions or eigenvectors can be a difficult mathematical concept. They are often presented as an advanced concept. However, eigenfunctions are actually a basic concept. They are relatively easy to understand if the detailed mathematics is avoided.

Eigenfunctions are the 'natural' functions of the system. For example, an eigenfunction of a swing is a sinusoidal back and forth motion. The swing naturally wants to go back and forth in a sinusoidal motion. The output of a system when the input is an eigenfunction is a scaled version of the input. The system does not change the shape of the eigenfunction, it just scales the magnitude. For example, if a pure tone is put into a good audio amplifier, the same tone comes out, only amplified.

The model implied by these descriptions is that there is a system that has an input and an output. In the first case, the system is a swing; in the second, the system is an audio amplifier. It should be noted that eigenfunctions are properties of systems. A more relevant example is where the system is a gamma camera, the input is the distribution of radioactivity from a patient and the output is an image.

The swing and the amplifier have sinusoids as eigenfunctions. In fact, a large class of systems have sinusoids as their eigenfunctions. All linear-time-invariant or linear-shift-invariant systems have sinusoids as their eigenfunctions. Time-invariant or shift-invariant mean that the properties of the systems do not change with time or with position. Real imaging systems are rarely linear-shift-invariant systems, but there is often a region over which they can be modelled as a linear-shift-invariant system, so the sinusoids will be very useful.

A common use of eigenfunctions in nuclear medicine is in measuring the modulation transfer function (MTF). The MTF is measured by imaging a spatial sinusoid and seeing how well the peaks and valleys of the sinusoid (the modulation) are preserved (transferred by the system). If they are completely preserved, the modulation transfer is 1. If the peaks and valleys are only half of the original, the modulation transfer is 0.5. The typical bar phantom is an approximation of a sinusoidal function.

The key point for this section is that a spatial sinusoid is an eigenfunction of the imaging system, at least over some region. The image has the same shape as the input, only the amplitude (modulation) of the sinusoid is altered. The basic properties of the system can be determined by measuring the modulation transfer of the eigenfunctions. Linearity means that the effect of the imaging system on any signal that is a combination of eigenfunctions can then be determined as a scaled sum of eigenfunctions.

12.3.1.4. Basis functions

A function, $f(t)$, can be made up of a sum of a number of other functions, $g_k(t)$. For example, a function can be made from a constant, a line passing through the origin, a parabola centred at the origin, etc. This function can be written as the sum of K powers of t:

$$f(t)=\sum_{k=0}^{K-1} a_k t^k \tag{12.1}$$

These functions are just polynomials. The terms t^k in the polynomial are basis functions, with coefficients a_k. Selecting different coefficients a_k, a large number of functions can be represented.

Usually, the powers of the independent variable t^k seem like they are the most important part of a polynomial, and in many ways they are. However, it will be useful to shift the focus from the powers of t to the coefficients a_k. To make a new polynomial, the key is to select the coefficients. From this viewpoint, the t^k terms are just placeholders.

Extending this point of view, the coefficients can be thought of as a discrete function $F[k]$. The polynomial function could be rewritten:

$$f(t) = \sum_{k=0}^{K-1} F[k] t^k \tag{12.2}$$

This polynomial function provides a method of transforming from the function of coefficients to the function of time. This process is called evaluation of the polynomial. The two functions, the function of coefficients and the function of time, are equivalent representations in the sense that it is possible to transform the function of the coefficients into the function of time using this transformation. It is also possible to select K points in $f(t)$ and transform them to $F[k]$. This process, called interpolation, is beyond the scope of this chapter.

For this discussion, the key point is that the coefficient and the time representations can be converted back and forth with the processes of evaluation and interpolation. $f(t)$ can be described as a function in the 'time domain'. $F[k]$ can be described as a discrete function in the 'coefficient domain'. These two very different functions represent the same thing in the sense that using evaluation or interpolation, it is possible to convert back and forth between the representations.

12.3.1.5. Fourier domain — 'k-space'

When learning about a new way of representation of a familiar concept, it is natural to think of the familiar concept as the 'real' representation and the new representation as 'artificial'. The time or space domains are the real domains, the frequency or the spatial-frequency domains are the artificial or transform domains. However, they are really equivalent representations. One way to understand this is to think about hearing and sight. The natural domain for sight is the space domain, but the natural domain for hearing is the frequency domain. In one case, the natural domain is frequencies and the other the natural domain is just the opposite. The equivalence of the two domains is called duality. Duality is a difficult concept. Electrical engineers, who work with Fourier transforms all the time, still think of the time domain as the real domain and the frequency domain as the transform domain.

One of the best examples of duality is MRI. Chemists used to working with nuclear magnetic resonance think of the time signal as the natural signal and the frequency signal as the transform signal. Imagers think of the image from a magnetic resonance imager as the natural signal and the spatial frequencies as the transform domain. However, due to the gradients, the frequency signal gives the spatial representation and the time signal is the spatial-frequency signal. There is no 'real' domain and 'transform' domain; it all depends on the point of view.

12.3.1.6. Fourier transform

The Fourier transform equations can be written compactly using complex exponentials:

$$F(\omega) = \int f(t) \mathrm{e}^{-\mathrm{i}\omega t} \, \mathrm{d}t \tag{12.3}$$

$$f(t) = \frac{1}{2\pi} \int F(\omega) \mathrm{e}^{\mathrm{i}\omega t} \, \mathrm{d}\omega \tag{12.4}$$

where i is $\sqrt{-1}$.

The first equation (Eq. (12.3)) from the time or space domain to the frequency domain is called Fourier analysis; the second equation (Eq. (12.4)), going from the frequency domain to the time or space domain, is called Fourier synthesis. Fourier analysis is analogous to polynomial interpolation; Fourier synthesis is analogous to polynomial evaluation. The relation of these equations to the sine and cosine transforms can be seen by substituting for the complex exponential using Euler's formula:

$$e^{\mathrm{i}\omega t} = \cos(\omega t) + \mathrm{i}\sin(\omega t) \tag{12.5}$$

These equations have been written in terms of time and frequency. Analogous equations can be written in terms of space and spatial frequency. For the 2-D case:

$$F(k_x, k_y) = \int f(x, y) \mathrm{e}^{-\mathrm{i}(k_x x + k_y y)} \, \mathrm{d}x \, \mathrm{d}y \tag{12.6}$$

$$f(x,y)=\frac{1}{2\pi}\int F(k_x,k_y)\mathrm{e}^{\mathrm{i}(k_x x-k_y y)}\,\mathrm{d}k_x\,\mathrm{d}k_y \tag{12.7}$$

The common use of the letter k with a subscript for the spatial frequency variable has led to the habit of calling the spatial frequency domain the 'k-space', especially in MRI. These equations are exactly analogous to the time and frequency equations with t replaced by x, y, and ω replaced by k_x, k_y. In the case of three dimensions, t is replaced by x, y, z, and ω by k_x, k_y, k_z.

The previous equations refer to continuous functions. The limits of integration of the integrals were not specified, but in fact, the limits are assumed to be $-\infty$ to $+\infty$. However, computer representation is digital. Computer representation is often thought of as discrete, not digital, since the accuracy of representation of numbers is often high, so that the quantification effects can be ignored (see Section 12.2.3.1). The Fourier transform equations in discrete form can be written as:

$$F[k]=\sum\nolimits_{n=0}^{N-1} f[n]\mathrm{e}^{-\mathrm{i}2\pi kn/N} \tag{12.8}$$

$$f[n]=\frac{1}{N}\sum\nolimits_{k=0}^{N-1} F[k]\mathrm{e}^{\mathrm{i}2\pi kn/N} \tag{12.9}$$

In image processing, the unit of n is often pixels, and the frequency unit k is often given as a fraction, cycles/pixel. The space variable n runs from 0 to $N-1$; the spatial frequency variable k runs from –0.5 cycles/pixel to (but not including) +0.5 cycles/pixel in steps of $1/N$.

12.3.1.7. Fourier transform as a correlation

Some understanding of how the Fourier transform pair works can be obtained by noting the analogy between these equations and a correlation. The correlation coefficient is written:

$$r=\frac{\sum x_i y_i}{\sqrt{\left(\sum x_i^2 \sum y_i^2\right)}} \tag{12.10}$$

where x and y are the two variables, and i indices over the number of samples. The denominator is just normalization, so that r ranges from –1 to 1. The action is in the numerator. It should be noted that the key feature of a correlation is that it is the sum of the products.

There is an analogy between the correlation equation and the Fourier transform equation. The index i is analogous to the variable t; x_i is analogous to $f(t)$; y_i is analogous to the sinusoid. The Fourier transform equation determines how much a function is 'like' the sinusoid, just as the correlation coefficient determines how much two variables are alike.

12.3.2. Sampling requirements

To represent a function, $f[n]$, with N points, N frequency values, $F[k]$, are needed; to represent a frequency domain signal, $F[k]$, with N points, N time or space values, $f[n]$, are needed. As a general heuristic, if a signal has N points in one representation, N points are needed in another representation. This heuristic assumes that the N points are independent.

A typical property of real systems is that they can only handle signal variations up to a particular maximum frequency. Thus, real signals often have a limited frequency content. The sampling theorem says that if a signal only contains frequencies up to a maximum of f, then it can be exactly reproduced from samples that are taken every $2/f$ per second. Sampling needs to be at twice the highest frequency. If the sampling is not fast enough, then the high frequencies appear to occur at lower frequencies. This process that occurs by sampling a frequency that is too low is called aliasing.

12.3.3. Convolution

Convolution can be used to describe any linear-time-invariant or linear-shift-invariant system. The input, the system and the output are each described by a function. The system function $h(t)$ is defined to be the output $g(t)$ when the input $f(t)$ is equal to a delta function $\delta(t)$.

In general, the input $f(t)$ can be conceptualized as a sequence of delta functions, $\delta(t-\tau)$, scaled by the input at time $f(\tau)$. Each delta function produces a component of the output that is the system function, $h(t-\tau)$, which has been shifted by a time equal to the time of the input and scaled by the magnitude of $f(\tau)$. As the system is time-invariant, the same response $h(t)$ can be used for an input at any time. As the system is linear, the inputs at different times can be considered separately and the separate outputs can be added. The formula for convolution is:

$$g(t) = \int h(t-\tau) f(\tau)\, \mathrm{d}\tau \tag{12.11}$$

12.3.3.1. Shift, scale, add

Convolution can be summarized as shift, scale, add — shift the system function by an amount equal to the time of the input, scale by the size of the input and add all of the resulting shifted and scaled system functions. Convolution can be used to describe a time–activity curve from a region of interest. The arterial concentration is the input, the system function is the time–activity curve after arterial injection of a bolus of activity, and the time–activity curve after intravenous administration is the convolution of the arterial concentration and the system function.

Convolution is not limited to 1-D examples. The system function for an Anger camera is the image obtained in response to a point source of activity. A point source can be modelled as a 2-D delta function. The image of a point source will be a blob where the size of the blob is related to the camera's resolution. The blob can be shifted to every point in the field of view (FOV); it can be scaled for the amount of activity at each location; and all of the blobs can be added up. This process — shift, scale, add — will produce the final output image.

The formula for 2-D convolution is:

$$g(x, y) = \int h(x-x', y-y', z-z') f(x', y')\, \mathrm{d}x'\, \mathrm{d}y' \tag{12.12}$$

In three dimensions:

$$g(x, y, z) = \int h(x-x', y-y', z-z') f(x', y', z')\, \mathrm{d}x'\, \mathrm{d}y'\, \mathrm{d}z' \tag{12.13}$$

12.3.3.2. Mapping convolution to multiplication

In Section 12.3.1.4, on basis functions, it was noted that interpolation could be used to transform from a set of values of a signal to a representation in terms of the coefficients of a polynomial function, and that evaluation could be used to transform from the coefficients of a polynomial function into values. The coefficients can be thought of as a function in the coefficient domain and the values can be thought of as a function in the value domain. The process of multiplying two polynomials involves keeping track of coefficients of the t^0s, t^1s, t^2s, etc. The t's are essentially placeholders.

In polynomial multiplication, the coefficients of the first polynomial are shifted by the exponent of the second polynomial, scaled by the coefficients of the second polynomial, and the products are added. The process — shift, scale, add — is convolution. Multiplying polynomials is complicated. However, the process is much simpler in the value domain. Two signals are multiplied in the value domain by simply multiplying their values. In the value domain, multiplication of two signals is implemented by multiplying their values; in the coefficient domain, multiplication of two signals is implemented by convolution. Transformation of the complicated process of convolution in one domain into the simple process of multiplication in the other domain is one of the key properties of representing signals in terms of a set of basis functions.

In an exactly analogous fashion, the Fourier transform maps convolution into multiplication. The Fourier transform of the convolution of two functions is equal to the product of the Fourier transforms of the functions. Symbolically:

$$\int h(t-\tau)f(\tau)\,\mathrm{d}\tau \leftrightarrow F(\omega)H(\omega) \tag{12.14}$$

where the symbol '$\leftrightarrow$' represents Fourier transformation, the left side shows the time domain operations, and the right side shows the Fourier domain operation. The complicated integral in the time domain is transformed into a simple multiplication in the frequency domain.

At first, it may seem that the Fourier transform operation is just as complicated as convolution. However, since the fast Fourier transform algorithm provides an efficient method for calculating the Fourier transform, it turns out that, in general, the most efficient method of performing convolution is to transform the two functions, multiply their transforms and do an inverse transform of the product. Calculation of convolutions is one of the major practical applications of the Fourier transform.

12.3.4. Filtering

The word 'filtering' refers to processing data in the Fourier domain by multiplying the data by a function, the filter. Conceptually, the idea is that the frequency components of a signal are altered. The data are thought of in terms of their frequency or spatial frequency content, not in terms of their time or space content. The most efficient process in general is (i) Fourier transform, (ii) multiply by a filter and (iii) inverse Fourier transform. Convolution performs this same operation in the time or space domain but, in general, is less efficient.

If, however, a filter can be represented in the time or space domain with only a small number of components that are non-zero, then filter implementation using convolution becomes more efficient. Many image processing filtering operations are implemented by convolution. The non-zero portion of the time or space representation of these filters is called a kernel. In this section, many examples of small kernels are given, which are used in image processing.

The act of multiplying a signal by a function in the time or space domain is sometimes called windowing. Multiplying a signal by a window changes the amplitude of the time or space components of a signal. Owing to duality, filtering and windowing are essentially the same operation, and the distinction is not always maintained.

Psychologists refer to the just noticeable difference as the smallest difference that can be perceived by humans. In vision, the just noticeable difference is a function of the spatial frequency (more exactly the angular frequency). Humans have limited spatial resolution and cannot detect variations at very high spatial frequency. At somewhat lower frequency, detail can be perceived, but only if it is very high contrast. As the frequency decreases, lower and lower contrast variations can be perceived. At very low spatial frequencies, sensitivity to low contrast again decreases. The just noticeable contrast sensitivity peaks at about 2 cycles per degree and falls off rapidly for higher and more slowly for lower angular frequencies. These characteristics are important in image filtering design, which often seeks to maximize transfer of the image information to a human.

12.3.4.1. One dimensional processing

A 1-D signal $f(t)$ can be filtered with a frequency domain signal $H(\omega)$ by transforming $f(t)$ to $F(\omega)$, multiplying the signals to obtain $G(\omega) = F(\omega)H(\omega)$ and then inverse transforming $G(\omega)$ to produce the result $g(t)$. Alternately, the time domain representation of $H(\omega)$, $h(t)$, can be convolved with $f(t)$:

$$g(t) = \int h(t-\tau) f(t)\, \partial t \qquad (12.15)$$

A good example of 1-D processing is a graphic equalizer. The sliders on a graphic equalizer represent $H(\omega)$. Each slider corresponds to a range of frequencies. The tones in the input signal are multiplied by the values represented by the slider to produce the output signal.

12.3.4.2. Two- and three dimensional processing

Two dimensional processing is exactly analogous to 1-D processing with the 1-D variables t and ω replaced with the 2-D variables x, y and k_x, k_y. For 3-D processing, t is replaced by x, y, z, and ω is replaced by k_x, k_y, k_z. Four-, five-, six-, etc. dimensional processing can be performed similarly.

A property of the Fourier transform equations is separability — the transform with respect to x can be performed first followed by the transform on y. It is sometimes convenient to implement a 2-D transform by first doing the transform on the rows, and then doing the transform on the columns. All that is needed is a 1-D Fourier transform subroutine, and any number of dimensions can be implemented just by managing the indexing.

12.3.5. Band-pass filters

Band-pass filters maintain a range of frequencies while eliminating all other frequencies. An ideal band-pass filter is exactly one for some range of frequencies and exactly zero for the remaining frequencies. There is a very sharp transition between the pass- and the stop-zones. A general heuristic is that a sharp edge in one domain will tend to create ripples in the other domain. Ideal band-pass filtering often creates ripples near sharp transitions in a signal. It is, therefore, common to make the transition from the pass-zone to the stop-zone more gradual.

12.3.5.1. Low-pass, noise suppression and smoothing filters

A low-pass filter is a type of band-pass filter that passes low frequencies and stops high frequencies. In an image, the high spatial frequencies are needed for detail. Zeroing the high spatial frequencies smooths the image. By suppressing the high frequencies compared to the lower frequencies, the image noise is often suppressed. Low-pass filters both smooth an image and decrease the noise in an image.

Information theory tells us that image processing can only lower the information in an image. (An exception is when processing includes a priori information.) Noise suppression filtering lowers the information carried by the noise. It is common to say that processing 'brings out' a feature, but actually what it is doing is suppressing the noise which is confusing to the human viewer. All of the information about the content exists in the original image; processing merely reduces the 'useless information' contained in the noise.

A somewhat similar effect can often be obtained by simply moving away from the image or by making an image smaller in size. Moving away makes the high frequency noise correspond to angular frequencies where vision has low contrast sensitivity. These operations are not entirely equivalent to low-pass filtering. They may also move the important content to a point where contrast sensitivity is reduced. Instead, it is best to view an image at a size where the content is at the maximum visual sensitivity and the noise is suppressed with a low-pass filter.

A common method of implementing a low-pass filter is as a convolution with a small kernel. For example, Table 12.3 shows a 3 × 3 smoothing kernel, also commonly called a 9-point smoothing kernel. Each point in the smoothed image is made up of the scaled sum of nine surrounding points. Also shown is a 5 × 5 smoothing kernel. Each point is made up of the scaled sum of 25 surrounding points. The top of Fig. 12.2 shows the effect of the 5 × 5 kernel on a circular

TABLE 12.3. COMMONLY USED KERNELS IN IMAGE PROCESSING

Smoothing 3 × 3		1	2	1	
		2	4	2	
		1	2	1	
Smoothing 5 × 5	1	2	3	2	1
	2	4	8	4	2
	4	8	16	8	4
	2	4	8	4	2
	1	2	3	2	1
Sharpening	–1	–1	–1	–1	–1
	–1	–1	–1	–1	–1
	–1	–1	25	–1	–1
	–1	–1	–1	–1	–1
	–1	–1	–1	–1	–1
Unsharp mask	0	0	–1	0	0
	0	–1	–2	–1	0
	–1	–2	17	–2	–2
	0	–1	–2	–1	0
	0	0	–1	0	0
X gradient	0	–1	0	1	0
	0	–1	0	1	0
	0	–1	0	1	0
	0	–1	0	1	0
	0	–1	0	1	0
Y gradient	0	0	0	0	0
	–1	–1	–1	–1	–1
	0	0	0	0	0
	1	1	1	1	1
	0	0	0	0	0

region of constant intensity. The effects are relatively small and are limited to the edges of the circle. The bottom of Fig. 12.2 shows the effect on a very noisy image of the same object. In this case, the effect is much more pronounced.

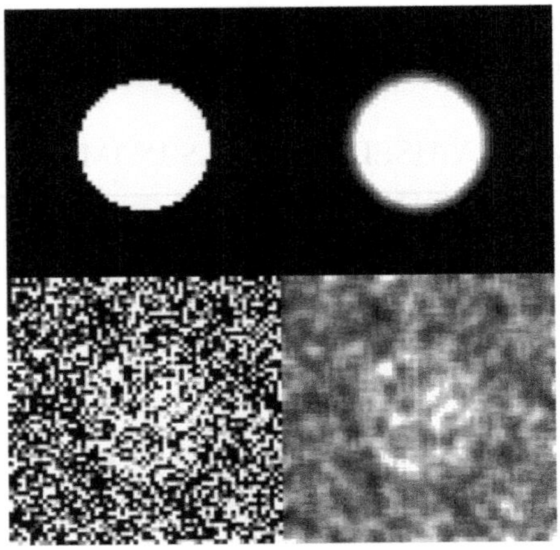

FIG. 12.2. The upper left quadrant shows a circular region of constant intensity; the upper right quadrant shows the effect of applying a 5-by-5 smoothing kernel; the lower left quadrant shows a very noisy version of the circular region; the bottom right quadrant shows the effect of applying a 5-by-5 smoothing kernel.

A common method of specifying a low-pass filter in the frequency domain is given by the Butterworth filter:

$$H(k) = \frac{1}{\sqrt{(1+(k/k_0)^{2n})}} \tag{12.16}$$

The Butterworth filter has two parameters, k_0 and n. The parameter k_0 is the cut-off frequency and n is the order. The filter reaches the value $1/\sqrt{2}$ when the spatial frequency k is equal to k_0. The parameter n determines the rapidity of the transition between the pass-zone and the stop-zone.

The filter is sometimes shown (Fig. 12.3) in terms of the square of the filter, $1/(1 + (k/k_0)^{2n})$. In that case, the filter reaches the value of half at the cut-off frequency. Confusion can arise since sometimes the filter itself is defined as the square value, i.e. without the square root. Furthermore, the filter is sometimes defined with an exponent of n instead of $2n$.

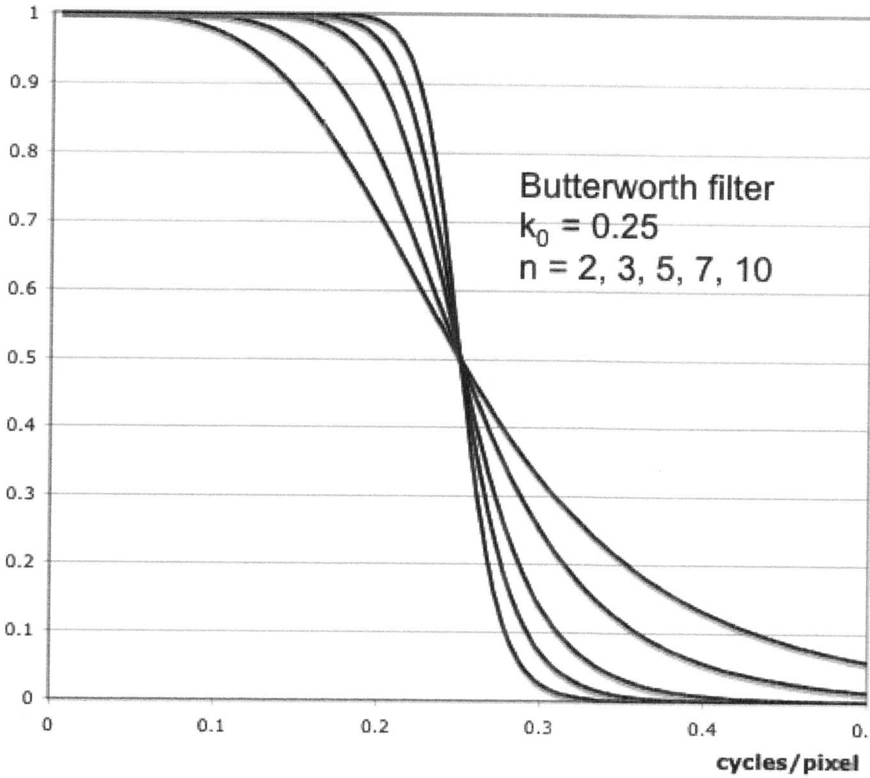

FIG. 12.3. This figure shows the square of the Butterworth filter in the spatial frequency domain with a cut-off equal to 0.25 cycles/pixel and n equal to 2, 3, 5, 7 and 10.

12.3.5.2. High-pass and edge-detection

Edges, and more generally detail, in an image are encoded with high spatial frequencies. Large regions of constant intensity are coded with low spatial frequencies. Thus, a high-pass filter will tend to emphasize the edges in an image.

A common method of implementing a high-pass filter is as a convolution with a small kernel. Table 12.3 shows a 5-by-5 sharpening kernel. The top of Fig. 12.4 shows the effect of applying this filter to a circular region with smooth edges. A high-pass filter will also emphasize pixel to pixel noise. The bottom of Fig. 12.4 shows the same circular object with a small amount of noise. Applying the high-pass filter to this image results in an image where the noise predominates. In general, a high-pass, edge-detection filter is used when the starting image has low noise.

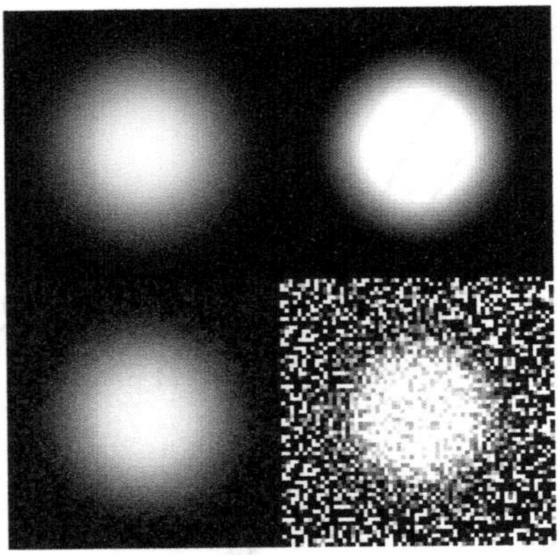

FIG. 12.4. The upper left quadrant shows a smooth circular region; the upper right quadrant shows the effect of applying the 5-by-5 sharpening kernel; the lower left quadrant shows a circular region with a small amount of noise; the bottom right quadrant shows the effect of the sharpening kernel on amplifying noise.

12.3.5.3. Unsharp masking and retinal ganglion cells

A band-pass filter, which passes intermediate frequencies and stops both low and high frequencies, will both smooth and edge enhance. It will smooth over a short range of pixels and it will enhance edges between regions at a slightly longer distance. As the low-pass and high-pass filters, this filter is often implemented in the space domain by convolution with a small kernel. Table 12.3 shows a band-pass kernel. The pixels near the centre of the kernel will be added together; those pixels will smooth the highest frequencies in the image. An annulus of pixels at a greater distance is subtracted from the centre pixels. This part of the kernel will emphasize edges that exist over this distance.

Two dimensional kernels with this form are sometimes called 'Mexican hat kernels' because the space of the kernel looks something like a Mexican hat. A smooth Mexican hat filter can be made from the difference of two Gaussians. This same operation is also called unsharp masking due to the similarity of a film processing method of that name.

High latitude means that images with a large dynamic range, with bright sunlight and shadows, can be displayed. Since film does not have the dynamic range of the human eye, the only way to obtain the highest latitude is to have low contrast. Even worse, a printed image, even a glossy image, has a dynamic range many orders of magnitude lower than the human eye. In an unsharp masked image, the intensity contrast over the range of the unsharp kernel is recovered from a low contrast original. The contrast between more distant objects is reduced. Although markedly altered, unsharp masked images often look verisimilar to the natural scene. In fact, they sometimes look 'better' than the natural scene. Unsharp masking is used extensively in 'coffee table books'.

In order to understand why these distorted images look verisimilar to the original scene, it is useful to consider the first ganglion cells in the retina. The first ganglion cells in the retina have a centre/surround organization. There is a small region on the retina where the cells have an excitatory input to a cell. There is a larger surrounding region where the cells have an inhibitory input to a cell. What is coded in the optic nerve is the relation of the small excitatory regions to the surrounding inhibitory regions. The ganglion cell processing is very much like unsharp masking. The brain then decodes these inputs to make a final impression of what the scene shows.

One important feature of human vision is its ability to discount illumination. To a very large extent, humans see the same colours under a wide variety of illuminations. The visual system discounts illumination. In the scene with bright sunlight and shadows, the human is aware of the illumination, but within a region, objects are still recognized even though the light reflecting from them may be orders of magnitude different depending on what part of the scene they are in. It is likely that this centre/surround mechanism is key for discounting illumination.

The important thing to remember in terms of image processing is that even though an unsharp masked image may appear to humans as verisimilar to the natural scene, it has been considerably altered from the original scene.

12.3.6. Deconvolution

Deconvolution is the process of undoing convolution. If we know the input, $f(t)$, to a system, $h(t)$, and the output, $g(t)$, from the system, then deconvolution is the process of finding $h(t)$ given $f(t)$ and $g(t)$. In the frequency or spatial frequency domain, deconvolution is straightforward — a simple division:

$$H(\omega)=\frac{G(\omega)}{F(\omega)} \tag{12.17}$$

There is a problem with this simple deconvolution. If $F(\omega)$ is zero for any ω, then $H(\omega)$ is infinite. Furthermore, since nuclear medicine deals with real systems that have noise, any time $F(\omega)$ is small, then any noise in $G(\omega)$ is amplified. The solution is to filter the simple deconvolution by a filter that depends on the power in $F(\omega)$.

Since signals usually have less power at high frequency or high spatial frequency, the first practical deconvolution was performed by simple high-pass filtering. A Metz filter (Section 12.3.7.2) can be used to emphasize the mid-frequency range while attenuating the high frequencies. If the statistical properties of the signal and noise are known or can reasonably be approximated, then a Wiener filter (Section 12.3.7.3) can be used.

12.3.7. Image restoration filters

Some filters attempt to restore degraded images. Degradations can come from many sources, but the most important for this book is degradations caused by imaging instruments such as gamma cameras. If the imaging system is linear-shift-invariant, then it can be modelled by convolution. A restoration filter needs to undo the effect of the imaging system, deconvolving for the system function. Even in the case where the system is not linear or shift-invariant, it is often possible to ameliorate the effects of a system with a filter.

12.3.7.1. Ramp

The effect of a projection/back projection is to smooth an object (see Section 12.3.5.1). In polar coordinates, the system function is given by the simple equation:

$$H(k,\theta)=\frac{1}{k} \tag{12.18}$$

It should be noted that the system function is not a function of θ. The higher frequencies in the image are reduced by a factor proportional to their spatial frequencies.

The obvious method of restoring the original image is to multiply it by a restoration filter given by:

$$G(k,\theta)=k \tag{12.19}$$

The restoration filter increases in magnitude in proportion to the spatial frequency. Owing to the shape of the filter, it is called a ramp filter. A ramp filter reverses the projection/back projection operation.

12.3.7.2. Metz

If the noise in the detected signal is uniform, then a restoration filter will alter the noise spectrum. For small restorations, the signal to noise may still be favourable, but for frequencies with large restorations, the noise may dominate. The best result is often to restore small effects completely, but restore large effects less. The Metz filter combines these two goals into a single filter. For any system function, $H(k, \theta)$, the Metz filter is given by:

$$G(k, \theta) = H^{-1}(k, \theta)(1-(1-H(k, \theta)^2)^x) \quad (12.20)$$

The system function is often written as MTF(k, θ) to emphasize that it is an MTF.

The first term in this equation, $1/H(k, \theta)$, reverses the effect of the system. When $H(k, \theta)$ is nearly one, the second term is about equal to one; when $H(k, \theta)$ is nearly zero, the second term is about equal to zero; at intermediate values, the second term transitions smoothly between these two values. In Fig. 12.5, a simulated system function is shown by the dotted lines. Four Metz filters with different X parameters are shown. At low frequencies, the Metz filter counteracts the effects of the imaging system. At high frequencies, it takes on the character of a low-pass filter. The transition between characters is controlled by the parameter X.

12.3.7.3. Wiener

Wiener filtering is used when the statistical properties of the signal and of the noise are known. A function known as the power spectral density, $S(\omega)$, gives the expected amount of power in a signal as a function of frequency. 'Power' means that the function is related to the square of the magnitude of a signal. The Wiener filter is given by:

$$H(\omega) = \frac{S_f(\omega)}{S_f(\omega) + S_n(\omega)} \quad (12.21)$$

where $S_f(\omega)$ is the power spectral density of the signal $f(t)$, and $S_n(\omega)$ is the power spectral density of the noise $n(t)$. Under the proper circumstances, it can be shown that the Wiener filter is the optimal estimate of a process $f(t)$ given the noisy data, $f(t) + n(t)$.

For those frequencies where the signal is much larger than the noise, the Wiener filter is equal to one. For those frequencies where the noise is much larger than the signal, the Wiener filter is equal to zero. The Wiener filter transitions smoothly from the pass-zone to the stop-zone when there is more noise in the data than signal. The Wiener filter provides a solid mathematical basis for the heuristic that has already been used several times.

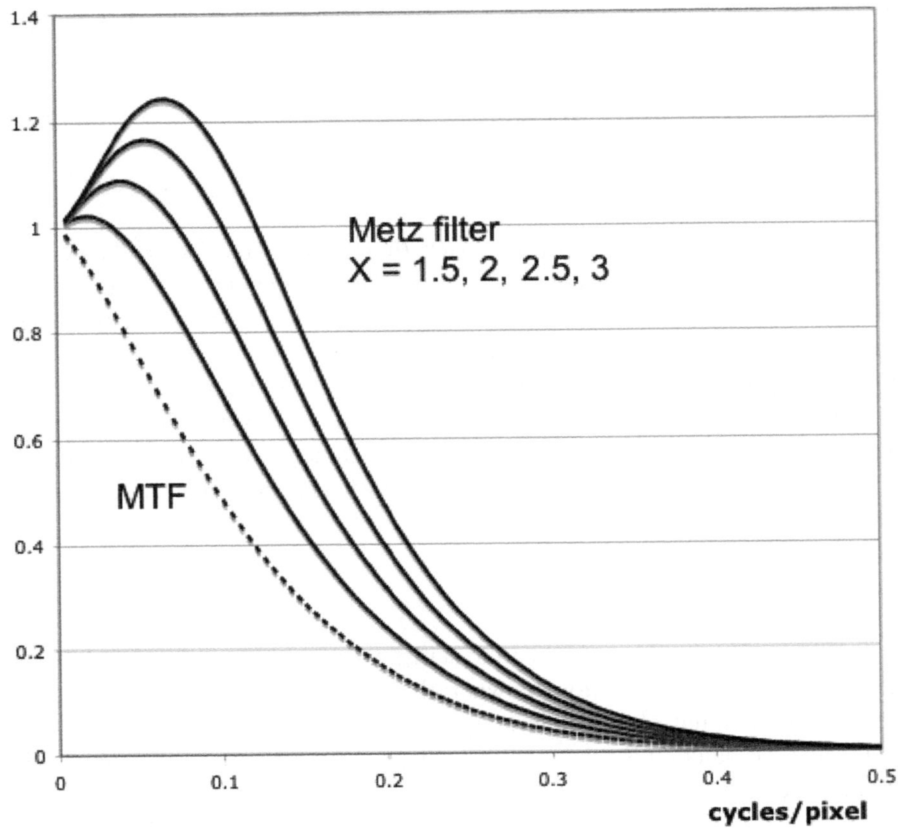

FIG. 12.5. The dotted line shows a simulated modulation transfer function (MTF). The four solid lines show Metz filters for X equal to 1.5, 2, 2.5 and 3.

12.3.8. Other processing

This chapter is predominantly concerned with linear-time-invariant or linear-shift-invariant signals. Fourier transform methods can be applied to these signals. Fourier transforms provide considerable insight and using the fast Fourier transform algorithm, they are very efficient computationally. However, the types of system that can be modelled are limited. Any real imaging system typically has an FOV. At the edge of the FOV, the system is no longer shift-invariant. Furthermore, attenuation, scatter and depth dependent resolution cannot be modelled using linear-shift-invariant systems.

12.3.8.1. Linear systems

Linear systems without being limited to time- or shift-invariance are much more powerful. Limited FOV, attenuation, scatter and depth dependant resolution can all be modelled. In addition, non-standard imaging geometries such as limited angle tomography can be modelled.

The relation between the input and the output of a linear system is equivalent to a system of simultaneous equations. An efficient method of writing a set of simultaneous equations is to use matrix algebra:

$$\boldsymbol{g} = \boldsymbol{H f} \tag{12.22}$$

where $\boldsymbol{f}$ and $\boldsymbol{g}$ are vectors and $\boldsymbol{H}$ is a matrix.

In the case of an image or especially a volume, the vectors $\boldsymbol{f}$ and $\boldsymbol{g}$ represent the whole image or volume. In the case of a 256 × 256 × 256 volume, each vector has 16 mega elements. The matrix, which is 16 mega elements by 16 mega elements, has 256 tera elements. It should be obvious that this simple expression hides a great deal of complexity. However, it does bring out the analogy between linear systems and linear-time-invariant or linear-shift-invariant systems.

12.3.8.2. Non-linear systems

The output of a linear system to the sum of two signals is equal to the sum of the outputs from each signal separately. Furthermore, the output of a system to a signal multiplied by a number is equal to the output due to the signal multiplied by the same number. Real systems are almost never linear. There is usually some size input for which real systems break down. Above that magnitude, increases in the input are not reflected in the output.

Non-linear systems are very difficult to deal with. Consequently, non-linear systems are often modelled by restricting their application to the range where the system is nearly linear. Other times, it is possible to transform the input or the output in order to make the system behave approximately as a linear system. There are methods of dealing with non-linear systems, but since they are usually very difficult, non-linear systems are rarely modelled directly.

12.4. DATA ACQUISITION

Modern nuclear medicine data acquisition devices have considerable computing power embedded in them that performs many tasks to improve images. These interesting and important computing functions are described in Chapter 11. This chapter will treat the camera as a black box and begin with the data that have already been obtained.

12.4.1. Acquisition matrix size and spatial resolution

For planar imaging, the FOV of a camera is divided into discrete elements that correspond to the pixels in a digital image. Most commonly, a rectilinear system is used and the digital image is analogous to a matrix (see Section 12.2.3.2). The size of the FOV and the spatial resolution of the camera are important in choosing the size of the matrix to be used. If the imaging device is modelled as a linear space-invariant system, then to avoid aliasing, the sampling should be twice the highest frequency that can be recorded by a camera (see Section 12.3.2). For example, if the FOV of the camera is 40 cm and the camera can detect frequencies up to 2 cycles/cm, then the matrix should have:

2 samples/cycle × 2 cycles/cm × 40 cm = 160 samples

For a normal large FOV Anger camera, a 256 × 256 matrix is more than sufficient for most imaging situations. For clinical imaging, a 128 × 128 matrix is often sufficient. For rapid dynamic images that are often very count limited, a 64 × 64 matrix can be used. For vertex to thigh imaging, a 1:4 matrix, e.g. 256 × 1024, will cover 160 cm. When including the legs, a non-power-of-two matrix may be most logical.

In the past, when memory and storage were more expensive, there was emphasis on using as small a matrix as possible. Currently, it makes more sense to emphasize that the use of a matrix is large enough to be sure that imaging is not limited.

12.4.2. Static and dynamic planar acquisition

Many of the processes of interest in nuclear medicine involve a changing pattern of distribution of a radiopharmaceutical in the body. In matrix-mode, a sequence of images collected over time allows visualization and measurement of the biodistribution of the radiopharmaceutical. Each image is often called a frame of data. Collection of a sequence of images over time is called a dynamic acquisition; collection of an image at a single point in time is called a static acquisition. Several static images may be grouped together as a multiple static acquisition; however, the difference between multiple static and dynamic acquisitions is largely conceptual. A dynamic acquisition that shows vascular delivery of the radiopharmaceutical is often called a flow study. A dynamic collection during renal scintigraphy is called a renogram.

The FOV of the gamma camera is divided into discrete rectangles that are generally square. Each rectangle corresponds to one pixel in the image and each is represented by one memory location. The size of the rectangles puts a limit on the resolution that is achievable, but the rectangle size is not the resolution. The resolution of a gamma camera is the overall ability to resolve adjacent sources of activity. For each scintillation, the gamma camera identifies the rectangle corresponding to the location of the scintillation and increments the corresponding memory location.

12.4.3. SPECT

SPECT is typically performed by acquiring a sequence of 2-D projection images from multiple different angles around the object. The 2-D projections can be represented by $p[x', z]$, where z is the axial position. The set of projections can be described by a 3-D function, $p(x', z, \theta)$, where θ is the position of the camera head.

12.4.3.1. Sinogram

The first step in reconstruction is to reformat the data in terms of axial position. The reformatted data can be described by a function $p'(x', \theta, z)$. For each axial position, the set of data, $p'(x', \theta)$, has one line of data from each of the projections. Each of the images, $p'(x', \theta)$, is called a sinogram. The x position in the sinogram is the same x position in the raw projection image. The other position θ is the projection angle. A point source will project onto all of the projection images at the same axial position z. The position of the point source in the θ direction will form a sine wave in the sinogram, hence the name sinogram.

The sinogram is not a particularly useful way of visualizing the data, but it is frequently available in nuclear medicine systems. It allows easy identification of a few artefacts. If there is patient movement during data acquisition, there will be discontinuity of the sine waves at the projection angle θ where the movement occurred. A detector problem will show up as straight lines of different intensity in the θ direction. Intermittent problems with the whole detector will result in straight lines in the x direction.

12.4.3.2. Sampling requirements in SPECT

Each of the sinograms $p'(x', \theta)$ needs to be reconstructed into an image $f(x, y)$ where x and y are coordinates with respect to the patient. The two types of reconstruction are Fourier reconstruction and iterative reconstruction. For an adequately sampled linear-shift-invariant system, Fourier reconstruction is exact in the absence of noise. Iterative reconstruction is approximate, but allows a much more realistic model of the system including attenuation, scatter, depth dependent collimator resolution and non-regular sampling.

A good heuristic is that in order to reconstruct N points in a slice, N data samples are needed. Sampling tangentially to the slice is reasonably straightforward; x' should have the same sampling rate as x and y. A good rule for Fourier reconstruction is that the number of angular samples over 180° is $\pi n/2$, where n is the samples along x'. This number is equal to the heuristic within a factor of $\pi/2$.

Resolution recovery is used in several nuclear medicine applications. Resolution recovery improves 'resolution' compared to the heuristic given above. The extra information required to improve 'resolution' is supplied by an a priori assumption, e.g. the edges in the image are relatively sharp. If the object does not meet the a priori assumption, then artefacts will be produced. As long as the assumptions are reasonable, resolution recovery can be useful.

Well designed collimators always involve a trade-off between resolution and sensitivity. A good heuristic is that for optimal detection, the collimator should have a resolution equal to the object to be detected. However, there may be more than one resolution object to be detected. For example, in cardiac imaging, if a low resolution initial image is needed for positioning followed by a high resolution image, then using resolution recovery for the higher resolution image can ameliorate use of a lower resolution collimator.

Each tomographic data sample includes a combination of the voxels in the object. For detectors that measure energy, so-called square-law detectors, the data samples will always be sums of object voxels. (This limitation is not true for MRI where the data have both an amplitude and a phase.) For tomographic data, square-law detectors always oversample low frequencies and reconstruction must amplify the high frequencies compared to the low frequencies. Typically, noise is relatively uniform in the projection space. Thus, after reconstruction, the high frequency noise is amplified with respect to the low frequency noise.

12.4.4. PET acquisition

The detectors in positron cameras detect single photons. However, the key feature of a positron camera is detection of pairs of photons in coincidence. Two photons detected in coincidence define an LOR. The LORs are then reconstructed into a 3-D image set. At that point, PET and SPECT data are the same.

Most PET cameras are made from individual crystals. The 'singles' data from the crystals can be recorded in a number of memory elements equal to the number N of crystals. $N \times N$ memory elements are needed for the LORs. (Some potential LORs do not traverse the imaging area.) The crystals in a PET camera make a 2-D array. In the usual cylindrical geometry, the two dimensions are the axial slice and the angular position around the ring of detectors for that slice. LORs are defined by two crystals and four dimensions. To limit the electronics, early PET cameras limited coincidences to a single slice. In that case, the slice position is the same for both detectors, and the LORs are 3-D, one axial and two angular position dimensions.

12.4.4.1. '2-D' and '3-D' volume acquisition

Modern PET cameras can collect oblique LORs, greatly increasing sensitivity. Some cameras have axial collimators that can be inserted or retracted while others do not. Imaging with axial collimation involves a 3-D object, a 3-D image and 3-D data. Strangely, this type of imaging is called '2-D'. Imaging without axial collimation again involves a 3-D object and a 3-D image, but in this case the data are 4-D. This type of imaging is called '3-D' imaging. Although these terms for imaging with and without axial collimation do not make sense, they are imbedded in the PET vernacular.

12.4.4.2. Time of flight

Time of flight (TOF) imaging measures the difference in arrival of the two photons at the two detectors. The difference in arrival time can be used to position the annihilation event along the LOR. This provides an additional dimension to the data. Non-axial collimation, TOF PET data are 5-D. The resolution along the LOR is relatively low. The speed of light is about 3×10^8 m/s, which means that in 1 ns photons travel 30 cm. Current detectors can detect differences in arrival of about 0.5 ns. Thus, improvements using TOF detection are important, but modest.

12.4.4.3. Sampling requirements in PET

Data sampling is determined by the LORs. Determination of the resolution of a PET camera is complicated, but once it is determined, then the size of the 3-D reconstructed matrix is the same as in SPECT imaging. Non-axial collimation PET adds oblique data with the potential of increasing the axial dimension. In practice, the axial resolution is similar for axial collimation and non-axial collimation PET. With cylindrical cameras, there are many more oblique angles that can be detected on the centre slice than on end slices. In non-axial collimated PET, the end slices are noisy. The non-uniform axial sampling is ameliorated in whole body imaging by overlapping the different bed positions, so that the data are more uniformly sampled.

12.4.5. Gated acquisition

Data acquisition can be gated to a physiological signal. Owing to count limitation, if the signal is repetitive, the data from multiple cycles are usually summed together. A gated, static acquisition will have three dimensions — two spatial and one physiological. A gated dynamic acquisition will have four dimensions — two spatial, one time and one physiological. A gated SPECT or PET acquisition will have a different four dimensions — three spatial and one physiological. A gated dynamic SPECT or PET acquisition will have five dimensions — three spatial, one time and one physiological. Count limitations will often limit the dimension and/or the elements per dimension.

12.4.5.1. Cardiac-synchronized

Gated cardiac studies were one of the early studies in nuclear medicine. Usually, the electrocardiogram is used to identify cardiac timing. The R-wave is relatively easy to detect; however, at least a little knowledge of electrocardiography is useful to know how to select electrode positions where the R-wave has a high amplitude.

The timing of events during the cardiac cycle changes with cycle length. The timing of contraction, systole, and the initial part of relaxation, diastole, change relatively little as cycle length changes. Most of the change in cycle length is a change in diastasis, the period between early relaxation and atrial contraction. Atrial contraction, atrial systole, has a relatively constant relation to the end of the cardiac cycle, i.e. to the next R-wave. Thus, for evaluation of systolic and early diastolic events, it is better to sum cycles using constant timing from the R-wave than to divide different cycles using a constant proportion of the cycle length.

However, with constant timing, different amounts of data are collected in later frames depending on the number of cycles that are long enough to reach those frames. Normalizing the data for the acquisition time for each frame can ameliorate this problem. Alternatively, systole and early diastole can be gated forwards in time from the preceding R-wave, and atrial systole can be gated backwards in time, joining the two during diastasis.

12.4.5.2. Respiratory-synchronized

Gating to respiration has been used in nuclear medicine both to study respiratory processes and to ameliorate the blurring caused by breathing. The signal to use for respiratory synchronization is not clear-cut. Successful synchronization has been based on different types of plethysmography, on chest or abdominal position changes, and on image data. Plethysmography, measurement of the air in the lungs, can be measured by integrating the rate of airflow or indirectly from changes in temperature of the air as it is breathed in and out. Corrections often need to be made for errors that build up from cycle to cycle. Electrical impedance across the chest changes with respiration and can also be used to detect lung volume changes. Chest and abdominal motion often correlate with the respiratory cycle, although detection of motion and changes in diaphragmatic versus chest wall breathing also pose problems. Although difficult to measure, respiratory gating is becoming more common, particularly in quantitative and high resolution PET applications.

12.4.6. List-mode

Instead of collecting data as a sequence of matrixes or frames, the position of each scintillation can be recorded sequentially in a list. Periodically, a special mark is entered into the list that gives the time and may include physiological data such as the electrocardiogram or respiratory phase.

List-mode data are generally much less efficient than matrix-mode data. For example, a 1 million-count frame of data collected into a 256 by 256 matrix will require 64 000 memory locations. In list-mode, the same data will require 1 million memory locations. The reason list-mode is so much less efficient is that it contains extra information, namely the order of arrival of each of the scintillations. However, this information is not interesting. List-mode is more efficient than matrix-mode when there is less than one event per pixel on average.

Since an image with one event per pixel is very noisy, it may be surprising that list-mode is useful at all. However, with gated images, multiple individual images are added to produce one output image. List-mode can be more efficient if it is necessary to keep the separate images that make up one gated image for post-processing.

12.5. FILE FORMAT

“The nicest thing about standards is that there are so many of them to choose from.” — Ken Olsen, founder of Digital Equipment Corp.

12.5.1. File format design

12.5.1.1. Interfaces

Standards have facilitated the rapid development of computers. Standards can be considered as part of a larger topic, interfaces. In general, an interface is a connection between an entity and the rest of the system. If each entity in the system has an interface and if the interface is the only allowed way for the other parts of the system to interact with an entity, then there is a great simplification of the overall system. Development of one part of a system depends only on that part and the interface. It does not depend on any other part of the system. Each part is greatly simplified because it is separate from the rest of the system. A system that is divided up into parts with well designed interfaces between the parts is called modular.

There was an early problem with development of complex systems. As more resources such as programmers were added to a task, very little additional output resulted. The problem was that as the task grew in complexity, more time was spent on communication between the programmers and less was spent on programming. At times, the output actually decreased as more resources were devoted to the task. The problem is that as a project becomes larger, the complexity tends to increase exponentially.

In a modular system, the details of each portion of the task are hidden from other parts of the task. Data hiding is a goal for good system design. Modularizing a task tends to linearize the complexity. As a task with complexity n becomes twice as large, the work becomes about $2n$, not n^2 or e^{2n} times as much.

File formats are a type of interface. They define how information will be transferred from one system to another. By having a well designed format, each system becomes separate. The goal of this section is to describe existing file formats and understand how they can simplify the design of a nuclear medicine information system.

12.5.1.2. Raster graphics

There are two general image formats — raster graphics and vector graphics. Vector graphics define an image in terms of components such as points, lines, curves and regions, e.g. polygons. Points, lines, curves and the boundaries of regions can have a thickness and a colour. The interiors of regions can have colours, gradients or patterns. For images that are made up of these types of component, the vector graphics description is very compact. Vector graphics can be smoothly expanded to any size. Modern type faces, which are defined in terms of vector graphics, can be expanded to any size while retaining their smooth outlines. Vector graphics are used extensively in gaming software. Vector graphics are not useful for nuclear medicine images, so this chapter will only describe raster graphics.

Raster graphics images are made up of a matrix of picture elements or pixels. Each pixel represents one small rectangle in the image. The primary data in nuclear medicine are numbers of counts. Counts are translated into a shade of grey or a colour for display. There are many ways the shade of grey or the colour of a pixel can be encoded. One factor is the number of bits that are used for each pixel. Grey scale images often have 8 bits or 1 byte per pixel. That allows 256 shades of grey, with values from 0 to 255.

A common way to encode colour images is in terms of their red, green and blue (RGB) components. If each of the colours is encoded with 8 bits, 24 bits or 3 bytes are needed per pixel. If each colour is encoded with 10 bits, 30 bits are needed per pixel. RGB encoding is typical in nuclear medicine, but there are other common encodings such as intensity, hue and saturation. For printing, images need to be encoded in terms of cyan, magenta and yellow. Use of more complicated encodings allows more vibrant images that use additional colours, e.g. black, orange, green, silver or gold.

12.5.1.2.1. Transparency

Graphic user interfaces often allow a composition of images where background images can be seen through a foreground image. In such cases, each pixel in an image may be given a transparency value that determines how much of the background will come through the foreground. A common format is 8 bits for each of the RGB colours and an 8-bit transparency value, giving a total of 32 bits per pixel.

12.5.1.2.2. Indexed colour

Typically, an image will include only a small number of the 16 million (2^{24}) possible colours in a 24 bit RGB palette. A common method of taking advantage of this is to use indexed colour. Instead of storing the colour in each pixel, an index value is stored. Typically, the index is 8 bits, allowing 256 colours to be specified. In the case where more than 256 colours are used, it is often possible to approximate the colour spectrum using a subset of colours. Sometimes, a combination of colours in adjacent pixels will help to approximate a broader spectrum. The algorithms for converting full spectrum RGB images into a limited palette are remarkably good. It often requires very careful inspection of a zoomed portion of the image to identify any difference.

For indexed colour, a colour palette is stored with the images. The colour palette has 256 colour values corresponding to the colours in the image. The 8 bit index value stored in a pixel is used to locate the actual 24 bit colour in the colour palette, and that colour is displayed in the pixel. Each pixel requires only 8 bits to store the index as opposed to 24 bits to store the actual colour. The colour palette introduces an overhead of 8 bits for each of the 256 colours, but since most images have tens of thousands or hundreds of thousands of pixels this overhead is small.

12.5.1.2.3. Compression

The information content of an image is often much smaller than the information capacity of an image format. For example, images often have large blank areas or areas that all have the same colour value. Therefore, it is possible to encode the image using less space. Improving efficiency for one class of images results in decreasing efficiency for another type of image. The trick is to pick an encoding which is a good match for the range of images that are typical of a particular application.

One of the simplest and easiest encodings to understand is run-length encoding. This method works well for images where large areas all have the same value, e.g. logos. Instead of listing each pixel value, a value and the number of times it is repeated are listed. If there are 50 yellow pixels in a row, rather than listing the value for yellow 50 times, the value '50' is followed by the value for yellow. Two values instead of 50 values need to be listed.

Another common encoding is called Lempel–Ziv–Welch (LZW) after its developers. It was originally under patent, but the patent expired on 20 June 2003, and it may now be used freely. The LZW algorithm is relatively simple but performs well on a surprisingly large class of data. It works well on most images and works particularly well on images with low information content, logos and line drawings. It is used in several of the file formats described below.

Both run-length encoding and LZW encoding are non-destructive or reversible or lossless coding; the original image can be exactly reconstructed from the encoded image. After destructive or irreversible or lossy coding, the original image cannot be exactly recovered from the coded image. However, much more efficient coding can be performed with destructive encoding. When non-destructive encoding results in reduction of the image size by a factor of 2–3, destructive encoding will often result in a reduction of the image size by a factor of 15–25. The trick is to pick an encoding system where the artefacts introduced by the coding are relatively minor.

The human visual system has decreased sensitivity to low contrast, very high resolution variations. Details, high resolution variations, are conspicuous only at high contrast. Discrete cosine transform (DCT) encoding takes advantage of this property of the visual system. DCT encoding is in the Joint Photographic Experts Group (JPEG) standard. The low spatial resolution content of the image is encoded with high fidelity, and the high spatial resolution content is encoded with low fidelity. The artefacts introduced tend to be low contrast detail. High contrast detail, and both low and high contrast features at low resolution are faithfully reproduced. Although there are artefacts introduced by the coding, they are relatively inconspicuous for natural scenes. For nuclear medicine images, the artefacts are more apparent on blown up images of text.

Wavelets are a generalized form of sinusoids. Some improvement in compression ratios at the same level of apparent noise can be obtained using wavelet-transform coding. The 'blocky' artefacts that degrade DCT images are not seen with wavelet-transform images. The JPEG 2000 standard uses wavelet-transform coding.

Non-destructive coding systems often make use of the fact that adjacent pixels are equal. The statistical noise in nuclear medicine images reduces the similarity of adjacent pixels, thus reducing the utility of non-destructive coding. Destructive coding may overcome this limitation, and since the statistical variations generally do not carry any significant information, image quality may not be greatly degraded.

12.5.2. Common image file formats

Common image file formats could be used for some or all of the raw nuclear medicine image data. In fact, they are rarely used for this purpose. However, secondary images, especially when used for distribution of image information, generally use these standard file formats. This section will sketch the format of these files, the advantages and disadvantages of the formats, and the typical uses of these formats.

12.5.2.1. Bitmap

Bitmap (BMP) is a general term that may refer to a number of different types of data. When used in reference to image formats, it may be used to mean a raster graphic as opposed to a vector graphic format, but it often refers to a Windows image format. The Windows file format is actually called a device independent bitmap (DIB). The external DIB file format is distinguished from various device dependent internal Windows BMPs. File name extensions .bmp and .dib are used for the BMP image file format. BMP may be used to imply an uncompressed format, but the DIB format defines several types of compression.

12.5.2.2. Tagged Image File Format

Aldus Corp. created the Tagged Image File Format (TIFF). Adobe Systems Inc., which merged with Aldus, now owns the copyright. The TIFF format can be used free of licensing fees. The TIFF format includes many types of image. Most commonly, it is used for 8 bit grey scale, 8 bit indexed colour or 24 bit RBG colour images. Images are typically compressed with the non-destructive LZW algorithm.

The TIFF format is often used for high quality single images that are non-destructively compressed. Although multiple other uses are possible, this application is often thought of as the strength of the TIFF format. The wide variety of options provided by the TIFF format is both a strength and a weakness. Few programs support all of these options, so that a TIFF image produced by one program may not be readable by another program. A particular weakness for nuclear medicine data is that the multiframe option is rarely supported.

12.5.2.3. Graphics Interchange Format

CompuServe, now a subsidiary of AOL, developed Graphics Interchange Format (GIF). GIF is a relatively simple indexed format with up to 8 bits per element. Up to 256 colours are selected from a 24 bit RGB palette. Pixels can be transparent, in which case the background colour is displayed. The 89a specification included multiple images that may be displayed as a cine. Compression is performed with the LZW algorithm.

As the GIF format is relatively simple yet quite useful, it is very commonly supported. All web browsers support this format. It is particularly good for storing low information content pictures such as logos and clip art. It is also good for images such as nuclear medicine images. Cine capability is widely supported, so it is very convenient for showing dynamic, gated and 3-D data.

12.5.2.4. Joint Photographic Experts Group

The Joint Photographic Experts Group developed the JPEG format in 1992. JPEG was approved in 1994 by the International Organization for Standardization as ISO 10918-1. The JPEG standard leaves some issues unspecified. The JPEG File Interchange Format (JFIF) clarifies these issues. To indicate that a JPEG image also follows this standard, it is sometimes called a JPEG/JFIF image, but generally JFIF is assumed. The JPEG standard is a reasonably simple, non-indexed grey scale and colour image format that allows adjustable destructive compression.

The JPEG coding tries to match the coding to human vision. It uses more precision for brightness than for hue. It uses more precision for low frequency data than for high frequency detail. The JPEG format is particularly good at compressing images of natural scenes, the type of images in routine photography. All web browsers support this format, and it is very widely used in general photography products. It is not a particularly good format for line drawings and logos; the GIF format is better for these types of image.

12.5.3. Movie formats

Multitrack movie formats allow audio and video to be blended together. Often, there are many audio tracks and many video tracks where the actual movie is a blending of these sources of information. A key part of a multitrack format is a timing track, which has information about how the tracks are sequenced and blended in a final presentation. However, in nuclear medicine, there is rarely a need for all of this capacity. Often, all that is needed is a simple sequence of images. A more complex movie format can be used for this type of data, but often a simpler format is easier to implement.

12.5.3.1. Image sequence formats

Of the image formats described so far, BMP, TIFF and GIF can be used to define an image sequence. An extension of the JPEG format, JPEG 2000, also allows image sequences. As with the multiframe version of BMP and TIFF formats, JPEG 2000 is relatively poorly supported. However, the multiframe version, 89a, of the GIF format is widely supported. It is supported by all web browsers and by almost all image processing and display programs. The GIF format is a logical choice for distribution of cine images.

12.5.3.2. Multitrack formats

There are several multitrack movie formats available. As newer formats that are still under development, they tend to be controlled by particular vendors. The AVI format is controlled by Microsoft, the QuickTime format is controlled by Apple, the RealVideo format is controlled by RealNetworks and the Flash format is controlled by Adobe.

12.5.4. Nuclear medicine data requirements

There are two types of information in a nuclear medicine study — image information and non-image information. As described in the previous section, there are several general purpose image and movie formats. Often, these formats lack capabilities that would be optimal for medical imaging. However, some formats, e.g. TIFF, are general enough to be used for the image portion of the study data. The advantage of using a widely accepted format is that it allows the great diversity of software available for general imaging to be adapted to medical imaging.

12.5.4.1. Non-image information

There is unique nuclear medicine information that must be carried along reliably with the images. This information includes: identification data, e.g. name, medical record number (MRN); study data, e.g. type of study, pharmaceutical; how the image was acquired, e.g. study date, view; etc. This information is sometimes called meta-information. Most general image formats are not flexible enough to carry this information along with the image.

12.5.4.1.1. American Standard Code for Information Interchange

Text information is usually and most efficiently coded in terms of character codes. Each character, including punctuation and spacing, is coded as a number of bits. Initially, 7 bits were used; 7 bits allowed $2^7 = 128$ codes, which were enough codes for the 26 small and 26 capital letters, plus a fairly large number of symbols and control codes. Using 1 byte (8 bits) for each character meant that the extra bit could be used for error checking using a scheme called parity. Error checking each character became more of an issue, so the American Standard Code for Information Interchange (ASCII) was extended to 8 bits, allowing the addition of 128 new codes. Characters in many of the Latin languages could be encoded using 8 bit ASCII.

12.5.4.1.2. Unicode

Computer usage has transcended national boundaries; it is now just as easy to communicate with the other side of the Earth, as to communicate within a single building. Internationalization has meant that a single server or client needs to be multilingual. Multilingual programming is now common. The ASCII code is not adequate for this task, so a multilingual code, Unicode, has superseded it. For example, the Java programming language specifies that programs be written in Unicode. Unicode uses 32 bits, and there are more than 100 000 character codes, including all common languages and many uncommon languages.

There are different ways to implement Unicode called, Unicode Transformation Format (UTF) encodings. One of the most common is a system called UTF-8, which uses a variable length coding system. Different numbers of bytes from one to four are used for different character codes. If the first byte in a code has a zero in the highest bit, then 1 byte is used. If the first byte in a code has a one in the highest bit, then subsequent bytes are used. The single byte codes are identical to the 7-bit ASCII codes. For a document that only includes the characters in the 7-bit ASCII system, encoding in ASCII and UTF-8 are identical. This results in an efficient encoding of Latin languages. In addition, any other Unicode character — Greek, Japanese or symbol — can be included occasionally in a predominantly Latin text while maintaining average coding efficiency.

12.5.4.1.3. Markup language

The difference between text editors and word processors is that the former largely edit the characters, while the latter define how the document will appear — the size of the text, the spacing, indentations, etc. For a word processor, the layout of the document as it is entered is the same as the layout when it is printed. In computer jargon, 'what you see is what you get' (WYSIWYG).

Markup has the advantage that it can be read by humans and can be edited with any text editor. Hypertext Markup Language (HTML), the language used by the World Wide Web, was originally a markup language. When HTML was first developed, text editors were used to writing it. Page layout capabilities were added, and now WYSIWYG editors are generally used. A markup language allows other types of information to be included. For example, one of the key features of HTML is that it includes hyperlinks to other HTML pages on the Internet.

12.5.4.1.4. Extensible Markup Language

Currently, a very popular and increasingly used method for encoding text information is a standard called Extensible Markup Language (XML). It provides a method of producing machine-readable textual data. For example, the Real Simple Syndication (RSS) standard is encoded with XML. A common use of the RSS standard is for web publishing. Almost all newspaper web sites use RSS to communicate with subscribers.

XML is not a markup language itself, but rather a general format for defining markup languages. It defines how the markup is written. A document is said to be 'well formed' if it follows the XML standard. If it is well formed, then the markup can be separated from the rest of the text. More information, provided by an XML schema, defines a particular markup language. The most recent version of HTML, XHTML, is fully compatible with the XML standard.

For nuclear medicine file formats, one of the key properties of XML is that it can be used to make text information readable by machine. For example, consider the following section of an XML document:

```
<patient>
    <name>
        <last>Parker</last>
        <first>Tony</fist>
    </name>
    <medical_record_number>10256892</medical_record_number>
</patient>
```

It would be straightforward for a computer to unambiguously determine the name and MRN from this document. Human readability and text editor processing of XML make it highly self-documenting. Owing to its wide acceptance in the computer world, XML is the most appropriate current format for storing non-imaging nuclear medicine information.

XML can be used to store numerical data as characters, e.g. <number>12.56</number>. The scalable vector graphics format for vector graphics is an XML language. However, XML is too inefficient for coding raster graphics image information.

12.5.4.2. Non-image information in general formats

(a) JPEG 2000: An extension of the JPEG format, JPEG 2000[1], has several very interesting capabilities. The JPEG 2000 standard is a general purpose image format; however, it was developed with medical imaging as a potential application. Since it uses wavelet-transform compression, the image quality at high compression ratios is considerably better than with DCT compression. JPEG 2000 allows multiple frames, so it could be used for cine; it has an advantage over GIF of providing good compression of information rich images such as natural scenes. However, the reason that it is included in this section is its capability to carry considerable meta-information in tight association with the image data. Although this format was defined nearly a decade ago, it has not found wide acceptance. The JPEG 2000 wavelet-transform coding is allowed in the Digital Imaging and Communications in Medicine (DICOM) standard, but mass market software, such as browsers, generally do not support this standard. It has

[1] http://www.jpeg.org/jpeg2000/

been included not as a solution that is currently practical, but rather as an example of a non-medical standard that has capabilities that could be applied to the particular requirements of nuclear medicine.

(b) Portable Document Format (PDF): PDF has not been used extensively in medicine, but it has many useful capabilities. It is intended for page layout. It can contain text, audio, images and movies. It is very widely supported, including support by all web browsers. Although uncommonly used in medicine, it has many properties that would make it an excellent format for distribution of image or cine information.

12.5.4.3. Image information

The image portion of a nuclear medicine study can be a sequence of static images, a dynamic series, a gated series, multiple dynamics series, raw data from a tomographic collection, reconstructed images, a dynamic series of tomographic collections, a dynamic series of gated tomographic collections, related tomographic datasets, curves from regions of interest, functional datasets derived from calculations on other datasets, etc. Thus, it should be clear that a rather general data format is needed. This section will present suggestions for general image format principles.

12.5.4.3.1. Types of data

The types of data in the last paragraph are not only long, but also incomplete; it is easy to think of other variations. A different data format for each type of data would add complexity without functionality. Therefore, it is logical to separate the interpretation of the data from the data. Whether the data represent a dynamic series or a set of static planar images does not have to be represented in the dataset. There is no necessity to have two dataset formats — one for a 128 × 128 image and another for a set of 128 curves with 128 points.

The interpretation of the data should be part of the non-image information, rather than part of the data. The data type should be determined by the format of the data elements. Separation of the interpretation from the data format will simplify processing and handling of the data. Adding two 128 × 128 images and adding two sets of 128 curves with 128 points is exactly the same operation; an add function should not concern itself with the interpretation of the data.

12.5.4.3.2. Data element

Several different data elements are needed. The raw data collection bins can generally be 8 or 16 bit unsigned integers, depending on whether a maximum of

255 or 65 535 counts per pixel are required. Since processed data may require a larger dynamic range, floating point or complex data may be appropriate in many circumstances. For some analysis, signed integers may be most appropriate. A region of interest can be represented as a single bit raster or with vector graphics.

The selection of a small number of data element formats would simplify the programming task. The trick is to limit the number of formats while maintaining all of the functionality. It would be logical to select some standard set of data element formats, especially a set of formats associated with a widely accepted data standard. However, it is not clear that such a standard exists. At a minimum, 16 bit integer and floating point formats are required. Probably signed and unsigned 8 and 16 bit formats should be included. It is not clear whether increasing complexity by including a vector graphics format is worth the added functionality.

12.5.4.3.3. Organization

A logical first level of organization of the image data is what will be called a 'dataset'. A dataset is an *n* dimensional set of data in which each of the data elements is the same format, e.g. 8 bit unsigned integer, 16 bit signed integer and IEEE 32 bit floating point data. The key characteristic is that the dataset is a sequence of identical elements. The dimensions do not need to be the same; 7 sets of 100 curves with 256 points is a 7 ×100 × 256 dataset.

The dataset should be the 'atom' of the data; there should not be a lower level of organization. The lower levels of organization will depend on the type of data and the dimensions. For example, a 256 × 256 × 256 tomographic volume could be considered as 256 axial slices of 256 × 256, but it would be equally valid to consider that dataset as 256 coronal slices of 256 × 256. The organization of a dataset consisting of 4-D PET lines of response will depend entirely on the reconstruction algorithm. The lower levels of organization depend on the non-image information and should not be part of the image data format.

12.5.5. Common nuclear medicine data storage formats

12.5.5.1. Interfile

The Interfile format has been used predominantly in nuclear medicine. The final version of Interfile, version 3, was defined in 1992. Although Interfile has been largely replaced by DICOM, it has some interesting properties. The metadata, encoded in ASCII, are readable and editable by any text editor. The lexical structure of the metadata was well defined, so that it is computer readable.

12.5.5.2. Digital Imaging and Communications in Medicine

DICOM is the most widely accepted radiological imaging format. DICOM began as ACR-NEMA, a collaboration between the American College of Radiology and the National Electrical Manufacturers Association. Version 3 of that standard changed the name to DICOM, in part to position the standard in a more international framework. DICOM is often thought of as a file format; however, the standard covers communication more broadly. It defines a transmission protocol, a query and retrieval standard, and workflow management. Unfortunately, it is overly complex, non-self-describing and has a heavy 2-D image bias (see Section 12.6.4).

12.6. INFORMATION SYSTEM

12.6.1. Database

Databases are one of the most common applications for computers. Web sites frequently depend heavily on a database. For example, search engines provide information from a database about the web and on-line retailers make extensive use of databases both for product searches and for information about customers and orders.

12.6.1.1. Table

Almost all of the information in databases is contained in tables. A table is a simple 2-D matrix of information. The rows in a table are called records and the columns are called fields. The rows refer to individual entities. In a customer table, the rows are customers. In an order table, the rows are orders. In a patient table, the rows are patients. In a study table, the rows are the studies. The columns are the attributes of an entity. In a patient database, the columns might be first name, middle name, last name, date of birth, MRN, etc. In an administer dose table, the columns might be radioisotope, radiopharmaceutical, administered dose, time of injection, etc.

A key concept is that tables are very simple — 2-D matrixes of data. The simplicity enables reliability. Almost all of the information in the database is in a simple format that is easy to back up and easy to transfer between databases.

There is usually one field in each table that is unique for each row. That unique field can be used for lookup. As that field allows access to the record, it is called an accession number. The accession number for a patient table might be the MRN; the accession number for a study table might be the study number;

the accession number for a customer table would be the customer number. Some databases use a combination of fields as the accession number, but the basic idea is the same, there must be something unique which identifies each row.

12.6.1.2. Index

An index provides rapid access to the records (rows) of a table. The index is separate from the table and is sorted to allow easy searching, often using a tree structure. The tables themselves are usually not sorted; the records are just added one after another. The index, not the table provides organization of the records for fast access. Indexes are one of the important technologies provided by a database vendor.

Indexes can be complicated and complex. However, there is no information stored in an index. An index can be rebuilt from the tables. In fact, when transferring data to a new database, the indexes are usually rebuilt. Indexes are important for efficiency, but the only information they contain is how to efficiently access the tables.

12.6.1.3. Relation

A key element of relational databases is relations. Relations connect one table to another table. Conceptually, the relations form a very important part of a database, and provide much of the complexity. However, the relations themselves are actually very simple. A relation between the patient table and the study table might be written:

patient.MRN = study.MRN

This says that the records in the patient table are linked to the records in the study table by the MRN field.

A physician thinks of the database in terms of patients who have studies that involve radiopharmaceuticals; however, a radiopharmacist thinks of isotopes and radiopharmaceuticals. The relations facilitate these different points of view. The physician can access the database in terms of patients and the radiopharmacist can access the database in terms of radiopharmaceuticals. The same tables are used; it is just the small amount of information contained in the relations that is different.

12.6.2. Hospital information system

A hospital information system is a large distributed database. Data come from clinical laboratories, nuclear medicine and financial systems, etc.

12.6.2.1. Admission, discharge, transfer

Most hospital information systems have an admission, discharge, transfer (ADT) database. The ADT system is the master patient identification system. Other systems use the ADT system for institution-wide, reliable, coordinated identification of each patient.

12.6.2.2. Information gateway

Often, the goal in a business or a division of a larger organization is to maintain a simple business model — a small product line, etc. The simple business model is often reflected in a simple database design. Medicine is very complex and the principle of a simple, well structured design is a dream. Generally, the different departments in the hospital have incompatible databases; even within one department, there may be incompatible systems. One method of ameliorating this problem is the development of Health Level Seven (HL7), a standard message format for communicating between systems in a hospital. However, the connections between systems still differ. If there are n systems, communication between them becomes a problem that grows proportionally to n^2.

An information gateway ameliorates this problem. The only task of an information gateway is to connect systems, translating messages so that they can be understood by other systems. Each system only needs to connect to the gateway, and the gateway communicates with all of the systems in the hospital. The growth in complexity tends to increase more like n than n^2.

12.6.3. Radiology information system

The radiology information system (RIS) supports scheduling, performing, reporting procedure results and billing. When nuclear medicine is a division of radiology, the RIS usually also functions as the nuclear medicine information system. However, nuclear medicine procedures have some unique characteristics, such as studies that extend over several days, which may not be well handled by a general purpose RIS.

12.6.4. Picture archiving and communication system

The image information from the imaging equipment is usually stored in a picture archiving and communication system (PACS) that is separate from but coordinated with the radiology/nuclear medicine information system. DICOM is the predominant standard for PACSs.

12.6.4.1. Study

The top level of organization in DICOM is the study. This level of organization comes from the organization of the health care system. Health care providers request services from radiology/nuclear medicine by a request for a consultation. Imaging or another service is performed and a report, ideally including image information, is returned to the provider. Each study is linked to a single patient, but a patient can have any number of separate studies.

12.6.4.2. Sequence

Sequence is the next level of organization in DICOM. Sequence comes from a sequence of images; for a volume of image data, sequence is the top level of organization. For example, it is common to collect a sequence of axial images. A more general name for this level of organization would be dataset.

12.6.4.3. Image

Originally, DICOM used a 2-D image as the basic atom. Other data structures are composed of a number of images. A considerable amount of metadata are included with each image, defining both the structure of the image, patient information, data collection information and relation of the image to other images. Somewhat more recently, a multiframe format was defined in DICOM. One file may contain information from a volume of data, from a time series, from a gated sequence, from different photopeaks, etc. Nuclear medicine tends to use the multiframe format much more commonly than other modalities. Describing the organization of a dataset in terms of multiple images is a particularly awkward feature of DICOM.

12.6.4.4. N-dimensional data

It is unfortunate that DICOM selected the image as the basic atom of organization (see Section 12.5.4.3.3). Even before it was introduced, it was apparent that an N-dimensional data model would be more appropriate. Nuclear medicine and MRI often dealt with 1-D curves, 2-D images, 3-D volumes, 4-D gated or dynamic volumes, etc. However, radiology tended to be film based, and volume data such as CT was anisotropic, so some of the early developers had an image based orientation.

12.6.5. Scheduling

Scheduling is a much more complicated task than it may initially seem. Several appointments in different departments may need to be scheduled at the same time. Sequencing of studies may be important, e.g. thyroid imaging should not be performed in proximity to iodinated contrast usage. Some studies require a prior pregnancy test or other laboratory values. Prior data need to be made available prior to performing some studies; for example, a prior electrocardiogram should be available before a myocardial stress study.

Since scheduling needs to be coordinated between departments and make use of hospital-wide information, it is logically a function of the hospital information system. However, scheduling involves many issues that are local to the modality, for example, availability of resources such as radiopharmaceuticals, staff and equipment. Often, scheduling is local, making use of humans to provide much of the hospital-wide coordination.

The scheduling system creates a 'worklist', a list of studies that need to be performed. Most modern imaging equipment can use a worklist provided by the RIS. When the technologist starts an imaging study, the appropriate patient is picked from the worklist. Selecting a patient from a list results in far fewer errors than re-entering all of the demographic identifier data for each study.

12.6.6. Broker

A broker is essentially an information gateway. The term 'information gateway' is generally used for a hospital-wide system. The term 'broker' is generally used when talking about a system within a radiology department. The broker handles incompatible details related to RIS, PACS and imaging devices that are local to radiology.

12.6.7. Security

Health information is private. Although the privacy of health information is often a relatively minor concern for the general public, it is a major concern for politicians, who make rules about the security needed for medical information. The theory of secure communication over insecure channels is thought to be a solved problem. However, vigilance is necessary, because the practical implementations of the theory often have security holes that hackers can exploit. Furthermore, with aggregation of private health information, the potential extent of a security breach becomes catastrophic. Security depends not only on computer systems, but also on humans who are often the weak link.

There needs to be a balance between the damage caused due to a security breach and the expense of the security measures. Often, relatively low damage situations are addressed with overly expensive systems, especially in terms of lost productivity. The dominant effect of security should not be to prevent authorized users from accessing information. Nuclear medicine tends to be a relatively low risk environment, so in most circumstances the balance should favour productivity over security.

BIBLIOGRAPHY

HUTTON, B.F., BARNDEN, L.R., FULTON, R.R., "Nuclear medicine computers", Nuclear Medicine in Clinical Diagnosis and Treatment (ELL, P.J., GAMBHIR, S.S., Eds), Churchill Livingston, London (2004).

LEE, K.H., Computers in Nuclear Medicine: A Practical Approach, Society of Nuclear Medicine, 2nd edn (2005).

PARKER, J.A., Image Reconstruction in Radiology, CRC Press, Boston, MA (1990).

PIANYKH, O.S., Digital Imaging and Communications in Medicine (DICOM): A Practical Introduction and Survival Guide, Springer, Berlin (2008).

SALTZER, J.H., KAASHOEK, M.F., Principles of Computer System Design: An Introduction, Morgan Kaufmann, Burlington (2009).

TODD-POKROPEK, A., CRADDUCK, T.D., DECONINCK, F., A file format for the exchange of nuclear medicine image data: a specification of Interfile version 3.3, Nucl. Med. Commun. **13** (1992) 673–699.

CHAPTER 13

IMAGE RECONSTRUCTION

J. NUYTS
Department of Nuclear Medicine and Medical Imaging Research Center,
Katholieke Universiteit Leuven,
Leuven, Belgium

S. MATEJ
Medical Image Processing Group,
Department of Radiology,
University of Pennsylvania,
Philadelphia, Pennsylvania,
United States of America

13.1. INTRODUCTION

This chapter discusses how 2-D or 3-D images of tracer distribution can be reconstructed from a series of so-called projection images acquired with a gamma camera or a positron emission tomography (PET) system [13.1]. This is often called an 'inverse problem'. The reconstruction is the inverse of the acquisition. The reconstruction is called an inverse problem because making software to compute the true tracer distribution from the acquired data turns out to be more difficult than the 'forward' direction, i.e. making software to simulate the acquisition.

There are basically two approaches to image reconstruction: analytical reconstruction and iterative reconstruction. The analytical approach is based on mathematical inversion, yielding efficient, non-iterative reconstruction algorithms. In the iterative approach, the reconstruction problem is reduced to computing a finite number of image values from a finite number of measurements. That simplification enables the use of iterative instead of mathematical inversion. Iterative inversion tends to require more computer power, but it can cope with more complex (and hopefully more accurate) models of the acquisition process.

13.2. ANALYTICAL RECONSTRUCTION

The (n-dimensional) radon transform maps an image of dimension n to the set of all integrals over hyperplanes of dimension (n – 1) [13.2]. Thus, in two dimensions, the radon transform of image Λ corresponds to all possible line integrals of Λ. In three dimensions, the radon transform contains all possible plane integrals.

The (n-dimensional) X ray transform maps an image of dimension n to the set of all possible line integrals. In all PET and in almost all single photon emission computed tomography (SPECT) applications, the measured projections can be well approximated as a subset of the (possibly attenuated) X ray transform, because the mechanical (SPECT) or electronic (PET) collimation is designed to acquire information along lines (the line of response (LOR), see Chapter 11). Consequently, reconstruction involves computing the unknown image Λ from (part of) its X ray transform. Figure 13.1 shows PET projections, which are often represented as a set of projections or a set of sinograms.

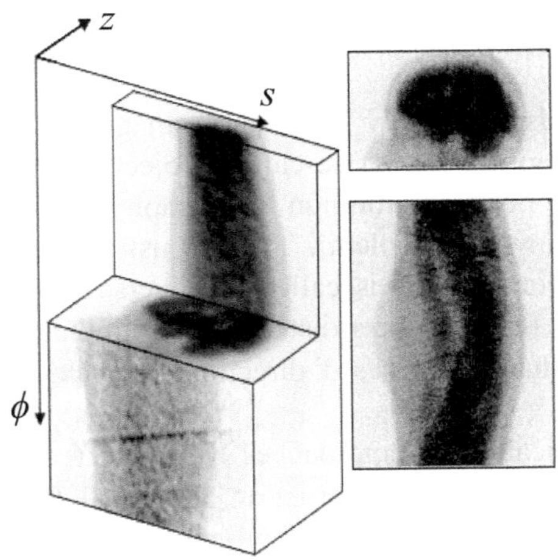

FIG. 13.1. The relation between projections and sinograms in parallel-beam projection. The parallel-beam (PET) acquisition is shown as a block with dimensions s, ϕ and z. A cross-section at fixed ϕ yields a projection; a cross-section at fixed z yields a sinogram.

An important theorem for analytical reconstruction is the central slice (or central section) theorem, which gives a relation between the Fourier transform of an image and the Fourier transforms of its parallel projections. Below, the central slice theorem for 2-D is found as Eq. (13.7) and the 3-D central section theorem as Eq. (13.29).

The direct Fourier method is a straightforward application of the central section theorem: it computes the Fourier transform of the projections, uses the central section theorem to obtain the Fourier transform of the image and applies the inverse Fourier transform to obtain the image. In practice, this method is rarely used; the closely related filtered back projection (FBP) algorithm is far more popular.

13.2.1. Two dimensional tomography

13.2.1.1. X ray transform: projection and back projection

In 2-D, the radon transform and X ray transform are identical. Mathematically, the 2-D X ray (or radon) transform of the image Λ can be written as follows:

$$\begin{aligned} \mathrm{Y}(s,\phi) &= \int_{-\infty}^{\infty} \int_{-\infty}^{\infty} \Lambda(x,y)\delta_{s=x\cos\phi+y\sin\phi}\, \mathrm{d}x\, \mathrm{d}y \\ &= \int_{-\infty}^{\infty} \Lambda(s\cos\phi - t\sin\phi,\ s\sin\phi + t\cos\phi)\, \mathrm{d}t \end{aligned} \tag{13.1}$$

where the δ function is unity for the points on the LOR (s, ϕ) and zero elsewhere. It should be noted that with the notation used here, $\phi = 0$ corresponds to projection along the y axis.

The radon transform describes the acquisition process in 2-D PET and in SPECT with parallel-hole collimation, if attenuation can be ignored. Assuming that $\Lambda(x, y)$ represents the tracer distribution at transaxial slice Z through the patient, then $\mathrm{Y}(s, \phi)$ represents the corresponding sinogram, and contains the z-th row of the projections acquired at angles ϕ. Figure 13.1 illustrates the relation between the projection and the sinogram.

The X ray transform has an adjoint operation that appears in both analytical and iterative reconstruction. This operator is usually called the back projection operator, and can be written as:

$$\begin{aligned} \mathrm{B}(x,y) &= \mathrm{Backproj}\ (\mathrm{Y}(s,\phi)) \\ &= \int_0^{\pi} \mathrm{d}\phi \int_{-\infty}^{\infty} \mathrm{Y}(s,\phi)\delta_{s=x\cos\phi+y\sin\phi}\, \mathrm{d}s \\ &= \int_0^{\pi} \mathrm{Y}(x\cos\phi + y\sin\phi, \phi)\, \mathrm{d}\phi \end{aligned} \tag{13.2}$$

The back projection is not the inverse of the projection, $B(x, y) \neq \Lambda(x, y)$. Intuitively, the back projection sends the measured activity back into the image by distributing it uniformly along the projection lines. As illustrated in Fig. 13.2, projection followed by back projection produces a blurred version of the original image. This blurring corresponds to the convolution of the original image with the 2-D convolution kernel $1/\sqrt{x^2+y^2}$.

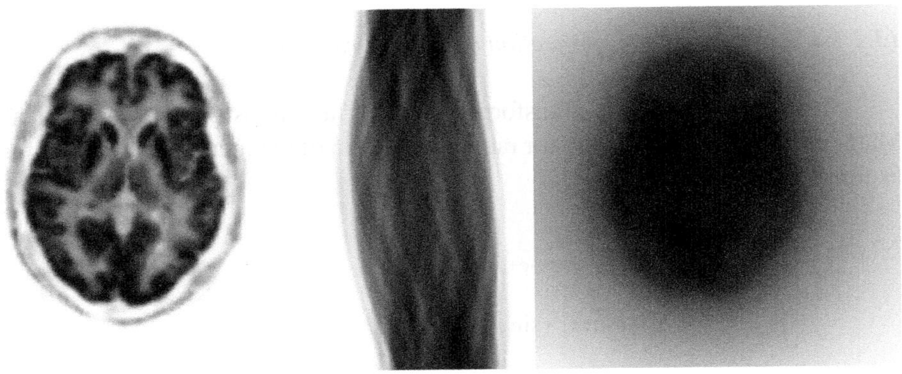

FIG. 13.2. The image (left) is projected to produce a sinogram (centre), which in turn is back projected, yielding a smoothed version of the original image.

13.2.1.2. Central slice theorem

The central slice theorem gives a very useful relation between the 2-D Fourier transform of the image and the 1-D Fourier transform of its projections (along the detector axis). Consider the projection along the y axis, $\phi = 0$, and its 1-D Fourier transform:

$$Y(s,0) = \int_{-\infty}^{\infty} \Lambda(s,t)\, dt \tag{13.3}$$

$$\begin{aligned}(\mathcal{F}_1 Y)(v_s,0) &= \int_{-\infty}^{\infty} Y(s,0) e^{-i2\pi v_s s}\, ds \\ &= \int_{-\infty}^{\infty} \int_{-\infty}^{\infty} \Lambda(s,t) e^{-i2\pi v_s s}\, dt\, ds\end{aligned} \tag{13.4}$$

and compare this to the 2-D Fourier transform of the image $\Lambda(x, y)$:

$$(\mathcal{F}_2\Lambda)(\nu_x,\nu_y)=\int_{-\infty}^{\infty}\int_{-\infty}^{\infty}\Lambda(x,y)\mathrm{e}^{-\mathrm{i}2\pi(\nu_x x+\nu_y y)}\,\mathrm{d}x\,\mathrm{d}y \tag{13.5}$$

Both expressions are equal if we set $\nu_y = 0$:

$$(\mathcal{F}_1\mathrm{Y})(\nu_s, 0) = (\mathcal{F}_2\Lambda)(\nu_x, 0) \tag{13.6}$$

$(\mathcal{F}_1\mathrm{Y})(\nu_s, 0)$ is the 1-D Fourier transform of the projection along the y axis and $(\mathcal{F}_2\Lambda)(\nu_x, 0)$ is a 'central slice' along the ν_x axis through the 2-D Fourier transform of the image. Equation (13.6) is the central slice theorem for the special case of projection along the y axis. This result would still hold if the object had been rotated or equivalently, the x and y axes. Consequently, it holds for any angle ϕ:

$$(\mathcal{F}_1\mathrm{Y})(\nu_s, \phi) = (\mathcal{F}_2\Lambda)(\nu_s\cos\phi, \nu_s\sin\phi) \tag{13.7}$$

13.2.1.3. Two dimensional filtered back projection

The central slice theorem (Eq. (13.7)) can be directly applied to reconstruct an unknown image $\Lambda(x, y)$ from its known projections $\mathrm{Y}(s, \phi)$. The 1-D Fourier transform of the projections provides all possible central slices through $(\mathcal{F}_2\Lambda)(\nu_x, \nu_y)$ if $\mathrm{Y}(s, \phi)$ is known for all ϕ in an interval with a length of at least π (Tuy's condition). Consequently, $(\mathcal{F}_2\Lambda)(\nu_x, \nu_y)$ can be constructed from the 1-D Fourier transform of $\mathrm{Y}(s, \phi)$. Inverse 2-D Fourier transform then provides $\Lambda(x, y)$.

However, a basic Fourier method implementation with a simple interpolation in Fourier space does not work well. In contrast, in the case of the FBP algorithm derived below, a basic real-space implementation with a simple convolution and a simple interpolation in the back projection works well. Inverse Fourier transform of Eq. (13.5) yields:

$$\Lambda(x,y)=\int_{-\infty}^{\infty}\int_{-\infty}^{\infty}(\mathcal{F}_2\Lambda)(\nu_x,\nu_y)\mathrm{e}^{\mathrm{i}2\pi(\nu_x x+\nu_y y)}\,\mathrm{d}\nu_x\,\mathrm{d}\nu_y \tag{13.8}$$

This can be rewritten with polar coordinates as:

$$\Lambda(x,y)=\int_{-\infty}^{\infty}\mathrm{d}\nu\int_{0}^{\pi}(\mathcal{F}_2\Lambda)(\nu\cos\phi,\nu\sin\phi)\mathrm{e}^{\mathrm{i}2\pi(x\nu\cos\phi+y\nu\sin\phi)}|\nu|\,\mathrm{d}\phi \tag{13.9}$$

Application of the central slice theorem (Eq. (13.7)) and reversing the order of integration finally results in:

$$\Lambda(x,y)=\int_0^{\pi}\mathrm{d}\phi\int_{-\infty}^{\infty}(\mathcal{F}_1\mathrm{Y})(\nu,\phi)|\nu|\mathrm{e}^{\mathrm{i}2\pi\nu(x\cos\phi+y\sin\phi)}\,\mathrm{d}\nu \tag{13.10}$$

which is the FBP algorithm. This algorithm involves the following steps:

(a) Apply 1-D Fourier transform to $\mathrm{Y}(s,\phi)$ to obtain $(\mathcal{F}_1\mathrm{Y})(\nu,\phi)$;
(b) Filter $(\mathcal{F}_1\mathrm{Y})(\nu,\phi)$ with the so-called ramp filter $|\nu|$;
(c) Apply the 1-D inverse Fourier transform to obtain the ramp filtered projections $\hat{\mathrm{Y}}(s,\phi)=\int(\mathcal{F}_1\Lambda)(\nu,\phi)|\nu|\mathrm{e}^{\mathrm{i}2\pi\nu s}\,\mathrm{d}\nu$;
(d) Apply the back-projection operator Eq. (13.2) to $\hat{\mathrm{Y}}(s,\phi)$ to obtain the desired image $\Lambda(x,y)$.

It should be noted that the ramp filter sets the DC component (i.e. the amplitude of the zero frequency) of the image to zero, while the mean value of the reconstructed image should definitely be positive. As a result, straightforward discretization of FBP causes significant negative bias. The problem is reduced with 'zero padding' before computing the Fourier transform with fast Fourier transform (FFT). Zero padding involves extending the sinogram rows with zeros at both sides. This increases the sampling in the frequency domain and results in a better discrete approximation of the ramp filter. However, a huge amount of zero padding is required to effectively eliminate the bias completely. The next paragraph shows how this need for zero padding can be easily avoided. It should be noted that after inverse Fourier transform, the extended region may be discarded, so the size of the filtered sinogram remains unchanged.

Instead of filtering in the Fourier domain, the ramp filtering can also be implemented as a 1-D convolution in the spatial domain. For this, the inverse Fourier transform of $|\nu|$ is required. This inverse transform actually does not exist, but approximating it as the limit for $\varepsilon \to 0$ of the well behaved function $|\nu|\mathrm{e}^{-\varepsilon|\nu|}$ gives [13.3, 13.4]:

$$\mathcal{F}^{-1}(|\nu|\mathrm{e}^{-\varepsilon|\nu|})=\frac{\varepsilon^2-(2\pi s)^2}{(\varepsilon^2+(2\pi s)^2)^2} \tag{13.11}$$

$$\approx-\frac{1}{(2\pi s)^2}\quad\text{for}\quad|s|\gg\varepsilon \tag{13.12}$$

In practice, band limited functions are always worked with, implying that the ramp filter has to be truncated at the frequencies $\nu=\pm1/(2\tau)$, where τ represents the sampling distance. The corresponding convolution kernel h then equals [13.3]:

$$h(s) = \mathcal{F}^{-1}(|\nu| b(\nu)) = \frac{1}{2\tau^2}\frac{\sin(\pi s/\tau)}{\pi s/\tau} - \frac{1}{4\tau^2}\left(\frac{\sin(\pi s/(2\tau))}{\pi s/(2\tau)}\right)^2 \tag{13.13}$$

with $b(\nu) = 1$ if $|\nu| \leq 1/(2\tau)$
$= 0$ if $|\nu| > 1/(2\tau)$

The kernel is normally only needed for samples $s = n\tau$: $h(n\tau) = 1/(4\tau^2)$ if $n = 0$, $h(n\tau) = 0$ if n is even and $h(n\tau) = -1/(n\pi\tau)^2$ if n is odd. The filter can either be implemented as a convolution or the Fourier transform can be used to obtain a digital version of the ramp filter. Interestingly, this way of computing the ramp filter also reduces the negative bias mentioned above. The reason is that this approach yields a non-zero value for the DC component [13.3]. When the filtering is done in the frequency domain, some zero padding before FFT is still recommended because of the circular convolution effects, but far less is needed than with straightforward discretization of $|\nu|$.

Although this is not obvious from the equations above, an algorithm equivalent to FBP is obtained by first back projecting $\mathrm{Y}(s, \phi)$ and then applying a 2-D ramp filter to the back projected image $\mathrm{B}(x, y)$ [13.4]:

$$\mathrm{B}(x,y) = \int_0^{\pi} \mathrm{Y}(x\cos\phi + y\sin\phi, \phi)\, \mathrm{d}\phi \tag{13.14}$$

$$(\mathcal{F}_2\Lambda)(\nu_x,\nu_y) = \sqrt{\nu_x^2+\nu_y^2}\,(\mathcal{F}_2\mathrm{B})(\nu_x,\nu_y) \tag{13.15}$$

This algorithm is often referred to as the ‘back project-then-filter’ algorithm.

FBP assumes that the projections $\mathrm{Y}(s, \phi)$ are line integrals. As discussed in Chapter 11, PET and SPECT data are not line integrals because of attenuation, detector non-uniformities, the contribution of scattered photons and/or random coincidences, etc. It follows that one has to recover (good estimates of) the line integrals by pre-correcting the data for these effects. However, a particular problem is posed by the attenuation in SPECT because, different from PET, the attenuation depends on the position along the projection line, precluding straightforward pre-correction. A detailed discussion of this problem is beyond the scope of this contribution, but three solutions are briefly mentioned here:

(a) If it can be assumed that the attenuation is constant inside a convex body contour, then FBP can be modified to correct for the attenuation. Algorithms have been proposed by Bellini, Tretiak, Metz and others; an algorithm is presented in Ref. [13.3].

(b) If the attenuation is not constant, an approximate correction algorithm proposed by Chang can be applied [13.5]. It is a post-correction method, applied to the image obtained without any attenuation correction. To improve the approximation, the attenuated projection of the first reconstruction can be computed, and the method can be applied again to the difference of the measurement and this computed projection.

(c) Finally, a modified FBP algorithm, compensating for non-uniform attenuation in SPECT, was found by Novikov in 2000. An equivalent algorithm was derived by Natterer [13.6]. However, because this algorithm was only found after the successful introduction of iterative reconstruction in clinical practice, it has not received much attention in the nuclear medicine community.

13.2.2. Frequency–distance relation

Several very interesting methods in image reconstruction, including Fourier rebinning, are based on the so-called frequency–distance relation, proposed by Edholm, Lewitt and Lindholm, and described in detail in Ref. [13.7]. This is an approximate relation between the orthogonal distance to the detector and the direction of the frequency in the sinogram. The relation can be intuitively understood as follows.

Consider the PET acquisition of a point source, as illustrated in Fig 13.3. Usually, the acquisition is described by rotating the projection lines while keeping the object stationary. However, here the equivalent description is considered, where projection is always along the y axis, and tomographic acquisition is obtained by rotating the object around the origin. Suppose that the point is located on the x axis when $\phi = 0$. When acquiring the parallel projections for angle ϕ, the point has polar coordinates (r, ϕ), with r the distance from the centre of the field of view (FOV) and ϕ the angle with the x axis. The distance to the x axis is $d = r\sin\phi$. The corresponding sinogram $Y(s, \phi)$ is zero everywhere, except on the curve $s = r\cos\phi$ (Fig. 13.3). The complete sinogram is obtained by rotating the point over 360° $\phi = -\pi\ldots\pi$. Consider a small portion of this curve, which can be well approximated as a tangential line segment near a particular point (s, ϕ),

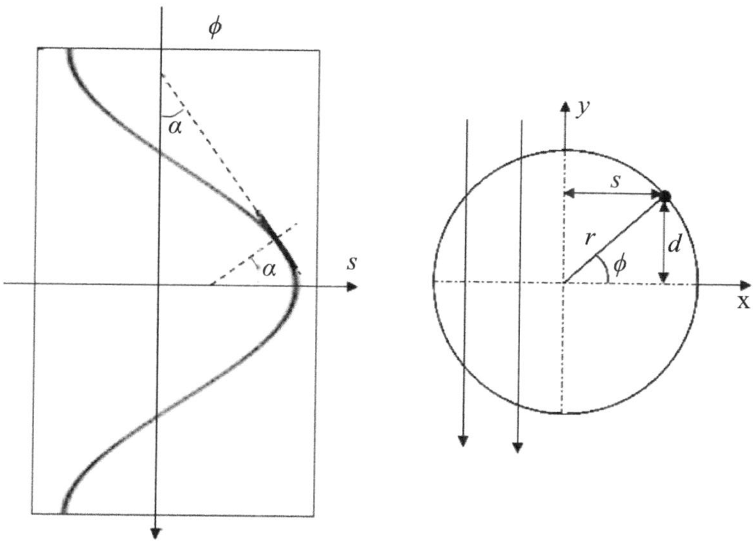

FIG. 13.3. The frequency–distance principle. Left: sinogram; right: vertical projection of a point located at polar coordinates (r, ϕ).

as illustrated in Fig. 13.3. In the 2-D Fourier transform of the sinogram, this line segment contributes mostly frequencies in the direction orthogonal to the line segment. This direction is represented by the angle α, given by:

$$\tan\alpha = \frac{\partial}{\partial\phi}(r\cos\phi) = -r\sin\phi = -d \tag{13.16}$$

Thus, in the 2-D Fourier transform $(\mathcal{F}g)$ (ν_s, ν_ϕ), the value at a particular point (ν_s, ν_ϕ) carries mostly information about points located at a distance $d = -\tan\alpha = -\nu_\phi/\nu_s$ from the line through the centre, parallel to the detector. This relation can be exploited to apply distance dependent operations to the sinogram. One example is distance dependent deconvolution, to compensate for the distance dependent blurring in SPECT. Another example is Fourier rebinning, where data from oblique sinograms are rebinned into direct sinograms.

13.2.3. Fully 3-D tomography

13.2.3.1. Filtered back projection

Owing to the use of electronic collimation, the PET scanner can simultaneously acquire information in a 4-D space of line integrals. These are the so-called LORs, where each pair of detectors in coincidence defines a single LOR. In this section, the discrete nature of the detection is ignored, since the analytical approach is more easily described assuming continuous data. Consider the X ray transform in 3-D, which can be written as:

$$\mathrm{Y}(\hat{\boldsymbol{u}}, \boldsymbol{s}) = \int_{-\infty}^{\infty} \Lambda(\boldsymbol{s} + t\hat{\boldsymbol{u}})\, \mathrm{d}t \tag{13.17}$$

where the LOR is defined as the line parallel to $\hat{\boldsymbol{u}}$ and through the point $\boldsymbol{s}$. The vector $\hat{\boldsymbol{u}}$ is a unit vector, and the vector $\boldsymbol{s}$ is restricted to the plane orthogonal to $\hat{\boldsymbol{u}}$, hence $(\hat{\boldsymbol{u}}, \boldsymbol{s})$ is 4-D.

Most PET systems are either constructed as a cylindrical array of detectors or as a rotating set of planar detector arrays and, therefore, have cylindrical symmetry. For this reason, the inversion of Eq. (13.17) is studied for the case where $\hat{\boldsymbol{u}}$ is restricted to the band Ω_{θ_0} on the unit sphere, defined by $|u_z| \leq \sin\theta_0$, as illustrated in Fig. 13.4. It should be noted that only half of the sphere is actually needed because $\mathrm{Y}(\hat{\boldsymbol{u}}, \boldsymbol{s}) = \mathrm{Y}(-\hat{\boldsymbol{u}}, \boldsymbol{s})$, but working with the complete sphere is more convenient.

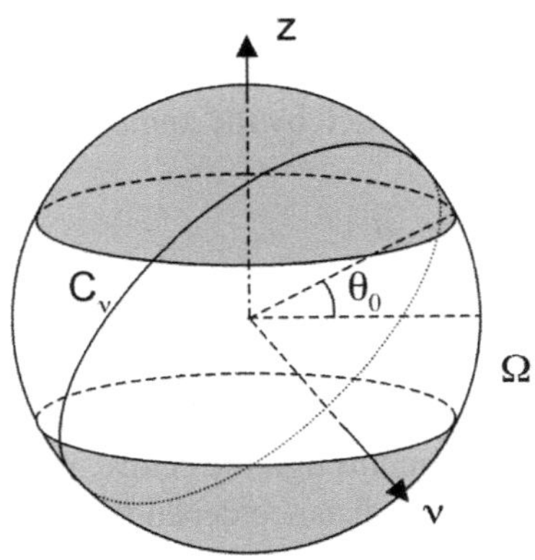

FIG. 13.4. Each point on the unit sphere corresponds to the direction of a parallel projection. An ideal rotating gamma camera with a parallel-hole collimator only travels through the points on the equator. An idealized 3-D PET system would also acquire projections along oblique lines; it collects projections for all points of the set Ω. The set Ω, defined by θ_0, is the non-shaded portion of the unit sphere. To recover a particular frequency v (of the Fourier transform of the object), at least one point on the circle C_v is required.

With $\theta_0 = 0$, the problem reduces to 2-D parallel projection (for multiple slices), which was shown to have a unique solution. It follows that with

$|\theta_0| > 0$, the problem becomes overdetermined, and there are infinitely many ways to compute the solution. This can be seen as follows. Each point of Ω corresponds to a parallel projection. According to the central slice theorem, this provides a central plane perpendicular to $\hat{\boldsymbol{u}}$ of the 3-D Fourier transform $\mathcal{L}(\boldsymbol{\nu})$ of $\Lambda(\boldsymbol{x})$. Thus, the set Ω_0 (i.e. all points on the equator of the unit sphere in Fig. 13.4) provides all planes intersecting the ν_z axis, which is sufficient to recover the entire image $\Lambda(\boldsymbol{x})$ via inverse Fourier transform. The set Ω_0 with $\theta_0 > 0$ provides additional (oblique) planes through $\mathcal{L}(\boldsymbol{\nu})$, which are obviously redundant. A simple solution would be to select a sufficient subset from the data. However, if the data are noisy, a more stable solution is obtained by using all of the measurements. This is achieved by computing $\mathcal{L}(\boldsymbol{\nu})$ from a linear combination of all available planes:

$$\mathcal{L}(\boldsymbol{\nu}) = \int_{\Omega_{\theta_0}} \mathcal{Y}(\hat{\boldsymbol{u}},\boldsymbol{\nu})\mathrm{H}(\hat{\boldsymbol{u}},\boldsymbol{\nu})\delta(\hat{\boldsymbol{u}},\boldsymbol{\nu})\,\mathrm{d}\hat{\boldsymbol{u}} \tag{13.18}$$

Here, $\mathcal{Y}(\hat{\boldsymbol{u}},\boldsymbol{\nu})$ is the 2-D Fourier transform with respect to $\boldsymbol{s}$ of the projection $\mathrm{Y}(\hat{\boldsymbol{u}},\boldsymbol{s})$. The Dirac function $\delta(\hat{\boldsymbol{u}},\boldsymbol{\nu})$ selects the parallel projections $\hat{\boldsymbol{u}}$ which are perpendicular to $\boldsymbol{\nu}$ (i.e. the points on the circle C_ν in Fig. 13.4). Finally, the filter $\mathrm{H}(\hat{\boldsymbol{u}},\boldsymbol{\nu})$ assigns a particular weight to each of the available datasets $\mathcal{Y}(\hat{\boldsymbol{u}},\boldsymbol{\nu})$. The combined weight for each frequency should equal unity, leading to the filter equation:

$$\int_{\Omega_{\theta_0}} \mathrm{H}(\hat{\boldsymbol{u}},\boldsymbol{\nu})\delta(\hat{\boldsymbol{u}},\boldsymbol{\nu})\,\mathrm{d}\hat{\boldsymbol{u}} = 1 \tag{13.19}$$

A solution equivalent to that of unweighted least squares (LS) is obtained by assigning the same weight to all available data [13.8]. This results in the Colsher filter which can be written as:

$$\begin{aligned}\mathrm{H}_\mathrm{C}(\hat{\boldsymbol{u}},\boldsymbol{\nu}) &= |\boldsymbol{\nu}|/(2\pi) && \text{if } \sin\psi \le \sin\theta_0 \\ &= |\boldsymbol{\nu}|/(4\arcsin(\sin\theta_0/\sin\psi)) && \text{if } \sin\psi > \sin\theta_0\end{aligned} \tag{13.20}$$

where ψ is the angle between ν and the z axis: $\nu_z/|\nu| = \cos\psi$. The direct Fourier reconstruction method can be applied here, by straightforward inverse Fourier transform of Eq. (13.18). However, an FBP approach is usually preferred, which can be written as:

$$\Lambda(\boldsymbol{x}) = \int_{\Omega_{\theta_0}} \mathrm{d}\hat{\boldsymbol{u}}\,\mathrm{Y}^\mathrm{F}(\hat{\boldsymbol{u}},\boldsymbol{x} - (\boldsymbol{x}\cdot\hat{\boldsymbol{u}})\hat{\boldsymbol{u}}) \tag{13.21}$$

Here, Y^F is obtained by filtering Y with the Colsher filter (or another filter satisfying Eq. (13.19): $Y^F(\hat{\boldsymbol{u}},\boldsymbol{s}) = \mathcal{F}^{-1}(H_C(\hat{\boldsymbol{u}},\boldsymbol{\nu})\mathcal{Y}(\hat{\boldsymbol{u}},\boldsymbol{\nu}))$. The coordinate $\boldsymbol{s} = \boldsymbol{x} - (\boldsymbol{x}\cdot\hat{\boldsymbol{u}})\hat{\boldsymbol{u}}$ is the projection of the point $\boldsymbol{x}$ on the plane perpendicular to $\hat{\boldsymbol{u}}$; it selects the LOR through $\boldsymbol{x}$ in the parallel projection $\hat{\boldsymbol{u}}$.

13.2.3.2. The reprojection algorithm

The previous analysis assumed that the acceptance angle θ_0 was a constant, independent of $\boldsymbol{x}$. As illustrated in Fig. 13.5, this is not the case in practice. The acceptance angle is maximum for the centre of the FOV, becomes smaller with increasing distance to the centre and vanishes near the axial edges of the FOV. In other words, the projections are complete for $\hat{\boldsymbol{u}}$ orthogonal to the z axis (these are the 2-D multislice parallel-beam projections) and are truncated for the oblique parallel projections. The truncation becomes more severe for more oblique projections (Fig. 13.5).

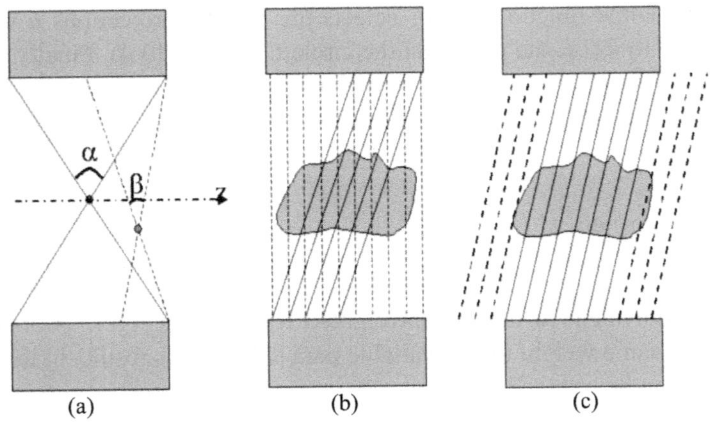

FIG. 13.5. An axial cross-section through a cylindrical PET system, illustrating that the acceptance angle is position dependent (a). Oblique projections are truncated (b). In the reprojection algorithm, the missing oblique projections (dashed lines) are computed from a temporary multislice 2-D reconstruction (c).

As the acceptance angle is position dependent, the required filtering is position dependent as well, and cannot be implemented as a shift-invariant convolution (or Fourier filter). Several strategies for dealing with this truncation have been developed. One approach is to subdivide the image into a set of regions, and then optimize a shift-invariant filter in each of the regions. The filter is determined by the smallest acceptance angle of the region, so some of the data will not be used. A good compromise between minimum data loss and practical implementation must be sought [13.9].

Another approach is to start with a first reconstruction, using the smallest acceptable angle over all positions $\boldsymbol{x}$ in the FOV. This usually means that only the parallel projections orthogonal to the z axis are used. The missing oblique projection values are computed from this first reconstruction (Fig. 13.4) and used to complete the measured oblique projections. This eliminates the truncation, and the 3-D FBP method of the previous section can be applied. This method [13.10] was the standard 3-D PET reconstruction method for several years, until it was replaced by the faster Fourier rebinning approach (see below).

13.2.3.3. Rebinning techniques

The complexity (estimated as the number of LORs) increases linearly with the axial extent for 2-D PET, but quadratically for 3-D PET. To keep the processing time acceptable, researchers have sought ways to reduce the size of the data as much as possible, while minimizing the loss of information induced by this reduction.

Most PET systems have a cylindrical detector surface: the detectors are located on rings with radius R, and the rings are combined in a cylinder along the z axis. The data are usually organized in sinograms which can be written as:

$$Y_P(s,\phi,z,\Delta_z)=\int_{-\infty}^{\infty}\mathrm{d}t\Lambda(s\cos\phi+t\hat{u}_x,s\sin\phi+t\hat{u}_y,z+t\hat{u}_z) \tag{13.22}$$

where $\hat{\boldsymbol{u}}$ is a unit vector in the direction of the LOR:

$$\hat{\boldsymbol{u}}=\boldsymbol{u}/\|\boldsymbol{u}\| \quad \text{with} \quad \boldsymbol{u}=(-\sin\phi,\cos\phi,\Delta_z/(2\sqrt{R^2-s^2}))$$

The parameter s is the distance between the LOR and the z axis. The LOR corresponds to a coincidence between detector points with axial positions $z-\Delta_z/2$ and $z+\Delta_z/2$. Finally, ϕ is the angle between the y axis and the projection of the LOR on the xy plane. The coordinates (s, ϕ, z) are identical to those often used in 2-D tomography. It should be noted that, in practice, $s \ll R$ and, as a result, the direction of the LOR, the vector $\hat{\boldsymbol{u}}$, is virtually independent of s. In other words, a set of LORs with fixed Δ_z can then be treated as a parallel projection with good approximation. LORs with $\Delta_z = 0$ are often called 'direct' LORs, while LORs with $\Delta_z \neq 0$ are called 'oblique'.

The basic idea of rebinning algorithms is to compute estimates of the direct sinograms from the oblique sinograms. If the rebinning algorithm is good, most of the information from the oblique sinograms will go into these estimates.

As a result, the data have been reduced from a complex 3-D geometry into a much simpler 2-D geometry without discarding measured signal. The final reconstruction can then be done with 2-D algorithms, which tend to be much faster than fully 3-D algorithms. A popular approach is to use Fourier rebinning, followed by maximum-likelihood reconstruction.

13.2.3.4. Single slice and multislice rebinning

The simplest way to rebin the data is to treat oblique LORs as direct LORs [13.11]. This corresponds to the approximation:

$$\mathrm{Y_P}(s,\phi,z,\Delta_z) \approx \mathrm{Y_P}(s,\phi,z,0) \qquad (13.23)$$

The approximation is only exact if the object consists of points located on the z axis, and it introduces mis-positioning errors that increase with increasing distance to the z axis and increasing Δ_z. Consequently, single slice rebinning is applicable when the object is small and positioned centrally in the scanner or when Δ_z is small. The axial extent of most current PET systems is too large to rebin an entire 3-D dataset with Eq. (13.23). However, single slice rebinning is used on all PET systems to reduce the sampling of the Δ_z dimension in the 3-D data, by combining sinograms with similar Δ_z. This typically reduces the data size by a factor of about ten, when compared to the finest possible sampling.

Application of Eq. (13.23) obviously causes blurring in the z direction, to a degree proportional to the distance from the z axis. However, it may also cause severe inconsistencies in the sinograms, producing blurring artefacts in the xy planes of the reconstructed images as well. Lewitt et al. [13.12] proposed distributing the oblique LOR values $\mathrm{Y_P}(s,\phi,z,\Delta_z)$ over all LORs with $z \in [z - \Delta_z R_\mathrm{f} / (2R), z + \Delta_z R_\mathrm{f} / (2R)]$, i.e. over all slices intersected by the LOR, and within an FOV with radius R_f. This so-called multislice rebinning reduces the inconsistencies in the sinograms, eliminating most of the xy blurring artefacts in the reconstruction. Unfortunately, the improvement comes at the cost of strong axial blurring. This blurring depends strongly on *z*, but it is found to be approximately independent of *x* and *y*. A *z*-dependent 1-D axial filter is applied to reduce this axial blurring [13.12]. Multislice rebinning is superior to single slice rebinning, but the noise characteristics are not optimal.

13.2.3.5. Fourier rebinning

Fourier rebinning [13.13] is based on the frequency–distance principle, which was explained previously. The Fourier rebinning method is most simply formulated when the projection is written as follows:

$$\mathrm{Y}(s,\phi,z,\delta)=\int_{-\infty}^{\infty}\mathrm{d}t\lambda(s\cos\phi-t\sin\phi,s\sin\phi+t\cos\phi,z+t\delta) \tag{13.24}$$

where

$\delta = \tan\theta$;
θ is the angle between the LOR and the xy plane;

and the integration variable t is the distance between the position on the LOR and the z axis.

It follows that:

$$\mathrm{Y}(s,\phi,z,\delta)=\frac{\mathrm{Y_P}(s,\phi,z,\Delta_z=2\delta\sqrt{R^2-s^2})}{\sqrt{1+\delta^2}} \tag{13.25}$$

$$\approx\frac{\mathrm{Y_P}(s,\phi,z,\Delta_z=2\delta R)}{\sqrt{1+\delta^2}} \tag{13.26}$$

where the approximation is valid whenever $s \ll R$. In this case, no interpolation is needed; it is sufficient to scale the PET data $\mathrm{Y_P}$ with the weight factor $\sqrt{1+\delta^2}$.

Fourier rebinning uses the frequency–distance principle to find the distance d corresponding to a particular portion of the oblique sinogram. As illustrated in Fig. 13.6, distance is used to locate the direct sinogram to which this portion should be assigned. Denoting the 2-D Fourier transform of Y with respect to s and ϕ as $\mathcal{Y}$, this can be written as:

$$\mathcal{Y}(\nu_s,\nu_\phi,z,\delta)\approx\mathcal{Y}(\nu_s,\nu_\phi,z-\delta\frac{\nu_\phi}{\nu_s},0) \tag{13.27}$$

This equation explains how to distribute frequency components from a particular oblique sinogram into different direct sinograms. Frequencies located on the line

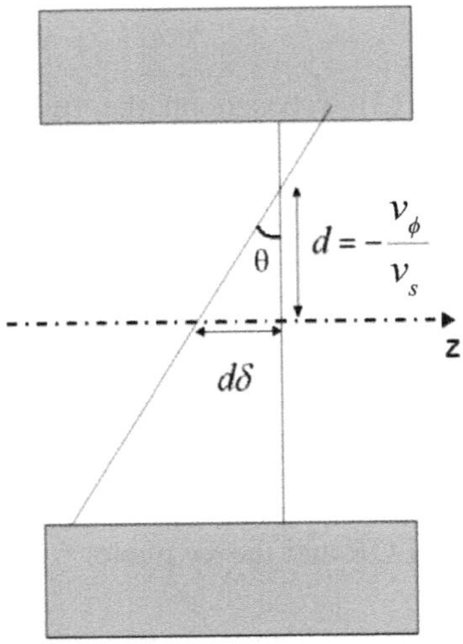

FIG. 13.6. Fourier rebinning: the distance from the rotation axis is obtained via the frequency–distance principle. This distance is used to identify the appropriate direct sinogram.

$\nu_\phi = \nu_s$ in the oblique sinogram z can be assigned to that same line in the direct sinogram $z + d\delta$.

The final rebinning algorithm (often called 'FORE') is obtained by averaging all of the available estimates of the direct sinogram:

$$\begin{aligned} \mathcal{Y}(\nu_s,\nu_\phi,z,0) &\approx \frac{1}{\delta_{\max}}\int_0^{\delta_{\max}} d\delta\, \mathcal{Y}(\nu_s,\nu_\phi,z+\delta\frac{\nu_\phi}{\nu_s},\delta) && \text{if} \quad |\nu_s| \gg 0 \\ &\approx \mathcal{Y}(0,0,z,0) && \text{if} \quad \nu_s \approx 0, \nu_\phi \approx 0 \\ &\approx 0 && \text{if} \quad |\nu_\phi / \nu_s| > R_f \end{aligned} \tag{13.28}$$

It should be noted that the rebinning expression is only valid for large ν_s. In the low frequency range, only the direct sinogram is used. The last line of Eq. (13.28) holds because the image $\Lambda(x, y, z)$ is assumed to be zero outside the FOV $\sqrt{x^2 + y^2} > R_f$.

A more rigorous mathematical derivation of the frequency–distance relation is given in Ref. [13.14]. Alternative derivations based on exact rebinning expressions are given in Ref. [13.13].

After Fourier rebinning, the resulting 2-D dataset can be reconstructed with any 2-D reconstruction algorithm. A popular method is the combination of Fourier rebinning with a 2-D statistical reconstruction algorithm.

13.2.3.6. Exact rebinning methods

Fourier rebinning is an approximate method, but was found to be sufficiently accurate for apertures up to $\theta_0 = 25°$, and it is, therefore, largely sufficient for most current PET systems. However, there is a tendency towards still larger acceptance angles, and a more exact Fourier rebinning algorithm may be needed in the future. An example of an 'exact' rebinning algorithm is FOREX [13.13]. It is exact in the sense that the rebinning expression is exact for the continuous 3-D X ray transform.

According to the central section theorem, the 2-D Fourier transform of a projection $Y(s, \phi, z, \delta)$ equals a cross-section through the 3-D Fourier transform of the image $\Lambda(x, y, z)$:

$$\mathcal{Y}_{13}(\nu_s,\phi,\nu_z,\delta) = \mathcal{L}(\nu_s \cos\phi + \nu_z \delta \sin\phi, \nu_s \sin\phi - \nu_z \delta \cos\phi, \nu_z) \tag{13.29}$$

The subscript of $\mathcal{Y}_{13}$ denotes a Fourier transform with respect to s and z. Defining:

$$\sigma = \arctan(\delta \nu_z / \nu_s)$$

$$\nu'_s = \nu_s \sqrt{1 + \delta^2 \nu_z^2 / \nu_s^2}$$

Equation (13.29) can be rewritten as:

$$\mathcal{Y}_{13}(\nu_s,\phi,\nu_z,\delta) = \mathcal{L}(\nu'_s \cos(\phi - \sigma), \nu'_s \sin(\phi - \sigma), \nu_z) \tag{13.30}$$

Taking the 1-D Fourier transform with respect to ϕ yields:

$$\mathcal{Y}_{123}(\nu_s,\nu_\phi,\nu_z,\delta) = \mathrm{e}^{-i\nu_\phi \sigma} \int_0^{2\pi} \mathrm{e}^{-i\nu_\phi \phi} \mathcal{L}(\nu'_s \cos\phi, \nu'_s \sin\phi, \nu_z)\, \mathrm{d}\phi \tag{13.31}$$

By comparing the expressions for $\mathcal{Y}_{123}(\nu_s,\nu_\phi,\nu_z,\delta)$ and $\mathcal{Y}_{123}(\nu_s,\nu_\phi,\nu_z,0)$, one finally obtains:

$$\mathcal{Y}_{123}(\nu_s,\nu_\phi,\nu_z,\delta) = e^{-i\nu_\phi \sigma} \mathcal{Y}_{123}(\nu'_s,\nu_\phi \nu_z,0) \tag{13.32}$$

A problem of FOREX is that it needs the 1-D Fourier transform along z, which cannot be computed for truncated projections. Similar as with 3-D filtered back projection, the problem can be avoided by completing the truncated projections with synthetic data. Fortunately, Eq. (13.32) can be used in both ways, and allows estimation of (missing) oblique sinograms from the available direct sinograms. The resulting algorithm is slower than FORE, but still considerably faster than 3-D FBP with reprojection.

13.2.4. Time of flight PET

In time of flight (TOF) PET, the difference in arrival time of the two detected photons is used to estimate the position of their emission along the LOR. The uncertainty in the time estimation results in a similar uncertainty in the position estimation, which can be well modelled as a Gaussian distribution. As a result, the TOF projections correspond to Gaussian convolutions along lines, rather than to line integrals, as illustrated in Fig. 13.7.

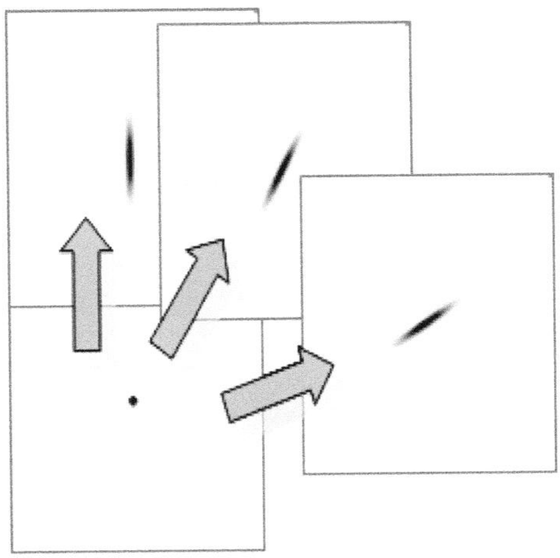

FIG. 13.7. Time of flight projection can be well modelled as a 1-D Gaussian convolution in the direction of the line of response.

The corresponding TOF back projection corresponds to convolving the measured data with the same 1-D Gaussians, followed by summation over all angles.

Recall from Eq. (13.2) that the regular projection followed by the regular back projection corresponds to a convolution with a blurring filter:

$$B_{\text{nonTOF}}(x,y)=\frac{1}{\sqrt{x^2+y^2}} \tag{13.33}$$

The Fourier transform of $1/\sqrt{x^2+y^2}$ equals $1/\sqrt{\nu_x^2+\nu_y^2}$. Consequently, this blurring can be undone by the ramp filter $\sqrt{\nu_x^2+\nu_y^2}$, which can be applied either before or after back projection (see Section 13.2.1).

If σ_{TOF} is the standard deviation of the TOF-blurring kernel, then TOF projection followed by TOF back projection corresponds to convolution with the blurring kernel:

$$B_{\text{TOF}}(x,y) = \frac{\text{Gauss}\,(x,y,\sqrt{2}\sigma_{\text{TOF}})}{\sqrt{x^2+y^2}} \tag{13.34}$$

$$= \frac{1}{\sqrt{x^2+y^2}}\frac{1}{2\sqrt{\pi}\sigma_{\text{TOF}}}\exp\left(-\frac{x^2+y^2}{4\sigma_{\text{TOF}}^2}\right) \tag{13.35}$$

It should be noted that the Gaussian in the equation above has a standard deviation of $\sqrt{2}\sigma_{\text{TOF}}$. This is because the Gaussian blurring is present in the projection and in the back-projection. The filter required in TOF PET FBP is derived by inverting the Fourier transform of B_{TOF}, and equals:

$$\text{TOF\_recon\_filter}\,(\nu) = \frac{1}{\exp(-2\pi^2\sigma_{\text{TOF}}^2\nu^2)I_0(2\pi^2\sigma_{\text{TOF}}^2\nu^2)} \tag{13.36}$$

where I_0 is the zero order modified Bessel function of the first kind.

This FBP expression is obtained by using the ‘natural’ TOF back projection, defined as the adjoint of the TOF projection. This back projection also appears in LS approaches, and it has been shown that with this back projection definition, FBP is optimal in an (unweighted) LS sense [13.15]. However, TOF PET data are redundant and different back projection definitions could be used; they would yield different expressions for $B_{\text{TOF}}(x, y)$ in Eq. (13.34) and, therefore, different TOF reconstruction filters.

Just as for non-TOF PET, exact and approximate rebinning algorithms for TOF PET have been derived to reduce the data size. As the TOF information limits the back projection to a small region, the errors from approximate rebinning are typically much smaller than in the non-TOF case.

13.3. ITERATIVE RECONSTRUCTION

13.3.1. Introduction

13.3.1.1. Discretization

In analytical reconstruction, it is initially assumed that the unknown object can be represented as a function $\Lambda(\vec{x})$ with $\vec{x} \in \mathbb{R}^3$, and that the acquired data can be represented as a function $Y(s, \theta)$ with $s \in \mathbb{R}^2$ and θ a unit vector in $\mathbb{R}^2$ or $\mathbb{R}^3$. The reconstruction algorithm is then derived by mathematical inversion (assuming some convenient properties for Λ and Y), and finally the resulting algorithm is discretized to make it ready for software implementation. In iterative reconstruction, one usually starts by discretizing the problem. This reduces the reconstruction problem to finding a finite set of unknown values from a finite set of equations, a problem which can be solved with numerical inversion. The advantage of numerical inversion is that only a model for the acquisition process is needed, not for its inverse. That makes it easier (although it may still be non-trivial) to take into account some of the undesired but unavoidable effects that complicate the acquisition, such as photon attenuation, position dependent resolution, gaps between the detectors and patient motion.

After discretization, the unknown image values and the known measured values can be represented as column vectors $\boldsymbol{\lambda}$ and $\boldsymbol{y}$. The PET or SPECT acquisition process is characterized by the system matrix $\boldsymbol{A}$ and an additive contribution $\bar{\boldsymbol{b}}$, and $\boldsymbol{n}$ is the measurement noise:

$$\boldsymbol{y} = \boldsymbol{A}\boldsymbol{\lambda} + \bar{\boldsymbol{b}} + \boldsymbol{n} \quad \text{or} \quad y_i = \sum_{j=1}^{J} A_{ij}\lambda_j + \bar{b}_i + n_i, \quad i = 1,\ldots,I \tag{13.37}$$

The symbol y_i denotes the number of photons measured at LOR i, where the index i runs over all of the sinogram elements (merging the three or four sinogram dimensions into a single index). The index j runs over all of the image voxels, and A_{ij} is the probability that a unit of radioactivity in j gives rise to the detection of a photon (SPECT) or photon pair (PET) in LOR i. The estimate of the additive contribution is denoted as $\bar{\boldsymbol{b}}$. This estimate is assumed to be noise-free and includes, for example, scatter and randoms in PET or cross-talk between different energy windows in multitracer SPECT studies. Finally, n_i represents the noise contribution in LOR i.

Image reconstruction now consists of finding $\boldsymbol{\lambda}$, given $\boldsymbol{A}$, $\boldsymbol{y}$ and $\bar{\boldsymbol{b}}$, and a statistical model for $\boldsymbol{n}$.

For further reading about this subject, the recent review paper on iterative reconstruction by Qi and Leahy [13.16] is an ideal starting point.

13.3.1.2. Objective functions

The presence of the noise precludes exact reconstruction. For this reason, the reconstruction is often treated as an optimization task: it is assumed that a useful clinical image can be obtained by maximizing a well chosen objective function. When the statistics of the noise are known, a Bayesian approach can be applied, searching for the image $\hat{\lambda}$ that maximizes the conditional probability on the data:

$$\begin{aligned}\hat{\lambda} &= \arg\max_{\boldsymbol{\lambda}} p(\boldsymbol{\lambda} \mid \boldsymbol{y}) \\ &= \arg\max_{\boldsymbol{\lambda}} \frac{p(\boldsymbol{y} \mid \boldsymbol{\lambda}) p(\boldsymbol{\lambda})}{p(\boldsymbol{y})} \qquad (13.38) \\ &= \arg\max_{\boldsymbol{\lambda}} p(\boldsymbol{y} \mid \boldsymbol{\lambda}) p(\boldsymbol{\lambda}) \\ &= \arg\max_{\boldsymbol{\lambda}} (\ln p(\boldsymbol{y} \mid \boldsymbol{\lambda}) + \ln p(\boldsymbol{\lambda})) \qquad (13.39)\end{aligned}$$

The second equation is Bayes' rule. The third equation holds because $\boldsymbol{y}$ does not depend on $\boldsymbol{\lambda}$, and the fourth equation is valid because computing the logarithm does not change the position of the maximum. The probability $p(\boldsymbol{y}|\boldsymbol{\lambda})$ gives the likelihood of measuring a particular sinogram $\boldsymbol{y}$, when the tracer distribution equals $\boldsymbol{\lambda}$. This distribution is often simply called the likelihood. The probability $p(\boldsymbol{\lambda})$ represents the a priori knowledge about the tracer distribution, available before PET or SPECT acquisition. This probability is often called the prior distribution. The knowledge available after the measurements equals $p(\boldsymbol{y}|\boldsymbol{\lambda})p(\boldsymbol{\lambda})$ and is called the posterior distribution. To keep things simple, it is often assumed that no prior information is available, i.e. $p(\boldsymbol{\lambda}|\boldsymbol{y}) \propto p(\boldsymbol{y}|\boldsymbol{\lambda})$. Finding the solution then reduces to maximizing the likelihood $p(\boldsymbol{y}|\boldsymbol{\lambda})$ (or its logarithm). In this section, maximum-likelihood algorithms are discussed. Maximum a posteriori (MAP) algorithms are discussed in Section 13.3.5, as a strategy to suppress noise propagation.

A popular approach to solve equations of the form of Eq. (13.37) is LS estimation. This is equivalent to a maximum-likelihood approach, if it is assumed that the noise is Gaussian with a zero mean and a fixed, position independent standard deviation σ. The probability to measure the noisy value y_i when the expected value was $\sum_j A_{ij}\lambda_j + \bar{b}_i$ then equals:

$$p_{\mathrm{LS}}(y_i \mid \sum_j A_{ij}\lambda_j + \bar{b}_i) = \frac{1}{\sqrt{2\pi}\sigma}\exp\left(-\frac{(y_i - (\sum_j A_{ij}\lambda_j + \bar{b}_i))^2}{2\sigma^2}\right) \tag{13.40}$$

As the noise in the sinogram is not correlated, the likelihood (i.e. the probability of measuring the entire noisy sinogram $\boldsymbol{y}$) equals:

$$p_{\mathrm{LS}}(\boldsymbol{y} \mid \boldsymbol{\lambda}) = p_{\mathrm{LS}}(\boldsymbol{y} \mid \boldsymbol{A\lambda} + \bar{\boldsymbol{b}}) = \prod_i p_{\mathrm{LS}}(y_i \mid \sum_j A_{ij}\lambda_j + \bar{b}_i) \tag{13.41}$$

It is more convenient to maximize the logarithm of p_{LS}; dropping constants, the objective function L_{LS} is finally obtained:

$$L_{\mathrm{LS}} = -\sum_i (y_i - (\sum_j A_{ij}\lambda_j + \bar{b}_i))^2 = -(\boldsymbol{y} - (\boldsymbol{A\lambda} + \bar{\boldsymbol{b}}))'(\boldsymbol{y} - (\boldsymbol{A\lambda} + \bar{\boldsymbol{b}})) \tag{13.42}$$

where the prime denotes matrix transpose. Setting the first derivatives with respect to λ_j to zero for all j gives:

$$\begin{aligned} &\boldsymbol{A}'(\boldsymbol{y} - \boldsymbol{A\lambda} - \bar{\boldsymbol{b}}) = 0 \\ &\boldsymbol{\lambda}(\boldsymbol{A}'\boldsymbol{A})^{-1}\boldsymbol{A}'(\boldsymbol{y} - \bar{\boldsymbol{b}}) \end{aligned} \tag{13.43}$$

provided that $\boldsymbol{A}'\boldsymbol{A}$ is non-singular. The operator $\boldsymbol{A}$ is the discrete projection; its transpose $\boldsymbol{A}'$ is the discrete back projection. Its analytical counterpart was given in Eq. (13.2) and illustrated in Fig. 13.1. The same figure shows that the operator $\boldsymbol{A}'\boldsymbol{A}$ behaves as a blurring filter.

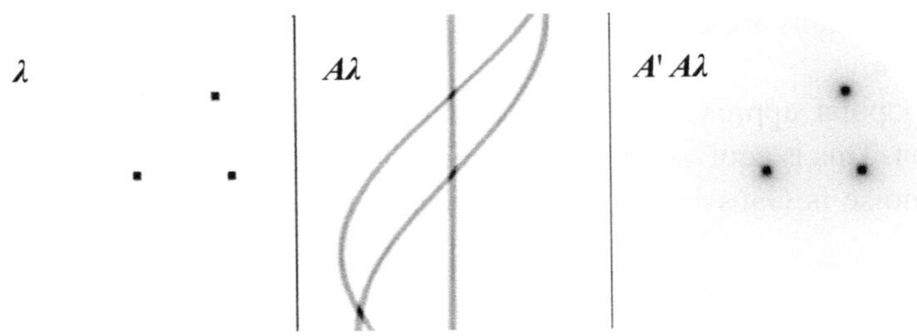

FIG. 13.8. The image of point sources is projected and back projected again along ideal parallel beams. This yields a shift-invariant blurring.

Figure 13.8 is similar, but illustrates A and $A'A$ on an image of three point sources, using ideal parallel-beam projection. The figure shows the resulting point spread functions of $A'A$ for each of the point sources. They are identical: for ideal parallel-beam projection, $A'A$ is shift-invariant, equivalent to a convolution. It follows that $(A'A)^{-1}$ is the corresponding shift-invariant deconvolution, which is easily computed via the Fourier transform. In this situation, LS reconstruction (Eq. (13.43)) is the discrete equivalent of the 'back project-then-filter' algorithm (Eq. (13.15)), applied to the data after pre-correction for $\bar{b}$.

Figure 13.9 illustrates A and $A'A$ for a projector that models the position dependent blurring of a typical parallel-beam SPECT collimator. The blurring induced by $A'A$ is now shift-variant — it cannot be modelled as a convolution and its inverse cannot be computed with the Fourier transform. For real life problems, direct inversion of $A'A$ is not feasible. Instead, iterative optimization is applied to find the maximum of Eq. (13.42).

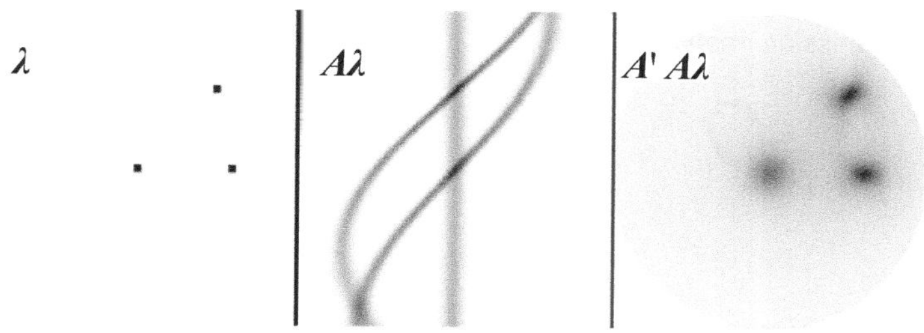

FIG. 13.9. The image of point sources is projected and back projected again with collimator blurring. This yields a shift-variant blurring.

It is known that the number of detected photons is subject to Poisson noise, not to uniform Gaussian noise. The Poisson distribution can be well approximated with a Gaussian distribution, where the variance of the Gaussian equals its mean. With this approximation, σ must be replaced by σ_i in Eq. (13.40) because now there is a different Gaussian distribution for every sinogram pixel i. Proceeding as before, the weighted least squares (WLS) objective function is:

$$\begin{aligned} L_{\mathrm{WLS}} &= -\sum_i \frac{(y_i - (\sum_j A_{ij}\lambda_j + \bar{b}_i))^2}{\sigma_i^2} \\ &= -(\boldsymbol{y} - (\boldsymbol{A\lambda} + \bar{\boldsymbol{b}}))' \boldsymbol{C}_y^{-1} (\boldsymbol{y} - (\boldsymbol{A\lambda} + \bar{\boldsymbol{b}})) \end{aligned} \tag{13.44}$$

where $\boldsymbol{C}_y$ is the covariance matrix of the data.

For emission tomography, it is a diagonal matrix (all covariances are zero) with elements $C_y[i, i] = \sigma_i^2$. The corresponding WLS reconstruction can be written as:

$$\lambda = (A' C_y^{-1} A)^{-1} A' C_y^{-1} (y - \bar{b}) \tag{13.45}$$

The operator $A' C_y^{-1} A$ is always shift-variant, even for ideal parallel-beam tomography. This is illustrated in Fig. 13.10. The noise-free sinogram $\bar{y}$ is computed for a particular activity distribution. Setting $C_y = \mathrm{diag}(\bar{y})$, the operator $A' C_y^{-1} A$ can be analysed by applying it to the image of a few point sources, called x in the figure. The image x is projected, the sinogram Ax is divided by $\bar{y}$ on a pixel basis and the result is back projected. Clearly, position dependent blurring is obtained. Consequently, iterative optimization must be used for WLS reconstruction.

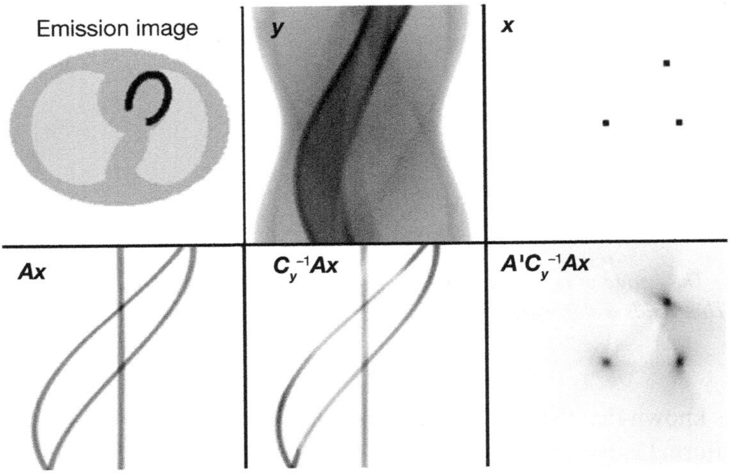

*FIG. 13.10. The operator $A' C_y^{-1} A$ is derived for a particular activity distribution (top left) and then applied to a few point sources **x**. Although ideal parallel-beam projection was used, shift-variant blurring is obtained.*

In practice, because there is only a noisy sinogram y, the noise-free sinogram $\bar{y}$ must be estimated to find C_y. There are basically two approaches. In the first approach, $\bar{y}$ is estimated from y, e.g. by smoothing y to suppress the noise. In the second approach, $\bar{y}$ is estimated as $A\lambda^{(k)} + \bar{b}$ during the iterative optimization, where $\lambda^{(k)}$ is the estimate of the reconstruction available at iteration k. A drawback of the first approach is that the noise on the data affects the weights, with a tendency to give higher weight when the noise contribution happens to be negative. A complication of the second approach is that it makes σ_i a function of $\boldsymbol{\lambda}$. In this case, the normalizing amplitude $1/(\sqrt{2\pi\sigma_i})$ of the Gaussians cannot be dropped, implying that an additional term $-\sum_i \ln\sigma_i$ should be added to Eq. (13.44).

It is possible to use the Poisson distribution itself, instead of approximating it with Gaussians. The probability of the noise realization y_i then becomes:

$$p_{\text{ML}}(y_i \mid \sum_j A_{ij}\lambda_j + \bar{b}_i) = \frac{e^{-(\sum_j A_{ij}\lambda_j + \bar{b}_i)} (\sum_j A_{ij}\lambda_j + \bar{b}_i)^{y_i}}{y_i!} \tag{13.46}$$

Proceeding as before, the log-likelihood function is:

$$\ln\left(\prod_i p_{\text{LS}}(y_i \mid \sum_j A_{ij}\lambda_j + \bar{b}_i)\right) = \sum_i y_i \ln(\sum_j A_{ij}\lambda_j + \bar{b}_i) - (\sum_j A_{ij}\lambda_j + \bar{b}_i) - \ln y_i!$$

$$L_{\text{ML}} = \sum_i y_i \ln(\sum_j A_{ij}\lambda_j + \bar{b}_i) - (\sum_j A_{ij}\lambda_j + \bar{b}_i) \tag{13.47}$$

It should be noted that the term $\ln y_i!$ can be dropped, because it is not a function of $\boldsymbol{\lambda}$. As L_{ML} is a non-linear function of $\boldsymbol{\lambda}$, the solution cannot be written as a product of matrixes. However, it is sometimes helpful to know that the features of the Poisson-objective function are often very similar to those of the WLS function (Eq. (13.44)).

13.3.2. Optimization algorithms

Many iterative reconstruction algorithms have been proposed to optimize the objective functions L_{WLS} and L_{ML}. Here, only two approaches are briefly described: preconditioned conjugate gradient methods and optimization transfer, with expectation maximization (EM) as a special case of the latter.

13.3.2.1. Preconditioned gradient methods

The objective function will be optimized when its first derivatives are zero:

$$\hat{y}_i = \sum_i A_{ij}\lambda_j + \bar{b}_i \tag{13.48}$$

$$\frac{\partial L_{\mathrm{WLS}}(\boldsymbol{\lambda})}{\partial \lambda_j} = \sum_i A_{ij} \frac{y_i - \hat{y}_i}{\sigma_i^2} \tag{13.49}$$

$$\frac{\partial L_{\mathrm{ML}}(\boldsymbol{\lambda})}{\partial \lambda_j} = \sum_i A_{ij} \frac{y_i - \hat{y}_i}{\hat{y}_i} \tag{13.50}$$

The optimization can be carried out by a steepest ascent method, which can be formulated as follows:

$$\begin{aligned} \boldsymbol{d}^k &= \nabla L(\boldsymbol{\lambda}^{k-1}) \\ \boldsymbol{\lambda}^k &= \boldsymbol{\lambda}^{k-1} + \alpha_k \boldsymbol{d}^k \\ \alpha_k &= \arg\max_{\alpha} L(\boldsymbol{\lambda}^{k-1} + \alpha \boldsymbol{d}^k) \end{aligned} \tag{13.51}$$

where the superscripts k and k–1 denote the iteration numbers and ∇L is the vector of the first derivatives of L with respect to λ_j.

Steepest gradient ascent is known to be suboptimal, requiring many iterations for reasonable convergence. To find a better update, it is required that after the update, the first derivatives of L are zero as intended. Approximating this with a first order Taylor expansion yields:

$$\begin{aligned} &\nabla L(\boldsymbol{\lambda}^{k-1} + \boldsymbol{p}^k) = 0 \\ &\nabla L(\boldsymbol{\lambda}^{k-1}) + \boldsymbol{H}\boldsymbol{p}^k \approx 0 \\ &\boldsymbol{p}^k \approx -\boldsymbol{H}^{-1} \nabla L(\boldsymbol{\lambda}^{k-1}) = -\boldsymbol{H}^{-1}\boldsymbol{d}^k \end{aligned} \tag{13.52}$$

where the Hessian $\boldsymbol{H}$ is the matrix of second derivatives of L. This is obviously a very large matrix, but its elements are relatively easy to compute:

$$\text{for WLS: } H_{jk} = -\sum_i \frac{A_{ij} A_{ik}}{\sigma_i^2} = -(\boldsymbol{A}'\boldsymbol{C}_y{}^{-1}\boldsymbol{A})[j,k] \tag{13.53}$$

$$\text{for ML: } H_{jk} = -\sum_i \frac{A_{ij} A_{ik} y_i}{\hat{y}_i^2} \approx \sum_i \frac{A_{ij} A_{ik}}{\hat{y}_i} \tag{13.54}$$

$$\approx -(\boldsymbol{A}'\boldsymbol{C}_y{}^{-1}\boldsymbol{A})[j,k] \quad \text{if} \quad \hat{\boldsymbol{y}} \approx \overline{\boldsymbol{y}} \tag{13.55}$$

For a Gaussian likelihood, Eq. (13.52) is in fact exact, and a single iteration would suffice. As shown before, however, it is usually impossible to compute $\boldsymbol{H}^{-1}$. Instead, approximations to the Hessian (or other heuristics) can be used to obtain a good $\boldsymbol{M}$ to derive a so-called preconditioned gradient ascent algorithm:

$$\boldsymbol{d}^k = \nabla L(\lambda^{k-1})$$
$$\lambda^k = \lambda^{k-1} + \alpha_k \boldsymbol{M}\boldsymbol{d}^k \tag{13.56}$$

To ensure that the convergence is preserved, the matrix $\boldsymbol{M}$ must be symmetric positive definite (it should be noted that $-\boldsymbol{H}^{-1}$ is symmetric positive definite, since $\boldsymbol{H}$ is symmetric negative definite, if $\boldsymbol{A}$ has maximum rank).

A simple way to obtain a reasonable $\boldsymbol{M}$ is to use only the diagonal elements of $\boldsymbol{H}$: $M_{ii} = -1/H_{ii}$ and $M_{ij} = 0$ if $i \neq j$. A more sophisticated approach is discussed in Ref. [13.17]: a circulant, i.e. shift-invariant approximation of the Hessian is proposed. Such an approximation is easily computed by fixing j at a particular location in the image in Eqs (13.53) or (13.54), which yields an image that can be considered as the point spread function of a convolution operator. This shift-invariant operator is then inverted via the Fourier transform, yielding a non-diagonal matrix $\boldsymbol{M}$. For cases where the true Hessian depends heavily on position, the approach could be repeated for a few well chosen positions j, applying linear interpolation for all other positions.

13.3.2.2. Conjugate gradient methods

Figure 13.11 shows the convergence of the steepest gradient ascent algorithm for a nearly quadratic function of two variables. In every iteration, the algorithm starts moving in the direction of the maximum gradient (i.e. perpendicular to the isocontour), and keeps moving along the same line until a maximum is reached (i.e. until the line is a tangent to the isocontour). This often leads to a zigzag line, requiring many iterations for good convergence.

The conjugate gradient algorithm is designed to avoid these oscillations [13.18]. The first iteration is identical to that of the steepest gradient ascent. However, in the following iterations, the algorithm attempts to move in a direction for which the gradient along the previous direction(s) remains the same (i.e. equal to zero). The idea is to eliminate the need for a new optimization along these previous directions. Let $\boldsymbol{d}_{\text{old}}$ be the previous direction and $\boldsymbol{H}$ the Hessian matrix (i.e. the second derivatives). It is now required that the new direction $\boldsymbol{d}_{\text{new}}$ be such that the gradient along $\boldsymbol{d}_{\text{old}}$ does not change. When moving in direction $\boldsymbol{d}_{\text{new}}$, the

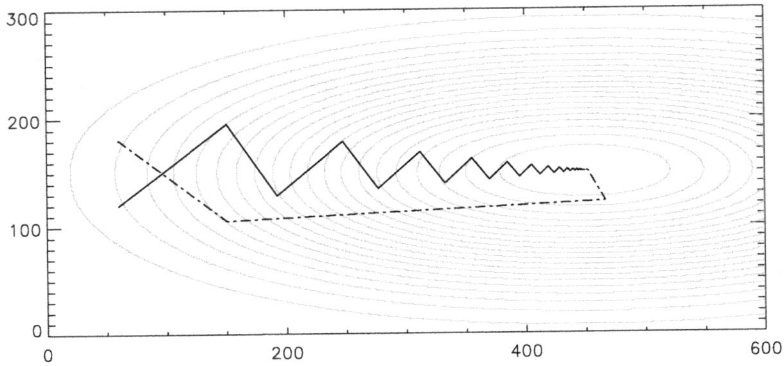

FIG. 13.11. The dotted lines are isocontours of the objective function. The solid line shows the convergence of the steepest gradient ascent algorithm, the dashed line the convergence of conjugent gradient ascent. It should be noted that the starting points are equivalent because of the symmetry. The objective function equals $-(a\,|\,x-x_0\,|^p + b\,|\,y-y_0\,|^p)$*, with p = 2.15.*

gradient will change (using a quadratic approximation) as $\boldsymbol{Hd}_{\text{new}}$. Requiring that the resulting change along $\boldsymbol{d}_{\text{old}}$ is zero yields the condition:

$$\boldsymbol{d}'_{\text{old}}\,\boldsymbol{Hd}_{\text{new}} = 0 \tag{13.57}$$

This behaviour is illustrated by the dashed line in Fig. 13.11: in the second iteration, the algorithm moves in a direction such that the trajectory cuts the isocontours at the same angle as in the starting point. For a quadratic function in n dimensions, convergence is obtained after no more than n iterations. As the function in Fig. 13.11 is not quadratic, more than two iterations are required for full convergence.

The new direction can be easily computed from the previous ones, without computation of the Hessian $\boldsymbol{H}$. The Polak–Ribiere algorithm is given by [13.18]:

$$\begin{aligned}
&\boldsymbol{g}_{\text{new}} = \nabla L(\boldsymbol{\lambda}_{\text{old}}) \\
&\gamma = \frac{(\boldsymbol{g}_{\text{new}} - \boldsymbol{g}_{\text{old}})'\boldsymbol{g}_{\text{new}}}{\boldsymbol{g}'_{\text{old}}\boldsymbol{g}_{\text{old}}} \\
&\boldsymbol{d}_{\text{new}} = \boldsymbol{g}_{\text{new}} + \gamma \boldsymbol{d}_{\text{old}} \\
&\alpha = \arg\max_{\alpha} L(\boldsymbol{\lambda}_{\text{old}} + \alpha \boldsymbol{d}_{\text{new}}) \\
&\boldsymbol{\lambda}_{\text{new}} = \boldsymbol{\lambda}_{\text{old}} + \alpha \boldsymbol{d}_{\text{new}}
\end{aligned} \tag{13.58}$$

This algorithm requires storage of the previous gradient $\boldsymbol{g}_{\text{old}}$ and the previous search direction $\boldsymbol{d}_{\text{old}}$. In each iteration, it computes the new gradient and search direction, and applies a line search along the new direction.

13.3.2.3. Preconditioned conjugate gradient methods

Both techniques mentioned above can be combined to obtain a fast reconstruction algorithm, as described in Ref. [13.17]. The preconditioned conjugate gradient ascent algorithm (with preconditioning matrix $\boldsymbol{M}$) can be written as follows:

$$\begin{aligned}
\boldsymbol{g}_{\text{new}} &= \nabla L(\boldsymbol{\lambda}_{\text{old}}) \\
\boldsymbol{p}_{\text{new}} &= \boldsymbol{M}\boldsymbol{g}_{\text{new}} \\
\gamma &= \frac{(\boldsymbol{g}_{\text{new}} - \boldsymbol{g}_{\text{old}})' \boldsymbol{p}_{\text{new}}}{\boldsymbol{g}'_{\text{old}} \boldsymbol{p}_{\text{old}}} \\
\boldsymbol{d}_{\text{new}} &= \boldsymbol{p}_{\text{new}} + \gamma \boldsymbol{d}_{\text{old}} \\
\alpha &= \arg\max_{\alpha} L(\boldsymbol{\lambda}_{\text{old}} + \alpha \boldsymbol{d}_{\text{new}}) \\
\boldsymbol{\lambda}_{\text{new}} &= \boldsymbol{\lambda}_{\text{old}} + \alpha \boldsymbol{d}_{\text{new}}
\end{aligned} \tag{13.59}$$

13.3.2.4. Optimization transfer

The log-likelihood function (Eq. (13.47)) can be maximized by setting its gradients (Eq. (13.50)) to zero for all $j = 1 \ldots J$. A problem is that each of these derivatives is a function of many voxels of $\boldsymbol{\lambda}$, which makes the set of equations very hard to solve. The idea of 'optimization transfer' is to replace the problematic log-likelihood function with another function $\Phi(\boldsymbol{\lambda})$ that leads to a simpler set of equations, usually one where the derivative with respect to λ_j is only a function of λ_j and not of the other voxels of $\boldsymbol{\lambda}$. That makes the problem separable into J 1-D optimizations, which are easily solved. Ideally, Φ and L should have the same optimum, but that is asking for too much. The key is to design $\Phi(\boldsymbol{\lambda})$ in such a way that maximization of $\Phi(\boldsymbol{\lambda})$ is guaranteed to increase $L(\boldsymbol{\lambda})$. This leads to an iterative algorithm, since new functions Φ will have to be designed and maximized repeatedly to maximize L. At iteration k, the surrogate function $\Phi(\lambda)$ needs to satisfy the following conditions (illustrated in Fig. 13.12):

$$\Phi(\lambda^{(k)}) = L(\lambda^{(k)}) \tag{13.60}$$

$$\Phi(X) \leq L(X) \tag{13.61}$$

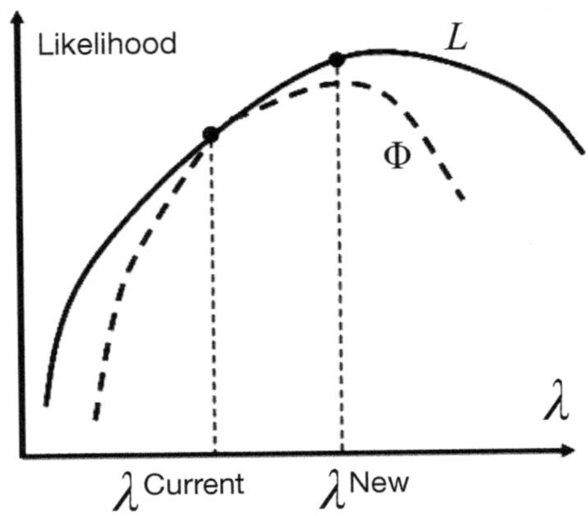

FIG. 13.12. Optimization transfer: a surrogate function is designed, which is equal to the likelihood in the current reconstruction, and less or equal everywhere else.

It follows that the new reconstruction image $\lambda^{(k+1)}$ which maximizes $\Phi(\lambda)$ has a higher likelihood than $\lambda^{(k)}$:

$$L(\lambda^{(k)}) = \Phi(\lambda^{(k)}) \leq \Phi(\lambda^{(k+1)}) \leq L(\lambda^{(k+1)}) \tag{13.62}$$

Several algorithms for maximum-likelihood and MAP reconstruction in emission and transmission tomography have been developed with this approach. De Pierro [13.19] has shown how the well known maximum-likelihood expectation-maximization (MLEM) algorithm can be derived using the optimization transfer principle. He also showed how this alternative derivation provides a natural way to extend it to an MAP algorithm.

13.3.3. Maximum-likelihood expectation-maximization

13.3.3.1. Reconstruction from sinogram data

There are many ways to derive the MLEM algorithm, including the original statistical derivation by Shepp and Vardi [13.20] (based on the work by Dempster et al. [13.21]) and the optimization transfer approach by De Pierro [13.19]. Only the EM recipe is given below.

Recall that we wish to find the image $\boldsymbol{\lambda}$ that maximizes the likelihood function L_{ML} of Eq. (13.47). The EM does this in a remarkable way. Instead of concentrating on L_{ML}, an alternative (different) likelihood function is derived by introducing a set of so-called 'complete data' x_{ij}, defined as the number of photons that were emitted at voxel j and detected in LOR i during the measurement. These unobserved data are 'complete' in the sense that they describe in more detail than the observed data y_i what happened during the measurement. These variables x_{ij} are Poisson distributed. Just as for the actual data y_i, one can write the log-likelihood function for observing the data x_{ij} while $\bar{x}_{ij} = A_{ij}\lambda_j$ were expected:

$$L_x(\boldsymbol{\lambda}) = \sum_i \sum_j x_{ij} \ln(A_{ij}\lambda_j) - A_{ij}\lambda_j \tag{13.63}$$

However, this likelihood cannot be computed, because the data x_{ij} are not available. The emission measurement only produces sums of the complete data, since:

$$y_i = \sum_j A_{ij}x_{ij} + b_i \tag{13.64}$$

where b_i represents the actual (also unobserved) additive contribution b_i in LOR i.

The EM recipe prescribes computing the expectation of L_x, based on the available data and on the current reconstruction $\boldsymbol{\lambda}^{(k)}$. Based on the reconstruction alone, one would write $E(x_{ij} \mid \boldsymbol{\lambda}^{(k)}) = A_{ij}\lambda_j^{(k)}$. However, it is also known that x_{ij} should satisfy Eq (13.64). It can be shown that this leads to the following estimate:

$$E(x_{ij} \mid \boldsymbol{\lambda}^{(k)}, \boldsymbol{y}) = \frac{y_i}{\sum_j A_{ij}\lambda_j^{(k)} + \bar{b}_i} A_{ij}\lambda_j^{(k)} \tag{13.65}$$

where $\bar{b}_i$ is the noise-free estimate of b_i, which is assumed to be available.

Inserting this in Eq. (13.63) produces the expectation of $L_x(\boldsymbol{\lambda})$ and completes the expectation (E) step. For the maximization (M) step, the first derivatives are simply set to zero:

$$\frac{\partial L_x(\boldsymbol{\lambda})}{\partial \lambda_j} = \sum_i \left(\frac{y_i}{\sum_j A_{ij}\lambda_j^{(k)} + \bar{b}_i} A_{ij}\lambda_j^{(k)} \frac{1}{\lambda_j} - A_{ij} \right) = 0 \tag{13.66}$$

This is easily solved for λ_j, yielding the new reconstruction $\lambda_j^{(k+1)}$:

$$\lambda_j^{(k+1)} = \frac{\lambda_j^{(k)}}{\sum_i A_{ij}} \sum_i A_{ij} \frac{y_i}{\sum_j A_{ij}\lambda_j^{(k)} + \bar{b}_i} \tag{13.67}$$

This is the well known MLEM algorithm for emission tomography.

It can be shown that this recipe has the wonderful feature that each new EM iteration increases the value of the likelihood L_{ML}. It should be noted that the complete data x_{ij} do not appear in Eq. (13.67); they are needed in the derivation but they do not need to be computed explicitly. This is very fortunate as there is a huge number of them.

An initial image $\boldsymbol{\lambda}^{(1)}$ is required to start the iterations. As experience (and theoretical analysis) has shown that higher spatial frequencies have slower convergence, and because smooth images are preferred, the initial image is usually chosen to be uniform, by setting $\lambda_j^{(1)} = C$ and $j = 1 \ldots J$, where C is a strictly positive constant.

The MLEM algorithm is multiplicative, implying that it cannot change the value of a reconstruction voxel, when the current value is zero. For this reason, the voxels in the initial image should only be set to zero if it is known a priori that they are indeed zero. The derivation of the MLEM algorithm uses the assumption that all y_i, all x_{ij} and all λ_j are non-negative. Assuming that $y_i \geq 0$ and $i = 1 \ldots I$, and considering that the probabilities A_{ij} are also non-negative, it is clear that when the initial image $\boldsymbol{\lambda}^{(1)}$ is non-negative, all subsequent images $\boldsymbol{\lambda}^{(k)}$ will be non-negative as well. However, when, for some reason, a reconstruction value becomes negative (e.g. because one or a few sinogram values y_i are negative), then convergence is no longer guaranteed. In practice, divergence is almost guaranteed in that case. Consequently, if the sinogram is pre-processed with a procedure that may produce negatives (e.g. randoms subtraction in PET), MLEM reconstruction will only work if all negative values are set to a non-negative value.

13.3.3.2. Reconstruction from list-mode data

The measured data y_i considered in the derivations above (so-called 'binned' data) represent the number of counts acquired within an individual crystal pair i (LOR i), that is, y_i represents the sum of those acquired events (indexed by m) that were assigned (histogrammed) to the i-th LOR: $y_i = \sum_{m \in i} 1$. However, in modern PET systems, the number of possible LORs within the FOV typically exceeds (often by many times) the number of events acquired in a clinical PET study. Consequently, the binned data are very sparse and it is more efficient to store and process each acquired event (with all of its relevant information) separately, in the so-called 'list-mode' format.

Modification of the maximum-likelihood algorithms is straightforward (whether MLEM or accelerated algorithms based on ordered subsets discussed later), as shown in works by Parra and Barrett [13.22], and by Reader et al. [13.23]. It should be noted that the same is not true about other algorithms, for example, algorithms with additive updates. The MLEM algorithm for the list-mode data can be obtained by replacing y_i in the MLEM equation (Eq. (13.67)) by the above mentioned sum over events, skipping the LORs with zero counts (which do not contribute to the MLEM sum), and combining the sum over LORs i with the sum over events m:

$$\lambda_j^{(k+1)} = \frac{\lambda_j^{(k)}}{\sum_{i \in \text{LORs}} A_{ij}} \sum_{m \in \text{event-list}} A_{i_m j} \frac{1}{\sum_j A_{i_m j} \lambda_j^{(k)} + \bar{b}_{i_m}} \tag{13.68}$$

where i_m represents the LOR index in which the m-th event has been recorded.

The main difference is that the MLEM sum is now evaluated (including calculations of the relevant forward and back projections) only over the list of the available events (in any order). However, it is important to mention here that the normalizing term in front of the sum (sensitivity matrix $\sum_i A_{ij}$) still has to be calculated over all possible LORs, and not only those with non-zero counts. This represents a challenge for the attenuated data (attenuation considered as part of the system matrix $\boldsymbol{A}$), since the sensitivity matrix has to be calculated specifically for each object and, therefore, it cannot be pre-computed. For modern systems with a large number of LORs, calculation of it often takes more time than the list-mode reconstruction itself. For this reason, alternative approaches (involving certain approximations) have been considered for the calculation of the sensitivity matrix, such as subsampling approaches [13.24] or Fourier based approaches [13.25].

13.3.3.3. Reconstruction of time of flight PET data

In the TOF case, the probability of a pair of photons arriving from a particular point along the LOR (as reported based on the difference of their detection times) is given by a Gaussian kernel having a width determined by the timing uncertainty of the detection system. In contrast, in the non-TOF case, the probability of detecting the event is approximately uniform along the LOR. Modification of iterative reconstruction algorithms (whether for binned or list-mode data) to account for the TOF is straightforward. Integrations along the LORs (the main component of the system matrix $\boldsymbol{A}$) just need to be replaced with the TOF kernel weighted integrations along the LORs. The forward projection (or back projection) in a certain direction can now be viewed, and performed, as a convolution of the image with a proper TOF kernel in the LOR direction (see Fig. 13.13). The rest of the algorithm, i.e. formulas derived in the previous subsections, stays exactly the same (only the form of the system matrix $\boldsymbol{A}$ is changed). Additional information provided by the TOF measurements, leading to more localized data, results in faster, and more uniform, convergence, as well as in improved signal to noise ratios in reconstructed images, as widely reported in the literature.

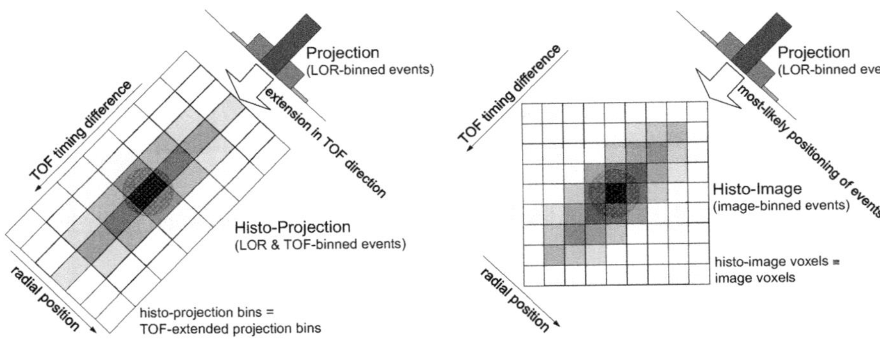

FIG. 13.13. Comparison of the data formats for binned time of flight (TOF) data (left: histo-projection for a 45° view) and for the DIRECT (direct image reconstruction for TOF) approach (right: histo-image for a 45° view). Histo-projections can be viewed as an extension of individual non-TOF projections into TOF directions (time bins), and their sampling intervals relate to the projection geometry and timing resolution. Histo-images are defined by the geometry and desired sampling of the reconstructed image. Acquired events and correction factors are directly placed into the image resolution elements of individual histo-images (one histo-image per view) having a one to one correspondence with the reconstructed image voxels.

The TOF mode of operation has some practical consequences (and novel possibilities) for the ways the acquired data are stored and processed. The list-mode format is very similar to the non-TOF case. The event structure is just slightly expanded by a few bits (5–8 bits/event) to include the TOF information, and the events are processed event by event as in the non-TOF case.

On the other hand, the binned data undergo considerable expansion when accommodating the TOF information. The projection (X ray transform) structures are expanded by one dimension, that is, each projection bin is expanded in the LOR direction into the set of time bins forming the so-called histo-projections (see Fig. 13.13 (left)). In practice, the effect of this expansion on the data size is not as bad as it appears, because the localized nature of TOF data allows decreased angular sampling (typically about 5–10 times) in both azimuthal and co-polar directions (views), while still satisfying angular sampling requirements. The resulting data size, thus, remains fairly comparable to the non-TOF case. During the reconstruction process, the histo-projection data are processed time-bin by time-bin (instead of projection line by line in the non-TOF case). It should be noted that hybrid approaches also exist between the two aforementioned approaches, in which the data are binned in the LOR space, but events are stored in list-mode for each LOR bin.

TOF also allows a conceptually different approach of data partitioning, leading to more efficient reconstruction implementations, by using the DIRECT (direct image reconstruction for TOF) approach utilizing so-called histo-images (see Fig. 13.13 (right)) [13.25]. In the DIRECT approach, the data are directly histogrammed (deposited), for each view, into image resolution elements (voxels) of desired size. Similarly, all correction arrays and data are estimated or calculated in the same histo-image format. The fact that all data and image structures are now in image arrays (of the same geometry and size) makes possible very efficient computer implementations of the data processing and reconstruction operations.

13.3.3.4. Reconstruction of dynamic data

Data acquired from an object dynamically changing with time in activity distribution, or in morphology (shape), or in both is referred to as dynamic data. An example of the first case would be a study looking at temporal changes in activity uptake in individual organs or tissues, so-called time–activity curves. An example of the second case would be a gated cardiac study providing information about changes of the heart morphology during the heart beat cycle (such as changes of the heart wall thickness or movements of the heart structures).

The dynamic data can be viewed as an expansion of static (3-D) data by the temporal information into 4-D (or 5-D) data. The dynamic data are usually subdivided (spread) into a set of temporal (time) frames. In the first application, each time frame represents data acquired within a certain sequential time subinterval of the total acquisition time. The subintervals can be uniform, or non-uniform with their durations adjusted, for example, to the speed of the change of the activity curves. In the second application, each time frame represents the total counts acquired within a certain stage (gate) of the periodic organ movement (e.g. gated based on the electrocardiogram signal). In the following, issues of the reconstruction of dynamic data in general are addressed. Problems related to the motion and its corrections are discussed in Section 13.3.6.4.

Once the data are subdivided (during acquisition) or sorted (acquired list-mode data) into the set of time frames, seemingly the most natural way of reconstructing them is to do it for each time frame separately. It should be noted that this is the only available option for the analytical reconstruction approaches, while the iterative reconstruction techniques can also reconstruct the dynamic data directly in 4-D (or 5-D). A problem with frame by frame reconstruction is that data in the individual time frames are quite noisy, since each time frame only has a fraction of the total acquired counts, leading to noisy reconstructions. Consequently, the resulting reconstructions often have to be filtered in the spatial and/or temporal directions to obtain images of any practical value. Temporal filtering takes into account time correlations between the signal components in the neighbouring time frames, while the noise is considered to be independent. Filtering, however, leads to resolution versus noise trade-offs.

On the other hand, reconstructing the whole 4-D (or 5-D) dataset together, while using this correlation information in the (4-D) reconstruction process via proper temporal (resolution) kernels or basis functions, can considerably improve those trade-offs as reported in the literature (similarly to the case of spatial resolution modelling). The temporal kernels (basis functions) can be uniform in shape and distribution, or can have a non-uniform shape (e.g. taking into account the expected or actual shape of the time–activity curves) and can be distributed on a non-uniform grid (e.g. reflecting count levels at individual frames or image locations). The kernel shapes and distributions can be defined, or determined, beforehand and be fixed during the reconstruction. During the reconstruction process, just the amplitudes of the basis functions are reconstructed. The algorithms derived in the previous subsections basically stay the same, where the temporal kernels can be considered as part of the system matrix $\boldsymbol{A}$ (comparable to including the TOF kernel in TOF PET). Another approach, more accurate but mathematically and computationally much more involved, is to iteratively build up the shape (and distribution) of the temporal kernels during the reconstruction in conjunction with the reconstruction of the emission activity (that is, the amplitude of the basis functions).

While iterative methods lead to a clear quality improvement when reconstructing dynamic data, thanks to the more accurate models of the signal and data noise components, for the quantitative dynamic studies their shortcoming is their non-linear behaviour, especially if they are not fully converged. For example, the local bias levels can vary across the time frames as the counts, local activity levels and object morphology change, which can lead to less accurate time–activity curves. On the other hand, analytical techniques which are linear and consequently do not depend on the count levels and local activity, might provide a more consistent (accurate) behaviour across the time frames in the mean (less bias of the mean), but much less consistent (less precise) behaviour in the variance due to the largely increased noise. It is still an open issue which of the two approaches provides more clinically useful results, and the discussions and research on this topic are still open and ongoing.

13.3.4. Acceleration

13.3.4.1. Ordered-subsets expectation-maximization

The MLEM algorithm requires a projection and a back projection in every iteration, which are operations involving a large number of computations. Typically, MLEM needs several tens to hundreds of iterations for good convergence. Consequently, MLEM reconstruction is slow and many researchers have studied methods to accelerate convergence.

The method most widely used is ordered-subsets expectation-maximization (OSEM) [13.26]. The MLEM algorithm (Eq. (13.67)) is rewritten here for convenience:

$$\hat{y}_i^{(k)} = \sum_j A_{ij}\lambda_j^{(k)} + \bar{b}_i \tag{13.69}$$

$$\lambda_j^{(k+1)} = \frac{\lambda_j^{(k)}}{\sum_i A_{ij}} \sum_i A_{ij} \frac{y_i}{\hat{y}_i^{(k)}} \tag{13.70}$$

where k is the iteration number and $\boldsymbol{\lambda}^{(1)}$ is typically set to a uniform, strictly positive image.

In OSEM, the set of all projections $\{1 \ldots I\}$ is divided into a series of subsets S_t, $t = 1 \ldots T$. Usually, these subsets are exhaustive and non-overlapping, i.e. every projection element i belongs to exactly one subset S_t. In SPECT and PET, the data $\boldsymbol{y}$ are usually organized as a set of (parallel- or fan-beam) projections, indexed by projection angle φ. Therefore, the easiest way to produce subsets of $\boldsymbol{y}$ is by assigning all of the data for each projection angle to exactly one of the subsets.

However, if the data $\boldsymbol{y}$ are stored in list-mode (see Section 13.3.2), the easiest way is to simply cut the list into blocks, assigning each block to a different subset.

The OSEM algorithm can then be written as:

$$
\begin{aligned}
&\text{initialize } \lambda_j^{\text{old}}, j = 1,\ldots J \\
&\text{for } k = 1,\ldots K \\
&\quad \text{for } t = 1,\ldots T \\
&\qquad \hat{y}_i = \sum_j A_{ij}\lambda_j^{\text{old}} + \bar{b}_i, \quad i \in \boldsymbol{S}_t \\
&\qquad \text{for } j = 1,\ldots J \\
&\qquad\quad \lambda_j^{\text{new}} = \frac{\lambda_j^{\text{old}}}{\sum_{i\in \boldsymbol{S}_t} A_{ij}} \sum_{i\in \boldsymbol{S}_t} A_{ij}\frac{y_i}{\hat{y}_i}
\end{aligned}
\tag{13.71}
$$

If all of the projections are combined into a single subset, the OSEM algorithm is identical to the MLEM algorithm. Otherwise, a single OSEM iteration k consists of T sub-iterations, where each sub-iteration is similar to an MLEM iteration, except that the projection and back projection are only done for the projections of the subset $\boldsymbol{S}_t$. If every sinogram pixel i is in exactly one subset, the computational burden of a single OSEM iteration is similar to that of an MLEM iteration. However, MLEM would update the image only once, while OSEM updates it T times. Experience shows that this improves convergence by a factor of about T, which is very significant.

Convergence is only guaranteed for consistent data and provided that there is subset balance, which requires:

$$\sum_{i\in \boldsymbol{S}_t} A_{ij} = \sum_{i\in \boldsymbol{S}_u} A_{ij} \tag{13.72}$$

where $\boldsymbol{S}_t$ and $\boldsymbol{S}_u$ are different subsets.

In practice, these conditions are never satisfied, and OSEM can be shown to converge to a limit cycle rather than to a unique solution, with the result that the OSEM reconstruction is noisier than the corresponding MLEM reconstruction. However, in many applications, the difference between the two is not clinically relevant.

The procedure is illustrated with a simple simulation in Fig. 13.14. As there was no noise and no attenuation, convergence of OSEM is guaranteed in this example. In more realistic cases, it may be recommended to have four or more projections in a single subset, to prevent excessive noise amplification at higher iteration numbers.

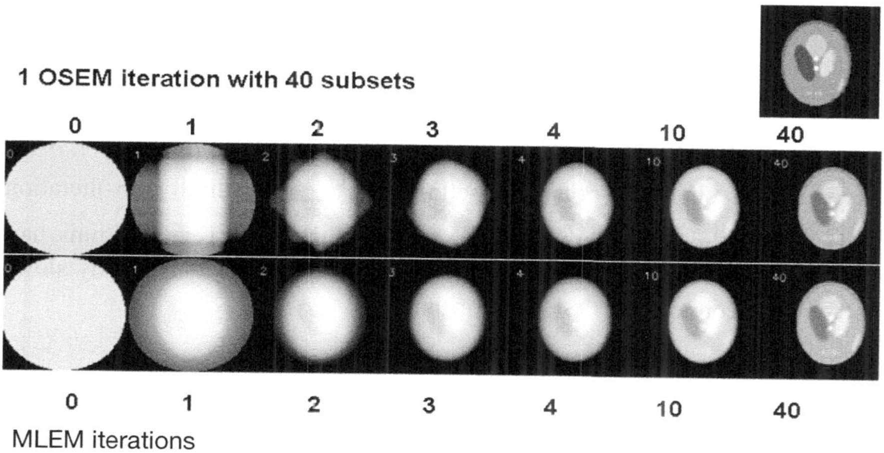

FIG. 13.14. A simulation comparing a single ordered-subsets expectation-maximization (OSEM) iteration with 40 subsets, to 40 maximum-likelihood expectation-maximization (MLEM) iterations. The computation time of the MLEM reconstruction is about 40 times longer than that of OSEM. In this example, there were only two (parallel-beam) projection angles per subset, which is clearly visible in the first OSEM iteration.

13.3.4.2. Refinements of the ordered-subsets expectation-maximization algorithm

As mentioned above, OSEM converges to a limit cycle: after many iterations, it starts cycling through a series of solutions rather than converging to the maximum-likelihood solution. When compared to the initial image (usually a uniform image), these series of solutions are 'relatively close' to the maximum-likelihood solution. Consequently, the convergence of OSEM is initially much faster but otherwise similar to that of MLEM; the better performance of MLEM only becomes noticeable at high iteration numbers. Thus, a simple solution to avoid the limit cycle is to gradually decrease the number of subsets: this approach preserves the initial fast convergence of OSEM, avoiding the limit cycle by returning to MLEM at high iteration numbers. A drawback of this approach is that convergence becomes slower each time the number of subsets is reduced. In addition, there is no theory available that prescribes how many sub-iterations should be used for each OSEM iteration.

Many algorithms have been proposed that use some form of relaxation to obtain convergence under less restrictive conditions than those of OSEM. As an example, relaxation can be introduced by rewriting the OSEM Eq. (13.71) in an additive way. Then, a relaxation factor α is inserted to scale the update term to obtain RAMLA (row-action maximum-likelihood algorithm [13.27]):

$$\lambda_j^{\text{new}} = \lambda_j^{\text{old}} + \alpha \lambda_j^{\text{old}} \sum_{i \in \boldsymbol{S}_t} A_{ij} \left(\frac{y_i}{\hat{y}_i} - 1 \right) \quad \text{with} \quad \alpha < \frac{1}{\max_t (\sum_{i \in \boldsymbol{S}_t} A_{ij})} \tag{13.73}$$

The relaxation factor α decreases with increasing iteration number to ensure convergence. It should be noted that setting $\alpha = 1 / \sum_{i \in \boldsymbol{S}_t} A_{ij}$ for all (sub-)iterations yields OSEM. Several alternative convergent block iterative algorithms have been proposed. They are typically much faster than MLEM but slightly slower than the (non-convergent) OSEM algorithm.

13.3.5. Regularization

MLEM maximizes the likelihood, by making the computed projections (from the current reconstruction) as similar as possible to the measured projections, where the similarity is measured based on the Poisson distribution. An upper limit of the likelihood would be obtained when the measured and calculated projections are identical. However, this is never possible, because Poisson noise introduces inconsistencies. Nevertheless, a large part of the noise is consistent, which means that it can be obtained as the projection of a (noisy) activity distribution. This part of the noise propagates into the reconstructed image, and is responsible for the so-called 'deterioriation' of the MLEM image at high iterations.

13.3.5.1. Stopping iterations early

An 'accidental' feature of the MLEM algorithm is its frequency dependent convergence: low spatial frequencies converge faster than higher frequencies. This is due to the low-pass effect of the back projection operation. This effect is easily verified for the reconstruction of the activity in a point source, if the MLEM reconstruction is started from a uniform image. The first iteration then yields the back projection of the point source measurement. As discussed in Section 13.2.1, this yields an image with intensity $\lambda(x,y) \propto 1/\sqrt{x^2 + y^2}$, if the point source was located at (0,0). Each iteration multiplies with a similar back projection, implying that after t iterations, the image intensity at (x, y) is proportional to $1/(x^2 + y^2)^{t/2}$, so that the peak at (0,0) becomes a bit sharper with every iteration. For more complicated objects, the evolution is more subtle.

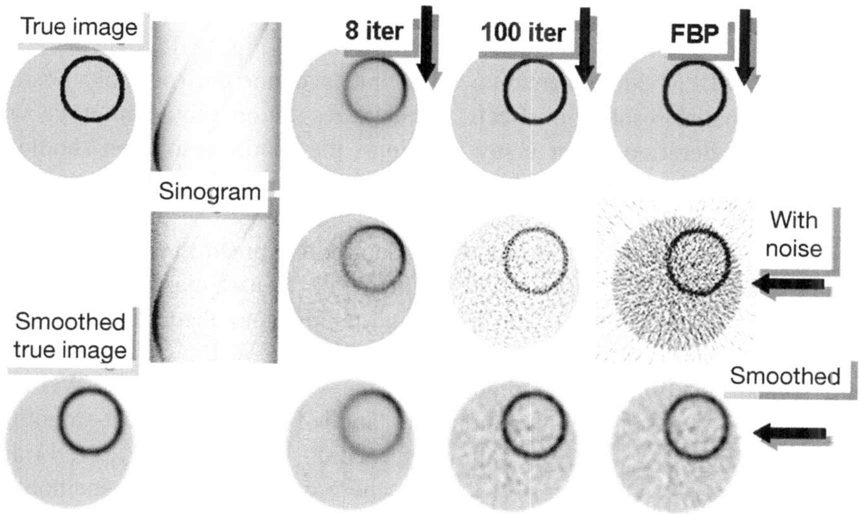

FIG. 13.15. Simulation study illustrating position dependent convergence in PET with attenuation. After 8 iterations (iter), convergence in highly attenuated regions is poor. After 100 iterations, good convergence is obtained, but with strong noise propagation. Post-smoothing yields a fair compromise between noise and nearly position independent resolution. FBP: filtered back projection.

It follows that reducing the number of iterations has an effect which is similar to reducing the cut-off frequency of a low-pass filter. However, the effect on the resolution is position dependent, as illustrated in Fig. 13.15. Attenuated PET projections of a highly radioactive uniform ring inside a less active disc were simulated with and without Poisson noise. After eight MLEM iterations, the reconstructed ring has non-uniform activity. In the centre of the phantom, convergence is slower, resulting in poorer resolution and poorer recovery of the activity in the ring. After 100 iterations, convergence is much better everywhere in the phantom, but for noisy data, there is very disturbing noise propagation.

If the image was acquired for detection (e.g. to see if there is a radioactive ring inside the disc or not), then the image produced after eight iterations is excellent. However, if the aim is quantification (e.g. analysing the activity distribution along the ring), then quantification errors can be expected at low iteration numbers.

13.3.5.2. Post-smoothed maximum-likelihood

The noise in the higher MLEM iterations is high frequency noise, and there are strong negative correlations between neighbouring pixels. As a result, a modest amount of smoothing strongly suppresses the noise at the cost of a mild loss of resolution. This is illustrated in the third row of Fig. 13.15.

If the MLEM implementation takes into account the (possibly position dependent) spatial resolution effects, then the resolution should improve with every MLEM iteration. After many iterations, the spatial resolution should be rather good, similar or even better than the sinogram resolution, but the noise will have propagated dramatically. It is assumed that the obtained spatial resolution corresponds to a position dependent point spread function which can be approximated as a Gaussian with a full width at half maximum (FWHM) of $F_{\mathrm{ML}}(x, y)$. Assume further that this image is post-smoothed with a (position independent) Gaussian convolution kernel with an FWHM of F_{p}. The local point spread function in the smoothed image will then have an FWHM of $\sqrt{(F_{\mathrm{ML}}(x,y))^2 F_{\mathrm{p}}^2}$. If enough iterations are applied and if the post-smoothing kernel is sufficiently wide, the following relation holds $F_{\mathrm{p}} \gg F_{\mathrm{ML}}(x, y)$ and, therefore, $\sqrt{(F_{\mathrm{ML}}(x,y))^2 + F_{\mathrm{p}}^2} \approx F_{\mathrm{p}}$. Under these conditions, the post-smoothed MLEM image has a nearly position independent and predictable spatial resolution. Thus, if PET or SPECT images are acquired for quantification, it is recommended to use many iterations and post-smoothing, rather than a reduced number of iterations, for noise suppression.

13.3.5.3. Smoothing basis functions

An alternative approach to counter noise propagation is to use an image representation that does not accomodate noisy images. Instead of representing the image with a grid of non-overlapping pixels, a grid of smooth, overlapping basis functions can be used. The two mostly used approaches are the use of spherical basis functions or 'blobs' [13.28] and the use of Gaussian basis functions or sieves [13.29].

In the first approach, the projector and back projector operators are typically adapted to work directly with line integrals of the basis functions. In the sieves approach, the projection of a Gaussian blob is usually modelled as the combination of a Gaussian convolution and projection along lines. The former approach produces a better approximation of the mathematics, while the latter approach yields a faster implementation.

The blobs or sieves are probably most effective when their width is very similar to the spatial resolution of the tomographic system. In this setting, the basis function allows accurate representation of the data measured by the tomographic system, and prevents reconstruction of much of the (high frequency) noise. It has been shown that using the blob during reconstruction is more effective than using the same blob only as a post-smoothing filter. The reason is that the post-filter always reduces the spatial resolution, while a sufficiently small blob does not smooth data if it is used during reconstruction.

If the blob or sieve is wider than the spatial resolution of the tomographic system, then its use during reconstruction produces Gibbs over- and undershoots, also known as 'ringing'. This effect always arises when steep edges have to be represented with a limited frequency range, and is related to the ringing effects observed with very sharp low-pass filters. For some imaging tasks, these ringing artefacts are a disadvantage.

13.3.5.4. Maximum a posteriori or penalized likelihood

Smoothing the MLEM image is not a very elegant approach: first, the likelihood is maximized, and then it is decreased again by smoothing the image. It seems more elegant to modify the objective function, such that the image that maximizes it does not need further processing. This can be done with a Bayesian approach, which is equivalent to combining the likelihood with a penalty function.

It is assumed that a good reconstruction image $\boldsymbol{\lambda}$ will be obtained if that image maximizes the (logarithm of the) probability $p(\boldsymbol{\lambda}|\boldsymbol{y})$ given by Eq. (13.39) and repeated here for convenience:

$$\hat{\boldsymbol{\lambda}} = \arg\max_{\boldsymbol{\lambda}}(\ln p(\boldsymbol{y}\,|\,\boldsymbol{\lambda}) + \ln p(\boldsymbol{\lambda})) \tag{13.74}$$

The second term represents the a priori knowledge about the tracer distribution, and it can be used to express our belief that the true tracer distribution is fairly smooth. This is usually done with a Markov prior. In a Markov prior, the a priori probability for a particular voxel, given the value of all other voxels, is only a function of the direct neighbours of that voxel:

$$p(\lambda_j\,|\,\lambda_k, \forall k \neq j) = p(\lambda_j\,|\,\lambda_k, k \in \boldsymbol{N}_j) \tag{13.75}$$

where $\boldsymbol{N}_j$ denotes the set of neighbour voxels of j.

Such priors are usually written in the following form:

$$P(\boldsymbol{\lambda}) = \ln p(\boldsymbol{\lambda}) = \sum_j \ln p(\lambda_j\,|\,\lambda_k, k \in \boldsymbol{N}_j) = -\beta \sum_j \sum_{k \in \boldsymbol{N}_j} E(\lambda_j \lambda_k) \tag{13.76}$$

where

the 'energy' function E is designed to obtain the desired noise suppressing behaviour and the parameter β is the weight assigned to the prior.

A higher weight results in smoother images, at the cost of a decreased likelihood, i.e. poorer agreement with the acquired data. In most priors, the expression is further simplified by making E a function of a single variable, the absolute value of the difference $|\lambda_j - \lambda_k|$.

Some popular energy functions $E(|\lambda_j - \lambda_k|)$ are shown in Fig. 13.16. A simple and effective one is the quadratic prior $E(x) = x^2$; an MAP reconstruction with this prior is shown in Fig. 13.17. Better preservation of strong edges is obtained with the Huber prior: it is quadratic for $|\lambda_j - \lambda_k| \leq \delta$ and linear for $|\lambda_j - \lambda_k| > \delta$, with a continuous first derivative at δ. Consequently, it applies less smoothing than the quadratic prior for differences larger than δ, as illustrated in Fig. 13.17. Even stronger edge tolerance is obtained with the Geman prior, which converges asymptotically to a constant for large differences, implying that it does not smooth at all over very large pixel differences.

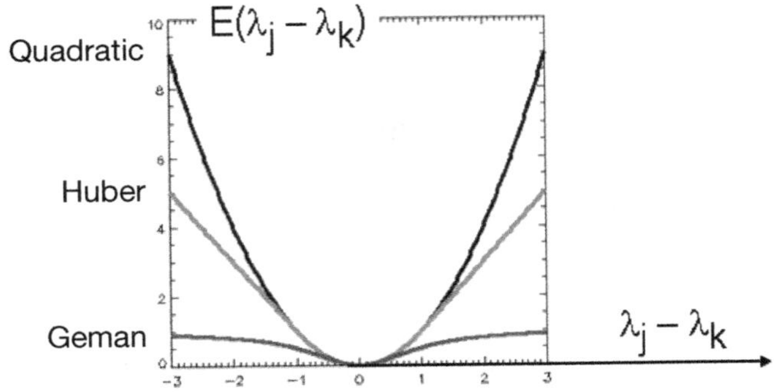

FIG. 13.16. The energy function of the quadratic prior, the Huber prior and the Geman prior.

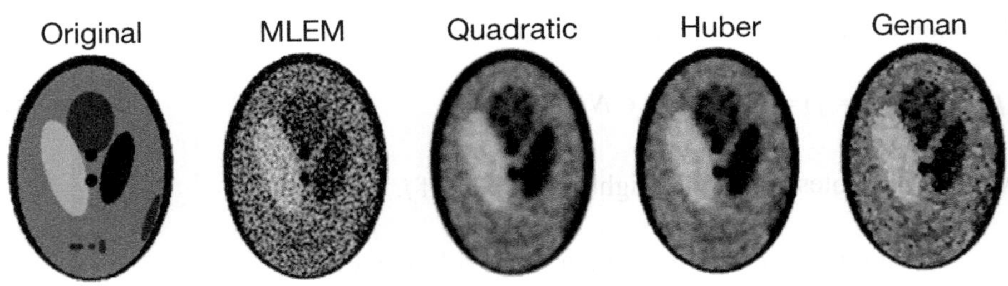

FIG. 13.17. Maximum-likelihood expectation-maximization (MLEM) and maximum a posteriori reconstructions of the Shepp–Logan phantom. Three different smoothing priors were used: quadratic, Huber and Geman. The latter smooth small differences quadratically, but are more tolerant for large edges.

It can be shown that the prior (Eq. (13.76)) is a concave function of λ if $E|\lambda_j - \lambda_k|$ is a convex function. Consequently, the quadratic and Huber energy functions yield a concave prior: it has a single maximum. In contrast, the Geman prior is not concave (see Fig. 13.16) and has local maximums. Such concave priors require careful initialization, because the final reconstruction depends on the initial image and on the behaviour of the optimization algorithm.

Figure 13.18 shows that MAP reconstructions produce position dependent spatial resolution, similar to MLEM with a reduced number of iterations. The reason is that the prior is applied with a uniform weight, whereas the likelihood provides more information about some voxels than about others. As a result, the prior produces more smoothing in regions where the likelihood is 'weaker', e.g. regions that have contributed only a few photons to the measurement due to high attenuation.

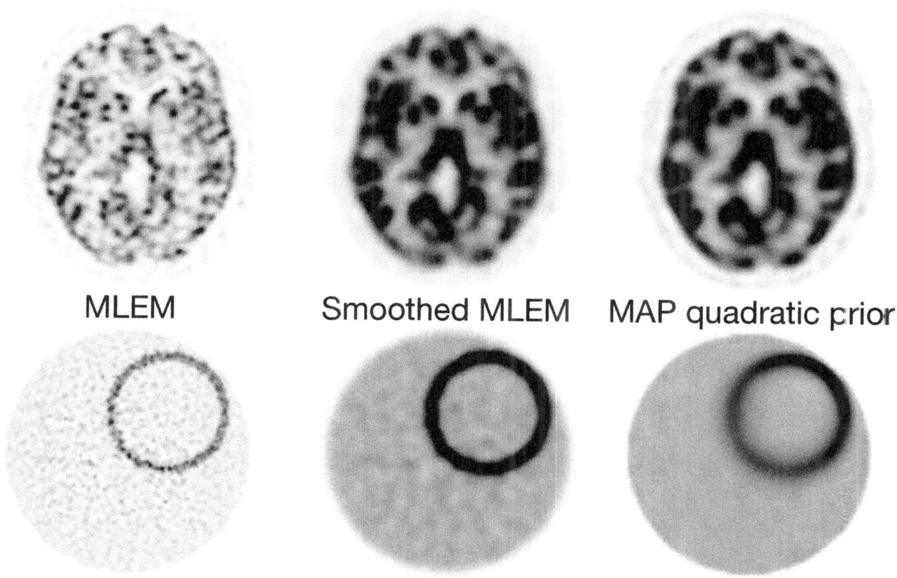

FIG. 13.18. Maximum-likelihood expectation-maximization (MLEM), smoothed MLEM and maximum a posteriori (MAP) (quadratic prior) reconstructions of simulated PET data of a brain and a ring phantom. The ring phantom reveals position dependent smoothing for MAP.

The prior can be made position dependent as well, to ensure that the balance between the likelihood and the prior is about the same in the entire image. In that case, MAP with a quadratic prior produces images which are very similar to MLEM images with post-smoothing: if the prior and smoothing are tuned to produce the same spatial resolution, then both algorithms also produce nearly identical noise characteristics.

Many papers have been devoted to the development of algorithms for MAP reconstruction. A popular algorithm is the so-called 'one step late' algorithm. Inserting the derivative of the prior P in Eq. (13.66) yields:

$$\frac{\partial(L_x(\boldsymbol{\lambda})+P(\boldsymbol{\lambda}))}{\partial\lambda_j}=\sum_i\left(\frac{y_i}{\hat{y}_i^{(k)}}A_{ij}\lambda_j^{(k)}\frac{1}{\lambda_j}-A_{ij}\right)+\frac{\partial P(\boldsymbol{\lambda})}{\partial\lambda_j}=0 \tag{13.77}$$

where $\hat{y}_i^{(k)}$ is the projection of the current reconstruction for detector i.

A problem with this equation is that $\partial P(\boldsymbol{\lambda})/\partial\lambda_j$ is itself a function of the unknown image $\boldsymbol{\lambda}$. To avoid this problem, the derivative of the prior is simply evaluated in the current reconstruction $\boldsymbol{\lambda}^{(k)}$. The equation can then be solved to produce the MAP update expression:

$$\lambda_j^{(k+1)}\frac{\lambda_j^{(k)}}{\sum_i A_{ij}-\left.\dfrac{\partial P(\boldsymbol{\lambda})}{\partial\lambda_j}\right|_{\boldsymbol{\lambda}^{(k)}}}\sum_i A_{ij}\frac{y_i}{\hat{y}_i^{(k)}} \tag{13.78}$$

Owing to the approximation, convergence is not guaranteed. The algorithm usually works fine, except with very high values for the prior.

The MLEM algorithm can be considered as a gradient ascent algorithm (see also Eq. (13.50)):

$$\lambda_j^{(k+1)}=\frac{\lambda_j^{(k)}}{\sum_i A_{ij}}\sum_i A_{ij}\frac{y_i}{\hat{y}_i^{(k)}} \tag{13.79}$$

$$=\lambda_j^{(k)}+\frac{\lambda_j^{(k)}}{\sum_i A_{ij}}\left.\frac{\partial \mathrm{L}_{\mathrm{ML}}(\boldsymbol{\lambda})}{\partial\lambda_j}\right|_{\boldsymbol{\lambda}^{(k)}} \tag{13.80}$$

Extensions to an MAP gradient ascent algorithm typically have the form:

$$\lambda_j^{(k+1)}=\lambda_j^{(k)}+S(\boldsymbol{\lambda}^{(k)})\left.\frac{\partial(L_{\mathrm{ML}}(\boldsymbol{\lambda})+P(\boldsymbol{\lambda}))}{\partial\lambda_j}\right|_{\boldsymbol{\lambda}^{(k)}} \tag{13.81}$$

where the key is to determine a good preconditioner S.

Several methods with (almost) guaranteed convergence have been based on the previously described optimization transfer method, by designing useful surrogate functions for both the likelihood and the prior.

13.3.6. Corrections

In typical emission data, the true events (having a Poisson character) are distorted and contaminated by a number of physical effects. To make the best use of the acquired data and of our knowledge of the acquisition system, these effects should be included in the reconstruction model. The distortion effects include resolution effects (such as detector resolution, collimator effects, and in PET also non-collinearity and positron range) and motion effects. The contamination effects can be divided, by their character and the way they are treated, into multiplicative and additive terms. The multiplicative factors include: attenuation of the annihilation photons by the object, the probability of the detector elements detecting an event once they are hit by the photon (detector normalization factors), coefficients accounting for the decay time and the geometrical restriction of directions/LORs for which true events are detected (axial acceptance angle, detector gaps). The additive terms include scattered and random (in the PET case) coincidences. Details on calculation of the correction factors and terms are discussed in other chapters. This chapter is limited to the discussion of their utilization within the reconstruction process.

The most straightforward approach is to pre-correct the data before reconstruction for the contamination effects (multiplying by multiplicative correction coefficients and subtracting the scatter and random estimates), so as to approximate the X ray transform (or attenuated X ray transform in the SPECT case) of the reconstructed object. For analytical reconstruction approaches (derived for the ideal X ray transform data), the data always have to be pre-corrected.

For the statistical reconstruction methods, derived based on the statistical properties of the data, an attempt is made to preserve the Poisson character of the data as much as possible by including the correction effects inside the reconstruction model. Theoretically, the most appropriate way is to include the multiplicative and scatter effects directly into the system matrix. The system matrix would have to include not only an accurate model of the direct data (true events) but also of the physical processes of the generation of the contamination scatter data. In a sense, the contamination would then become valid data, bringing extra information to our model and, thus, adding valid (properly modelled) counts to the image. However, inclusion of the scatter model into the system matrix tremendously increases the number of non-zero elements of the system matrix, i.e. the matrix is not sparse anymore, and consequently the system is more ill-posed (the contamination data are typically quite noisy) and computationally exceedingly expensive, and, thus, not feasible for routine clinical use.

The more practical, and commonly used, approach is to include correction effects as multiplicative factors and additive terms within the forward projection model of the iterative reconstruction approaches:

$$\boldsymbol{y} = \boldsymbol{A\lambda} + \boldsymbol{b} \tag{13.82}$$

where the effects directly influencing the direct (true) data are included inside the system matrix $\boldsymbol{A}$ and will be discussed in the following, while the additive terms $\boldsymbol{b}$ (including scatter and randoms) will be discussed separately in Section 13.3.6.2 on additive terms.

13.3.6.1. Factors affecting direct events — multiplicative effects

In the PET case, the sequence of the physical effects (described in previous chapters) that occur as the true coincident events are generated and detected can be described by the following factorization of the system matrix $\boldsymbol{A}$ as discussed in detail in Ref. [13.30]:

$$\boldsymbol{A} = \boldsymbol{A}_{\text{det.sens}}\boldsymbol{A}_{\text{det.blur}}\boldsymbol{A}_{\text{att}}\boldsymbol{A}_{\text{geom}}\boldsymbol{A}_{\text{tof}}\boldsymbol{A}_{\text{positron}} \tag{13.83}$$

where

$\boldsymbol{A}_{\text{positron}}$ models the positron range;

$\boldsymbol{A}_{\text{tof}}$ models the timing accuracy for the TOF PET systems (TOF resolution effects, as discussed in Section 13.3.3.3);

$\boldsymbol{A}_{\text{geom}}$ is the geometric projection matrix, the core of the system matrix, which is a geometrical mapping between the source (voxel j) and data (projection bin i, defined by the LOR, or its time bin in the TOF case); the geometrical mapping is based on the probability (in the absence of attenuation) that photon pairs emitted from an individual image location (voxel) reach the front faces of a given crystal pair (LOR);

$\boldsymbol{A}_{\text{att}}$ is a diagonal matrix containing attenuation factors on individual LORs;

$\boldsymbol{A}_{\text{det.blur}}$ models the accuracy of reporting the true LOR positions (detector resolution effects; discussed in Section 13.3.6.2);

and $\boldsymbol{A}_{\text{det.sens}}$ is a diagonal matrix modelling the probability that an event will be reported once the photon pair reaches the detector surface — a unique multiplicative factor for each detector crystal pair (LOR) modelled by normalization coefficients, but can also include the detector axial extent and detector gaps.

In practice, the attenuation operation A_{att} is usually moved to the left (to be performed after the blurring operation). This is strictly correct only if the attenuation factors change slowly, i.e. they do not change within the range of detector resolution kernels. However, even if this is not the case, a good approximation can be obtained by using blurred (with the detector resolution kernels) attenuation coefficients. In this case, the multiplicative factors $A_{det.sens}$ and A_{att} can be removed from the system matrix A and applied only after the forward projection operation as a simple multiplication operation (for each projection bin). The rest of the system matrix (except $A_{positron}$, which is object dependent) can now be pre-computed, whether in a combined or a factorized form, since it is now independent of the reconstructed object. On the other hand, the attenuation factors A_{att} (and $A_{positron}$, if considered) have to be calculated for each given object.

In the SPECT case, the physical effects affecting the true events can be categorized and factorized into the following sequence:

$$A = A_{det.sens} A_{det.blur} A_{geom,att} \tag{13.84}$$

where

$A_{det.sens}$ includes multiplicative factors (such as detector efficiency and decay time);

$A_{det.blur}$ represents the resolution effects within the gamma camera (the intrinsic resolution of the system);

and $A_{geom,att}$ is the geometric projection matrix, also including the collimator effects (such as the depth dependent resolution) and the depth and view dependent attenuation factors.

For gamma cameras, the energy and linearity corrections are usually performed in real time, and the remaining (detector efficiency) normalization factors are usually very close to one and can be, for all practical purposes, ignored or pre-corrected. Similarly, the theory says that the decay correction should be performed during the reconstruction, because it is different for each projection angle. However, for most tracers, the decay during the scan is very modest, and in practice it is usually either ignored or done as a pre-correction. The attenuation component is object dependent and needs to be recalculated for each reconstructed object. Furthermore, its calculation is much more computationally expensive than in the PET case, since it involves separate calculations of the attenuation factors for each voxel and for each view. This is one of the reasons why the attenuation factors have often been ignored in SPECT. More details on the inclusion of the resolution effects into the system matrix are discussed in Section 13.3.6.3.

13.3.6.2. Additive contributions

The main additive contaminations are scatter (SPECT and PET) and random events (PET). The simplest possibility of dealing with them is to subtract their estimates ($\bar{s}$ and $\bar{r}$) from the acquired data. While this is a valid (and necessary) pre-correction step for the analytical reconstructions, it is not recommended for statistical approaches since it changes the statistical properties of the data, causing them to lose their Poisson character. As the maximum-likelihood algorithm is designed for Poisson distributed data, its performance is suboptimal if the data noise is different from Poisson. Furthermore, subtraction of the estimated additive terms from the noisy acquired data can introduce negative values into the pre-corrected data, especially for low count studies. The negative values have to be truncated before the maximum-likelihood reconstruction, since it is not able to correctly handle the negative data. This truncation, however, leads to a bias in the reconstruction.

On the other end of the spectrum of possibilities, would be considering the scatter and randoms directly in the (full) system model, that is, including a complete physical model of the scatter and random components into a Monte Carlo calculation of the forward projection. However, this approach is exceedingly computationally expensive and is not feasible for practical use. A practical and the most common approach for dealing with the additive contaminations is to add their estimate ($\bar{\boldsymbol{b}} = \bar{\boldsymbol{s}} + \bar{\boldsymbol{r}}$) to the forward projection in the matrix model of the iterative reconstruction, i.e. the forward model is given by $\boldsymbol{A\lambda} + \bar{\boldsymbol{b}}$, as considered in the derivation of the MLEM reconstruction (Eq. (13.67)).

Special treatment has to be considered for clinical scanners in which the random events ($\boldsymbol{r}$, estimated by delayed coincidences) are on-line subtracted from the acquired data ($\boldsymbol{y}$, events in the coincidence window — prompts). The most important characteristic of the Poisson data is that their mean equals their variance: mean(y_i) = var(y_i). However, after the subtraction of the delays from the prompts (both being Poisson variables), the resulting data (γ) are not Poisson anymore, since mean(γ_i) = mean($y_i - r_i$) = mean(y_i) – mean(r_i), while var(γ_i) = var($y_i - r_i$) = var(y_i) + var(r_i). To regain the main characteristic of the Poisson data (at least of the first two moments), the shifted Poisson approach can be used, utilizing the fact that adding a (noiseless) constant value to the Poisson variable changes the mean but preserves the variance of the result. To modify the mean of the subtracted data γ to be equal to their variance (i.e. var(y_i) + var(r_i)), we need to add to the subtracted data an estimate (of the mean) of the randoms ($\overline{r}$) multiplied by two. This gives $\text{mean}(\gamma_i + 2\overline{r}_i) = \text{mean}(y_i - r_i - 2\overline{r}_i) = \text{mean}(y_i) + \text{mean}(r_i)$, which is equal to $\text{var}(\gamma_i + 2\overline{r}_i) = \text{var}(y_i) + \text{var}(y_i)$. The MLEM algorithm using the shifted Poisson model can then be written as:

$$\lambda_j^{(k+1)} = \frac{\lambda_j^{(k)}}{\sum_i A_{ij}} \sum_i A_{ij} \frac{\gamma_i + 2\overline{r}_i}{\sum_j A_{ij}\lambda_j^{(k)} + \overline{s}_i + 2\overline{r}_i} \tag{13.85}$$

It is worthwhile mentioning here that even in the shifted Poisson case, the negative values in the subtracted data and consequent truncation leading to the bias and artefacts cannot be completely avoided. However, the chance of the negative values decreases since the truncation of the negative values is being performed on the 'value-shifted' data $(\gamma_i + 2\overline{r}_i)$. Examples of reconstructions from data with a subtracted additive term, using the regular MLEM algorithm and using MLEM with the shifted Poisson model, are shown in Fig. 13.19. As the counts were relatively high in this simulation, the subtraction did not produce

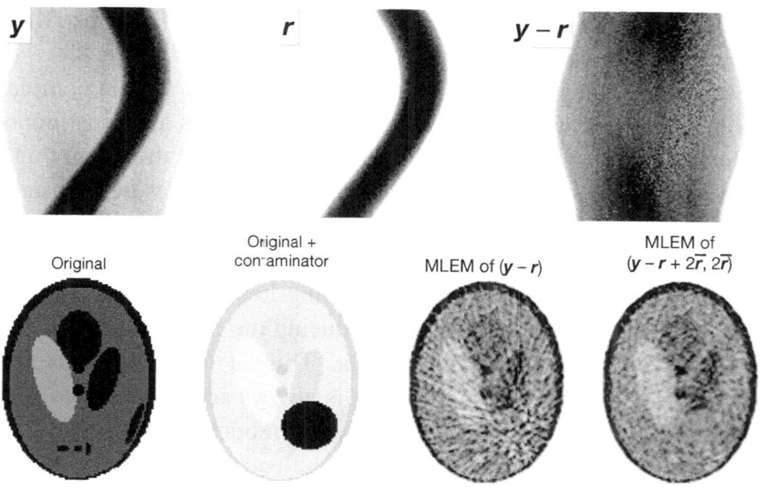

*FIG. 13.19. Illustration of (exaggerated case of) reconstructions from contaminated data **y** from which the additive contamination term **r** was subtracted (both data and contamination term are Poisson). The top row shows the sinograms. The increased noise level in the contaminated area in the sinogram (**y** – **r**) should be noted. The bottom row shows the true image without and with the contaminator, the maximum-likelihood expectation-maximization (MLEM) reconstruction from the subtracted data (**y** – **r**) and the shifted Poisson MLEM reconstruction, in which the estimated (noiseless) additive term $2\overline{\mathbf{r}}$ is added to the subtracted data and forward projection as given by Eq. (13.85).*

negatives. MLEM of $(\boldsymbol{y}-\boldsymbol{r})$ creates streaks because the reliability of the subtracted data is overestimated.

It should be noted that in the reconstruction model (as well as in the pre-correction approaches) the estimates of the scatter and randoms have to be treated in the same way as the estimates of the true events in the forward projection, including consideration of the normalized or un-normalized events, attenuation corrected or uncorrected data, gaps in the data, etc. Various challenges exist for the scatter and randoms estimations in general, such as modelling of the out of FOV scatter. This is addressed in Chapter 11.

13.3.6.3. Finite spatial resolution

There are a number of physical and geometrical effects and limitations (such as positron range, non-collinearity, depth of interaction, size of detector crystal elements, inter-crystal scatter, collimator geometry, etc.) affecting PET and SPECT resolution as described in more detail in Chapter 11. To get the most out of the acquired data and to correct for the resolution degradation, these effects have to be properly modelled in the system matrix of statistical reconstruction, as considered in the components ($\boldsymbol{A}_{\text{det.blur}}$, $\boldsymbol{A}_{\text{geom}}$, $\boldsymbol{A}_{\text{positron}}$) of the factorized system matrix outlined in Section 13.3.6.1. This step does not influence the mathematical definition of the reconstruction algorithm (such as MLEM, as given by Eq. (13.67)); only the form of its system matrix is changed.

However, this step has very practical consequences for the complexity of the algorithm implementation, for computational demands and most importantly for the quality of the reconstructed images. By including the resolution effects into the reconstruction model, a larger fraction of the data is being used for the reconstruction within each point of the space, with the true signal component becoming more consistent, while the noise components becoming less consistent with the model. Thus, the resolution modelling helps twice, by improving the image resolution while at the same time reducing the image noise, as illustrated in Fig. 13.20 for simulated SPECT data. This is quite different from the filtering case, where the noise suppression is always accompanied by resolution deterioration. On the other hand, the resolution modelling has a price in terms of a considerable increase in the computational load (both in space/memory and time) since the system matrix is much less sparse, that is, it contains a larger proportion of non-zero elements. This not only leads to more computational load per iteration, but also to a slower convergence of the iterative reconstruction and, consequently, to the need for more iterations.

Resolution effects can be subdivided into the effects dependent on the particular object, such as the positron range, and the effects influenced by the scanner geometry, design and materials (which can be determined beforehand for the given scanner). The positron range depends on the particular attenuation structures in which the the positrons annihilate, and also varies from isotope to isotope. Furthermore, the shape of the probability function (kernel) of the positron annihilation abruptly changes at the boundaries of two tissues, such as at the boundary of the lungs and surrounding soft tissues, and, thus, it strongly depends on the particular object's morphology and is quite challenging to model accurately. In general, the positron range has a small effect (compared to the other effects) for clinical scanners, particularly for studies using ^{18}F-labelled tracers, and can often be ignored. However, for small animal imaging and for other tracers (such as ^{82}Rb), the positron range becomes an important effect to be considered.

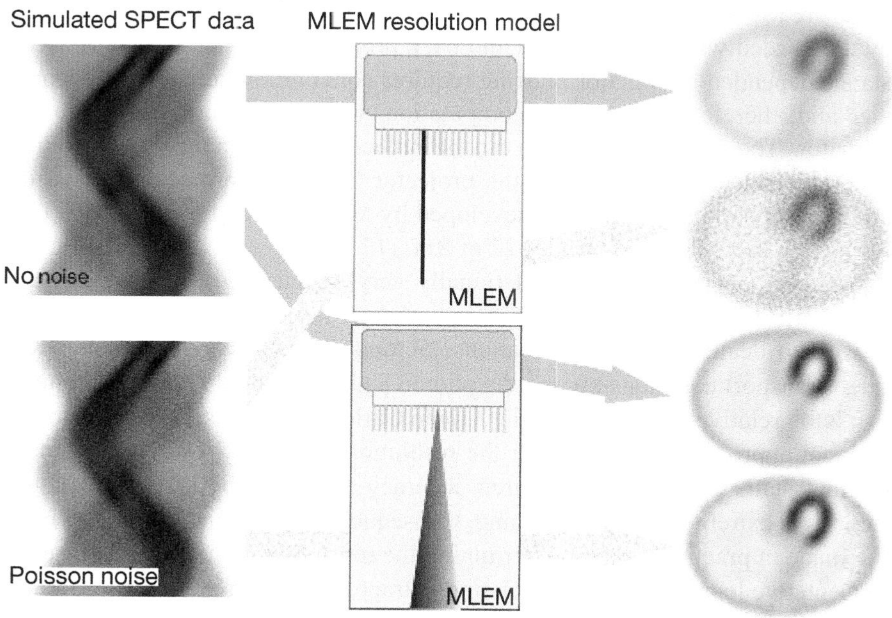

FIG. 13.20. Examples of the effects of resolution modelling within statistical iterative reconstruction. Data were simulated for a SPECT system with depth dependent resolution. It is clearly seen that using the proper resolution model within statistical reconstruction (lower two images on the right) not only improves the resolution of the images, but also helps to efficiently suppress the noise component.

There is a whole spectrum of approaches to determine and implement the scanner dependent resolution models. Only the main ones are addresssed. The simplest, but least accurate, approach is to approximate the system resolution model by a spatially invariant resolution kernel, usually a spherically symmetric Gaussian, with the shape (FWHM) estimated from point source measurements at one or more representative locations within the given scanner. This approach typically provides satisfactory results within the central FOV of large, whole body PET scanners. However, for PET systems with smaller ring diameters (relative to the reconstruction FOV), such as animal systems, and for SPECT systems with depth dependent resolution (and in particular with non-circular orbits), it is desirable to use more accurate spatially variant resolution models.

The second category is using analytically calculated resolution functions (usually spatially variant anisotropic kernels) for each location (LOR) as determined based on analytical models of physical effects affecting the resolution. This approach is usually limited to simple analytical models representing (or approximating) only basic physical characteristics of the system. The resolution kernels are usually calculated in real time during the reconstruction process when they are needed within the forward and back projection calculations. In SPECT, distance dependent collimator blurring requires convolution kernels that become wider and, therefore, need more computation, with increasing distance to the collimator. The computation time can be reduced considerably by integrating an incremental blurring step into the projector (and back projector), based on Gaussian diffusion. This method, developed by McCarthy and Miller in 1991, is described in more detail in chapter 22 of Ref. [13.5].

A more accurate but computationally very demanding approach is using Monte Carlo simulations of the resolution functions based on a set of point sources at various (ideally all) image locations. Setting up an accurate mathematical model (transport equations tracing the photon paths through the detector system/crystals) is relatively easy within the Monte Carlo simulations, compared to the analytical approach of determining the resolution function. However, to obtain sufficient statistics to get the desired accuracy of the shape of the resolution functions is extremely time consuming. Consequently, simplifications often have to be made in practice, such as determining the resolution kernels only at a set of representative locations and interpolating/extrapolating from them the resolution kernels at other locations.

The most accurate but also most involved approach is based on experimental measurements of the system response by measuring physical point sources at a set of image locations within the scanner. This is a tedious and very time consuming process, involving point sources with long half-life isotopes and usually requiring the use of accurate robotic stages to move the point source. Among the biggest challenges is to accumulate a sufficient number of counts to obtain an accurate point spread function, even at a limited number of locations. Consequently, the actual resolution kernels used in the reconstruction model are often estimated by fitting analytical functions (kernels) to the measured data, rather than directly using the measured point spread functions.

At the conclusion of this subsection, it is worth making the following general comment. In the light of the resolution modelling possibilities discussed above, one might wonder whether it is worth spending energy and resources on building new PET and SPECT systems with improved resolution properties. However, although it has been shown in the literature that proper system models lead to improved reconstructed image quality, they can never fully recover information that has been lost through resolution effects and other instrumentation limitations. Furthermore, due to the increased level of modelling, the system matrix becomes more dense, and consequently the inverse problem (reconstruction) becomes more ill-posed, thus making it impossible to attain perfect recovery for the realistic data. There is no doubt that improved instrumentation as well as novel and more accurate reconstruction models play an important role in improving image quality and quantitative accuracy, and eventually increasing the general clinical utility of emission tomography systems.

13.3.6.4. Motion corrections

Owing to the relatively long acquisition times, motion effects, caused by patient movement and organ motion and deformation, cannot be avoided in emission tomography. In the following, all of these effects are covered under the simple term 'motion'. With the continuous improvements of PET and SPECT technology, leading to improved spatial resolution, signal to noise ratio, image quality and accuracy of quantitative studies, corrections for motion effects become more important. In fact, artefacts caused by motion are becoming the single most important factor for image degradation, especially in PET or PET/computed tomography (CT) imaging of the upper torso region. For example, motion effects can lead to the loss of small lesions by blurring them out completely in regions with strong motion (such as near the lower lung wall), or to their misplacement into the wrong anatomical region (e.g. into the liver from the lungs, or vice versa). Motion correction has become an important research topic; however, a thorough discussion of this topic is out of the scope of this chapter and interested readers are referred to the literature on this topic. In the following, the main concepts of motion correction as dealt with within the reconstruction process are outlined.

The two main sources of motion related artefacts in emission studies are the motion during the emission scan and the discrepancy (caused by the motion) between the attenuation and emission data. The motion during the emission scan means that the emission paths (LORs) through the object (as considered in the system matrix) change during the scan time. If this time dependent change is not accounted for, the system model becomes inconsistent with the data, which results in artefacts and motion blurring in the reconstructed images. On the other hand, the transmission scan (CT) is relatively short and can usually be done in a breath-hold mode. Consequently, the attenuation image is usually motion-free and captures only one particular patient position and organ configuration (time frame). If the attenuation factors obtained from this fixed-time position attenuation image are applied to the emission data acquired at different time frames (or averaged over many time frames), this leads to artefacts in the reconstructed images, which tend to be far more severe in PET than in SPECT. This is, for example, most extremely pronounced at the bottom of the lungs which can typically move several centimetres during the breathing cycle, causing motion between two regions with very different attenuation coefficients.

Emission data motion: Correction approaches for motion during the emission scan are discussed first. The first step is subdividing the data (in PET, typically list-mode data) into a sufficient number of time frames to ensure that the motion within each frame is small. For the organ movement, the frames can be distributed over a period of the organ motion (e.g. breathing cycle). For the patient motion, the frames would be typically longer and distributed throughout the scan time. Knowledge about the motion can be obtained using external devices, such as cameras with fiducial markers, expansion belts or breathing sensors for respiratory motion, the electrocardiogram signal for cardiac motion, etc. There are also a limited number of approaches for estimating the motion directly from the data.

Once the data are subdivided into the set of the frames, the most straightforward approach is to reconstruct data independently in each frame. The problem with this approach is that the resulting images have a poor signal to noise ratio because the acquired counts have been distributed into a number of individual (now low count) frames. To improve the signal to noise ratio, the reconstructed images for individual frames can be combined (averaged) after they are registered (and properly deformed) to the reference time frame image. However, for statistical non-linear iterative reconstruction algorithms, this is not equivalent to (and typically of a lower quality than) the more elaborate motion correction approaches, taking into account all of the acquired counts in a single reconstruction, as discussed below.

For rigid motion (e.g. in brain imaging), the events on LORs (LOR_i) from each time frame, or time position, can be corrected for motion by translation (using affine transformations) into the new LORs (LOR_i) in the reference frame (see Fig. 13.21 (top right, solid line)), in which the events would be detected if there were no motion. Reconstruction is then done in a single reference frame using all acquired counts, leading to a better signal to noise ratio in the reconstructed images. Care has to be taken with the detector normalization factors so that the events are normalized using the proper factors (N_i) for the LORs on which they were actually detected (and not into which they were translated).

Attenuation factors are obtained on the transformed lines ($att_{i'}$) through the attenuation image in the reference frame. Care also has to be given to the proper treatment of data LORs with events being translated into, or out of, the detector gaps or detector ends. This is important, in particular for the calculation of the sensitivity matrix, which then becomes a very time consuming process.

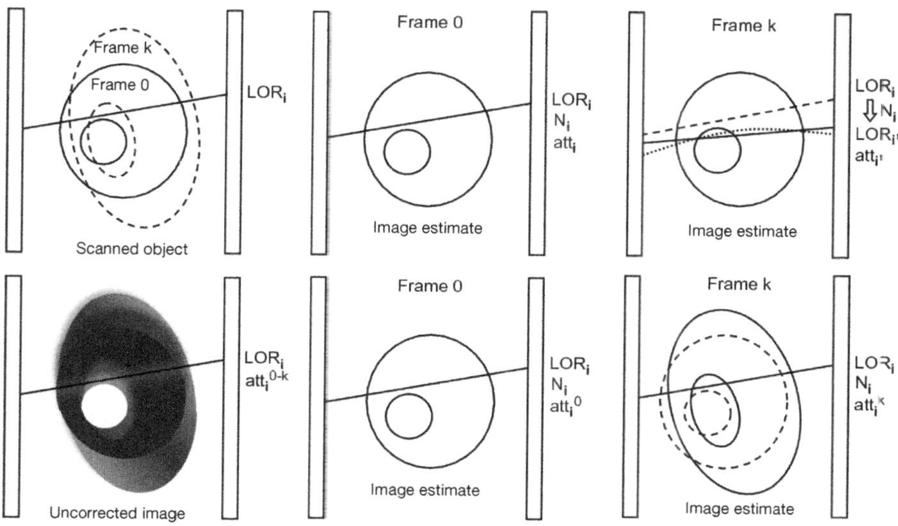

FIG. 13.21. Illustration of motion corrections for events acquired within line of response LOR_i with corresponding normalization N_i and attenuation att_i factors. Left top: positions and shapes of the object in the reference time frame 0 and frame k. Left bottom: illustration of blurring in the reconstruction combining events from all frames without motion correction (attenuation factors are also averaged over the whole range of the frames att_i^{0-k}). Middle column: processing within the reference time frame. Right top: LOR based motion correction for frame k — the LOR_i (dashed line) has to be transformed to the $LOR_{i'}$ (solid line for rigid motion, dotted line for non-rigid motion) which represents the paths that the photons would travel through the reference object if there were no motion. It should be noted that although the LORs are transformed, the normalization factors are used for the crystal pairs (LORs) in which the events were detected (N_i), while the used attenuation factors are for the transformed paths ($att_{i'}$). Right bottom: image based motion correction, including image morphing of the estimated image from the reference frame (dashed lines) into the given frame (solid lines).

For non-rigid (elastic) motion, which is the case for most of the practical applications, the motion correction procedures become quite involved. There are two basic possibilities. The first approach is to derive the transformations of individual paths of events (LORs) from each frame into the reference frame (see Fig. 13.21 (top right, dotted line)). For the non-rigid motion, the transformed paths through the reference object frame are not straight lines anymore, thus leading to very large computational demands for the calculations of the forward and back projection operations. The same care for normalization, gaps and detector ends has to be taken as above.

The second, more efficient, approach involves morphing the image estimate (of the reference image) into the frame for which current events (LORs) are being processed (see Fig. 13.21 (bottom right, solid line)). It should be noted that some pre-sorting of the data is considered, so that events from each frame are processed together (using a common image morphing operation). Here, the acquired LORs (LOR_i) and their normalization coefficients (N_i) are directly used without modification. However, the sensitivity matrix still needs to be carefully calculated, taking into consideration update and subset strategy, e.g. including the morphing operation if subset data involve several frames. This is, however, a simpler operation than in the LOR based case since the morphing is done in the image domain. This image based approach is not only more efficient, but also better reflects/models the actual data acquisition process during which the acquired object is being changed (morphed).

Attenuation effects: In the following, it is considered that either attenuation information for each time frame is available, for example, having a sequence of CT scans for different time positions, or there is knowledge of the motion and tools to morph a fixed-time position CT image to represent attenuation images at individual time frames. It is further considered that tools are available to obtain the motion transformation of data and/or images between the individual time frames. If the emission data are stored or binned without any motion gating, they represent motion-blurred emission information over the duration of the scan. Using attenuation information for them for a fixed time position is not correct. It would be better to pre-correct those data using proper attenuation factors for each frame, but then the statistical properties (Poisson character) are lost due to the pre-correction. A good compromise (although not theoretically exact) is to use motion-blurred attenuation factors during the pre-correction or the reconstruction process.

For data stored in multiple time frames, separate attenuation factors (or their estimates) are used for each frame, such that they reflect attenuation factors (for each LOR) at that particular time frame. For the case when there are multiple CT images, this is simply obtained by calculation (forward projection) of the attenuation coefficients for each frame from the representative CT image for that frame. For the case when there is only one CT image, attenuation factors have to be calculated on the modified LORs (for each time frame) in the LOR based corrections, or to morph the attenuation image for each frame and then calculate the attenuation factors from the morphed images in the image based corrections.

13.4. NOISE ESTIMATION

13.4.1. Noise propagation in filtered back projection

The pixel variance in an image reconstructed with FBP can be estimated analytically, by propagating the uncorrelated Poisson noise in the data through the reconstruction operation. The FBP algorithm can be written as:

$$\Lambda(x,y)=\int_0^{\pi}\mathrm{d}\phi\int_{-\infty}^{\infty}\mathrm{Y}(x\cos\phi+y\sin\phi-s)\mathrm{h}(s)\,\mathrm{d}s \tag{13.86}$$

where h(s) is the convolution kernel, combining the inverse Fourier transform of the ramp filter and a possible low-pass filter to suppress the noise.

The variance on the measured sinogram $\boldsymbol{Y}(s, \phi)$ data equals its expectation $\bar{\boldsymbol{Y}}(s, \phi)$; the covariance between two different sinogram values $\boldsymbol{Y}(s, \phi)$ and $\boldsymbol{Y}(s', \phi')$ is zero. Consequently, the covariance between two reconstructed pixel values $\Lambda(x, y)$ and $\Lambda(x', y')$ equals:

$$\mathrm{covar}(\Lambda(x,y),\Lambda(x',y'))=\int_0^{\pi}\mathrm{d}\phi\int_{-\infty}^{\infty}\bar{\mathrm{Y}}(x\cos\phi+y\sin\phi-s)\,\mathrm{d}s$$
$$\mathrm{h}(s)\mathrm{h}(s+(x'-x)\cos\phi+(y'-y)\sin\phi) \tag{13.87}$$

This integral is non-zero for almost all pairs of pixels. As h(s) is a high-pass filter, neighbouring reconstruction pixels tend to have fairly strong negative correlations. The correlation decreases with increasing distance between (x, y) and (x', y'). The variance is obtained by setting $x = x'$ and $y = y'$, which produces:

$$\mathrm{var}(\Lambda(x,y))=\int_0^{\pi}\mathrm{d}\phi\int_{-\infty}^{\infty}\bar{\mathrm{Y}}(x\cos\phi+y\sin\phi-s)|\mathrm{h}(s)|^2\,\mathrm{d}s \tag{13.88}$$

Figure 13.22 shows the variance image of the FBP reconstruction of a simulated PET sinogram of a heart phantom. The image was obtained by reconstructing 400 sets of noisy PET data. The figure also shows a noise-free and one of the noisy FBP images. The noise creates streaks that extend to the edge of the image. As a result, the variance is non-zero in the entire image.

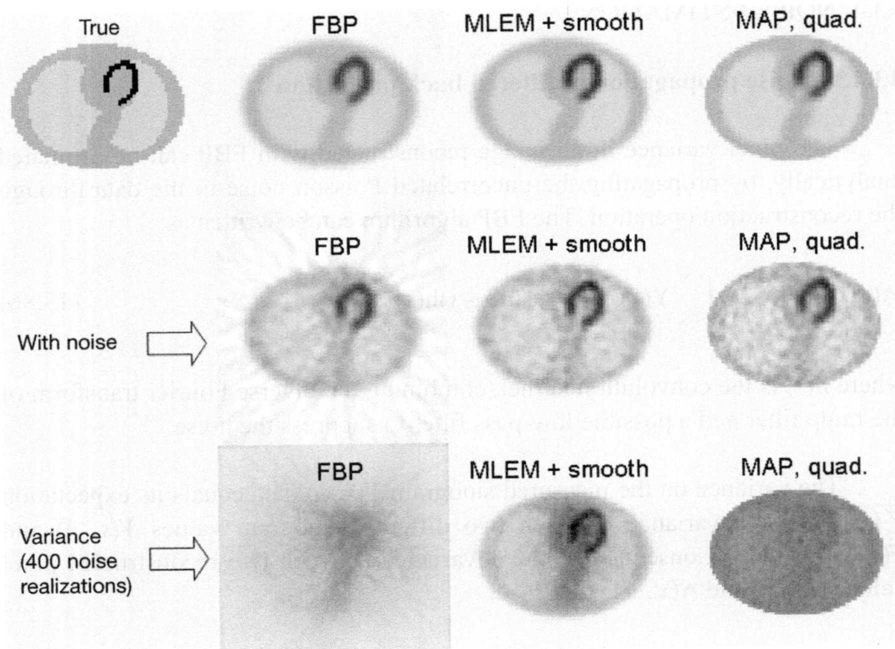

FIG. 13.22. Simulated PET reconstructions of a heart phantom. Reconstructions were done with filtered back projection (FBP), maximum-likelihood expectation-maximization (MLEM) with Gaussian post-smoothing and with maximum a posteriori (MAP) using a quadratic prior. For each algorithm, a noise-free and a noisy reconstruction are shown, and also the pixel variance obtained from 400 independent Poisson noise realizations on the simulated PET data. All reconstructions (first two rows) are shown on the same grey value scale. A second scale was used to display the three variance images. The noisy FBP image contains negative pixels (displayed in white with this scale).

13.4.2. Noise propagation in maximum-likelihood expectation-maximization

The noise analysis of MLEM (and MAP) reconstruction is more complicated than that for FBP because these algorithms are non-linear. However, the MLEM algorithm has some similarity with the WLS algorithm, which can be described with matrix operations. The WLS reconstruction was described previously; Eq. (13.45) is repeated here for convenience (the additive term was assumed to be zero for simplicity):

$$\boldsymbol{\lambda} = (\boldsymbol{A}'\boldsymbol{C}_y^{-1}\boldsymbol{A})^{-1}\boldsymbol{A}'\boldsymbol{C}_y^{-1}\boldsymbol{y} \tag{13.89}$$

$\boldsymbol{C}_y$ is the covariance of the data, which is defined as $\boldsymbol{C}_y = E(\boldsymbol{y}-\bar{\boldsymbol{y}})(\boldsymbol{y}-\bar{\boldsymbol{y}})'$, where E denotes the expectation and $\bar{\boldsymbol{y}}$ is the expectation of $\boldsymbol{y}$.

The covariance of the reconstruction is then:

$$\begin{aligned} C_\lambda &= E(\lambda-\bar{\lambda})(\lambda-\bar{\lambda})' \\ &= (A'C_y^{-1}A)^{-1}A'C_y^{-1}E(y-\bar{y})(y-\bar{y})'C_y^{-1}A(A'C_y^{-1}A)^{-1} \\ &= (A'C_y^{-1}A)^{-1} \end{aligned} \quad (13.90)$$

This matrix gives the covariances between all possible pixel pairs in the image produced by WLS reconstruction. The projection $\boldsymbol{A}$ and back projection $\boldsymbol{A}'$ have a low pass characteristic. Consequently, the inverse $(\boldsymbol{A}'\boldsymbol{C}_y^{-1}\boldsymbol{A})^{-1}$ acts as a high-pass filter. It follows that neighbouring pixels of WLS reconstructions tend to have strong negative correlations, as is the case with FBP. Owing to this, the MLEM variance decreases rapidly with smoothing.

Figure 13.22 shows mean and noisy reconstructions and variance images of MLEM with Gaussian post-smoothing and MAP with a quadratic prior. For these reconstructions, 16 iterations with 8 subsets were applied. MAP with a quadratic prior produces fairly uniform variance, but with a position dependent resolution. In contrast, post-smoothed MLEM produces fairly uniform spatial resolution, in combination with a non-uniform variance.

REFERENCES

[13.1] LEWITT, R.M., MATEJ, S., Overview of methods for image reconstruction from projections in emission computed tomography, Proc. IEEE Inst. Electr. Electron. Eng. **91** (2003) 1588–1611.

[13.2] NATTERER, F., The Mathematics of Computerized Tomography, Society for Industrial and Applied Mathematics (SIAM), Philadelphia, PA (1986).

[13.3] KAK, A.C., SLANEY, M., Principles of Computerized Tomographic Imaging, Society for Industrial and Applied Mathematics (SIAM), Philadelphia, PA (1988).

[13.4] BARRETT, H.H., MYERS, K.J., Foundations of Image Science, John Wiley and Sons, Hoboken, NJ (2004).

[13.5] WERNICK, M.N., AARSVOLD, J.N. (Eds), Emission Tomography, The Fundamentals of PET and SPECT, Elsevier Academic Press (2004).

[13.6] NATTERER, F., Inversion of the attenuated radon transform, Inverse Probl. **17** (2001) 113–119.

[13.7] XIA, W., LEWITT, R.M., EDHOLM, P.R., Fourier correction for spatially variant collimator blurring in SPECT, IEEE Trans. Med. Imaging **14** (1995) 100–115.

[13.8] DEFRISE, M., CLACK, R., TOWNSEND, D.W., Image reconstruction from truncated, two-dimensional, parallel projections, Inverse Probl. **11** (1996) 287–313.

[13.9] DEFRISE, M., KUIJK, S., DECONINCK, F., A new three-dimensional reconstruction method for positron cameras using plane detectors, Phys. Med. Biol. **33** (1988) 43–51.

[13.10] KINAHAN, P.E., ROGERS, J.G., Analytic three-dimensional image reconstruction using all detected events, IEEE Trans. Nucl. Sci. **NS-36** (1990) 964–968.

[13.11] DAUBE-WITHERSPOON, M.E., MUEHLLEHNER, G., Treatment of axial data in three-dimensional PET, J. Nucl. Med. **28** (1987) 1717–1724.

[13.12] LEWITT, R.M., MUEHLLEHNER, G., KARP, J.S., Three-dimensional image reconstruction for PET by multi-slice rebinning and axial image filtering, Phys. Med. Biol. **39** (1994) 321–339.

[13.13] DEFRISE, M., et al., Exact and approximate rebinning algorithms for 3D PET data, IEEE Trans. Med. Imaging **16** (1997) 145–158.

[13.14] DEFRISE, M., A factorization method for the 3D X-ray transform, Inverse Probl. **11** (1995) 983–994.

[13.15] TOMITANI, T., Image reconstruction and noise evaluation in photon time-of-flight assisted positron emission tomography, IEEE Trans. Nucl. Sci. **NS-28** (1981) 4582–4589.

[13.16] QI, J., LEAHY, R.M., Iterative reconstruction techniques in emission computed tomography, Phys. Med. Biol. **51** (2006) R541–578.

[13.17] FESSLER, J.A., BOOTH, S.D., Conjugate-gradient preconditioning methods for shift variant PET image reconstruction, IEEE Trans. Image Process. **8** (1999) 688–699.

[13.18] PRESS, W.H., FLANNERY, B.P., TEUKOLSKY, S.A., VETTERLING, W.T., Numerical Recipes, The Art of Scientific Computing, Cambridge University Press (1986).

[13.19] DE PIERRO, A.R., A modified expectation maximization algorithm for penalized likelihood estimation in emission tomography, IEEE Trans. Med. Imaging **14** (1995) 132–137.

[13.20] SHEPP, L.S., VARDI, Y., Maximum likelihood reconstruction for emission tomography, IEEE Trans. Med. Imaging **MI-1** (1982) 113–122.

[13.21] DEMPSTER, A.P., LAIRD, N.M., RUBIN, D.B., Maximum likelihood from incomplete data via the EM algorithm, J. R. Stat. Soc. Series B Stat. Methodol. **39** (1977) 1–38.

[13.22] PARRA, L., BARRETT, H.H., List-mode likelihood: EM algorithm and image quality estimation demonstrated on 2-D PET, IEEE Trans. Med. Imaging **17** 2 (1998) 228–235.

[13.23] READER, A.J., ERLANDSSON, K., FLOWER, M.A., OTT, R.J., Fast accurate iterative reconstruction for low-statistics positron volume imaging, Phys. Med. Biol. **43** 4 (1998) 835–846.

[13.24] QI, J., Calculation of the sensitivity image in list-mode reconstruction, IEEE Trans. Nucl. Sci. **53** (2006) 2746–2751.

[13.25] MATEJ, S., et al., Efficient 3-D TOF PET reconstruction using view-grouped histo images: DIRECT — Direct Image Reconstruction for TOF, IEEE Trans. Med. Imaging **28** (2009) 739–751.

[13.26] HUDSON, M.H., LARKIN, R.S., Accelerated image reconstruction using ordered subsets of projection data, IEEE Trans. Med. Imaging **13** (1994) 601–609.

[13.27] BROWNE, J., DE PIERRO, A.R., A row-action alternative to the EM algorithm for maximizing likelihoods in emission tomography, IEEE Trans. Med. Imaging **15** (1996) 687–699.

[13.28] DAUBE-WITHERSPOON, M.E., MATEJ, S., KARP, J.S., LEWITT, R.M., Application of the row action maximum likelihood algorithm with spherical basis functions to clinical PET imaging, IEEE Trans. Nucl. Sci. **48** (2001) 24–30.

[13.29] SNYDER, D.L., MILLER, M.I., THOMAS, L.J., Jr., POLITTE, D.G., Noise and edge artefacts in maximum-likelihood reconstructions for emission tomography, IEEE Trans. Med. Imaging **MI-6** (1987) 228–238.

[13.30] LEAHY, R.M., QI, J., Statistical approaches in quantitative positron emission tomography, Stat. Comput. **10** (2000) 147–165.

CHAPTER 14

NUCLEAR MEDICINE IMAGE DISPLAY

H. BERGMANN
Center for Medical Physics and Biomedical Engineering,
Medical University of Vienna,
Vienna, Austria

14.1. INTRODUCTION

The final step in a medical imaging procedure is to display the image(s) on a suitable display system where it is presented to the medical specialist for diagnostic interpretation. The display of hard copy images on X ray film or photographic film has largely been replaced today by soft copy image display systems with cathode ray tube (CRT) or liquid crystal display (LCD) monitors as the image rendering device. Soft copy display requires a high quality display monitor and a certain amount of image processing to optimize the image both with respect to the properties of the display device and to some psychophysiological properties of the human visual system. A soft copy display system, therefore, consists of a display workstation providing some basic image processing functions and the display monitor as the intrinsic display device. Display devices of lower quality may be used during intermediate steps of the acquisition and analysis of a patient study. Display monitors with a quality suitable for diagnostic reading by the specialist medical doctor are called primary devices, also known as diagnostic devices. Monitors with lower quality but good enough to be used for positioning, processing of studies, presentation of images in the wards, etc. are referred to as secondary devices or clinical devices.

Nuclear medicine images can be adequately displayed even for diagnostic purposes on secondary devices. However, the increasing use of X ray images on which to report jointly with images from nuclear medicine studies, such as those generated by dual modality imaging, notably by positron emission tomography (PET)/computed tomography (CT) and single photon emission computed tomography (SPECT)/CT, requires display devices capable of visualizing high resolution grey scale images at diagnostic quality, i.e. primary display devices. Both grey scale and colour display devices are used, the latter playing an important role in the display of processed nuclear medicine images and in the display of overlaid images such as from registered dual modality imaging studies.

Owing to the advances of picture archiving and communication systems (PACSs), the location of a display device which used to be adjacent to the gamma camera or PET scanner now widely varies, and can, for example, be in special reporting rooms, patient wards and operating rooms. An important requirement for display devices connected to a PACS is that, regardless of the display device and the imaging modality used, the display of an image should be consistent in appearance and presentation among the monitors and printers used at different locations and under different environmental lighting conditions. A special standard arising from this need was created specifying the requirements to display grey scale images on different display devices [14.1]. For colour displays, similarity in visual appearance is achieved by using an industry standard colour management system (CMS) [14.2]. The requirements for inclusion of a CMS into a PACS are addressed in Ref. [14.3].

The technology of soft copy display devices is presently experiencing a rapid transition from CRT based monitors to LCD technologies, the latter potentially offering better image quality, improved stability, reduced weight and reduced costs.

Regardless of the technology, display hardware needs to operate at maximum performance and needs to deliver consistent and reproducible results. Quality assurance of a display device, both at the time of installation and at regular intervals during its lifespan, ensures stability and optimum performance, and should be considered an important component of the quality system in nuclear medicine.

14.2. DIGITAL IMAGE DISPLAY AND VISUAL PERCEPTION

Emissive display devices such as CRTs or LCDs display a soft copy of an image. Hard copies consist of transmissive film or of reflective prints, both produced by printing devices.

The digital image to be displayed is represented as a rectangular matrix of integer values. Each matrix element corresponds to a pixel of the image. Regardless of the modality, the original intensity values as produced by the acquisition device are integer values with a pixel depth of between 8 to 16 bits. For display, the original images are stored in the memory of a computer, the display workstation. The soft copy display system of a workstation consists of the monitor which produces the visible image on the screen, and of the associated display controller, also known as a graphics card or video card, which holds the intensity values and carries out the conversion of the digital values to the analogue signals which control the electronics of the display device.

Originally, there is no colour in the image, but pseudo-colours are routinely used in nuclear medicine images to improve the visibility or to emphasize special features within an image. Another common use of colour is with overlay display techniques of pairs of registered images, e.g. originating from dual modality acquisitions such as PET/CT or SPECT/CT. Colour display devices, therefore, play an increasingly important role in the visualization of medical images.

14.2.1. Display resolution

The resolution of a digital display device is commonly described in terms of the number of distinct pixels in each dimension that can be displayed. Displays are designed in such a way that individual pixels have an approximately quadratic shape. The basic question arising is how many pixels are needed for adequate visualization. The desirable lower limit of pixel size would be reached if the human eye would not be able to distinguish between the individual pixels which make up the screen. Assuming a spatial resolution of the human eye of 1 arc minute and an average reading distance for an X ray image of 65 cm, a human observer would not be able to discern two adjacent pixels as being different for pixels smaller than about 0.18 mm. A modern 3 megapixel (MP) colour LCD monitor with a screen diagonal of 540 mm (~21 in) has pixel matrix dimensions of 1536 × 2048, resulting in a pixel size of 0.21 mm which is close to the resolution limit of the eye, so that individual pixels are almost indistinguishable. Monitors of this quality are already used routinely in radiology as primary display devices except for mammography. A 5 MP LCD monitor of the same screen size would have pixel matrix dimensions of 2048 × 2560 and linear pixel dimensions of 0.17 mm. Monitors with such high resolution are accepted as primary devices for reading digital mammography images. Limits of display resolution can be overcome by magnification and interpolation techniques included as standard features in the display workstation.

LCD monitors are composed of a large number of liquid crystals. This number represents the 'native' resolution of the display device since each pixel can be addressed individually to change brightness and colour. The display of an image is best if each pixel of the image maps to a pixel of the screen. If the mapping requires interpolation of the image pixels to the screen pixels, the image loses sharpness. A CRT display, in contrast, can change screen pixel size without loss of image sharpness by changing deflection and modulation frequencies. Several display resolution values can, therefore, be used with equal sharpness.

A display issue specific to nuclear medicine arises from the fact that the matrix dimensions of the original acquired or reconstructed image are much smaller than the display resolution would permit to display. The small original image size is due to both the poor spatial resolution of a nuclear medicine imaging device and its noise characteristics. As a rule of thumb, the sampling size used for a nuclear medicine image preserves the information content in the image when it is approximately one third of the full width at half maximum (FWHM) of the spatial resolution of the acquisition system. Thus, a scintillation camera equipped with, for example, a high resolution collimator, a system spatial resolution of 8 mm FWHM and a field of view of 540 mm × 400 mm would require a matrix size of at most 256 × 256 pixels to preserve the information content transmitted by the camera. Even a current commercial off the shelf (COTS) display device, on the other hand, has minimum pixel dimensions of 1024 × 1280. Displaying the original image matrix at native pixel resolution would result in an image too small for visual interpretation. It is, therefore, essential for the image to be magnified to occupy a reasonable sized part of the available screen area. Straightforward magnification using a simple interpolation such as the 'nearest neighbour' interpolation scheme would result in a block structure with clearly visible square elements which strongly interfere with the intensity changes created by the true structure of the object generating artefacts in the interpretation. Magnification is necessary and can be done without artefact generation by a suitable interpolation algorithm that generates smooth transitions between screen pixels while preserving the intensity variations within the original image. It is the task of the display workstation to provide software for visualizing this type of image.

14.2.2. Contrast resolution

This refers to the number of intensity levels which an observer can perceive for a given display. It is referred to as perceived dynamic range (PDR).

Brightness refers to the emitted luminance on screen and is measured in candelas per square metre (cd/m^2). The maximum brightness of a monitor is an important quality parameter. Specifications of medical display devices also include the calibrated maximum brightness which is lower but is recommended for primary devices to ensure that the maximum luminance can be kept constant during the lifespan of the display device. Typical values for a primary device LCD monitor are 700 and 500 cd/m^2 for the maximum and the calibrated maximum luminance, respectively.

The dynamic range of a display monitor is defined as the ratio between the highest and the lowest luminance (brightness) a monitor is capable of displaying. The dynamic range is highest if measured in the absence of ambient light. It is then called contrast ratio (CR = L_H/L_L) and is the figure usually quoted by vendors in the specifications. A typical CR of a grey scale primary LCD monitor is 700:1, measured in a dark reading room. If luminance values are measured with ambient light present, which is the scenario in practice, CR is replaced by the luminance ratio (LR = L'_H/L'_L), which is the ratio of the highest and the lowest luminance values including the effect of ambient light. It can be considerably smaller than the CR, since the effect of ambient lighting is added as a luminance L_{amb} to both the minimum and the maximum luminances. The CR is related to the PDR, but its potential usefulness as a predictor of monitor performance suffers from the lack of standardized measurement procedures and the effect of ambient light. The dark room CR is a common performance parameter quoted by manufacturers.

The PDR is the number of intensity levels an observer can actually distinguish on a display. It can be estimated based on the concept of just noticeable differences (JNDs). The JND is the luminance difference of a given target under given viewing conditions that the average human observer can just perceive. The measured JND depends strongly on the conditions under which the experiment is performed, for example, on the size, shape and position of the target. The PDR is defined as the number of JNDs across the dynamic range of a display. The PDR for grey scale displays has been assessed in Ref. [14.4] to be around a hundred. The number of intensity values which a pixel of the digital image can hold is much higher. The pixel of an image matrix is usually 1 or 2 bytes deep. It is an integer number between 256 and 65 536, and is given by the pixel depth of the image matrix. It is a further task of the display system to scale the original intensity values to a range compatible with the performance of the human observer. It is common to use 256 intensity values to control the brightness of a display device as this is sufficient to produce a sequence of brightness levels perceived as continuous by the human observer.

Colour displays using pseudo-colour scales can extend the PDR to about 1.5 times that of a grey scale display. This has been demonstrated for a heated-object scale which has the additional advantage of producing a 'natural' image [14.4]. Owing to the enormous number of possible colour scales and the fact that the majority of them produce 'unnatural' images, the concept of JNDs, while valid in principle, cannot be transferred directly to colour displays.

14.3. DISPLAY DEVICE HARDWARE

14.3.1. Display controller

The display controller, also known as a video card or graphics card, is the interface between the computer and the display device. Its main components are the graphical processing unit (GPU), the video BIOS, the video memory and the random access memory digital to analogue converter (RAMDAC). The GPU is a fast, specialized processor optimized for graphics and image processing operations. The video memory holds the data to be displayed. The capacity of current COTS graphics cards ranges from 128 MB to 4 GB which is sufficient to store even a large sequence of images. This permits the rapid change between images on the screen for, for example, cine display, rapid browsing through a sequence of tomographic slices or the use of overlay techniques to display simultaneously additional information such as text, markers or regions of interest without having to modify the image data. The output signals generated by the RAMDAC depend on the monitor type. For CRT monitors, the RAMDAC generates analogue positioning and intensity signals and additional control signals for the deflection and synchronization of the electron beam. The output of the graphics card is via a video graphics array connector. For current LCD displays, the graphics card provides standardized digital output signals via the digital visual interface connector which avoids image distortion and electrical noise, and can be configured to use the native spatial resolution of the LCD display directly.

Lookup tables (LUTs) constitute a crucial part of the video memory because they play an important role in the implementation of intensity transformations and in the display of colours. An LUT contains the digital values which are converted by the RAMDAC to the analogue intensity values driving the monitor. These values are called digital driving levels (DDLs). The range of DDLs is typically 8 bit, i.e. from 0 to 255. For medical displays, the range can be larger, up to 12 bit. The maximum value contained in an LUT produces the maximum brightness the screen can display. Since there is some latitude in the maximum brightness a screen can display, it is adjusted by the monitor hardware controls or firmware, and in practice follows manufacturer's recommendations to ensure optimum performance with respect to both image quality and life expectancy of the monitor. The luminance values generated by the sequence of available DDLs (e.g. 0…255) produce the characteristic curve of the display device.

The intensity values of an image stored in video memory are mapped to the values in an LUT. The mapping transformation associates each intensity with an LUT index. In the case of a colour display, a triple of DDLs, one each for the three primary colours red, green and blue, is used for a pixel. The LUT consisting of triples of primary DDL values is referred to as a colour lookup table (CLUT). For a medical image, which by nature is a grey scale image, a pseudo-colour image is generated by a mapping transformation which associates the intensity value of an image pixel to an appropriate index of a CLUT.

14.3.2. Cathode ray tube

The CRT is a vacuum tube containing an electron gun and a fluorescent screen (Fig. 14.1). The electron gun produces electrons which are focused by a system of magnetic and electrostatic lenses into a pencil beam. The electron beam is accelerated by a positive high voltage applied to the anode towards the fluorescent screen. The screen is covered with a crystalline phosphorescent coating that produces a visible light spot when hit by the electron beam. The light distribution of the spot is given by a 2-D Gaussian function. The directional dependence of the intensity of the emitted light follows Lambert's cosine law, implying that the apparent luminance of the screen is independent of the viewing angle. Using the deflecting coils attached to the collar of the tube, the beam is made to scan the screen area in a rectangular pattern. At the same time, the intensity of the electron beam is controlled by the control grid, thereby producing different light intensities. The digital image matrix containing the numerical intensity values is visualized by synchronizing scanning motion and intensity modulation given by the LUT to produce an intensity pattern, the visual image, on the screen. The elements of the matrix occupy a rectangular grid on the screen, with the luminance of the centre of each grid point corresponding to the LUT value of the corresponding matrix element.

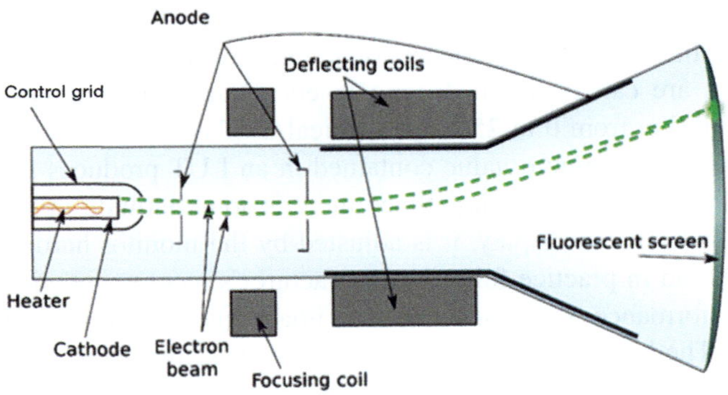

FIG. 14.1. Diagram of a cathode ray tube.

Colour CRTs use three different phosphors which emit red, green and blue light, respectively. The phosphors are packed together in clusters called 'triads' or in stripes. Colour CRTs have three electron guns, one for each primary colour. Each gun's beam reaches the dots of exactly one type of phosphor. A mask close to the screen absorbs electrons that would otherwise hit the wrong phosphor. The triads or stripes are so small that the intensities of the primary colours merge in the eye to produce the desired colour.

14.3.3. Liquid crystal display panel

An LCD display panel consists of a rectangular array of liquid crystal cells in front of a light source, the backlight. Each cell acts as a tiny light valve which transmits the backlight to an extent determined by a voltage applied to the liquid crystal. The image on the screen is formed by applying voltages to each cell separately, thereby modulating the light intensity into the desired intensity pattern.

A typical liquid crystal cell (Fig. 14.2) consists of a liquid crystal in twisted nematic phase between two glass plates, G, coated with alignment layers (not shown) that precisely twist the liquid crystal by 90° when no external field is present (left diagram). Light from the backside is polarized by polarizer P_1 and rotated by the crystal structure. The second polarizer P_2, set at 90° to P_1, then permits the light to pass. If a voltage is applied to the two transparent electrodes, E1 and E2, the nematics realign themselves (right diagram) and the polarized light is blocked by P_2. Partial realignment is achieved by varying the voltage and permits the transmitted intensity to vary.

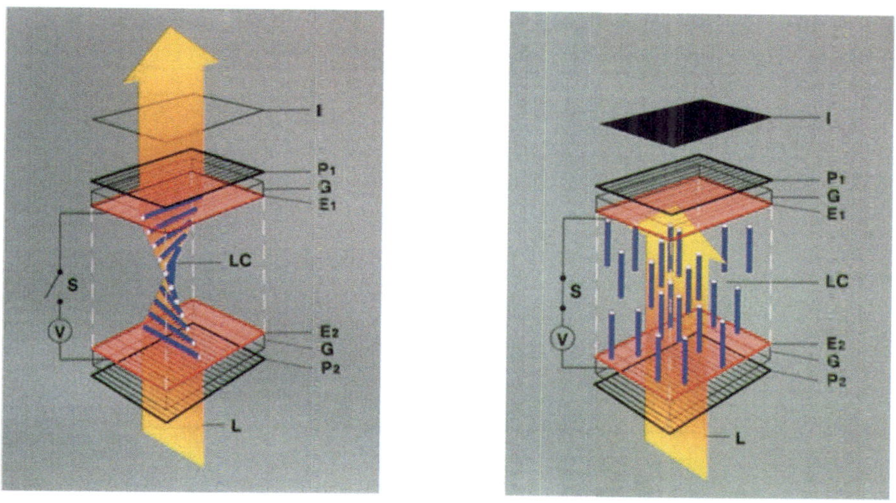

FIG. 14.2. Illustration of construction and operation of a single pixel of a twisted nematic liquid crystal cell. No voltage applied = OFF state (left diagram); voltage applied = ON state (right diagram) (courtesy of M. Schadt).

In colour LCDs, each individual pixel is divided into three cells, or subpixels, which are coloured red, green and blue, respectively, by additional filters. Each subpixel can be controlled independently, so that thousands or millions of possible colours can be obtained for each pixel.

An active matrix LCD is the predominant type of flat panel display. It is the standard display device used as general computer displays, in notebooks and increasingly as high quality displays for medical imaging. The active matrix design permits switching each pixel individually by applying a row and column addressing scheme. It is implemented by thin film transistor technology which supplies each pixel of the display with its own transistor. The circuit is made by depositing a thin film of silicon on the glass surface where the transistors are fabricated. The transistors take up only a small fraction of the surface and the rest of the silicon film is etched away to let the light pass through (Fig. 14.3).

LCD displays using twisted nematic liquid crystals exhibit a strong dependence of display brightness and colour on viewing angle. Developments in technology have considerably improved the angular viewing response. The preferred technique used for medical display devices is currently the in-plane switching (IPS) technology. IPS aligns the crystals horizontally and applies the voltage to realign the liquid crystal structure to both ends of the cell. Non-uniformity of both luminance and luminance ratio of an LCD is expressed as a function of viewing angle (from the normal to the display surface) for horizontal and vertical directions separately.

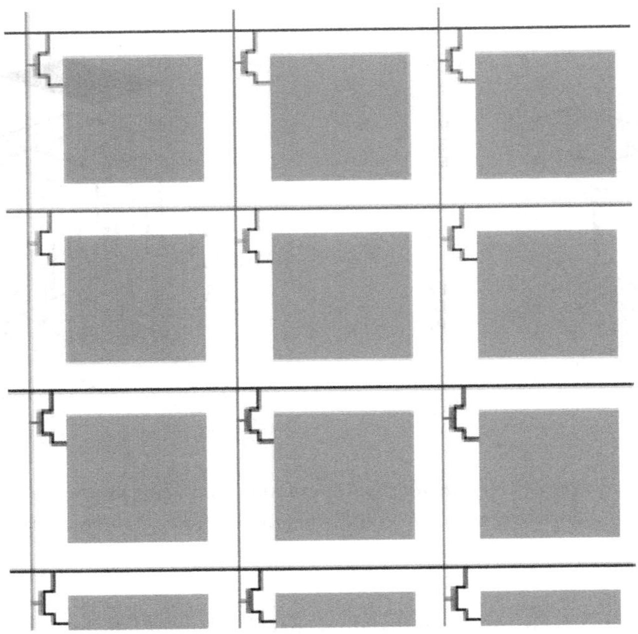

FIG. 14.3. A diagram of the pixel layout. Each liquid crystal pixel is connected to a transistor which provides the voltage that controls the brightness. The pixel is addressed by a row–column scheme.

14.3.4. Hard copy devices

Although reporting is increasingly performed using soft copy displays and PACS facilities, there is still a need for hard copies of images, such as for use in the operating room or to be sent to referring physicians. Hard copy images of diagnostic quality are printed on X ray film and use high resolution laser printing and dry processing. In nuclear medicine, where the original images of a diagnostic procedure often use pseudo-colours for the final display of the results, even cheap mainstream colour printers may adequately be used. The long term storage properties of hard copy media are not an issue when used in connection with a PACS, where images are stored in electronic form. Therefore, technologies such as dry laser film, thermal printers, colour laser printers or inkjet printers are all acceptable hardcopy output devices.

The spatial resolution of a printer is given in dots per inch (dpi), a measure of spatial printing density. It is defined as the number of individual dots that can be placed within the span of 1 in (2.54 cm).

14.3.4.1. Film laser imager

Dry laser imagers print X ray images on transparent film with the same quality as available with conventional X ray film. Spatial resolution is up to 650 dpi, adequate for diagnostic quality output for all imaging procedures, including mammography. Contrast resolution depends on film quality and can reach a D_{max} of up to 4.0.

14.3.4.2. Colour printer

Mainstream colour laser printers produce cheap grey scale and colour output of images with a spatial resolution of typically 600–1200 dpi. For normal paper, the CR is low. Image quality can be improved by using special paper with a smooth surface for improved toner gloss and sharpness.

Inkjet printers can have a spatial resolution of up to 9600 × 2400 dpi. With special photo paper, excellent image quality equivalent to colour photographs can be achieved. The stability of the printout is known to be somewhat fragile, with the image fading within a couple of years even under optimal storage conditions.

14.4. GREY SCALE DISPLAY

Nuclear medicine images do not require the same high quality grey scale displays as is necessary for the display of X ray images. This can be attributed to the fact that the nuclear medicine image is a low count image with considerable statistical fluctuations, making the comparison of tiny intensity differences meaningless. A main difference to diagnostic X ray reporting is the fact that colour in the image was recognized early as a helpful technique to improve diagnostic reading and a tradition of visualization in colour has been established. Therefore, images and analysis of results, in particular curves and functional or metabolic images, are preferably displayed using colours. The display is usually done on workstations with special nuclear medicine software and using current COTS colour LCD screens as standard display devices. Typical screen sizes are from 20 to 24 in and native display resolutions from 1024 × 1280 pixels to 1200 × 1600 pixels. Depending on the capabilities of the nuclear medicine display workstation's software, several monitors can be used simultaneously.

The need to be able to perform concurrent diagnostic reporting on X ray images, generated by dual mode acquisition techniques such as PET/CT and SPECT/CT, and the inclusion of images from other modalities via PACS in the reporting session, requires the use of grey scale display devices of diagnostic quality (primary devices) at the nuclear medicine display workstation.

Both CRT and LCD display devices are available with spatial resolution and CRs satisfying the requirements for a primary device. LCD displays are rapidly replacing CRT displays for several reasons:

— LCDs typically have about twice the brightness of CRTs. An overall brighter image is less sensitive to changes in the level of ambient light and is preferred for reporting.
— LCD monitors exhibit no geometric distortion.
— LCDs have a weight that is about one third of that of a comparable CRT.
— LCDs are less prone to detrimental ageing effects.
— LCDs are less expensive.

Furthermore, high quality colour LCD devices can be used as grey scale primary devices, which is not feasible for a colour CRT monitor.

14.4.1. Grey scale standard display function

Today's ubiquity of PACSs enables deployment of display devices at all locations where access to medical images is needed. The main challenge when using different display devices in a PACS is to ensure that an image presented to an observer appears identical irrespective of the display device used, be it a CRT based or LCD based soft copy display, or hard copy displays, such as film laser printers or paper printers. The Digital Imaging and Communications in Medicine (DICOM) grey scale standard display function (GSDF) offers a strategy that ensures that a medical image displayed or printed at any workstation or printing system for which the GSDF is implemented has the same visual appearance, within the capabilities of the particular display device [14.1]. This implies that a display device complying with the GSDF must be standardized and calibrated, and that a scheme of regular quality control is required for the display systems on the PACS. Colour display systems may also be used for the purpose of displaying grey scale images if calibrated to the GSDF [14.3].

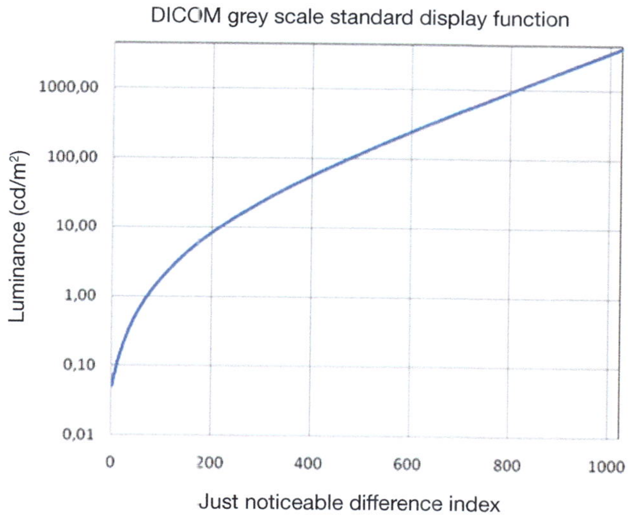

FIG. 14.4. Digital Imaging and Communications in Medicine (DICOM) grey scale standard display function.

The visual appearance of a native image as produced by the acquisition device (gamma camera, PET scanner, CT scanner) depends, if no corrections are applied, on the characteristic curve of the particular display device used at a display workstation. An image displayed using the characteristic curves inherent to a particular device could be significantly different in visual perception from the optimal presentation. Part 14 of the DICOM standard [14.1] standardizes the display of grey scale images. It does so by introducing the GSDF which can be seen as a universal characteristic curve (Fig. 14.4). The GSDF is based on human contrast sensitivity. It covers a luminance range from 0.05 to 4000 cd/m^2. The minimum luminance is the lowest that can be used in practice with a CRT display, whereas the maximum luminance is slightly above the luminance of a very bright unattenuated light box used for the examination of mammography X ray films, so that it covers the range of luminance values of all display devices in current use. Human contrast sensitivity is non-linear within this range. Perceptual similarity of a displayed image is achieved by linearizing the GSDF with respect to contrast sensitivity. This is done by introducing the JND index. One step in the JND index corresponds to a luminance difference that is just noticeable, regardless of the mean luminance level. The DICOM standard contains the standard GSDF as a tabulation of luminance (brightness) against the JND index. Table 14.1 shows the first and the last few JND indices of the tabulation. It can be seen clearly that the relative changes of luminance need to be much larger on the dark side of the curve than on the bright side to achieve a JND.

TABLE 14.1. TABULATED JUST NOTICEABLE DIFFERENCE INDICES OF THE GREY SCALE STANDARD DISPLAY FUNCTION

Just noticeable difference	Luminance (cd/m^2)
1	0.0500
2	0.0547
3	0.0594
4	0.0643
—	—
—	—
—	—
1021	3941.8580
1022	3967.5470
1023	3993.4040

Note: The relative difference between the luminance of consecutive just noticeable difference indices is much higher for low indices (~9%) than for high indices (~0.6%).

An individual display device with a luminance range L_{min}–L_{max} and a DDL range of, for example, 8 bits exhibits a characteristic luminance curve as a function of the DDL as shown in Fig. 14.5. The device specific characteristic curve will usually not match the corresponding segment of the GSDF. A transformation is needed, which is implemented as an LUT. The LUT will map a DDL D_s which should produce the standard luminance value L_s to the value D_m, so that for input level D_s the transformed value D_m will produce the correct luminance as required by the GSDF. The transformation LUT correcting for the deviations of the characteristic curve of a specific display system from the GSDF may be implemented directly in the display device or in the video memory of the display controller. The result of the transformation is that the modified DDLs operating the display will generate a characteristic curve identical to the GSDF.

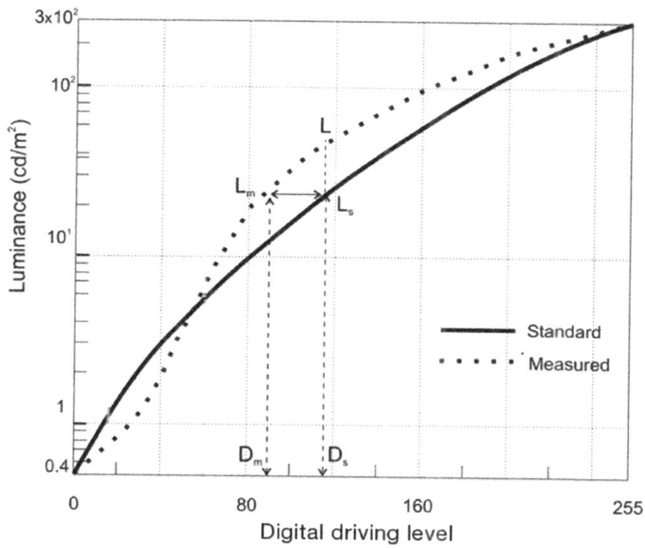

FIG. 14.5. Mapping of digital driving level D_s to the value D_m, so that for input level D_s which should produce the standard luminance value L_s the transformed value D_m will produce the correct luminance as given by the grey scale standard display function.

14.5. COLOUR DISPLAY

The human eye can distinguish millions of different colours. The full range of colours the average human can see is given by the spectrum of sunlight. Each colour can be characterized by three coordinates representing the colour as a mixture of three primary colours in a colour space.

One of the first colour spaces introduced in 1931 is the International Commission on Illumination (CIE) xyz colour space [14.5]. It is derived from a model of human colour perception and uses three tri-stimulus values to compose a colour. It is designed in such a way that one of the three coordinates defines luminance (brightness); the other two coordinates represent the chromaticity (hue). This leads to the well known CIE 1931 chromaticity diagram with its typical horseshoe shape in which all colours the human visual system can perceive are represented as a function of two coordinates, x and y (Fig. 14.6). The third coordinate, representing brightness, would only change the saturation of a colour, so, for example, varying the brightness coordinate for the chromaticity 'white' would run through all levels of grey from black to the maximum white a display device can render.

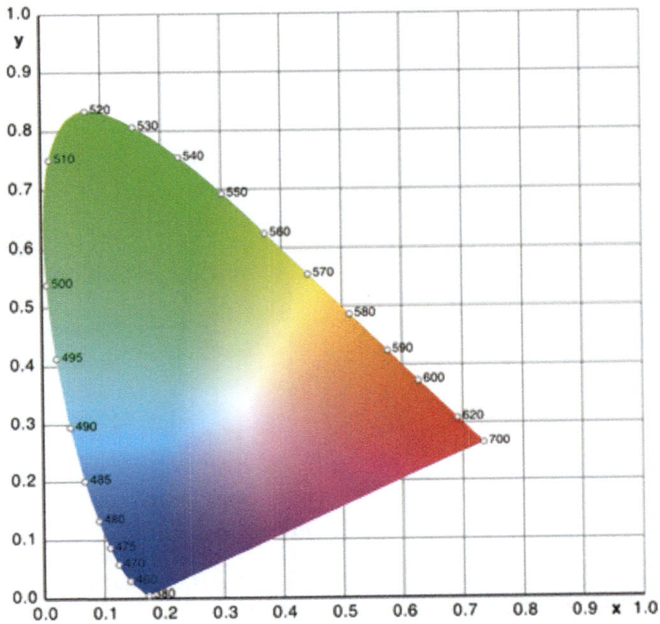

FIG. 14.6. International Commission on Illumination xy chromaticity diagram. The outer curved boundary is the spectral locus, with wavelengths shown in nanometres.

Another frequently used colour space is the red, green, blue (RGB) space, a natural colour space for a CRT or LCD colour monitor. It uses as coordinates the intensities of the red, green and blue primary colours to generate a colour pixel (Fig. 14.7).

The colour space used for hard copy printers is the cyan, magenta, yellow, key (black) (CMYK) space.

The quality of the colour image depends on the colour depth (the range of colour intensities) with which each subpixel contributes. Colour quality increases with subpixel depth. A common classification of the display controller's ability to reproduce colours is: 8 bit colour (can display 256 colours), 15/16 bit colour (high colour: can display 65 536 colours), 24 bit colour (true colour: can display 16 777 216 colours) and 30/36/48 bit colour (deep colour: can typically display over a billion colours).

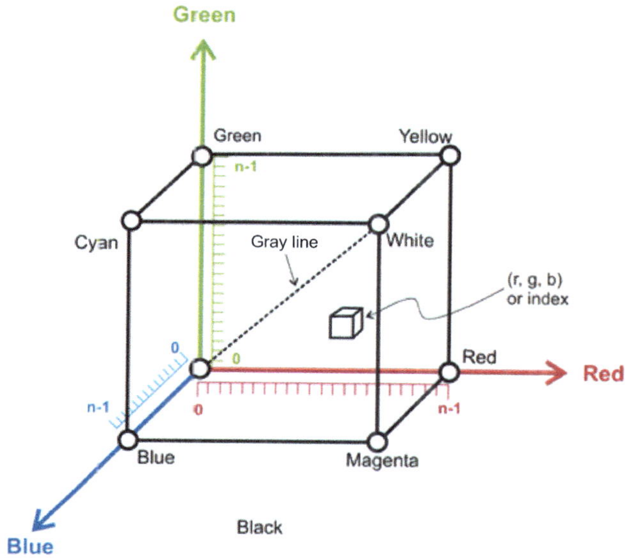

FIG. 14.7. Red, green, blue colour cube with a grey line as diagonal. The number of (r, g, b) voxels, i.e. colours available, depends on the bit depth of each coordinate. A depth of 8 bits for each component would result in 16 777 216 colours.

A nuclear medicine display controller can usually handle true colour pixel depths, with 8 bits available for each primary colour.

Colour was utilized in nuclear medicine already at an early stage of the development of digital displays. Since the original image data contain no colour information, the allocation of colour to an image pixel can be freely chosen. The allocation takes the form of a CLUT. Conceptually, the CLUT is an array structure containing the colour coordinates for each colour included in the table. A colour is defined by three values representing the intensities of the red, green and blue subpixels. Each pixel intensity in the image maps to an array index of the LUT, so that each intensity is associated with a particular colour. This is accomplished by a transformation algorithm. The transformation is usually carried out by the GPU of the display controller. The CLUT is stored in the memory of the graphics card. The LUT is usually much smaller in size than the image. Usual CLUTs contain 64–256 elements, respectively colours. An advantage of a CLUT is that colours can be changed by changing the LUT, resulting in better display performance. It is worthwhile noting that for a real world colour image, the colour of each pixel is determined by the image itself and cannot be arbitrarily associated with a colour such as is the case for pseudo-colour display. Thus, the quality of a real world image increases the larger the number of colours that can be reproduced. Using a CLUT for a colour image of the real world implies a loss of quality, as can easily be seen on images on the Internet which use CLUTs with typically 64 colours to save on image size. The addition of colour information to native nuclear medicine and X ray images always results in a pseudo-colour image, with the colours chosen by the user.

A modern nuclear medicine system typically uses 16–32 different CLUTs. The choice of colours is a complex issue. A continuous colour scale can be achieved if the individual components vary slowly and continuously. Pseudo-colour can be used to increase the PDR relative to grey scale; other CLUTs may emphasize regions with a specific intensity as, for example, in the case when performing a Fourier analysis of the beating heart to highlight amplitude and phase information.

14.5.1. Colour and colour gamut

As with grey scale images, it is expected that a colour image displayed on a PACS display device has the same colour appearance regardless of the type or the individual characteristics of the display device. Fortunately, the problem of producing digital colour images with the same perception of colours regardless of the display device, including display monitors and hard copy printers, has already been resolved by the printing industry and the photographic industry.

Since each colour is a unique entity, it is to be expected that unambiguous transformations exist between the coordinates representing the colour in different colour spaces. Such transformations are indeed available and are the basis of a CMS. The purpose of a CMS is to produce a colour image that is perceived as being the same by a human observer regardless of the output device used.

The gamut or colour gamut is defined as the entire range of colours a particular display device can reproduce. The gamut depends on the type of display and on design characteristics. The number of vertices of the gamut is given by the number of primary colours used to compose a colour. In the case of an LCD or a CTR monitor, the three primary colours, red, green and blue, are used to produce a colour. For a printer, the colours of several inks or dyes can be mixed to produce a colour on paper. Most printers can create dots in a total of six colours which are cyan, yellow, magenta, red (which combines yellow and magenta), green (yellow plus cyan) and blue (cyan plus magenta). Typical gamuts for an LCD monitor and for a printer are shown in Fig. 14.8. It is obvious that the monitor can display colours unavailable to the printer and vice versa. The International Color Consortium (ICC) has published procedures including colour transformations (CMS) that ensure that a colour image that is displayed on, for example, a monitor has the same appearance on, for example, a colour printout [14.6]. The system is based on describing the colour properties of a colour display device by an ICC colour profile. The colour profile contains manufacturer provided or, preferably, the measured gamut of the individual display device in a format which permits transformation of the colours between the representation on the display device and an intermediary colour space, for which the CIE xyz space or the CIE lab space are used. The intermediary colour space acts as a colour reference and is called the profile connection space (PCS). The PCS is utilized by DICOM [14.3], analogous to the GSDF for grey scale displays, as a reference space for the transformation of colours from one colour display device to another. Unlike the GSDF, it does not, however, claim to linearize perceived contrast. The colours of an individual display device can be transformed with the help of the ICC profiles to any other display device, including hard copy colour printers while maintaining the same visual perception of the colours. For colours available on one device but not on the other, the PCS substitutes colours similar in perception to the missing ones. In order to make the PCS work, it is necessary that all display systems involved have their ICC profiles available. The DICOM standard formalizes the information required for colour management by adding the necessary tags to the data dictionary, so that in a medical PACS the colour transformations are carried out by the CMS in a manner transparent to the user.

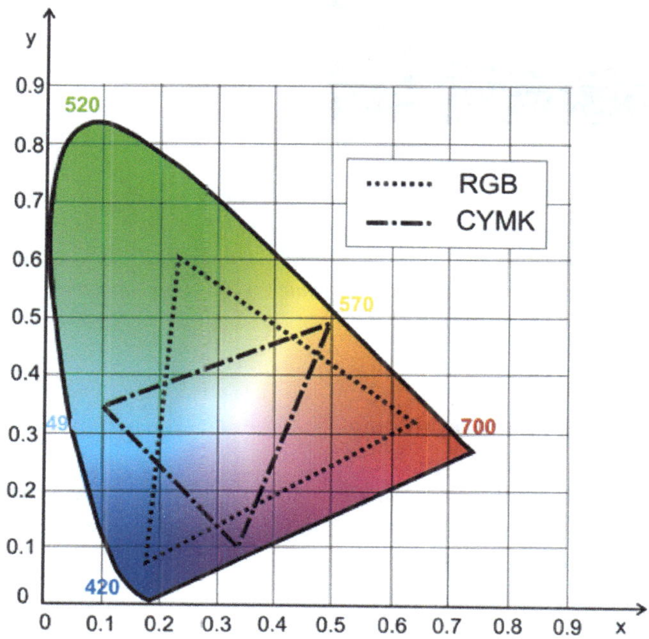

FIG. 14.8. Typical gamuts for a liquid crystal display monitor and a colour printer. The large non-overlapping areas of the colours which cannot be reproduced by the other device and must be substituted by similar colours should be noted.

14.6. IMAGE DISPLAY MANIPULATION

14.6.1. Histograms

The intensity histogram of an image represents the distribution of the grey values in an image. It is obtained by dividing the range of grey values into intervals of equal width, the bins, and calculating the number of pixels with intensity values falling into each bin. The number of bins to store the frequencies can be freely chosen, but the most informative displays are obtained with bin numbers between 32 and 256. The graphical display of the histogram transmits a rough idea about the distribution of intensities (Fig. 14.9).

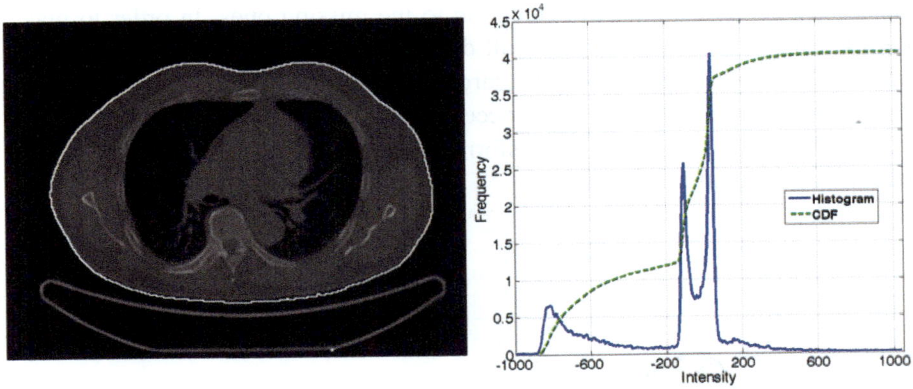

FIG. 14.9. Transverse CT slice at the height of the heart (left), with the corresponding intensity histogram, using only the pixels within the region surrounding the trunk. The number of bins is 256. Even when excluding all background pixels, the unequal distribution of intensities is seen, especially the lack of high intensity values which is in agreement with the small proportion of bony structure in the image.

14.6.2. Windowing and thresholding

The most basic intensity transformations used in image display are to transform a pixel intensity I to a grey scale intensity value r within the available grey scale range Q of the display monitor:

$$r = T(I) \tag{14.1}$$

Q is normally in the range 0...255. The transformations do not take into account the values of surrounding pixels; each pixel is processed independently.

Windowing and thresholding are linear intensity transformations of that type. They provide an easy method to emphasize contrast and visibility of structures in areas of interest by only mapping intensity values within an intensity window defined by a threshold and a window width to the available range of brightness values. Values below the threshold are set to black; values above the upper level are set to maximum display intensity. Thus, for an intensity threshold level t and a window width w, the pixel intensity I is transformed into the grey scale value r according to:

$$r = \begin{cases} \beta I, t \leq I \leq t + w \\ 0, I < t \\ Q, I > t + w \end{cases} \tag{14.2}$$

with $\beta = Q/W$.

Windowing and thresholding may be hardware implemented, i.e. the values may be changed by turning knobs on the monitor or the console or, more frequently, by software implementation using mouse movements, sliders or the arrows on the keyboard of the display workstation. The diagnostic value offered

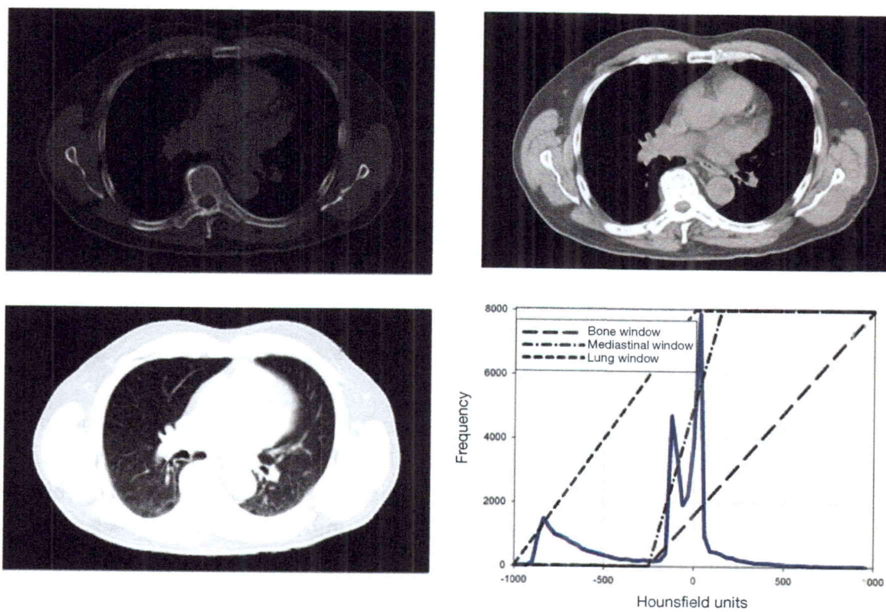

FIG. 14.10. CT slice from Fig. 14.9 with typical lung and mediastinal windowing (upper row from left to right), a bone window (bottom left) and a histogram with corresponding linear window functions (bottom right).

by windowing and thresholding can be appreciated when using typical CT windows for bone, mediastinum and lung. This is shown for a transaxial CT slice through the chest, together with the range of intensity values actually visualized out of the total histogram (Fig. 14.10).

Simple algorithms may be successfully used to perform automatic windowing and thresholding using histogram data, such as minimum and maximum intensity or setting the threshold and window width in such a way that a small percentage of the lowermost and uppermost intensity values are discarded. Suitable values are between 0.5 and 2%.

14.6.3. Histogram equalization

Image intensity values may utilize the range of display intensities inefficiently. The CT slice of Fig. 14.9 is a typical example of a medical image and demonstrates that most of the intensity values are the Hounsfield units for soft tissue and the lung.

Histogram equalization aims at utilizing each grey scale level available for display with the same frequency. If all intensity values were present in equal numbers in the image, the histogram would be flat, and the corresponding cumulative density function would increase linearly. A redistribution of intensity values s to approximately equally distributed grey scale intensity values r can be achieved using the transformation:

$$r_{\text{eq}} = \text{CDF}(I) \times (Q - 1)/(M \times N) \tag{14.3}$$

where

CDF(I) is the cumulative density function of the original image;
Q is the available range of grey scale values;

and the image size is $M \times N$ pixels.

For more details, see Ref. [14.7]. Figure 14.11 demonstrates the effect of histogram equalization using the standard algorithm of the image processing software package ImageJ [14.8] on the CT slice of Fig. 14.9. In the processed image, the structures of both the bronchi of the lung and the ribs are visualized in the same image without underflow or overflow and with approximately the same information content as in the three windowed images of Fig. 14.10 together. The drawbacks of the method are that the visual appearance of an image depends on the shape of the histogram and may, therefore, be significantly different between patients, and the fact that the resulting intensity data can no longer be used to extract quantitative information. The latter is nicely shown by the range of intensity values in the histogram of Fig. 14.11, which no longer shows the familiar range of CT numbers.

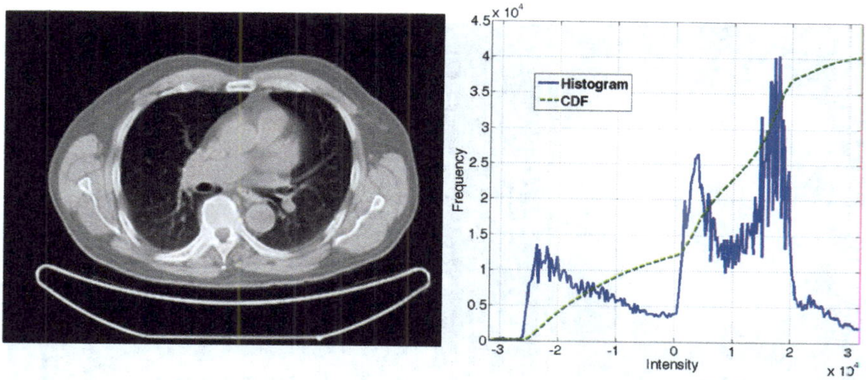

FIG. 14.11. CT slice from Fig. 14.9 after histogram equalization with a corresponding intensity histogram, using only the pixels within the region surrounding the trunk. The number of bins is 256. The cumulative density function is now approximately linear. The intensity values are no longer related to Hounsfield units. Owing to the better distribution of the intensity values, all of the structures of interest, including the bone and the bronchi of the lung, are visualized simultaneously.

14.7. VISUALIZATION OF VOLUME DATA

14.7.1. Slice mode

An image volume dataset consists of a series of adjacent image slices through the patient's body. The slices can be displayed sequentially with manual stepping through the images or automatically as a movie, or they can be displayed simultaneously as a montage of several images. Specialized viewing software offers easy ways to manipulate the display further and permits, for example, zooming and panning. Panning consists of quickly moving around a zoomed image too large to be displayed completely on the screen by utilizing the mouse, a joystick or a scroll wheel.

From the original slices, orthogonal views can easily be calculated by rearranging the pixel matrix. Presenting the orthogonal views simultaneously on the screen facilitates the anatomical location of structures. Slices with oblique orientations can also be calculated. In myocardial SPECT and PET, reorientation along the long and the short axes of the left ventricle are the standard display for visualization (Fig. 14.12). A substantial gain in anatomical information can be achieved by using 'linked' cursors. The technique consists of moving a cursor in one image while a second cursor is simultaneously moved by software to identical locations in the other views.

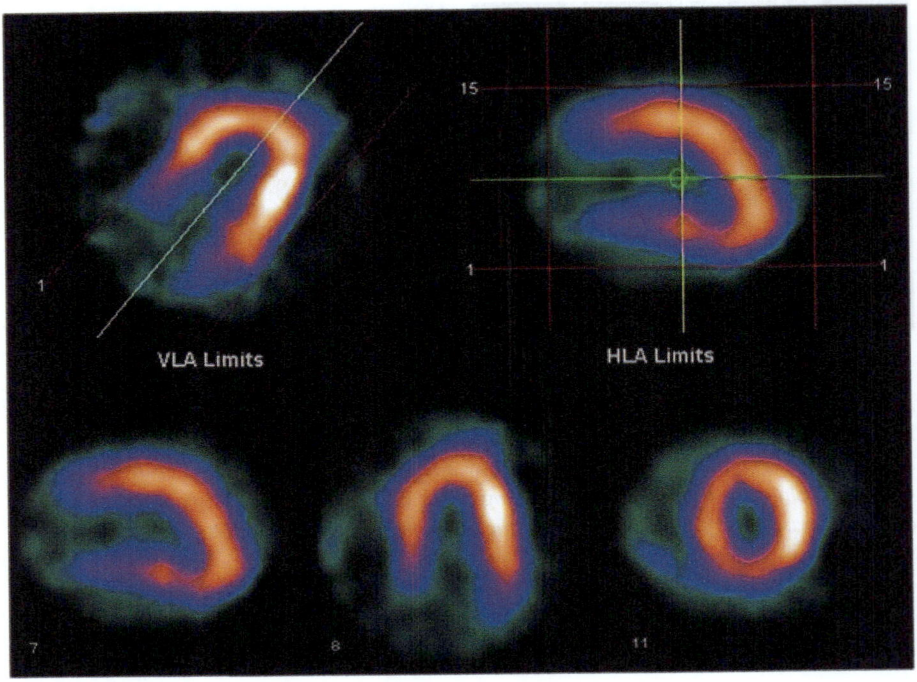

FIG. 14.12. Orthogonal views of myocardial perfusion SPECT with orientation of the slices along the long axis of the heart. The upper row shows an original transaxial slice through the myocardium with the white line indicating the long axis (left) and a sagittal slice (right). The bottom row shows the reoriented views with the vertical and horizontal slices through the long axis, and a slice perpendicular to the long axis (from left to right). (Courtesy of B. König, Hanuschkrankenhaus, Vienna.)

14.7.2. Volume mode

Volume mode display refers to techniques which extract the information about structures in the 3-D image dataset by selecting intensity information directly from the volume dataset and projecting the selected values on the display screen. The ray casting technique projects a line from a viewpoint through the data starting from a pixel on the display screen. It calculates the value of interest using the image intensities along its path (Fig. 14.13). Less frequently used in nuclear medicine are splatting techniques consisting of the projection of possibly modified image voxels on the display screen; these techniques will not be considered further here. Details can be found, for example, in Ref. [14.9]. The dominant ray casting geometry in nuclear medicine applications and in dual mode imaging is parallel projection. Perspective projection is predominantly used in virtual endoscopy and is not yet used routinely in dual mode imaging.

14.7.2.1. Transmission type volume rendering

Maximum intensity projection (MIP) consists of projecting the maximum intensity value encountered along the trajectory of the ray through the data volume on the corresponding screen pixel. It improves the visualization of small isolated hot areas by enhancing the contrast (Fig. 14.14). MIP is successfully employed for lesion detection in PET oncological whole body studies. Its efficiency for the detection of lesions is further increased by displaying the MIP projections as a sequence of projection angles in cine mode.

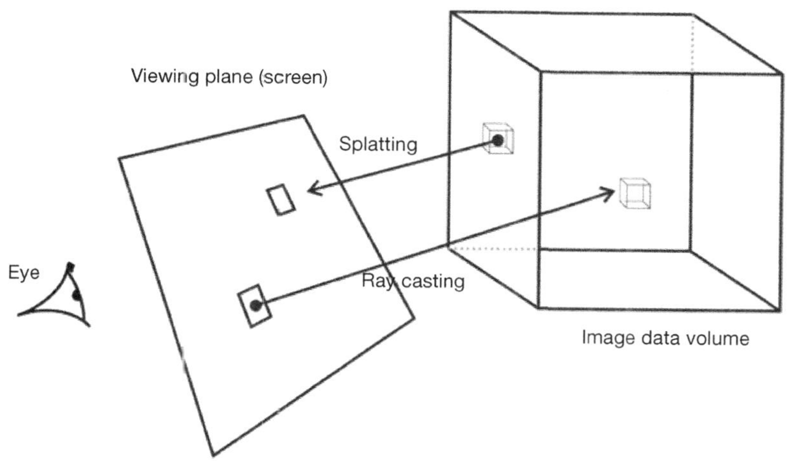

FIG. 14.13. Principle of ray casting and splatting geometry. In ray casting, the ray collects intensity transformation throughout its trajectory. A voxel is usually hit outside its centre which has to be corrected for by interpolation. Splatting starts from the centre of a voxel and distributes its intensity on several screen pixels.

Summed voxel projection produces the rendered image by summing all intensities along a ray trajectory. If applied to a CT volume, it is known as a digitally rendered radiograph. If central projection geometry is used, the projection image simulates a conventional X ray image. Digitally rendered radiographs of CT data are used in radiotherapy for the positioning and registration of patients. Tomographic datasets from nuclear medicine displayed as digitally rendered radiographs may be used to compare lesion extensions with planar X ray images of the patient.

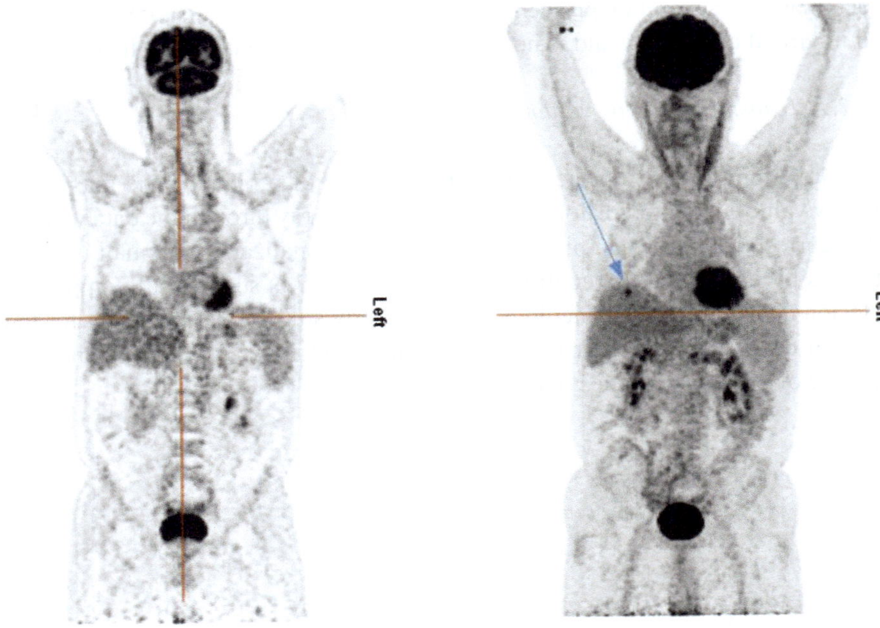

FIG. 14.14. A maximum intensity projection (right) compared to a standard coronal slice. A solitary lesion is clearly visible in the maximum intensity projection image (arrow) while it is missing in the coronal standard view.

14.7.2.2. Reflection type volume and surface rendering

The purpose of surface and volume rendering is to visualize structures in volume datasets as objects and display them in a pseudo-3-D mode. Volume rendering techniques extract information about objects directly from the 3-D volume data. They start by casting rays through the image volume data and by processing the intensities of the voxels along the ray trajectory. Depending on the handling of the intensities, different types of 3-D display can be generated.

Three dimensional rendering utilizes standard computer graphics techniques, such as illumination, shading and the application of textures, to produce a realistic appearance of anatomical structures and tumours. This is useful for the visualization of complex anatomical relationships, can improve the orientation of surgeons and resolve ambiguities of localization.

For registered images originating from different imaging techniques, such as images from magnetic resonance and from PET, anatomical and functional data can be displayed simultaneously, thereby taking advantage of the excellent morphological resolution of one modality and of functional, blood flow or metabolic information of the other modality. Such combined images are capable of displaying the spatial relationships between different objects, such as, for example, the surface of the left ventricle together with the location of the coronary arteries, or the surface of the brain grey matter rendered from a magnetic resonance study combined with the blood flow obtained by an HMPAO (hexamethylpropyleneamine oxime) SPECT study.

Surface rendering traditionally starts from a sequence of contours extracted from the object of interest. The surface is obtained by fitting a mosaic of triangles followed by illumination and shading. The relatively small number of parameters needed to describe an object permits real time visualization of transformations, useful, for example, for interactive surgical planning. Analytical descriptions of the objects can also be generated, which can be used by other programs, such as CAD/CAM, or which can be used to produce physical models of the objects of interest using lithographic techniques.

Three dimensional surfaces are generated by specifying an intensity threshold. The method is closely related to the generation of contours. When a ray encounters the threshold value along its trajectory, the location of that voxel is interpreted as a surface point of the structure of interest. The appearance of a 3-D image is improved further by utilizing illumination and shading techniques. In order to apply these effects, additional knowledge about the orientation of the surface element is required for which gradient techniques with various levels of sophistication are employed.

Voxel gradient shading is the most successful technique to produce illuminated and shaded surfaces. It calculates a gradient vector from a voxel neighbourhood and renders a realistic pseudo-3-D image by calculating diffuse reflective illumination from an external light source and applying shading. Noise in the surface is reduced by smoothing (Fig. 14.15, middle).

Volume compositing can be considered a generalization of voxel gradient shading. It aims at visualizing internal structures beyond the limit given by a threshold by using information from all voxels along a ray. The method consists of calculating a gradient for each voxel along the ray, thereby assigning a surface to each voxel, and applying lighting and shading to that surface. The light transmitted and reflected by each voxel is then collected into the pixel value on-screen by assigning opacities to each voxel and summing the results along the ray (Fig. 14.15, right). Volume compositing is by far the most complex and time consuming rendering method. The results are similar to the voxel gradient method. Under favourable conditions and through careful selection of the rendering parameters, volume compositing can visualize several objects simultaneously in the rendered image.

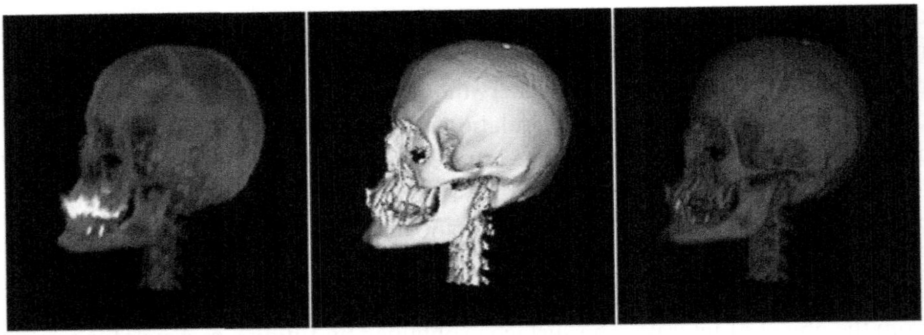

FIG. 14.15. Surface of a skull from CT image data using (from left to right) maximum intensity projection, voxel gradient shading and volume compositing for rendering. Rendered images were produced with ANALYZE© 9.0.

14.7.3. Polar plots of myocardial perfusion imaging

Myocardial perfusion imaging is a tomographic technique using a myocardial perfusion tracer such as the potassium analogue ^{201}Tl or ^{99m}Tc-MIBI (methoxyisobutylisonitrile) to produce SPECT images of the perfusion of the left ventricle. Myocardial perfusion is reduced or absent in ischaemic and infarcted areas. The size and severity of perfusion defects is of high diagnostic and prognostic value. Therefore, myocardial perfusion imaging is among the most frequent nuclear medicine investigations. Unfortunately, visual interpretation of the tomographic slices is difficult and suffers from high inter-observer variability due to the poor spatial resolution of SPECT studies in general and to the added blurring of the images by the motion of the heart during acquisition. Display methods tailored to a more reliable detection of perfusion defects were, therefore, developed shortly after the introduction of myocardial perfusion SPECT.

The initial visual presentation of the tomographic images makes use of a coordinate system natural to the anatomy of the left ventricle. One coordinate axis passes through the long axis of the heart; the other two are perpendicular to the long axis and to each other (Fig. 14.16). The standard display consists of slices perpendicular to the long axis, the short axis slices and two sets of slices parallel to the long axis. The left ventricle has an annular shape in the short axis slices. The annuli can be aggregated into one image using a polar map representation [14.10]. In a first step, each annulus is reduced to a circumferential profile. The methods of choosing the circumferential profile vary, for example, using the maximum intensity at each angular step only or taking a mean intensity, and the thickness of the annulus can also be considered at each angular step. Thereafter, all of the profiles are arranged into one image, starting with the profile representing the apex at the innermost position, with each following profile surrounding the previous one. The resulting display is referred to as a bull's-eye display or polar map (Fig. 14.17). The latter name refers to the fact that the intensity along a given annulus can easily be handled by a polar coordinate system. The intensities displayed for each annulus correspond to the myocardial perfusion in that slice or a segment thereof. Absolute perfusion values cannot be derived from the intensities. The method to obtain an estimate of the degree of hypo-perfusion and of the location of the perfusion defect consists of comparing the relative intensity values in different segments of the annuli to the maximally perfused segments of the same patient, and then comparing the pattern of relative perfusion of the individual study with normal perfusion patterns. This permits an estimate of both the degree and extent of the perfusion defects as well as a good anatomical allocation to the coronary arteries causing the hypo-perfusion.

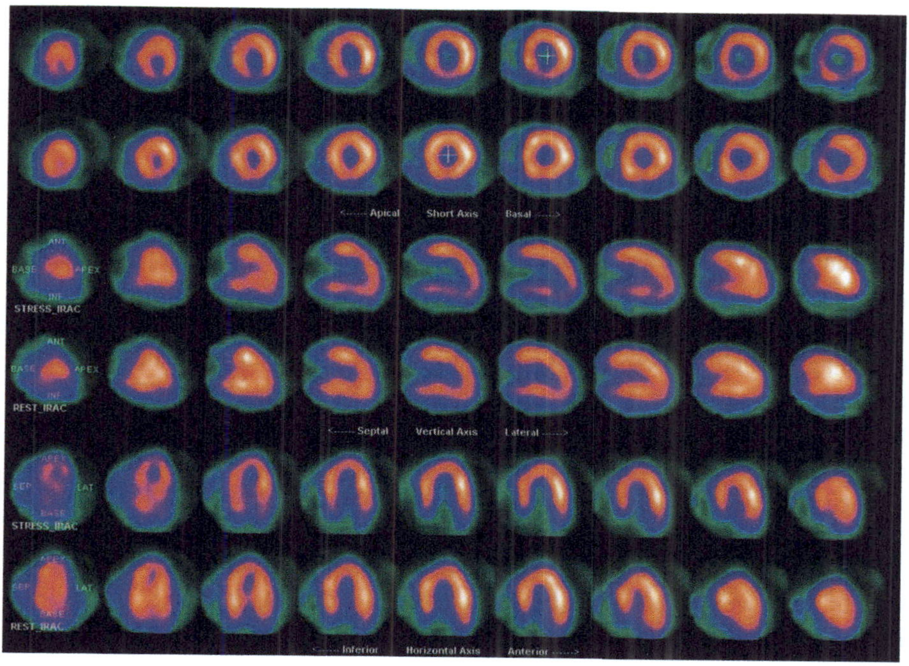

FIG. 14.16. Results of a stress–rest perfusion study with ^{99m}Tc-MIBI (methoxyisobutylisonitrile) with the orientation of slices adapted to the long axis of the heart. Images show hypo-perfusion of the inferior wall. The upper rows are stress images; the lower rows are images at rest. (Courtesy of B. König, Hanuschkrankenhaus, Vienna.)

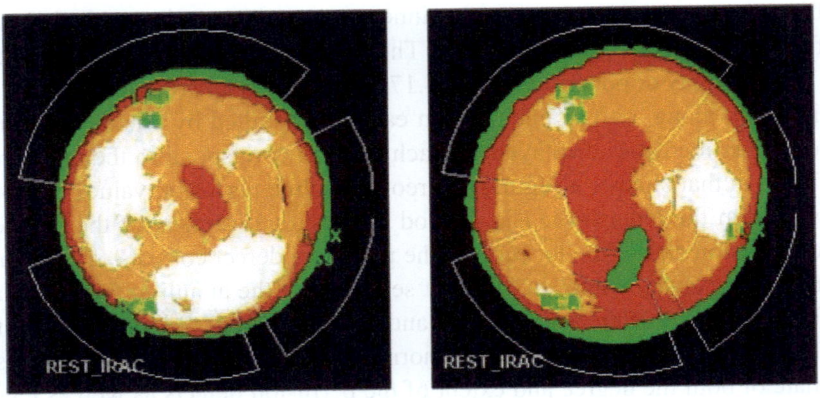

FIG. 14.17. Bull's-eye displays of myocardial SPECT perfusion studies. Normal perfusion (left) and hypo-perfusion of inferior wall (right). The colours indicate the degree of perfusion, with white — normal, orange — acceptable, red — hypo-perfused, and green — no perfusion. Also indicated are the perfusion areas for the main coronary vessels LAD, LCX and RCA. (Courtesy of B. König, Hanuschkrankenhaus, Vienna.)

14.8. DUAL MODALITY DISPLAY

Several techniques have been developed to display registered images originating from different modalities, such as from a PET/CT study. The simplest technique is to display the images belonging together side by side. Anatomical information can be gained easily by using the linked cursor technique. The CT image which has superior spatial resolution is, thus, used to determine the anatomical location of a lesion visible in the PET image. The linked cursor technique, while providing precise anatomical information, is impractical if several lesions are present in the image as often is the case in oncological studies. In such situations, alpha blending is helpful, which combines both the CT and the PET image into one composite image.

Alpha blending consists of adding the images pixelwise with different weight. The weight is called the transparency factor α, with $0 \leq \alpha \leq 1$. The composite pixel I_{CS} is given by:

$$I_{CS}(m, n) = \alpha \times I_{BG}(m, n) + (1 - \alpha) \times I_{FG}(m, n) \tag{14.4}$$

where

I_{BG} is the intensity of the background pixel;

and I_{FG} is the intensity of the foreground pixel.

When using the native grey scale images for both modalities, it is difficult to distinguish clearly which intensity comes from which modality. The composite display becomes much easier to interpret if one of the images uses a CLUT. In this case, the formula has to be applied to each colour component separately:

$$R_{CS}(m,n) = \alpha \times R_{BG}(m,n) + (1-\alpha) \times I_{FG}(m,n) \tag{14.5}$$

$$G_{CS}(m,n) = \alpha \times G_{BG}(m,n) + (1-\alpha) \times I_{FG}(m,n) \tag{14.6}$$

$$B_{CS}(m,n) = \alpha \times B_{BG}(m,n) + (1-\alpha) \times I_{FG}(m,n) \tag{14.7}$$

where

R, G and B are the colour components of the background image;

and I is the grey value of the foreground image.

In PET/CT alpha blending, the background image is usually the PET image whereas the CT image as the foreground image retains the grey scale (Fig. 14.18). The composite display can be further improved by changing thresholds and windows for each modality separately and interactively. For CT, the traditional windows are usually employed.

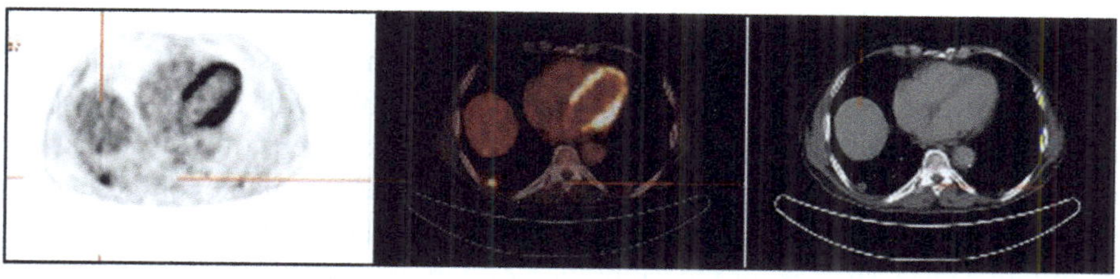

FIG. 14.18. PET/CT fused image display with a PET image on the left showing a hot lesion at the border between the lung and rib cage. The fused image in the middle shows the location inside the lung close to the pleura; the CT image on the right confirms and permits close inspection of the position. Linked cursors point to the position of the lesion. The transparency factor α is 0.5.

14.9. DISPLAY MONITOR QUALITY ASSURANCE

Several standards define the performance parameters of display systems (see, for example, Refs [14.11, 14.12]). The Report of task group 18 of the American Association of Physicists in Medicine offers an exhaustive up to date set of performance parameters for current medical display monitors, together with procedures and accompanying test images to assess performance [14.13]. It contains limits of acceptance for all parameters, distinguishing between primary devices and secondary devices.

In addition to quality assurance aimed at the individual display device, a major component of quality assurance is to ensure a consistent display of the image at all display workstations of a PACS, including different ambient light conditions. This is resolved by including the calibration and validation of the DICOM GSDF into the quality assurance framework.

Quality control at regular intervals of a display device is required because the performance may change over time, due to ageing processes of the display device, both for CRT and LCD displays, and due to changes in environmental lighting with time.

14.9.1. Acceptance testing

The purpose of acceptance testing is to ensure that the performance of purchased equipment complies with the specifications established in the sales contract. The user should clearly specify in the contract the required performance, the test procedures to assess the performance parameters and the limits of acceptability. Reference [14.13] lists a set of performance parameters which completely characterize the performance of a soft copy display device. These are summarized in Table 14.2.

For each of the parameters, several tests at various levels of sophistication are described in detail. Most of the parameters can be assessed visually by analysing the test images listed in Table 14.2, possibly with additional use of templates on transparency sheets, such as for the assessment of distortions. An exhaustive set of test images has been made electronically available for these tests, both in Joint Photographic Experts Group (JPEG) and DICOM format [14.3]. For quantitative tests, such as for the calibration of luminance characteristic curves, of the DICOM GSDF and of chromaticity values, luminance meters and colorimeters with computer readout of the measured values and special software are necessary.

14.9.2. Routine quality control

In order to make sure that a display system meets the expected performance during its economic lifetime, assessment of performance parameters at regular intervals is necessary. Reference [14.13] recommends that a subset of the tests in Table 14.2, namely geometric distortions, reflection, luminance response, luminance dependencies, resolution and chromaticity, be performed at monthly to quarterly intervals, depending on the monitor's stability. Tests for

TABLE 14.2. PERFORMANCE PARAMETERS OF DISPLAY MONITORS AND EQUIPMENT FOR MEASUREMENT
(according to Ref. [14.13])

Test	Major required tools	
	Equipment	Patterns
Luminance response	Luminance and illuminance meters	TG18-LN TG18-CT TG18-MP
Luminance dependencies	Luminance meter	TG18-UNL TG18-LN TG18-CT
Reflection	Luminance and illuminance meters	TG18-AD
Resolution	Luminance meter, magnifier	TG18-QC TG18-CX TG18-PX
Geometric distortions	Flexible ruler or transparent template	TG18-QC
Noise	None	TG18-AFC
Veiling glare	Baffled funnel, telescopic photometer	TG18-GV TG18-GVN TG18-GQs
Chromaticity	Colorimeter	TG18-UNL80

geometric distortions and for resolution are more important for CRTs, whereas the dependence of luminance on the viewing angle is important only for LCD displays.

In addition, Ref. [14.13] recommends that a daily check prior to clinical work be performed by the user. It consists of evaluating anatomical test images or a suitable geometrical test image such as a TG18-QC test image (Fig. 14.19) to verify adequate display performance. The instructions for assessing the quality of a display device when using the TG18-QC test pattern are given in Box 14.1.

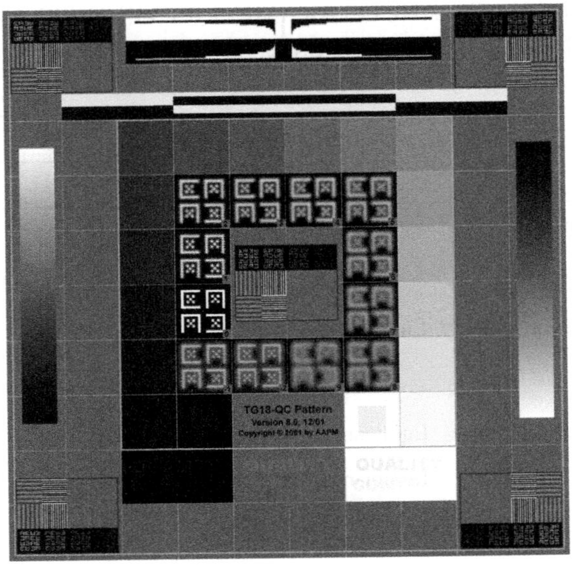

FIG. 14.19. Test pattern TG18-QC suitable for daily quality control of display monitor performance using visual inspection [14.13].

Box 14.1. Instructions for visual assessment of image quality using the TG18-QC test pattern as part of daily quality control by the user [14.23]

1. General image quality and artefacts: Evaluate the overall appearance of the pattern. Note any non-uniformities or artefacts, especially at black-to-white and white-to-black transitions. Verify that the ramp bars appear continuous without any contour lines.
2. Geometric distortion: Verify that the borders and lines of the pattern are visible and straight and that the pattern appears to be centered in the active area of the display device. If desired, measure any distortions (see section 4.1.3.2).
3. Luminance, reflection, noise, and glare: Verify that all 16 luminance patches are distinctly visible. Measure their luminance using a luminance meter, if desired, and evaluate the results in comparison to GSDF (section 4.3.3.2). Verify that the 5% and 95% patches are visible. Evaluate the appearance of low-contrast letters and the targets at the corners of all luminance patches with and without ambient lighting.
4. Resolution: Evaluate the Cx patterns at the center and corners of the pattern and grade them compared to the reference score (see section 4.5.3.1). Also verify the visibility of the line-pair patterns at the Nyquist frequency at the centre and corners of the pattern, and if desired, measure the luminance difference between the vertical and horizontal high-modulation patterns (see section 4.5.3.1).

The most frequent influence on reporting comes from changes in ambient lighting. An increase in the level of ambient light results in poorer discrimination of structures in the darker parts of the image. This has to be compensated for by adapting the GSDF to the current light level. Modern medical display monitors, therefore, provide automatic measurement and recalibration features. A typical, high performance LCD display used as a primary device includes a luminance meter covering a small area of the monitor to continuously control the display characteristic curve and the maximum brightness level. A second photometer records the average ambient lighting. With such an arrangement, it is possible to adjust the GSDF continuously to the DICOM required luminance values, taking into account the changes in L'_{min} and L'_{max}, the minimum and maximum luminance values including the luminance L_{amb} added by ambient light.

As an annual quality control of a display device, the TG18 working group recommends performing all tests carried out during acceptance [14.13].

REFERENCES

[14.1] NATIONAL ELECTRICAL MANUFACTURERS ASSOCIATION, Digital Imaging and Communications in Medicine (DICOM), Part 14: Grayscale Standard Display Function, Rosslyn, VA (2003).

[14.2] INTERNATIONAL COLOR CONSORTIUM, Color Management, UK (2003), http://www.color.org/slidepres2003.pdf

[14.3] NATIONAL ELECTRICAL MANUFACTURERS ASSOCIATION, Digital Imaging and Communications in Medicine (DICOM), Supplement 100: Color Softcopy Presentation State Storage SOP Classes, DICOM Standards Committee WG1DNEMA, Rosslyn, VA (2005).

[14.4] PIZER, S.M., CHAN, F.H., Evaluation of the number of discernible levels produced by a display, INSERM **88** (1979) 561–580.

[14.5] SMITH, T., GUILD, J., The C.I.E. colorimetric standards and their use, Trans. Opt. Soc. **33** (1931) 73–134.

[14.6] INTERNATIONAL COLOR CONSORTIUM, The Role of ICC Profiles in a Colour Reproduction System (2004).

[14.7] SONKA, M., HLAVAC, V., BOYLE, R., Image Processing, Analysis, and Machine Vision, Brooks/Cole Publishing Company, Pacific Grove, CA (1999).

[14.8] ImageJ, A public domain Java image processing program, Version 1.32b (1997).

[14.9] BIRKFELLNER, W., et al., Wobbled splatting — a fast perspective volume rendering method for simulation of X-ray images from CT, Phys. Med. Biol. **50** (2005) N73–N84.

[14.10] GARCIA, E.V., et al., Quantification of rotational thallium-201 myocardial tomography, J. Nucl. Med. **26** (1985) 17–26.

[14.11] DIN V 6868-57. DIN V 6868-57: Sicherheit der Bildqualität in röntgendiagnostischen Betrieben, Teil 57: Abnahmeprüfung an Bildwiedergabegeräten, Normenausschuß Radiologie (NAR) im DIN Deutsches Institut für Normung e.V. (2001).

[14.12] VIDEO ELECTRONIC STANDARDS ASSOCIATION, Flat Panel Display Measurement Standard, Version 2, Milpitas, CA (2001).

[14.13] AMERICAN ASSOCIATION OF PHYSICISTS IN MEDICINE, Task Group 18 (TG18), Assessment of Display Performance for Medical Imaging Systems, AAPM On-Line Report No. 03, College Park, MD (2005).

CHAPTER 15

DEVICES FOR EVALUATING IMAGING SYSTEMS

O. DEMIRKAYA, R. AL-MAZROU
Department of Biomedical Physics,
King Faisal Specialist Hospital and Research Centre,
Riyadh, Saudi Arabia

15.1. DEVELOPING A QUALITY MANAGEMENT SYSTEM APPROACH TO INSTRUMENT QUALITY ASSURANCE

A quality management system (QMS) has three main components:

(a) Quality assurance (QA);
(b) Quality improvement;
(c) Quality control (QC).

The aim of a QMS is to ensure that the deliverables meet the requirements set forth by the users. The deliverables can be, in general, all the services provided in a nuclear medicine department, and the diagnostic imaging services in particular. In this section, the primary focus is the diagnostic imaging equipment and images produced by them.

15.1.1. Methods for routine quality assurance procedures

QA is a systematic programme for monitoring and evaluation of the process of production. It is an all-encompassing management plan to ensure the reliability of the production system. QA in diagnostic imaging, however, can help minimize the uncertainties and errors in equipment performance by supervising the entire image production process. This, in turn, will guarantee that the images generated are of diagnostic quality. QA can also help identify and rectify the problems, errors and malfunctioning and drifting of the performance earlier. Moreover, a QA programme can help the standardization of the image production process across centres and, thus, allows comparison of clinical results with other centres. This is especially imperative in multicentre clinical trials. A QA programme in nuclear medicine involves all aspects of nuclear medicine, including minimizing the exposure to personnel, patients and the public; preparation, safety, sterility and administration of radiopharmaceuticals; patient handling; and ensuring the diagnostic quality of images produced.

QC is the process by which the performance level of a product is measured and then compared against the existing standards or tolerance values. QC activities are a subset of QA activities. QA focuses on the processes while QC focuses on the product.

QC with regard to imaging systems may entail:

— A series of performance measurements to assess the quality of the imaging system;
— Keeping the record of measurements;
— Monitoring the accuracy and precision of the results;
— Taking corrective actions in case the performance measurements are outside the tolerance levels or above the predetermined action levels.

The items above require:

— Defining the performance parameters to be measured;
— Preparing written procedures as to how and by whom the measurements should be carried out;
— Establishment of the frequency of performance tests and expected results in the form of tolerance and action levels;
— Training the persons who perform these measurements;
— Designing record forms (preferably electronic) to keep the measurement values;
— Logging and reporting all of the problems and actions taken.

Tolerance levels define the range within which the results are acceptable while action levels define the range beyond which a corrective action is required. The upper level of the tolerance range may concur with the lower level of the action range. If the performance of the system is just outside the tolerance range, an immediate corrective action may not always be needed and the imaging system can still be used for patient scanning, but a close monitoring of the system performance is critical in the following tests. Record keeping is critical and essential for trending the performance parameters to monitor the system and to intervene, when necessary, in an effective and timely manner.

Phantoms are indispensable tools for QC measurements. They are utilized to evaluate diagnostic imaging systems, as well as for other reasons in radiation protection, radiobiology and radiotherapy. The phantoms can be hot (containing a known amount of radioactivity) or cold (containing no radioactivity) for primarily measuring radiation interaction. Phantoms used in nuclear medicine are usually injected with a radioisotope simulating a particular organ or tissue structure containing a particular radiopharmaceutical, while X ray computed tomography (CT) QC phantoms are employed to measure CT values of water and/or other materials by simulating different tissue types. IAEA Human Health Series Nos 1 and 6 [15.1, 15.2] include an extensive discussion of the QA for positron emission tomography (PET) and PET/CT systems, and single photon emission computed tomography (SPECT) systems, respectively.

The International Commission on Radiation Units and Measurements (ICRU) in ICRU Report 48 [15.3] defines the phantom as a material object that includes one or more tissue substitutes and is used to simulate radiation interaction in the body. Furthermore, “any material that simulates a body tissue in its interaction with ionizing radiation is termed a tissue substitute” [15.4].

The ICRU distinguishes the ‘physical phantoms’ from what are usually called ‘software phantoms’ by defining them as ‘phantoms’ and ‘computational models’, respectively. In this chapter, the convention of the ICRU is followed for consistency in the terminology and to avoid any potential misunderstanding that the other naming conventions may lead to.

According to Ref. [15.3], phantoms can be grouped under three categories with respect to their primary usage: dosimetric, calibration and imaging phantoms. The dosimetric phantoms are used to measure absorbed dose while calibration phantoms are employed to calibrate a particular photon detection system such as a PET scanner to convert the number of detected photons to actual activity per tissue volume. An imaging phantom is used for assessing image quality or characterizing imaging systems. The ICRU further defines three subcategories under the above three functional categories. These are body, standard and reference phantoms. Body phantoms are built in the shape of a body and consist of multiple tissue substitutes or organs. These phantoms are more often referred to as anthropomorphic phantoms as they simulate the human body. The anthropomorphic torso phantom, which is discussed later in the chapter, consisting of liver, heart, spine and lung inserts, is used in nuclear medicine for testing image quality and is an example of this category.

In this chapter, physical phantoms or simply phantoms and computational models that have applications in nuclear medicine are discussed. Throughout this chapter, many commercial phantoms are mentioned and are pictured in figures for ease of understanding. This does not, however, constitute an endorsement of these commercial products.

15.2. HARDWARE (PHYSICAL) PHANTOMS

The use of phantoms dates back to the beginning of the 20th century. In the 1920s, water tanks and wax blocks were often used for X ray experiments and, to this day, these materials are still in use in certain applications. In the 1960s, more reliable tissue substitutes and sophisticated phantoms began to appear.

Today, phantoms are used in performing numerous tasks within the field of diagnostic imaging and radiation therapy. This includes testing the performance of imaging equipment, measuring radiation dosage during therapy, teaching interventional image guided procedures and servicing equipment in the field.

Hardware phantoms are the indispensable tools for medical physicists to enquire about or characterize medical imaging systems. These phantoms provide the means to determine, not only qualitatively but also quantitatively, the performance characteristics of medical imaging systems.

As compared to computational models, physical phantoms may be advantageous in that data are acquired with an actual scanner and contain the effect of the parameters that impact on the entire photon detection process. One major disadvantage of physical phantoms, however, is the difficulty of simulating the change of the activity in an organ in time. Although phantoms that simulate cardiac motion, for instance, are available commercially or are being developed by researchers in various institutions, in general, phantoms simulating physiological processes such as breathing are difficult to build and are not widely available.

In this section, the hardware phantoms that are used to measure the performance characteristics of gamma cameras and PET scanners are discussed. Some of these phantoms are also known as test phantoms. Their physical characteristics are reviewed along with a brief description of their purpose of use. Some practical suggestions are also provided about the preparation of the phantoms that require injection of radioactivity. Although the focus of this section is primarily on discussing the phantoms themselves, the positioning and data acquisition requirements are also addressed. The analysis of the acquired phantom data is not the subject of this chapter. For the analysis of gamma camera and SPECT performance test data, please see Ref. [15.5] in which the test methods suggested by the National Electrical Manufacturers Association (NEMA) are discussed. The authors have also developed a software application and made it publicly available free of charge [15.5].

15.2.1. Gamma camera phantoms

The gamma camera is the most widely used diagnostic imaging system available in nuclear medicine departments. Owing to their physical characteristics, gamma cameras require very close attention and, therefore, more frequent and a larger number of tests than any other diagnostic imaging modality in radiology. One of the important QC tests that has to be carried out daily on every gamma camera is the uniformity test. This test shows the current status of the gamma camera and allows monitoring of any possible deterioration in the performance of the camera. It can also signal whether there has been any malfunctioning in the detector elements, such as the photomultiplier tubes or the crystal, since the last QC test was conducted. These assessments can be performed qualitatively or quantitatively by a computer program.

15.2.1.1. Point source holders

This phantom is used to hold point sources that are employed in intrinsic uniformity, resolution and linearity measurements. It is made up of lead and its main purpose is to shield the walls, ceiling and personnel, and collimate the γ radiation to the detector. Figure 15.1 shows a picture of a source holder. Copper plates (1–2 mm thick) should be placed in front of the source holder to act as absorbers and stop the low energy photons. When placed on the floor, source holder height can be adjusted such that the point source is directed to the centre of the detector under investigation.

FIG. 15.1. Point source holders in a slanted position so that they can point to the detectors from the floor.

15.2.1.2. ^{57}Co flood sheets

Gamma cameras should also be tested extrinsically (collimator in place) using a ^{57}Co sheet source. The cost of ^{57}Co sheet sources is relatively high and they should be replaced every 2 years. It should be noted that new sheet sources may contain ^{56}Co and ^{58}Co impurities. These radionuclides have a shorter half-life (77.234 and 70.86 d, respectively) than that of ^{57}Co (271.74 d) and emit high energy γ rays (>500 keV). If the impurities result in non-uniformities, the sources can also be left to decay for a while before being used. It is advisable to place the sheet source at a distance of 5–10 cm from the collimator during the scan. Figure 15.2 shows a commercial ^{57}Co flood source.

FIG. 15.2. Picture of a ^{57}Co flood source.

15.2.1.3. Fillable flood phantoms

Although ^{57}Co flood sources are more convenient and easy to use, their higher cost may be a factor affecting accessibility. If one cannot have access to ^{57}Co flood sources, then a fillable water phantom source is a good alternative. These phantoms are available commercially but they can also be manufactured in a machine shop from Perspex. The commercial ones are available in different dimensions for different detector sizes. It is necessary to be careful in filling these phantoms to prevent bubble formation, contamination of the outside surface of the phantom or the workplace, and/or bulging of the phantom in the centre. Bulging of the phantom and air bubbles formed in the phantoms can affect the uniformity of the captured image. Depending on the size and volume of the phantom, around 370 MBq (10 mCi) of ^{99m}Tc activity will be sufficient to give a count rate of 20 kcounts/s in the image. The acquisition of the image is performed in the same way as the ^{57}Co flood sources.

15.2.1.4. Slit phantom

Slit phantoms are used to measure the intrinsic resolution of a gamma camera detector. The phantom is made of a 3 mm thick lead mask consisting of 1 mm wide parallel slits that are 30 mm apart. Slit phantoms, which are usually manufactured by the gamma camera vendors, vary in size to fit perfectly to particular detectors. They are made in pairs to measure the intrinsic resolution in the X and Y directions (see Figure 15.3). These masks are placed in the closest possible proximity to the crystal covering its entire area. Measurement is performed using a ^{99m}Tc point source centred at a distance more than five times the largest dimension of the useful field of view (UFOV) of the crystal. The activity of the point source is adjusted, so that the count rate is less than 20 kcounts/s.

FIG. 15.3. Top: picture of the slit phantom designed for a cardiac camera whose field of view is smaller than that of a typical gamma camera. Bottom: acquired images of the slit phantoms for a typical gamma camera to measure the resolution in the Y (left image) and X (right image) directions. The white vertical and horizontal lines denote the image of 1 mm slits.

15.2.1.5. Dual-line source phantom and scattering medium

This phantom, suggested by NEMA NU 1-2007 [15.6], is used to measure the extrinsic resolution of the system with and without a scattering medium. It consists of two parallel line sources 1 mm in internal diameter and with a centre to centre distance of 5 cm. The line sources are built so that they are positioned 10 cm above the collimator. Figure 15.4 shows a simple custom built, dual-line source phantom. The capillary tube shown as dark lines in the figure is commercially available but a butterfly IV line can also be utilized.

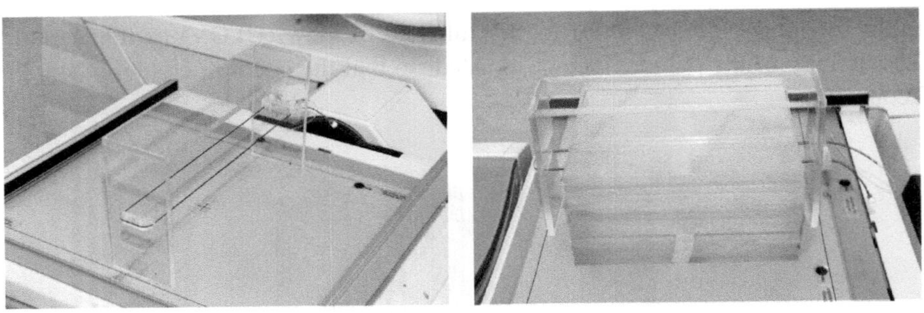

FIG. 15.4. A custom built, dual-line source phantom. On the left is the phantom positioned on the detector, and on the right the same line sources are immersed in a scattering medium consisting of sheets of Perspex.

The line is filled with ^{99m}Tc activity solution with a concentration of about 550 MBq/mL (15 mCi/mL) to achieve an adequate count rate when used with the scattering medium. When measuring X and Y resolutions, the lines are placed parallel to the Y and X directions, respectively. In both cases, one of the lines should be positioned in the centre of the field of view (FOV). The acquired image should have at least 1000 counts in the peak channel of the line spread function.

To measure the extrinsic resolution with scatter, the dual-line source is embedded into Perspex sheets, 10 cm of which are placed between the collimator and the line sources and 5 cm placed above the lines as seen in Fig. 15.4. The Perspex sheets under the sources create a scattering medium and the ones above a backscattering medium. For a perfect contact between the sheet and the line sources, it is recommended to make two grooves, through which the lines run, in one of the sheets to insert the two lines.

15.2.1.6. Bar phantom

The second most frequent QC test in nuclear medicine is the resolution test performed with bar phantoms. Bar phantoms can be used to measure, semi-quantitatively (i.e. visually), the extrinsic and the intrinsic resolution of a gamma camera. Images of bar phantoms can also be useful for the qualitative evaluation of the gamma camera linearity which is normally measured by the slit phantom.

Bar phantoms are made of lead strips embedded into plastic and typically arranged in four quadrants. The lead strips are radio-opaque, while plastic strips are radio-lucent. Each quadrant has strips of different thickness. The rectangular bar phantom image shown in Fig. 15.5 (middle) has four quadrants with strip sizes of 2.0, 2.5, 3.0 and 3.5 mm, while the image on the right has four quadrants with strips of sizes 3.2, 4.6, 6.3 and 10 mm. In the images of the bar phantoms, displayed in grey colour maps, white lines correspond to the plastic strips while black lines correspond to lead strips. In a bar phantom, the strips are separated with the same distance as the strip width.

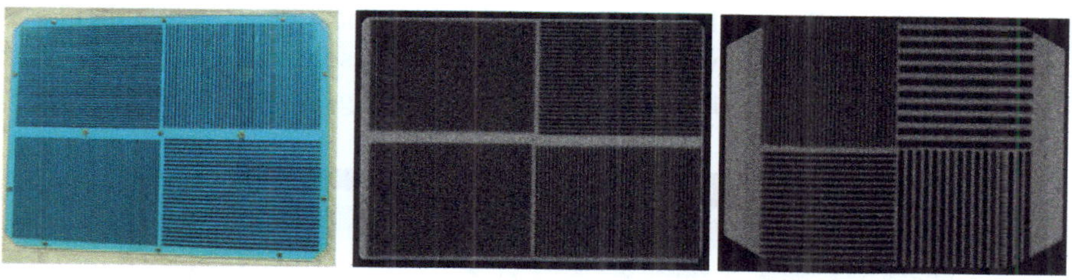

FIG. 15.5. Left: picture of a typical four-quadrant rectangular bar phantom. Middle: image of the left bar phantom acquired by an ECAM gamma camera. Right: image of a bar phantom acquired with an ADAC FORTE gamma camera. Both images were acquired at a matrix size of 512 × 512 and with a total count of 10 Mcounts.

In routine QC tests, normally performed weekly or biweekly, bar phantoms are used for the visual assessment of the extrinsic resolution (collimator mounted), together with a flood source discussed in the previous section. Normally, a low energy high resolution, parallel-hole collimator is used during this test. The bar phantom is first placed directly on the collimator and the flood source is placed on top of the bar phantom. Since the gamma camera resolution is dependent on the distance from the detector, operators should make sure that the bar phantom and the collimator are in direct contact with each other. A 10 Mcount image of the bar phantom is normally acquired and evaluated visually to check the detector resolution and linearity.

When used for determining the intrinsic resolution, the bar phantom is again placed on the detector without the collimator in place, and a ^{99m}Tc point source is placed at a distance five times the largest dimension of the crystal away from the bar phantom. As a rule of thumb, the intrinsic resolution of a detector in terms of the full width at half maximum (FWHM) of the line spread function can be approximately determined as FWHM ≈ $1.7S_b$, where S_b is the size of the smallest resolvable bars.

15.2.1.7. Dual-line phantom for whole body imaging

This phantom is used to test the whole body resolution of a gamma camera system. It consists of two parallel line sources which are 1 mm in internal diameter and 10 cm centre to centre. Figure 15.6 shows a custom built, dual-line phantom. The line is usually filled with ^{99m}Tc activity with a concentration of about 370 MBq/mL (10 mCi/mL) to achieve an adequate count rate. During the testing, the line sources are placed at a distance of 10 cm from both collimators. When measuring the perpendicular resolution, the lines should be placed parallel to the bed direction with one of them being in the centre of the bed. When measuring the parallel resolution, the lines should be positioned perpendicular to the direction of the bed movement. The whole body resolution is calculated from the FWHMs of the line profiles extracted from the image of the dual-line sources.

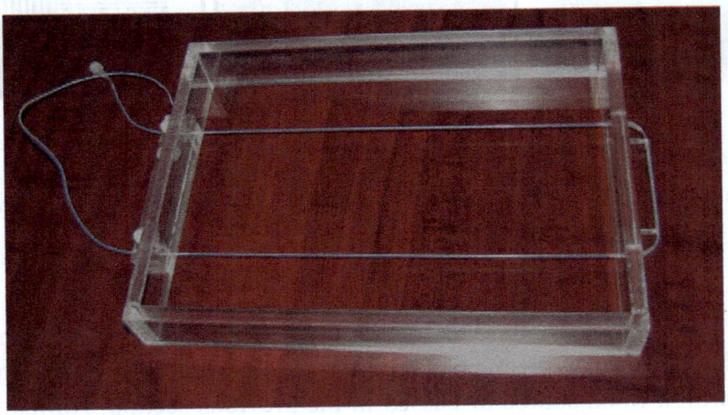

FIG. 15.6. Dual-line phantom for whole body resolution tests.

15.2.1.8. Planar sensitivity phantom

In a planar sensitivity test, the accuracy of the response of the detector to a radioactive source of known activity is measured for the particular collimator. It is suggested to use a Petri dish containing around 3 mm of water homogeneously mixed with a suitable activity (around 40 MBq) of ^{99m}Tc. The activity should be drawn into a syringe and then measured accurately in the dose calibrator. After injecting the activity in the Petri dish, the residual activity in the syringe should be measured. The residual activity is subtracted from the initial activity to determine the net activity injected into the dish. This dish should be placed at a distance of 10 cm from the face of the collimator. It is recommended to acquire two images. The average count, in units of counts per megabecquerel per second or counts per minute per microcurie, is determined to measure the planar sensitivity of the system.

15.2.1.9. Multiple window spatial registration phantom: lead-lined point source holders

A multiple window spatial registration test measures the camera's ability to position photons of different energies. In this section, the phantom is discussed, as described in Ref. [15.6], together with its preparation and the measurement procedures. The details of the test conditions and test phantoms can be found in Ref. [15.6]. A schematic drawing of the lead phantom is given in Fig. 15.7. As suggested by NEMA, nine of these lead-lined source holders are placed on the surface of the detector. The relative position of each holder is shown in the drawing. Plastic vials, as seen in Fig. 15.7, can be used to hold the actual activity of ^{67}Ga (~7–11 MBq (200–300 μCi) in each). Other acquisition parameters and camera settings are given in Table 15.1.

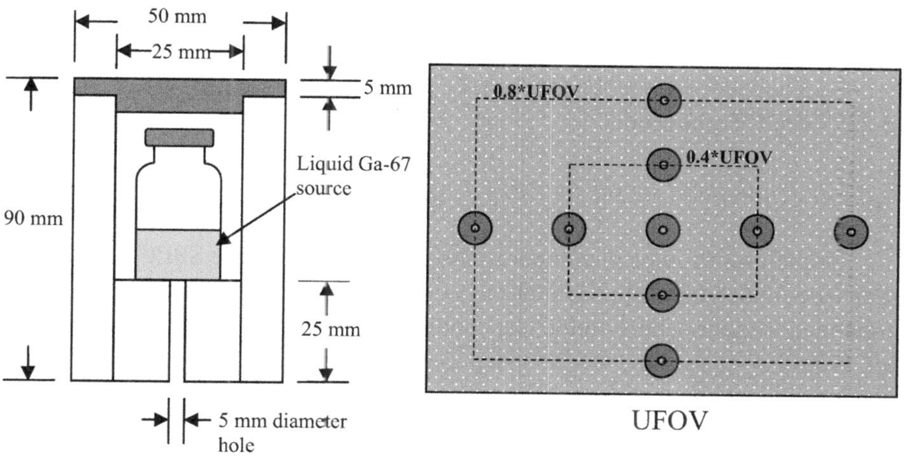

FIG. 15.7. Multiple window spatial registration phantom lead-lined point source holders. On the right is the top view of the point sources or source holders placed on the detector crystal. The locations of the point sources are determined by multiplying the dimensions of the useful field of view (UFOV) by 0.4 and 0.8. On the left is the cross-sectional view of the source holder together with the source vial.

Images of nine (or four) point sources of ^{67}Ga are acquired normally at three different photopeak energy windows (the three photopeaks for ^{67}Ga are 93, 185 and 296 keV).

The aim of the subsequent calculation is to find the centroids of these points in the image acquired at different energy windows and to compare the displacement between the point source images acquired at different energy windows. The maximum displacement between the centroids of point sources is the performance parameter indicating the error in multiple window spatial registration. The details of the calculation of this performance parameter can be found in Ref. [15.6].

TABLE 15.1. IMAGE ACQUISITION AND CAMERA SETTINGS

Radionuclide	^{67}Ga
Activity	~7–10 MBq in each source
Total counts	1000 counts in the peak pixel of each point source
Energy window	15%
Count rate	<10 kcounts/s
Pixel size	<2.5 mm
Matrix size	~1024 × 1024

15.2.2. SPECT phantoms

15.2.2.1. Triple-point source for SPECT resolution

Triple-point source phantoms are used for measuring the SPECT resolution in air (i.e. under no scatter conditions) or measuring centre of rotation (COR) alignment. Further details on the test conditions and the phantom can be found in Ref. [15.6].

For this purpose, thin-walled glass capillary tubes with an internal diameter of less than 2 mm are used. These point sources can be prepared as follows. First, a ^{99m}Tc solution of high concentration (about 5.5 GBq/mL) is prepared in a small (1 mL) syringe. Then, drops of small sizes are created on the surface of a clean plastic. These small drops can be drawn up into the capillary tubes by the principle known as capillary action by simply touching them. It may take a few trials to get a small size drop. At the end, the capillary tubes should be sealed on both ends with a capillary tube sealer such as Critoseal®. The point sources should be made as spherical as possible, that is, their transaxial and axial extents should be similar in length. Their maximum dimension (the axial extent of the activity) should not exceed 2 mm. The activity in the point sources should not vary more than 10%. The point sources should be suspended in air and positioned in accordance with the suggestions in Ref. [15.6] (Fig. 15.8). An alternative practical solution to suspend the point sources in air is to mark the positions of the point sources on a thin paper attached to a polystyrene (widely known as Styrofoam) sheet, and use this as a source holder. The scatter caused by the holder should be negligible.

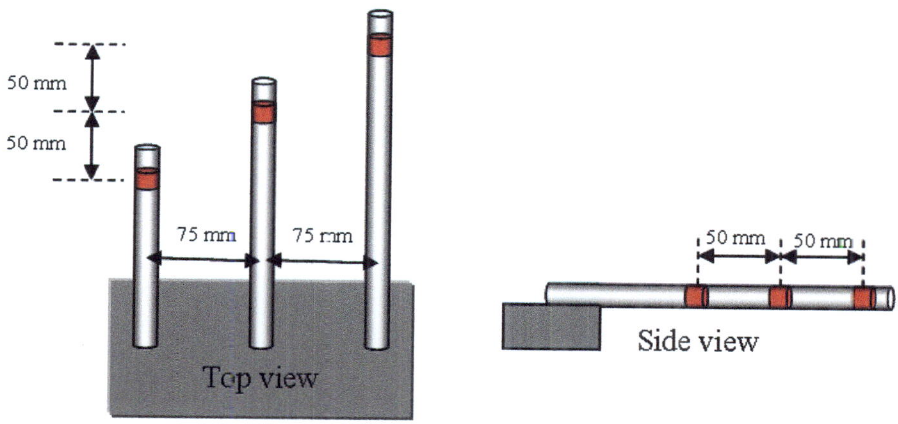

FIG. 15.8. Top and side views of the position of the point sources as suggested by the National Electrical Manufacturers Association.

15.2.2.2. Triple-line source phantom for SPECT resolution

The SPECT resolution with scatter is measured using the triple-line source phantom. This performance test is normally performed as part of the acceptance testing and annual testing. As described in Ref. [15.6], this phantom consists of a cylinder made of plastic (lucite or Perspex) with three line sources oriented along the axial direction (see Fig. 15.9). The cylinder is filled with water to create a scattering medium. The line sources are available either as inserts of ^{57}Co lines or hollow metal tubes to be filled with ^{99m}Tc solution. Here, the latter is discussed (Fig. 15.10). The inner diameter of the line sources is less than 2 mm. Both ends of the line sources are available for injecting the activity and are normally closed with small caps after the injection.

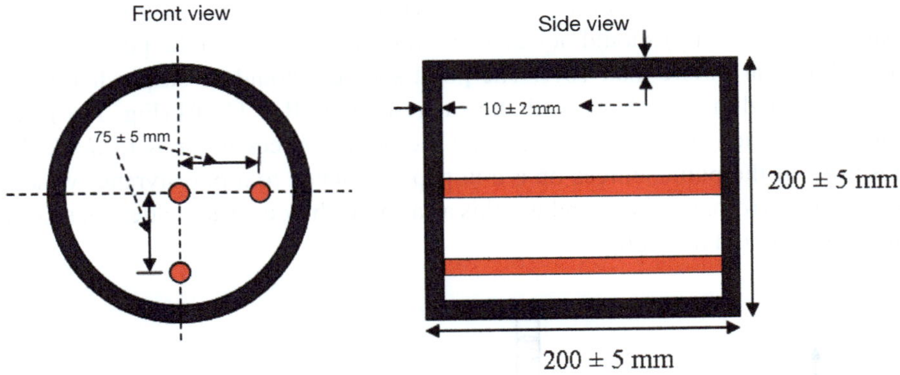

FIG. 15.9. Schematic drawing of the front and side views of a triple-line source phantom.

FIG. 15.10. A commercial triple-line source phantom with three line sources inside. The tank is filled with water to simulate a scattering medium.

The line sources should all be emptied of the decayed solution left from the previous test using two empty syringes attached at both ends of the line source. During the injection of each line source, two syringes are attached to both ends, one empty and one with activity of a concentration around 300–500 MBq/mL. While pushing the plunger of the syringe with the activity, that of the empty syringe should also be pulled very slowly until ^{99m}Tc solution appears from the other end. The filled line source should be securely sealed from both ends with the original caps, ensuring that there is no leak. It should also be ensured that the entire line source is uniformly filled.

During measurement, according to Ref. [15.6], the centre line source should be on the axis of rotation centred in the FOV within ±5 mm. The pixel size should be small enough (<FWHM/3) to prevent aliasing. Since this test has to be carried out with the collimator, a high resolution collimator is the best choice. During data acquisition, a 0–360° range should be evenly covered. Some of the acquisition parameters and camera settings are given in Table 15.2.

TABLE 15.2. ACQUISITION PARAMETERS AND CAMERA SETTINGS FOR THE SPECT RESOLUTION WITH SCATTER TEST

Radionuclide	^{99m}Tc
Count rate (kcounts/s)	<20
Total kilocounts per view	100
Scan time/view	~5 s at 20 kcounts/s
Energy window	15%
Collimator	Low energy high resolution
Radius of rotation	150 ± 5 mm
Total number of views	≥120
Pixel size	<2.5 mm

After the measurement, the resolution parameters should be calculated according to the method set forth in Ref. [15.6].

15.2.2.3. Volume sensitivity and detector to detector variation measurement phantom

Volume sensitivity is the total system sensitivity to a uniform concentration of activity in a specific cylindrical phantom. Factors such as detector configuration, collimator type, radionuclide, energy window setting and source configuration will impact the volume sensitivity in SPECT. Detector to detector sensitivity variation is the relative difference in sensitivity of the individual detector heads in a tomographic mode. The data acquired in a volume sensitivity test are directly used to calculate this performance parameter as well.

The volume sensitivity in SPECT is measured using a cylindrical phantom with an inner diameter and a length of 200 ± 5 mm (see Ref. [15.6]). The recommended wall thickness is 10 ± 2 mm. The volume of the phantom has to be accurately measured to accurately calculate the source concentration. The phantom is filled with water uniformly mixed with a known amount of activity (approximately 350 MBq) of ^{99m}Tc. The activity amount should be such that the count rate at the photopeak energy window is 10 000 ± 2000 counts/s. The following parameters have to be accurately determined and recorded to calculate the volume sensitivity:

— Volume of the phantom;
— Pre- and post-injection syringe activity to determine net injected activity;
— Elapsed time half way through the SPECT acquisition;
— Total scan time.

Further details of the measurement and calculations can be found in Ref. [15.6].

15.2.2.4. Total performance test phantoms

Image quality measures or overall SPECT system performance, such as noise, tomographic uniformity, contrast and lesion detectability, are measured using total performance phantoms. These phantoms are commercially available and are not so easy to build in an institutional workshop. There are several commercial phantoms for this purpose. Some of the phantoms that are frequently used to assess the performance of a SPECT system are discussed. It should be noted that these phantoms can also be used to evaluate PET systems.

15.2.2.5. Carlson phantom

The Carlson phantom (designed and developed by R.A. Carlson, Hutzel Hospital, Detroit, MI, USA, and J.T. Colvin, Texas Oncology PA, Dallas, TX, USA) in this category is frequently used for evaluating the tomographic uniformity, image contrast, noise and linearity. The main source tank (see Fig. 15.11) is made of acrylic with dimensions: 20.32 cm inside diameter, 21.59 cm outside diameter and 30.48 cm length. The phantom comes with various inserts, which are demonstrated and described in Fig. 15.11, to evaluate the performance parameters noted above. The thick plastic screws on the top lid allow easy filling and draining of the tank with water. The ^{99m}Tc solution injected inside the tank serves as the background activity, which may vary between 300 and 550 MBq, depending on the collimator used [15.7].

There is an insert or section for each performance measure. The SPECT uniformity is assessed using the uniform section of the phantom. The non-uniformities in the gamma camera can result in severe ring or bull's-eye artefacts. These artefacts can be checked for by looking at the uniform transverse slices. The amount of noise can be quantitatively calculated from the uniform section.

15.2.2.6. Jaszczak circular and elliptical phantoms

Similar to the Carlson phantom, Jaszczak elliptical and circular phantoms are used to evaluate the overall performance of SPECT systems after a repair or preventive maintenance, or during acceptance testing or quarterly testing. In addition to the purposes above, these phantoms can be used in evaluating the impact of reconstruction filters on resolution, as well as for other purposes in research studies.

Jaszczak phantoms consist of a main cylinder or tank made of acrylic with several inserts (see Fig 15.12). They are manufactured and sold by Data Spectrum Corporation (NC, USA). Jaszczak phantoms, which may have circular or elliptical tanks, come in several different flavours. The cylinders of all models of the circular flanged phantoms have the same physical specifications: 21.6 cm inside diameter, 18.6 cm inside height and 3.2 cm wall thickness. The principal differences between the different models of the flanged cylindrical Jaszczak phantoms are the diameters of the rods and solid sphere inserts. The circular phantom has flanged and flangeless models. The latter is recommended by the American College of Radiology for accreditation of nuclear medicine departments. These different models are designed to test a range of systems, from low resolution to ultra-high resolution, which has rods and spheres smaller than the others.

All Jaszczak phantoms have six solid spheres and six sets of cold rods. In flanged models, the sizes of the spheres vary. The number of rods in each set depends on the size of the rod in that set as different models of the phantom have rods of different sizes. In flangeless models, the diameters of the spheres are 9.5, 12.7, 15.9, 19.1, 25.4 and 31.8 mm, while the rod diameters are 4.8, 6.4, 7.9, 9.5, 11.1 and 12.7 mm. Both solid spheres and rod inserts mimic cold lesions in a hot background. Spheres are used to measure the image contrast while the rods are used to investigate the image resolution in SPECT systems.

15.2.2.7. Anthropomorphic torso phantoms

Anthropomorphic torso phantoms are used in testing gamma cameras in SPECT mode to evaluate data acquisition, attenuation correction and image reconstruction methods. They normally simulate or model the upper torso of the body (from the heart down to the diaphragm) of an average male or female patient. These phantoms consist of a body-shaped (elliptical) cylinder with fillable inserts for organs such as the heart, lungs and liver (see Fig. 15.13).

Phantom	Description
Hot lesions 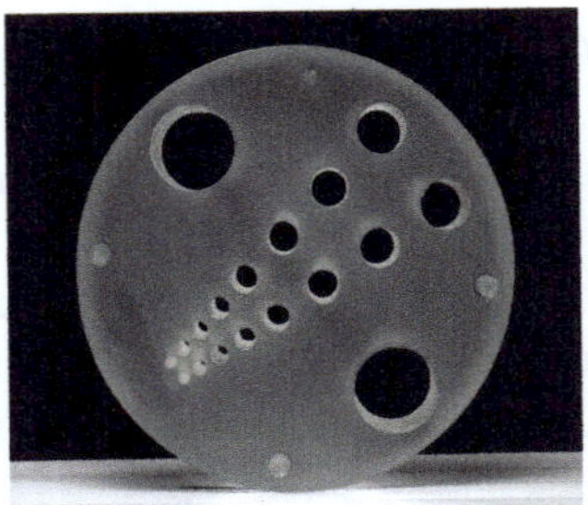	Eight pairs of holes drilled through a solid acrylic block, with diameters of 4.7, 5.9, 7.3, 9.2, 11.4, 14.3, 17.9 and 22.3 mm, model hot lesions with the background activity injected.
Cold rods and spheres 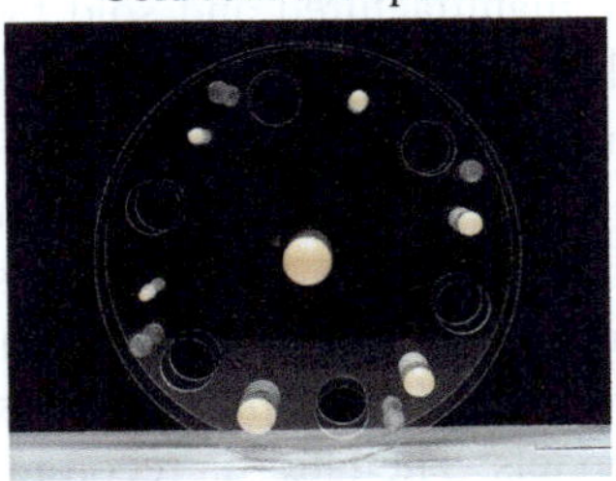	Seven rods, with diameters of 5.9, 7.3, 9.2, 11.4, 14.3, 17.9 and 22.3 mm, simulate cold lesions. Each rod is 25% larger in diameter than the preceding one. Seven solid spheres of the same diameters as rods, the centre one being the largest, are attached to the rods.
Linearity/uniformity section 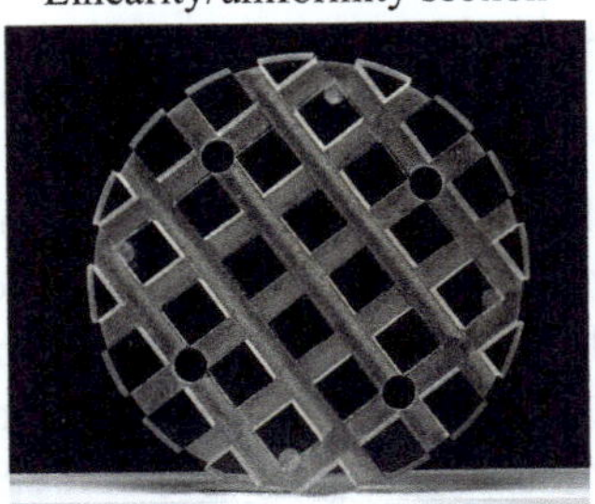	Crossed grid of cut out channels, again in an acrylic block, can be used to assess the linearity. The region where only background activity is available is used to evaluate the tomographic or SPECT uniformity.
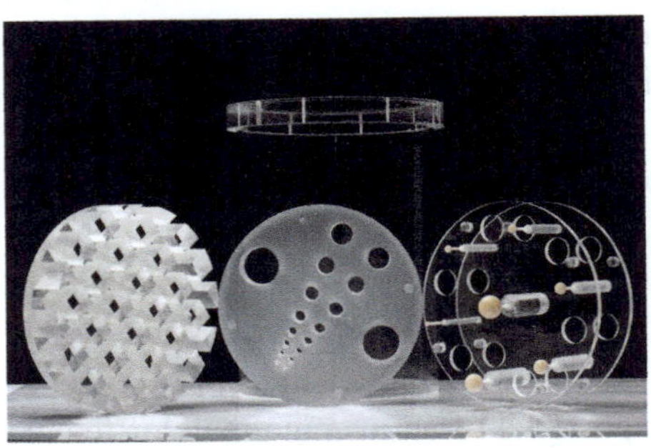	Picture of the Carlson phantom tank together with all three inserts.

FIG. 15.11. Carlson phantom and its inserts.

FIG. 15.12. Jaszczak phantom used for verifying image quality (phantom by Data Spectrum Corporation, USA).

Defects can also be added to the heart insert. Lung inserts are filled with Styrofoam beads and water to emulate lung tissue density. The phantoms can be used to evaluate non-uniform attenuation correction methods including CT based attenuation correction in SPECT/CT systems and scatter compensation methods. When used with the optional cardiac insert, cardiac SPECT data acquisition and reconstruction methods may also be evaluated.

Filling the inserts with different distributions of radioactivity is not as easy as filling other phantoms because of the multiple organs and the organ to background ratios that need to be adjusted. To set the concentration ratios, the volumes of the organ inserts need to be measured accurately a priori. For a simulation of a 1110 MBq (30 mCi) sestamibi stress study, the injected activity concentrations, as suggested in Ref. [15.8], are given in Table 15.3.

Torso phantoms can be integrated with the fillable breast phantom, which is also commercially attainable. These breast phantoms allow the inclusion of inserts to simulate breast lesions that can be employed to evaluate lesion detectability.

The volumes in the second column in Table 15.3 are the measured volumes of the torso phantom inserts.

TABLE 15.3. SUGGESTED ACTIVITY CONCENTRATIONS AND MEASURED VOLUMES OF INSERTS FOR THE ANTHROPOMORPHIC TORSO PHANTOM

Section	Volume (mL)	Activity concentration (kBq/mL)	Total activity (MBq)
Heart	117	250	30
Tissue	8620	25	225
Liver	1177	150	175
Lungs		0	0

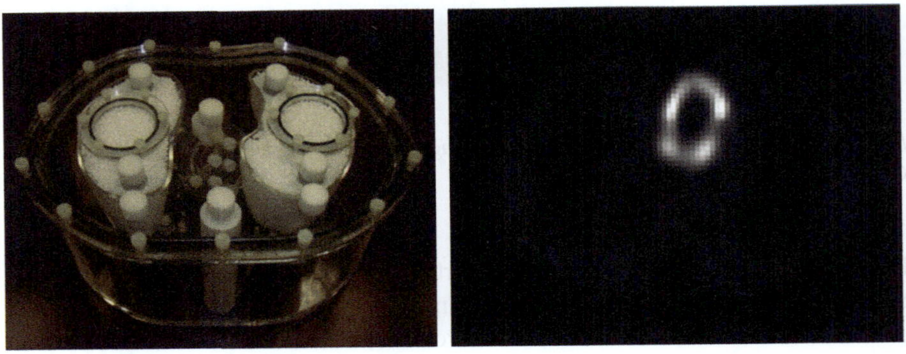

FIG. 15.13. A commercial anthropomorphic phantom and a transaxial slice cutting through the heart and lungs from its image acquired by a SPECT/CT system.

15.2.2.8. Hoffman brain phantom

This phantom, developed by Hoffman et al. [15.9], provides an anatomically accurate simulation of the radioactivity distribution in normal brain. Using this phantom, cerebral blood flow and metabolic activity in the brain can be simulated. It can be used in both PET and SPECT systems to optimize/investigate imaging acquisition protocols, to evaluate attenuation and scatter correction methods, and to measure the performance of imaging systems. It consists of a water fillable cylinder (i.e. a single-fillable chamber) containing 19 separate layers each 6.4 mm thick (see Fig. 15.14). The fillable water volume is about 1.2 L. Water freely permeates between layers to simulate concentration ratios of 4:1:0 between grey, white and ventricle, respectively, in normal brain. The 2-D version, consisting of a single slice, and a 3-D version of the phantom are available commercially.

FIG. 15.14. Three dimensional Hoffman phantom with a water fillable cylinder and layers of inserts (phantom by Data Spectrum Corporation, USA).

15.2.2.9. Defrise phantoms

These phantoms are designed for measuring the performance of small animal imaging systems (both SPECT and PET). They can be used to investigate image quality or resolution. Figure 15.15 shows the hot spot phantom manufactured by Data Spectrum Corporation, USA. This phantom is a miniaturized version of the image quality phantoms mentioned in the previous sections. The phantoms are available in different sizes for imaging systems with different FOVs.

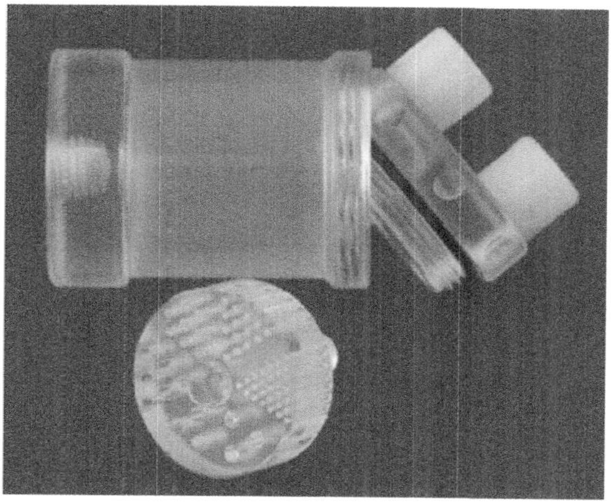

FIG. 15.15. Defrise hot spot phantom manufactured by Data Spectrum Corporation, USA.

15.2.3. PET phantoms

15.2.3.1. National Electrical Manufacturers Association image quality phantom

Measuring image quality in an objective manner has been one of the most difficult tasks in PET. Image quality in PET can be determined by calculating performance parameters, such as uniformity, noise, lesion contrast, spatial resolution, and the accuracy of the attenuation and scatter correction techniques. In this section, the NEMA image quality (IQ) phantom is described. This phantom (known as the NEMA IEC (International Electrotechnical Commission) body phantom) was originally recommended in IEC standards and was then adopted by NEMA. In addition to the above performance parameters, the image registration accuracy between the PET and CT gantries in a PET/CT scanner can be assessed. This phantom is commercially available from Data Spectrum Corporation, USA. The IQ phantom consists of four main parts:

(a) Fillable spheres: The six fillable spheres are used for measuring hot and cold lesion contrast. The inner diameters of the six spheres are 10, 13, 17, 22, 28 and 37 mm. The two largest spheres (28 and 37 mm) are filled with water to mimic cold lesions, while the rest are injected with ^{18}F activity with lesion to background ratios of 4:1 and 8:1 to mimic hot lesions. The spheres are attached to the cover or top lid through capillary stems. The

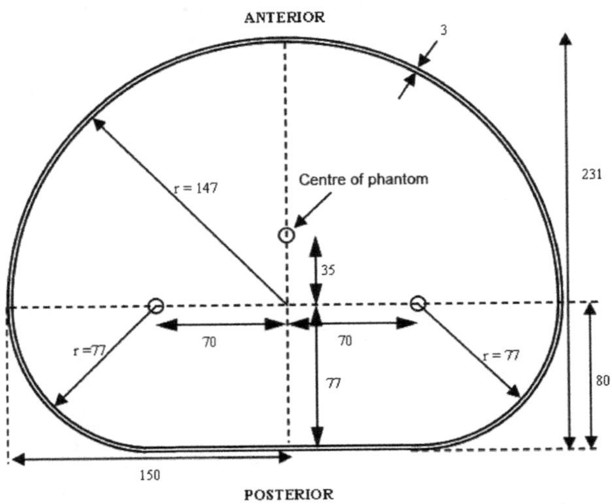

FIG. 15.16. Cross-section of the body part of the International Electrotechnical Commission image quality phantom made of acrylic. The dimensions are given in millimetres (reproduced with permission).

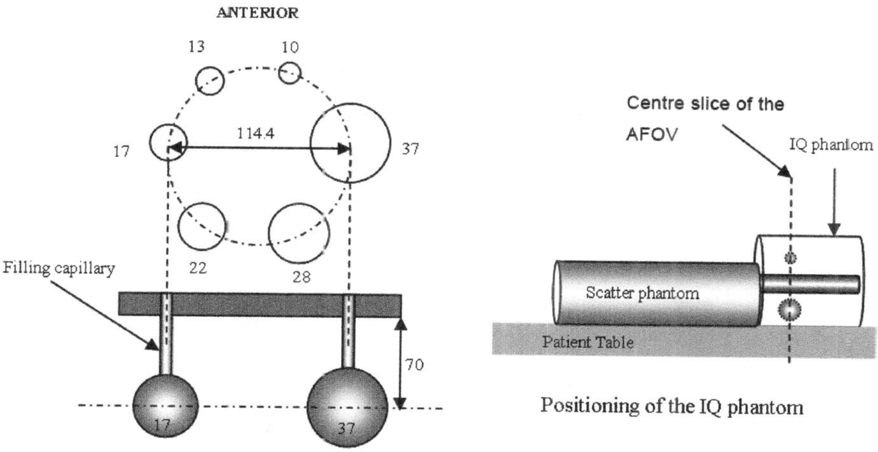

FIG. 15.17. Transaxial (top left) and coronal (bottom left) cross-sectional view of the image quality (IQ) phantom through the centres of fillable spheres. Sphere diameters and the other dimensions are given in millimetres (reproduced with permission). On the right is the schematic drawing demonstrating the positioning of the IQ phantom together with the scatter phantom.

filling is also done through the capillaries without removing the cover lid. Filler screws for each fillable part inside the body phantom allow easy access. A picture of the phantom is shown in Fig. 15.18.

(b) Cylindrical insert: A cylindrical section that is filled with a mixture of polystyrene beads and water to mimic lung (the density of which is around 0.3 ± 0.1 g/mL) attenuation is placed axially in the centre of the phantom with the same length as the body phantom. The outside diameter of the insert is about 5 cm.

(c) Phantom preparation: Table 15.4 shows the measured volumes of the various inserts and the torso cavity of the IQ phantom. It is suggested that all the volumes be measured upon acquiring a new IQ phantom. The activities used to fill the phantom should be measured using a calibration time that corresponds to the planned PET acquisition time, taking into account the time necessary for the preparation and positioning of the phantom for this test. Table 15.4 shows the typical activity concentrations that may be prepared and injected into the background and the hot spheres in order to have the proper activity concentration at the time of the scan (supposed to be performed 45 min after phantom preparation). It should be noted that the activity concentration ratio in the table is 8:1. A 4:1 activity concentration ratio can be easily obtained by doubling the amount of activity in the background.

FIG. 15.18. National Electrical Manufacturers Association/International Electrotechnical Commission image quality phantom.

TABLE 15.4. MEASURED VOLUMES OF THE NATIONAL ELECTRICAL MANUFACTURERS ASSOCIATION/INTERNATIONAL ELECTROTECHNICAL COMMISSION IMAGE QUALITY PHANTOM AND SUGGESTED ACTIVITIES FOR A CONCENTRATION RATIO OF 8:1

Phantom section	Volume (mL)	Typical activity (MBq)	Activity concentration at time of preparation (kBq/mL)	Activity concentration at time of scan (kBq/mL)
Torso cavity	9700	n.a.		
Four hot spheres	Different sizes	n.a.	56	42.4
Two cold spheres	Different sizes	n.a.		
Lung insert	353	n.a.		
Background (torso – all inserts)	9286	65	7	5.3

Note: n.a.: not applicable. The scan is supposed to be performed 45 min after phantom preparation. For a description of the phantom, please see: http://www.spect.com/pub/NEMA_IEC_Body_Phantom_Set.pdf

There are different suggestions as to how to prepare the IQ phantom. The following is a summary of one possible approach:

— The NEMA recommends an activity concentration for the background of 5.3 kBq/mL at the time of the scan, assuming that a normal 70 kg patient injected with 370 MBq of activity will have a similar background activity in the body (370 MBq/70 000 mL, ~5.3 kBq/mL).
— The amount of time to fill and position the phantom must be estimated to determine the amount of activity at the time of preparation of the phantom. A typical time frame for this process would be 45 min.
— Two separate activities of 65 MBq are prepared and one of them is injected into the background. This results in a background activity concentration of 65 MBq/9286 mL = 7 kBq/mL. The activity concentration will be reduced to ~5.3 kBq/mL after 45 min (time of scanning).
— Another solution for the hot spheres with an activity concentration of ~56 kBq/mL is prepared separately. ^{18}F activity of 5.6 MBq can be injected into 100 mL of cold (non-radioactive) water to obtain this concentration. If the measured activity is slightly more or slightly less, the volume of the cold water can be adjusted accordingly to achieve the intended concentration.
— The ^{18}F activity is injected into the torso cavity (i.e. background), which is already filled with cold water, and then the hot spheres are filled with the prepared ^{18}F solution.
— After having acquired the images of the phantom for the ratio of 8:1, the previously prepared activity of 65 MBq is added to the background in order to obtain an activity concentration ratio of 4:1.
— The phantom is acquired for the 4:1 ratio one half-life (~110 min) after the first scan, when the activity concentration in the background will be ~5.3 kBq/mL.

One of the disadvantages of the above filling method is the difficulty of mixing the background activity uniformly into the cold water; however, the method obviates the need for removal of the top lid with attached spheres and refilling of the background in each experiment.

15.2.3.2. National Electrical Manufacturers Association scatter phantom

The scatter phantom, whose specifications were set forth by NEMA guidelines (NEMA NU 2-2007 [15.10]), is used to measure the count rate performance of PET scanners in the presence of scatter. In other words, it is used to measure the amount of scatter in terms of the scatter fraction, the effect of dead time and the random events generated at different levels of source activity.

The phantom consists of a solid, 70 cm long, polyethylene cylinder with an outer diameter of 203 ± 3 mm and a line source insert. The line source insert is made of a clear polyethylene tube at least 80 cm in length, and with inner and outer diameters of 3.2 ± 0.2 and 4.8 ± 0.2 mm, respectively. The volume of the line source is approximately 6 mL. The solid cylinder comes in four segments for ease of fabrication and handling. During the assembly, these four segments should be tightly fitted to prevent the formation of scatter-free air gaps in between them. A hole (6.4 ± 0.2 mm in diameter) is drilled along the central axis of the cylinder at a radial distance of 45 ± 1 mm (see Fig. 15.19) to insert the aforementioned line source. The scatter phantoms are commercially available.

The line source insert should be uniformly filled with a ^{18}F solution. The amount of activity is usually recommended by the manufacturer and should be in the central 700 ± 5 mm part of the insert. The line source should be inserted such that the activity region remains completely within the 70 cm long phantom. Further detail about phantom preparation and data acquisition can be found in Ref. [15.10].

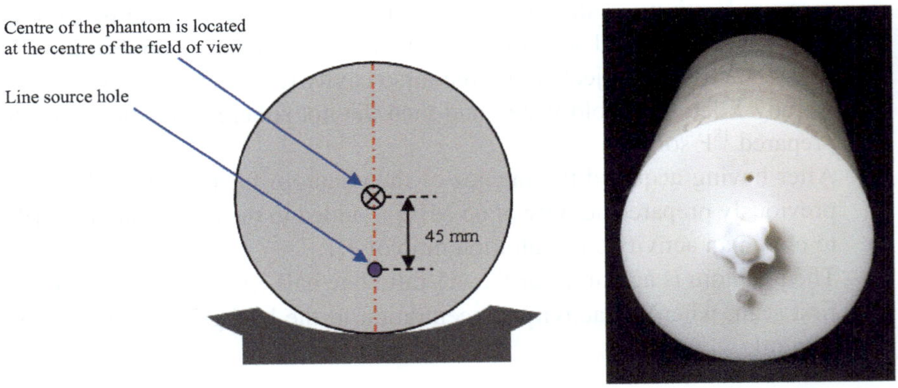

FIG. 15.19. Positioning of the scatter phantom on the patient bed: transaxial view (left); picture of the National Electrical Manufacturers Association scatter phantom (right).

15.2.3.3. National Electrical Manufacturers Association sensitivity phantom

Sensitivity is the number of counts per unit time per unit of radioactivity concentration within the FOV. To be able to compare different PET scanners, the sensitivity performance measure should be independent of factors such as scatter, attenuation, count losses and random events. Therefore, in PET, unlike SPECT, the sensitivity is measured using a special phantom developed by Bailey et al. [15.11] and later adapted by NEMA. The NEMA sensitivity phantom allows the determination of the attenuation-free sensitivity.

The sensitivity phantom consists of five concentric aluminium sleeves (70 mm in length), each with a wall thickness of 1.25 mm. The inner diameters of the five tubes are 3.9, 7, 10.2, 13.4 and 16.6 mm. The line source, made from clear polyethylene, is filled uniformly with ^{18}F in solution and inserted into the smallest sleeve and suspended in air within the FOV of the scanner. The line source is filled with activity, such that the dead time losses are less than 1% and the random events are less than 5% of the true rate. Figure 15.20 shows the sensitivity phantom: the five aluminium sleeves and the tube. The figure also shows the positioning of the phantom during the scan. In this case, a shower curtain rod and the point source holder from the scanner vendor are used to suspend the phantom within the FOV. A sling can be constructed from tape to hang the phantom in that position as well. It should be noted that the centre of the aluminium sleeves should coincide with the centre of the AFOV of the scanner.

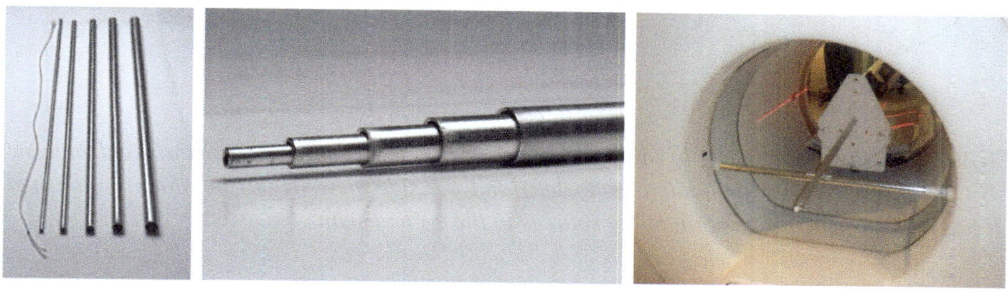

FIG. 15.20. Pictures of the National Electrical Manufacturers Association sensitivity phantom: positioning of the phantom within the gantry (right). A spring tensioned shower curtain rod and the point source holder are used to suspend the phantom within the field of view. The aluminium sleeves (left and centre) should coincide with the centre of the axial field of view.

15.2.3.4. Triple-point source phantom for spatial resolution

Hematocrit or capillary tubes are commonly used to create point sources for measuring the spatial resolution of PET scanners. The inner and outer diameters of these tubes should be less than 1 and 2 mm, respectively. The axial extent of the activity in the tube should be no more than 1 mm. As for the NEMA NU 2-2007 guidelines [15.10], three point sources should be positioned as shown in Fig. 15.21. It should be noted that the central point source is positioned 1 cm above the centre of the FOV.

A high concentration of ^{18}F activity in a solution should be prepared such that neither the dead time losses nor random events exceed 5% of the total event rate. The actual activity concentration to be used, more than approximately 200 MBq/mL, is normally provided by the manufacturer. The preparation of the point sources in hematocrit tubes is undertaken as discussed in Section 15.2.2.1. The point sources are positioned and the data are acquired at the centre of the FOV as well as at a distance a quarter of the FOV away from the centre (see Fig. 15.21). Figure 15.22 shows a point source holder with capillary tubes mounted on it.

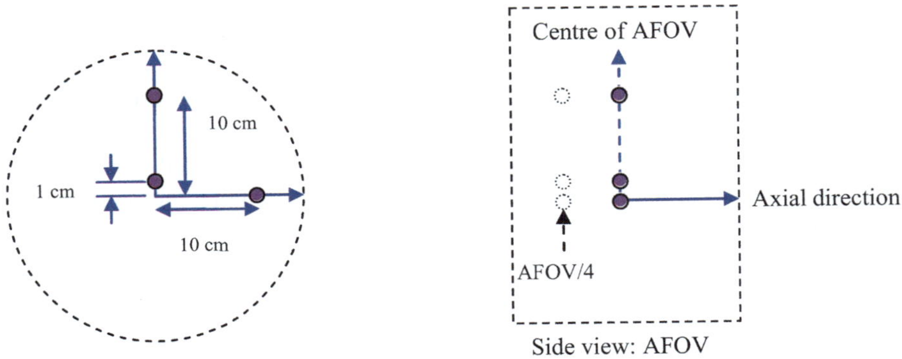

FIG. 15.21. Positioning of the three point sources in the centre of the axial field of view (AFOV). The view into the gantry bore (left) and the side view (right) in which the dashed circles denote the axial position of the sources in the second scan.

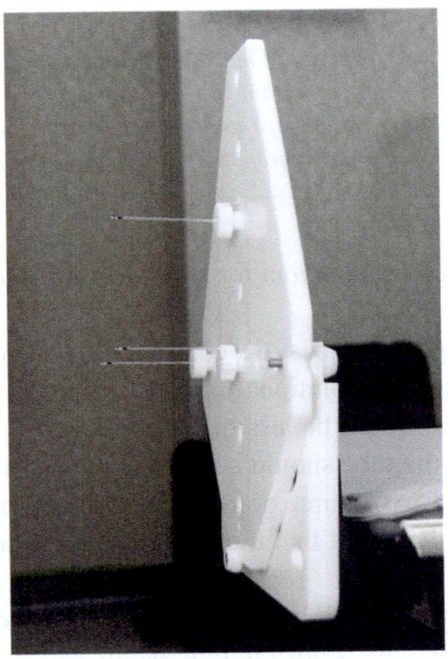

FIG. 15.22. Three capillary point sources mounted on a point source holder used in PET to measure spatial resolution.

15.3. COMPUTATIONAL MODELS

Computational models can be categorized in three groups:

(a) Mathematical models;
(b) Voxelized computational models;
(c) Hybrid computational models.

This enumeration also reflects the progress of the development as the listing is from simple to more realistic and sophisticated. This order of classification also reflects the chronological order of the development process of the computational models.

Mathematical models, also known as stylized models, simulate the organs with geometric primitives such as ellipsoids, cylinders, spheres and rectangular ellipsoids. These rather simple, geometrically well defined shapes representing the organs or structures in the body are defined using the surface equations of these primitives. The mathematical models were very early models and crude in their representation of organs. The well known models, which have been adopted by the Medical Internal Radiation Dose Committee of the Society of Nuclear Medicine and have been used for many years in dose calculations, are mathematical models. Although for a while these models served the purpose, the need for a more realistic definition of organs, and, therefore, a more realistic representation of the body, has always been there.

The advent of tomographic imaging technology, particularly X ray CT and magnetic resonance imaging (MRI) made it possible to obtain high resolution images of the body. In voxelized models, also known as tomographic models, the organs are defined by the structures segmented from high resolution tomographic images such as X ray CT and MRI. The segmented structures consist of volumetric image elements called voxels, each of which is assigned a value indicating the organ to which it belongs. The smaller the voxel dimensions, the more realistic the surface of organs can look. Depending on the dimension of the voxels, it may be challenging to define thin or small structures such as skin.

The Visible Human Project initiated and conducted by the United States National Library of Medicine has played a significant role in the development of voxelized models. As part of this project, CT and MRI images and cryosection photographs of a 38 year old male cadaver were made available in the public domain. To produce the colour photographic images of the cryosections, the cadaver was frozen and sliced into 1 mm thin sections and photographed at a resolution of 2048 pixels × 1216 pixels. This project has led to the development of many voxel based computational models [15.12–15.14]. The construction of voxel based models is a lengthy and tedious process and requires several steps.

First, high resolution images of the body need to be acquired. Then, the individual organs and structures are segmented from the high resolution images. The segmentation is the most challenging task as the boundaries between organs and tissues are often not well defined. Researchers, therefore, resort to tedious manual or semi-automated segmentation methods. Obtaining CT scans of desired pixel resolution or dimension and slice thickness may result in a significant amount of exposure to ionizing radiation; thus, it is difficult to recruit healthy subjects for this purpose. As a result, some of the voxel models have been constructed from medical images of patients. For example, the Zubal phantom [15.15] was created from CT scans of a patient by manual segmentation. These limitations on pixel dimensions and slice thickness have made cadavers an attractive choice for building voxel based models. In voxelized models, the surface of the organs are jagged, piece-wise continuous and, therefore, not smooth. Other issues, such as shifting of internal organs and non-rigid transformations in organ shape during the scan in the supine position, may limit the generality of these models.

Hybrid models combine the best of both worlds. Surfaces of the segmented structures in voxelized models are defined by mathematical formulations used to define irregularly shaped surfaces such as 3-D B-spline surfaces.

A group of researchers developed a series of 3-D and 4-D computational models. Their first model, the mathematical cardiac torso phantom, was a mathematical model based on simple geometric primitives but also used cut-planes and overlaps to create complex biological shapes to be used in nuclear medicine research. This model also included a beating heart based on gated MRI patient data and a respiratory model based on known respiratory mechanics. With this model, emission and transmission data could be simulated. The following models, 4-D NCAT and cardiac torso (XCAT) (see Fig. 15.23), were based on the visible human CT dataset. The organ shapes, i.e. surfaces, were reconstructed using the primitive non-uniform rational B-spline (NURBS) surfaces. The 4-D models use cardiac and respiratory motions developed using 4-D tagged MRI data and 4-D high resolution respiratory-gated CT data, respectively. These models, from the hybrid class, can successfully model not only the anatomy but also physiological functions such as respiratory and cardiac motion.

Such 4-D models can be used to accurately simulate SPECT and PET images of the torso and can be particularly helpful for optimizing image acquisition protocols and image reconstruction algorithms, and understanding the various effects of these complex motions on the acquired PET or SPECT images. These models are also accessible free of charge for academic research.

These models have been widely used in internal absorbed dose calculations in nuclear medicine or in calculation of dose distribution from external sources in radiation therapy and in studying issues pertinent to imaging systems and their performance characteristics. They have also been quite helpful in the optimization of image acquisition protocols and reconstruction methods.

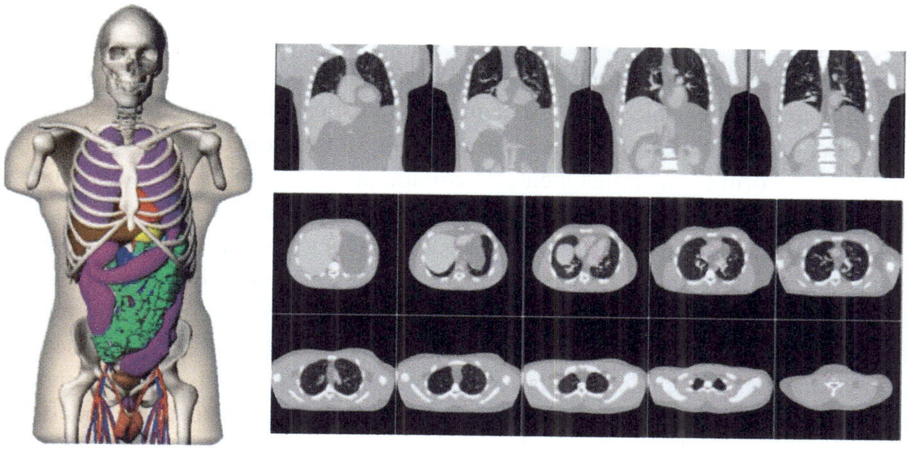

FIG. 15.23. Left: initial extension of the 4-D XCAT anatomy. Right: simulated chest X ray CT images from the extended 4-D XCAT. Coronal (top row) and transaxial (bottom two rows) reconstructed slices are shown (reproduced with permission from P. Segars).

Since anatomy and physiological functions are accurately known, they can serve as gold standards. Computational models may be preferred because the use of physical phantoms leads to unnecessary occupational exposure to radiation, and the preparation and repetition of the experiments using physical phantoms can be lengthy and time consuming.

An ideal model should be able to conform, reasonably well, to the size and shape of the object being represented. Currently, as personalized medicine is the strong driving impetus for most current research in many pertinent fields, personalized modelling should be the aim in computational model development research.

15.3.1. Emission tomography simulation toolkits

15.3.1.1. SimSET

SimSET, first released in 1993 and developed at the University of Washington, is a simulation package that can simulate PET and SPECT emission tomography systems using Monte Carlo simulations. It can model the photon interaction process as well as the imaging detector geometries. SimSET allows the use of a different object description such as a Zubal phantom to simulate a whole body phantom. SimSET is freely available for use.

15.3.1.2. GATE

Owing to the limitations of SimSET regarding the modelling of complex detector geometries, the need for a more sophisticated emission tomography simulator arose. To meet this need, a group of physicists from different institutions around the world formed the OpenGate collaboration. Out of this collaboration, a simulation toolkit (GATE) for nuclear medicine applications was developed and has been available since 2001. GATE uses the existing libraries of Geant4, which is a comprehensive simulation toolkit that simulates the interaction of particles as they traverse through matter. GATE is unique and superior in that it can model time dependent phenomena such as source and detector movement and source decay kinetics. It includes validated geometry modelling tools that can model complex scanner geometries. It also includes the description and models of several commercially available PET and SPECT scanners. GATE can simulate CT scans and can perform dose calculations. GATE is also freely available for use.

15.4. ACCEPTANCE TESTING

15.4.1. Introduction

As discussed in Ref. [15.16], gamma cameras are evaluated at different levels of testing before being made ready for clinical use. The first set of tests is carried out in the factory before shipment. Manufacturers test gamma cameras to check whether the performance parameters meet the specifications quoted to customers. NEMA has published several guidelines that describe the methods to measure the performance parameters of gamma cameras and PET systems [15.6, 15.10]. These guidelines provide standardized criteria for manufacturers to measure and report the performance of their scanner. The IEC has also published several technical reports [15.17–15.19] describing the tests to be performed during acceptance testing which reflects as closely as possible the clinical settings in which gamma cameras and PET systems are operated. Most manufacturers quote the performance of their systems according to NEMA guidelines [15.6].

The second level of testing is the acceptance testing performed after the scanner arrives at the site. These tests should be performed by the user or a third party, usually a qualified medical physicist, to determine whether the system performs according to the manufacturer's specifications and free of any deficiencies, flaws or defects.

The baseline performance of the equipment will also be established. These data provide guidance in the determination of the optimal operating parameters for routine use and ensure that the imaging equipment meets regulatory requirements for radiation safety [15.1].

These tests are usually very involved and require sophisticated phantoms and dedicated software to calculate the performance parameters. Several national and international agencies have set forth a range of tests, to be performed during acceptance testing, that are easier (than the NEMA tests) to conduct. The American Association of Physicists in Medicine (AAPM) is one of these agencies that has produced several publications for testing gamma cameras during acceptance and routine testing. The reports, AAPM 6, AAPM 9 and AAPM 22 [15.20–15.22], describe methods to perform acceptance testing on analogue, computer-aided and SPECT capable gamma cameras, respectively. These reports describe tests similar to those of NEMA. Moreover, the IAEA has published several books describing the methods to perform tests on gamma cameras during acceptance, reference and routine testing. Among them are TECDOC-317 and TECDOC-602 [15.16, 15.23]. The IAEA has recently published guidelines for QC and QA tests for PET and PET/CT scanners [15.1].

During acceptance testing, the user should also conduct reference tests which constitute the third level of testing. These tests reflect the performance of the system under clinical settings, are easy to perform and can be performed within an acceptable time frame. These tests, in addition to some other acceptance tests, will establish the baseline performance characteristics for routine QC tests. The results of routine tests are compared against the results of these tests.

Routine tests constitute the fourth level of testing. These are the tests performed on a regular basis by users. Depending on the variability (in time) of the performance parameter and its impact on image quality, test frequencies may range from daily to annual. Several guidelines have been published on routine tests, describing them and specifying their frequencies and the tolerance limits. The IEC published standards 61675-2 and 61948-2 [15.18, 15.24] for gamma camera routine testing including SPECT, and standard TR 61948-3 [15.25] for PET routine testing. The AAPM has also published Report No. 52 [15.7] which describes methods for measuring the quantification of SPECT performance.

Several issues regarding acceptance testing should be considered. Some phantoms are used during acceptance testing and after major repair only, and may be included in the purchasing contract. The manufacturer may also lend their customers these phantoms during the period of acceptance testing. The slit phantom used for testing linearity and intrinsic resolution of gamma cameras is a typical example. The calculation of performance parameters from the image data in PET and the gamma cameras and SPECT systems may require sophisticated software applications; thus, in such a case, the manufacturer must provide the calculation software. The documentation for the acceptance test procedures may be made available by the vendor. If needed, the recommendation of the manufacturer should be followed, for instance, with regard to the amount of activity required for each test. In multimodality imaging systems, additional tests, which are not discussed in the existing guidelines, such as the accuracy of image registration and attenuation correction, must also be conducted.

Before starting acceptance testing, the following additional issues should be considered:

- An accurate dose calibrator is an essential part of acceptance testing and must, therefore, be available.
- The required amount of radioactivity has to be arranged before starting acceptance testing, so that the acceptance testing procedure does not experience any interruption.
- Proper calibration of the imaging system prior to acceptance testing is of paramount importance. Any major erroneous calibration or lack of calibration may result in an increase in commissioning cost and undue delays in acceptance testing.
- The order of the tests that will be conducted must be arranged so that any malfunctioning or improper calibration can be discovered early on. This will minimize the number of tests that must be repeated after recalibration of the system.
- If the medical physicist is not familiar with the system, a vendor representative who knows how to operate the scanner and how to run the calculation software should be present during acceptance testing.
- All the required phantoms discussed in earlier sections of this chapter should be made ready and prepared in advance.

15.4.2. Procurement and pre-purchase evaluations

When an institution decides to buy an imaging system, the administration should start the planning properly by defining the purpose(s) for acquiring the system and form a committee of a team of professionals to take on all of the responsibilities from purchasing to setting up the system.

The purchasing committee should include the following professionals, as defined in Ref. [15.1]:

- Nuclear medicine and radiology physicians;
- A medical physicist with experience in nuclear medicine;

— If buying SPECT/CT or PET/CT, a medical physicist experienced in diagnostic radiological physics should be included;
— A medical physicist experienced in radiation therapy if the system will be used in radiation therapy planning;
— An administrator from the radiology department;
— A radiation protection expert;
— A person qualified in radiochemistry or radiopharmacy, if in-house production of radiopharmaceuticals;
— A nuclear medicine technologist;
— A hospital management expert;
— A bioengineering expert in imaging systems.

The role of this committee is to:

— Choose the location;
— Set the specifications of the system;
— Prepare the tender documents;
— Choose the proper system;
— Supervise the installation process;
— Supervise the acceptance and commissioning procedure.

This committee should start by choosing the proper space to host the system. This location should be inside a radiation-controlled area, with the door of the room opening to a closed vicinity (not to a public corridor). The room should be wide enough to host the scanner, give accessibility to patient stretchers and provide free space to maintenance engineers. If possible, the room should be far away from MRI scanners to avoid any interference from their magnetic field. If buying SPECT/CT and the CT sub-component will be used as a stand alone system occasionally, it is advisable to have the scanner as close as possible to the radiology CT scanner to act as a backup system when needed. This is also true for the PET/CT systems if there is no on-site cyclotron. For scanners inside institutions having a cyclotron, it is advisable to have the scanner as close as possible to the cyclotron. This will allow the quick transfer of isotopes with very short half-lives using dedicated lines or manual means.

The process of setting the specifications starts by agreeing on the purpose for which the scanner will be used. Again, based on the applications that the system will be used for, the different add-on components to be ordered will be decided.

After defining all of the components, the specifications of the scanner and each component should be set. To define a suitable specification, the committee members should know what suitable systems are available that may meet their needs. After studying these systems, the required specifications should be set with the aim of not excluding any available system initially. As a good practice, one or several Excel work-sheets should be developed. The work-sheet(s) should list all of the specifications, hardware, performance parameters, imaging table, standard software, optional software, etc. Under each category, a list of different specifications in that category should be listed with their limits. Examples of hardware specifications are crystal(s) dimension and shape, number of photomultiplier tubes, bore diameter and head movement ranges. Examples of performance specifications are resolution, uniformity, dead time, SPECT specifications, noise equivalent count rate and sensitivity. Examples of imaging table specifications are pallet thickness, attenuation factor, scan range and speed, minimum and maximum floor clearance, and weight limits. Knowing all of the software that comes with the system on the acquisition and processing stations and the optional ones available is necessary at this stage. The work-sheet(s) will be distributed to all vendors as a soft copy, so that the answers from each will be rearranged in one sheet to allow easy comparison of each specification between vendors.

The tender should be prepared by the committee members and should follow the institution's local regulations. It should include a summary of the terms and conditions of the new equipment purchase deal. The following items may be requested in a tender:

- Name and model of the equipment.
- Terms of pricing; way of payment, site preparation, accessories, etc.
- Application specialist training.
- System upgrade conditions.
- Equipment references; short list of current users of similar system, local or international.
- Training of staff.
- Equipment warranty.
- Scheduling installation process and way of coordination.
- Responsibility of site preparation, including removal of old equipment.
- User and engineering manuals and equipment specifications (NEMA and others).
- Acceptance testing to be performed by a medical physicist (the system should comply with NEMA or local specifications).
- Commitments of the vendor to provide maintenance, and spare parts readiness.
- Specifications of local civil work and materials used.

Other steps that may assist the committee in the evaluation stage are:

- — Site visits: Manufacturers take the prospective customers to their reference sites to evaluate the systems and listen to the users.
- — Evaluation of the clinical and phantom images provided by the manufacturers: It is recommended that this be carried out on a common imaging workstation for an objective comparison of different imaging systems because each imaging workstation may process images differently before displaying them on the screen. The medical physicist has to facilitate the unbiased and blind comparison of the clinical images by the nuclear medicine physicians.
- — Surveying centres with similar systems through a written questionnaire can also be very effective and beneficial.
- — Inviting the vendor representatives to present their product in detail.

After thorough evaluation of all systems, the committee decides on the most appropriate system upon considering the cost and other factors such as the availability of a good maintenance service in the region.

After the system is chosen, the committee should supervise the installation process. It should help the vendor representative to finalize all of the paper work and get the access permits to the location. The system should be installed completely with all the accessories and software ordered.

The local medical physicist or a private consultant should perform the acceptance testing on the system. The committee should facilitate and make available all of the necessary resources to the medical physicist to complete the task and get the system ready for clinical use.

15.4.3. Acceptance testing as a baseline for regular quality assurance

As mentioned in Section 15.4.1, the medical physicist should produce reference tests during acceptance testing. Tests should be acquired that are easy to perform with less sophisticated procedures and that can be conducted within an acceptable period by the user. These tests should reflect the performance of the system in the working environment. The results of the routine tests should be compared against the results of these reference tests.

For example, the medical physicist may acquire a five or ten million counts uniformity image as a reference image for the system uniformity test during the acceptance period. This is less sophisticated than the usual 30 million counts uniformity image acquired for the acceptance testing. Another example is acquiring a 10 million counts image for the bar phantom during the acceptance testing and considering it a reference image. Some of the results of acceptance testing would be considered reference values to be used during routine testing. The multiple window spatial registration, maximum count rate and system spatial resolution values are examples of these tests.

15.4.4. What to do if the instrument fails acceptance testing

During acceptance testing, most of the performance parameters of the system should be tested and compared with the manufacturer's specifications. These specifications should be required during tendering and be provided with the system. If any of the test results do not meet the specifications, the analyses should be re-evaluated carefully. Following this, the test should be repeated again, paying close attention to any possible mistakes made during the acquisition and processing of the data. The analysis should also be carried out carefully, making sure that an accurate method has been followed.

If the problem persists, the engineer should be called to rectify the problem and then it becomes the responsibility of the vendor to resolve the issue. The engineer should look for any malfunction in the system, repair it and then recalibrate the scanner. All of the required calibrations after this repair should be performed and the system should be ready for testing. The medical physicist should not start acceptance testing if the system still needs more calibration, as some calibrations may readjust some parameters.

If two or three tests of the same parameter fail, the vendor should either replace the affected part (if it was not done before) or replace the system. The latter procedure should be the last option to be taken, as it will affect the routine work of the clinic. The vendor should compensate the clinic and the medical physicist (if a third party) for unnecessary delays.

15.4.5. Meeting the manufacturer's specifications

The verification of performance specifications is one of the key reasons for performing acceptance testing. Acceptance testing should follow the local recommendations (in the institute or country) or one of the international bodies' recommendations. As was discussed earlier, for both PET and gamma cameras, there are a number of guidelines set forth by various international bodies or agencies (NEMA, IEC, AAPM and IAEA reports) as to what kind of tests should be carried out.

Currently, there are task groups that have been formed by the AAPM working on a new set of guidelines because the existing guidelines need additional sets of performance tests to evaluate the hybrid systems as a whole, and some modifications for the recently emerged new technologies are necessary.

The results of these tests should meet the specifications set by the manufacturer, as they are usually one of the main reasons for selecting a particular system. If one or more test results do not meet the manufacturer's specifications, the test should be repeated carefully. In the case of similar results, the vendor engineer should rectify the problem at hand and then repeat the calibrations if necessary.

REFERENCES

[15.1] INTERNATIONAL ATOMIC ENERGY AGENCY, Quality Assurance for PET and PET/CT Systems, IAEA Human Health Series No. 1, IAEA, Vienna (2009).

[15.2] INTERNATIONAL ATOMIC ENERGY AGENCY, Quality Assurance for SPECT Systems, IAEA Human Health Series No. 6, IAEA, Vienna (2009).

[15.3] INTERNATIONAL COMMISSION ON RADIATION UNITS AND MEASUREMENTS, Phantoms and Computational Models in Therapy Diagnosis and Protection, ICRU Rep. 48, Bethesda, MD (1992).

[15.4] INTERNATIONAL COMMISSION ON RADIATION UNITS AND MEASUREMENTS, Phantoms and Computational Models in Therapy Diagnosis and Protection, ICRU Rep. 44, Bethesda, MD (1992).

[15.5] DEMIRKAYA, O., AL MAZROU, R., Performance test data analysis of scintillation cameras, IEEE Trans. Nucl. Sci. **54** (2007) 1506–1515.

[15.6] NATIONAL ELECTRICAL MANUFACTURERS ASSOCIATION, Performance Measurements of Gamma Cameras, Standards Publication NU 1-2007, NEMA (2007).

[15.7] GRAHAM, L.S., FAHEY, F.H., MADSEN, M.T., VAN ASWEGEN, A., YESTER, M.V., Quantitation of SPECT Performance: Report of Task Group 4, Nuclear Medicine Committee (AAPM Report No. 52), Med. Phys. **22** 4 (1995) 401–409.

[15.8] NICHOLS, K.J., et al., Instrumentation quality assurance and performance, J. Nucl. Cardiol. **13** (2006) 25–41.

[15.9] HOFFMAN, E.J., CUTLER, P.D., DIGBY, W.M., MAZZIOTTA, J.C., 3-D phantom to simulate cerebral blood flow and metabolic images for PET, IEEE Trans. Nucl. Sci. **37** (1990) 616–620.

[15.10] NATIONAL ELECTRICAL MANUFACTURERS ASSOCIATION, Performance Measurements of Positron Emission Tomography, Standards Publication NU 2-2007, NEMA (2007).

[15.11] BAILEY, D.L., JONES, T., SPINKS, T.J., A method for measuring the absolute sensitivity of positron emission tomographic scanners, Eur. J. Nucl. Med. **18** (1991) 374–379.

[15.12] XU, X.G., CHAO, T.C., BOZKURT, A., VIP-Man: an image-based whole-body adult male model constructed from color photographs of the visible human project for multi-particle Monte Carlo calculations, Health Phys. **78** (2000) 476–486.

[15.13] ZAIDI, H., XU, X.G., Computational anthropomorphic models of the human anatomy: the path to realistic Monte Carlo modeling in radiological sciences, Annu. Rev. Biomed. Eng. **9** (2007) 471–500.

[15.14] CAON, M., Voxel-based computational models of real human anatomy: a review, Radiat. Environ. Biophys. **42** (2004) 229–235.

[15.15] ZUBAL, I.G., et al., Computerized three-dimensional segmented human anatomy, Med. Phys. **21** (1994) 299–302.

[15.16] INTERNATIONAL ATOMIC ENERGY AGENCY, Quality Control of Nuclear Medicine Instruments, IAEA-TECDOC-602, IAEA, Vienna (1991).

[15.17] INTERNATIONAL ELECTROTECHNICAL COMMISSION, Medical Electrical Equipment — Characteristics and Test Conditions of Radionuclide Imaging Devices — Anger Type Gamma Cameras, 3rd edn, IEC 60789, IEC, Geneva (2005).

[15.18] INTERNATIONAL ELECTROTECHNICAL COMMISSION, Radionuclide Imaging Devices — Characteristics and Test Conditions — Part 2: Single Photon Emission Computed Tomographs, Edn 1.1, IEC 61675-2, IEC, Geneva (2005).

[15.19] INTERNATIONAL ELECTROTECHNICAL COMMISSION, Radionuclide Imaging Devices — Characteristics and Test Conditions — Part 3: Gamma Camera Based Whole Body Imaging Systems, 1st edn, IEC 61675-3, IEC, Geneva (1998).

[15.20] AMERICAN ASSOCIATION OF PHYSICISTS IN MEDICINE, Scintillation Camera Acceptance Testing & Performance Evaluation, Report No. 6, AAPM, College Park, MD (1980).

[15.21] AMERICAN ASSOCIATION OF PHYSICISTS IN MEDICINE, Computer-aided Scintillation Camera Acceptance Testing, Report No. 9, AAPM, College Park, MD (1982).

[15.22] AMERICAN ASSOCIATION OF PHYSICISTS IN MEDICINE, Rotating Scintillation Camera SPECT Acceptance Testing and Quality Control, Report No. 22, AAPM, College Park, MD (1987).

[15.23] INTERNATIONAL ATOMIC ENERGY AGENCY, Quality Control of Nuclear Medicine Instruments, IAEA-TECDOC-317, IAEA, Vienna (1984).

[15.24] INTERNATIONAL ELECTROTECHNICAL COMMISSION, Nuclear Medicine Instrumentation — Routine Tests — Part 2: Scintillation Cameras and Single Photon Emission Computed Tomography imaging, IEC TR 61948-2, IEC, Geneva (2001).

[15.25] INTERNATIONAL ELECTROTECHNICAL COMMISSION, Nuclear Medicine Instrumentation — Routine Tests — Part 3: Positron Emission Tomographs, IEC TR 61948-3, IEC, Geneva (2005).

CHAPTER 16

FUNCTIONAL MEASUREMENTS IN NUCLEAR MEDICINE

M.J. MYERS
Institute of Clinical Sciences,
Imperial College London,
London, United Kingdom

16.1. INTRODUCTION

The strength of nuclear medicine lies in using the tracer method to acquire information about how an organ is or is not functioning as it should. This modality, therefore, focuses on physiological organ function for diagnoses and not on anatomical information such as X ray computed tomography (CT) or magnetic resonance imaging.

The three aspects involved in the process are: (i) choice of radioactive tracer, (ii) method of detection of the emissions from the tracer, and (iii) analysis of the results of the detection. The radioactive tracers on which nuclear medicine (or molecular imaging as it is increasingly being called) is based are designed to participate in or 'trace' a chosen function of the body. Their distribution is then found by detecting and locating the emissions, usually γ photons, of the radioactive tracer. The tracer may be involved in a metabolic process, such as iodine in the thyroid, or it may take part in a physiological process because of its physical make-up, such as macroaggregate of albumin (MAA) in the lungs.

A number of methods of detection can be used. One is imaging with a gamma camera or positron emission tomography (PET) scanner in a number of modes: static (showing an accumulated or integrated function), dynamic (showing the variation of the function with time), whole body and tomographic (single photon emission computed tomography (SPECT) and PET analysis). Another is simple counting over areas of the body which can also be static or dynamic. Yet another is through laboratory analysis of blood samples. Imaging often provides a rough anatomical distribution of the function but, more importantly, a quantitative idea of the extent of the function in the whole functional unit or in component parts such as the right and left kidney. The anatomical picture has little of the detail of the other modalities but may be a more direct tool in assessing pathology since it provides primary information rather than displaying the anatomical consequences of pathology such as changes in density. The images created can be directly related to the uptake of the radiopharmaceutical, either in terms of counts or, with more sophisticated processing, with activity in becquerels. As parametric images, they can also represent a parameter such as the distribution of the ratio of ventilation V to perfusion Q, the V:Q ratio. The results can be acquired and displayed as 2-, 3- or 4-D images with time as the last dimension. In addition, because of the unique property of nuclear medicine to be able to detect specific radionuclides with different energy γ emissions, a number of functions can be followed simultaneously. Thus, the ventilation of the lung traced by ^{81m}Kr, a 190 keV γ emitter, can be investigated at the same time as the perfusion of the lung traced by the 140 keV γ emitting ^{99m}Tc-MAA.

16.2. NON-IMAGING MEASUREMENTS

'Non-imaging' measurements refer to the analysis of data from radionuclide procedures that are not derived from interpreting normal and pathological patterns of uptake of tracer in images from gamma cameras and PET scanners. Images may be used for non-imaging measurements but only to provide regions of interest (ROIs) for subsequent quantification of function. A common example of this is in the investigation of glomerular filtration in the kidney using a tracer such as ^{51}Cr-EDTA which can be measured from timed blood samples without using images for information about morphological changes.

16.2.1. Renal function measurements

16.2.1.1. General discussion

Renal haemodynamic functions can be divided into measurements of renal blood flow and glomerular filtration. The first depends on the supply of blood to the cortical and extramedullary nephrons which are the functional unit of the kidney. The second class of function depends on the transfer of fluids across the glomerulus. A number of radioactive tracers may be used depending on the function to be studied, the most common being ^{99m}Tc labelled diethylenetriaminepentaacetic acid (DTPA), dimercaptosuccinic acid (DMSA) and mercaptoacetyltriglycine (MAG3).

16.2.1.2. Glomerular filtration rate plasma clearance

Calculation of glomerular filtration rate (GFR) is used, for example, in the general assessment of renal function and the monitoring of renal function in patients undergoing therapy with nephritic drugs. Radioisotope measurements depend on the assessment of plasma clearance with time as seen with blood sampling of a tracer that is handled exclusively by glomerular filtration and does not enter blood cells. The most common radiopharmaceutical used is ^{51}Cr-EDTA, though ^{99m}Tc-DTPA and ^{125}I-iodothalamate are also seen.

GFR is obtained by constructing the clearance curve from one or a series of timed measurements of plasma activity. In the multi-sample method, the expected multi-exponential curve is defined accurately with samples taken at 10, 20 and 30 min, and 2, 3 and 4 h, or approximated with samples taken at about 2 and 4 h or even at only one time point between 2 and 4 h. As taking multiple samples over a period of hours may be impractical, further simplification of the process to the taking of a single sample is attractive. An empirical relationship between the apparent volume of distribution and the GFR has been derived and validated to allow this at a less precise but acceptable accuracy.

The object of the measurements is to construct the total area under the plasma clearance curve. It is sufficient for accuracy to assume a bi-exponential curve with a fast and slow component between times of 10 min and 4 h, ignoring any initial very fast components. The zero time intercepts and rate constants for the fast and slow components are C_{10} and α, and C_{20} and β, respectively. The equation for GFR is:

$$\text{GFR} = \frac{\text{injected activity}}{\text{total area under plasma clearance curve}} = \frac{Q_0}{A} = \frac{Q_0}{\frac{C_{10}}{\alpha} + \frac{C_{20}}{\beta}} \tag{16.1}$$

where

injected activity has units of megabecquerels (MBq);
C_{10} and C_{20} are in count rates that are converted into megabecquerels per millilitre (MBq/mL);

and α and β have units of min^{-1}, so that the GFR has units of millilitres per minute (mL/min).

As the contribution to the whole area from the fast component is relatively small and can be approximated without too much loss of accuracy, the equation can be simplified to:

$$\text{GFR} = \frac{Q_0}{C_{20}/\beta} \tag{16.2}$$

This produces an estimate of GFR that is obviously too small, though with poor renal function the approximation has less of an effect. A correction factor to convert the approximate GFR to the 'true' GFR can be used. Although this correction factor depends on the renal function, a figure of 1.15 can be used in most cases.

GFR will vary with body size and is conventionally normalized to a standard body surface area of 1.73 m^2, though other normalization variables such as the extracellular fluid volume have been suggested. The calculation for GFR requires measurement of the activity injected into the patient as well as the activity in the post-injection syringe, in a standard and in the blood samples. A number of methods are available in practice. These are based on the difference in weights of the pre- and post-injection syringe or on measurement of a fixed volume in a dose calibrator or on known dilutions. The counts recorded by the well counter measuring the small activities in the blood samples also have to be calibrated in terms of megabecquerels per count rate (MBq/count rate).

16.2.1.3. Effective renal plasma flow measurements

Renal plasma flow, often referred to as renal blood flow, has been investigated in the past using ^{131}I or ^{123}I labelled ortho-iodohippurate (hippuran) or para-amino hippurate (PAH).

Hippuran is almost completely excreted by the renal tubules and extracted on its first pass through the renal capillary system. As the extraction is not quite 100%, the renal function measured is called the effective renal plasma flow (ERPF). A modern variation is to use the ^{99m}Tc labelled tubular agent MAG3 as the ^{99m}Tc label is more available than ^{123}I. However, the extraction fraction of MAG3 at below 50% is inferior to that of hippuran, so the measurements of ERPF are simply estimates.

The ERPF measurement is very much the same as that of GFR in that a known activity of the radiopharmaceutical is injected and blood samples are taken at intervals. The timing of the intervals is, however, earlier than for GFR and occurs at 5 min intervals and then at 30, 50, 70 and 90 min. The resulting two-exponential time–activity curve is plotted from which the function is given as:

$$\text{ERPF} = \frac{Q_0}{\dfrac{C_{10}}{\alpha} + \dfrac{C_{20}}{\beta}} \tag{16.3}$$

using the symbols for GFR as above.

16.2.2. ^{14}C breath tests

The ^{14}C urea breath test is used for detecting *Helicobacter pylori* infection in cases of, for example, patients with duodenal and other ulcers following and monitoring anti-*H. pylori* treatment. The test is based on the finding that the bacterium *H. pylori* produces the enzyme urease in the stomach. As urease is not normally found in that organ, its occurrence can, therefore, denote the existence of *H. pylori* infection. The activity of ^{14}C used in the test is very small, about 37 kBq, and the effective dose for this is also low, at less than 3 μSv.

To carry out the test, ^{14}C urea is administered orally in the form of a capsule. The urease in the stomach converts the urea to ammonia and ^{14}C carbon dioxide which is exhaled and can be detected in breath samples using a liquid scintillation counter. The counter can measure minute quantities of the β emitting ^{14}C. One or two samples are usually collected after 10–30 min using a straw and balloon technique. Also counted are a known standard sample and a representative background sample. Counting is done either locally or by sending the samples to a specialized laboratory. The net disintegrations per minute (dpm) registered by the counter are compared with standard values to assess the degree of infection. dpm is given by:

$$\text{dpm} = \frac{(S-B)\times S_{\text{t}}}{(S_{\text{t}}-B)} \tag{16.4}$$

where

S is sample counts per minute;
B is blank counts per minute;

and S_{t} is standard counts per minute.

A non-radioactive test employing ^{13}C in place of ^{14}C is also used. Carbon-13 is measured by (non-radio)isotope ratio mass spectrometry. For ^{13}C, a baseline sample is required before the capsule is swallowed to compare with the post-capsule sample.

16.3. IMAGING MEASUREMENTS

These include static image acquisition and analysis for quantitative assessment of uptake, e.g. thyroid uptake measurement with organ delineation and background subtraction. Furthermore, time–activity curves can be derived from dynamic 2-D imaging and quantitative parameters assessed from images, e.g. renal function, gastric function, gall bladder emptying, gastrointestinal transit and oesophageal transit. Time–activity curves can also be derived from dynamic 3-D imaging with quantitative parameters assessed from physiologically triggered images, such as cardiac ejection fraction measurement.

16.3.1. Thyroid

Measurement of thyroid function is one of the oldest techniques in nuclear medicine, though the original radioisotope used, ^{131}I, has been replaced by ^{99m}Tc and ^{123}I, and the collimated sodium iodide crystal uptake probe has almost exclusively been replaced by the gamma camera.

Tests on the thyroid consist of both imaging the morphology of the organ and assessing its 'uptake'. Uptake consists of measuring the activity taken up by the gland of an ingested or intravenously administered activity of radioactive iodine or intravenously administered ^{99m}Tc-pertechnetate. The uptake mechanisms are different for the two radioisotopes. Iodine is both trapped and organified by thyroid follicular cells, a process more like the true thyroid function, whereas the pertechnetate is simply trapped.

The tests allow assessment of functionality of thyroid lesions and nodules, and investigations of thyroiditis and ectopic tissue. They can confirm a diagnosis of an excess of circulating thyroid hormones, as in Graves' disease and toxic nodular goitre, and lead to a more quantitative approach to treatment of hyperthyroidism with ^{131}I.

Uptake of tumours secondary to thyroid cancer that have disseminated through the body following surgery may also be assessed. In this case, whole body images are made using ^{131}I and a scanning gamma camera. The tumours can be located and, in some cases, a quantitative measurement of uptake made, thus allowing the effectiveness of treatment to be monitored.

The choice of radiopharmaceutical used can be dictated by the availability and cost of the preferred isotope ^{123}I and ^{99m}Tc. Each is suitable for use with a gamma camera. Iodine-123 has a half-life of 13 h and a γ emission of 160 keV. Technetium-99m has a half-life of 6 h and an energy of 140 keV. However, the cyclotron produced ^{123}I is not readily available and is expensive, while ^{99m}Tc-pertechnetate is readily available from the standard hospital molybdenum generator. The timing of the uptake of the two isotopes is also different. While iodine uptake is usually performed 18–24 h after injection or 2–6 h after ingestion, ^{99m}Tc is measured after 20 min.

Uptake measurement with a scintillation probe involves counting over the neck, over a 'thyroid phantom' and over the thigh of the patient to simulate counts in the neck that do not arise from the thyroid. The thyroid phantom (standardized by, for example, the IAEA) consists of a small source of known activity in a plastic cylinder offering the same attenuation as a neck. This acts as the standard. Counts are obtained at a distance of about 25 cm from the collimator face to offset any inverse square errors due to different locations of the thyroid. The percentage uptake U is then calculated from the formula:

$$U = \frac{N - T}{C_{\mathrm{a}}} \times 100 \tag{16.5}$$

where

N is the counts in the neck;
T is the counts in the thigh;

and C_{a} is the administered counts corrected for background.

The administered counts can be measured directly with an isotope calibrator before the test or can be related to the activity in the thyroid phantom. Corrections for decay are made throughout.

Similar measurements can be made with the gamma camera in place of the probe, except that corrections for neck uptake can be made using ROIs over the delineated thyroid and regions away from the gland. The camera can be fitted with a pinhole collimator in order to provide a degree of magnification, although the image is somewhat distorted and is subject to distance errors, in which case a fixed distance (perhaps employing a spacer) is used. Images obtained with a parallel-hole collimator may appear smaller but are not prone to distance effects, though subject to the same attenuation. Anterior views are generally enough but a lateral view may also be used to locate ectopic tissue.

Quantification of the uptake is achieved in two ways. By calibrating the camera with a known activity in a suitable phantom, the activity injected into the patient can be measured, or by measuring the injection directly in the syringe before administration. Each will yield the sensitivity of the camera in terms of counts per megabecquerel and allow the activities seen in the thyroid glands and background to be calculated. The process is often an automatic one performed by the camera computer software that delineates the outlines of the thyroid lobe(s) and establishes a suitable background region used to correct for the presence of activity in tissue overlying and underlying the thyroid tissue correction. It is important that a local normal range is established and that the calibration of the camera in terms of counts per megabecquerel is subject to a quality assurance programme.

A simple check on the quantification is to perform the test on a syringe of activity as if the syringe is the thyroid gland, i.e. imaging the syringe as the activity to be injected and re-imaging it as the activity taken up by the thyroid. An 'uptake' of 100% would be expected.

16.3.2. Renal function

16.3.2.1. General discussion

The study of renal function has been a mainstay of nuclear medicine for many decades and is an established efficient technique for, among other functions, assessing renal perfusion, quantifying divided or individual kidney function, and studying output efficiency and obstruction.

Two aspects of renal function are exploited: (i) glomerular filtration, i.e. the transfer of fluids across the glomerulus, investigated by measuring the clearance of ^{99m}Tc-DTPA ('pentetate'); and (ii) tubular secretion, investigated by measuring the clearance of ^{99m}Tc-MAG3 ('tiatide').

16.3.2.2. Renal function measurements

The basis of measurements of most renal functions is time–activity curves obtained by imaging the kidneys using a gamma camera. Views are obtained at different times and over different periods after administration of one of a number of radiotracers and often after some intervention. The result is usually a curve, called a 'renogram', showing the rise and fall of counts in each kidney. This curve is corrected for non-renal background counts using computer software, a feature lacking in old counting methods using probes. Figure 16.1 shows a series of images obtained after administration of ^{99m}Tc-MAG3 (left), the ROIs drawn over the two kidneys as well as the background region (top right), and the renogram curves for the right kidney (RK) and left kidney (LK) (bottom right). The analysis programmes, supported by commercial software providers, allow the calculation of a number of renal function parameters, including relative perfusion, relative function, mean and minimum transit times, and outflow efficiency.

There is an overabundance of different methods of analysis of the curves to create a plethora of indices of function, some relying on the definition of aspects of the renograms that have little basis in renal physiology. It is, therefore, prudent to understand what is happening to the tracer as it traverses the kidneys, and how it appears in the images and renogram. In fact, after adjustment for the contributions of activity in the renal vasculature, the corrected curve displaying a relatively fast rise and subsequent slower fall in activity can be described by two distinct phases. The first spans the time of injection to the end of the minimum transit time when the kidney contains the sum of all of the tracer extracted from the blood and has, therefore, been termed the sum phase. The second starts at the end of the first and reflects the net activity left after loss from the kidney and has been called the difference phase. Other terms such as ‘vascular spike’ and ‘secretory phase’ may not reflect purely renal function and are, therefore, not very helpful.

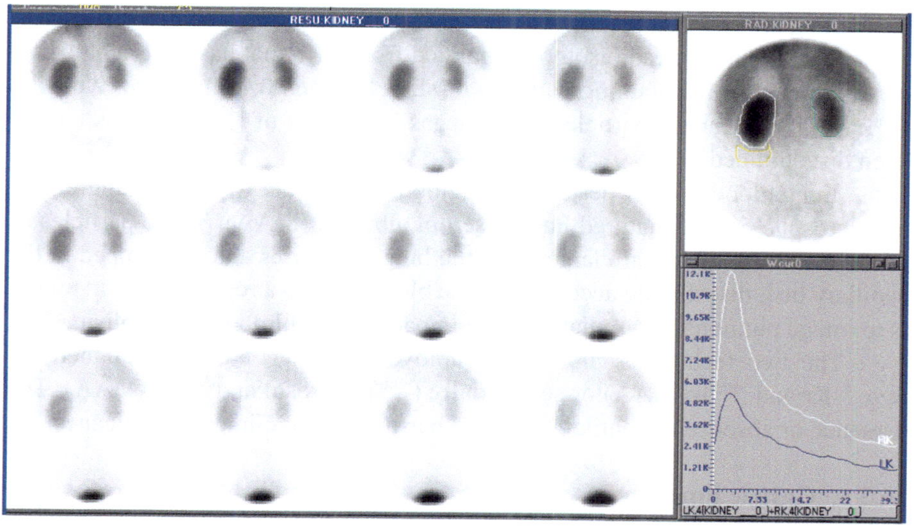

FIG. 16.1. Dynamic renal flow study after administration of ^{99m}Tc-MAG3 (left), regions of interest drawn over the kidneys and background (top right), and corresponding renogram curves for the right kidney (RK) and left kidney (LK) (bottom right).

What is helpful is often a quantitative comparison of the two kidneys with derivation of a relative function. This may be calculated from Patlak plots or from the uptake slope or the integral of the renogram curves. The programme can calculate the relative perfusion and function from the retention functions. The outflow curves for both kidneys and the value of the outflow at any selected time can also be displayed. In the case of assessment of kidney transplants, other aspects can be used to calculate relevant parameters.

A number of radioactive tracers may be used, depending on the function to be studied. These functions are measured by ^{99m}Tc labelled DTPA, DMSA and MAG3. Renal blood flow (or renal plasma flow) has been measured from clearance of hippurate from the plasma.

16.3.3. Lung function

The functions of the lung that are most investigated using nuclear medicine techniques are regional ventilation and pulmonary blood flow or perfusion, leading to studies of the ventilation perfusion ratio. Other studies cover intrapulmonary vascular shunting, pulmonary granulocyte kinetics and lung tissue permeability.

Lung air flow or ventilation imaging is carried out either with a gas such as ^{81m}Kr (a 13 s half-life radioisotope generated from ^{81}Rb that has a 4.6 h half-life) or with an aerosol containing particles of sizes between 0.1 and 2 μm, typically ^{99m}Tc-DTPA or carbon particles ('Technegas'). The use of ^{133}Xe gas has largely been discontinued.

Lung blood flow or perfusion imaging is carried out with macroaggregates or microspheres of denatured human serum albumin (MAA). These particles of average size 20–40 μm are larger than the lung capillaries and are trapped in the capillary bed, distributing according to the blood flow to a region. Their main use is to image pulmonary vascular disease, in particular, pulmonary embolism.

The two techniques are often employed together, either simultaneously (e.g. ^{81}Kr and ^{99m}Tc-MAA) or sequentially (^{99m}Tc aerosol and ^{99m}Tc-MAA). The presence or absence of ventilation and/or perfusion is of clinical significance.

16.3.4. Gastric function

Nuclear medicine allows a full, non-invasive and quantitative assessment of the way the oesophagus moves both solid and liquid meals to the stomach, how the stomach handles these meals and how they transit through the gastrointestinal tract.

16.3.4.1. Stomach emptying of solid and liquid meals

As the choice of both solid and liquid test meals determines the standard values used as criteria for evaluating the function, a 'standard meal' has been agreed. Solid meals are based on preparations including eggs (into which ^{99m}Tc sulphur colloid has been mixed), toast and water.

Anterior and posterior dynamic images are obtained at suitable intervals of time following ingestion of the meal and are repeated for the same positioning at hourly intervals for up to 4 h. The two views are obtained simultaneously or sequentially, depending on the availability of a dual or single headed gamma camera. The reason for the two view approach is to obtain a geometric mean of the activity in the field of view that accounts for the movement of activity between the anterior and posterior surfaces of the body. Relying on a simple anterior view leads to artefacts due to differential attenuation of the ^{99m}Tc γ rays.

The data are analysed by drawing ROIs around the organs of interest (stomach and parts of the gastrointestinal tract) and creating a decay corrected time–activity curve. An assessment of the gastric emptying function is made from standard values. An alternative way of expressing the result is through the half-emptying time.

16.3.4.2. Analysis of colonic transit

Colonic transport analysis can be performed using ^{111}In labelled non-absorbable material, such as DTPA or polystyrene micropellets administered orally. Indium-111 is chosen rather than the more accessible ^{99m}Tc because of its longer half-life (2.7 d) and the possibility of imaging over a longer time since images are taken at, for example, 6, 24, 48 and 72 h.

As with the stomach procedures, a geometric mean parametric image of anterior and posterior views may be used in the quantification, if this facility is available in the nuclear medicine software. A geometric centre of the activity (also called centre of mass) may be tracked over time by defining particular segments in the colon, perhaps 5–11 in number (e.g the ascending, transverse, descending, rectosigmoid and excreted stool), multiplying the individual segment counts by weighting factors from 1 to 5, respectively, and summing the resulting numbers. In addition to time–activity curves for the individual segments, the rate of movement of the geometric centre as a function of time can be assessed by plotting this as a time position profile. A colonic half-clearance time may be calculated and compared with historical normal control values of colonic transport.

16.3.4.3. Oesophageal transit

This function is studied by imaging the transit of a bolus of radiolabelled non-metabolized material such as ^{99m}Tc sulphur colloid. Either the whole oesophagus may be included in an ROI and a time–activity curve generated for the whole organ, or a special display may be generated, whereby the counts in successive regions of the oesophagus are displayed on a 2-D space–time map called a condensed image. The counts in the regions are displayed in the y direction as colour or grey scale intensities corresponding to the count rate against time along the x axis (Fig. 16.2). The result is a pictorial idea of the movement of the bolus down the oesophagus.

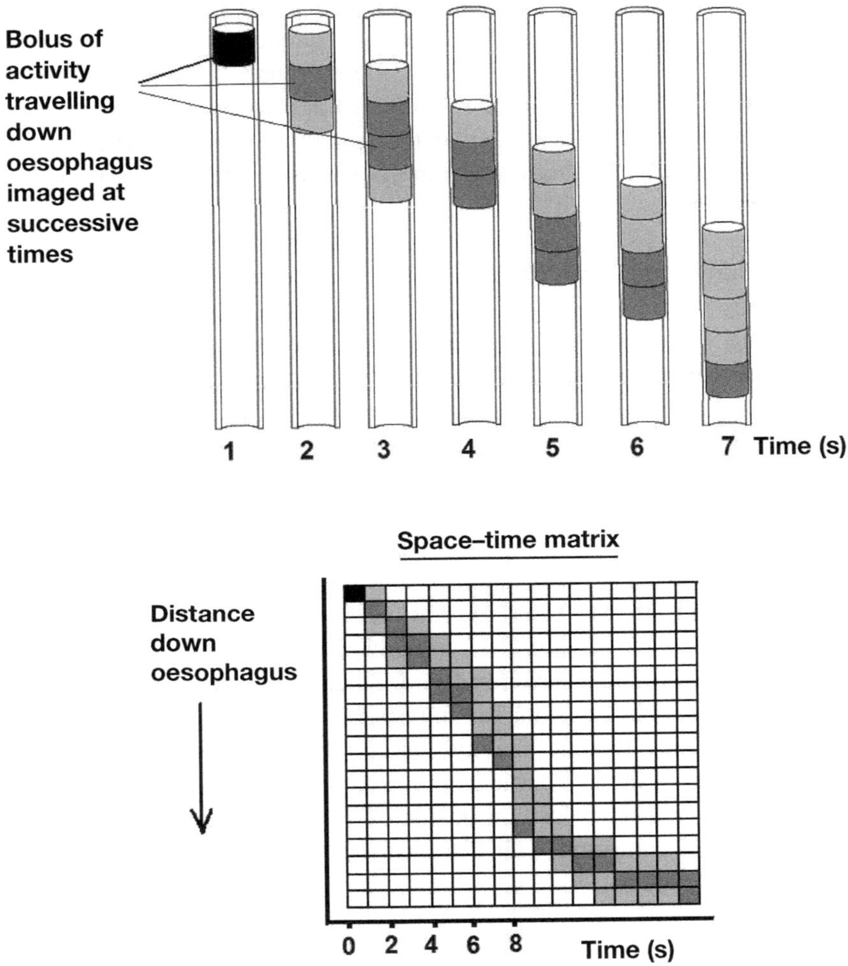

FIG. 16.2. Oesophageal transit is imaged as a space–time matrix. As the bolus of radioactivity passes down the oesophagus, the counts from successive regions of interest, represented on a grey scale, are placed in consecutive positions in the matrix in the appropriate time slot. A normal transit will be shown as a movement of the bolus down and to the right in the matrix. Retrograde peristalsis will be shown as a movement to the right and upwards.

16.3.4.4. Gall bladder ejection function

The gall bladder is investigated using hydroxy iminodiacetic acid labelled with ^{99m}Tc. This is injected and can be imaged using a gamma camera after excretion by the liver into the bile and as it passes through the gall bladder and bile ducts. The gall bladder is then made to contract and empty by injecting a hormone called cholecystokinin and the imaging of the gall bladder continued, the whole test taking between 1 and 2 h. The amount of the radiolabel that leaves the gall bladder is assessed by the difference in counts in the ROI over the emptied gall bladder divided by the counts from the ROI over the full gall bladder. Expressed as a percentage, this gives the ejection fraction. An ejection fraction above 50% is considered as normal and an ejection fraction below about 35% as abnormal, suggesting, for example, chronic acalculous cholecystitis.

16.3.5. Cardiac function

16.3.5.1. General discussion

The two main classes of cardiac function are blood flow in the myocardium and in the blood pool and ventricles. Images are acquired in both planar and tomographic modes, and the data may be acquired dynamically over sequential time periods or as a gated study triggered by the electrocardiogram (ECG), or as part of a first-pass study. The information is presented on a global or regional basis as conventional or parametric images, or as curves from which quantitative parameters may be derived. A range of pharmaceutical agents labelled with single and positron emitting isotopes are used.

Cardiac functions that may be investigated with nuclear medicine techniques run into many dozens, though relatively few are covered by standard clinical practice and some are confined to research. A list of cardiac functions would, therefore, include myocardial perfusion, myocardial metabolism of glucose and fatty acids, myocardial receptors, left ventricular ejection fraction, first-pass shunt analysis, wall motion and thickness, stroke volumes, cardiac output and its fractionation, circulation times, systolic emptying rate, diastolic filling rate, time to peak filling or emptying rate, and regional oxygen utilization.

Despite the usefulness of nuclear cardiology procedures, a worldwide survey has shown a wide variation in their use and availability. There is a high application rate in the United States of America (where cardiology accounts for about half of nuclear medicine procedures) and Canada, less in western Europe and Japan, and low application elsewhere such as in the Russian Federation, Asia and some parts of South America. One reason for this pattern of use may be the degree of access to training for physicians. Another reason may be that gated SPECT imaging and analysis requires a high level of instrumentation and software.

However, the procedures that can be carried out in any particular department will depend very much on the nuclear medicine software provided by the supplier of the gamma camera or from a specialized nuclear software supplier.

A commercial suite of programmes will usually only offer a limited selection of functional analysis. Typically, these include blood pool gated planar or SPECT analysis for ventricular volumes and ejection fractions, and cardiac perfusion analysis of gated SPECT images acquired under stress/rest conditions. The sophistication of the programmes may also be limited in the extent of automation of the preparation of cardiac images and their analysis. Whereas in the earlier days of nuclear medicine a programmer could be asked to tailor programmes to suit the need of the practitioner, and a high level language or macro-language might be provided to allow this, the fashion nowadays is for a fixed set of programmes dictated by the commercial supplier to satisfy a perceived general need. A typical omission might, therefore, be list-mode data acquisition with the consequent inability to trigger on selected parts of the ECG to omit ectopic heart beats.

In addition to the aspect of the software provided is the choice of radiopharmaceutical available. A wide choice is theoretically possible, depending on the particular function to be explored. Thus, for example, myocardial receptors can be investigated using ^{123}I-MIBG (metaiodobenzylguanidine) or ^{11}C-hydroxyephedrine, myocardial glucose using ^{18}F-FDG (fluorodeoxyglucose) and fatty acid metabolism using ^{123}I-heptadecanoic acid or BMIPP (β-methyl-p-iodophenylpentadecanoic acid). SPECT techniques use ^{201}Tl, ^{99m}Tc-sestamibi and other perfusion agents. PET viability studies can employ ^{13}N-ammonia, ^{18}F-FDG and ^{11}C-acetate.

In terms of which gamma camera radiopharmaceutical is best to use, there may be limitations in the availability or the licensing of a ^{201}Tl or a ^{99m}Tc product. As far as PET imaging goes, it is obvious that the non-research use of short lived positron emitters such as ^{11}C or ^{13}N that would be useful in the study of myocardial function would depend on the proximity of a cyclotron as well as the funding to allow the production of the labelled products. It would also presuppose the existence of an expensive PET/CT scanner for the procedures.

16.3.5.2. First-pass angiography

First-pass studies typically involve the acquisition of about 2000 frames of data with a duration of about 50 ms following the bolus injection of autologous red blood cells labelled in vivo or in vitro with ^{99m}Tc as they pass through the right ventricle for the first time. A time–activity curve derived from an ROI over the ventricle shows a curve that rises to a peak and then falls off, the curve also showing a saw tooth pattern corresponding to the filling and emptying of the left ventricle during the cardiac cycle. The first-pass curve is illustrated in Fig. 16.3(a) alongside the volume curve in Fig. 16.3(b), also used to derive the ejection fraction obtained from a multiple-gated acquisition (MUGA) study.

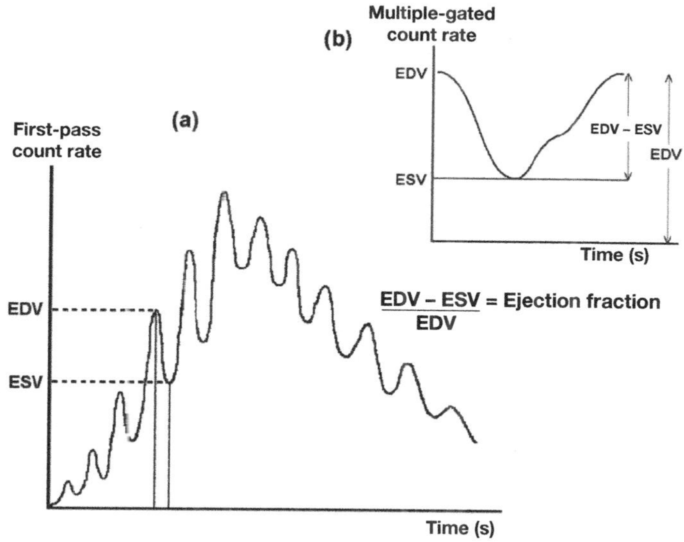

FIG. 16.3. Curves obtained from (a) first-pass angiography and (b) a multiple-gated acquisition study, showing how end diastolic volume (EDV) and end systolic volume (ESV), used to determine ejection fraction, are represented in each technique.

By suitable positioning of the gamma camera in the right anterior oblique view, this saw tooth can be used as an estimate of ejection fraction in the right ventricle, a parameter that is only amenable to analysis in the first pass before uptake in adjoining structures. The ejection fraction is derived from the ratio of the peak of any saw tooth (the end diastolic volume (EDV)) minus the value of the next trough (the end systolic volume (ESV)) to the EDV. The ejection fraction for the left ventricle would be assessed from the curve obtained by viewing in the left anterior oblique position. Although this parameter is more reliably obtained from a MUGA study, the first-pass procedure is much quicker and may be suitable for patients who cannot tolerate the much longer MUGA study. First-pass kinetics also provide a measure of left to right cardiac shunts (where some activity goes from the left ventricle through a septal defect to the right ventricle and, thus, recirculates between the heart and the lungs) and the pulmonary systemic flow ratio (Q_p:Q_s), as well as systolic emptying and diastolic filling rates and ventricular volumes.

16.3.5.3. The multiple-gated acquisition scan

The MUGA scan traces heart muscle activity from the distribution of the administered radiopharmaceutical, allowing the calculation of the left ventricular ejection fraction and demonstrating myocardial wall motion. It may be obtained while the patient is at rest and after physically or pharmacologically induced stress.

The same radiopharmaceutical preparation is used as for the first-pass study and a bolus is injected. The gamma camera views the patient in the left anterior oblique position so as to best separate the projections of the two ventricles. Dynamic images of the left ventricle in a beating heart are acquired at the same time as the ECG and the results stored. A trigger ('gating signal') corresponding to the R-wave marks the start of each heart cycle and the start of each sequence of images. The time period between successive R-waves (R–R interval) is divided into time intervals and the beat by beat left ventricular images corresponding to each time interval are each integrated into a single combined 'gated' image that provides a stop motion image of the heart at intervals in its cycle. As any one frame would not have enough data to provide sufficient counts and would, therefore, be statistically unreliable, many frames at the same interval are superimposed on each other. The signal for the start of each sequence is derived from an ECG monitor connected to the patient that provides a short electronic pulse as it detects the peak of an R-wave. Usually, about 32 equally time-spaced frames ('multiple gates') are used and these are defined between each R–R interval. Beats within 10% of the mean length are accepted. The result is a series of images of the heart at end diastole and at end systole, and at stages in between (Fig. 16.4).

The image at end diastole when the heart has filled with blood contains the maximum number of counts, and the end systolic image the least number. A direct relationship is made between the number of counts in a region of the ventricle and its volume.

For each frame, the computer, starting with an initial rough outline provided by the operator, defines the boundary of the left ventricle. Depending on the computer programme used, a different method of edge detection may be employed. This may be based on an isocount contour or on a maximum slope normal to the edge. As there is interference with the ventricular image from activity seen in pulmonary blood, the computer will also define a suitable background ROI close to the wall and correct the ventricular image at each stage. The plot of the counts in the ventricular region against time forms the volume curve that starts with the EDV, falls to the ESV level and rises again. The ejection fraction is calculated as before from the standard formula:

$$\mathrm{EF} = \frac{\mathrm{EDV} - \mathrm{ESV}}{\mathrm{EDV}} \tag{16.6}$$

Simple differentiation of the volume curve provides indices of the ventricular filling and emptying rates. Indeed, if the cardiac output is known, the rates can be quantified in terms of millilitres per minute. Other information can be obtained from phase/amplitude analysis of the sequence of images.

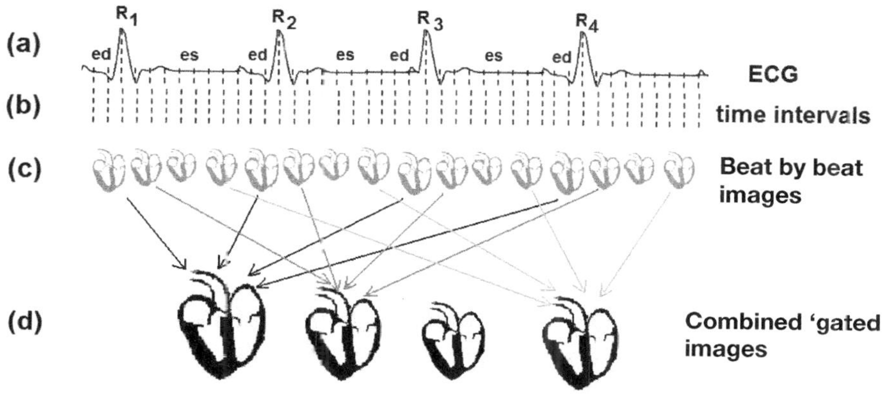

FIG. 16.4. The multiple-gated acquisition scan. The sequence of these gated images shows the heart cycle with higher counts and better statistics than the individual images, allowing better interpretation of the data than in Fig. 16.3(b).

16.3.5.4. Myocardial perfusion imaging

Myocardial perfusion imaging allows a regional assessment of blood flow to the heart and demonstrates areas of ischaemic myocardium where the blood flow is diminished when the patient undergoes a stress test.

Imaging follows administration of specific radiopharmaceuticals such as ^{201}Tl-chloride, ^{99m}Tc-tetrofosmin or ^{99m}Tc-sestamibi after the patient has been subjected to physical exercise or, if this is not suitable, to pharmacological stress with vasodilators to raise the heart rate and stimulate the myocardium. This induces a difference of uptake of radiolabel between normal and ischaemic myocardial tissue that can be imaged and localized after planar and/or ECG gated tomographic imaging. It is also possible to use positron emitting radiopharmaceuticals such as the cyclotron produced 10 min half-life ^{13}N-ammonia and generator produced 75 s half-life ^{82}Rb, but there are the usual limitations on applicability of PET, such as availability of the product or imaging system.

Once a stable ECG pattern is observed, the patient is imaged using a gamma camera operating in SPECT mode. ECG gating is applied throughout and produces sets of 16 images at each acquisition angle. The stress images and their analysis may be compared with similar ones obtained with the patient at rest. A number of protocols involving different times of examination and different administrations of radioactivity have been devised to carry out the stress/rest examinations in one rather than two days, given the potential long washout periods involved. Imaging can be performed 'early' (at about 15 min) following injection of ^{201}Tl or ^{99m}Tc-sestamibi at rest or after the stress test and/or 'delayed' (after 1–4 h or after 24 h) after injection at rest or under stress of the longer lived ^{201}Tl. These protocols give rise to different types of image. In general, the imaging properties of ^{99m}Tc give superior images, though ^{201}Tl is superior from a physiological viewpoint as it is a better potassium analogue.

Conventional cardiac SPECT imaging may be carried out with a single or double headed gamma camera using circular or elliptical orbits, the latter allowing closer passes over the patient and, consequently, better resolution. Attenuation correction may be performed on the emission images using an isotope or CT X ray source. However, there is still debate as to the usefulness of attenuation correction since this technique may give rise to artefacts due to mismatch of the emission and transmission images on the fused images.

16.3.5.5. Technical aspects of SPECT and PET imaging

Thallium is not an ideal gamma camera imaging radionuclide, as it emits rather low energy characteristic X rays between 69 and 80 keV that are easily attenuated (and, therefore, lost) in the body. The attenuation, by breasts and the chest wall, varies for the different projections around the body and gives rise to artefacts in the perfusion images if not corrected. The higher γ energy of ^{99m}Tc (140 keV) while still liable to attenuation, allows better collection of data from the heart and less variation in the attenuation. Although SPECT/CT has still not become a viable option (as has PET/CT where the two modalities have become inseparable), this would be a better option for attenuation correction than the rather cumbersome isotope attenuation correction devices that have been used in the past.

Scattering of the γ photons before detection in the camera also leads to problems in that their origin might be misplaced and loss from deeper structures occurs. Recently, software to reduce the effects of scattering by modelling its behaviour within the field of view has become available. Another source of degradation of the image quality is the loss of resolution with distance from the collimator face. Although 'high' resolution collimators are usually chosen for ^{99m}Tc imaging, the basic resolution of the camera at the level of the heart is rather poor. Again, software techniques to model this behaviour and correct for it have become available.

Gamma camera images, unlike X ray ones, are always subject to lack of counts and are, therefore, prone to statistical errors. Using a double headed rather than a single headed system is, therefore, an advantage. There is still discussion on the best angle between the heads and this may vary between less than 90° and 180°. Scanning the patient with the collimator as close to the source of activity as possible also ensures the best resolution, so a non-circular orbit is usually chosen. Owing to the lack of accessible counts with ^{201}Tl (its relatively long half-life and high effective dose reducing the activity that can be administered), a general purpose collimator is used, which is more efficient but less accurate than the high resolution collimator used with ^{99m}Tc.

PET imaging based on the simultaneous detection of opposed annihilation γ radiation from the original positron radiotracer is intrinsically tomographic and invariably comes with anatomical land marking and attenuation correction from the CT. It is also more sensitive and more accurate (because of its better resolution) than SPECT and uptake of the radiopharmaceuticals can be quantified absolutely. In theory, the use of ^{13}N labelled ammonia and ^{18}F-FDG can differentiate more about the state of the myocardium, its blood flow and metabolism, than the SPECT tracers.

Processing of the data starts with filtered back projection of the data using standard filters such as a Butterworth or Hanning filter with appropriate cut-off spatial frequencies and order, or an iterative reconstruction may be performed. Attenuation correction may be applied. This stage produces a set of contiguous slices parallel to the transverse axis of the patient which can be combined to populate a 3-D matrix of data. As the heart lies at an angle to this body axis, a process of reorientation is performed. From the original matrix, the data that lay parallel to the axes of the heart itself can be selected to form vertical long axis (parallel to the long axis of the left ventricle and perpendicular to the septum), horizontal long axis (parallel to the long axis of the left ventricle and to the septum) and short axis (perpendicular to the long axis of the left ventricle) slices through the myocardium of the particular patient, as shown in Figs 16.5 and 14.12.

Cardiac processing software, working on features extracted from the shape of the myocardium, allows easy automatic alignment which may also be operator guided. The reoriented sections form three sets of images that are displayed in a standard format to show, for example, the apex and heart surfaces at each stage of gating of the heart cycle (Fig. 16.4). The display may take several forms including the simultaneous display of many sections in each of the three axes both at rest and after stress, or as a moving image of the beating heart. The algorithms used to carry out this process differ with the provider of the computer software and are recognized under specific names.

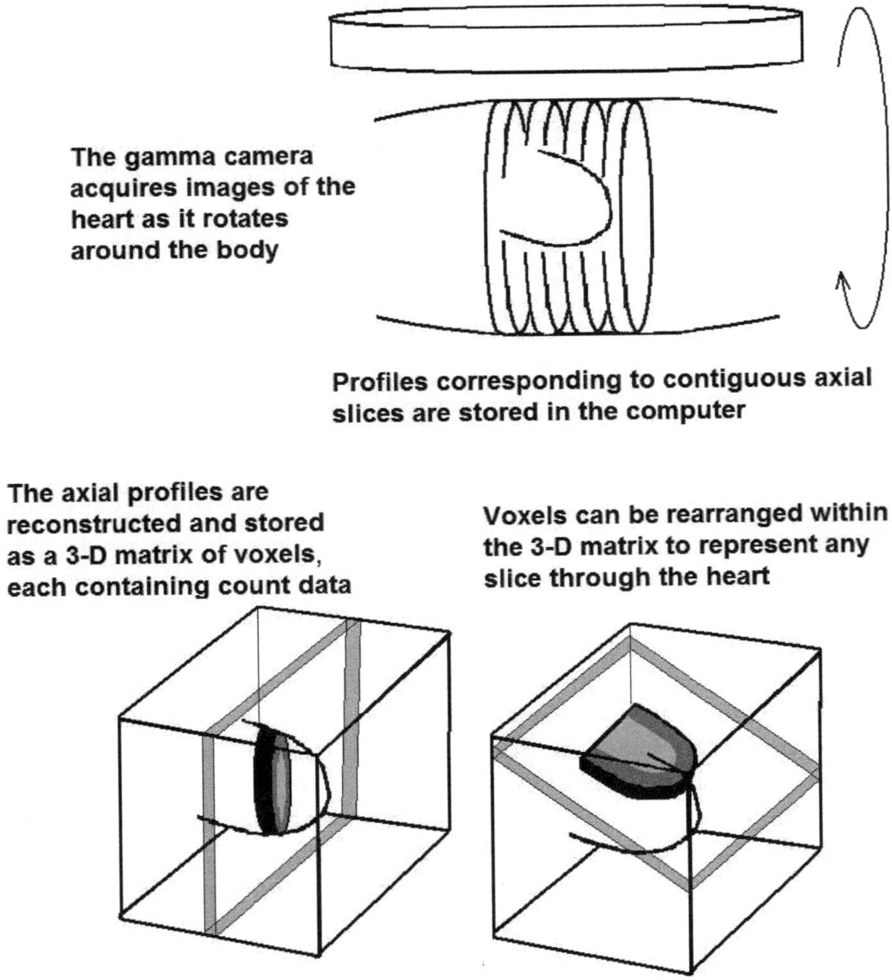

FIG. 16.5. Principle of rearrangement of acquired axial tomographic images into slices aligned with the axes of the heart.

Presentation of the slice data is often as a so-called polar diagram or bull's-eye display illustrated in Fig. 14.17. This allows the 3-D information about the myocardium, which would be difficult to interpret easily, to be depicted as a simple, 2-D, colour coded, semi-quantitative image. The process is often described as imagining the myocardial surface as the peel of half an orange which is flattened out to form the polar diagram. This is divided into accepted segments and values, and colours associated with each segment. Again, there is a variation of the exact form and mathematical basis of the polar diagram in the commercial products available. This results in different looking maps that, although individually validated, are not directly comparable. It would, therefore, be prudent for one software package to be standardized at any one reporting centre. The results from a particular study can be compared with a reference image derived from a so-called normal database to allow a better estimation of the extent of the defects. However, it is often difficult to match the population being examined with the available validated normal data, for example, in terms of gender and ethnicity.

BIBLIOGRAPHY

PETERS, A.M., MYERS, M.J., Physiological Measurements with Radionuclides in Clinical Practice, Oxford University Press, Oxford (1998).

ZIESSMAN, H.A., O'MALLEY, J.P., THRALL, J.H., Nuclear Medicine — The Requisites, 3rd edn, Mosby-Elsevier (2006).

FURTHER READING

Recommended methods for investigating many of these functions may be found on the web sites of the American Society of Nuclear Medicine (www.snm.org), the British Nuclear Medicine Society (www.bnms.org.uk) and the International Committee for Standardization in Haematology (http://www.islh.org/web/published-standards.php).

CHAPTER 17

QUANTITATIVE NUCLEAR MEDICINE

J. OUYANG, G. EL FAKHRI
Massachusetts General Hospital
and Harvard Medical School,
Boston, United States of America

Planar imaging is still used in clinical practice although tomographic imaging (single photon emission computed tomography (SPECT) and positron emission tomography (PET)) is becoming more established. In this chapter, quantitative methods for both imaging techniques are presented. Planar imaging is limited to single photon. For both SPECT and PET, the focus is on the quantitative methods that can be applied to reconstructed images.

17.1. PLANAR WHOLE BODY BIODISTRIBUTION MEASUREMENTS

Planar whole body imaging is almost always carried out by translating the patient and bed in the z direction between opposed heads of a dual head standard scintillation camera, typically in the anterior and posterior positions. The resulting images are 2-D projections of the 3-D object being studied. The attenuation of photons varies with the distance and the material the photons have to travel through the object before reaching the detector. One approach to compensate for attenuation in planar imaging is to perform conjugate counting with the geometric mean, which consists of acquiring data from opposite views and combining them into a single dataset. Figure 17.1 shows an imaging object with uniform attenuation viewed by two gamma detectors placed in opposite directions. A point source in the object has attenuation depth d_1 and d_2 to detectors 1 and 2, respectively. The projected counts P_1 and P_2 measured on detectors 1 and 2, respectively, are:

$$P_1 = I_0 \exp(-\mu d_1), \quad P_2 = I_0 \exp(-\mu d_2) \tag{17.1}$$

where

I_0 is the 'unattenuated' number of counts that would be detected (in the absence of attenuation);

and μ is the attenuation coefficient in the object.

These two conjugate views are combined using the geometric mean P_G defined as:

$$P_G = \sqrt{P_1 P_2} = I_0 \exp(-\mu D/2) \quad (17.2)$$

where D is the total thickness and $D = d_1 + d_2$.

The geometric mean depends on the total thickness, but not on the source depths. This result is exact only for a point source. Corrections can be made for extended sources [17.1]. The geometric mean using conjugate views is a popular quantification method for planar imaging. The arithmetic mean (average of opposite views) has also been proposed previously, but yields inferior results to the geometric mean approach described here.

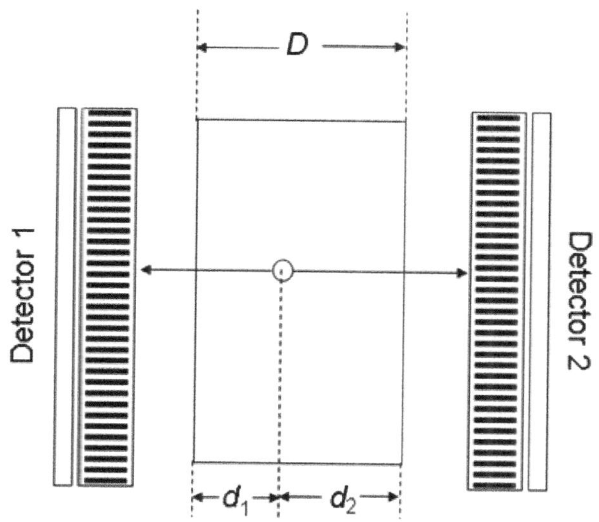

FIG. 17.1. Illustration of conjugate viewing with the geometric mean for attenuation compensation.

17.2. QUANTITATION IN EMISSION TOMOGRAPHY

17.2.1. Region of interest

Quantitative measurement of tracer uptake often requires the definition of a region of interest (ROI), where the tracer uptake is to be quantified. A reliable although time consuming way to define an ROI is to draw the boundary of the volume of interest on every slice containing the organ or tumour of interest. It is difficult and time consuming to manually draw the boundary of an ROI because the activity profile at the edge of the area of interest is not abrupt, but changes slowly, and, therefore, deciding on a threshold to determine such a boundary is not always straightforward. The manually drawn ROIs are, therefore, generally not reproducible. Alternative semi-automatic and automatic methods use edge detection techniques, which include count threshold, isocountours, maximum slope or maximum count gradient, to improve reproducibility. Finally, another approach that has been successfully used to determine organs or volumes with a specific time–activity behaviour in a dynamic acquisition is factor analysis of dynamic sequences [17.2, 17.3].

17.2.2. Use of standard

When performing quantification from projections in the clinic, it can be helpful to image a standard activity (known measured activity) along with the patient (i.e. in the same projection). The standard (usually a small flask of the radiotracer) provides a conversion between radioactivity concentration (MBq/mL) and counts in the projections. It should be noted that the use of standards does not guarantee accurate absolute quantification because the standard activity is not affected by scatter, attenuation and partial volume effects in the same way as the activity distribution in the patient.

17.2.3. Partial volume effect and the recovery coefficient

Ideally, the activity intensity within a region in a reconstructed image should be proportional to the actual activity level in the region if scatter, attenuation, randoms (PET only) and dead time corrections are properly applied, and if it is assumed that there is very little noise. However, the partial volume effect (PVE) significantly affects the quantification based on the size of the object of interest. The PVE includes two different phenomena. One is the image blurring effect caused by the finite spatial resolution. The blurring results in spillover between regions. The image of a hot region, such as a tumour, appears larger and dimmer. This limited resolution effect is theoretically described by a convolution operation. The other PVE phenomenon is the so-called tissue fraction effect caused by the fact that the boundaries of certain voxels do not match the underlying activity distribution. The net PVE effect is the reduced contrast between the object and the surrounding areas, as well as the reduced absolute uptake in a hot region. For tumour imaging, the PVE can affect both the tumour apparent uptake and tumour apparent size.

The PVE is dependent on the size and shape of the region, the activity distribution in the surrounding background, image spatial resolution, pixel size and how uptake value is measured. The PVE correction is complicated by the fact that not only activity from inside the region spills out but also activity from outside the region spills in. As these two activity dependent flows are not usually balanced, it is difficult to predict the overall PVE effect.

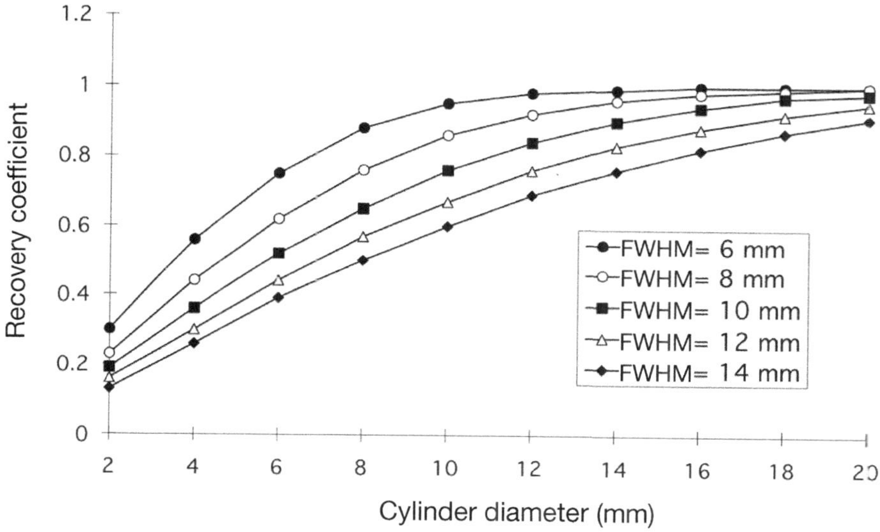

FIG. 17.2. Recovery coefficient versus cylinder diameter for different full widths at half maximum (FWHM) when imaging a cylinder filled with radioactivity in a 'cold' background.

The ratio of the apparent concentration to true concentration is called the recovery coefficient (RC). If the spatial resolution of the system can be characterized at the location where the object of interest is (this is usually the case in both SPECT and PET as the dependence of spatial resolution on location is known), and if the size and shape of a region are known (i.e. from a different modality such as computed tomography or magnetic resonance imaging), the RC can be pre-calculated and then applied to the measured concentration value in the region. Figure 17.2 illustrates RC versus cylinder size for different SPECT or PET spatial resolutions in the case of a cylindrical region which contains a uniform level of radioactivity, surrounded by a 'cold' (i.e. has no radiotracer uptake) background. An RC close to one is obtained when the size of the object is more than twice the full width at half maximum. As the RC depends on the activity concentration in the region and in the surrounding background, the RC will be different if the surrounding background is also 'hot' (i.e. has radiotracer uptake). The RC correction method is a very simple method commonly used for PVE correction in nuclear medicine [17.4]. However, the RC method can only be used to correct the spillover between two structures. Geometric transfer matrix is an approach that can account for the spillover among any number of structures [17.5]. Deconvolution is another approach to perform PVE correction without any assumption regarding tumour size, shape, homogeneity or background activity. These three correction methods are used to correct the uptake value in a region. It is also possible to model PVE in image reconstruction to obtain PVE corrected images.

17.2.4. Quantitative assessment

Reconstructed images can be assessed qualitatively and quantitatively. Qualitative interpretation is based on visualization that identifies regions with abnormal patterns of uptake of the injected radiopharmaceutical as compared to the known variants of radiotracer distribution. Quantitative assessment can be either relative or absolute:

— Target to background contrast: The target to background contrast is the ratio between the concentration within the target region and the concentration within the surrounding background. Therefore, contrast is considered a relative quantification metric.
— Radiotracer concentration: The radiotracer concentration (Bq/mL) is the amount of radioactivity per unit volume within a defined ROI. Sometimes, radiotracer concentration is converted into a different metric. For example, the standardized uptake value (SUV) is the radiotracer concentration normalized by injected dose and patient weight, and is mainly used to assess tumour glucose utilization for fluorodeoxyglucose (FDG) PET. Therefore, SUV is a semi-quantitative metric.
— Kinetic parameters: A time sequence of PET measurements, i.e. dynamic quantitative PET, makes it possible to measure tracer kinetics that describe the interaction between the tracer and physiological processes. For example, a water based tracer can be used to measure blood flow; a glucose based tracer, such as FDG, can be used to measure metabolic rate. This is the most accurate and absolute quantification metric that can be derived from PET measurements. Usually, absolute quantification is best achieved with PET because all projections are acquired simultaneously and, therefore, dynamic imaging can be more easily performed than with rotating SPECT cameras.

Clinical studies are generally analysed using qualitative and semi-quantitative (e.g. contrast, SUV) information. This is also the case for clinical trials and drug development.

17.2.4.1. Relative quantification using contrast ratio

Image contrast is the ratio of the signal level of a target relative to the signal level in the surrounding background. The contrast ratio (CR) is defined as:

$$\mathrm{CR} = \frac{C_\mathrm{T} - C_\mathrm{B}}{C_\mathrm{B}} \tag{17.3}$$

where C_T and C_B are the mean concentration within the defined target and background region, respectively.

17.2.4.2. Relative quantification using the standardized uptake value

The SUV [17 6] is a widely used semi-quantitative metric to measure tumour glucose utilization in, mostly, PET imaging. SUV (g/mL) is defined as:

$$\mathrm{SUV} = \frac{C_\mathrm{i}(\mathrm{kBq/mL})}{\mathcal{A}(\mathrm{kBq})/W(\mathrm{g})} \tag{17.4}$$

where

C_i is either the mean (for $\mathrm{SUV_{mean}}$) or maximum (for $\mathrm{SUV_{max}}$) decay-corrected activity concentration (kBq/mL) within the defined region in the image (or tissue);

$\mathcal{A}$ is the injected activity (kBq);

and W is the patient weight (g).

It is normally assumed that the density of tissue is equivalent to 1.0 g/mL, such that the units effectively cancel and the SUV becomes a dimensionless measure. The primary use of the SUV is to quantify activity in an ROI independent of administered activity and patient weight. It has been shown that the SUV may correlate with the metabolic rate of FDG in different tumour types, especially when normalized for plasma glucose levels [17.7]. $\mathrm{SUV_{max}}$ instead of $\mathrm{SUV_{mean}}$ is often used because it is less sensitive to PVEs and it avoids including necrotic or other non-tumour elements. However, $\mathrm{SUV_{max}}$ has lower reproducibility and a larger bias than $\mathrm{SUV_{mean}}$ as it is computed over a smaller number of voxels [17.8].

To address this issue, a further term has been introduced, 'SUV_{peak}', which is defined as the mean SUV value in a group of voxels surrounding the voxel with the highest activity concentration in the tissue. The SUVs will be affected by the level of image noise, which is affected by the reconstructed activity concentration, the parameters chosen in the reconstruction algorithm and a myriad of other factors. SUV_{peak} is intended to be a more robust measure than SUV_{max}.

The primary drawback with the SUV is that it is affected by many sources of variability [17.9]. In addition to mathematical factors (e.g. ROI, noise and PVE) that can alter the accuracy of the SUV, there are a number of biological factors that can variably and unpredictably impact the SUV. Firstly, the SUV calculation is based on the total administered dose. If a portion of the radiotracer becomes interstitially infiltrated during intravenous injection, it is not routine practice to correct for the activity that is trapped at the injection site and, therefore, failing to circulate through the body. As a result, the calculated SUV can be artificially low, because the total administered dose used in calculating the SUV is greater than the actual dose reaching its intended intravascular target. Secondly, the glucose avidity of tissues in the body is dependent on numerous factors such as the presence of diabetes, insulin level and glucose level (the latter two fluctuate widely depending on the patient's most recent meal). If a patient is diabetic, glucose is metabolized poorly by normal tissues, therefore leaving more glucose available in the bloodstream to be metabolized by abnormally glycolytic tissues such as tumour and infection, theoretically resulting in an artificially elevated SUV. Conversely, if a diabetic patient has just received a dose of insulin, the opposite effect may occur, lowering the SUV. In non-diabetic patients, a patient who has recently eaten will have high glucose and insulin levels, with a similar effect of lowering the SUV. Thirdly, FDG is cleared from the body through urinary excretion. Patients with impaired renal function will extract FDG more slowly from the bloodstream, leaving more FDG available for metabolism by both normal and hypermetabolic tissues. Finally, body mass is also a parameter used in calculating the SUV. In patients with a large number of ascites, or other significant '3rd-spacing' processes, the body mass of the patient will be elevated by the presence of fluid that is neither intravascular nor capable of uptake of radiotracer. Therefore, the denominator used in calculating the SUV will be artificially large due to overestimation of the size of the patient, theoretically causing an artificially low SUV to be calculated. Another factor that plays a key role in standardization of tumour uptake is the reproducibility of the measurement, and it has been previously shown that the correlation between uptake measurements made in an identical manner at different clinical sites was relatively low and significantly different to the standard reference measurement. It is not uncommon that the SUV can vary 50% because of one or more such effects. Therefore, the SUV should be properly corrected for all of these effects.

Another useful metric to assess tumour response to therapy is total lesion glycolysis (TLG) defined as:

$$\mathrm{TLG} = \mathrm{SUV}_{\mathrm{mean}} \times V \tag{17.5}$$

where V is the tumour volume (mL) that can be obtained using 3-D contour software.

TLG provides a measurement of the total FDG uptake in the tumour region, which reflects total rather than average tumour metabolism.

17.2.4.3. Absolute quantification using kinetic modelling

Dynamic imaging consists of acquiring data as a series of time frames that capture the time–activity curve in each voxel over time, making it possible to quantify tracer kinetics in vivo. Using a given radiotracer, the interaction of the radiotracer with the body's physiological, biochemical or pharmocokinetic processes can be monitored. For example, glucose metabolism can be assessed by FDG, a glucose analogue radiotracer. With an understanding of the underlying physiological factors that control the tissue radioactivity levels, mathematical models, known as kinetic models, can be constructed with one or more parameters that describe the distribution of radiotracers as a function of time in the body and fit the time–activity curves in each voxel in the organ of interest. Kinetic models used in nuclear medicine are based on compartments within a volume or space within which the radiotracer becomes uniformly distributed almost instantly, i.e. contains no significant concentration gradients. In other words, compartmental modelling describes systems that vary in time but not in space. More complicated kinetic models that include spatial gradients are generally not applicable to nuclear medicine because of limited spatial resolution.

For a single-tissue compartmental model illustrated in Fig. 17.3, the rate of change in tracer concentration in a tissue is:

$$\frac{\mathrm{d}C_{\mathrm{t}}(t)}{\mathrm{d}t} = K_1 C_{\mathrm{a}}(t) - k_2 C_{\mathrm{t}} \tag{17.6}$$

where

$C_{\mathrm{t}}(t)$ is the tracer concentration in the tissue;
$C_{\mathrm{a}}(t)$ is the tracer concentration in the blood;

and K_1 and k_2 are the first order rate constants for the flux into the tissue and out of the tissue, respectively.

The solution for the above equation is:

$$C_t(t) = K_1 C_a(t) \times \exp(-k_2 t) \tag{17.7}$$

As the blood is the tracer delivery compartment, $C_a(t)$ is also known as the input function. The input function is directly measured from blood samples or images of the blood, typically using the left ventricular blood pool. The tracer concentration in the tissue $C_t(t)$ is usually obtained from ROI analysis of a dynamic series of images. Knowing both $C_a(t)$ and $C_t(t)$, regression analysis can be applied to solve both K_1 and k_2. These kinetic parameters can be used to interpret the underlying physiology. K_1 is closely related to blood flow when the extraction fraction is large, but is more related to permeability when the extraction fraction is small. The ratio K_1/k_2 is the equilibrium volume of distribution that is defined as C_t/C_a when the net tracer flux between compartments is zero after sufficient time. By plotting the kinetic parameters as a function of the spatial coordinates, parametric images of the radiotracer can be constructed.

Other applications of kinetic modelling in PET imaging include measurement of glucose metabolic rate [17.10], and measurements of receptors and neurotransmitters [17.11], etc.

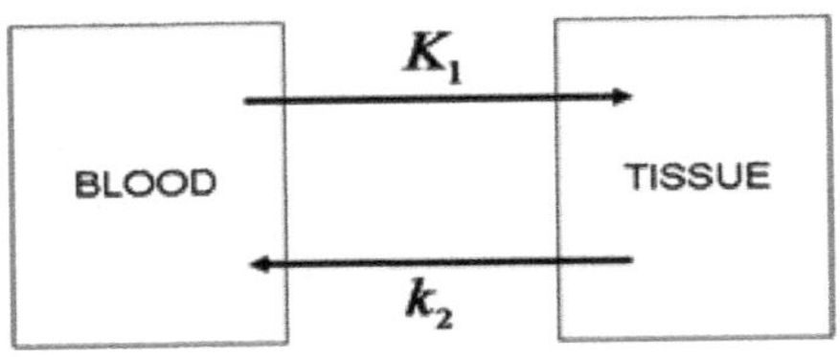

FIG. 17.3. Single-tissue compartmental model that describes the tracer exchange between blood and tissue.

17.2.5. Estimation of activity

Bias, precision and accuracy are three important statistical concepts for quantitative nuclear medicine. Bias is the difference between a population mean of the measurements or test results and an accepted reference or true value.

The bias is mainly due to faulty measuring devices or procedures. One common bias measure is defined as:

$$\text{BIAS} = \frac{1}{N}\sum_{i=1}^{N}(x_i - t) \tag{17.8}$$

where

N is the total number of measurements;
x_i is the measured value for the ith measurement;

and t is the true value.

Random error is a variable deviation and arises from the fluctuations in experimental conditions, such as Poisson noise. Random error is also called variance, but it is also often defined as the inverse, namely precision, referring to the absence of random error. Precision can be quantified by the variance defined as:

$$\sigma^2 = \frac{1}{N}\sum_{i=1}^{N}(x_i - \bar{x})^2 \tag{17.9}$$

where

$$\bar{x} = \frac{1}{N}\sum_{i=1}^{N}x_i \tag{17.10}$$

It is important to note that the calculation of precision does not require knowledge of the true value. Therefore, precision alone cannot be used to evaluate the performance of a measurement.

Bias and precision can be combined to assess the performance of a measurement. Less biased and more precise measurements yield more accurate estimations. Accuracy is, thus, defined as the overall difference between the measured value and the true value. Accuracy can be quantified by the mean square error (MSE) [17.12] defined as:

$$\text{MSE} = \frac{1}{N}\sum_{i=1}^{N}(x_i - t)^2 \tag{17.11}$$

It can be shown that [17.13]:

$$\text{MSE} = \sigma^2 + \text{BIAS}^2 \tag{17.12}$$

To have the same scale as the mean value, precision is also quantified by standard deviation, defined as the square root of variance. Similarly, accuracy is also quantified as root MSE defined as the square root of MSE.

17.2.6. Evaluation of image quality

Image quality is a concept that has received a lot of attention recently in an effort to better define what image quality is. A good surrogate for image quality is image utility, i.e. the usefulness of an image for a particular detection or quantification task, rather than measures of image properties, such as resolution, contrast or stationarity of the point spread function, or of image fidelity, such as normalized MSE. Measures of quantitative accuracy, precision and root MSE are, of course, very useful when first assessing a new system or a quantification method; however, for more rigorous evaluation or for definitive optimization of data acquisition strategies, reconstruction techniques or image processing procedures, it is recommended to carry out an objective assessment of image quality based on detection or quantification tasks. Performance metrics for task based estimation or detection tasks can be viewed as measures of image utility which are the most clinically relevant bases on which to evaluate or optimize imaging systems.

The most conclusive assessment of image quality is based on human-observer studies. However, such studies are not routinely performed clinically because they are time and resource consuming. Instead, a numerical (or mathematical) observer is often used. It is beyond the scope of this chapter to detail the different numerical observers used in SPECT and PET (for a review, see Refs [17.13, 17.14]).

One of the simplest numerical observer methods is the non-prewhitening matched technique, which is the optimal observer when images have uncorrelated noise. Assuming N noise realizations of target-present image $\boldsymbol{S}$ and N noise realizations of target-absent image $\boldsymbol{B}$, the non-prewhitening signal to noise ratio (SNR_{NPW}) can be calculated as:

$$\text{SNR}_{\text{NPW}} = \frac{\left|\langle \boldsymbol{S}\cdot\boldsymbol{T}\rangle - \langle \boldsymbol{B}\cdot\boldsymbol{T}\rangle\right|}{\sqrt{1/2[\sigma^2(\boldsymbol{S}\cdot\boldsymbol{T}) + \sigma^2(\boldsymbol{B}\cdot\boldsymbol{T})]}} \tag{17.13}$$

where $\boldsymbol{T}$ is the target matched filter.

More sophisticated numerical observers include the channelized Hotelling observer that is a better surrogate for a human-observer under less ideal situations.

REFERENCES

[17.1] SORENSON, J.A., "Quantitative measurement of radioactivity in vivo by whole-body counting", Instrumentation of Nuclear Medicine (HINE, G.J., SORENSON, J.A., Eds), Academic Press, New York **2** (1974) 311–348.

[17.2] BAZIN, J.P., DI PAOLA, R., GIBAUD, B., ROUGIER, P., TUBIANA, M., "Factor analysis of dynamic scintigraphic data as a modelling method. An application to the detection of the metastases", Information Processing in Medical Imaging (DI PAOLA, R , KAHN, E., Eds), INSERM, Paris (1980) 345–366.

[17.3] HOUSTON, A.S., The effect of apex-finding errors on factor images obtained from factor analysis and oblique transformation, Phys. Med. Biol. **29** (1984) 1109–1116.

[17.4] AVRIL, N., et al., Breast imaging with fluorine-18-FDG PET: quantitative image analysis, J. Nucl. Med. **38** (1997) 1186–1191.

[17.5] ROUSSET, O.G., MA, Y., EVANS, A.C., Correction for partial volume effects in PET: Principle and validation, J. Nucl. Med. **39**(5) (1998) 904–911.

[17.6] ZASADNY, K.R., WAHL, R.L., Standardized uptake values of normal tissues at PET with 2-[fluorine-18]-fluoro-2-deoxy-D-glucose: variations with body weight and a method for correction, Radiology **189** (1993) 847–850.

[17.7] FLANAGAN, F.L., DEHADASHTI, F., SIEGEL, B.A., PET in breast cancer, Semin. Nucl. Med. **XXVIII** (1998) 290–302.

[17.8] KRAK, N.C., et al., Effects of ROI definition and reconstruction method on quantitative outcome and applicability in a response, Eur. J. Nucl. Med. **32** (2005) 294–301.

[17.9] KEYES, J.W., Jr., SUV: Standard uptake or silly useless value? J. Nucl. Med. **36** (1995) 1836–1839.

[17.10] PHELPS, M.E., et al., Tomographic measurement of local cerebral glucose metabolic rate in humans with [18]-fluoro-2-deoxy-D-glucose: validation of method, Ann. Neurol. **6** (1979) 371–388.

[17.11] MINTUM, M.A., RAICHLE, M.E., KILBOURN, M.R., WOOTEN, G.F., WELCH, M.J., A quantitative model for the in vivo assessment of drug binding sites with positron emission tomography, Ann. Neurol. **15** (1984) 217–227.

[17.12] BURKHOLDER, D.L., "Point estimation", International Encyclopedia of Statistics: A Review (KRUSKAL, W.H., TANUR, J.M., Eds), J. Am. Stat. Assoc. **88** (1978) 251–259.

[17.13] BARRETT, H.H., MYERS, K.J., Foundations of Image Science, John Wiley & Sons, NJ (2004).

[17.14] ABBEY, C.K., BARRETT, H.H., Human- and model-observer performance in ramp-spectrum noise; effects of regularization and object variability, J. Opt. Soc. Am. A Opt. Image Sci. Vis. **18** (2001) 473–488.

CHAPTER 18

INTERNAL DOSIMETRY

C. HINDORF
Department of Radiation Physics,
Skåne University Hospital,
Lund, Sweden

18.1. THE MEDICAL INTERNAL RADIATION DOSE FORMALISM

18.1.1. Basic concepts

The Committee on Medical Internal Radiation Dose (MIRD) is a committee within the Society of Nuclear Medicine. The MIRD Committee was formed in 1965 with the mission to standardize internal dosimetry calculations, improve the published emission data for radionuclides and enhance the data on pharmacokinetics for radiopharmaceuticals [18.1]. A unified approach to internal dosimetry was published by the MIRD Committee in 1968, MIRD Pamphlet No. 1 [18.2], which was updated several times thereafter. Currently, the most well known version is the MIRD Primer from 1991 [18.3]. The latest publication on the formalism was published in 2009 in MIRD Pamphlet No. 21 [18.4], which provides a notation meant to bridge the differences in the formalism used by the MIRD Committee and the International Commission on Radiological Protection (ICRP) [18.5]. The formalism presented in MIRD Pamphlet No. 21 [18.4] will be used here, although some references to the quantities and parameters used in the MIRD Primer [18.3] will be made. All symbols, quantities and units are presented in Tables 18.1 and 18.2.

The MIRD formalism gives a framework for the calculation of the absorbed dose to a certain region, called the target region, from activity in a source region. The absorbed dose D is calculated as the product between the time-integrated activity $\tilde{A}$ and the S value:

$$D = \tilde{A} \cdot S \tag{18.1}$$

The International System of Units unit of absorbed dose is the joule per kilogram (J/kg), with the special name gray (Gy) (1 J/kg = 1 Gy).

The time-integrated activity equals the number of decays that take place in a certain source region, with units Bq · s, while the S value denotes the absorbed dose rate per unit activity, expressed in Gy · (Bq · s)$^{-1}$ or as a multiple thereof, for example, in mGy · (MBq · s)$^{-1}$. The time-integrated activity was named the cumulated activity in the MIRD Primer [18.3] and the absorbed dose rate per unit activity was named the absorbed dose per cumulated activity (or the absorbed dose per decay). A source or a target region can be any well defined volume, for example, the whole body, an organ/tissue, a voxel, a cell or a subcellular structure. The source region is denoted r_S and the target region r_T:

$$D(r_T) = \tilde{A}(r_S) \cdot S(r_T \leftarrow r_S) \tag{18.2}$$

The number of decays in the source region, denoted the time-integrated activity, is calculated as the area under the curve that describes the activity as a function of time in the source region after the administration of the radiopharmaceutical ($A(r_S, t)$). The activity in a region as a function of time is commonly determined from consecutive quantitative imaging sessions, but it could also be assessed via direct measurements of the activity on a tissue biopsy, a blood sample or via single probe measurements of the activity in the whole body. Compartmental modelling is a theoretical method that can be used to predict the activity in a source region in which measurements are impossible.

$$\tilde{A}(r_S) = \int A(r_S, t)\,\mathrm{d}t \tag{18.3}$$

The time-integration period T_D, for which the time-integrated activity in the source region is determined, is commonly chosen from the time of administration of the radiopharmaceutical until infinite time, e.g. 0 to ∞ (Eq. (18.4)). However, the integration period should be matched to the biological end point studied in combination with the time period in which the relevant absorbed dose is delivered.

$$\tilde{A}(r_S, T_D) = \int_0^{T_D} A(r_S, t)\,\mathrm{d}t \tag{18.4}$$

The time-integrated activity coefficient $\tilde{a}$ is defined as the time-integrated activity divided by the administered activity A_0, as can be seen in Eq. (18.5), and has the unit of time. The time-integrated activity coefficient was named the 'residence time' in the MIRD Primer [18.3]. Figure 18.1 further demonstrates the concept. The area under the curve describing the activity as a function of time equals the area for the rectangle ($\int_0^{T_D} A(r_S,t)\,dt = \tilde{a}(r_S)\cdot A_0$) and the time-integrated activity coefficient can be described as an average time that the activity spends in a source region.

$$\tilde{a}(r_S) = \frac{\tilde{A}(r_S)}{A_0} \tag{18.5}$$

The S value is defined according to Eq. (18.6), which includes the energy emitted E, the probability Y for radiation with energy E to be emitted, the absorbed fraction ϕ and the mass of the target region $M(r_T)$. The absorbed fraction is defined as the fraction of the energy emitted from the source region that is absorbed in the target region and equals a value between 0 and 1. The absorbed fraction is dependent on the shape, size and mass of the source and target regions, the distance and type of material between the source and the target regions, the type of radiation emitted from the source and the energy of the radiation:

$$S = \frac{EY\phi}{M(r_T)} \tag{18.6}$$

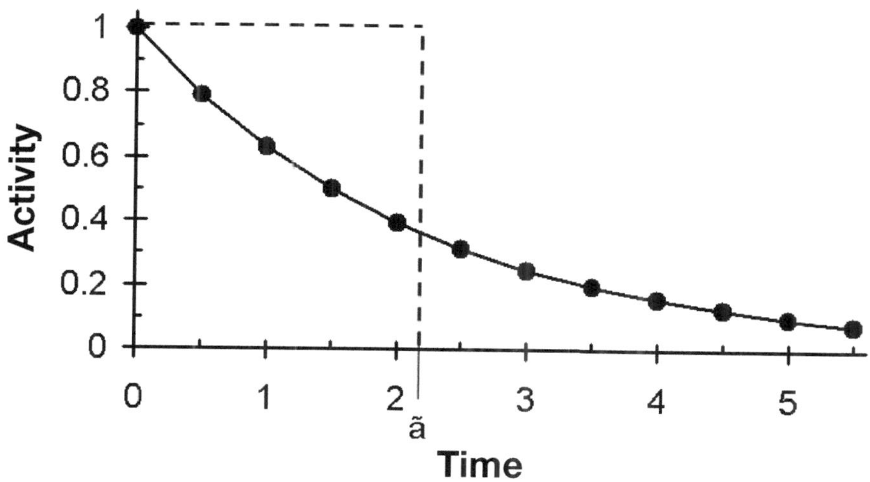

FIG. 18.1. The time-integrated activity coefficient (the residence time in the MIRD Primer [18.3]) is calculated as the time-integrated activity divided by the injected activity, which gives an average time the activity spends in the source region.

The product of the energy emitted E and its probability to be emitted Y is denoted Δ, the mean energy emitted per decay of the radionuclide. The full formalism also includes a summation over all of the transitions i per decay:

$$S(r_{\mathrm{T}} \leftarrow r_{\mathrm{S}}) = \sum_{i} \frac{\Delta_i \phi(r_{\mathrm{T}} \leftarrow r_{\mathrm{S}}, i)}{M(r_{\mathrm{T}})} \tag{18.7}$$

The absorbed fraction divided by the mass of the target region is named the specific absorbed fraction Φ:

$$\Phi(r_{\mathrm{T}} \leftarrow r_{\mathrm{S}}, E_i) = \frac{\phi(r_{\mathrm{T}} \leftarrow r_{\mathrm{S}}, E_i)}{M(r_{\mathrm{T}})} \tag{18.8}$$

The mass of both the source and target regions can vary in time, which means that the absorbed fraction will change as a function of time after the administration, and the full time dependent version of the internal dosimetry nomenclature must be applied (Eq. (18.9)). This phenomenon has been noted in the clinic for tumours, the thyroid and lymph nodes, and can significantly influence the magnitude of the absorbed dose.

$$\Phi(r_{\mathrm{T}} \leftarrow r_{\mathrm{S}}, E_i, t) = \frac{\phi(r_{\mathrm{T}} \leftarrow r_{\mathrm{S}}, E_i, t)}{M(r_{\mathrm{T}}, t)} \tag{18.9}$$

The total mean absorbed dose to the target region $D(r_{\mathrm{T}})$ is given by summing the separate contributions from each source region r_{S} (Eq. (18.10)). The self-absorbed dose commonly gives the largest fractional contribution to the total absorbed dose in a target region. The self-absorbed dose refers to when the source and target regions are identical, while the cross-absorbed dose refers to the case in which the source and the target regions are different from each other.

$$D(r_{\mathrm{T}}) = \sum_{r_{\mathrm{S}}} \tilde{A}(r_{\mathrm{S}}) S(r_{\mathrm{T}} \leftarrow r_{\mathrm{S}}) \tag{18.10}$$

The full time dependent version of the MIRD formalism can be found in Eq. (18.11), where $\dot{D}$ denotes the absorbed dose rate:

$$D(r_{\mathrm{T}}, T_{\mathrm{D}}) = \sum_{r_{\mathrm{S}}} \int_0^{T_{\mathrm{D}}} \dot{D}(r_{\mathrm{T}}, t)\,\mathrm{d}t = \sum_{r_{\mathrm{S}}} \int_0^{T_{\mathrm{D}}} A(r_{\mathrm{S}}, t) S(r_{\mathrm{T}} \leftarrow r_{\mathrm{S}}, t)\,\mathrm{d}t \tag{18.11}$$

TABLE 18.1. EXPLANATION OF SYMBOLS USED IN THE MEDICAL INTERNAL RADIATION DOSE FORMALISM

Symbol	Parameter
R	Type of radiation
r_S	Source region
r_T	Target region
T_D	Integration period

TABLE 18.2. EXPLANATION OF THE SYMBOLS USED TO REPRESENT QUANTITIES IN THE MEDICAL INTERNAL RADIATION DOSE FORMALISM

Symbol	Quantity	Unit
$\tilde{A}(r_S, T_D)$	Time-integrated activity	Bq · s
$\tilde{a}(r_S, T_D)$	Time-integrated activity coefficient	s
$D(r_T)$	Absorbed dose to the target region r_T	Gy
$\dot{D}$	Absorbed dose rate	Gy/s
Δ_i	Mean energy of the *i*th transition per nuclear transformation	J (Bq · s)$^{-1}$ or MeV (Bq · s)$^{-1}$
E_i	Mean energy of the *i*th transition	J or MeV
$M(r_T, t)$	Mass of target region	kg
$S(r_T \leftarrow r_S, t)$	Absorbed dose rate per unit activity	mGy (MBq · s)$^{-1}$
t	Time	s
Y_i	Number of *i*th transitions per nuclear transformation	(Bq · s)$^{-1}$
$\phi(r_T \leftarrow r_S, E_i, t)$	Absorbed fraction	Dimensionless
$\Phi(r_T \leftarrow r_S, E_i, t)$	Specific absorbed fraction	kg^{-1}

18.1.2. The time-integrated activity in the source region

The physical meaning of the time-integrated activity in the source region would be the number of decays in the source region during the relevant time period. The time-integrated activity was named the cumulated activity in the MIRD Primer [18.3].

The activity as a function of time $A(t)$ can often be described by a sum of exponential functions (Eq. (18.12)), where j denotes the number of exponentials, A_j the initial activity for the jth exponential, λ the decay constant for the radionuclide, λ_j the biological decay constant and t the time after the administration of the radiopharmaceutical. The sum of the j coefficients A_j gives the total activity in the source region at the time of administration of the radiopharmaceutical ($t = 0$):

$$A(r_{\mathrm{T}},t) = \sum_j A_j \cdot \mathrm{e}^{-t(\lambda+\lambda_j)} \tag{18.12}$$

The decay constant λ equals the natural logarithm of 2 (ln 2 = 0.693) divided by the half-life. The decay constant in an exponential function matches the slope of the curve it describes (in a linear-log plot of the function).

$$\lambda = \frac{\ln 2}{T_{1/2}} \tag{18.13}$$

The physical half-life $T_{1/2}$ and the biological half-life $T_{1/2,j}$ can be combined into an effective half-life $T_{1/2,\mathrm{eff}}$ according to Eq. (18.14). The effective half-life is always shorter than both the biological and the physical half-lives alone.

$$\frac{1}{T_{1/2,\mathrm{eff}}} = \frac{1}{T_{1/2,j}} + \frac{1}{T_{1/2}} \tag{18.14}$$

The cumulated activity for the relevant time period is commonly calculated as the time integral of an exponential function (Eq. (18.15)). However, other functions could be used, with trapezoidal or Riemann integration (Fig. 18.2). The trapezoidal and the Riemann methods could be reproduced with a higher accuracy than the integration of an exponential, depending on how well the exponential fit could be performed.

$$\tilde{A} = \int_0^{\infty} A(r_{\mathrm{S}},0)\,\mathrm{e}^{-t(\lambda+\lambda_j)}\,\mathrm{d}t = \frac{A(r_{\mathrm{S}},0)}{\lambda+\lambda_j} \tag{18.15}$$

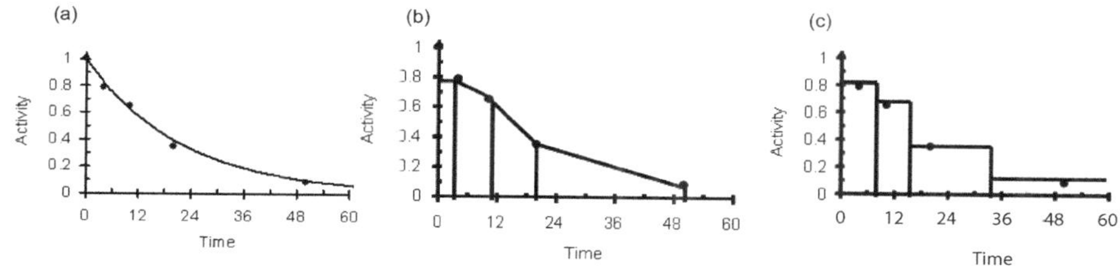

FIG. 18.2. Demonstration of different methods to calculate the time-integrated activity using a fit to an exponential function (a), trapezoidal integration (b) and Riemann integration (c).

Relevant biological data need to be acquired to perform absorbed dose calculations with accuracy. The shape of the fitted curve, which describes the activity as a function of time after the administration of the radiopharmaceutical, can be strongly influenced by the number and timing of the individual activity measurements (see Fig. 18.3). Three data points per exponential phase should be considered the minimum data required to determine the pharmacokinetics, and data points should be followed for at least two to three effective half-lives.

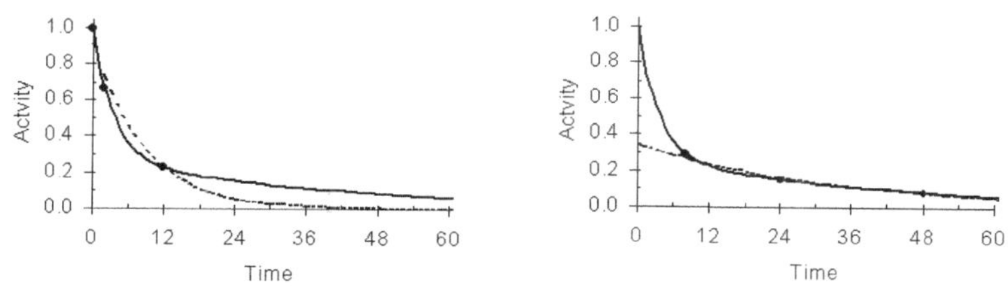

FIG. 18.3. Two examples of the possible influences of curve fitting caused by the number and the timing of activity measurements. The solid line gives the real activity versus time; the dotted line represents the exponential curve fitted to the measurements, which are shown as black dots.

The extrapolation from time zero to the first measurement of the activity in the source region, and the extrapolation from the last measurement of the activity in the source region to infinity, can also strongly influence the accuracy in the time-integrated activity (see Fig. 18.4).

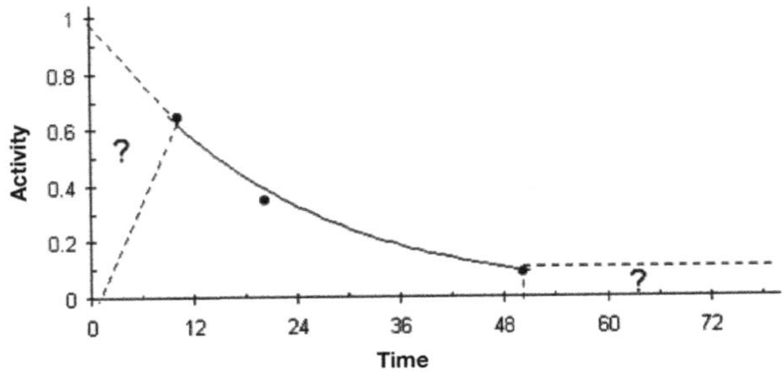

FIG. 18.4. Extrapolation before the first and after the last measurement point.

18.1.3. Absorbed dose rate per unit activity (*S* value)

The *S* value for a certain radionuclide and source–target combination is generated from Monte Carlo simulations in a computer model of the anatomy.

The first models were analytical phantoms, in which the anatomy was described by analytical equations. A coordinate system was introduced and simple geometrical shapes such as spheres or cylinders were placed in the coordinate system to represent important structures of the anatomy. Several analytical phantoms exist: adult man, non-pregnant woman, pregnant woman for each trimester of pregnancy, children (from the newborn and up to 15 years of age) as well as models of the brain, kidneys and unit density spheres.

Voxel based phantoms are the second generation of phantoms used for the calculation of *S* values. These phantoms offer the possibility of more detailed models of the anatomy. Voxel based phantoms can be based on the segmentation of organs from tomographic image data, such as computed tomography (CT) images.

The third generation of phantoms is created using non-uniform rational B-spline (NURBS). NURBS is a mathematical model used in computer graphics to represent surfaces. NURBS provides a method to represent both geometrical shapes and free forms with the same mathematical representation, and the surfaces are flexible and can easily be rotated and translated. This means that movements in time, such as breathing and the cardiac cycle, can be included, allowing for 4-D representations of the phantoms [18.6].

Anatomical phantoms for the calculation of *S* values for use in pre-clinical studies on dogs, rats and mice have also been developed.

A common assumption in radionuclide dosimetry is that radiation emissions can be divided into penetrating (p) or non-penetrating (np) and that the absorbed fractions for these two types can be set to equal 0 and 1, respectively ($\phi_p \approx 0$ and $\phi_{np} \approx 1$). Electrons are often considered non-penetrating and photons as penetrating radiation, but this is an oversimplification. The validity of the assumption is very dependent on the energy of the radiation in combination with the size of the source region and must, therefore, be assessed on a case by case basis, as is evident from Fig. 18.5. For electrons, the absorbed fraction is greater than 0.9 if the mass of the unit density sphere is greater than 10 g and the electron energy is lower than 1 MeV. This means that the approximation of electrons as non-penetrating radiation is good at an organ level for humans, but as the mass decreases, the approximation ceases to be valid. For photons, the absorbed fraction is less than 0.1 if the mass of the sphere is less than 100 g and if the photon energy is larger than 50 keV. The approximation of considering photons as penetrating radiation is valid in most pre-clinical situations, but as the mass increases, the approximation becomes inappropriate.

The self absorbed S values can be scaled by mass according to the following equation:

$$S(r_{\mathrm{T}} \leftarrow r_{\mathrm{T}},\mathrm{scaled}) \approx S(r_{\mathrm{T}} \leftarrow r_{\mathrm{T}},\mathrm{tabulated}) \cdot \frac{M(r_{\mathrm{T}},\mathrm{tabulated})}{M(r_{\mathrm{T}},\mathrm{scaled})} \tag{18.16}$$

This is a useful method to adjust the S value found in a table to the true weight of the target region. When scaling an S value, the absorbed fraction is considered to be constant in the interval of scaling. The change in the S value is then set equal to the change in mass of the target. It should be noted that linear interpolation should never be performed in S value tables (Fig. 18.6).

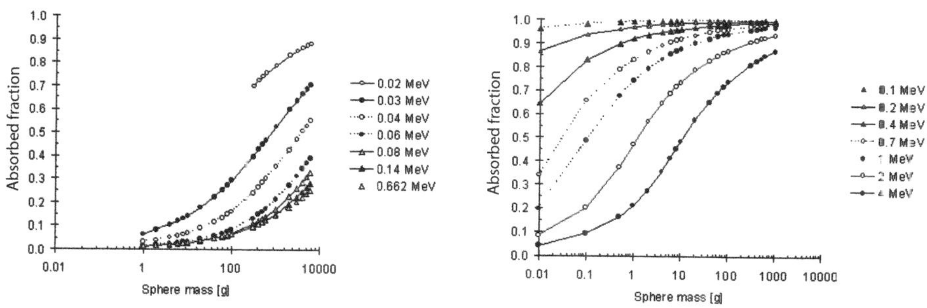

FIG. 18.5. Absorbed fraction for unit density spheres as a function of the mass of the spheres for mono-energetic photons (left) and electrons (right) (data from Ref. [18.7]).

A more sophisticated and probably more accurate way of recalculation of the S value is by separating the total S value into two parts: one for penetrating and one for non-penetrating radiation (S_{p} and S_{np}, respectively). If the absorbed fraction for non-penetrating radiation is assumed to be equal to 1 ($\phi_{\mathrm{np}} = 1$), the S value for penetrating radiation can be calculated (Eqs (18.17)–(18.19)).

The absorbed fractions for photons are relatively constant, so the S value for penetrating radiation can be scaled by mass.

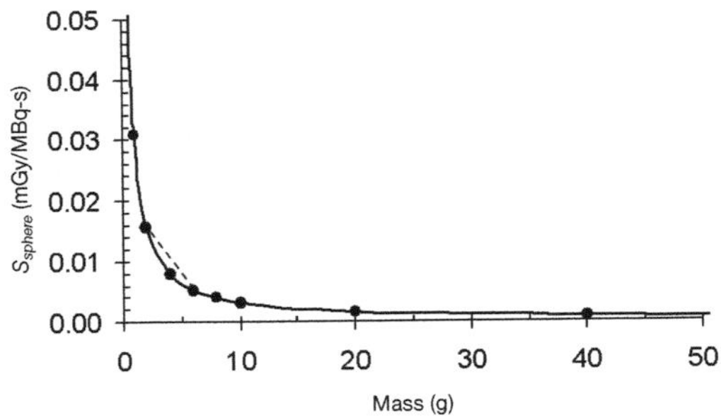

FIG. 18.6. Linear interpolation in S value tables gives S values that are too large. In this particular case for a unit density sphere of ^{131}I, linear scaling would give an S value that is significantly greater than when scaling according to mass is performed.

$$S = S_{\mathrm{p}} + S_{\mathrm{np}} = S_{\mathrm{p}} + \frac{\Delta_{\mathrm{np}}}{m} \tag{18.17}$$

$$S_{\mathrm{p}} = (S - \frac{\Delta_{\mathrm{np}}}{m}) \cdot \frac{m_{\mathrm{phantom}}}{m_{\mathrm{true}}} \tag{18.18}$$

$$S_{\mathrm{recalculated}} = (S - \frac{\Delta_{\mathrm{np}}}{m}) \cdot \frac{m_{\mathrm{phantom}}}{m_{\mathrm{true}}} + \frac{\Delta_{\mathrm{np}}}{m} \tag{18.19}$$

The absorbed fractions for photons and electrons vary according to the initial energy and the volume/mass of the target region and, thus, the suitability of the recalculation will also vary, as was discussed in the previous section (Fig. 18.5).

The principle of reciprocity means that the S value is approximately the same for a given combination of source and target regions, i.e. $S(r_{\mathrm{T}} \leftarrow r_{\mathrm{S}})$ is equal to $S(r_{\mathrm{S}} \leftarrow r_{\mathrm{T}})$. The reciprocity principle is only truly valid under ideal conditions, in regions with a uniformly distributed radionuclide within a material that is either (i) infinite and homogenous or (ii) absorbs the radiation without scatter. The ideal conditions are not present in the human body, although the reciprocity principle can be seen in S value tables for human phantoms as the numbers are almost mirrored along the diagonal axis of the table.

S values for a sphere of a certain volume and material should be scaled according to density if the material in the sphere is different from the material in the phantom (Eq. (18.20)). The technique can be applied when an *S* value for a unit density sphere is used for the calculation of the absorbed dose to a tumour made up of bone or lung. However, it should be noted that an *S* value with the correct mass could be chosen instead of scaling the *S* value for the correct volume by the density.

$$S_{\text{volume, material X}} = S_{\text{volume, material Y}} \cdot \frac{\phi_{\text{material Y}}}{\phi_{\text{material X}}} \tag{18.20}$$

18.1.4. Strengths and limitations inherent in the formalism

Two assumptions are automatically made when the MIRD formalism is applied:

(a) The activity distribution in the source region is assumed to be uniform;
(b) The mean absorbed dose to the target region is calculated.

These assumptions are approximations of the reality. The strengths of the MIRD implementation are its simplicity and ease of use. The limitation of these assumptions is that the absorbed dose may vary throughout the region.

It is important to note that the MIRD formalism does not set any restrictions on either the volume or the shape of the source or target, as long as uniformity can be assumed. This means that the source and target volumes could be defined so that the condition of uniformity is met.

The absorbed dose *D* is defined by the International Commission on Radiation Units and Measurements as the quotient of the mean energy imparted $\mathrm{d}\bar{\varepsilon}$ and the mass d*m* [18.8]:

$$D = \frac{\mathrm{d}\bar{\varepsilon}}{\mathrm{d}m} \tag{18.21}$$

The absorbed dose is defined at a point, but it is determined from the mean specific energy and is, thus, a mean value. This is more obvious from an older definition of absorbed dose, where it is defined as the limit of the mean specific energy as the mass approaches zero [18.9]:

$$D = \lim_{m \to 0} \bar{z} \tag{18.22}$$

The dosimetric quantity that considers stochastic effects and is, thus, not based on mean values, is the specific energy z. The specific energy represents a stochastic distribution of individual energy deposition events ε divided by the mass m in which the energy was deposited [18.10]:

$$z = \frac{\varepsilon}{m} \tag{18.23}$$

The unit of the specific energy is joules per kilogram and its special name is gray. Its relevance is especially important in microdosimetry which is the study of energy deposition spectra within small volumes corresponding to the size of a cell or cell nucleus.

The energy imparted to a given volume is the sum of all energy deposits ε_i in the volume:

$$\varepsilon = \sum_i \varepsilon_i \tag{18.24}$$

The energy deposit is the fundamental quantity that can be used for the definition of all other dosimetric quantities. Each energy deposit is the energy deposited in a single interaction i:

$$\varepsilon_i = \varepsilon_{\text{in}} - \varepsilon_{\text{out}} + Q \tag{18.25}$$

where

- ε_{in} is the kinetic energy of the incident ionizing particle;
- ε_{out} is the sum of the kinetic energies of all ionizing particles leaving the interaction;

and Q is the change in the rest energies of the nucleus and of all of the particles involved in the interaction.

If the rest energy decreases, Q has a positive value and if the rest energy increases, it has a negative value. The unit of energy imparted and energy deposited is joules or electronvolts. The summation of the energy deposits to receive the energy imparted may be performed for one or more events, which is a term denoting the energy imparted from statistically correlated particles, such as a proton and its secondary electrons.

The absorbed dose is a macroscopic entity that corresponds to the mean value of the specific energy per unit mass, but is defined at a point in space.

When considering an extended volume such as an organ in the body, then for the mean absorbed dose to be a true representation of the absorbed dose to the target volume, either radiation equilibrium or charged particle equilibrium must exist. Radiation equilibrium means that the energy entering the volume must equal the energy leaving the volume for both charged and uncharged radiation. The conditions under which radiation equilibrium are present in a volume containing a distributed radioactive source are [18.11]:

— The radioactive source must be uniformly distributed;
— The atomic composition of the medium must be homogeneous;
— The density of the medium must be homogeneous;
— No electric or magnetic fields may disturb the paths of the charged particle.

Charged particle equilibrium always exists if radiation equilibrium exists. However, charged particle equilibrium can exist even if the conditions for radiation equilibrium are not fulfilled.

If only charged particles are emitted from the radioactive source (as is the case for β emitters such as ^{90}Y and ^{32}P), charged particle equilibrium exists if radiative losses are negligible. Radiative losses increase with increasing electron energy and with an increase in the atomic number of the medium. The maximum β energy for pure β emitters commonly used in nuclear medicine (e.g. ^{90}Y, ^{32}P and ^{89}Sr) is less than 2.5 MeV and the ratio of the radiative stopping power to the total stopping power is 0.018 and 0.028 for skeletal muscle and cortical bone, respectively, for an electron energy of 2.5 MeV. This would imply that the radiative losses can be neglected in internal dosimetry and charged particle equilibrium can be assumed.

If both charged and uncharged particles (photons) are emitted (as is the case with most radionuclides used in nuclear medicine), charged particle equilibrium exists if the interaction of the uncharged particles within the volume is negligible. A negligible number of interactions means that the photon absorbed fraction is low. Photon absorbed fractions as a function of mass can be seen in Fig. 18.5, but it should be pointed out that the relative photon contribution for a radionuclide is also dependent on the energy and the probability of emission of electrons. For example, the photon contribution to the absorbed dose cannot be disregarded for ^{111}In in a 10 g sphere, where the photons contribute 45% to the total S value.

18.1.4.1. Non-uniform activity distribution

The activity distribution is seldom completely uniform over the whole tissue. This effect was theoretically investigated on a macroscopic level by Howell et al. [18.12] by introducing activity distributions that varied as a function of the radius of a sphere. The non-uniformity in the activity distribution can be overcome by redefining the source region into a smaller volume. This is a feasible approach until the activity per unit volume becomes small enough to cause a break-down of both radiation and charged particle equilibrium.

Redistribution of the radioactive atoms over time is responsible for creating non-uniformities of the absorbed dose distribution over time. This effect is handled indirectly in the MIRD formalism, which utilizes the concept of cumulated activity, defined as the total number of decays during the time of integration. However, in most practical applications of MIRD dosimetry, heterogeneities of source distribution within organs are neglected.

18.1.4.2. Non-uniform absorbed dose distribution

If the activity of an α or β emitting radionuclide is uniformly distributed within a sphere, then the absorbed dose distribution will be uniform from the centre of the sphere out to a distance from the rim corresponding to the range of the most energetic particle emission. If the radius of the sphere is large relative to the particle emission ranges, then radiation equilibrium will be established except at the rim and the mean absorbed dose will give a representative value of the absorbed dose. If the radius of the sphere is of the same order as the range of the emitted electrons, significant gradients in the absorbed dose distribution will be formed at the borders of the sphere. As a rule of thumb, it can be assumed that the absorbed dose at the border of the sphere will be half of the absorbed dose at the centre. If the sphere is small compared to the range of the electrons, charged particle equilibrium is never established and the absorbed dose distribution will never be uniform inside the sphere. For α emitting radionuclides, the absorbed dose is uniform for almost all sized spheres, except within 70–90 μm from the rim, corresponding to the α particle range.

Interfaces between media, such as soft tissue/bone or soft tissue/air, will cause non-uniformity in the absorbed dose distribution due to differences in backscatter. This can be significant when estimating the contribution of absorbed dose to the stem cells in the bone marrow from backscatter off the bone surfaces. For planar geometry, the maximum increase in absorbed dose was 9%, as determined by Monte Carlo simulations of ^{90}Y. Experimental measurements with ^{32}P showed a maximal increase of 7%, in close agreement with the theoretical estimates. For a spherical interface with a 0.5 mm radius of curvature, the absorbed dose to the whole sphere showed a maximum increase for 0.5 MeV electrons of as much as 12%.

Non-uniformities in the absorbed dose distribution will also be caused by the cross-absorbed dose, i.e. when one organ is next to another such as lung and heart. In human subjects, the separation between organs is sufficiently great that the cross-absorbed dose results from penetrating photon radiation only. It is important to note that the cross-organ absorbed dose from high energy β emitters, such as ^{90}Y and ^{32}P, can be significant in preclinical small animal studies used to study radiation toxicity. The importance of the cross-absorbed dose in comparison to the self-absorbed dose strongly depends on both the *S* value and the relative size of the time-integrated activity within the source and the target regions.

The MIRD formalism as such is equally applicable to any well defined source and target region combinations [18.13, 18.14]. Depending on the volume and dimensions of the regions, different types of emitted radiation will be of different importance. To conclude the above discussion, a number of factors causing non-uniformity in the absorbed dose distribution have been identified:

— Edge effects due to lack of radiation equilibrium;
— Lack of radiation equilibrium and charged particle equilibrium in the whole volume (high energy electrons emitted in a small volume);
— Few atoms in the volume, causing a lack of radiation equilibrium and introduction of stochastic effects;
— Temporal non-uniformity due to the kinetics of the radiopharmaceutical;
— Gradients due to hot spots;
— Interfaces between media causing backscatter;
— Spatial non-uniformity in the activity distribution.

18.2. INTERNAL DOSIMETRY IN CLINICAL PRACTICE

18.2.1. Introduction

Internal dosimetry is performed with different purposes, which would require different levels of accuracy in the calculated absorbed dose, depending on the subgroup:

— Dosimetry for diagnostic procedures utilized in nuclear medicine;
— Dosimetry for therapeutic procedures (radionuclide therapy);
— Dosimetry in conjunction with accidental intake of radionuclides.

The dosimetry for a diagnostic procedure is performed to optimize the procedure concerning radiation protection consistent with the requirements of an accurate diagnostic test. This is an optimization of a clinical procedure applicable to all persons. The most relevant would, therefore, be to utilize the mean pharmacokinetics for the radiopharmaceutical for the calculation of the time-integrated activity and *S* values based on a reference man phantom. The ICRP has published the absorbed dose per injected activity for most radiopharmaceuticals used for diagnostic procedures in the clinic in Publication 53 [18.5], with updates published in Publications 80 and 106 [18.15, 18.16].

The purpose of performing dosimetry for a patient that receives radionuclide therapy is to optimize the treatment so as to achieve the highest possible absorbed dose to the tumour, consistent with absorbed dose limiting toxicities. Thus, individualized treatment planning should be performed that takes into account the patient specific pharmacokinetics and biodistribution of the therapeutic agent.

The procedure to apply after an accidental intake of radionuclides must be decided on a case by case basis. The procedure to apply will depend on the level of activity, which radionuclide, the number of persons involved, whether the dosimetry is performed retrospectively or as a precaution, and whether there is a possibility to perform measurements after the intake.

18.2.2. Dosimetry on an organ level

Dosimetry on an organ level could be performed from activity quantification using either 2-D or 3-D images. Two dimensional images may include whole body scans or spot views covering the regions of interest. Three dimensional single photon emission computed tomography (SPECT) is mostly a limited field of view study that includes only the essential structures of interest. The advantage of 3-D tomographic methods is that they avoid the problems associated with corrections for activity in overlying and underlying tissues (e.g. muscle, gut and bone), and corrections for activity in partly overlapping tissues (e.g. liver and right kidney). Three dimensional positron emission tomography (PET) is emerging as a powerful dosimetric tool because of the greater ease and accuracy of radiotracer quantification with this modality.

S value tables for human phantoms can be found in MIRD Pamphlet No. 11 [18.17], in the OLINDA EXM software [18.18] and on the RADAR web site (www.doseinfo-radar.com). OLINDA/EXM stands for organ level internal dose assessment/exponential modelling, and is a software for the calculation of absorbed dose to different organs in the body. OLINDA includes *S* values for most radionuclides and for ten different human phantoms (adult and children at different ages as well as pregnant and non-pregnant female phantoms). Tumours are not included in the phantoms, although the *S* values for unit density spheres provided in the software could be applied for the calculation of the self-absorbed dose to the tumour. OLINDA also includes a module for biokinetic analysis, allowing the user to fit an exponential equation to the data entered on the activity in an organ at different time points. *S* values can be scaled by mass within

OLINDA, thus allowing for a more patient specific dosimetry to be performed. MIRDOSE [18.19] is the predecessor of OLINDA/EXM.

When calculating the absorbed dose with the MIRD formalism and using tabulated S values for a phantom, for example, the reference man, it is assumed that the patient's anatomy is the same as that of the phantom. To employ the MIRD scheme and yet make the dosimetry more patient specific, the S values can be scaled to the mass of each patient's target organ. Owing to the inverse relation between the absorbed dose and the mass of the target region, scaling can have a considerable influence on the result. The organ mass can be estimated from CT, magnetic resonance imaging or ultrasound images, provided that the anatomical size equals the functional size (the volume/mass of the organ that is actually physiologically functioning and has an activity uptake).

$$S_{\text{patient}} \approx S_{\text{phantom}} \cdot \frac{m_{\text{phantom}}}{m_{\text{patient}}} \tag{18.26}$$

Since it requires a great deal of work to determine the mass of every organ for each patient, it was suggested that the S values might be scaled to the total mass of the patient. This is a more crude method, assuming that the organ size follows the total mass of the body. The lean body weight of the patient should be used to avoid unrealistic values of the organ mass and, thus, the S values due to obese or very lean patients.

$$S_{\text{patient}} \approx S_{\text{phantom}} \cdot \frac{m_{\text{TB,phantom}}}{m_{\text{TB,patient}}} \tag{18.27}$$

Tumours are not included in reference man phantoms. However, S values could be used for spheres of the correct mass to get an approximation of the self-absorbed dose to the tumour. The drawback with this method is that neither the contribution from the cross-absorbed dose from activity in normal organs to the tumour nor the cross-absorbed dose from activity in the tumour to normal organs can be included in the calculations.

18.2.3. Dosimetry on a voxel level

The activity in an image could be quantified on a voxel level, to display the activity present in each voxel. Images that display the activity distribution at different points in time after injection may be co-registered to each other to allow for an exponential fit on a voxel by voxel basis. A parametric image that gives the time-integrated activity (the total number of decays) on a voxel level can, thus, be calculated. Parametric images that display the biological half-life for each voxel could also be produced by this technique.

The registration of the images acquired at different points in time after the administration becomes essential for the accuracy that can be achieved in the calculation of the time-integrated activity on a voxel level. Another important factor that determines the accuracy in the time-integrated activity and, thus, in the absorbed dose, is the acquired number of counts per voxel (a random error), the accuracy in the attenuation correction (systematic error) and the calibration factor that translates the number of counts to the activity (random and systematic errors). Multimodality imaging such as SPECT/CT and PET/CT facilitates the interpretation of the images as the CT will provide anatomical landmarks to support the functional images, which could change from one acquisition to the next.

A dose point kernel describes the deposited energy as a function of distance from the site of emission of the radiation. Figure 18.7 displays a dose point kernel for 1 MeV mono-energetic electrons. Convolution of a dose point kernel and the activity distribution from an image acquired at a certain time after the injection gives the absorbed dose rate. Dose point kernels provide a tool for fast calculation of the absorbed dose on a voxel level. However, the main drawback is that a dose point kernel is only valid in a homogenous medium, where it is commonly assumed that the body is uniformly unit density soft tissue.

Monte Carlo simulations that use the activity distribution from a functional image (PET or SPECT) and the density distribution from a CT image avoid the problem of non-uniform media, although full Monte Carlo simulations are time consuming. EGS (electron gamma shower), MCNP (Monte Carlo N-particle transport code), Geant and Penelope are commonly used Monte Carlo codes.

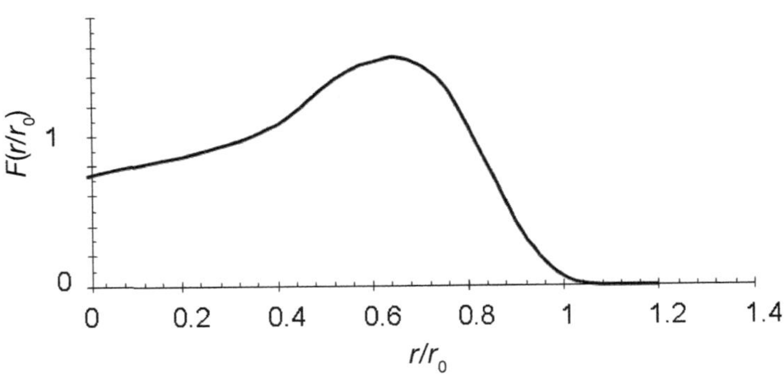

FIG. 18.7. A scaled dose point kernel for 1 MeV electrons [18.20]. r/r_0 expresses the distance scaled to the continuous slowing down approximation range of the electron and $\int_0^\infty F(r/r_0, E_0)\,\mathrm{d}(r/r_0) = 1$.

The concept of dose–volume histograms (DVHs), extensively used to describe the tumour and organ dose distribution in external beam radiotherapy, can be used to display the non-uniformity in the absorbed dose distribution from radionuclide procedures. A differential DVH shows the fraction of the volume that has received a certain absorbed dose as a function of the absorbed dose, while a cumulative DVH shows the fraction of the volume that has received an absorbed dose less than the figure given on the x axis. A truly uniform absorbed dose distribution would produce a differential DVH that shows a single sharp (δ function) peak and a step function on a cumulative DVH. Since the mean absorbed dose in internal dosimetry may be a poor representation of the absorbed dose to the tissue, as discussed above, the use of DVHs might be used to assist the correlation between absorbed dose and biological effect.

REFERENCES

[18.1] STELSON, A.T., WATSON, E.E., CLOUTIER, R.J., A history of medical internal dosimetry, Health Phys. **69** (1995) 766–782.

[18.2] LOEVINGER, R., BERMAN, R.M., A schema for absorbed-dose calculations for biologically-distributed radionuclides, MIRD Pamphlet No. 1, J. Nucl. Med. **9** Suppl. 1 (1968) 7–14.

[18.3] LOEVINGER, R., BUDINGER, T.F., WATSON, E.E., MIRD Primer for Absorbed Dose Calculations (Revised Edition), The Society of Nuclear Medicine, MIRD, Reston, VA (1991).

[18.4] BOLCH, W.E., ECKERMAN, E.F., SGOUROS, G., THOMAS, S.R., A generalized schema for radiopharmaceutical dosimetry — standardization of nomenclature, MIRD Pamphlet No. 21, J. Nucl Med. **50** (2009) 477–484.

[18.5] INTERNATIONAL COMMISSION ON RADIOLOGICAL PROTECTION, Radiation Dose to Patients from Radiopharmaceuticals, Publication 53, Pergamon Press, Oxford (1987).

[18.6] SEGARS, W.P., TSUI, B.M., FREY, E.C., JOHNSON, G.A., BERR, S.S., Development of a 4-D digital mouse phantom for molecular imaging research, Mol. Imaging Biol. **6** (2004) 149–159.

[18.7] STABIN, M.G., KONIJNENBERG, M.W., Re-evaluation of absorbed fractions for photons and electrons in spheres of various sizes, J. Nucl. Med. **41** (2000) 149–160.

[18.8] INTERNATIONAL COMMISSION ON RADIATION UNITS AND MEASUREMENTS, Fundamental Quantities and Units for Ionizing Radiation, Rep. 60, ICRU, Bethesda, MD (1998).

[18.9] INTERNATIONAL COMMISSION ON RADIATION UNITS AND MEASUREMENTS, Radiation Quantities and Units, Rep. 33, ICRU, Bethesda, MD (1983).

[18.10] INTERNATIONAL COMMISSION ON RADIATION UNITS AND MEASUREMENTS, Microdosimetry, Rep. 36, ICRU, Bethesda, MD (1983).

[18.11] ATTIX, F.H., Introduction to Radiological Physics and Radiation Dosimetry, John Wiley & Sons, New York (1986).

[18.12] HOWELL, R.W., RAO, D.V., SASTRY, K.S.R., Macroscopic dosimetry for radioimmunotherapy: Nonuniform activity distributions in solid tumours, Med. Phys. **16** (1989) 66–74.

[18.13] HOWELL, R.W., The MIRD schema: From organ to cellular dimensions, J. Nucl. Med. **35** (1994) 531–533.

[18.14] KASSIS, I.E., The MIRD approach: Remembering the limitations, J. Nucl. Med. **33** (1992) 781–782.

[18.15] INTERNATIONAL COMMISSION ON RADIOLOGICAL PROTECTION, Radiation Dose to Patients from Radiopharmaceuticals (Addendum to ICRP Publication 53), Publication 80, Pergamon Press, Oxford and New York (1998).

[18.16] INTERNATIONAL COMMISSION ON RADIOLOGICAL PROTECTION, Radiation Dose to Patients from Radiopharmaceuticals (Addendum 3 to ICRP Publication 53), Publication 106, Elsevier (2008).

[18.17] SNYDER, W.S., FORD, M.R., WARNER, G.G., WATSON, S.B., MIRD Pamphlet No. 11, S, Absorbed Dose per Unit Cumulated Activity for Selected Radionuclides and Organs, The Society of Nuclear Medicine, Reston, VA (1975).

[18.18] STABIN, M.G., SPARKS, R.B., CROWE, E., OLINDA/EXM: The second-generation personal computer software for internal dose assessment in nuclear medicine, J. Nucl. Med. **46** (2005) 1023–1027.

[18.19] STABIN, M.G., MIRDOSE: Personal computer software for internal dose assessment in nuclear medicine, J. Nucl. Med. **37** (1996) 538–546.

[18.20] BERGER, M., Improved point kernels for electron and beta-ray dosimetry, NBSIR 73-107, National Bureau of Standards (1973).

CHAPTER 19

RADIONUCLIDE THERAPY

G. FLUX[1], YONG DU[2]
Joint Department of Physics[1] and Nuclear Medicine[2],
Royal Marsden Hospital and Institute of Cancer Research,
Surrey, United Kingdom

19.1. INTRODUCTION

Cancer has been treated with radiopharmaceuticals since the 1940s. The radionuclides originally used, including ^{131}I and ^{32}P, are still in use. The role of the physicist in radionuclide therapy encompasses radiation protection, imaging and dosimetry. Radiation protection is of particular importance given the high activities of the unsealed sources that are often administered, and must take into account medical staff, comforters and carers, and, as patients are discharged while still retaining activity, members of the public. Regulations concerning acceptable levels of exposure vary from country to country. If the administered radiopharmaceutical is a γ emitter, then imaging can be performed which may be either qualitative or quantitative. While a regular system of quality control must be in place to prevent misinterpretation of image data, qualitative imaging does not usually rely on the image corrections necessary to determine the absolute levels of activity that are localized in the patient. Accurate quantitative imaging is dependent on these corrections and can permit the distribution of absorbed doses delivered to the patient to be determined with sufficient accuracy to be clinically beneficial.

Historically, the majority of radionuclide therapies have entailed the administration of activities that are either fixed, or may be based on patient weight or body surface area. This follows methods of administration necessarily adopted for chemotherapy. However, given that in vivo imaging is possible for many radiopharmaceuticals and that the mechanism of therapy is the delivery of a radiation absorbed dose, the principles of external beam radiation therapy apply equally to radionuclide therapies. These are summarized in European Directive 97/43:

> "For all medical exposure of individuals for radiotherapeutic purposes exposures of target volumes shall be individually planned; taking into account that doses of non-target volumes and tissues shall be as low as

reasonably achievable and consistent with the intended radiotherapeutic purpose of the exposure".

In this directive, the term 'radiotherapeutic' specifically includes nuclear medicine for therapy.

Radionuclide therapy is a rapidly expanding cancer treatment modality, both in terms of the number and range of procedures given, and many new radiopharmaceuticals are now entering the market. At present, there is a paucity of guidelines governing levels of activity to administer and these vary widely according to local protocols. The application of internal dosimetry to therapeutic procedures will allow the data to be collected on which to establish the evidence necessary to optimize treatment protocols.

For many therapy procedures, dosimetry studies have been conducted. These have demonstrated that a wide range of absorbed doses are delivered both to target tissues and to normal tissues from the administration of fixed activities due to variations in uptake and retention of a radiopharmaceutical. It is likely that, in conjunction with patient variations in radiosensitivity, this accounts for the variable response seen with radionuclide therapy.

Recent advances in the quantification of single photon emission computed tomography and positron emission tomography data, and increased research into patient specific rather than model based dosimetry, have led to the possibility of personalizing patient treatments according to individual biokinetics.

19.2. THYROID THERAPIES

19.2.1. Benign thyroid disease

Benign thyroid disease (typically hyperthyroidism or thyrotoxicosis) is most commonly caused by Graves' disease, an autoimmune disease affecting the whole thyroid gland and causing it to swell. Thyroid toxic nodules, consisting of abnormal thyroid tissue, are also responsible for overactive thyroid glands. Iodine-131 NaI (radioiodine) has been used since the 1940s to treat hyperthyroidism successfully.

There is a long standing wide acceptance of radioiodine as a treatment for hyperthyroidism, particularly for patients with solitary toxic adenoma, although treatment protocols vary, and there is limited evidence to compare long term results from surgery, anti-thyroid drugs or radioiodine. Guidelines are available from the European Association of Nuclear Medicine (EANM) and the American Thyroid Association, as well as from individual countries including Germany and the United Kingdom.

19.2.1.1. Treatment specific issues

Standard administrations can vary from 200 to 800 MBq, depending on the patient situation and local practice. While persistence of symptoms will result from inadequate treatment, entailing further administrations, it can be argued that patients should not receive more activity than is necessary to render them euthyroid. Therefore, in common with other radionuclide therapies, a major issue is that of personalized treatment based on patient specific dosimetry. This necessitates determination of the thyroid volume and calculation of the activity required to administer a fixed absorbed dose based on a tracer study. A range of methods have been followed to determine the thyroid volume, using, for example, ultrasound, ^{123}I-NaI or ^{124}I-NaI, and there is some discrepancy in the reported absorbed doses required to achieve euthyroidism. Further research work would certainly benefit patients.

Radiation protection advice must be given to a patient undergoing radioiodine treatment although standard treatments can usually be conducted on an out-patient basis.

19.2.2. Thyroid cancer

Thyroid cancer accounts for less than 0.5% of all cancers and there are 28 000 new cases diagnosed each year in Europe and the United States of America. Papillary and follicular thyroid cancer account for 80–90% of cases, with the remainder being anaplastic carcinomas, medullary carcinomas, lymphomas and rare tumours. Increased risk is associated with benign thyroid disease, radiation therapy to the neck and poor diet. As with benign thyroid disease, thyroid cancer has also been treated with radioiodine for over sixty years and in conjunction with total or near total thyroidectomy is widely used for an initial ablation of residual thyroid tissue. Up to 20% of cases may present with metastatic disease, usually to the lungs or bones although also to liver and brain. Treatment for distant metastases usually involves further and often higher administrations of radioiodine. This treatment is the most common application of radionuclides for therapy and is very successful, with complete response rates of 80–90%. Nevertheless, the disease can prove fatal in a higher proportion of patients that are most at risk, which include the young and the elderly.

19.2.2.1. Treatment specific issues

There are a number of controversies concerning the treatment of thyroid cancer with radioiodine. These include the extent of a low iodine diet prior to administration, levels of activity to administer for ablation or for therapy, and the time interval between ablation and the determination of success, which itself is subject to debate. The issue that most affects the physicist is that of standardized versus personalized treatments, which has been debated since the early 1960s. Fixed activities given for ablation can vary from 1100 to 4500 MBq, and those given for subsequent therapy procedures can be in excess of 9000 MBq. Published guidelines report the variation in fixed activities but do not make recommendations concerning these levels.

It has been conclusively demonstrated in a number of dosimetry studies that patients administered fixed activities of radioiodine receive absorbed doses to remnant tissue, residual disease and to normal organs that can vary by several orders of magnitude. This potentially has important consequences, as it implies that, in many cases, patients may be receiving less absorbed dose than is required for a successful ablation or therapy, while in other cases patients will receive absorbed doses to malignant and normal tissues that are excessively higher than necessary. Undertreatment will result in further administrations of radioiodine with the risk of dedifferentiation over time, so that tumours become less iodine avid. Overtreatment can result in unnecessary toxicity which can take the form of sialadenitis and pancytopenia. Radiation pneumonitis and pulmonary fibrosis have been seen in patients with diffuse lung metastases, and there is a risk of leukaemia in patients receiving high cumulative activities. Personalized treatments were first explored in the 1960s with patients administered activities required to deliver a 2 Gy absorbed dose to the blood and constraints regarding radioactive uptake levels at 48 h. Further approaches have been taken, based on whole body absorbed doses, which can be considered a surrogate for absorbed doses to the red marrow.

Dosimetry for thyroid ablations presents a different set of challenges to that performed for therapies. In the former case, the small volume of remnant tissue can render delineation inaccurate, which subsequently impinges on the accuracy of the dose calculation. Therapies of metastatic disease can involve larger volumes, although heterogeneous uptake is frequently encountered and lung metastases in particular require careful image registration and attenuation correction (Fig. 19.1).

A current issue is that of 'stunning', whereby a tracer level of activity may mitigate further uptake for an ablation or therapy. This phenomenon would have consequences for individualized treatment planning, although at present its extent and indeed its existence is being contested. However, it is not infrequent that a greater extent of uptake may be seen from a larger therapy administration than from a tracer administration (Fig. 19.1).

While subject to national regulations, patients receiving radioiodine treatment frequently require in-patient monitoring until retention of activity falls to levels acceptable to allow contact with family members and the public. It is, therefore, necessary for the physicist to give strict advice on radiation protection, taking into account the patient's home circumstances.

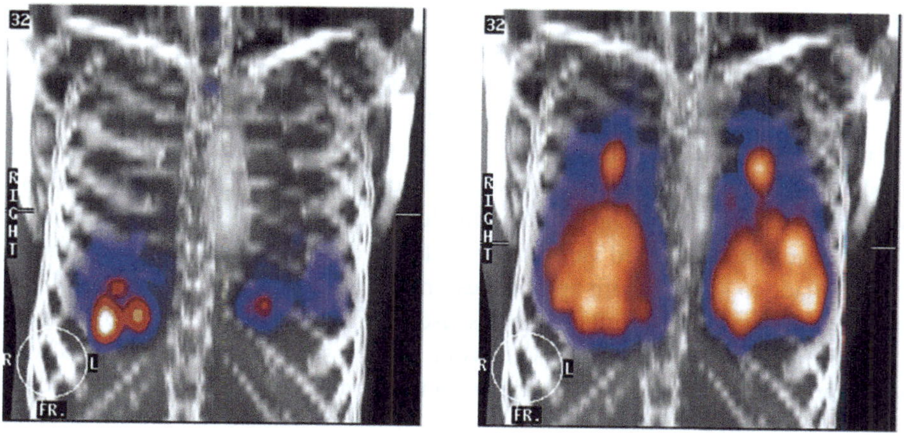

FIG. 19.1. *Absorbed dose maps resulting from a tracer administration of 118 MBq ^{131}I-NaI (left) and, subsequently, 8193 MBq ^{131}I-NaI (right) for therapy (maximum absorbed dose: 90 Gy). The absorbed doses were calculated using 3-D dosimetry on a voxel by voxel basis*

19.3. PALLIATION OF BONE PAIN

Bony metastases arise predominantly from prostate and breast cancer. Bone pain is experienced by up to 90% of patients with castration resistant prostate cancer in the later phases of their disease. Radiopharmaceuticals have been established as an effective agent for bone pain palliation for almost 70 years, with ^{89}Sr first being used in 1942. A wide range of radiopharmaceuticals have been used to treat bone metastases and there are two commercially available products, ^{89}Sr chloride (Metastron) and ^{153}Sm-lexidronam (Quadramet) that have received US Food and Drug Administration (FDA) approval (in 1993 and 1997, respectively). A number of other radiopharmaceuticals have been used, including ^{32}P, ^{186}Re-HEDP, ^{188}Re-HEDP, ^{117m}Sn and ^{177}Lu-EDTMP. More recently, the α emitter ^{223}Ra has undergone evaluation in randomized phase III clinical trials and has also received FDA approval.

In the case of ^{89}Sr and ^{153}Sm, administered activities tend to be standardized according to the manufacturer's guidelines. However, administered activities for other agents vary widely according to local protocols, and published guidelines are largely concerned with procedure. Re-treatments are generally considered to be beneficial, subject to recovery of haematological toxicity, and recommendations for the timing of these have been made by both the EANM and the IAEA. However, no trials have yet been performed to assess the optimal timing or levels of administration.

19.3.1. Treatment specific issues

The main issue concerning the use of radiopharmaceuticals for the treatment of bone pain is that of determining the ideal treatment protocol, including the optimal radionuclide to use, and whether this should be standardized or could be modified on an individual patient basis. In practice, local logistics and availability will have a strong impact on the radionuclide of choice. It is of particular interest that the radionuclides used vary widely in terms of their β emissions. Arguments can be made to support both approaches, in that the longer range β emitters may be more likely to target all of the disease, while the shorter range β emitters (and particularly an α emitter) will avoid unnecessary toxicity. There is also a wide range of physical half-lives between these radionuclides and there is some evidence to suggest that the longer lived ^{89}Sr can produce a response that takes longer to occur but that is longer lasting.

Dosimetry for bone pain palliation is challenging due to the difficulties of assessing the distribution of uptake in newly formed trabecular bone and its geometrical relation to viable red marrow and to disease. Nevertheless, models have been developed to address this interesting problem and a statistically significant correlation has been demonstrated between whole body absorbed doses and haematological toxicity. Dosimetry for other radionuclides is highly dependent on the imaging properties of these radionuclides, although it could potentially be used to increase administered activities in individual patients.

19.4. HEPATIC CANCER

Hepatocellular carcinoma is a major cause of cancer deaths. In recent years, primary and secondary liver cancers have been treated with a range of radionuclides administered intra-arterially, based on the fact that while the liver has a joint blood supply, tumours are supplied only by the hepatic artery. The advantage of this approach is that treatments can be highly selective and can minimize absorbed doses delivered to normal organs, including healthy liver. This procedure requires interventional radiology as the activity must be administered directly into the common, right or left hepatic artery via an angiographic catheter under radiological control and so is a prime example of the multidisciplinary nature of radionuclide therapy. Prior to administration, a diagnostic level of ^{99m}Tc-macroaggregate of albumin (MAA) is given to ascertain the likelihood of activity shunting to the lung. This is usually evaluated semi-quantitatively.

To date, two commercial products have received FDA approval, classified as medical devices rather than as drugs. Both use ^{90}Y. Theraspheres comprise ^{90}Y incorporated into small silica beads and SIR-Spheres consist of ^{90}Y incorporated into resin. Lipiodol, a mixture of iodized ethyl esters of the fatty acids of poppy seed oil, has also been explored for intra-arterial administration. Lipiodol has been radiolabelled with both ^{131}I and ^{188}Re, the latter having the benefit of superior imaging properties, a longer β path length and fewer concerns for radiation protection due to the shorter half-life.

19.4.1. Treatment specific issues

As with other therapies, outstanding issues include the optimal activity to administer, which is usually based on patient weight or body surface area, arteriovenous shunting observed prior to treatment and the extent of tumour involvement. However, there have been examples of treatments planned according to estimated absorbed doses delivered to the normal liver and this treatment offers the potential for individualized treatment planning based on potential toxicity. Radiobiological consequences have been considered by conversion of absorbed doses to biologically effective doses and there are tentative conclusions that multiple treatments may deliver higher absorbed doses to tumours while minimizing absorbed doses to normal liver.

A particular issue of this treatment concerning the physicist is that of imaging, due to the need to ascertain lung uptake from the pre-therapy ^{99m}Tc-MAA scan, and the possibility of bremsstrahlung imaging as a basis for calculation of absorbed doses delivered at therapy.

19.5. NEUROENDOCRINE TUMOURS

Neuroendocrine tumours (NETs) arise from cells that are of neural crest origin and usually produce hormones. There are several types of neuroendocrine cancer, including phaeochromocytoma, which originates in the chromaffin cells of the adrenal medulla, and paraganglioma, which develops in extra-adrenal ganglia, often the abdomen. Carcinoid tumours are slow growing and arise mainly in the appendix or small intestine although they can also be found in the lung, kidney and pancreas. Medullary thyroid cancer is a special case of an NET that arises from the parafollicular cells of the thyroid gland, which produce calcitonin. For the purposes of radionuclide therapy, NETs tend to be considered as one malignancy and similar radiopharmaceutical treatments are administered, although due in part to differences in radiosensitivity and proliferation, response is variable between diseases. NETs are frequently treated with radiopharmaceuticals. The mean age at diagnosis is around 60 years although tumours may present at any age.

There are two main mechanisms by which NETs are targeted with radiopharmaceuticals. NETs have been treated with the noradrenaline analogue ^{131}I-MIBG (metaiodobenzylguanidine) for over twenty years, although this is still largely considered an experimental treatment. Although generally considered to be a palliative treatment, high uptake can be achieved and complete responses have been seen. More recently, a number of peptide analogues of somatostatin have been developed, radiolabelled with ^{90}Y, ^{111}In or ^{177}Lu, that offer a range of treatment options. Guidelines for radionuclide therapy of NETs have been produced by the EANM and the European Neuroendocrine Tumour Society, and focus mainly on procedural aspects. Recommendations are not given for administered activities, and these can vary from 3700 to 30 000 MBq of ^{131}I-MIBG and cumulated activities of 12 000–18 000 MBq of ^{90}Y-DOTATOC. Administrations are often repeated, although there are no standardized protocols for the intervals between therapies.

To date, there have been almost no studies directly comparing the relative merits of the different radiopharmaceuticals available for the treatment of neuroendocrine cancer. Key considerations are largely related to the relative path lengths of the radionuclides used, their imaging properties and toxicity. For example, ^{111}In-octreotide therapy readily lends itself to imaging due to dual emission peaks at 173 and 247 keV, and relies on internalization due to radiation delivered by Auger emissions, whereas ^{90}Y labelled analogues can cause irradiation over 1 cm although they can only be imaged with bremsstrahlung scintigraphy. ^{90}Y labelled analogues and ^{131}I-MIBG can cause myelosuppression, thus the administration of higher activities may require stem cell support. Kidney toxicity can be another activity-limiting factor for the somatostatin analogues.

19.5.1. Treatment specific issues

The range of activities administered and the increasingly available range of radiopharmaceuticals developed for the treatment of NETs is indicative of the main issue facing this treatment, which is to determine the optimal treatment protocol. As with other therapies, the issue of personalized versus standardized treatments is at the forefront of this debate, with administrations based on fixed activities modified according to patient weight or, in some cases, based on absorbed whole body doses. It has been shown that a wide range of absorbed doses are delivered to both tumours and to normal organs from fixed activities. Dosimetry for high activities of ^{131}I-MIBG has been the subject of extensive research in recent years, and must deal with problems resulting from camera dead time, photon scatter and attenuation. Image based dosimetry of ^{90}Y labelled pharmaceuticals has been performed using low levels of ^{111}In given either prior to therapy or included with the therapy administration. More recently, bremsstrahlung imaging has been developed to enable dosimetry to be performed directly.

19.6. NON-HODGKIN'S LYMPHOMA

Of the malignancies arising from haematological tissues, non-Hodgkin's lymphoma is most commonly targeted with radiopharmaceuticals. Various forms of lymphoma are classified into high grade or low grade, depending on the rate of growth. Lymphomas are inherently radiosensitive, express a number of antigens and can be successfully treated with radioimmunotherapy (RIT) using monoclonal antibodies radiolabelled usually with either ^{131}I or ^{90}Y. A number of radiolabelled monoclonal antibodies have been developed and two, ^{90}Y-Ibritumomab Tiuxitan (Zevalin) and ^{131}I-Tositumomab (Bexxar), have received FDA approval. Both target the B-cell specific CD 20 antigen and have been used successfully in a number of clinical trials. Both agents have demonstrated superior therapeutic efficacy to prior chemotherapies in various clinical settings.

As with chemotherapy, RIT is more successful when administered at an early stage of disease. Clinical trials are ongoing to determine how to more effectively integrate RIT into the current clinical management algorithm in lymphoma patients.

19.6.1. Treatment specific issues

Internal dosimetry has been applied to a number of studies using RIT with varying results and conclusions. Initial dosimetry trials for Zevalin found that while absorbed doses were delivered to tumours and to critical organs that varied by at least tenfold, these did not correlate with toxicity or response, and that at the levels of activity prescribed, the treatment was deemed to be safe, obviating the need for individualized dose calculations, and treatment now tends to be administered based on patient weight. However, FDA approval incorporated the need for biodistribution studies to be performed prior to therapy using the antibody radiolabelled with ^{111}In as a surrogate for ^{90}Y under the assumption that the tracer kinetics would translate into clinical therapy and a number of studies are now concerned with assessing the biodistribution and dosimetry based on bremsstrahlung imaging (Fig. 19.2).

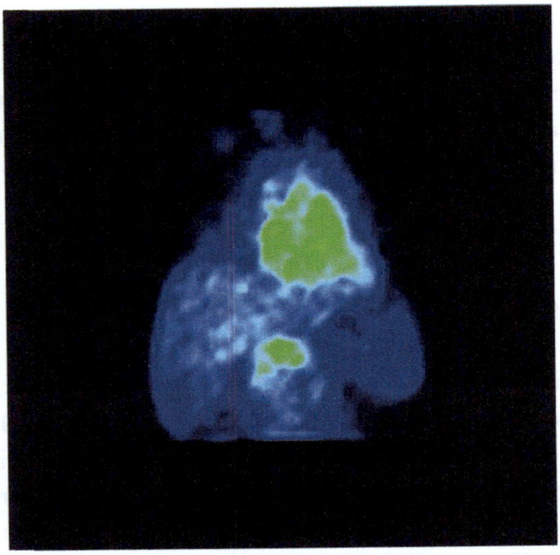

FIG. 19.2. An absorbed dose map (maximum dose: 39 Gy) resulting from 3-D dosimetry of bremsstrahlung data acquired from treatment of non-Hodgkin's lymphoma with ^{90}Y-Ibritumomab Tiuxitan (Zevalin).

In contrast, studies using ^{131}I-Tositumomab (Bexxar) have demonstrated that at least bone marrow toxicity is significantly related to dosimetry. As a result, ^{131}I-Tositumomab is one of the few radionuclide therapies (or indeed radiotherapy procedures) based on individualizing absorbed doses delivered to the critical organ, which in this case is the bone marrow. To this end, the level of administered activity is determined according to a whole body absorbed dose of 0.75 Gy, calculated from a series of three whole body scintigraphy scans.

19.7. PAEDIATRIC MALIGNANCIES

Cancer in children is rare, with an incidence of less than 130 per million and an overall relative survival rate of 57%. Leukaemia and lymphoma account for nearly 50% of cases. Radionuclide therapy for children and young people is correspondingly rare and entails particular scientific and logistical challenges that justify consideration independently of adult treatments. Issues of in-patient care predominantly arise due to the combination of increased nursing requirements and radiation protection considerations. Radiation protection must also play a large role in decisions to allow children to leave hospital, as they frequently have siblings at home.

19.7.1. Thyroid cancer

The ablation and therapy of thyroid cancer with radioiodine is performed for children, who are considered a high risk group. There is commonly a significantly higher incidence of metastatic disease in children than in adults. Fatalities can be as high as 25%, often occurring after many years of repeated radioiodine treatments and consequently high cumulated activities. Thus, potential late toxicity in children from radionuclide therapy needs to be considered.

19.7.2. Neuroblastoma

Neuroblastoma is a malignancy of the neuroendocrine system specific to children and young people. Neuroblastoma is inherently radiosensitive and has been treated with ^{131}I-MIBG since the 1980s, particularly for primary refractory or relapsed patients. Although treatments are generally intended to be palliative, complete responses have been reported. More recently, there has been interest in treating neuroblastoma with radiolabelled peptides, such as ^{177}Lu-Dotatate.

19.7.2.1. Treatment specific issues

While it is generally recognized that protocols for administering radioiodine to children should be modified from those applied to adults, there is little agreement on how such modifications should be determined and, in practice, these can be based on body weight, surface area or age. The EANM guidelines on radioiodine therapy of differentiated thyroid cancer support the principle of individual treatment of thyroid cancer in children and German procedure guidelines advocate administration based partly on the 24 h uptake of a tracer activity prior to ablation.

Despite the small number of centres that have treated children with ^{131}I-MIBG for the treatment of neuroblastoma, there has been a very wide variation in treatment protocols. While many treatments have relied on fixed activities (which have generally ranged from 3700 to 7400 MBq), substantial research and development into quantitative imaging and internal dosimetry of ^{131}I has led to a higher degree of dosimetry based personalized treatments than has been the case for adult therapies. This has led in particular to administered activities being calculated based on whole body absorbed doses which have been shown to correlate with haematological toxicity. Further complications, particularly relating to toxicity and radiation protection, can be caused by an increasing trend towards higher activities, possibly administered with stem cell support and concomitant chemotherapy which can act as a radiosensitizer.

19.8. ROLE OF THE PHYSICIST

The physicist is responsible for a wide range of tasks in nuclear medicine and must perform duties that include the procurement and maintenance of imaging equipment and associated computer systems; responsibility for radiation protection and interpretation; and implementation of national legislation. This comprehensive and challenging role is exemplified in the treatment of cancer and benign disease with radiopharmaceuticals that can entail levels of radioactivity far exceeding those used for diagnostic purposes.

Quality control of scintillation cameras is fundamental to good clinical practice in radionuclide therapy to ensure that the diagnostic information used as a basis for treatment is accurate. This relies on the development of and adherence to a strict procedure of well defined protocols and procedures that must be performed regularly.

Radiation protection pertaining to unsealed sources used for radionuclide therapy entails a greater degree of responsibility than that commonly encountered in diagnostic imaging. Staff are potentially exposed to high levels of radiation of γ, β and α emissions which, therefore, requires careful handling, dispensing and administration of therapeutic radiopharmaceuticals. Therapy procedures can involve staff groups that usually do not encounter high levels of radiation, so that extra precautions are needed. This particularly applies to the care of patients, which may be provided to some extent by family members and carers as well as by nurses, and the scanning of patients following administrations of high activities by radiographers and technicians. Careful monitoring of the involved staff must be performed at all times and the physicist must be aware of national regulations.

There is an increasing opportunity for development of a number of related areas in radionuclide therapy which predominantly involve the physicist. Foremost among these is quantitative imaging. While the clinical viewpoint of radionuclide therapy is focused on the indication and on the treatment, imaging concerns only matters that, to some extent, may be independent of these. Thus, for example, quantitative imaging of ^{131}I uptake in the abdomen will follow the same procedure whether this results from ^{131}I-MIBG treatment of an NET or an ^{131}I-radiolabelled monoclonal antibody for the treatment of lymphoma. Optimization of imaging of bony metastases with a given agent will follow similar procedures whether the metastases arise from prostate or breast cancer. Quantitative imaging is, therefore, predominantly focused on the radionuclide, independently of its formulation, and on the extent and localization of uptake as well as the imaging equipment employed for the purpose. Quantitative imaging must take into account a number of factors that are often of little concern in diagnostic imaging. Scatter is a significant impediment to accurate quantitative imaging, particularly where high energy emitters such as ^{131}I are used for therapy. Corrections can be applied with relative ease by assessing and subtracting the scatter contribution from one or more energy windows placed adjacent to the peak energy window. Attenuation correction is essential to quantitative imaging and can be performed using a variety of methods. These can range from a straightforward approach that assumes the patient consists entirely of water, to more sophisticated methods that take into account the electron density on a voxel by voxel basis. Dead time corrections are frequently overlooked in the imaging of patients undergoing radionuclide therapy as these seldom require consideration for diagnostic scanning. However, this is an issue of paramount importance that will severely inhibit accurate quantification if ignored. Again, this is a particularly significant factor when using high activities of ^{131}I, and it is essential that each camera is characterized accordingly prior to image processing and analysis.

Accurate image quantification enables further avenues of research and development that have only recently begun to emerge as substantial areas of study. Pharmacokinetic analysis, derived from sequential scanning, can allow inter- and intra-patient variations in uptake and retention to be calculated which can aid understanding and optimization of a radiopharmaceutical. This is particularly relevant for new products. Accurate analysis is dependent on the acquisition of sufficient statistics and data which must include the number and timing of scans. Inherent errors and uncertainties should be considered where possible.

Quantitative imaging and analysis facilitate the accomplishment of accurate internal dosimetry calculations, which are of paramount importance to patient specific treatment planning and which are now becoming mandatory for new products and necessary for the acquisition of the evidence base on which treatments should be performed. Although isolated studies into image based, patient specific dosimetry have been performed for many years, this remains a newly emerging field for which no standardized protocols or guidelines exist. This puts greater responsibility on the physicist whose role must be to advise on image acquisition. This invariably entails balancing scientific requirements with local resource restrictions. As there is only limited software available for dosimetry calculations at present, it may prove necessary to develop software or spreadsheets to perform absorbed dose calculations.

Essentially, the end point of such calculations should lead to a prediction of absorbed doses to the tumour and critical organs from a given administration and confirmation of the absorbed doses delivered after the therapy has been performed. However, interpretation and understanding of the biological relevance of these absorbed doses is not straightforward. Radiobiology for radionuclide therapy has not been developed at the rate seen for external beam radiotherapy (EBRT) but is now attracting more attention. There are both biological and physics aspects to radiobiology, with the latter largely concerned with constructing models to explain physiological phenomena. To some extent, these models may be adapted from those formulated for EBRT, which are predominantly based on the so-called linear quadratic model. This model is largely predicated on the assumption that cellular radiation damage can be considered separately according to single or double strand DNA (deoxyribonuleic acid) breaks and while this has a validity for radionuclide therapy, a number of confounding factors should be taken into account that can accommodate the relatively low but continuous absorbed dose rates delivered by radioactive uptake and emerging evidence to suggest that DNA is not the only target causing cell death. Radiobiology for radionuclide therapy is likely to become more complicated as radiopharmaceuticals are administered with concomitant chemotherapy or with EBRT and as new factors are discovered, such as the bystander effect, in which unirradiated cells can be killed if they are in proximity to cells that have been irradiated, and hyper-radiosensitivity, which has indicated excessive sensitivity to very low levels of irradiation.

In summary, it is likely that the varied role enjoyed by the physicist will become more complicated as radionuclide therapies are increasingly subject to accountability and as the field expands. Radionuclide therapy is the only cancer treatment modality that allows imaging of the therapeutic drug in situ. It is the duty of the physicist to capitalize on this by providing the information necessary to enable optimal and cost effective treatment.

19.9. EMERGING TECHNOLOGY

In the coming years, a number of factors will affect the development of radionuclide therapy and, in particular, the role of the physicist. New imaging technology has emerged in recent years in the form of hybrid scanners, which will have a significant impact on improving the accuracy of dosimetry from radionuclide therapies.

A range of new radiopharmaceuticals is now emerging and previously used radiopharmaceuticals are being revisited. In particular, there is currently a growing interest in α emitters, although these have been used in some form for many decades. The physical properties of α particles have distinct advantageous and disadvantageous implications for radionuclide therapy. The short range of emissions (10–100 μm in soft tissue) means that uniform uptake of a radiopharmaceutical is critical to a successful treatment as there is little radiation crossfire. However, the high linear energy transfer ensures that radiation damage resulting from uptake in a cell is likely to be lethal and that cells immediately adjacent are also likely to be killed. Examples of alpha therapy to date include ^{211}At for direct infusion into resected gliomas, antibodies radiolabelled with

^{213}Bi or ^{225}Ac for the treatment of leukaemia and ^{223}Ra for the treatment of bone metastases. Dosimetry for α emitters remains largely unexplored and is subject to a number of challenges due to the difficulty of localization and the need to take into account the emissions of daughter products, which may not remain at the initial site of uptake.

The introduction of more stringent regulatory procedures will increase the need for accurate internal dosimetry. The FDA now requires dosimetric evaluation of new radiopharmaceuticals, and it is becoming commonplace for new uses of existing agents to also be the subject of a phase I/II clinical trial to ascertain absorbed doses delivered to critical organs. As brief palliative effects translate into longer lasting survival, critical organ dosimetry will become more important to ensure minimization of unnecessary late toxicity. However, it should be noted that basic dosimetry calculations aimed at estimating critical organ absorbed doses are not necessarily sufficient to ensure an optimal treatment protocol, which must take tumour dosimetry into account.

The increasing trend towards accountability and evidence based medicine will require adherence to strict radiation protection procedures for patients, families and staff, and it may become necessary to assess exposure, particularly to family members, with greater accuracy.

Scientific developments are likely to proceed rapidly, and many are now within the reach of departments that have only basic research facilities, since accurate absorbed dose calculations can be obtained from careful imaging procedures and a relatively simple spreadsheet. Individualization of absorbed dose calculations can then be achieved. Dosimetry based treatment planning will also become an essential element of patient management as options of chemotherapy or EBRT administered concomitantly with radiopharmaceuticals are explored. The practice of internal dosimetry itself continues to evolve and can be divided into categories that require different approaches. In addition to image based dosimetry, these include whole body dosimetry, blood based dosimetry and model based dosimetry. A particular focus at present is on red marrow dosimetry, as this is the absorbed dose limiting organ for many therapies.

Multi-centre prospective data collection is critical to the development of this field, and international networks will be required to accrue a sufficient number of patient statistics to enable the formulation of agreed and standardized treatment protocols.

19.10. CONCLUSIONS

Nuclear medicine physics is playing an increasingly important role in the service and management of radionuclide therapies. In addition to the tasks traditionally associated with nuclear medicine, which primarily involve the maintenance of imaging and associated equipment, radiation protection and administration of national regulations, there is a growing requirement for patient specific treatment planning which requires quantitative imaging, internal dosimetry and radiobiological considerations.

While radionuclide therapy is usually performed within nuclear medicine, it is often to be found within clinical or medical oncology, or within endocrinology. In effect, radionuclide therapy requires a multidisciplinary approach that involves diverse groups of staff. The adoption of treatments based on individualized biokinetics, obtained from imaging and external retention measurements, places the physicist more centrally within this network, as is seen in EBRT.

There is currently the need for increased training in this field, and due to the relatively low numbers of patients treated even at specialist centres, multi-centre networks will facilitate the exchange of expertise and the gathering of prospective data necessary to advance the field. As this hitherto overlooked area of cancer management expands, the scientific opportunities available to the nuclear medicine physicist will also increase.

BIBLIOGRAPHY

BUFFA, F.M., et al., A model-based method for the prediction of whole-body absorbed dose and bone marrow toxicity for Re-186-HEDP treatment of skeletal metastases from prostate cancer, Eur. J. Nucl. Med. Mol. Imaging **30** (2003) 1114–1124.

CREMONESI, M., et al., Radioembolisation with Y-90-microspheres: dosimetric and radiobiological investigation for multi-cycle treatment, Eur. J. Nucl. Med. Mol. Imaging **35** (2008) 2088–2096.

DU, Y., HONEYCHURCH, J., JOHNSON, P., GLENNIW, M., ILLIDGE, T., Microdosimetry and intratumoral localization of administered ^{131}I labelled monoclonal antibodies are critical to successful radioimmunotherapy of lymphoma, Cancer Res. **67** (2003) 1335–1343.

GAZE, M.N., et al., Feasibility of dosimetry-based high-dose I-131-meta-iodobenzylguanidine with topotecan as a radiosensitizer in children with metastatic neuroblastoma, Cancer Biother. Radiopharm. **20** (2005) 195–199.

LASSMANN, M., LUSTER, M., HANSCHEID, H., REINERS, C., Impact of I-131 diagnostic activities on the biokinetics of thyroid remnants, J. Nucl. Med. **45** (2004) 619–625.

MADSEN, M., PONTO, J., Handbook of Nuclear Medicine, Medical Physics Publishing, Madison, WI (1992).

NINKOVIC, M.M., RAICEVIC, J.J., ADROVIC, F., Air kerma rate constants for gamma emitters used most often in practice, Radiat. Prot. Dosimetry **115** (2005) 247–250.

STOKKEL, M.P., HANDKIEWICZ JUNAK, D., LASSMANN, M., DIETLEIN, M., LUSTER, M., EANM procedure guidelines for therapy of benign thyroid disease, Eur. J. Nucl. Med. Mol. Imaging **37** (2010) 2218–2228.

CHAPTER 20

MANAGEMENT OF THERAPY PATIENTS

L.T. DAUER
Department of Medical Physics,
Memorial Sloan Kettering Cancer Center,
New York, United States of America

20.1. INTRODUCTION

The basic principles of radiation protection and their implementation as they apply to nuclear medicine are covered in general in Chapter 3. This chapter will look at the specific case of nuclear medicine used for therapy. In addition to the standards discussed in Chapter 3, specific guidance on the release of patients after radionuclide therapy can be found in the IAEA's Safety Reports Series No. 63 [20.1].

When the patient is kept in hospital following radionuclide therapy, the people at risk of exposure include hospital staff whose duties may or may not directly involve the use of radiation. This can be a significant problem. However, it is generally felt that it can be effectively managed with well trained staff and appropriate facilities. On the other hand, once the patient has been released, the groups at risk include members of the patient's family, including children, and carers; they may also include neighbours, visitors to the household, co-workers, those encountered in public places, on public transport or at public events, and finally, the general public. It is generally felt that these risks can be effectively mitigated by the radiation protection officer (RPO) with patient-specific radiation safety precaution instructions.

20.2. OCCUPATIONAL EXPOSURE

20.2.1. Protective equipment and tools

Protective clothing should be used in radionuclide therapy areas where there is a likelihood of contamination. The clothing serves both to protect the body of the wearer and to help to prevent the transfer of contamination to other areas. Protective clothing should be removed prior to going to other areas such as staff rooms. The protective clothing may include laboratory gowns, waterproof gloves, overshoes, and caps and masks for aseptic work. When β emitters are handled, the gloves should be thick enough to protect against external β radiation (perhaps double gloves should be utilized, when appropriate).

In radionuclide therapy nuclear medicine, most of the occupational exposures come from ^{131}I, which emits 356 keV photons. The attenuation by a lead apron at this energy is minimal (less than a factor of two) and is unlikely to result in significant dose reductions and may not justify the additional weight and discomfort of wearing such protective equipment. Typically, thicker permanent or mobile lead shielding may be more effectively applied for those situations which warrant its use. The RPO should determine the need and types of shielding required for each situation.

20.2.2. Individual monitoring

Individual monitoring, as discussed in Chapter 3, needs to be considered during the management of radionuclide therapy patients. In addition to general advice (see Chapter 3) on persons most likely to require individual monitoring in nuclear medicine, consideration needs to be given to nursing or other staff who spend time with therapy patients.

20.3. RELEASE OF THE PATIENT

Protection of the patient in therapeutic nuclear medicine is afforded through the application of the principles of justification and optimization — the principle of dose limitation is not applied to patient exposures. A discussion of these principles is given in Chapter 3. However, a patient that has undergone a therapeutic nuclear medicine procedure is a source of radiation that can lead to the exposure of other persons that come into the proximity of the patient. External irradiation of the persons close to the patient is related to the radionuclide used, its emissions, half-life and biokinetics, which can be important with some radionuclides. Excretion results in the possibility of contamination of the patient's environment and of inadvertent ingestion by other persons.

The system of radiation protection handles, in different ways, people that may be exposed by therapeutic nuclear medicine patients. If the person is in close proximity because their occupation requires it, then they are subject to the system of radiation protection for occupationally exposed persons. If the person, other than occupationally, is knowingly and voluntarily providing care, comfort and support to the patient, then their exposure is considered part of medical exposure, and they are subject to dose constraints as discussed in Chapter 3. If the person is simply a member of the public, including persons whose work in the nuclear medicine facility does not involve working with radiation, then their exposure is part of public exposure and that is discussed in the next section.

While precautions for the public are rarely required after diagnostic nuclear medicine procedures, some therapeutic nuclear medicine procedures, particularly those involving ^{131}I, can result in significant exposure to other people, especially those involved in the care and support of patients. Hence, members of the public caring for such patients in hospital or at home require individual consideration.

20.3.1. The decision to release the patient

Patients do not need to be hospitalized automatically after all radionuclide therapies. Relevant national dose limits must be met and the principle of optimization of protection must be applied, including the use of relevant dose constraints. The decision to hospitalize or release a patient should be determined on an individual basis. In addition to residual activity in the patient, the decision should take many other factors into account. Hospitalization will reduce exposure to the public and relatives, but will increase exposure to hospital staff. Hospitalization often involves a significant psychological burden as well as monetary and other costs that should be analysed and justified.

Medical practitioners shall determine whether the patient is willing and is physically and mentally able to comply with appropriate radiation safety precautions in the medical facility, should medical confinement be necessary, or at home after release. For some patients, hospitalization during and following treatment may be necessary and appropriate. The medical practitioners can determine that such patients may need to remain hospitalized beyond the period of time dictated by other dose constraint or clinical criteria. For example, incontinent patients or ostomy patients may require extended hospitalization to ensure safe collection and disposal of radioactively contaminated body wastes. Where the social system and infrastructure is such that there may be contamination risks from discharged patients, it may be necessary to hospitalize the patient or extend the normal hospitalization time, to avoid risk to the environment or other persons [20.1].

The decision to hospitalize or release a patient after therapy should be made on an individual basis considering several factors including residual activity in the patient, the patient's wishes, family considerations (particularly the presence of children), environmental factors, and existing guidance and regulations. The nuclear medicine physician has the responsibility to ensure that no patient who has undergone a therapeutic procedure with unsealed sources is discharged from the nuclear medicine facility until it has been established by either a medical physicist or by the facility's RPO that the activity of radioactive substances in the body is such that the doses that may be received by members of the public and family members would meet national criteria, including compliance with relevant dose limits and the application of relevant dose constraints. Iodine-131 typically results in the largest dose to medical staff, the public, caregivers and relatives. Other radionuclides used in therapy are usually simple β emitters (e.g. ^{32}P, ^{89}Sr and ^{90}Y) that pose much less risk.

The modes of exposure to other people are: external exposure, internal exposure due to contamination, and environmental pathways. The dose to adults from patients is mainly due to external exposure. Internal contamination of family members is most likely in the first seven days after treatment. In most circumstances, the risks from internal contamination of others are less significant than those from external exposure [20.1]. In general, contamination of adults is less important than external exposure. However, contamination of infants and children with saliva from a patient could result in significant doses to the child's thyroid [20.2]. Therefore, it is important to avoid contamination (particularly from saliva) of infants, young children and pregnant women owing to the sensitivity of fetal and paediatric thyroids to cancer induction [20.1, 20.3]. Written instructions to the patient concerning contact with other persons and relevant precautions for radiation protection must be provided as necessary (see Section 20.3.2).

The day to day management of hospitalization and release of patients should be the responsibility of the licensee. In applying dose constraints, registrants and licensees should have a system to measure or estimate the activity in patients prior to discharge and assess the dose likely to be received by members of the household and members of the public. The result should be recorded. A method to estimate the acceptable activity of radiopharmaceuticals for patients on discharge from hospitals is to calculate the time integral of the ambient dose equivalent rate and compare it with the constraints for patient comforters, or for other persons who may spend time close to the patient. For this calculation, either a simple conservative approach based on the physical half-life of the radionuclide or a more realistic one, based on patient-specific effective half-life, can be used. The assumptions made in these calculations with regard to time (occupancy factors) and distance should be consistent with the instructions given to patients and comforters at the time the patient is discharged from hospital. In the calculation of the effective half-life, the behaviour of ^{131}I can be modelled using two components for the biological half-life: the extra-thyroidal (i.e. existing outside the thyroid) iodine and thyroidal iodine following uptake by thyroid tissue. The assumptions used often err on the side of caution; it is sometimes felt that they significantly overestimate the potential doses to carers and the public. Examples of such calculations are found in the literature [20.4, 20.5]. Further guidance on radiation protection following radionuclide therapy can be found in Ref. [20.1] (especially in annex II).

When deciding on the appropriate discharge activity for a particular patient, the licensee should take into account the transport and the living conditions of the patient, such as the extent to which the patient can be isolated from other family members and the requirement to dispose safely of the patient's contaminated excreta. Special consideration shall be given to the case of incontinent patients. In some cases, such as for the elderly or children, it may be necessary to discuss the precautions to be taken with other family members.

Additional guidance on specific release considerations depending on various radionuclide therapies can be found in annex V of Ref. [20.1].

20.3.2. Specific instructions for releasing the radioactive patient

Current recommendations regarding release of patients after therapy with unsealed radionuclides vary widely around the world. However, the decision to release a patient is based on the assumption that the risk can be controlled when the patient returns to their home. This is generally achieved by combining an appropriate release criterion with well tailored instructions and information for the patient that will allow them to deal effectively with the potential risk [20.1].

When required, the patient or legal guardian shall be provided with written (and perhaps a verbal explanation of) instructions with a view to the restriction of doses to persons in contact with the patient as far as reasonably achievable, and information on the risks of ionizing radiation. It is important to develop effective communication methods. The IAEA gives example information/leaflet information in Safety Reports Series No. 63 [20.1]. Specific instructions should include items such as instructions to patients concerning the spread of contamination, minimization of exposure to family members, cessation of breast-feeding, and conception after therapy. The amount of time that each precaution should be implemented should be determined based on an estimate of the activity in patients prior to discharge and an assessment of the dose likely to be received by carers and comforters or members of the public under various precaution formulations as compared to the appropriate dose constraints. Procedures for advising carers and comforters should be in place in consultation with the RPO. Registrants and licensees should ensure that carers and comforters of patients during the course of treatment with radionuclides (e.g. ^{131}I for hyperthyroidism and thyroid carcinoma; ^{89}Sr, ^{186}Re for pain palliation) receive sufficient written instructions on relevant radiation protection precautions (e.g. time and proximity to the patient). Example methodologies for evaluating precaution time requirements have been published [20.5, 20.6].

Female patients should be advised that breast-feeding is contraindicated after therapeutic administration of radionuclides, and females as well as males should be advised concerning the avoidance of conception after therapeutic administrations. The IAEA's Safety Reports Series No. 40 [20.7] recommends cessation of breast-feeding for a patient given 5550 MBq (150 mCi) of ^{131}I-NaI. Following treatment with a therapeutic activity of a radionuclide, female patients should also be advised to avoid pregnancy for an appropriate period. The International Commission on Radiological Protection (ICRP) suggests that women should not become pregnant for some time after radionuclide therapy (e.g. 6 months for radioiodine, the most common radionuclide used) [20.2]. Various shorter or longer times for this and other radionuclides are given in ICRP Publication 94 [20.8] and Ref. [20.7] which identifies periods of 3, 4 and 24 months for ^{32}P, ^{131}I and ^{89}Sr treatments, respectively. Some practitioners use a 6–12 month gap for ^{131}I, with a view to providing further confidence in this regard. Table 13 of Ref. [20.1] gives additional information on precaution times for female avoidance of conception for specific radionuclide therapies [20.1].

The administration of therapeutic doses of relatively long lived radionuclides in ionic chemical forms to males is also a possible source of concern because of the appearance of larger quantities of these radionuclides in ejaculate and in sperm. It is widely recommended in practice, on the basis of prudence, that male patients take steps to avoid fathering children during the months immediately following therapy [20.1]. However, there is no strong evidence base to support this view. Some have suggested that it may be prudent to advise sexually active males who have been treated with ^{131}I (iodide), ^{32}P (phosphate) or ^{89}Sr (strontium chloride) to avoid fathering children for a period of 4 months after treatment, a period suggested as it is longer than the life of a sperm cell [20.9].

Patients travelling after radioiodine therapy rarely present a hazard to other passengers if travel times are limited to a few hours. Travel for 1–2 h immediately post-treatment in a private automobile large enough for the patient to maintain a distance of 1 m or greater from the other vehicle occupant(s) is generally permissible. A case by case analysis is necessary to determine the actual travel restrictions for each patient, especially for longer trips and for travel by public transport.

Current international security measures, such as those in place at airports and border crossing points, can include extremely sensitive radiation detectors. It is quite possible that patients treated with γ emitting radionuclides could trigger these alarms, particularly in the period immediately following discharge. Environmental or other radiation detection devices are able to detect patients who have had radioiodine therapy and some diagnostic procedures for several weeks after treatment [20.10, 20.11]. Triggering of an alarm does not mean that a patient is emitting dangerous levels of radiation — the detectors are designed to detect levels of radioactivity far below those of concern to human health. The security authorities are well aware of this possibility, and if a patient is likely to travel soon after discharge, the hospital or the patient's doctor should provide a written statement of the therapy and radionuclide used, for the patient to carry. The IAEA gives an example of a credit card-style card that might be given to a patient at the time of discharge (see Fig. 20.1) [20.1]. Personnel operating such detectors should be specifically trained to identify and deal with nuclear medicine patients. Records of the specifics of therapy with unsealed radionuclides should be maintained at the hospital and given to the patient along with written precautionary instructions [20.2].

(a)

(b)

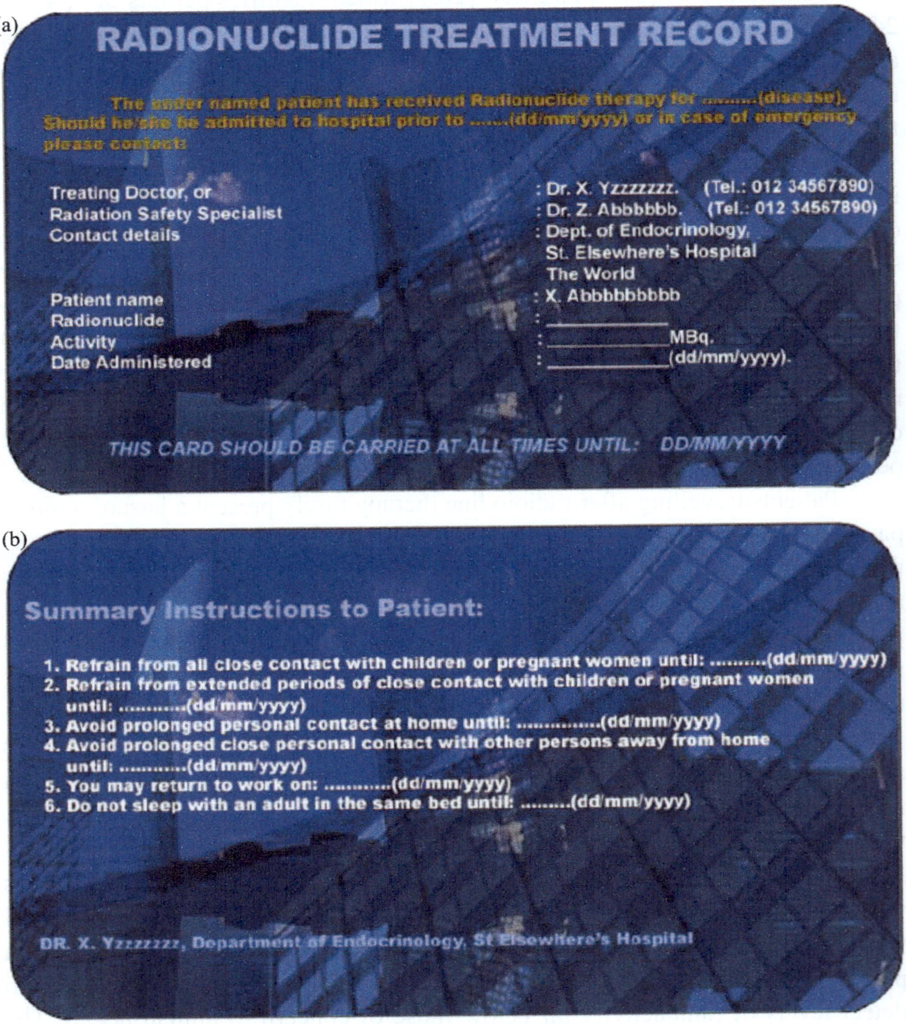

FIG. 20.1. Example of a credit card-style card that might be given to a patient at the time of discharge: (a) front side; (b) rear side [20.1].

20.4. PUBLIC EXPOSURE

The registrant or licensee is responsible for controlling public exposure resulting from a nuclear medicine practice [20.6]. The presence of members of the public in and near the nuclear medicine facility shall be considered when designing the shielding and flow of persons in the facility. Exposure to members of the general public from released patients also occurs, but this exposure is almost always very small. The unintentional exposure of members of the public in waiting rooms and on public transport is usually not high enough to require special restrictions on nuclear medicine patients, except for those being treated with radioiodine [20.3] who should receive patient-specific instructions for limiting public exposure [20.8, 20.12]. In addition, exposure of those immediately involved with the patient and the general population can occur through environmental pathways including sewerage, discharges to water, incinerated sludge or cremation of bodies. From the point of view of the individual doses involved, this is of relatively minor significance [20.1].

20.4.1. Visitors to patients

Arrangements should be made to control access of visitors (with special emphasis on controlling access of pregnant visitors or children) to patients undergoing radionuclide therapy and to provide adequate information and instruction to these persons before they enter the patient's room, so as to ensure appropriate protection. Registrants and licensees should also take measures for restricting public exposure to contamination in areas accessible to the public.

20.4.2. Radioactive waste

Registrants and licensees are responsible for ensuring that the optimization process for measures to control the discharge of radioactive substances from a source to the environment is subject to dose constraints established or approved by the regulatory body [20.13–20.15]. Chapter 3 gives specific recommendations for managing radioactive waste within the hospital facility.

For diagnostic patients, there is no need for collection of excreta and ordinary toilets can be used. For therapy patients, there are very different policies in different countries, but, in principle, the clearance criteria should follow a dilution and decay methodology. Much of the activity initially administered is eventually discharged to sewers. Storing a patient's urine after therapy appears to have minimal benefit as radionuclides released into modern sewage systems are likely to result in doses to sewer workers and the public that are well below public dose limits [20.8]. Once a patient has been released from hospital, the excreted radioactivity levels are low enough to be discharged through the toilet in their home without exceeding public dose limits. The guidelines given to patients will protect their family, carers and neighbours, provided the patient follows these guidelines.

20.5. RADIONUCLIDE THERAPY TREATMENT ROOMS AND WARDS

The following aims should be considered in the design of radionuclide therapy treatment rooms and wards: optimizing the exposure to external radiation and contamination, maintaining low radiation background levels to avoid interference with imaging equipment, meeting pharmaceutical requirements, and ensuring safety and security of sources (locking and control of access).

Typically, rooms for high activity patients should have separate toilet and washing facilities. The design of safe and comfortable accommodation for visitors is important. Floors and other surfaces should be covered with smooth, continuous and non-absorbent surfaces that can be easily cleaned and decontaminated. Secure areas should be provided with bins for the temporary storage of linen and waste contaminated with radioactive substances.

20.5.1. Shielding for control of external dose

Radiation sources used in radiopharmaceutical therapy have the potential to contribute significant doses to medical personnel and others who may spend time within or adjacent to rooms that contain radiation sources. Meaningful dose reduction and contamination control can be achieved through the use of appropriate facility and room design. Shielding should be designed using source related source constraints for staff and the public. The shielding should be designed using the principles of optimization of protection and taking into consideration the classification of the areas within it, the type of work to be done and the radionuclides (and their activity) intended to be used. It is convenient to shield the source, where possible, rather than the room or the person. Structural shielding is, in general, not necessary for most of the areas of a nuclear medicine department. However, the need for wall shielding should be assessed in the design of a therapy ward to protect other patients and staff, and in the design of rooms housing sensitive instruments (e.g. well counters and gamma cameras) to keep a low background.

Special consideration should be given to avoiding interference with work in adjoining areas, such as imaging or counting procedures, or where fogging of films stored nearby can occur. Imaging rooms are usually not controlled areas.

Placing radiopharmaceutical therapy patients in unshielded hospital rooms may expose persons in adjacent areas to levels that might cause dose constraints to be exceeded. Vacating adjacent rooms or areas or installing shielding (e.g. permanent poured concrete, solid concrete block, steel plates, lead sheets or portable shielding devices) may be necessary to ensure dose constraints are maintained in adjacent areas. Table 20.1 gives typical shielding effectiveness values for ^{131}I. Exposure rate or dose rate measurements should be taken after each radionuclide therapy administration, or worst case scenario evaluations documented to confirm that these are below levels that could cause a dose constraint to be exceeded.

TABLE 20.1. TYPICAL SHIELDING EFFECTIVENESS VALUES FOR ^{131}I

Material	Half-value layer	Tenth-value layer
Lead [20.16]	3.0 mm	11 mm
Concrete [20.17]	5.5 cm	18 cm

For permanent shielding evaluations, the design effective dose rate *P* (in millisieverts per year or millisieverts per week) in a given occupied area is derived by selecting a source related dose constraint, with the condition that the individual effective doses from all relevant sources will be well below the prescribed effective dose constraints for persons occupying the area to be shielded. Table 20.2 gives typical values for design effective dose in occupied areas adjacent to a radionuclide therapy room [20.18]. A critical review of conservative assumptions should be performed, so as to achieve a balanced decision and avoid accumulation of over-conservative measures that may go far beyond optimization.

It is preferable that patient treatment rooms be for individual patients and adjacent to each other. If this is not possible, appropriate shielding between one patient and another is required. When required, shielding should be provided for nurses and visitors of radionuclide therapy patients, for which movable shields may be used within patient rooms. When required, prior to each treatment, movable shields should be placed close to the patient's bed in such a way that exposure of the nurses caring for the patient is minimized. This is achieved by anticipating the nurse's tasks, positions and movements throughout the room.

TABLE 20.2. TYPICAL VALUES FOR DESIGN EFFECTIVE DOSE *P* IN OCCUPIED AREAS ADJACENT TO A RADIOTHERAPY TREATMENT ROOM

	Annual effective dose (mSv/a)	Weekly effective dose (mSv/week)
Occupational worker	10	0.2
Member of the public	0.5	0.01

20.5.2. Designing for control of contamination

Floors and other surfaces should be covered with smooth, continuous and non-absorbent surfaces that can be easily cleaned and decontaminated. The floors should be finished in an impermeable material which is washable and resistant to chemical change, curved to the walls, with all joints sealed and glued to the floor. The walls should be finished in a smooth and washable surface, for example, painted with washable, non-porous paint.

Control of access is required to source storage, preparation areas and rooms for hospitalized patients undergoing radionuclide therapy. A separate toilet room for the exclusive use of therapy patients is recommended. A sign requesting patients to flush the toilet well and wash their hands should be displayed to ensure adequate dilution of excreted radioactive materials and to minimize contamination. The facilities shall include a sink as a normal hygiene measure. Bathrooms designated for use by nuclear medicine patients should be finished in materials that are easily decontaminated. Hospital staff should not use patient washing facilities, as it is likely that the floors, toilet seats and sink tap handles will frequently be contaminated.

The design of safe and comfortable accommodation for visitors is important. Shielding should be designed using source related dose constraints for staff and the public. Secure areas should be provided with bins for the temporary storage of linen and waste contaminated with radioactive substances.

20.6. OPERATING PROCEDURES

General advice on operating procedures in a nuclear medicine facility is given in Chapter 3. Management of radionuclide therapy patients should be planned and performed in a way that minimizes the spread of contamination in air and on surfaces. Work with unsealed sources should be restricted to a minimum number of locations.

20.6.1. Transport of therapy doses

Specific radiation safety considerations for the radiopharmacy are addressed in Chapter 9. Radiopharmaceuticals need to be transported within the facility in shielded, spill-proof containers if warranted by the type of radionuclide and amount of activity. The shielding should be such that external doses are maintained as low as reasonably achievable (ALARA). The facility RPO should be consulted in designing or evaluating the appropriateness of shielding and transport methods.

20.6.2. Administration of therapeutic radiopharmaceuticals

Administration is normally by the oral route, intravenous injection (systemic) or instillation of colloidal suspensions into closed body cavities (intracavitary). Shielded syringes should be utilized during the intravenous administration of radiopharmaceuticals as necessary to ensure that extremity doses are maintained below occupational dose constraints. Absorbent materials or pads should be placed underneath an injection or infusion site. The facility RPO should be consulted to determine the necessity of other protective equipment (e.g. shoe covers, step-off-pads, etc.) for particular radiopharmaceutical therapies.

For oral administrations of therapeutic radiopharmaceuticals, the radioactive material should be placed in a shielded, spill-proof container. Care should be taken to minimize the chance for splashing liquid or for dropping capsules. Appropriate long-handled tools should be utilized when handling unshielded radioactive materials. For intravenous administrations by bolus injections, when dose rates warrant, the syringe should be placed within a syringe shield (plastic for β emitting radionuclides to minimize bremsstrahlung, high Z materials for photon-emitting radionuclides) with a transparent window to allow for visualization of the material in the syringe. For intravenous administrations by slower drip or infusions, the activity container should be placed within a suitable shield. For high energy photons, a significant thickness of lead or other high Z material may need to be evaluated. In addition, consideration should be given for shielding pumps and lines.

Procedures for administering a therapeutic radiopharmaceutical shall include considerations to ensure as complete a delivery as possible of the prescribed therapeutic activity. Any retention of material in syringes, tubing, filters or other equipment utilized for administration should be analysed. Where appropriate, equipment should be flushed or rinsed with isotonic saline (or another physiological buffer) for parenteral administration or water for oral administrations. All materials utilized in administrations shall be considered as medical and radioactive waste, and should be labelled with the radionuclide, a radiation precaution sticker, and stored and or disposed of in a manner consistent with local regulations.

20.6.3. Error prevention

Care should be exercised in avoiding administration of a therapeutic radiopharmaceutical to the wrong patient. In addition, prior to administration, the following should be verified:

— The dose on the radiopharmaceutical label matches the prescription;
— Identification of the patient by two independent means;
— Identity of the radionuclide;
— Identity of the radiopharmaceutical;
— Total activity;
— Date and time of administration;
— Patients have been given information about their own safety.

The therapeutic radiopharmaceutical, activity, the date and time of administration, and verification of the initial and residual assay should be entered in some form in the patient's medical record.

Pregnancy is a strong contraindication to unsealed radionuclide therapy, unless the therapy is life-saving. This advice is all the more valid for radioiodine therapy and for other radionuclides with the potential to impart radiation doses to the fetus in the range of a few millisieverts. Therefore, where treatment is likely or anticipated, the patient should be advised to take appropriate contraceptive measures in the time prior to therapy [20.1]. Some radiopharmaceuticals, including ^{131}I as iodide and ^{32}P as phosphate, rapidly cross the placenta, so that the possibility of pregnancy should be carefully excluded before administration. Before any procedure using ionizing radiation, it is important to determine whether a female patient is pregnant. The feasibility and performance of medical exposures during pregnancy require specific consideration owing to the radiation sensitivity of the developing embryo/fetus [20.3]. Some procedures and some radiopharmaceuticals (e.g. radioiodides) can pose increased risks to the embryo/fetus. The ICRP has given detailed guidance in Publications 84 [20.19] and 105 [20.2]. Radiation risks after prenatal radiation exposure are discussed in detail in ICRP Publication 90 [20.20].

20.6.4. Exposure rates and postings

Values of ambient dose equivalent from the patient should be determined. This information will assist in deriving appropriate arrangements for entry by visitors and staff. Following the administration of the therapeutic radiopharmaceutical to the patient, anterior exposure rates at the surface of and 1 m from the patient should be measured at the level of the patient's umbilicus (or other location as appropriate for the type of nuclear medicine administered), using a calibrated radiation monitor (e.g. a portable ionization chamber). Typically, these initial measurements are to be taken within 1 h of administration of the radiopharmaceutical therapy.

Rooms with radiotherapy patients should be controlled areas. A sign such as that recommended by the International Organization for Standardization (ISO) [20.21] should be posted on doors to the patient's room and radioactive material storage areas as an indicator of radiation (see Fig. 20.2).

FIG. 20.2. (a) International Organization for Standardization (ISO) radiation symbol; (b) New IAEA/ISO radiation warning symbol.

It should be noted, however, that the ISO radiation symbol is not intended to be a warning signal of danger but only of the existence of radioactive material. A new symbol has been launched by the IAEA and the ISO to help reduce needless deaths and serious injuries from accidental exposure to large radioactive sources [20.22]. It will serve as a supplementary warning to the trefoil, which has no intuitive meaning and little recognition beyond those educated in its significance. The new symbol is intended for IAEA category 1, 2 and 3 sources [20.23] defined as dangerous sources capable of death or serious injury, including food irradiators, teletherapy machines for cancer treatment and industrial radiography units. The symbol is to be placed on the device housing the source, as a warning not to dismantle the device or to get any closer. It will not be visible under normal use, only if someone attempts to disassemble the device. For radionuclide therapy applications, the new symbol will not be located on building access doors, transport packages or containers. Rather, the

ISO radiation symbol should be utilized to notify individuals of the existence of radioactive material.

Facilities may also consider placing a 'radioactive precautions' wristband on the patient's wrist if the patient is to remain in medical confinement. In addition, for those patients remaining in medical confinement, the patient should be resurveyed each day at the point of maximal uptake of the radiopharmaceutical. The exposure rate or dose rate measured can then be used in determining the activity remaining in the patient as well as developing appropriate release instructions for the patient (see Section 20.3).

20.6.5. Patient care in the treating facility

Medical practitioners should exercise their clinical duties consistent with patient safety and good quality medical care. Unless otherwise specified by the facility RPO, nurses, physicians and other health care personnel are to perform all routine duties, including those requiring direct patient contact, in a normal manner. However, medical practitioners should avoid lingering near the patient unnecessarily and should spend as little time as necessary in close proximity to radioactive materials or patients treated with radiopharmaceuticals and remain at distances appropriate for the exposure rate or dose rate measurements from such materials and patients. When necessary, portable shielding should be used to reduce radiation levels to medical practitioners.

Ward nurses should be informed when a patient may pose a radioactive hazard, and advice and training should be provided. The training should include radiation protection and specific local rules, in particular, for situations where there is a risk of significant contamination from, for example, urine, faeces or vomiting. Appropriate training should also be given to night staff. In the case of high activity patients, only essential nursing should be carried out. Other nursing should be postponed for as long as possible after administration, to take full advantage of the reduction of activity by decay and excretion. In addition, there should be minimum handling of contaminated bed linen, clothing, towels, crockery, etc. during the initial period and the instructions on how long these precautions should be maintained should be documented.

The nursing staff should be familiar with the implications of the procedure, the time and date of administration, and any relevant instructions to visitors. Values of ambient dose equivalent at suitable distances should be determined. This information will assist in deriving appropriate arrangements for entry by visitors and staff. These arrangements should be made in writing in the local rules.

20.6.6. Contamination control procedures

Work procedures should be formulated so as to minimize exposure from external radiation and contamination, to prevent spillage from occurring and, in the event of spillage, to minimize the spread of contamination. All manipulation for dispensing radioactive materials should be carried out over a drip tray, in order to minimize the spread of contamination due to breakages or spills.

Persons working with unsealed sources or nursing high activity patients should wash their hands before leaving the work area. Patients treated with high activity should use designated toilets. Simple precautions such as laying plastic backed absorbent paper on the floor around the toilet bowl and instructions to flush the toilet after each use will help to minimize exposure to external radiation and contamination.

Particular attention and measures to limit spread of contamination are required in the case of incontinent patients and, in cases of oral administration, if there are reasons for believing that the patient may vomit. Contaminated bedding and clothing should be changed promptly and retained for monitoring. Crockery and cutlery may become contaminated. Local rules should specify washing up and segregation procedures, except for disposable crockery and cutlery.

Where possible, a radionuclide therapy patient that requires medical confinement should be placed in a private hospital room with a private toilet and sink. The use of disposable plastic-backed absorbent pads or plastic sheeting taped in place in the areas most likely to be contaminated, such as the floor around the toilet and sink, may be appropriate for a facility. In all cases, consideration of the ALARA principle should be maintained. Removal of loose contaminated items from the patient's room should be done on a daily basis.

In the event of a large volume spill of blood, urine or vomitus, medical practitioners or staff should cover the spill with an absorbent material and immediately contact the facility radiation safety service for appropriate cleanup assistance and specific instructions. After such a spillage, the following actions should be taken:

(a) The RPO should immediately be informed and directly supervise the cleanup;
(b) Absorbent pads should be thrown over the spill to prevent further spread of contamination;
(c) All people not involved in the spill should leave the area immediately;
(d) All people involved in the spill should be monitored for contamination when leaving the room;
(e) If clothing is contaminated, it should be removed and placed in a plastic bag labelled 'radioactive';

(f) If contamination of skin occurs, the area should be washed immediately;
(g) If contamination of an eye occurs, it should be flushed with large quantities of water.

Upon discharge and release of the patient, all remaining waste and contaminated items should be removed and segregated into bags for disposable items and launderable items. All radioactively contaminated waste items should be labelled with the radionuclide and a radiation precaution sticker, and be stored and or disposed of in a manner consistent with local regulations. The patient's room should be surveyed and checked for removable contamination utilizing appropriate survey equipment (e.g. a Geiger–Müller counter or scintillation survey meter). Where necessary, wipe tests should be performed. Facility procedures should address applicable criteria for removable radioactive contamination. Contamination monitoring is required for:

— All working surfaces (including the interior of enclosures), tools, equipment, the floor and any items removed from this area. Monitoring is also required during the maintenance of contained workstations, ventilation systems and drains.
— Protective and personal clothing, and shoes, particularly when leaving an area that is controlled due to the risk of contamination (monitors should be available near the exit).
— Clothing and bedding of therapy patients.

20.7. CHANGES IN MEDICAL STATUS

If the medical condition of a patient deteriorates such that intensive nursing care becomes necessary, urgent medical care is a priority and should not be delayed. However, the advice of the RPO should be sought immediately. In the event of a deterioration in the patient's medical condition, frequent or continual monitoring of the patient may be necessary (e.g. septic shock, pulmonary oedema, stroke or myocardial infarction). In some cases, the patient may need to be transferred to intensive, special care or cardiac care units. It is possible that patients in these units are in close proximity to each other with little or no shielding available. As such, radionuclide therapy patients may present a radiation hazard to other patients or medical practitioners. The nuclear medicine physician and the RPO shall be notified of the transfer to a special unit as soon as possible or prior to the transfer. The RPO shall determine whether portable shielding is necessary to reduce doses to other patients or medical practitioners, whether specific personnel monitoring is necessary, and whether specific radiation precautions are necessary to keep radiation exposures ALARA.

20.7.1. Emergency medical procedures

Life-saving efforts shall take precedence over consideration of radiation exposures received by medical personnel. This is particularly important for therapy patients containing large amounts of radioactivity. Medical personnel should, therefore, proceed with emergency care (e.g. when a patient has suffered a stroke), while taking precautions against the spread of contamination and minimizing external exposure. The staff should avoid direct contact with the patient's mouth, and all members of the emergency team should wear protective gloves. Medical staff should be informed and trained on how to deal with radioactive patients. Rehearsals of the procedures should be held periodically.

The only exceptional, life-saving situations are those medical emergencies involving immediate care of patients in the case of strokes or similar situations, when large amounts of radioactive material have been incorporated (of the order of 2 GBq of ^{131}I) [20.7].

20.7.2. The radioactive patient in the operating theatre

Radiation protection considerations should not prevent or delay life-saving operations in the event that surgery on a patient is required. The following precautions should be observed:

— The operating room staff should be notified;
— Operating procedures should be modified under the supervision of the RPO to minimize exposure and the spread of contamination;
— Protective equipment may be used as long as efficiency and speed are not affected;
— Rotation of personnel may be necessary if the surgical procedure is lengthy;
— The RPO should monitor all individuals involved;
— Doses to members of staff should be measured as required.

The RPO should consider whether personnel monitoring is required. The number of persons in the operating theatre should be minimized, and operating personnel should only remain in the operating room for the minimum amount of time consistent with surgical objectives. If it is estimated that the circulating blood or the area of the body to be treated surgically contains a significant quantity of the radiopharmaceutical, the RPO and the surgeon should discuss the procedures to be performed to keep radiation exposure to surgical personnel

ALARA. The spread of radioactive contamination can be minimized through the use of typical primary precautions used in the operating theatre. Radioactive material can be kept off of surgeons through the use of gloves (the use of double gloves may be appropriate). If an injury to surgical staff such as a cut or puncture occurs, radioactive contamination of the skin or wound may occur. The RPO should be consulted to evaluate contamination and any possible radiation hazard, including the possibility of internal intakes. Any specimens sent for pathological examination should be monitored for contamination. Tools and other equipment from the surgery should be monitored for radioactive contamination, decontaminated as necessary, and stored for radioactive decay or treated as radioactive waste in accordance with local regulations.

20.7.3. Radioactive patients on dialysis

The care of patients receiving radiopharmaceutical therapy and who are on dialysis may require additional consideration. In general, for systemic treatments, these patients will not biologically clear radioactive materials as quickly as typical patients since the clearance is highly dependent on the schedule of the dialysis session. It may be necessary to reduce or otherwise adjust the activity required for a therapy. The decision as to the activity required for such patients should be based on either a trace trial administration of activity and the observed elimination rate, or a careful review of the available literature for similar patient administrations. Typically, the largest amount of radioactivity will be eliminated during the first dialysis session following radiopharmaceutical therapy.

The RPO should assess the radiation exposures likely to be received by medical practitioners during the sessions. In such cases, no significant contamination of dialysis machines has been reported [20.1]. The materials, tubing, filters and waste containers used during the sessions should be checked by radiation safety staff to evaluate whether these need to be considered low level radioactive waste and managed in accordance with facility and local regulations (see Section 20.4.2). There may be slight contamination of disposable items such as liners and waste bags, which may require storage for some time, in the case of ^{131}I. In most cases, however, no special precautions will be required and the dialysis and radiation safety staff will advise patients on how to deal with disposables.

20.7.4. Re-admission of patients to the treating institution

If a patient who still contains a therapeutic amount of radioactive material is re-admitted to the treating institution, the RPO shall be notified as soon as possible after re-admission. Patient medical charts should include information on dates of cessation of radiation precautions (perhaps in electronic chart systems which could provide useful triggers for the needed precautions in the event of re-admission). The RPO shall monitor the patient and specify any required precautions to be followed by medical practitioners. Where required, radiation precaution tags should be placed on the patient, the patient's room and chart.

20.7.5. Transfer to another health care facility

Some patients may need to be transferred to another health care facility (i.e. another hospital, skilled nursing facility, nursing home or hospice, etc.) following therapy treatments. In such a case, care must be taken that, in addition to practical measures and advice to ensure safety of other staff, any legal requirements relevant to the second institution are also complied with [20.1]. Patients transferred to another health care facility should meet the criteria for unrestricted clearance. However, the possibility for the generation of low level radioactive waste should be examined by the RPO of the treating facility and any issues should be discussed with the facility accepting the patient transfer. In the rare event that a patient being transferred to another health care facility does not meet the criteria for unrestricted clearance, the RPO shall ensure that the facility accepting the patient transfer has an appropriate registration or licence that would allow acceptance of the patient with therapeutic amounts of radioactive materials on board. The RPO should provide radiation safety information and precautions, if any, for the patient and for the receiving health care facility.

20.8. DEATH OF THE PATIENT

Therapeutic amounts of radioactive materials are typically not administered to critically ill patients unless there are circumstances where the palliative use of radioactive materials in terminal patients will significantly improve the quality of life of the patient. However, should the patient die in the period immediately following therapy, special consideration may need to be given to the treatment of the corpse. To facilitate this, the patient should be given a small card with details of their treatment and contact details for a radiation protection specialist/medical physicist associated with the department responsible for the therapy (see Fig. 20.1). Additional specific guidance for the death of a radionuclide therapy patient has been developed by the IAEA [20.1].

Areas of concern arise with respect to embalming, burial or cremation of the corpse and the conduct of autopsy examinations. National regulations, some quite dated, are available for some or all of these in many countries, but there is a lack of international recommendations. Practice tends to be guided by an untidy mixture of custom, professional guidance and national regulation [20.1]. Recent reports have emphasized the need to be sensitive to the wishes of the deceased and their family when decisions about the disposal of the corpse are being made. This may be particularly important if the possibility of retaining some organs for radiation protection reasons is being considered [20.1].

The authorities in many countries now place limits on the radioactivity that may be present in the corpse before autopsy, embalming, burial or cremation. No special precautions are required for direct burial or cremation, without embalming, provided the activity involved is not in excess of national limits. No special precautions are required for embalming if activities do not exceed the levels mentioned in table 14 of Ref. [20.1] for autopsy. If the activities are greater, then a corpse should not normally be embalmed, but if embalming is required an RPO should be consulted.

20.8.1. Death of the patient following radionuclide therapy

In cases where the death occurs in a hospital, access to the room occupied by the deceased should be controlled until the room has been decontaminated and surveyed. Radioactive bodies should be identified as potential hazards by a specified form of identifier. Identification of the possibility that a body may contain radioactive substances relies on information provided in the patient records, the information card (Fig. 20.1) or information gleaned from relatives or others. A body bag may need to be used to contain leakage of radioactive substances. To minimize external radiation risk, the corpse may need to be retained in a controlled area.

In the event that a patient dies within the treating health care facility while still containing a therapeutic quantity of radioactive material, the treating medical practitioner and the RPO shall be notified immediately. Depending on the number of days that have elapsed between radiopharmaceutical treatment and death, the radiation hazard may have been reduced considerably, and precautions minimized. In the rare event that large quantities of radiopharmaceuticals are still within the body, the RPO shall identify specific radiation precautions as necessary, depending on the type of radionuclide and measured exposure rates or dose rates. Nursing staff should be provided with instructions informing them that the normal procedure of pressing down on the abdomen of a corpse must not be performed due to the radiation and/or contamination levels that may result [20.1]. The RPO shall notify the morgue prior to the arrival of the body, and the RPO should discuss radiation safety precautions with morgue personnel, as required.

In most cases, if the patient has already been released from the treating facility, no special precautions are generally necessary for handling the body.

20.8.2. Organ donation

If organ donation is being considered, the RPO shall determine necessary precautions for operating theatre personnel who will harvest the organ(s). Unless the organ is directly involved in the treatment regime, it is unlikely that the donated organ will contain an amount of radioactive material to cause significant damage to the organ or deliver a radiation dose to the recipient sufficient to nullify the donation. However, the nuclear medicine physician and RPO should be prepared to estimate such quantities and doses.

20.8.3. Precautions during autopsy

The dose constraints applying to pathology staff responsible for the conduct of autopsy examinations will be either those for the general public or those for radiation workers, depending on the training and classification of the staff concerned. However, it is almost inevitable that some members of the pathology staff will be classified as members of the public from a radiation protection point of view. These constraints and the radiation safety procedures to be applied in practice should be determined in close consultation with the RPO from the department in which the therapy was administered. Where the possibility that the corpse may be radioactive arises, a proposed autopsy should be suspended until the situation is clarified to the greatest extent possible and a risk assessment has been undertaken by the RPO. This should establish the type, nature and location of the radioactive material used and when the therapy occurred. Any reporting or notifications required by law and/or good practice should also be undertaken. Additional points to consider for autopsies of radioactive bodies are noted in annex IV of Ref. [20.1].

Although it is rare that the body of a patient will be sent for autopsy shortly after administration of a therapeutic radiopharmaceutical, if death occurs within 24–48 h post-administration, a considerable amount of activity may be present in blood and urine. In these cases, the RPO or radiation safety staff should supervise the autopsy. Any residual activity in tissue samples should be evaluated prior to releasing the samples to the pathology laboratory. If death occurred more than 48 h post-administration, there will typically be little, if any, activity in the blood or urine. In these cases, activity may only be present in residual treated areas or metastatic disease sites. The staff dose may be reduced by deferring the autopsy where necessary and practical. Finally, Singleton et al. [20.24] conclude that:

> "provided that appropriate precautions are implemented, determined through consultation with a qualified expert in radiation protection and by

completion of risk assessment, the radioactive autopsy can be undertaken safely and in compliance with relevant legislative requirements."

Unsealed radioactive substances may be present in a particular body cavity or organ, or they may have concentrated after systemic administration (e.g. ^{131}I in the thyroid gland). Drainage of the cavity or excision of the organ will reduce exposure if undertaken at the start of the autopsy. In addition, care should be given with respect to organs with significant activity. In cases where the patient had received a dose of β emitting colloid or spheres (e.g. ^{32}P chromic phosphate into a body cavity or ^{90}Y microspheres into the liver), significant activity may be present in the cavity fluid or in the embolized organ. Beta radiation sources may provide a significant dose to the hands because they will be in close contact with body tissues and fluids [20.5]. Autopsy and pathology staff should wear standard protective clothing (i.e. gloves, lab coats, eye protection, etc.) and personnel monitoring should be considered. For β emitters, double surgical gloves may be helpful in reducing skin exposures. An intake of airborne material inadvertently released during cutting or movement of radioactive tissue or organs can be prevented by wearing eye protection and a face mask.

20.8.4. Preparation for burial and visitation

Funeral directors will need to be advised of any necessary precautions, and notification of the relevant national competent authorities may be required. It is essential that funeral directors and ministers of religion do not overreact to the risks associated with the radioactive corpse. Careful communication is needed to ensure that adequate controls are implemented without compromising dignity. Situations where the wishes of the next of kin have to be significantly disrupted should be rare [20.1].

In cases where the body will be prepared for burial without autopsy or embalming, if the RPO believes that the potential dose likely to be received by the personnel preparing the body will not exceed the appropriate dose constraint, burial can proceed. In rare cases where dose constraints may be exceeded, the RPO should provide radiation precaution information (e.g. restricting the time spent near the body). In cases where the body will be prepared for burial by embalming, the RPO should notify the morgue or funeral home that the body contains therapeutic quantities of radioactive material and should provide them with precautions to minimize radiation exposure and radioactive contamination. Embalming is conducted by injecting an embalming fluid into the body and flushing body fluids into the drain. Embalming staff should wear standard protective clothing (i.e. gloves, lab coats, eye protection, etc.) and personnel monitoring should be considered. Careful cleaning of equipment in the usual manner will remove radioactive contamination [20.5].

In most cases, no precautions will be necessary during visitation. If the possibility exists that there are measurable dose rates at 30 cm from the body, the family and funeral home should be given appropriate precautions, as necessary to provide assurance that dose constraints are met.

20.8.5. Cremation

A proportion of the activity retained will appear in cremated remains and may be sufficient, particularly in the case of long lived radionuclides, to require controls to be specified. The main concern is in respect to the scattering of ashes, although contact dose rates with the container may have to be considered if cremation takes place shortly after administration [20.1].

Assuming that the body has been prepared in accordance with the recommendations in the preceding sections, no additional handling precautions are necessary in transporting the body to the crematorium. The crematorium personnel should be informed by the treating facility or family that the body might contain radioactive material. Crematorium personnel may contact the treating facility if they need additional guidance on handling the body. Such guidance should include methods to minimize radiation exposure, contamination of the retort and especially methods to minimize radioactive ash particles. Crematorium employees may receive external exposure from the radioactive body or from contamination of the crematorium or internal exposure from inhalation of radioactive particles while handling the ashes [20.25]. Bodies that contain γ emitting radionuclides may result in some external exposure to employees of the crematorium. No precautions are necessary as long as there is minimal time required to handle the body at the crematorium (a likely assumption). Cremation of non-volatile radionuclides might result in contamination of the retort. As the most significant hazard from this contamination is inhalation of ash particles during cleaning of the retort, it is appropriate for workers who clean the retort to wear dust masks and protective garments.

The most likely hazard to the general population in the vicinity of the crematorium is the inhalation of radioactive material emitted with the stack gases. Each crematorium should maintain records of the type and activity in bodies cremated, when known.

The potential for effective doses from cremation of bodies containing ^{131}I should be evaluated and some have suggested that if a crematorium were to handle bodies that contain ^{131}I and do not exceed 100 GBq in a single year, the effective dose to individuals in the surrounding population would not likely exceed 0.1 mSv [20.5]. It, therefore, appears that no specific radiation hazard would exist even if a crematorium were to handle several bodies per year containing ^{131}I.

REFERENCES

[20.1] INTERNATIONAL ATOMIC ENERGY AGENCY, Release of Patients After Radionuclide Therapy, Safety Reports Series No. 63, IAEA, Vienna (2009).

[20.2] INTERNATIONAL COMMISSION ON RADIOLOGICAL PROTECTION, Radiological Protection in Medicine, Publication 105, Elsevier, Oxford (2008).

[20.3] INTERNATIONAL COMMISSION ON RADIOLOGICAL PROTECTION, Recommendations of the ICRP, Publication 103, ICRP, Elsevier, Oxford (2008).

[20.4] NUCLEAR REGULATORY COMMISSION, Consolidated Guidance about Materials Licensees, Rep. NUREG-1556, Vol. 9, Office of Standards Development, Washington, DC (1998).

[20.5] NATIONAL COUNCIL ON RADIATION PROTECTION AND MEASUREMENT, Management of Radionuclide Therapy Patients, Rep. No. 155, Bethesda, MD (2006).

[20.6] ZANZONICO, P.B., SIEGEL, J.A., ST. GERMAIN, J., A generalized algorithm for determining the time of release and the duration of post-release radiation precautions following radionuclide therapy, Health Phys. **78** (2000) 648–659.

[20.7] INTERNATIONAL ATOMIC ENERGY AGENCY, Applying Radiation Safety Standards in Nuclear Medicine, Safety Reports Series No. 40, IAEA, Vienna (2005).

[20.8] INTERNATIONAL COMMISSION ON RADIOLOGICAL PROTECTION, Release of Patients After Therapy with Unsealed Sources, Publication 94, Pergamon Press, Oxford (2004).

[20.9] STRAUSS, J., BARBIERI, R.L. (Eds), Yen and Jaffe's Reproductive Endocrinology, 6th edn, Saunders, Elsevier, Philadelphia, PA (2009).

[20.10] DAUER, L.T., WILLIAMSON, M.J., ST. GERMAIN, J., STRAUSS, H.W., Tl-201 stress tests and homeland security, J. Nucl. Cardiol. **14** (2007) 582–588.

[20.11] DAUER, L.T., STRAUSS, H.W., ST. GERMAIN, J., Responding to nuclear granny, J. Nucl. Cardiol. **14** (2007) 904–905.

[20.12] INTERNATIONAL COMMISSION ON RADIOLOGICAL PROTECTION, Radiological Protection in Medicine, Publication 73, Pergamon Press, Oxford (1996).

[20.13] INTERNATIONAL ATOMIC ENERGY AGENCY, Applications of the Concepts of Exclusion, Exemption and Clearance, IAEA Safety Standards Series No. RS-G-1.7, IAEA, Vienna (2004).

[20.14] INTERNATIONAL ATOMIC ENERGY AGENCY, Regulatory Control of Radioactive Discharges to the Environment, IAEA Safety Standards Series No. WS-G-2.3, IAEA, Vienna (2000).

[20.15] INTERNATIONAL ATOMIC ENERGY AGENCY, Management of Waste from the Use of Radioactive Material in Medicine, Industry, Agriculture, Research and Education, IAEA Safety Standards Series No. WS-G-2.7, IAEA, Vienna (2005).

[20.16] DELACROIX, D., GUERRE, J.P., LEBLANC, P., HICKMAN, C., Radionuclide and Radiation Protection Data Handbook, Oxford University Press, Oxford (2002).

[20.17] SCHLEIEN, B., BIRKY, B., SLABACK, L., Handbook of Health Physics and Radiological Health, 3rd edn, Williams and Wilkins, Baltimore, MD (1998).

[20.18] PODGORSAK, E.B. (Ed.), Radiation Oncology Physics: A Handbook for Teachers and Students, IAEA, Vienna (2005).

[20.19] INTERNATIONAL COMMISSION ON RADIOLOGICAL PROTECTION, Pregnancy and Medical Radiation, Publication 84, Pergamon Press, Oxford (2000).

[20.20] INTERNATIONAL COMMISSION ON RADIOLOGICAL PROTECTION, Biological Effects after Prenatal Irradiation (Embryo and Fetus), Publication 90, Pergamon Press, Oxford (2003).

[20.21] INTERNATIONAL ORGANIZATION FOR STANDARDIZATION, Basic Ionizing Radiation Symbol, ISO 361, Geneva (1975).

[20.22] INTERNATIONAL ORGANIZATION FOR STANDARDIZATION, Ionizing-radiation Warning — Supplementary Symbol, ISO 21482, Geneva (2007).

[20.23] INTERNATIONAL ATOMIC ENERGY AGENCY, Categorization of Radioactive Sources, IAEA Safety Standards Series No. RS-G-1.9, IAEA, Vienna (2005).

[20.24] SINGLETON, M., START, R.D., TINDALE, W., RICHARDSON, C., CONWAY, M., The radioactive autopsy: safe working practices, Histopathology **51** (2007) 289–304.

[20.25] WALLACE, A.B., BUSH, V., Management and autopsy of a radioactive cadaver, Australas. Phys. Eng. Sci. Med. **14** (1991) 119–124.

Appendix I

ARTEFACTS AND TROUBLESHOOTING

E. BUSEMANN SOKOLE
Department of Nuclear Medicine,
Academic Medical Center,
Amsterdam, Netherlands

N.J. FORWOOD
Department of Nuclear Medicine,
Royal North Shore Hospital,
Sydney, Australia

I.1. THE ART OF TROUBLESHOOTING

I.1.1. Basics

Troubleshooting refers to the process of recognizing and identifying the cause of an artefact, a malfunction or a problem in an instrument. The problem could be immediately obvious, for example, the instrument does not work at all or a particular component stops working (such as the computer, the mechanism for whole body scanning or the automatic mechanism for collimator exchange). The malfunction could also be less obvious, and be recognized only by an abnormality in the expected result (such as the pattern formed by a defective photomultiplier tube (PMT) in the gamma camera clinical or quality control (QC) image or an unexpected calibration result in a radionuclide dose calibrator). Such an abnormality is generally referred to as an artefact, in particular, when observed in images.

The malfunctioning of an instrument can occur at any time. It might become evident from a routine QC test. However, it is especially stressful when it occurs during a patient investigation. In such a situation, the first lines of action are to minimize the distress to the patient that a problem has occurred, to remain calm and clear headed, to immediately try to identify the problem and correct it, if possible, and to decide whether the investigation can be continued, either on the same instrument or another similar one, or whether the investigation must be rescheduled. An action flow chart is useful in the decision making process. Such a flow chart is shown in Fig. I.1 for actions following a QC test.

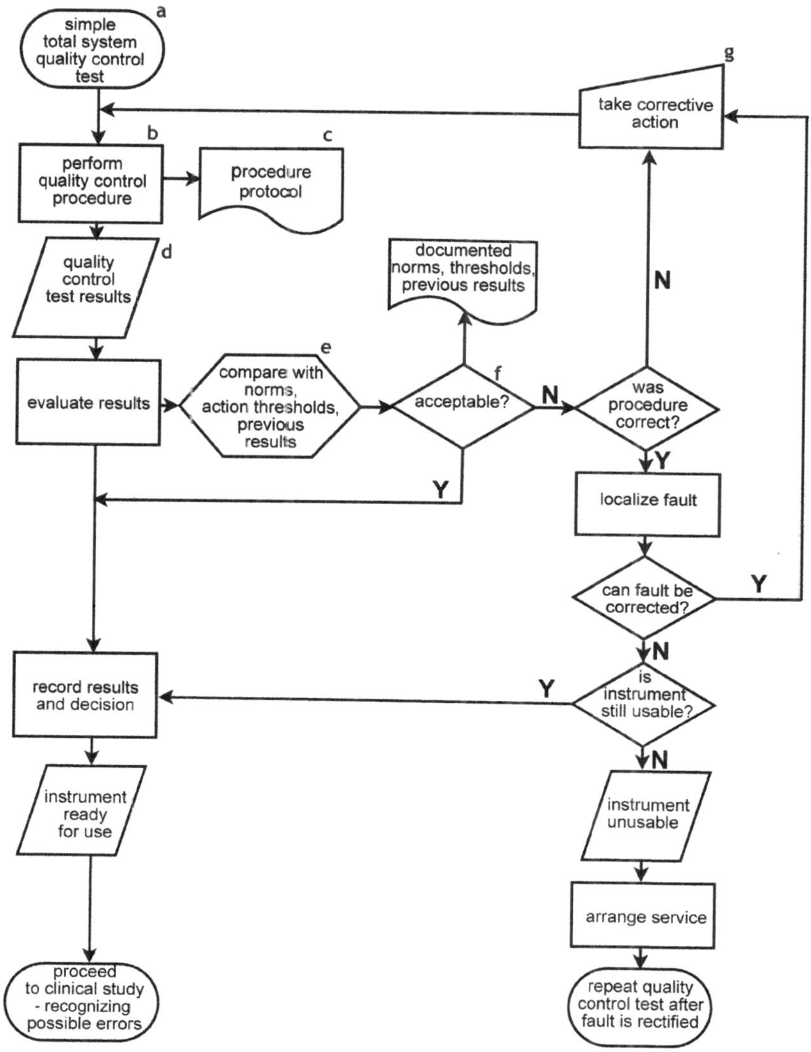

FIG. I.1. Decision tree suggested for performance, evaluation and follow-up of a quality control test. The symbols indicate: a — start or end; b — process to be performed; c — protocol; d — intermediate results; e — checks required; f — decision to be made; g — action taken. Question answer: Y — yes; N — no.

Good communication and teamwork between the personnel of the department are essential, especially when the consequences of an instrument failure or malfunction may involve the action of different disciplines (e.g. taking care of the patient, undertaking first line troubleshooting and problem solving, decision making if the problem cannot be immediately solved, patient rescheduling).

Available qualified personnel who understand the basic functioning of the instrumentation and the digital environment (e.g. the imaging or measuring instrument, computers, peripherals, network, picture archiving and communication system) are desirable. Up to date protocols for instrumentation function and set-up, instrument calibration, work procedures, clinical studies, QC tests (including action thresholds) and phantom preparation are necessary for creating and maintaining uniform methods within the department. These are an important reference during troubleshooting.

Illustrative examples of artefacts shown in this appendix are restricted to only a few examples. The IAEA Quality Control Atlas for Scintillation Camera Systems is a further valuable resource with many more examples of gamma camera artefacts.

I.1.2. Processes in troubleshooting

Troubleshooting in a clinical environment requires first and foremost immediate action, clear thinking and resources to a network of qualified personnel, in order to minimize down time. The following processes are suggestions to assist when setting up a troubleshooting system within the department:

(a) Identify a qualified person within the department who will have the responsibility for that day to be called upon as first-line support should a problem occur. This person could be the physicist, technologist or technical engineer. They should be called upon immediately when a problem is signalled and be responsible for communication regarding the problem and decisions made.

(b) When a problem is signalled during a patient's nuclear medicine investigation, make every effort to localize the problem as soon as possible. Decide whether the problem can be solved immediately, so that the investigation can be continued (possibly without moving the patient beneath an imaging system), or whether the problem solving will take more time and the patient has to leave the room and return to the waiting room until further decisions have been made.

   (i) Example A: An example of necessary fast action is when the computer or the computer network halts during acquisition of a planar dynamic gamma camera study started immediately after injection of the radiopharmaceutical (e.g. renography). If at all possible, the problem should be solved immediately and the dynamic study continued or restarted. Depending on the type of study, the diagnostic value of the study may still be salvaged, without requiring the patient to return on another day and receive another radioactive injection. The problem of patching the dynamic study between the first and second parts with missing data may then be tackled afterwards. The nuclear medicine physician should decide whether the interrupted study still has diagnostic value.

(ii) Example B: A problem during an electrocardiogram (ECG) gated cardiac study may be related to an inappropriate ECG signal to the computer, simply requiring the repositioning of the ECG leads (e.g. a negative R-wave instead of the correct positive R-wave).

Caution: If the data acquisition computer and/or imaging system is to be shut down and restarted, the patient must first be removed from the patient pallet.

(c) If the identity of the problem is not obvious or the problem cannot be solved immediately, a decision must be made as to whether the instrument is totally unusable or usable with limitations until repaired. A partially performed clinical investigation should, if possible and appropriate, be repeated on another similar instrument in the department. This also applies to the other investigations scheduled for that instrument for patients already administered with radioactivity. Any change in instrumentation or protocol must also be noted in the patient record.

Caution: Caution should be exercised regarding the comparative validity of quantitative data when a study is performed on another instrument (e.g. cardiac studies assessing left ventricular ejection fraction).

(d) The problem may need to be solved by an in-house service, a telephone consultation with the service centre of the vendor or by a vendor's service visit, which should be initiated as soon as possible. To assist with localizing the problem, as much information as possible should be documented regarding the circumstances at the time of malfunction, such as other activities being performed (e.g. data processing, data transfer over a computer network, temperature and humidity, power stability, nearby surrounding activities, time of day). A digital photograph of the situation and any error message display on the instrument monitors can be a helpful tool for troubleshooting. An example from a digital log book (created in house using File Maker Pro software) is shown in Fig. I.2.

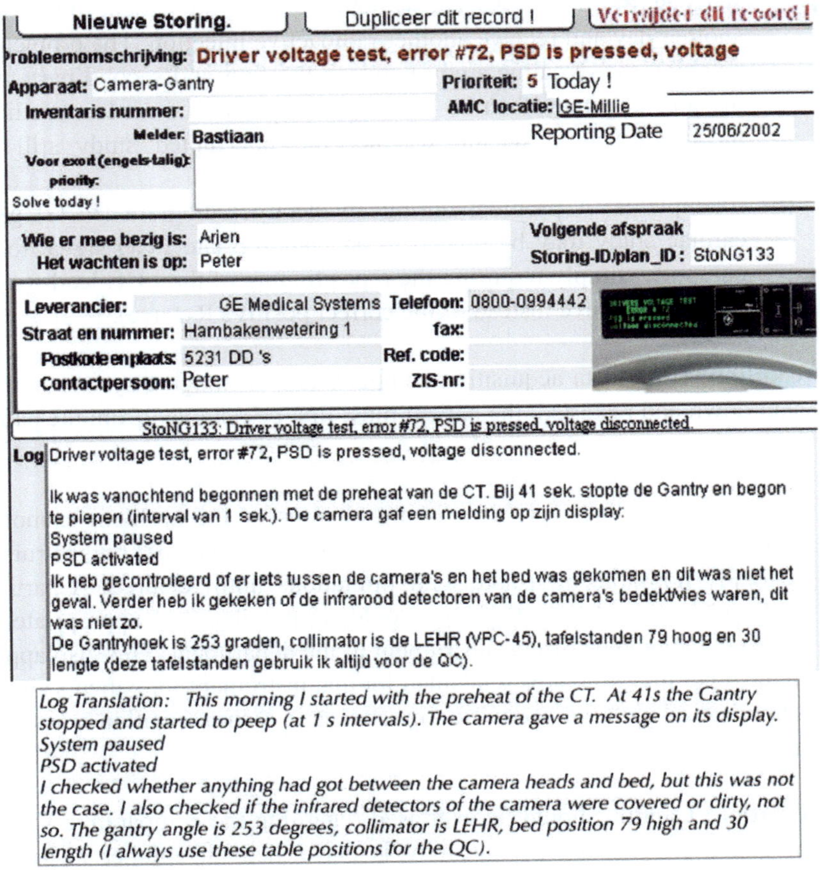

FIG. I.2. Example of the digital documentation (in Dutch, with log translation) of a gamma camera gantry error. This is one item from the digital database (developed in house) of a troubleshooting log. The error report includes the instrument type, the date of problem report, action priority, room location, name of responsible person, company information, log describing the problem, first-line actions taken and their results. A photograph is included of the gantry error message readout.

(e) Enter all problems and as much related data as possible into the log book specific for the instrument. The solutions should also be documented. A well documented and maintained digital record can be especially useful for assisting with troubleshooting by a search for a previous similar problem or if a repeat problem occurs at a later date. This log book should be started at installation and maintained throughout the lifetime of the instrument, together with preventive maintenance reports and any major modification. Such a log book can also be linked to the QC results and records.

(f) A problem may manifest itself during a routine QC test (such as an artefact from a malfunctioning PMT observed in a routine QC uniformity image). A calibration procedure may fail or show values that are outside the acceptable range. A decision must then immediately be made regarding the acceptability of continuing to use that instrument, whether the instrument can be used with recognized limitations, or whether it must be taken out of use until the problem is solved. This decision making should be communicated with the other responsible staff members. The problem and follow-up actions must be documented in the log book. Two examples of artefacts discovered at the time of routine QC testing of flood-field uniformity are shown in Figs I.3 and I.4.

At the moment of discovering a problem in a QC test, it is uncertain when the malfunction causing the artefact or calibration failure first occurred. The assumption is that it may have occurred at any time between the current and previous QC test. The clinical studies prior to the current QC test should, therefore, be carefully reviewed in order to ascertain when the artefact first occurred, and if the artefact in the clinical images might

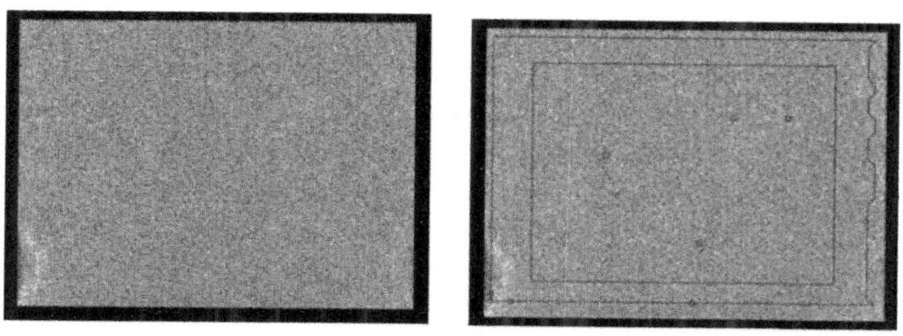

NEMA		Uniformity(%)	Specifications(%)
UFOV	Integral	9.53	<3.6
	Differential	6.4	<3.0
CFOV	Integral	3.72	
	Differential	2.79	

FIG. I.3. Weekly system uniformity image (left) from one detector of a dual head gamma camera. The image was obtained with all corrections activated (linearity, energy, uniformity), low energy high resolution collimator, ^{57}Co flood source, symmetric energy window over 122 keV, 256 × 256 matrix, 4 million counts. On the lower left side, there is an irregular hot semicircular area. The National Electrical Manufacturers Association (NEMA) uniformity quantification in the useful field of view (UFOV; right image outer rectangle) confirms that this non-uniformity is outside of specifications. The problem was a loss of gel between the border photomultiplier tube and crystal. Once the gel was replaced, the uniformity was restored. Note: This camera required a service. The defect affected imaging at the edge of the field of view. This detector could, therefore, still be used for imaging within the central field of view (CFOV; right image, inner rectangle) area, for which the NEMA differential uniformity values were satisfactory (planar and whole body imaging, and with caution SPECT imaging).

NEMA		Head 1 Uniformity (%)	Head 2 Uniformity (%)
UFOV	Integral	3.54	20.94
	Differential	2.08	20.00
CFOV	Integral	3.11	15.79
	Differential	2.08	14.64

FIG. I.4. Routine system uniformity images from a dual head gamma camera. The images were obtained with all corrections activated, low energy all-purpose collimators, ^{57}Co flood source, symmetric energy window over 122 keV, 64 × 64 matrix, 16 Mcounts. The uniformity was quantified with the National Electrical Manufacturers Association (NEMA) integral and differential uniformity parameters in both the useful field of view (UFOV) and central field of view (CFOV). The values from detector head 1 were within the expected limits. Detector head 2 shows a gross non-uniformity pattern corresponding to the photomultipliers. This pattern was due to a failure of the electronic correction due to bad electrical contacts of the circuit boards. After re-seating the relevant circuit boards, the problem was solved and uniformity was restored as shown in a follow-up test (not shown here). Note: The non-uniformity of head 2 was extensive and, thus, imaging with this detector had to be suspended until the problem was solved.

have resulted in an incorrect diagnostic report. If it appears that the artefact may have compromised the images and report, a decision must be made whether to recall the patient and redo the study after the problem has been solved, or redo the study on another instrument. An example of an artefact not discovered until the following QC test is shown in Fig. I.5. The nuclear medicine physician should be informed and consulted.

(g) Particular care must be exercised at all times to be alert to artefacts in clinical images, abnormal quantitative readings and data analysis results. An obvious problem is present when an organ uptake measurement is >100%. Constant alertness is an ongoing process, which should be an integral part of daily practice for all members of the nuclear medicine team.

If the same or a very similar abnormal pattern is observed in successive clinical images from different patients, then the abnormal pattern may be caused by a malfunction in the instrument rather than metabolic dysfunction in the patients. If such a situation is suspected, the problem should first be investigated before further patients are injected and imaged. Troubleshooting may involve not only investigating the instrument, but also checking the radiopharmaceutical quality and integrity of the radioactive administration, etc. It often requires a QC test to assess the situation. An example of uniformity artefacts in lung perfusion studies that were not immediately related to instrument artefacts is shown in Fig. I.6.

The acquired image data for an investigation should always be reviewed carefully before the patient is allowed to leave the department. An artefact or inadequate data may require that the data acquisition be repeated.

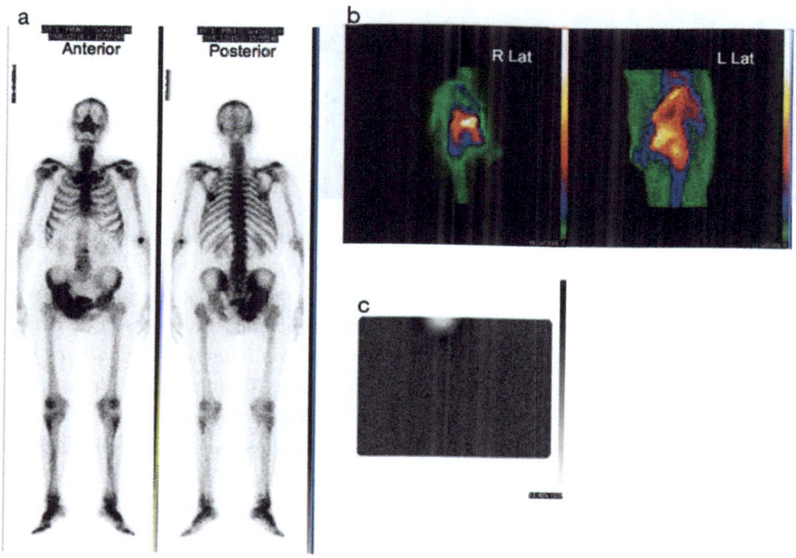

FIG. I.5. A ^{99m}Tc phosphonate bone scan obtained with a dual head gamma camera. (a) Anterior and posterior whole body images. (b) R lateral and L lateral static images of the left knee. (c) Routine system uniformity quality control image of detector 1 taken 2 d later. The photomultiplier tube artefact is at the upper border of the field of view. Note: The bone scan was reported without noticing the malfunctioning photomultiplier tube of detector 1, which was only discovered at the following routine QC test. On review, the effect of the photomultiplier tube artefact was not discernible in the anterior whole body scan made with detector 1 (a), but was visible in the R lateral static of the bone scan using a colour table and high contrast that highlighted low count areas. This example illustrates alertness to an unexpected malfunction. This camera required a service but could still be used with caution for planar imaging within a limited part of the detector. Owing to the nature of the clinical bone study and the location of the photomultiplier tube artefact, this study was not repeated. The gamma camera required a service.

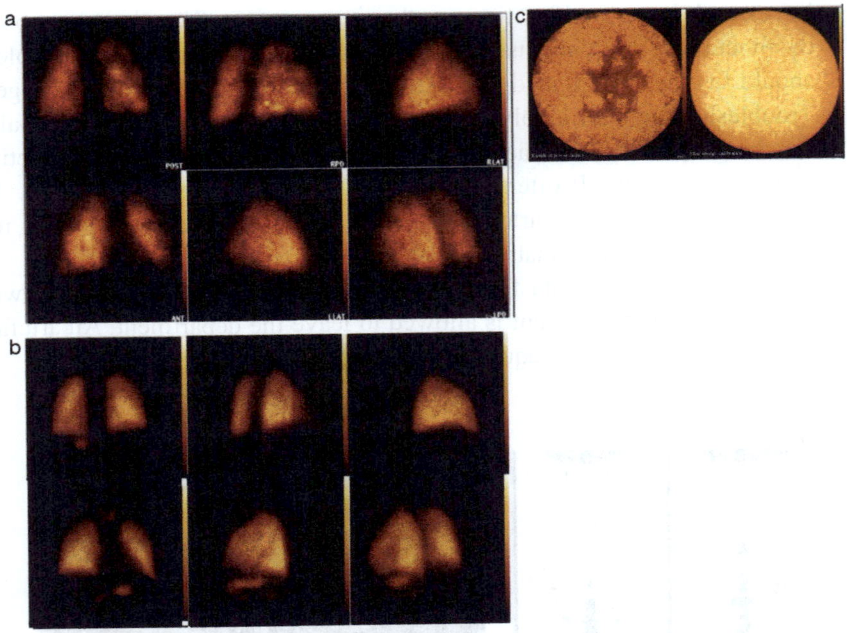

FIG. I.6. A clinical lung perfusion study using ^{99m}Tc macroaggegates. (a) Images of the lungs were obtained with camera 1 (top row images: posterior, right posterior oblique, right lateral; bottom row images: anterior, left lateral, left posterior oblique). The irregular pattern of hot and cold areas was not at first recognized as a camera problem. After two subsequent patients demonstrated the same patchy pattern in their lung perfusion images, the clinicians reviewing the studies questioned the results and troubleshooting was initiated. Quality control of the radiopharmaceutical was acceptable. A uniformity quality control test of the gamma camera was made and this revealed gross non-uniformity ((c) — left image). All patients were recalled and re-imaged on camera 2. (b) Lung perfusion images obtained on camera 2 (with the same image order as in (a)). Camera 1 was retuned, which restored uniformity ((c) — right image). Further investigation revealed that there had been a power disruption during the night. This had corrupted the energy correction values, and explained the reason for the non-uniformity. Note: A routine quality control uniformity test had not been performed at the start of the day's clinical imaging. If this had been done, the problem would have been identified immediately. An uninterruptible power supply was later installed in order to prevent a similar future occurrence. (For more details, see the IAEA Quality Control Atlas for Scintillation Camera Systems.)

(h) After a problem has been solved, the instrument should be tested for correct functioning before being released for clinical use. If computer software or hardware has been changed, reboot and restart the system to ensure that the system works after a power down:

  (i) Be aware of any changes in hardware or software that could affect quantitative results. Validate the results.

(ii) Changes to hardware may require QC testing before the instrument is released for clinical use.

(i) Be aware of an intermittent or repetitive problem. Creative testing and dedicated persistence is required to locate the cause of such a problem. Even after a problem appears to be solved, it may still be present because of instability in a component. For example, this has been the case with electronic grounding, cable connections, cable breaks and a fluctuating power supply. An example of the effect of voltage instability in a single photon emission computed tomography (SPECT) study is shown in Fig. I.7. Continued alertness is always required, as well as repeated QC testing.

I.1.3. Troubleshooting remedies

Various first-line troubleshooting tactics can be useful before resorting to contacting the service. If a service contract is available for an instrument, it should be clear where the responsibilities and limits lie. Some general hints to be considered are given below, although the circumstances are left to the discretion of the troubleshooter. The troubleshooting section of the instruction manual of the instrument should also be consulted.

In the event of failure or instability of an instrument or component (to be carried out by a qualified person or the service engineer):

(a) Check electrical power, circuit breakers, fuses, cables and cable connections, fans;

(b) For accessible batteries, check the level of battery power within the instrument that regulates a specific function;

(c) Check for dust, and cleanliness of sensors and metal contacts.

Computer and network:

(a) If a program halts, exit and restart the program. If this is unsuccessful, shut down and restart the computer. (If restarting the data acquisition computer, make sure that the patient is not on the imaging table.)

(b) For a suspected communications failure between the instrument and the computer, shut down and restart the computer. If this does not solve the problem, shut down both the computer and the instrument, and, after about 30 s, restart the instrument and then the computer. Be careful to follow a correct startup procedure and make sure that the patient is not on the imaging table.

(c) If peripheral equipment (such as a printer) stops functioning or produces an error message, shut down and restart that equipment.

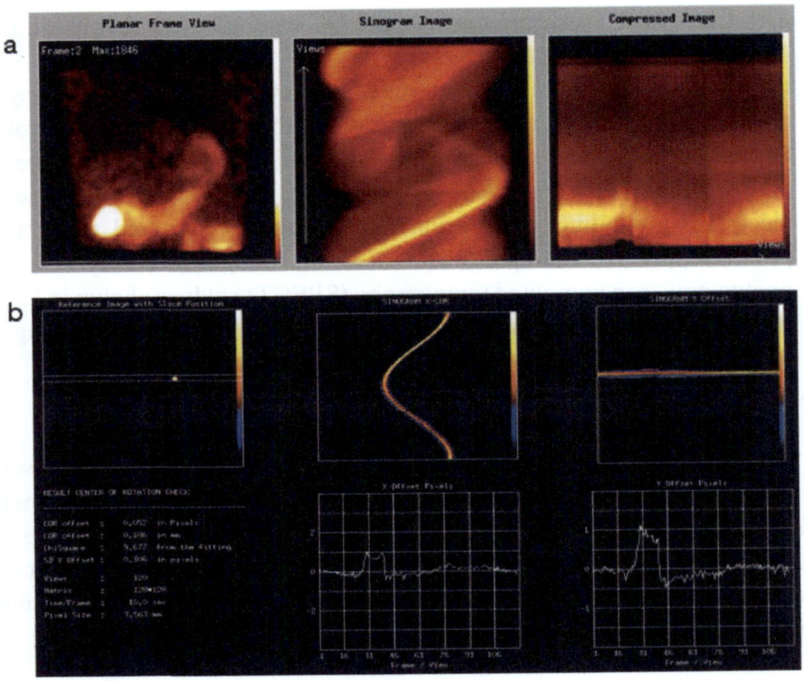

FIG. I.7. (a) Quality control images over the whole field of view of the acquisition data of a SPECT myocardial perfusion study (left — one projection image, middle — sinogram over the whole field of view (X), right — linogram over the whole field of view (Y)). The images were obtained from a 3-detector SPECT system (120° rotation per head, starting with head 1, and a 360° total rotation). The linogram shows an upwards shift in the images from head 1 towards the finish of the 120° rotation (first third of the dataset). In order to clarify the situation, a point source was placed off-axis and imaged with the same data acquisition parameters. (b) In order to test the system, a SPECT acquisition was made of a point source placed off-axis. Quality control images of this acquisition (same image order as above), and their quantitative offset analysis (lower row). Offsets are seen in both X and Y in detector 1 data, identified clearly by the jump in offset in both X and Y on the quantitative analysis. The problem was due to a decrease in voltage to the signal board of detector head 1 at certain projection angles. This was found to be due to instability in the power cable connected to the signal board. The problem was resolved only after replacing the cable. Note: This problem was difficult to locate. A problem was signalled in patient studies by the reporting nuclear medicine physician who observed upwards motion in the quality control review of the patient SPECT data from successive patients. Initially, a corrupted centre of rotation calibration was considered to be the cause. However, the problem repeated itself, and was again recognized on subsequent clinical and point source SPECT acquisitions. It took much persistence from the department to keep on testing the system and several visits by the service engineer before the problem was found. The upwards shift was only seen in one detector head, thus pointing to a problem with the camera and not movement of the patient, which would have been seen in the acquired data of all three heads, at the same time frames.

(d) Check the computer network and communications.
(e) Note: Before instrument re-use, a simple QC check may be necessary to ensure that the instrument is functioning correctly.

Error and artefact in results:

(a) Check the radiopharmaceutical and injection quality, study parameters and instrument settings, and patient positioning before investigating instrument malfunction. Consult the department procedure manuals.
(b) If observed in a QC test, check that the test method was correct.
(c) Initiate appropriate supplementary QC tests, as appropriate.

I.2. IMAGE ARTEFACTS

I.2.1. Recognizing image artefacts and their underlying causes

Pattern recognition is the essential ingredient of interpretation of nuclear medicine images. Thus, recognizing an artefact is also an essential part of pattern recognition. Image artefacts manifest themselves in different ways as a result of different factors. Relating an artefact to the underlying problem causing the artefact is a developing process of understanding the instrument and how it should be used. A particular instrument may show characteristic artefact patterns that repeat and become familiar over time.

An artefact may also be caused by incorrect instrument settings or be patient related. Thus, a daily component of troubleshooting is to review the acquired data before each patient is allowed to leave the department.

Some causes of problems encountered in gamma camera images are given below for different performance parameters. Related images can be found in the IAEA Quality Control Atlas for Scintillation Camera Systems, which is an extensive resource of different types of image artefact that may be encountered in planar, whole body and SPECT imaging modes of a gamma camera system. Examples given in the Atlas include results from QC tests as well as clinical examples.

Problems encountered in gamma camera images for different performance parameters:

(a) Distortion of the energy spectrum photopeak shape, loss of energy resolution: A change may be caused by:
   (i) Poor tuning or energy calibration.
   (ii) Malfunctioning PMT or preamplifier.

(iii) Inadequate or unstable electrical grounding.
(iv) Instability in electrical contact in the detector power supply.
(v) Deteriorating detector material.
(vi) Interference from nearby radionuclides.

(b) Decrease in detector sensitivity: The decrease in count response may be related to:
(i) Incorrect centering of the photopeak window.
(ii) Change in the PMT tuning values or gain values.
(iii) Malfunctioning PMT.
(iv) Deteriorating detector material.

(c) Poor image uniformity of a gamma camera: Image uniformity may be affected by a variety of problems, for example:
(i) Deterioration in detector properties, so that energy and/or linearity corrections no longer correspond.
(ii) Inadequate energy correction for radionuclides other than ^{99m}Tc.
(iii) Offset in centering of the image with respect to the image matrix and corrections.
(iv) Poor tuning of the PMTs.
(v) Malfunctioning or defective PMT(s).
(vi) Loss of optical coupling between PMT and light guide, light guide and crystal surface, or PMT and crystal surface.
(vii) Asymmetrical or erroneous position of the energy window on the photopeak.
(viii) Defects in the collimator (extrinsic uniformity).
(ix) Radioactive contamination on the collimator or detector crystal.
(x) Crystal hydration.
(xi) Broken detector crystal (due to impact or thermal changes).
(xii) Improper QC procedure, including errors due to phantom preparation, e.g. size of a point source, flood source filling, source positioning.

A test of flood field uniformity is the basic most sensitive QC test for the gamma camera. This QC test should be considered as a first troubleshooting QC test when an image artefact is encountered or suspected (see example in Fig. I.8). However, if there is suspicion of a local artefact, relating to a possible PMT malfunction, in the patient's images, it may be the simplest to first repeat an image after moving the patient within the field of view (FOV); if the artefact moves with the patient, the problem is patient oriented; if the artefact remains in the same location, then it is instrument oriented. Further investigations can then be performed (e.g. by making a uniformity QC test). An example of the effect of a defective PMT artefact in static bone scans is shown in Fig. I.9.

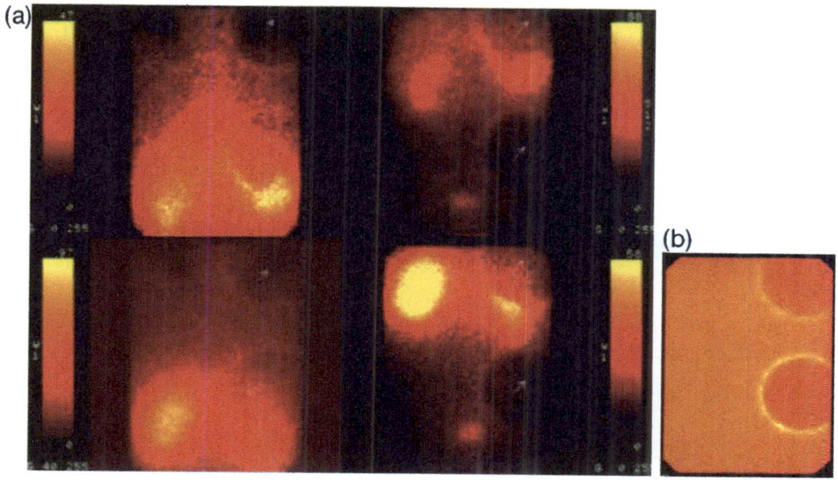

FIG. I.8. (a) A clinical ^{111}In somatostatin receptor study obtained with a single detector gamma camera. The upper and lower abdomen in the anterior view (top two images), and the upper and lower abdomen in the posterior view (bottom two images). Each image of the clinical study showed two large, diffuse, circular colder areas (indicated by arrows). (b) Uniformity image obtained after the clinical study, which shows two large cold areas with a hot border, each due to a defective photomultiplier tube. The non-uniformities in this example were large enough to be recognized in the patient's images at the time of imaging. The images could, therefore, be repeated on another gamma camera system. The problem occurred intermittently, but the fault was never localized. The camera was finally replaced. (Example 2.2.8.6 in the IAEA Quality Control Atlas for Scintillation Camera Systems.)

The appearance of an artefact in the uniformity QC test is dependent on the problem. If the non-uniformity is diffuse or unclear, a sensitive troubleshooting method is to make two further uniformity QC images with asymmetrically positioned energy windows: with the energy window set asymmetrically over the lower half of the photopeak and with the energy window set asymmetrically over the upper half of the photopeak. Non-uniformities are highlighted in such asymmetric images, with cold areas in the one image corresponding to hot areas in the other image. Asymmetric images highlight, for example, poor tuning, problems with an energy correction map, ADC (analogue to digital converter) problems and crystal hydration. Figure I.10 shows an example of the early appearance of extensive non-uniformity patterns in images made with asymmetrical energy windows. These artefacts were attributed by the service engineer to separation of the light pipe from the crystal, which could not be rectified by service and implied replacement of the whole detector. Six months later, these artefacts became evident in the clinically used symmetrical energy

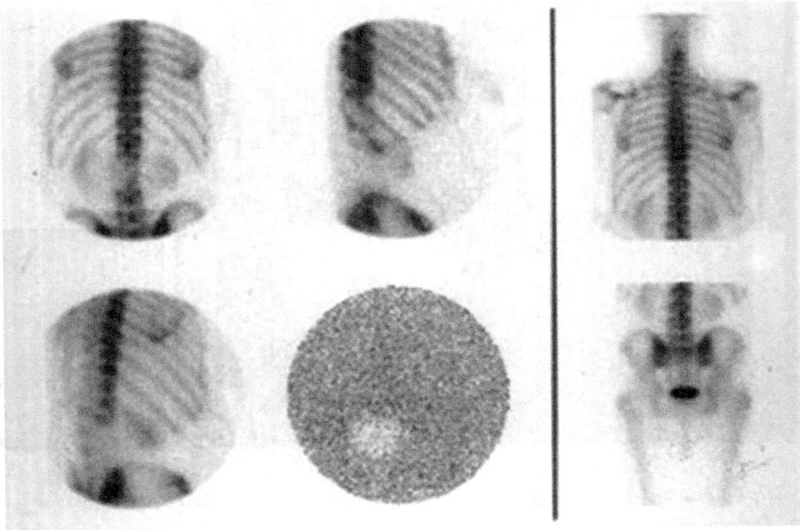

FIG. I.9. Static images of the skeleton after administration of ^{99m}Tc phosphonate. Images obtained on camera A (top left — posterior, top right — right anterior, bottom left — right anterior oblique) show an area of apparent decreased activity in the lower spine that is especially evident in the posterior and right anterior oblique views. As the cold area in the lower spine was unusual, it was not considered to indicate pathology but to be an artefact. The posterior skeleton was, therefore, imaged on camera B, and these images show a normal ^{99m}Tc-phosphonate distribution in the lower spinal column. Subsequently, a uniformity image was obtained on camera A that demonstrated a defective photomultiplier tube that corresponded to the area of decreased activity in the skeleton images. Camera A required servicing before further clinical images were performed. Note: The study was reviewed before the patient left the department. If a second camera had not been available, the patient could have been shifted so as to image the lower skeleton in another part of the camera field of view. (Example 2.2.8.5 in the IAEA Quality Control Atlas for Scintillation Camera Systems.)

window. Early observance of such a situation can assist with initiating a replacement plan. An unusual and unexpected discovery of crystal hydration in a new camera 3 months after installation is shown in Fig. I.11. In this situation, detector replacement was required but this was within the guarantee period. A dramatic example of hydration and poor tuning is shown in Fig. I.12.

(d) Poor image spatial resolution and image contrast: Poor planar image spatial resolution can be caused by:

   (i) Too large a distance between the patient and collimator.

   (ii) Poor linearity corrections of the detector: As visual evaluation of linearity is subjective and difficult to assess, linearity should be

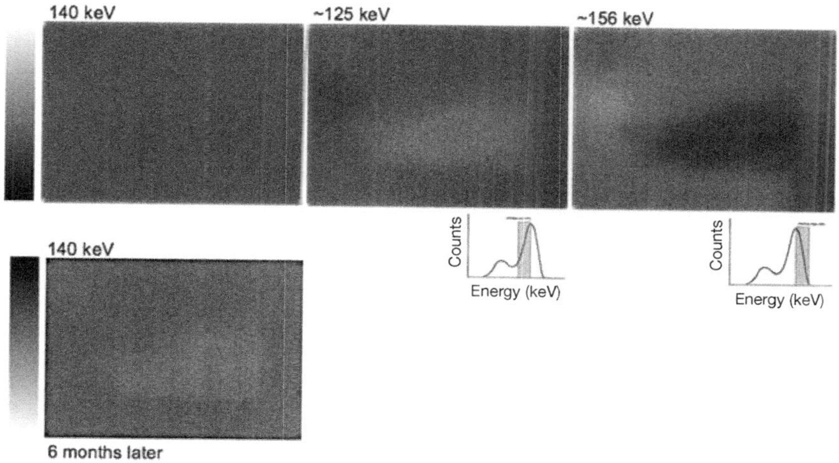

FIG. I.10. Top row: A series of routine intrinsic uniformity images obtained with ^{99m}Tc, uniformity correction turned off, with the energy window set symmetrically (left), asymmetrically low (middle) and asymmetrically high (right) over the photopeak. The bottom image is a repeat intrinsic uniformity image obtained 6 months later on the same gamma camera. Each image was made with 5 Mcounts in a 256 × 256 matrix. The asymmetrical low image shows a large diffuse rectangular hotter central area that corresponds on the asymmetrical high image to a colder central area. Six months later, the same central colder area is now visible on the image obtained with the symmetrical photopeak. This artefact was caused by a separation of the light pipe from the crystal. This problem could not be fixed and the whole detector required replacement. In this particular case, the detector replacement was covered in the service contract. It would otherwise have been a very expensive repair. Note: The colour scale used in these images is reversed between the first images (0–100% counts = black to white) and the image made 6 months later (0–100% counts = white to black). Recording the colour scale together with the images is essential, not only for quality control images but also for clinical images. The colour scale and any colour enhancement should always be taken into consideration when reviewing images.

quantified if software is available. Figure I.13 shows the results of a 6-monthly QC test of spatial resolution and linearity using a slit phantom, where the quantified National Electrical Manufacturers Association linearity values were outside of the specifications in both the X and Y directions, indicating that a new linearity map was necessary.

(iii) Poor multiple energy window registration.

A decrease in image contrast in acquired images may be caused by:

(i) Incorrect position of the energy window and may be an operator error: For example, this can occur if a test or calibration has been performed

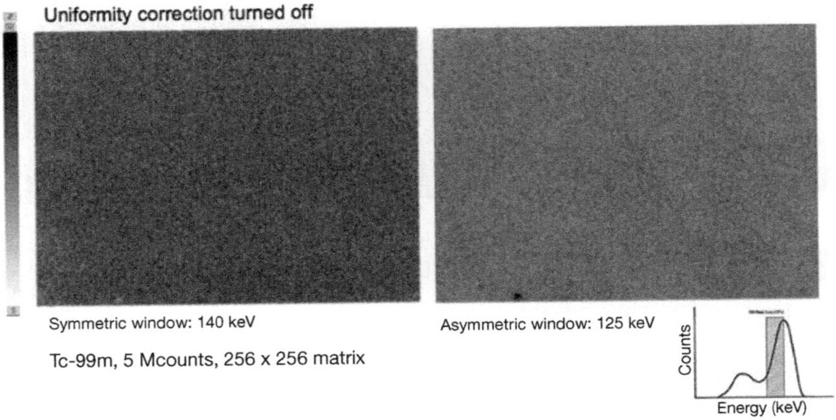

FIG. I.11. Intrinsic uniformity images obtained at 3 months after installation of a new gamma camera. The images were obtained with the uniformity correction turned off (but linearity and energy corrections activated), using a ^{99m}Tc point source, 256 × 256 matrix, 5 Mcounts total. The left image was obtained with the ^{99m}Tc energy window set symmetrically over the photopeak. It shows a suspicious small cold spot on the lower border. In order to investigate this further, the energy window was offset on the lower half of the photopeak (see diagram). The image obtained with this window setting (right) shows a distinct hot spot at the same location as the cold spot on the left images, as well as two other small hot spots close by. This is the result of crystal hydration. The detector can still be used at this moment in time, because the hydrated areas are at the edge of the field of view. However, hydration will continue to develop. The detector required replacement. In this case, the problem was discovered soon after installation within the guarantee period, so that replacement could be made under the guarantee. Note: If this situation is observed in an older gamma camera, a replacement strategy for the detector must be planned. The development of hydration requires close monitoring by weekly or monthly asymmetric uniformity images until replacement takes place.

with a ^{57}Co source, and inadvertently the energy window not been reset to ^{99m}Tc.

(ii) Performing an automatic ‘peaking’ procedure with the patient as the radioactive source: Owing to the large additional scatter component in the photopeak, the window will automatically adjust too low over the photopeak. The clinical image will, thereby, include unnecessary scatter in the image.

In SPECT, a decrease in resolution and contrast may be related to:

(i) The imaging technique (e.g. excessively large radius of rotation, poor choice of acquisition and reconstruction parameters).

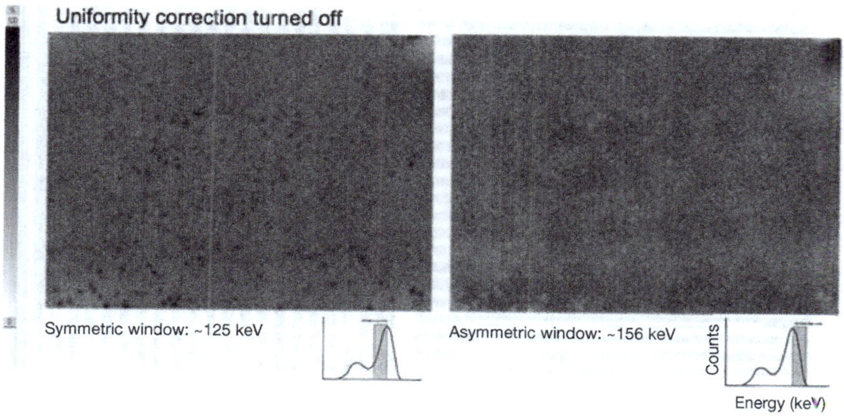

FIG. I.12. Periodic intrinsic uniformity images obtained with ^{99m}Tc, the uniformity correction turned off, and asymmetric energy windows set low (left) and high (right) over the photopeak (each image: 5 Mcounts, 256 × 256 matrix). The images demonstrate extreme crystal hydration over the whole field of view: small hot spots in the low asymmetric window correspond to small cold spots in the high asymmetric window. The asymmetric images also show some poor tuning (especially in the top right corner). The extent of the hydration indicates that this detector requires replacement.

(ii) Inadequate instrument calibration (offset in centre of rotation (in X or Y directions), detector tilt in Y direction, poor alignment of multiple detector heads, inadequate uniformity correction).
(iii) Artefacts in acquired data (missing projections, PMT artefact).
(iv) Artefacts in reconstructed data (e.g. ring artefacts from non-uniformity).

Troubleshooting artefacts observed in SPECT images may simply involve a test SPECT acquisition of a point source placed off-axis (see Fig. I.7) (e.g. for assessing problems of motion between detector heads), or might be more intensive involving a test SPECT acquisition of a cylindrical bottle or a SPECT phantom (e.g. assessing ring artefacts observed in clinical images). The recalibration of the uniformity correction map or the centre of rotation and head alignment (and head alignment in multiple detector systems) may solve the problem. However, a follow-up QC test following recalibration is always required, in order to check that the problem has been solved. Figure I.14 is an example where a recalibration of head alignment was insufficient, and a remaining problem of detector head tilt required a service visit. Further awareness is still required if the problem is an intermittent fault and not solved by recalibration alone (as in the situation of the example in Fig. I.7).

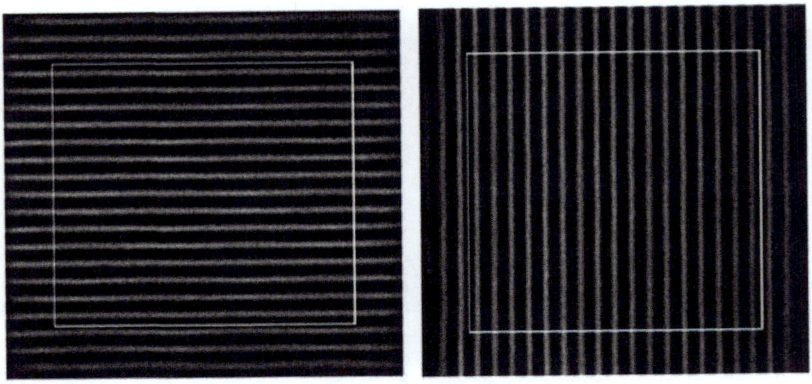

NEMA (mm)	Mean FWHM	Mean FWTM	Abs Dev	Max LineDev	Mean FWHM	Mean FWTM	Abs Dev	Max LineDev
UFOV	3.81	6.89	1.70	0.46	3.69	6.78	0.67	0.28
CFOV	3.61	7.07	0.80	0.30	3.61	6.98	0.54	0.25
Specifications: UFOV			0.40	0.20			0.40	0.20

FIG. I.13. Routine 6-monthly quality control test of intrinsic linearity and spatial resolution of a small field of view gamma camera, using a slit phantom with 1 cm spacing between slits, ^{99m}Tc point source, 1024 × 1024 matrix (pixel size: 0.29 mm). The acquired images were quantified within the indicated rectangles. The spatial resolution was within specification. However, linearity (absolute deviation: Abs Dev; maximum line deviation: Max LineDev) was out of specifications in both the X and Y directions. The linearity correction maps needed recalibration. NEMA: National Electrical Manufacturers Association; FWHM: full width at half maximum; FWTM: full width at tenth maximum; UFOV: useful field of view; CFOV: central field of view.

(e) Clinical investigations: A review of data acquired is essential before processing and quantification:

- (i) For SPECT data, a cine, sinogram and linogram can suffice to review the clinical data for patient movement, missing projections, instability (an artefact that appears in only some projections) and inadequate continuity of data from multiple detectors. If the review reveals such errors, then the study may need to be repeated.

Note that an artefact due to patient movement will be imaged by all detectors at the same time, so that the movement artefact will repeat. If a 'movement' artefact appears only in one detector of a multiple detector system, then the problem is probably due to the instrument (as an example, see Fig. I.7).

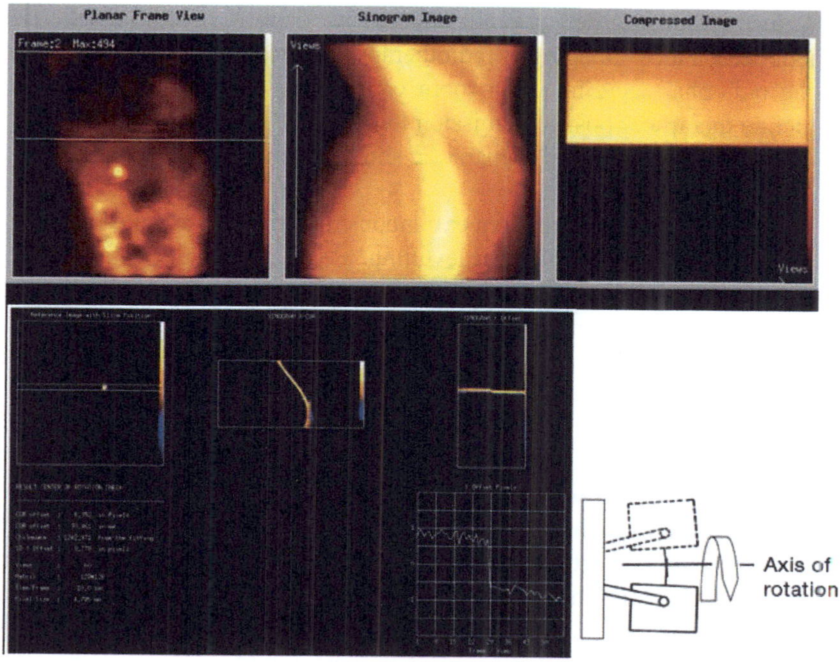

FIG. I.14. Top: Quality control images of a SPECT myocardial perfusion study using a dual head gamma camera in a 90° configuration, 180° rotation (90° per head), 128 × 128 matrix, 4.8 mm pixel size. Images: left — one projection of the acquired data, middle — sinogram (X) of a profile over the myocardium (shown in the left image), right — linogram (Y) for the same profile. There is a discontinuity between detector 1 and 2 visible on both the sinogram and linogram. Bottom: After recalibration of the centre of rotation and head alignment (first trouble-shooting technique applied), a test acquisition was made of a ^{99m}Tc point source placed off-axis (left image projection). The sinogram (middle) shows no discontinuity, whereas the linogram (right) shows both discontinuity and slope in the data — particularly evident in the quantified offset graph. The problem was a detector head tilt in both heads, which required a service.

I.3. MINIMIZING PROBLEMS

Problems can occur at any time, but taking appropriate precautions can minimize the likelihood.

I.3.1. Siting and room preparation

The location of an instrument and good preparation prior to installation are vital first steps. Particular care is needed to choose a location suitable for the specific instrument in order to avoid interference from X rays, magnetic fields (magnetic resonance imaging), radiotherapy machines, radioactive sources from the radiopharmacy, injection room or radioactive patients (such as from radionuclide therapy or positron emission tomography (PET)).

Considerations for room preparation should include the following:

(a) Necessary wall shielding from extraneous radiation and magnetic fields.
(b) Floor weight support and floor levelling.
(c) Continuous stable electrical power supply: Consider connecting the instrument to an emergency power supply and installing an uninterruptible power supply (UPS) (see Sections 7.6.2 and I.3.2). Consider the electrical conditioning, grounding and safety.
(d) Sufficient strategically placed power outlets for peripherals.
(e) Lighting and switches (to exclude electrical interference with equipment).
(f) Window placement with respect to the instrument position to avoid drafts and influence from direct sunlight. (Particular care is needed for the gamma camera with respect to exposing the crystal to a sudden temperature change such as might happen during collimator change and intrinsic QC measurements.)
(g) Stable air conditioning with respect to temperature (maximum, minimum, fluctuating temperatures) and humidity (non-condensing): Consideration should be given to these aspects not only during working hours, but also outside of working hours, including at weekends. A major hazard to a gamma camera crystal is a rapid change of temperature: a rule of thumb is that the temperature should not change more than 4°C over 1 h.
(h) Dust free environment: It is generally not possible to achieve a dust free environment in a hospital. However, maximizing a dust free environment should be aimed for, especially for the computers and picture archiving and communication system.
(i) Positioning of the instrument within the room to minimize interference from external radioactive sources.

I.3.2. Electrical power conditioning

The stability and correct voltage level of the electrical supply is crucial to reducing the likelihood of obtaining an instrument malfunction. This may be achieved by use of a surge protector, a constant voltage transformer or a UPS (battery backup), the choice being dependent on the type of equipment and local environmental requirements.

The UPS is essential where power failures or major power dips and surges occur, in order to avoid the disastrous failure of instruments and computers and the breakdown of components. Even if the instrument's regular power supply is connected to an emergency power supply, the interval between failure of the regular power supply and the initiation of the emergency power supply may produce a power dip sufficient for the instrument to halt or a circuit breaker to switch off (particularly disruptive when occurring during a patient study; see also Fig. I.6).

The investment for a UPS must be considered when requisitioning and purchasing the instrument. A UPS should be connected to all sensitive instruments that are required for daily routine patient care. The UPS specification is dependent on the instrument's power requirement and the local power situation. The UPS may be needed to ensure that, in the event of a power failure during working hours, the instrument can be manually shut down in the correct way. The UPS may be needed only to bridge the gap between a regular power failure and the switch to the emergency power supply.

The electrical conditioning includes appropriate grounding, and shielding of cables, especially for signal cables and data transmission cables.

Electrostatic disturbances can be minimized by adequate humidity control (air conditioners), and antistatic work surfaces and floor covering.

I.3.3. Regular preventive maintenance

Regular preventive maintenance and a service contract can help not only to minimize the chance of an unexpected problem occurring, but also to minimize the down time when a problem has occurred. The expense of a service contract can be considered as an insurance policy. A service contract should ensure fast response and priority access to spare parts. The service may include remote computer login and access. In-house access to trained personnel is also essential for first-line troubleshooting of electrical failures, computer and network failures, and mechanical failures. Without any access to appropriate support, a problem can take a considerable time to be solved, add extra expenses, and become a major obstacle to high quality and continuous patient care.

I.3.4. Acceptance testing and routine quality control testing

Thorough and careful acceptance testing is the first step towards ensuring that an instrument is performing according to specifications and as expected for clinical use. Any problems or suspected problems encountered at this early stage require instant rectification as the instrument is still under the guarantee period (see the example of crystal hydration observed at 3 months in Fig. I.11). Replacement of any defective component must be initiated. The collimator is particularly sensitive to damage from transport and must be tested carefully at

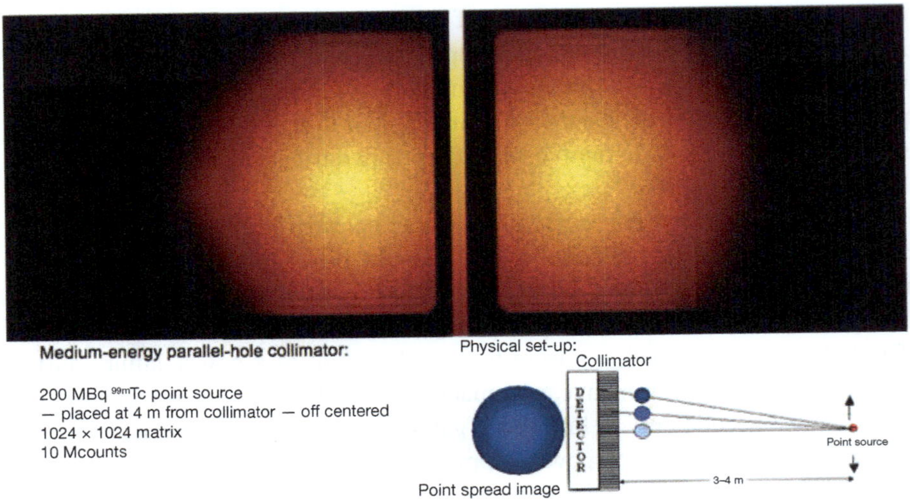

FIG. I.15. Acceptance testing of a medium energy collimator using a distant point source of ^{99m}Tc (acquisition parameters given above). The point source was positioned first to image the right part (left image) and then the left part of the collimator (right image). There are vertical discontinuities evident, probably from the manufacturing process. This collimator was replaced within the guarantee period. Note: This is a sensitive method for testing a collimator for hole angulation problems and for any suspected damage. A large distance between the source and collimator is essential. This test supplements a high count system uniformity test.

acceptance and at any time when damage is observed or suspected. An example of a defective collimator discovered at acceptance testing is shown in Fig. I.15.

Routine QC tests are performed on the gamma camera and SPECT system in order to assess performance of the instrument at a specific moment in time. They are intended to reassure the user that performance up to that moment is satisfactory. Monitoring the results of successive QC tests can indicate a stable functioning condition, deterioration or an impending problem. A database of results is recommended. The visual as well as quantitative results must always be reviewed together. Figure I.16 illustrates a situation in which the quantitative uniformity value appears to be acceptable, whereas the image shows there is a PMT problem. Routine QC tests are valuable in troubleshooting, and should neither be underestimated nor neglected. The results of QC tests can highlight the underlying problems.

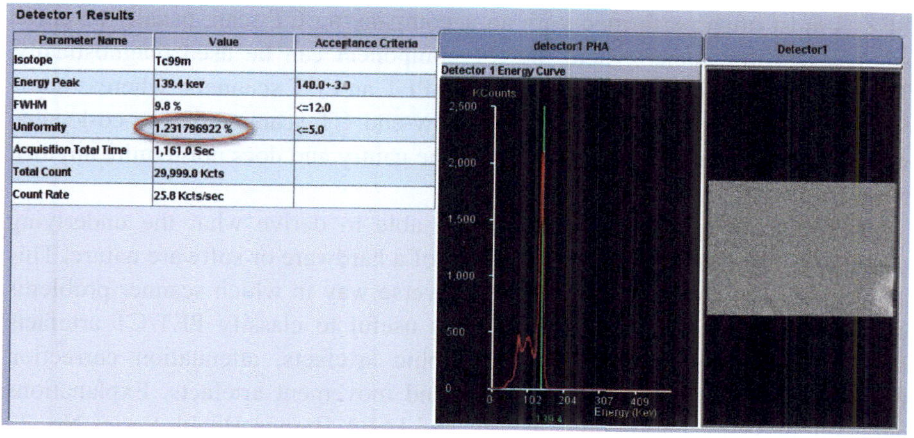

FIG. I.16. Routine intrinsic uniformity image with quantification. The image was obtained with a ^{99m}Tc point source, symmetrical energy window over 140 keV, 30 Mcounts. The image shows a hot semicircular area in the lower right field of view. The quantification indicates that the uniformity is acceptable. However, not indicated in the results, the uniformity calculation refers only to the central field of view. The artefact was due to a malfunctioning photomultiplier tube. Note: This camera required a service but could continue to be used with caution because of the lateral location of the defect. This example illustrates that it is essential to review together both image and quantification, and to understand the parameters provided in the results.

I.4. IMAGE ARTEFACTS IN PET/CT

PET and combined PET/computed tomography (CT) require users to develop skills in recognizing a range of artefacts which are quite distinct from those which may be seen in SPECT or SPECT/CT. PET and SPECT reconstruction do have aspects in common, resulting in analogous methods of recognition; however, the artefacts themselves may appear quite different due to intrinsically different modes of acquisition. PET scanners usually employ a fixed full-ring detection system, unlike SPECT which has a rotating gamma camera, thus eliminating the need for a centre of rotation correction and its associated artefacts. In a typical PET system, a ring of detectors surrounds the patient, each of which simultaneously and independently acquires data. In addition, there is no collimator used in 3-D PET, leading to a vast increase in scanner sensitivity such that acquisition times are generally shorter and whole body scans are the norm. The use of very many individual detectors in PET implies that cameras with minor defects can be tolerated unlike in SPECT, where a defect has a variable impact depending on its location (greater impact towards the centre of the FOV) but may still be usable if the defect is towards the edge of the FOV.

PET is most often performed with an accompanying CT scan, usually acquired using a hybrid scanner where the CT component can be used diagnostically. This requires bed translation between the PET and CT scanners, whereas some SPECT/CT scanners use an integrated low-end CT scanner that is co-located with the gamma camera detectors within the gantry and does not require any bed translation.

The medical physicist needs to be able to derive what the underlying problem is from the artefact, whether it is of a hardware or software nature. This task can be difficult owing to the very diverse way in which scanner problems can present themselves as artefacts. It is useful to classify PET/CT artefacts into the following categories: tomographic artefacts, attenuation correction (AC) artefacts, co-registration artefacts and movement artefacts. Explanations and examples of each are given below, and IAEA Human Health Series No. 27 provides further information and examples.

I.4.1. Tomographic artefacts

Tomographic artefacts are those which appear when some fundamental aspect of the tomographic system performs below specification or else fails entirely. Problems in the tomograph lead to systematic image abnormalities that occur regardless of the type of acquisition being performed. One such abnormality is due to incorrect normalization. Figure I.17 demonstrates an artefact that was created when the normalization of the tomograph had been corrupted. Normalization corrects for the sensitivity difference between different lines of response (LORs). Sensitivity differences are caused by both a geometric distortion, which needs to be measured only once at the factory, and by detector efficiency variations that can change with time and must be periodically recalibrated. Normalization errors occur in the projection space and appear as circular defects in the transaxial reconstructed space. In the example of Fig. I.17(a), the artefact is seen to be repeated in each bed position of the whole body acquisition, indicating that there was a problem with the tomograph itself.

The daily quality assurance routine is a good way to detect unexpected sudden normalization errors. Some quality assurance routines involve scanning a cylindrical phantom filled uniformly with radioactivity (e.g. ^{68}Ge, ^{18}F). Such a cylindrical phantom is designed to be large enough to cover many of the potential LORs in the system. The artefacts due to normalization errors seen in the clinical images of Fig. I.17(a) were clearly visible in the uniformity QC image, as shown in Fig. I.17(b).

Geometry is often the key in diagnosing tomographic artefacts, as can be seen in Fig. I.18, where there has been failure of a detector block leading to a distinctive pattern in the sinogram of the QC uniformity image. Detector block

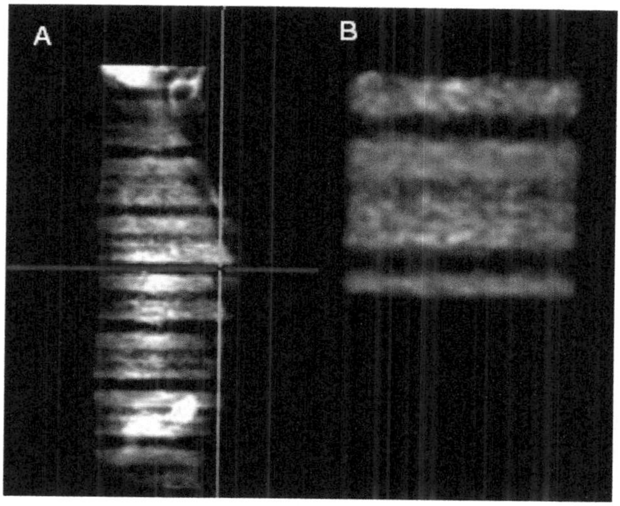

FIG. I.17. (a) Clinical whole body images obtained on a PET system in which the normalization correction was corrupted, but not known at the time. The sagittal view shows a pattern of repetitive cold horizontal stripes at consistent locations within each of the bed positions. The periodic nature of the artefact is a sign that the problem is associated with the system rather than this particular patient or acquisition. (b) Sagittal view of a uniformity quality control check of the PET system acquired using a uniformly filled cylindrical phantom. The image shows cold stripes indicative of errors in the normalization table. This quality control image was obtained after the clinical images revealed the artefacts shown in (a). The quality control image shows several cold streaks which indicate that the problem is most likely a corrupted normalization file. Normalization was recalibrated before further patient acquisitions were performed. (Courtesy of the Department of Nuclear Medicine, Monte Tabor São Rafael, Brazil.)

failure may not contraindicate the clinical use of a PET system since modern scanners have many detectors and the absence of one block may have little statistical impact. Examination of the sinogram (also in clinical images) is a good way to test for block failure, as it appears as a distinctive diagonal streak on the sinogram.

I.4.2. Attenuation correction artefacts

AC artefacts occur when the CT AC algorithm leads to a hot or cold spot in the attenuation corrected reconstructed data. AC effectively increases the counts in each voxel in proportion to the total attenuation along all LORs that pass through that voxel. When the CT image shows a highly attenuating material in a group of voxels, then the total counts along all lines of response that pass through those voxels are increased, and the group of voxels appears hot. This is

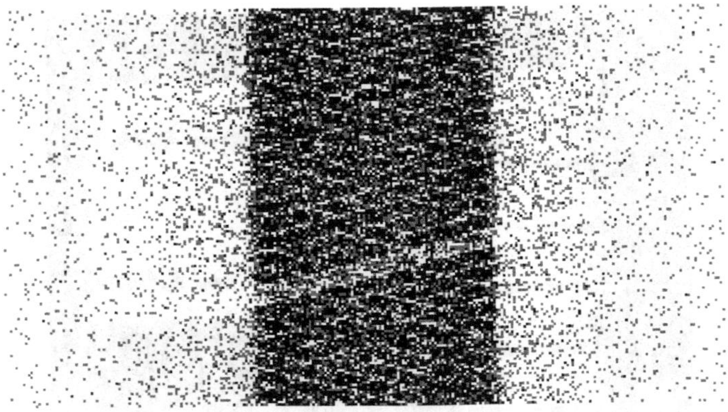

FIG. I.18. Sinogram from a PET system that has a detector block failure. The diagonal streak that is clearly visible is the pattern created when one detector block has failed and causes many lines of response to be zero. The failed detector block creates several blank lines of response at every projection angle at incremental radial positions and the result is a diagonal streak. With only one streak visible, and the fact that the streak is several pixels wide, it would be appropriate to assume that a whole detector block has failed. The noticeable width in the streak occurs because each detector block contains many individual detector elements. Multiple simultaneous detector block failure is unlikely in a system which has regular quality assurance tests. This system is still acceptable for clinical use because there are many detector blocks in a PET system and the loss of one block results in a drop in sensitivity of only ~0.5%.

particularly noticeable where the patient has metal implants or has taken contrast media. The attenuation of metal and contrast at the CT energy does not relate linearly to the attenuation at annihilation photon energy. In this situation, the AC is overestimated and a hot spot appears erroneously at the point where the metal or contrast media is found. Figure I.19 demonstrates a contrast artefact leading to a hot spot that appears cold on the corresponding non-AC PET image. The non-AC images are often a key component in recognizing metal or contrast based artefacts, but the presence of streaks in the CT image is also a warning sign. The non-AC images should always be reviewed whenever any dubious finding is suspected in the AC images.

Another AC artefact is due to truncation, where the CT and PET FOVs are not the same size, so that parts of the anatomy outside the CT FOV are not corrected for by the AC algorithm. This often occurs when the patient's arms (which are raised above the head during the acquisition) are outside the CT FOV and a cold stripe appears across the patient's head in the AC images. In Fig. I.20, a cold stripe is prominent in the AC images but not visible in the non-AC images. Some PET/CT systems include software that can reconstruct the truncated CT data to increase the FOV and, thereby, reduce the severity of the artefact.

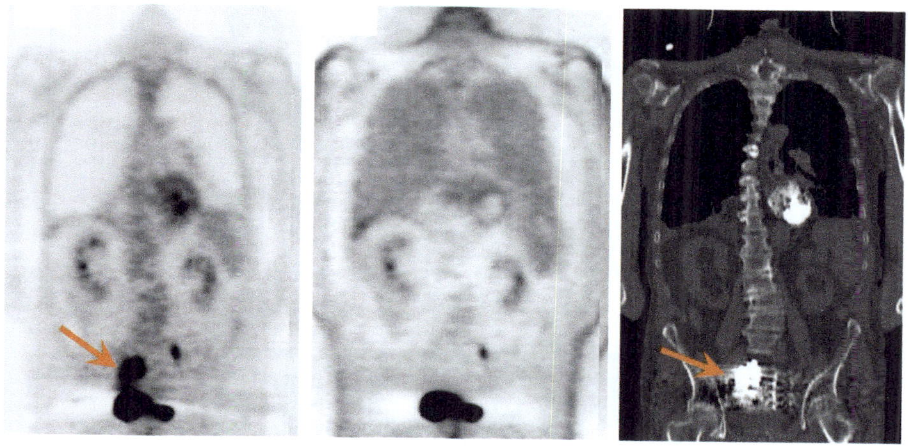

FIG. I.19. (a) CT attenuation corrected image of a patient showing a focal hot spot (indicated by the arrow). (b) Non-attenuation corrected image. The hot spot is no longer visible. (c) CT image showing a high density artefact from barium contrast pooled in the bowel. The artefact appears to be a region of high attenuation and the reconstruction algorithm overcompensates and creates a false hot spot. On the non-attenuation corrected image, the hot spot is entirely gone. These high density material artefacts are very common and the user should always examine the non-attenuation corrected image to check for the presence of such artefacts. Clues can also be found by examination of the corresponding location on the CT. (Courtesy of the Department of Nuclear Medicine, Memorial Sloane Kettering Cancer Center, New York.)

I.4.3. Co-registration and motion artefacts

Problems in co-registration in PET/CT are common and can be due to a system error or caused by movement of the patient. The system must be tested and recalibrated periodically, whereby a transform matrix is created to co-register the PET and CT data. Small errors in co-registration can often be seen in the head where the brain does not correctly fit inside the skull. Errors in the co-registration can occur either suddenly or gradually and can be a sign that there is a problem with the mechanism that controls the bed motion. Regular QC is required.

Alignment errors originating from patient motion are problematic and can have an effect on the medical interpretation of the image. The effect of an alignment error is demonstrated in Fig. I.21, where the patient's head has moved during data acquisition causing a misalignment between the PET and CT data. In the attenuation corrected images, the cortical uptake appears asymmetrical, but it can be seen from the fused PET and CT image that this was caused by a mis-registration error.

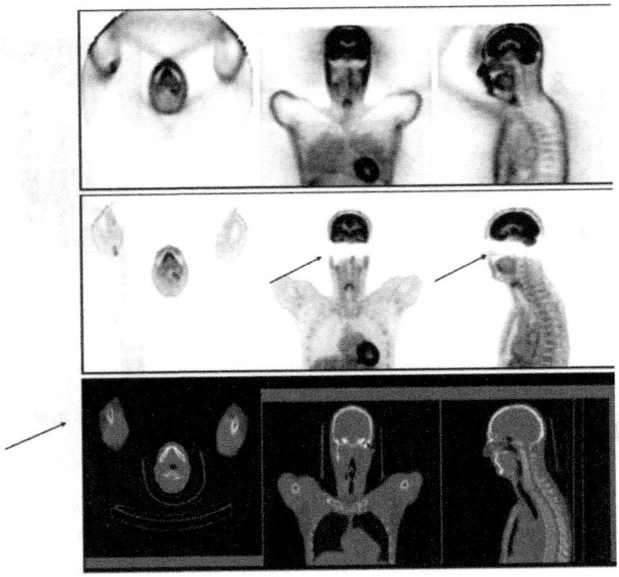

FIG. I.20. Clinical [^{18}F]-fluorodeoxyglucose images of the head and thorax from a PET/CT system. Top row: non-attenuation and scatter corrected PET images; middle row: the corresponding slices with attenuation and scatter correction; bottom row: CT images. The patient's arms have been truncated and extend beyond the CT field of view (arrow, bottom row). Although the truncation was relatively localized, using these CT data to correct for attenuation and scatter produced more extensive errors. This is shown by the cold band in corrected coronal and sagittal images of the head (arrows, middle row). The extent of this artefact may be due to an error in scaling of the scatter correction. The non-attenuation corrected images do not show the cold band. This example also demonstrates the essential value of comparing images with and without attenuation correction. (Courtesy of R. Boellaard, Department of Nuclear Medicine and PET Research, VU University Medical Centre, Amsterdam, Netherlands.)

Patient motion artefacts can be easily missed and lead to an incorrect diagnosis. In whole body PET scans, it is possible for the patient to move during the acquisition so that some part of the anatomy is accurately registered between PET and CT, while in another part of the anatomy the registration is poor. When not noticed by the operator and the reporting doctor, these artefacts can be misinterpreted as pathological uptake or be mistaken for a problem with the system. An example of this is shown in Fig. I.22, where the patient moved towards the end of the scan, causing an erroneous hot spot to appear in the wrong place.

Another movement artefact often seen is due to respiratory motion. PET images are acquired over many respiratory cycles, such that the final image is an average activity distribution across the respiratory cycle. The CT images are

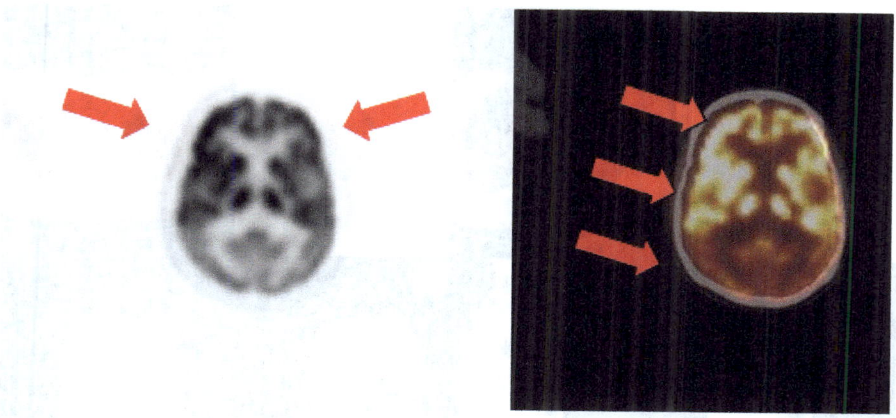

FIG. I.21. Apparent asymmetrical uptake of ^{18}F-fluorodeoxyglucose in the brain (left) caused by the slight misalignment of the PET and CT images (right). The non-attenuation corrected image did not show this asymmetry. These images form part of a whole body acquisition, which commenced at the thigh and moved up towards the head, which was the last part of the scan. Movement in the lower part of the body was not evident.

acquired far more quickly and, thus, demonstrate blurring over only a small component of the respiratory cycle. This can create a mis-registration and blurring in the AC PET images at the boundary of lung and liver. This kind of artefact can have significant clinical implications when a tumour is found near the border between the lungs and the liver. Figure I.23 shows an example where a lesion in the liver is incorrectly located in the lungs. This is due to the CT being acquired during full inspiration (the patient has taken a deep breath, pushing the diaphragm down and displacing the liver caudally), as opposed to the PET which is time averaged over regular tidal breathing, resulting in a severe mis-registration between the functional and anatomical location of the lesion. Conversely, Fig. I.24 shows an example where a tumour is incorrectly located in the liver due to a respiratory motion artefact.

Respiratory motion artefacts can also be seen in the CT image itself where the liver is not correctly rendered during reconstruction of the CT data because the patient is breathing during acquisition. This can be seen in Fig. I.25 where there is a characteristic artefact repeated along the axis of motion, leading to unclear definition of organ boundaries.

I.5. IMPORTANCE OF REGULAR QUALITY CONTROL

Regular QC procedures vary between scanner vendors; however, daily QC often requires checking the gantry status (voltages, temperatures, etc.) and

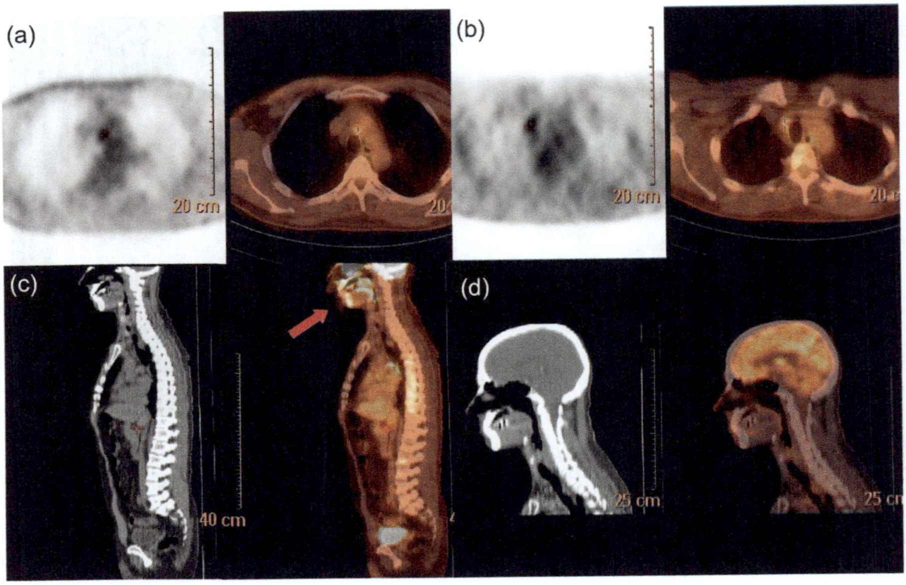

FIG. I.22. (a) A whole body PET/CT scan shows a point of focal uptake in the upper torso. (b) Separate head and neck acquisition of the same patient — the focal hot spot seems to have moved to a different location. (c) Whole body fused images show good registration between PET and CT in the bladder, spine and heart, and so it was assumed that the images were correctly registered; however, closer inspection of the head shows a clear mis-registration (indicated by the red arrow). (d) Fusion between PET and CT in the separate head and neck view shows good registration. The operator did not closely inspect the head and neck portion of the whole body view since there was a separate head and neck acquisition; however, the patient moved during the scan, probably by rotating the head. Had there not been a separate head and neck view, the doctor would have reported the focal uptake as metastatic cancer. This image was reported to the staff physicist as a problem with the system; however, it was in fact operator related. (Courtesy of the Nuclear Medicine Department, St. Vincent's Hospital, Darlinghurst, Australia.)

generation of the normalization from a high count emission image to ensure image quality (including checking a number of parameters, such as block noise and efficiency, scatter ratio, time alignment, etc.). The system vendor should provide a daily QC package that automates all of the above requirements so that QC can be performed quickly by the operator before the commencement of scanning. The daily QC procedure should produce a report indicating any unsatisfactory results which require further attention and allow for systematic monitoring of the scanner.

The characteristics of PET/CT systems allow for quantitative imaging that can display the absolute concentration of tracer in the subject. The ramification

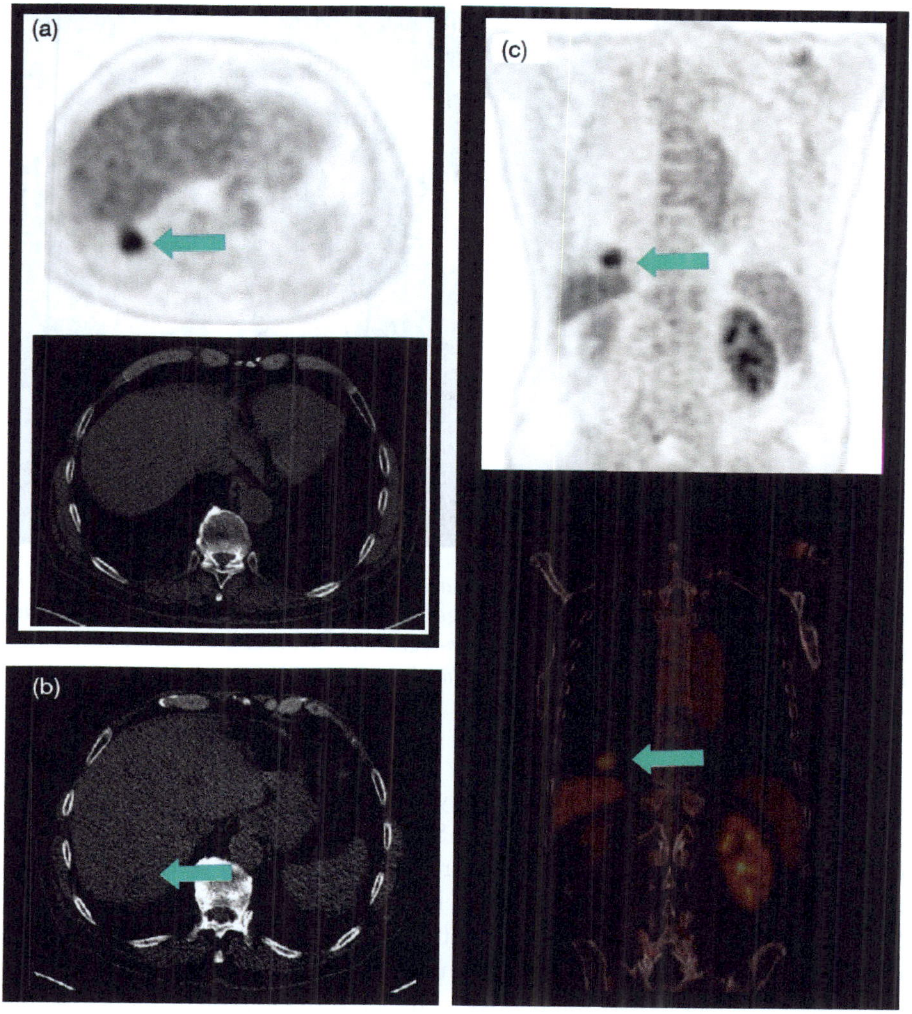

FIG. I.23. (a) Transaxial PET and CT images show a focal lesion, which appears to be in the lung. (b) The CT image shows the same lesion appearing to be in the liver. (c) Coronal view of PET and fused PET/CT where the lesion appears largely displaced from the liver. This problem occurs because the CT acquisition is very brief and captures the liver at one point in the respiratory cycle (in this case, full inspiration such that the diaphragm has displaced the liver caudally), while the PET acquisition is much longer and averaged over the whole respiratory cycle. The activity of the lesion is underestimated due to an attenuation correction artefact. This artefact stems from the fact that lung tissue is less attenuating than liver tissue and so the reconstruction process under-corrects for the attenuation of the signal from the lesion when it is assumed to be in the lung. The lesion demonstrates intense uptake and so is still highly visible despite the attenuation correction artefact.

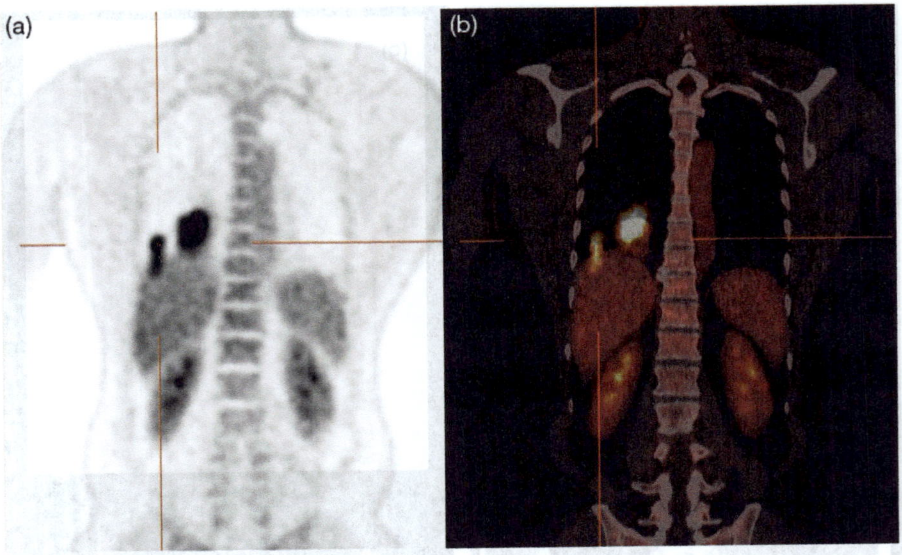

FIG. I.24. (a) Coronal PET image showing an area of focal uptake that appears to be both in the liver and in the lung, as well as another larger area that appears to be entirely in the lung. (b) Fused coronal PET/CT image showing a mis-registration in the larger lesion. Both of these lesions are entirely contained in the lung and the elongated appearance of the smaller lesions is an artefact created by respiratory motion.

of this for the physicist is the necessity to regularly perform a check of the standardized uptake value (SUV). The SUV is a quantitative parameter often quoted in clinical PET reporting that represents the uptake of activity in a lesion relative to background or healthy tissue (which should have an SUV = 1). The SUV measure is highly dependent on patient preparation, scanning protocol and reconstruction technique, and should be used with caution. It also relies heavily on accurate scanner calibration relative to the department dose calibrator, which allows PET data to be quantitative. Despite these difficulties, this index is often used by physicians for indicating abnormal uptake and, in particular, monitoring patient response during treatment by comparison of the SUV at baseline to the SUV during or after therapy. As such, the physicist must verify that such measures are accurate. A monthly check of the scanner SUV should be performed using a phantom of known volume, which, if activity is homogeneous and the scanner and dose calibrator are correctly calibrated, should produce an SUV = 1. Erroneously low SUVs may indicate that the physicist needs to recalibrate the dose calibrator and PET scanner, through the calculation of a new calibration factor.

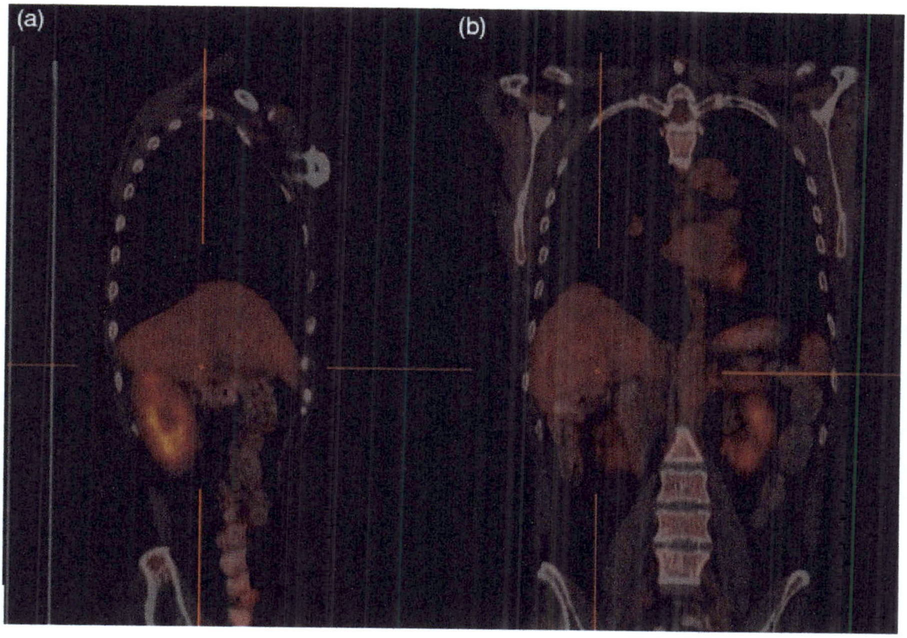

FIG. I.25. (a) A sagittal image shows a step-like artefact in the liver. (b) The same step-like artefact is seen on the coronal views. This problem is caused by the patient breathing during the CT acquisition and distorting the size of the liver. These artefacts are very common and can be compensated for by using a breath-hold technique.

This discussion of artefacts in PET/CT is by no means exhaustive and the reader is referred to the IAEA's PET/CT Atlas on Quality Control and Image Artefacts for a more comprehensive set of examples. As with all other instruments of the department, developing expertise in PET and PET/CT is essential for trouble-shooting and recognizing artefacts. Regular QC is a crucial factor in reducing artefacts due to the tomographic and/or CT system. Users of PET and PET/CT should be alert at all times to unexpected artefacts. A comparison between attenuation and scatter corrected images with non-corrected images should be part of routine clinical practice.

BIBLIOGRAPHY

BUSEMANN SOKOLE, E., PŁACHCÍNSKA, A., BRITTEN, A., Acceptance testing for nuclear medicine instrumentation, Eur. J. Nucl. Med. Mol. Imaging **37** (2010) 672–681.

BUSEMANN SOKOLE, E., et al., Routine quality control recommendations for nuclear medicine instrumentation, Eur. J. Nucl. Med. Mol. Imaging **37** (2010) 662–671.

INTERNATIONAL ATOMIC ENERGY AGENCY (Vienna)
Handbook on Care, Handling and Protection of Nuclear Medicine Instruments (1997).

IAEA Quality Control Atlas for Scintillation Camera Systems (2003).

Quality Assurance for SPECT Systems, IAEA Human Health Series No. 6 (2009).

Quality Assurance for PET and PET/CT Systems, IAEA Human Health Series No. 1 (2009).

PET/CT Atlas on Quality Control and Image Artefacts, IAEA Human Health Series No. 27 (2014).

Appendix II

RADIONUCLIDES OF INTEREST IN DIAGNOSTIC AND THERAPEUTIC NUCLEAR MEDICINE

Radionuclide	Half-life	Principal radiations emitted	Energy of emission[a] (MeV)	Abundance (in same order as emissions)
^{11}C	20.5 min	β^+ ($\gamma\pm$)	0.39	100%
^{13}N	9.97 min	β^+ ($\gamma\pm$)	0.49	100%
^{15}O	122.2 s	β^+ ($\gamma\pm$)	0.74	100%
^{18}F	109.8 min	β^+ ($\gamma\pm$)	0.24	96.9%
^{32}P	14.2 d	β^-	0.695	100%
^{51}Cr	27.7 d	γ	0.32	9%
^{57}Co	271.7 d	γ	0.122	86%
^{62}Cu	9.7 min	β^+	2.93 (max.)	97%
^{64}Cu	12.8 h	β^+ ($\gamma\pm$), β^-	0.28 (β^+) 0.19 (β^-)	17% (β^+) 39% (β^-)
^{67}Cu	62 h	β^-, γ	0.91 (γ_1) 0.93 (γ_2) 0.184 (γ_3) 0.121 (β^-_1) 0.154 (β^-_2) 0.189 (β^-_3)	7% (γ_1) 16% (γ_2) 49% (γ_3) 56% (β^-_1) 23% (β^-_2) 20% (β^-_3)
^{67}Ga	78 h	γ	0.093 0.184 0.300 0.394	38% 24% 16% 4%
^{68}Ga	68 min	β^+ ($\gamma\pm$)	0.74	88%
^{81m}Kr	13.1 s	γ	0.191	100%
^{82}Rb	75 s	β^+ ($\gamma\pm$)	1.4	96%
^{89}Sr	50.5 d	β^-	0.585	100%
^{89}Zr	78.4 h	β^+ ($\gamma\pm$) γ	0.897 (max.) 0.909	22.3% 99%

Radionuclide	Half-life	Principal radiations emitted	Energy of emission[a] (MeV)	Abundance (in same order as emissions)
^{90}Y	64 h	β^-	0.93	100%
^{99m}Tc	361.2 min	γ	0.1405	89%
^{111}In	67.4 h	γ	0.172 0.247	89% 94%
^{123}I	13 h	γ	0.159	97%
^{124}I	4.2 d	β^+ ($\gamma\pm$), γ	0.97 (β^+_1) 0.69 (β^+_2) 0.603 (γ_1) 1.69 (γ_2)	11% (β^+_1) 12% (β^+_2) 62% (γ_1) 0% (γ_2)
^{125}I	60.2 d	γ	0.036	7%
^{131}I	8.04 d	β^-, γ	0.19 (β^-) 0.364 (γ_1) 0.637 (γ_2)	90% (β^-) 83% (γ_1) 7% (γ_2)
^{131}Cs	9.7 d	γ	0.353	100%
^{133}Xe	5.25 d	β^-, γ	0.10 (β^-_1) 0.081 (γ_1)	100% (β^-_1) 37% (γ_1)
^{153}Sm	1.95 d	β^-, γ	0.20 (β^-_1) 0.23 (β^-_2) 0.27 (β^-_3) 0.070 (γ_1) 0.103 (γ_2)	32% (β^-_1) 50% (β^-_2) 18% (β^-_3) 5% (γ_1) 28% (γ_2)
^{166}Ho	26.8 h	β^-	0.70 (β^-_1) 0.65 (β^-_2)	50% (β^-_1) 49% (β^-_2)
^{169}Er	9.4 d	β^-	0.10	100%
^{177}Lu	6.73 d	β^-, γ	0.15 (β_1) 0.12 (β^-_2)	79% (β^-_1) 9% (β^-_2)
^{186}Re	3.78 d	β^-, γ	1.07 (β^-_1) 0.93 (β^-_2) 0.140 (γ)	74% (β^-_1) 21% (β^-_2) 9% (γ)
^{188}Re	17.0 h	β^-, γ	0.79 (β^-) 0.73 (β^-_2)	70% (β^-_1) 26% (β^-_2)
^{198}Au	2.7 d	β^-, γ	0.32 (β^-) 0.40 (γ)	99% (β^-) 96% (γ)

Radionuclide	Half-life	Principal radiations emitted	Energy of emission[a] (MeV)	Abundance (in same order as emissions)
^{201}Tl	73 h	γ, X	0.167 (γ) 0.070 (X_1) 0.080 (X_2)	8% (γ) 74% (X_1) 20% (X_2)
^{211}At	7.14 h	α	5.868	42%[b]
^{212}Bi	1.01 h	α, β^-	6.1 (α) 2.25 (β^-)	34% (α) 55% (β^-)
^{213}Bi	45.6 min	α	5.5–5.9 (α)	100% (α)
^{223}Ra	11.4 d	α, γ	5.5–5.7 (α) 0.154 (γ)	97% (α)[c] 6% (γ)

[a] Average energy of β emission.

[b] Astatine-211 undergoes a branched decay — 42% corresponds to direct α emission. The other 58% decays via electron capture to ^{212}Po, which also decays by α emission. As such, the net effect is one α emission per decay.

[c] Radium-223 has a complex decay chain, and when it comes to rest, it results in the emission of four α particles per decay.

ABBREVIATIONS

AAPM	American Association of Physicists in Medicine
AC	attenuation correction; alternating current
ACR	American College of Radiology
ADC	analogue to digital converter
ADT	admission, discharge, transfer
AFOV	axial field of view
ALARA	as low as reasonably achievable
ANSTO	Australian Nuclear Science and Technology Organisation
APD	avalanche photodiode
ASCII	American Standard Code for Information Interchange
BED	biological effective dose
BGO	bismuth germanate
BMIPP	β-methyl-p-iodophenylpentadecanoic acid
BMP	bitmap
BSS	Basic Safety Standards
CDR	collimator–detector response
CDF	cumulative density function
CERN	European Organization for Nuclear Research
CF	calibration factor
CFD	constant fraction discriminator
CFOV	central field of view
CIE	International Commission on Illumination
CLUT	colour lookup table
CMOS	complementary metal oxide semiconductor
CMS	colour management system
CMYK	cyan, magenta, yellow, key (black)
COTS	commercial off the shelf
cpm	counts per minute
cps	counts per second
CR	contrast ratio
CRT	cathode ray tube
CT	computed tomography
CZT	cadmium zinc telluride
DC	direct current
DCT	discrete cosine transform
DDL	digital driving level

DIB	device independent bitmap
DICOM	Digital Imaging and Communications in Medicine
DMSA	dimercaptosuccinic acid
DNA	deoxyribonucleic acid
dpi	dots per inch
DPM	disintegrations per minute
DRL	diagnostic reference level
DSB	double strand break
DTPA	diethylenetriaminepentaacetic acid
DVH	dose–volume histogram
EANM	European Association of Nuclear Medicine
EBRT	external beam radiotherapy
ECG	electrocardiogram
EDV	end diastolic volume
EGF	epidermal growth factor
EGS	electron gamma shower
e–h	electron–hole
EM	expectation maximization
ERPF	effective renal plasma flow
ESV	end systolic volume
FBP	filtered back projection
FDA	Food and Drug Administration
FDG	fluorodeoxyglucose
FFT	fast Fourier transform
FORE	final rebinning algorithm
FOV	field of view
FWHM	full width at half maximum
FWTM	full width at tenth maximum
GFR	glomerular filtration rate
GIF	Graphics Interchange Format
GPU	graphical processing unit
GSDF	grey scale standard display function
GSO	gadolinium oxyorthosilicate
HMPAO	hexamethlypropyleneamine oxime
HPMT	hybrid photomultiplier tube
HTML	Hypertext Markup Language
HU	Hounsfield unit

HVL	half-value layer
ICC	International Color Consortium
ICRP	International Commission on Radiological Protection
ICRU	International Commission on Radiation Units and Measurements
IEC	International Electrotechnical Commission
IEEE	Institute of Electrical and Electronics Engineers
IPS	in-plane switching
IQ	image quality
ISO	International Organization for Standardization
JFIF	JPEG File Interchange Format
JND	just noticeable difference
JPEG	Joint Photographic Experts Group
LCD	liquid crystal display
LET	linear energy transfer
LOR	line of response
LQ	linear–quadratic
LR	luminance ratio
LS	least squares
LSF	line spread function
LSO	lutetium oxyorthosilicate
LUT	lookup table
LZW	Lempel–Ziv–Welch
MAA	macroaggregate of albumin
MAG3	mercaptoacetyltriglycine
MAP	maximum a posteriori
MCNP	Monte Carlo N-particle transport code
MCP	microchannel plate
MFP	mean free path
MIBG	metaiodobenzylguanidine
MIBI	methoxyisobutylisonitrile
MIP	maximum intensity projection
MIRD	medical internal radiation dose
MLEM	maximum-likelihood expectation-maximization
MR	magnetic resonance
MRI	magnetic resonance imaging
MRN	medical record number
MSE	mean square error

MTF	modulation transfer function
MUGA	multiple-gated acquisition
MWPC	multiwire proportional chamber
NEA	negative electron affinity
NECR	noise equivalent count rate
NEMA	National Electrical Manufacturers Association
NET	neuroendocrine tumour
NHEJ	non-homologous end joining
NIST	National Institute of Standards and Technology
NPL	National Physical Laboratory
NTCP	normal tissue complication probability
NURBS	non-uniform rational B-spline
OER	oxygen enhancement ratio
OSEM	ordered-subsets expectation-maximization
PACS	picture archiving and communication system
PAH	para-amino hippurate
PAHO	Pan American Health Organization
PCS	profile connection space
PDF	Portable Document Format
PDR	perceived dynamic range
PET	positron emission tomography
PMT	photomultiplier tube
POPOP	para-phenylene-phenyloxazole
PSF	point spread function
PSRF	point source response function
PV	plasma volume
PVE	partial volume effect
QA	quality assurance
QC	quality control
QMS	quality management system
RAMDAC	random access memory digital to analogue converter
RAMLA	row-action maximum-likelihood algorithm
RBE	relative biological effectiveness
RC	recovery coefficient
RGB	red, green, blue
RIS	radiology information system

RIT	radioimmunotherapy
ROI	region of interest
RPO	radiation protection officer
RPP	radiation protection programme
RSS	Real Simple Syndication
SI	International System of Units
SiPM	silicon photomultiplier tube
SNR	signal to noise ratio
SPECT	single photon emission computed tomography
SPR	scatter to primary ratio
SSB	single strand break
SUV	standardized uptake value
TCP	tumour conrol probability
TDC	time to digital converter
TEW	triple energy window
TIFF	Tagged Image File Format
TOF	time of flight
TLG	total lesion glycolysis
TVL	tenth-value layer
UFOV	useful field of view
UNSCEAR	United Nations Scientific Committee on the Effects of Atomic Radiation
UPS	uninterruptable power supply
UTF	Unicode Transformation Format
WHO	World Health Organization
WLS	weighted least squares
WYSIWYG	what you see is what you get
XML	Extensible Markup Language

SYMBOLS

Roman symbols

a	year (unit of time)
a	acceleration; area; specific activity
$\tilde{a}$	time integrated activity coefficient
A	ampere (SI unit of current)
A	atomic mass number
Å	ångström (unit of distance: 1 Å = 10^{-10} m)
$\tilde{A}$	cumulated activity
$\mathcal{A}$	activity
b	barn (unit of cross-section)
Bq	becquerel (SI unit of activity)
c	speed of light
C	capacity; coulomb (SI unit of charge)
°C	degree Celsius (unit of temperature)
C	activity concentration; counts
cd	luminous intensity: candela
Ci	curie (unit of activity: 1 Ci = 3.7×10^{10} Bq)
d	day (unit of time)
d	depth; distance
D	dose; thickness
$\dot{D}$	dose rate
dB	decibel
e	electron charge
E	effective dose; electric field; energy
E_{ab}	absorbed energy
E_B	binding energy
E_K	kinetic energy
E_{tr}	transferred energy
E_γ	photopeak energy
$E(\tau)$	committed effective dose
$\bar{E}_{ab}$	mean absorbed energy
$\bar{E}_{tr}$	mean transferred energy
eV	electronvolt

F	farad (SI unit of capacitance)
F	Fano factor; force
G	gravitational constant
Gy	gray (SI unit of dose)
h	hour (unit of time)
h	Planck's constant
H_p	personal dose equivalent
H_T	equivalent dose
$H_T(\tau)$	committed equivalent dose
Hz	unit of frequency
I	electric current; intensity
j	current density
J	joule (SI unit of energy)
K	kelvin (SI unit of thermodynamic temperature)
K	kerma
kg	kilogram (SI unit of mass)
l	length
L	litre (unit of volume)
L	luminance
m	metre (SI unit of length)
m	mass
m_a	atomic mass in atomic mass units u
m_e	electron rest mass; positron rest mass
m_n	neutron rest mass
m_p	proton rest mass
M	nuclear mass
mol	amount of substance: mole
N	newton (SI unit of force)
N	number of counts; number of neutrons in an atom
N_a	number of atoms
N_A	Avogadro's number
N_{el}	number of electrons
N_{ph}	number of photons

p	momentum; probability
P	power; pressure
Pa	pascal (SI unit of pressure)
Q	disintegration energy; electric charge; perfusion; reaction energy
r	correlation coefficient; radius
R	roentgen (unit of exposure)
R	counting rate; dose rate; radius; random coincidence count rate
R_{E}	fractional energy resolution
R_0	nuclear radius constant
s	second (unit of time)
s	septal thickness; sample standard deviation
s_{col}	collision stopping power
s_{rad}	radiation stopping power
s_{tot}	total stopping power
s^2	sample variance
S	absorbed dose rate per unit activity; sensitivity
Sv	sievert (unit of equivalent dose and unit of effective dose)
t	time
T	tesla (SI unit of magnetic field strength)
T	temperature
$T_{1/2}$	half-life
u	atomic mass unit
V	volt (unit of voltage)
V	ventilation; voltage; volume
w	width
w_{R}	radiation weighting factor
w_{T}	tissue weighting factor
W	watt (SI unit of power)
W	mean energy to produce an information carrier; weight
$x_{1/2}$	half-value layer
$x_{1/10}$	tenth-value layer
$\bar{x}$	mean free path
$\bar{x}_{\mathrm{e}}$	experimental mean

$\bar{x}_t$	true mean
X	exposure
z	specific energy
Z	atomic number
Z_{eff}	effective atomic number

Greek symbols

α	alpha particle
α	linear radiosensitivity constant
β	beta particle
β	quadratic radiosensitivity coefficient
γ	gamma ray
Γ	air kerma rate constant
ε_0	electric constant (permittivity of vacuum)
ε_T	total energy imparted by radiation to the tissue or organ
$\bar{\varepsilon}$	mean energy imparted
η	quantum efficiency; refractive index
κ	pair production attenuation coefficient
$_a\kappa$	pair production atomic attenuation coefficient
λ	radioactive decay constant; wavelength
μ	linear attenuation coefficient; mobility of charged carriers
μ_{ab}	energy absorption coefficient
μ_m	mass attenuation coefficient
μ_{tr}	energy transfer coefficient
$_a\mu$	atomic attenuation coefficient
$_e\mu$	electronic attenuation coefficient
μ_0	magnetic constant (permeability of vacuum)
ν	photon frequency
ν_e	electronic neutrino
$\bar{\nu}_e$	electronic antineutrino

ρ	mass density
σ	cross-section; standard deviation
σ_C	Compton attenuation coefficient
σ_e	experimental standard deviation
σ_{eF}	experimental fractional standard deviation
σ_F	fractional standard deviation
σ_P	percentage standard deviation
σ_R	Rayleigh attenuation coefficient
${}_a\sigma_R$	Rayleigh atomic attenuation coefficient
${}_a\sigma_C$	Compton atomic attenuation coefficient
${}_e\sigma_C$	Compton electronic attenuation coefficient
σ^2	variance
σ_e^2	experimental variance
τ	dead time; mean life
${}_a\tau$	photoelectric atomic attenuation coefficient
υ	velocity
ω	angular frequency; fluorescence yield
Ω	ohm (SI unit of electrical resistance); solid angle

CONTRIBUTORS TO DRAFTING AND REVIEW

Al-Mazrou, R.	King Faisal Specialist Hospital and Research Centre, Saudi Arabia
Bailey, D.L.	Royal North Shore Hospital and University of Sydney, Australia
Bergmann, H.	Medical University of Vienna, Austria
Busemann Sokole, E.	Academic Medical Center, Netherlands
Carlsson, S.T.	Uddevalla Hospital, Sweden
Dale, R.G.	Imperial College London, United Kingdom
Daube-Witherspoon, M.E.	University of Pennsylvania, United States of America
Dauer, L.T.	Memorial Sloan Kettering Cancer Center, United States of America
Demirkaya, O.	King Faisal Specialist Hospital and Research Centre, Saudi Arabia
Du, Yong	Royal Marsden Hospital and Institute of Cancer Research, United Kingdom
El Fakhri, G.	Massachusetts General Hospital and Harvard Medical School, United States of America
Flux, G.	Royal Marsden Hospital and Institute of Cancer Research, United Kingdom
Forwood, N.J.	Royal North Shore Hospital, Australia
Frey, E.C.	Johns Hopkins University, United States of America
Hindorf, C.	Skåne University Hospital, Sweden
Humm, J.L.	Memorial Sloan Kettering Cancer Center, United States of America
Kesner, A.L.	International Atomic Energy Agency
Le Heron, J.C.	International Atomic Energy Agency
Lodge, M.A.	Johns Hopkins University, United States of America

Lötter, M.G.	University of the Free State, South Africa
Lundqvist, H.O.	Uppsala University, Sweden
Matej, S.	University of Pennsylvania, United States of America
Myers, M.J.	Imperial College London, United Kingdom
Nuyts, J.	Katholieke Universiteit Leuven, Belgium
Ott, R.J.	Royal Marsden Hospital and Institute of Cancer Research, United Kingdom
Ouyang, J.	Massachusetts General Hospital and Harvard Medical School, United States of America
Palm, S.	International Atomic Energy Agency
Parker, J.A.	Harvard Medical School, United States of America
Podgorsak, E.B.	McGill University, Canada
Poli, G.L.	International Atomic Energy Agency
Smart, R.C.	St. George Hospital, Australia
Soni, P.S.	Bhabha Atomic Research Centre, India
Stephenson, R.	Rutherford Appleton Laboratory, United Kingdom
Todd-Pokropek, A.	University College London, United Kingdom
van Aswegen, A.	University of the Free State, South Africa
Van Eijk, C.W.E.	Delft University of Technology, Netherlands
Willowson, K.P.	University of Sydney, Australia
Wondergem, J.	International Atomic Energy Agency
Zanzonico, P.B.	Memorial Sloan Kettering Cancer Center, United States of America

Consultants Meetings

Vienna, Austria: 1–5 September 2008, 14–16 April 2009, 17–19 May 2010, 7–11 March 2011

IAEA
International Atomic Energy Agency

ORDERING LOCALLY

In the following countries, IAEA priced publications may be purchased from the sources listed below or from major local booksellers.

Orders for unpriced publications should be made directly to the IAEA. The contact details are given at the end of this list.

AUSTRALIA
DA Information Services
648 Whitehorse Road, Mitcham, VIC 3132, AUSTRALIA
Telephone: +61 3 9210 7777 • Fax: +61 3 9210 7788
Email: books@dadirect.com.au • Web site: http://www.dadirect.com.au

BELGIUM
Jean de Lannoy
Avenue du Roi 202, 1190 Brussels BELGIUM
Telephone: +32 2 5384 308 • Fax: +32 2 5380 841
Email: jean.de.lannoy@euronet.be • Web site: http://www.jean-de-lannoy.be

CANADA
Renouf Publishing Co. Ltd.
5369 Canotek Road, Ottawa, ON K1J 9J3, CANADA
Telephone: +1 613 745 2665 • Fax: +1 643 745 7660
Email: order@renoufbooks.com • Web site: http://www.renoufbooks.com

Bernan Associates
4501 Forbes Blvd., Suite 200, Lanham, MD 20706-4391, USA
Telephone: +1 800 865 3457 • Fax: +1 800 865 3450
Email: orders@bernan.com • Web site: http://www.bernan.com

CZECH REPUBLIC
Suweco CZ, spol. S.r.o.
Klecakova 347, 180 21 Prague 9, CZECH REPUBLIC
Telephone: +420 242 459 202 • Fax: +420 242 459 203
Email: nakup@suweco.cz • Web site: http://www.suweco.cz

FINLAND
Akateeminen Kirjakauppa
PO Box 128 (Keskuskatu 1), 00101 Helsinki, FINLAND
Telephone: +358 9 121 41 • Fax: +358 9 121 4450
Email: akatilaus@akateeminen.com • Web site: http://www.akateeminen.com

FRANCE
Form-Edit
5 rue Janssen, PO Box 25, 75921 Paris CEDEX, FRANCE
Telephone: +33 1 42 01 49 49 • Fax: +33 1 42 01 90 90
Email: fabien.boucard@formedit.fr • Web site: http://www.formedit.fr

Lavoisier SAS
14 rue de Provigny, 94236 Cachan CEDEX, FRANCE
Telephone: +33 1 47 40 67 00 • Fax: +33 1 47 40 67 02
Email: livres@lavoisier.fr • Web site: http://www.lavoisier.fr

L'Appel du livre
99 rue de Charonne, 75011 Paris, FRANCE
Telephone: +33 1 43 07 50 80 • Fax: +33 1 43 07 50 80
Email: livres@appeldulivre.fr • Web site: http://www.appeldulivre.fr

GERMANY
Goethe Buchhandlung Teubig GmbH
Schweitzer Fachinformationen
Willstätterstrasse 15, 40549 Düsseldorf, GERMANY
Telephone: +49 (0) 211 49 8740 • Fax: +49 (0) 211 49 87428
Email: s.dehaan@schweitzer-online.de • Web site: http://www.goethebuch.de

HUNGARY
Librotade Ltd., Book Import
PF 126, 1656 Budapest, HUNGARY
Telephone: +36 1 257 7777 • Fax: +36 1 257 7472
Email: books@librotade.hu • Web site: http://www.librotade.hu

INDIA

Allied Publishers
1st Floor, Dubash House, 15, J.N. Heredi Marg, Ballard Estate, Mumbai 400001, INDIA
Telephone: +91 22 2261 7926/27 • Fax: +91 22 2261 7928
Email: alliedpl@vsnl.com • Web site: http://www.alliedpublishers.com

Bookwell
3/79 Nirankari, Delhi 110009, INDIA
Telephone: +91 11 2760 1283/4536
Email: bkwell@nde.vsnl.net.in • Web site: http://www.bookwellindia.com

ITALY

Libreria Scientifica "AEIOU"
Via Vincenzo Maria Coronelli 6, 20146 Milan, ITALY
Telephone: +39 02 48 95 45 52 • Fax: +39 02 48 95 45 48
Email: info@libreriaaeiou.eu • Web site: http://www.libreriaaeiou.eu

JAPAN

Maruzen Co., Ltd.
1-9-18 Kaigan, Minato-ku, Tokyo 105-0022, JAPAN
Telephone: +81 3 6367 6047 • Fax: +81 3 6367 6160
Email: journal@maruzen.co.jp • Web site: http://maruzen.co.jp

NETHERLANDS

Martinus Nijhoff International
Koraalrood 50, Postbus 1853, 2700 CZ Zoetermeer, NETHERLANDS
Telephone: +31 793 684 400 • Fax: +31 793 615 698
Email: info@nijhoff.nl • Web site: http://www.nijhoff.nl

Swets Information Services Ltd.
PO Box 26, 2300 AA Leiden
Dellaertweg 9b, 2316 WZ Leiden, NETHERLANDS
Telephone: +31 88 4679 387 • Fax: +31 88 4679 388
Email: tbeysens@nl.swets.com • Web site: http://www.swets.com

SLOVENIA

Cankarjeva Zalozba dd
Kopitarjeva 2, 1515 Ljubljana, SLOVENIA
Telephone: +386 1 432 31 44 • Fax: +386 1 230 14 35
Email: import.books@cankarjeva-z.si • Web site: http://www.mladinska.com/cankarjeva_zalozba

SPAIN

Diaz de Santos, S.A.
Librerias Bookshop • Departamento de pedidos
Calle Albasanz 2, esquina Hermanos Garcia Noblejas 21, 28037 Madrid, SPAIN
Telephone: +34 917 43 48 90 • Fax: +34 917 43 4023
Email: compras@diazdesantos.es • Web site: http://www.diazdesantos.es

UNITED KINGDOM

The Stationery Office Ltd. (TSO)
PO Box 29, Norwich, Norfolk, NR3 1PD, UNITED KINGDOM
Telephone: +44 870 600 5552
Email (orders): books.orders@tso.co.uk • (enquiries): book.enquiries@tso.co.uk • Web site: http://www.tso.co.uk

UNITED STATES OF AMERICA

Bernan Associates
4501 Forbes Blvd., Suite 200, Lanham, MD 20706-4391, USA
Telephone: +1 800 865 3457 • Fax: +1 800 865 3450
Email: orders@bernan.com • Web site: http://www.bernan.com

Renouf Publishing Co. Ltd.
812 Proctor Avenue, Ogdensburg, NY 13669, USA
Telephone: +1 888 551 7470 • Fax: +1 888 551 7471
Email: orders@renoufbooks.com • Web site: http://www.renoufbooks.com

United Nations
300 East 42nd Street, IN-919J, New York, NY 1001, USA
Telephone: +1 212 963 8302 • Fax: 1 212 963 3489
Email: publications@un.org • Web site: http://www.unp.un.org

Orders for both priced and unpriced publications may be addressed directly to:

IAEA Publishing Section, Marketing and Sales Unit, International Atomic Energy Agency
Vienna International Centre, PO Box 100, 1400 Vienna, Austria
Telephone: +43 1 2600 22529 or 22488 • Fax: +43 1 2600 29302
Email: sales.publications@iaea.org • Web site: http://www.iaea.org/books

13-21541

Printed and bound by CPI Group (UK) Ltd, Croydon, CR0 4YY

13/09/2026

14961051-0001